D0358441

Methods in Enzymology

Volume 310

BIOFILMS

METHODS IN ENZYMOLOGY

EDITORS-IN-CHIEF

John N. Abelson Melvin I. Simon

DIVISION OF BIOLOGY
CALIFORNIA INSTITUTE OF TECHNOLOGY
PASADENA, CALIFORNIA

FOUNDING EDITORS

Sidney P. Colowick and Nathan O. Kaplan

Methods in Enzymology

Volume 310

Biofilms

EDITED BY

Ron J. Doyle

DEPARTMENT OF MICROBIOLOGY AND IMMUNOLOGY
SCHOOLS OF DENTISTRY AND MEDICINE
UNIVERSITY OF LOUISVILLE
LOUISVILLE, KENTUCKY

ACADEMIC PRESS
San Diego London Boston New York Sydney Tokyo Toronto

This book is printed on acid-free paper. 

Academic Press
A Harcourt Science and Technology Company
525 B Street, Suite 1900, San Diego, California 92101-4495, USA
http://www.academicpress.com

Academic Press Limited
24-28 Oval Road, London NW1 7DX, UK
http://www.hbuk.co.uk/ap/

International Standard Book Number: 0-12-182211-7

PRINTED IN THE UNITED STATES OF AMERICA
99 00 01 02 03 04 MM 9 8 7 6 5 4 3 2 1

Table of Contents

Section I. Molecular Biology of Biofilm Bacteria

Section II. Microscopic Methods of Biofilm Formation and Physiology

Section III. Flow and Steady-State Methods

Section IV. Biofilms in Archaea

Section V. Physical Methods

Section VI. Physiology of Biofilm-Associated Microorganisms

Section VII. Substrate for Biofilm Development

Section VIII. Antifouling Methods

Contributors to Volume 310

Article numbers are in parentheses following the names of contributors.
Affiliations listed are current.

DONALD G. AHEARN (40), *Georgia State University, Atlanta, Georgia 30302-4010*

DAVID G. ALLISON (19), *School of Pharmacy and Pharmaceutical Sciences, University of Manchester, Manchester M13 9PL, United Kingdom*

ATSUO AMANO (36), *Division of Special Care Dentistry, Osaka University Faculty of Dentistry, Suita-Osaka, 565-0871 Japan*

JENS BO ANDERSEN (2), *Department of Microbiology, Technical University of Denmark–DTU, DK-2800 Lyngby, Denmark*

ROXANNA N. ANDERSEN (24), *Oral Infection and Immunity Branch, National Institute of Dental and Craniofacial Research, National Institutes of Health, Bethesda, Maryland 20892-4350*

NEIL ANDERSON (42), *The Royal Group of Hospitals and Dental Hospital Health and Social Services Trust, Belfast BT12 6BA, United Kingdom*

C. ARNOSTI (30), *Department of Marine Science, University of North Carolina, Chapel Hill, North Carolina 27599*

GEORGE S. BAILLIE (48), *Division of Infection and Immunity, Institute of Biomedical and Life Sciences, University of Glasgow, Glasgow G12 8QQ, United Kingdom*

RICHARD BOLD (34), *Biotechnology Research Department, Orange County Water District, Fountain Valley, California 92728-8300*

CHRISTOPHE J. P. BOONAERT (28), *Unité de Chimie des Interfaces, Université Catholique de Louvain, Louvain-la-Neuve 1348, Belgium*

ROYA N. BORAZJANI (40), *Georgia State University, Atlanta, Georgia 30302-4010*

GEORGE H. W. BOWDEN (17), *Department of Oral Biology, University of Manitoba, Winnipeg, Manitoba, Canada R3E OW2*

DAVID J. BRADSHAW (22), *Research Division, Centre for Applied Microbiology and Research (CAMR), Salisbury, Wiltshire SP4 0JG, United Kingdom*

BERIT K. BREDTVED (26), *Enzyme Development and Application, Novo Nordisk A/S, DK-2880 Bagsvaerd, Denmark*

R. G. BROWN (46), *Building Services Research and Information Association, Bracknell, Berkshire RG1R 7AH, United Kingdom*

VOLKER S. BRÖZEL (1), *Department of Microbiology and Plant Pathology, University of Pretoria, Pretoria 0002, South Africa*

ROBERT A. BURNE (33), *Department of Microbiology and Immunology, and Center for Oral Biology, University of Rochester Medical Center, Rochester, New York 14642*

HENK J. BUSSCHER (31, 38), *Department of Biomedical Engineering, University of Groningen, 9797kZ Groningen, The Netherlands*

BJARKE BAK CHRISTENSEN (2), *Department of Microbiology, Technical University of Denmark–DTU, DK-2800 Lyngby, Denmark*

JOSEPH J. COONEY (47), *Environmental, Coastal and Ocean Sciences, University of Massachusetts, Boston, Massachusetts 02125-3393*

J. WILLIAM COSTERTON (49), *Center for Biofilm Engineering, Montana State University, Bozeman, Montana 59717-3980*

MARTIN D. CURRAN (42), *Regional Histocompatibility and Immunogenetic Laboratory, Belfast City Hospital, Belfast BT9 7TS, United Kingdom*

RICHARD I. DAVIS (42), *Belfast City Hospital, Belfast BT9 7BL, United Kingdom*

TERESA R. DE KIEVIT (8), *Department of Microbiology and Immunology, University of Rochester School of Medicine and Dentistry, Rochester, New York 14642*

MARIA R. DIAZ-TORRES (32), *Genencor International Inc., Palo Alto, California 94304*

GEORGE H. DIBDIN (23), *Department of Oral and Dental Science, University of Bristol, Bristol BS1 2LY, United Kingdom*

L. JULIA DOUGLAS (48), *Division of Infection and Immunity, Institute of Biomedical and Life Sciences, University of Glasgow, Glasgow G12 8QQ, United Kingdom*

YVES F. DUFRÊNE (28), *Unité de Chimie des Interfaces, Université Catholique de Louvain, Louvain-la-Neuve 1348, Belgium*

CHARLES E. EDMISTON, JR. (15), *Surgical Microbiology Research Laboratory, Department of Surgery, Medical College of Wisconsin, Milwaukee, Wisconsin 53226*

JAMES G. ELKINS (44), *Department of Molecular Genetics, Biochemistry, and Microbiology, University of Cincinnati College of Medicine, Cincinnati, Ohio 45267-0524*

THERESA A. FASSEL (15), *Core Electron Microscope Unit, The Scripps Research Institute, La Jolla, California 92037-1027*

SUSANA FORTUN (49), *Lucent Technologies, Naperville, Illinois 60566*

TAKASHI FUKUOKA (43), *Biological Research Laboratories, Sankyo Company, Limited, Aoba-ku, Yokohama 225-0024, Japan*

MANAL M. GABRIEL (40), *Biology Department, Georgia State University, Atlanta, Georgia 30302-4010*

PETER GILBERT (19), *School of Pharmacy and Pharmaceutical Sciences, University of Manchester, Manchester M13 9PL, United Kingdom*

MICHAEL GIVSKOV (2), *Department of Microbiology, Technical University of Denmark–DTU, DK-2800 Lyngby, Denmark*

BIRGIT GIWERCMAN (16), *Department of Clinical Microbiology, University Hospital (Rigshospitalet), University of Copenhagen, DK-2200 Copenhagen, Denmark*

DARLA GOERES (45), *Center for Biofilm Engineering, Montana State University, Bozeman, Montana 59717-3980*

SEAN P. GORMAN (41, 42), *Medical Devices Group, School of Pharmacy, The Queen's University of Belfast, Belfast BT9 7BL, United Kingdom*

BOVRT GOTTENBOS (38), *Department of Biomedical Engineering, University of Groningen, 9797kZ Groningen, The Netherlands*

E. PETER GREENBERG (3), *Department of Microbiology, University of Iowa, Iowa City, Iowa 52242*

SHIGEYUKI HAMADA (36, 37), *Department of Oral Microbiology, Osaka University Faculty of Dentistry, Suita-Osaka, 565-0871 Japan*

MARTY HAMILTON (45), *Center for Biofilm Engineering, Montana State University, Bozeman, Montana 59717-3980*

PATRICIA L. HARTZELL (25), *Department of Microbiology, Molecular Biology, and Biochemistry, University of Idaho, Moscow, Idaho 83844-3052*

DANIEL J. HASSETT (44), *Department of Molecular Genetics, Biochemistry, and Microbiology, University of Cincinnati College of Medicine, Cincinnati, Ohio 45267-0524*

JOANNA HEERSINK (45), *Center for Biofilm Engineering, Montana State University, Bozeman, Montana 59717-3980*

CHRISTINE HEINEMANN (31), *Department of Microbiology and Immunology, University of Western Ontario, London, Ontario, Canada N6A 4V2*

MARK C. HERZBERG (7), *Department of Preventive Sciences, School of Dentistry, University of Minnesota, Minneapolis, Minnesota 55455*

NIELS HØIBY (16), *Department of Clinical Microbiology, University Hospital (Rigshospitalet), University of Copenhagen, DK-2200 Copenhagen, Denmark*

BRIAN C. HOSKINS (20), *Department of Petroleum and Geosystems Engineering, University of Texas, Austin, Texas 78712*

BARBARA H. IGLEWSKI (8), *Department of Microbiology and Immunology, University of Rochester School of Medicine and Dentistry, Rochester, New York 14642*

KENNETH ISHIDA (34), *Biotechnology Research Department, Orange County Water District, Fountain Valley, California 92728-8300*

CHARLOTTE JOHANSEN (26), *Enzyme Development and Application, Novo Nordisk A/S, DK-2880 Bagsvaerd, Denmark*

DAVID S. JONES (41), *Medical Devices Group, School of Pharmacy, The Queen's University of Belfast, Belfast BT9 7BL, United Kingdom*

RYAN N. JORDAN (29), *Center for Biofilm Engineering, Montana State University, Bozeman, Montana 59717-3980*

SHIGETADA KAWABATA (37), *Department of Oral Microbiology, Osaka University Faculty of Dentistry, Suita-Osaka, 565-0871 Japan*

KAREN KAZMERZAK (24), *Oral Infection and Immunity Branch, National Institute of Dental and Craniofacial Research, National Institutes of Health, Bethesda, Maryland 20892-4350*

ARSALAN KHARAZMI (16), *Department of Clinical Microbiology, University Hospital (Rigshospitalet), University of Copenhagen, DK-2200 Copenhagen, Denmark*

ROBIN D. KIRKEGAARD (18), *Center for Environmental Biotechnology, University of Tennessee, Knoxville, Tennessee 37932-2575*

TOM KNOELL (34), *Biotechnology Research Department, Orange County Water District, Fountain Valley, California 92728-8300*

TETSUFUMI KOGA (43), *Biological Research Laboratories, Sankyo Company, Limited, Aoba-ku, Yokohama 225-0024, Japan*

PAUL E. KOLENBRANDER (24), *Oral Infection and Immunity Branch, National Institute of Dental and Craniofacial Research, National Institutes of Health, Bethesda, Maryland 20892-4350*

ROBERTO KOLTER (6), *Department of Microbiology and Molecular Genetics, Harvard Medical School, Boston, Massachusetts 02115*

DARREN R. KORBER (1), *Department of Applied Microbiology and Food Science, University of Saskatchewan, Saskatoon, Saskatchewan, Canada S7N 5A8*

CHRISTOPHER LAPAGLIA (25), *Department of Microbiology, Molecular Biology, and Biochemistry, University of Idaho, Moscow, Idaho 83844-3052*

JOHN R. LAWRENCE (9, 10), *National Water Research Institute, Saskatoon, Saskatchewan, Canada S7N 3H5*

PHILIPPE LEJEUNE (4), *Laboratoire de Génétique Moléculaire des Microorganismes et des Interactions Cellulaires, CNRS UMR 5577, Institut National des Sciences Appliquées de Lyon, 69621 Villeurbanne, France*

JOHN T. LISLE (13), *Department of Microbiology, Montana State University, Bozeman, Montana 59717*

V. J. LUCAS (46), *Centre for Applied Microbiology and Research (CAMR), Salisbury, Wiltshire SP4 0JG, United Kingdom*

JU-FANG MA (44), *Department of Molecular Genetics, Biochemistry, and Microbiology, University of Cincinnati College of Medicine, Cincinnati, Ohio 45267-0524*

RAYNARD MACDONALD (1), *Department of Microbiology and Plant Pathology, University of Pretoria, Pretoria 0002, South Africa*

PAUL D. MAJORS (20), *Department of Petroleum and Geosystems Engineering, University of Texas, Austin, Texas 78712*

WERNER MANZ (5), *FG Microbial Ecology, Technical University Berlin, D-10587 Berlin, Germany*

ROBERT E. MARQUIS (33), *Department of Microbiology and Immunology, and Center for Oral Biology, University of Rochester Medical Center, Rochester, New York 14642*

PHILIP D. MARSH (22), *Research Division, Centre for Applied Microbiology and Research (CAMR), Salisbury, Wiltshire SP4 0JG, United Kingdom, and Leeds Dental Institute, Leeds LS2 9LU, United Kingdom*

TIMOTHY R. MCDERMOTT (44), *Department of Molecular Genetics, Biochemistry, and Microbiology, University of Cincinnati College of Medicine, Cincinnati, Ohio 45267-0524*

GORDON A. MCFETERS (13), *Department of Microbiology, Montana State University, Bozeman, Montana 59717*

ROBERT J. C. MCLEAN (20), *Department of Biology, Southwest Texas State University, San Marcos, Texas 78666*

BRUCE R. MCLEOD (49), *College of Graduate Studies, Montana State University, Bozeman, Montana 59717-2580*

BRIAN S. MILLER (32), *Genencor International Inc., Palo Alto, California 94304*

JACK H. MILLSTEIN (25), *Department of Microbiology, Molecular Biology, and Biochemistry, University of Idaho, Moscow, Idaho 83844-3052*

MARC W. MITTELMAN (39), *Altran Corporation, Boston, Massachusetts 02210*

SØREN MOLIN (2), *Department of Microbiology, Technical University of Denmark–DTU, DK-2800 Lyngby, Denmark*

SØREN MØLLER (26), *Enzyme Research, Novo Nordisk A/S, DK-2880 Bagsvaerd, Denmark*

NICOLA S. MORRIS (35), *Cardiff School of Biosciences, Cardiff University, Cardiff CF1 3TL, Wales, United Kingdom*

ICHIRO NAKAGAWA (36), *Department of Oral Microbiology, Osaka University Faculty of Dentistry, Suita-Osaka, 565-0871 Japan*

THOMAS R. NEU (9, 10), *Department of Inland Water Research, UFZ Centre for Environmental Research, 39114 Magdeburg, Germany*

DIANNE K. NEWMAN (6), *Department of Microbiology and Molecular Genetics, Harvard Medical School, Boston, Massachusetts 02115*

ALEX TOFTGAARD NIELSEN (2), *Department of Microbiology, Technical University of Denmark–DTU, DK-2800 Lyngby, Denmark*

TANJA NIEPEL (1), *Department of Microbiology, GBF–Gesellschaft für Biotechnologische Gesellschaft mbH, 38124 Braunschweig, Germany*

JAMES R. NIXON (42), *Withers Orthopaedic Centre, Musgrave Park Hospital, Belfast BT9 7JB, United Kingdom*

GEORGE A. O'TOOLE (6), *Department of Microbiology and Molecular Genetics, Harvard Medical School, Boston, Massachusetts 02115*

ROBERT J. PALMER, JR. (2, 11, 12, 18, 24), *Biofilm Imaging Facility, Center for Environmental Biotechnology, University of Tennessee, Knoxville, Tennessee 37932-2575*

MATTHEW R. PARSEK (3), *Department of Microbiology, University of Iowa, Iowa City, Iowa 52242*

SHEILA PATRICK (42), *Department of Microbiology and Immunobiology, School of Clinical Medicine, The Queen's University of Belfast, Belfast BT9 7BL, United Kingdom*

DONALD PHIPPS (14), *Biotechnology Research Department, Orange County Water District, Fountain Valley, California 92728-8300*

BETSEY PITTS (45), *Center for Biofilm Engineering, Montana State University, Bozeman, Montana 59717-3980*

LESLIE A. PRATT (6), *Department of Microbiology and Molecular Genetics, Harvard Medical School, Boston, Massachusetts 02115*

CLAIRE PRIGENT-COMBARET (4), *Laboratoire de Génétique Moléculaire des Microorganismes et des Interactions Cellulaires, CNRS UMR 5577, Institut National des Sciences Appliquées de Lyon, 69621 Villeurbanne, France*

E. QUINTERO (30), *Department of Biology Science, Universidad Santa Maria La Antigua, Panama*

ROBERT G. QUIVEY, JR. (33), *Department of Microbiology and Immunology, and Center for Oral Biology, University of Rochester Medical Center, Rochester, New York 14642*

GORDON RAMAGE (42), *Department of Microbiology and Immunobiology, and School of Pharmacy, The Queen's University of Belfast, Belfast BT12 6BN, United Kingdom*

GREGOR REID (31), *Lawson Research Institute, and Department of Microbiology and Immunology, University of Western Ontario, London, Ontario, Canada N6A 4V2*

HARRY F. RIDGWAY (14, 34), *Biotechnology Research Department, Orange County Water District, Fountain Valley, California 92728-8300*

A. D. G. ROBERTS (46), *Centre for Applied Microbiology and Research (CAMR), Salisbury, Wiltshire SP4 0JG, United Kingdom*

GRISEL RODRIGUEZ (14, 34), *Biotechnology Research Department, Orange County Water District, Fountain Valley, California 92728-8300*

M. A. ROPER (46), *Building Services Research and Information Association, Bracknell, Berkshire RG1R 7AH, United Kingdom*

PAUL G. ROUXHET (28), *Unité de Chimie des Interfaces, Université Catholique de Louvain, Louvain-la-Neuve 1348, Belgium*

JANA SAFARIK (34), *Biotechnology Research Department, Orange County Water District, Fountain Valley, California 92728-8300*

WOLFGANG SAND (27), *Abteilung Mikrobiologie, Institut für Allgemeine Botanik und Botanischer Garten, Universität Hamburg, D-22609 Hamburg, Germany*

E. SEAGREN (30), *Department of Civil Engineering, University of Maryland, College Park, Maryland 20742-0001*

MUKUL M. SHARMA (20), *Department of Petroleum and Geosystems Engineering, University of Texas, Austin, Texas 78712*

ROBERT B. SIMMONS (40), *Georgia State University, Atlanta, Georgia 30302-4010*

CLAUS STERNBERG (2), *Department of Microbiology, Technical University of Denmark–DTU, DK-2800 Lyngby, Denmark*

PHILIP S. STEWART (13, 49), *Department of Chemical Engineering and Center for Biofilm Engineering, Montana State University, Bozeman, Montana 59717*

DAVID J. STICKLER (35), *Cardiff School of Biosciences, Cardiff University, Cardiff CF1 3TL, Wales, United Kingdom*

PAUL STURMAN (45), *Center for Biofilm Engineering, Montana State University, Bozeman, Montana 59717-3980*

RUEY-JING TANG (47), *Environmental, Coastal and Ocean Sciences, University of Massachusetts, Boston, Massachusetts 02125-3393*

LIN TAO (7), *Department of Oral Biology, College of Dentistry, University of Illinois, Chicago, Illinois 60612-7213*

MICHAEL M. TUNNEY (41, 42), *Medical Devices Group, School of Pharmacy, and Department of Microbiology and Immunobiology, School of Clinical Medicine, The Queen's University of Belfast, Belfast BT9 7BL, United Kingdom*

HENNY C. VAN DER MEI (31, 38), *Department of Biomedical Engineering, University of Groningen, 9797kZ Groningen, The Netherlands*

MARTINE VELRAEDS (31), *Laboratory for Materia Technica, University of Groningen, 9797kZ Groningen, The Netherlands*

HENRY VON RÈGE (27), *Abteilung Mikrobiologie, Institut für Allgemeine Botanik und Botanischer Garten, Universität Hamburg, D-22609 Hamburg, Germany*

DIANE WALKER (45), *Center for Biofilm Engineering, Montana State University, Bozeman, Montana 59717-3980*

JAMES T. WALKER (46), *Centre for Applied Microbiology and Research (CAMR), Salisbury, Wiltshire SP4 0JG, United Kingdom*

PAULA I. WATNICK (6), *Department of Microbiology and Molecular Genetics, Harvard Medical School, Boston, Massachusetts 02115*

VALERIE B. WEAVER (6), *Department of Microbiology and Molecular Genetics, Harvard Medical School, Boston, Massachusetts 02115*

RONALD WEINER (30), *Department of Cell Biology and Molecular Genetics, University of Maryland, College Park, Maryland 20742-0001*

DAVID C. WHITE (11, 18), *Center for Environmental Biotechnology, University of Tennessee, Knoxville, Tennessee 37932-2575*

MARVIN WHITELEY (20), *Department of Microbiology, College of Medicine, University of Iowa, Iowa City, Iowa 52242*

MICHAEL WILSON (21), *Department of Microbiology, Faculty of Clinical Sciences, and Eastman Dental Institute, University College London, WC1X 8LD, United Kingdom*

JULIAN WIMPENNY (23), *Cardiff School of Biosciences, Cardiff University, Cardiff CF1 3TL, Wales, United Kingdom*

CAROLE WINTERS (35), *Cardiff School of Biosciences, Cardiff University, Cardiff CF1 3TL, Wales, United Kingdom*

GIDEON M. WOLFAARDT (1), *Department of Microbiology, University of Stellenbosch, 7600 Stellenbosch, South Africa*

ROSEMARY WU (24), *Oral Infection and Immunity Branch, National Institute of Dental and Craniofacial Research, National Institutes of Health, Bethesda, Maryland 20892-4350*

HIROSHI YASUDA (43), *Biological Research Laboratories, Sankyo Company, Limited, Aoba-ku, Yokohama 225-0024, Japan*

NICK ZELVER (45), *Center for Biofilm Engineering, Montana State University, Bozeman, Montana 59717-3980*

MANFRED S. ZINN (18), *Center for Environmental Biotechnology, University of Tennessee, Knoxville, Tennessee 37932-2575*

Preface

In the past 30–40 years, studies on microbial adhesion have spawned new journals, books, and thousands of journal articles. The studies have ranged from phenomenological, such as determination of adhesin specificities, to molecular biological, such as regulation of a particular adhesin phenotype. In some cases, adhesion has led to colonization. In turn, colonization has often led to biofilm development. Biofilm research is now an established field, depending on all the modern tools of molecular biology and physical instrumentation.

This is the first volume devoted solely to biofilm research methods. Its goal is to provide a contemporary source book for virtually any kind of experimental approach involving biofilms. It is certain that new approaches to biofilm studies will be developed in the future. This volume, however, will be a starting place for future developments.

A special effort was made to include bioengineering, molecular, genetic, microscopic, chemical, continuous culture, and physical methods in the volume. Only when directly relevant to biofilms were methods on adhesion considered.

I thank Atha Carter and Jan Powars for their superb assistance in preparation of the volume. I also thank the world's leaders in biofilm research for contributing to the volume. I depended on guidance from Shirley Light of Academic Press for editorial advice.

RON J. DOYLE

METHODS IN ENZYMOLOGY

Volume I. Preparation and Assay of Enzymes
Edited by Sidney P. Colowick and Nathan O. Kaplan

Volume II. Preparation and Assay of Enzymes
Edited by Sidney P. Colowick and Nathan O. Kaplan

Volume III. Preparation and Assay of Substrates
Edited by Sidney P. Colowick and Nathan O. Kaplan

Volume IV. Special Techniques for the Enzymologist
Edited by Sidney P. Colowick and Nathan O. Kaplan

Volume V. Preparation and Assay of Enzymes
Edited by Sidney P. Colowick and Nathan O. Kaplan

Volume VI. Preparation and Assay of Enzymes (*Continued*)
Preparation and Assay of Substrates
Special Techniques
Edited by Sidney P. Colowick and Nathan O. Kaplan

Volume VII. Cumulative Subject Index
Edited by Sidney P. Colowick and Nathan O. Kaplan

Volume VIII. Complex Carbohydrates
Edited by Elizabeth F. Neufeld and Victor Ginsburg

Volume IX. Carbohydrate Metabolism
Edited by Willis A. Wood

Volume X. Oxidation and Phosphorylation
Edited by Ronald W. Estabrook and Maynard E. Pullman

Volume XI. Enzyme Structure
Edited by C. H. W. Hirs

Volume XII. Nucleic Acids (Parts A and B)
Edited by Lawrence Grossman and Kivie Moldave

Volume XIII. Citric Acid Cycle
Edited by J. M. Lowenstein

Volume XIV. Lipids
Edited by J. M. Lowenstein

Volume XV. Steroids and Terpenoids
Edited by Raymond B. Clayton

Volume XVI. Fast Reactions
Edited by Kenneth Kustin

VOLUME XVII. Metabolism of Amino Acids and Amines (Parts A and B)
Edited by HERBERT TABOR AND CELIA WHITE TABOR

VOLUME XVIII. Vitamins and Coenzymes (Parts A, B, and C)
Edited by DONALD B. MCCORMICK AND LEMUEL D. WRIGHT

VOLUME XIX. Proteolytic Enzymes
Edited by GERTRUDE E. PERLMANN AND LASZLO LORAND

VOLUME XX. Nucleic Acids and Protein Synthesis (Part C)
Edited by KIVIE MOLDAVE AND LAWRENCE GROSSMAN

VOLUME XXI. Nucleic Acids (Part D)
Edited by LAWRENCE GROSSMAN AND KIVIE MOLDAVE

VOLUME XXII. Enzyme Purification and Related Techniques
Edited by WILLIAM B. JAKOBY

VOLUME XXIII. Photosynthesis (Part A)
Edited by ANTHONY SAN PIETRO

VOLUME XXIV. Photosynthesis and Nitrogen Fixation (Part B)
Edited by ANTHONY SAN PIETRO

VOLUME XXV. Enzyme Structure (Part B)
Edited by C. H. W. HIRS AND SERGE N. TIMASHEFF

VOLUME XXVI. Enzyme Structure (Part C)
Edited by C. H. W. HIRS AND SERGE N. TIMASHEFF

VOLUME XXVII. Enzyme Structure (Part D)
Edited by C. H. W. HIRS AND SERGE N. TIMASHEFF

VOLUME XXVIII. Complex Carbohydrates (Part B)
Edited by VICTOR GINSBURG

VOLUME XXIX. Nucleic Acids and Protein Synthesis (Part E)
Edited by LAWRENCE GROSSMAN AND KIVIE MOLDAVE

VOLUME XXX. Nucleic Acids and Protein Synthesis (Part F)
Edited by KIVIE MOLDAVE AND LAWRENCE GROSSMAN

VOLUME XXXI. Biomembranes (Part A)
Edited by SIDNEY FLEISCHER AND LESTER PACKER

VOLUME XXXII. Biomembranes (Part B)
Edited by SIDNEY FLEISCHER AND LESTER PACKER

VOLUME XXXIII. Cumulative Subject Index Volumes I–XXX
Edited by MARTHA G. DENNIS AND EDWARD A. DENNIS

VOLUME XXXIV. Affinity Techniques (Enzyme Purification: Part B)
Edited by WILLIAM B. JAKOBY AND MEIR WILCHEK

VOLUME XXXV. Lipids (Part B)
Edited by JOHN M. LOWENSTEIN

Volume XXXVI. Hormone Action (Part A: Steroid Hormones)
Edited by Bert W. O'Malley and Joel G. Hardman

Volume XXXVII. Hormone Action (Part B: Peptide Hormones)
Edited by Bert W. O'Malley and Joel G. Hardman

Volume XXXVIII. Hormone Action (Part C: Cyclic Nucleotides)
Edited by Joel G. Hardman and Bert W. O'Malley

Volume XXXIX. Hormone Action (Part D: Isolated Cells, Tissues, and Organ Systems)
Edited by Joel G. Hardman and Bert W. O'Malley

Volume XL. Hormone Action (Part E: Nuclear Structure and Function)
Edited by Bert W. O'Malley and Joel G. Hardman

Volume XLI. Carbohydrate Metabolism (Part B)
Edited by W. A. Wood

Volume XLII. Carbohydrate Metabolism (Part C)
Edited by W. A. Wood

Volume XLIII. Antibiotics
Edited by John H. Hash

Volume XLIV. Immobilized Enzymes
Edited by Klaus Mosbach

Volume XLV. Proteolytic Enzymes (Part B)
Edited by Laszlo Lorand

Volume XLVI. Affinity Labeling
Edited by William B. Jakoby and Meir Wilchek

Volume XLVII. Enzyme Structure (Part E)
Edited by C. H. W. Hirs and Serge N. Timasheff

Volume XLVIII. Enzyme Structure (Part F)
Edited by C. H. W. Hirs and Serge N. Timasheff

Volume XLIX. Enzyme Structure (Part G)
Edited by C. H. W. Hirs and Serge N. Timasheff

Volume L. Complex Carbohydrates (Part C)
Edited by Victor Ginsburg

Volume LI. Purine and Pyrimidine Nucleotide Metabolism
Edited by Patricia A. Hoffee and Mary Ellen Jones

Volume LII. Biomembranes (Part C: Biological Oxidations)
Edited by Sidney Fleischer and Lester Packer

Volume LIII. Biomembranes (Part D: Biological Oxidations)
Edited by Sidney Fleischer and Lester Packer

Volume LIV. Biomembranes (Part E: Biological Oxidations)
Edited by Sidney Fleischer and Lester Packer

VOLUME LV. Biomembranes (Part F: Bioenergetics)
Edited by SIDNEY FLEISCHER AND LESTER PACKER

VOLUME LVI. Biomembranes (Part G: Bioenergetics)
Edited by SIDNEY FLEISCHER AND LESTER PACKER

VOLUME LVII. Bioluminescence and Chemiluminescence
Edited by MARLENE A. DELUCA

VOLUME LVIII. Cell Culture
Edited by WILLIAM B. JAKOBY AND IRA PASTAN

VOLUME LIX. Nucleic Acids and Protein Synthesis (Part G)
Edited by KIVIE MOLDAVE AND LAWRENCE GROSSMAN

VOLUME LX. Nucleic Acids and Protein Synthesis (Part H)
Edited by KIVIE MOLDAVE AND LAWRENCE GROSSMAN

VOLUME 61. Enzyme Structure (Part H)
Edited by C. H. W. HIRS AND SERGE N. TIMASHEFF

VOLUME 62. Vitamins and Coenzymes (Part D)
Edited by DONALD B. MCCORMICK AND LEMUEL D. WRIGHT

VOLUME 63. Enzyme Kinetics and Mechanism (Part A: Initial Rate and Inhibitor Methods)
Edited by DANIEL L. PURICH

VOLUME 64. Enzyme Kinetics and Mechanism (Part B: Isotopic Probes and Complex Enzyme Systems)
Edited by DANIEL L. PURICH

VOLUME 65. Nucleic Acids (Part I)
Edited by LAWRENCE GROSSMAN AND KIVIE MOLDAVE

VOLUME 66. Vitamins and Coenzymes (Part E)
Edited by DONALD B. MCCORMICK AND LEMUEL D. WRIGHT

VOLUME 67. Vitamins and Coenzymes (Part F)
Edited by DONALD B. MCCORMICK AND LEMUEL D. WRIGHT

VOLUME 68. Recombinant DNA
Edited by RAY WU

VOLUME 69. Photosynthesis and Nitrogen Fixation (Part C)
Edited by ANTHONY SAN PIETRO

VOLUME 70. Immunochemical Techniques (Part A)
Edited by HELEN VAN VUNAKIS AND JOHN J. LANGONE

VOLUME 71. Lipids (Part C)
Edited by JOHN M. LOWENSTEIN

VOLUME 72. Lipids (Part D)
Edited by JOHN M. LOWENSTEIN

Volume 73. Immunochemical Techniques (Part B)
Edited by John J. Langone and Helen Van Vunakis

Volume 74. Immunochemical Techniques (Part C)
Edited by John J. Langone and Helen Van Vunakis

Volume 75. Cumulative Subject Index Volumes XXXI, XXXII, XXXIV–LX
Edited by Edward A. Dennis and Martha G. Dennis

Volume 76. Hemoglobins
Edited by Eraldo Antonini, Luigi Rossi-Bernardi, and Emilia Chiancone

Volume 77. Detoxication and Drug Metabolism
Edited by William B. Jakoby

Volume 78. Interferons (Part A)
Edited by Sidney Pestka

Volume 79. Interferons (Part B)
Edited by Sidney Pestka

Volume 80. Proteolytic Enzymes (Part C)
Edited by Laszlo Lorand

Volume 81. Biomembranes (Part H: Visual Pigments and Purple Membranes, I)
Edited by Lester Packer

Volume 82. Structural and Contractile Proteins (Part A: Extracellular Matrix)
Edited by Leon W. Cunningham and Dixie W. Frederiksen

Volume 83. Complex Carbohydrates (Part D)
Edited by Victor Ginsburg

Volume 84. Immunochemical Techniques (Part D: Selected Immunoassays)
Edited by John J. Langone and Helen Van Vunakis

Volume 85. Structural and Contractile Proteins (Part B: The Contractile Apparatus and the Cytoskeleton)
Edited by Dixie W. Frederiksen and Leon W. Cunningham

Volume 86. Prostaglandins and Arachidonate Metabolites
Edited by William E. M. Lands and William L. Smith

Volume 87. Enzyme Kinetics and Mechanism (Part C: Intermediates, Stereochemistry, and Rate Studies)
Edited by Daniel L. Purich

Volume 88. Biomembranes (Part I: Visual Pigments and Purple Membranes, II)
Edited by Lester Packer

Volume 89. Carbohydrate Metabolism (Part D)
Edited by Willis A. Wood

Volume 90. Carbohydrate Metabolism (Part E)
Edited by Willis A. Wood

VOLUME 91. Enzyme Structure (Part I)
Edited by C. H. W. HIRS AND SERGE N. TIMASHEFF

VOLUME 92. Immunochemical Techniques (Part E: Monoclonal Antibodies and General Immunoassay Methods)
Edited by JOHN J. LANGONE AND HELEN VAN VUNAKIS

VOLUME 93. Immunochemical Techniques (Part F: Conventional Antibodies, Fc Receptors, and Cytotoxicity)
Edited by JOHN J. LANGONE AND HELEN VAN VUNAKIS

VOLUME 94. Polyamines
Edited by HERBERT TABOR AND CELIA WHITE TABOR

VOLUME 95. Cumulative Subject Index Volumes 61–74, 76–80
Edited by EDWARD A. DENNIS AND MARTHA G. DENNIS

VOLUME 96. Biomembranes [Part J: Membrane Biogenesis: Assembly and Targeting (General Methods; Eukaryotes)]
Edited by SIDNEY FLEISCHER AND BECCA FLEISCHER

VOLUME 97. Biomembranes [Part K: Membrane Biogenesis: Assembly and Targeting (Prokaryotes, Mitochondria, and Chloroplasts)]
Edited by SIDNEY FLEISCHER AND BECCA FLEISCHER

VOLUME 98. Biomembranes (Part L: Membrane Biogenesis: Processing and Recycling)
Edited by SIDNEY FLEISCHER AND BECCA FLEISCHER

VOLUME 99. Hormone Action (Part F: Protein Kinases)
Edited by JACKIE D. CORBIN AND JOEL G. HARDMAN

VOLUME 100. Recombinant DNA (Part B)
Edited by RAY WU, LAWRENCE GROSSMAN, AND KIVIE MOLDAVE

VOLUME 101. Recombinant DNA (Part C)
Edited by RAY WU, LAWRENCE GROSSMAN, AND KIVIE MOLDAVE

VOLUME 102. Hormone Action (Part G: Calmodulin and Calcium-Binding Proteins)
Edited by ANTHONY R. MEANS AND BERT W. O'MALLEY

VOLUME 103. Hormone Action (Part H: Neuroendocrine Peptides)
Edited by P. MICHAEL CONN

VOLUME 104. Enzyme Purification and Related Techniques (Part C)
Edited by WILLIAM B. JAKOBY

VOLUME 105. Oxygen Radicals in Biological Systems
Edited by LESTER PACKER

VOLUME 106. Posttranslational Modifications (Part A)
Edited by FINN WOLD AND KIVIE MOLDAVE

VOLUME 107. Posttranslational Modifications (Part B)
Edited by FINN WOLD AND KIVIE MOLDAVE

VOLUME 108. Immunochemical Techniques (Part G: Separation and Characterization of Lymphoid Cells)
Edited by GIOVANNI DI SABATO, JOHN J. LANGONE, AND HELEN VAN VUNAKIS

VOLUME 109. Hormone Action (Part I: Peptide Hormones)
Edited by LUTZ BIRNBAUMER AND BERT W. O'MALLEY

VOLUME 110. Steroids and Isoprenoids (Part A)
Edited by JOHN H. LAW AND HANS C. RILLING

VOLUME 111. Steroids and Isoprenoids (Part B)
Edited by JOHN H. LAW AND HANS C. RILLING

VOLUME 112. Drug and Enzyme Targeting (Part A)
Edited by KENNETH J. WIDDER AND RALPH GREEN

VOLUME 113. Glutamate, Glutamine, Glutathione, and Related Compounds
Edited by ALTON MEISTER

VOLUME 114. Diffraction Methods for Biological Macromolecules (Part A)
Edited by HAROLD W. WYCKOFF, C. H. W. HIRS, AND SERGE N. TIMASHEFF

VOLUME 115. Diffraction Methods for Biological Macromolecules (Part B)
Edited by HAROLD W. WYCKOFF, C. H. W. HIRS, AND SERGE N. TIMASHEFF

VOLUME 116. Immunochemical Techniques (Part H: Effectors and Mediators of Lymphoid Cell Functions)
Edited by GIOVANNI DI SABATO, JOHN J. LANGONE, AND HELEN VAN VUNAKIS

VOLUME 117. Enzyme Structure (Part J)
Edited by C. H. W. HIRS AND SERGE N. TIMASHEFF

VOLUME 118. Plant Molecular Biology
Edited by ARTHUR WEISSBACH AND HERBERT WEISSBACH

VOLUME 119. Interferons (Part C)
Edited by SIDNEY PESTKA

VOLUME 120. Cumulative Subject Index Volumes 81–94, 96–101

VOLUME 121. Immunochemical Techniques (Part I: Hybridoma Technology and Monoclonal Antibodies)
Edited by JOHN J. LANGONE AND HELEN VAN VUNAKIS

VOLUME 122. Vitamins and Coenzymes (Part G)
Edited by FRANK CHYTIL AND DONALD B. MCCORMICK

VOLUME 123. Vitamins and Coenzymes (Part H)
Edited by FRANK CHYTIL AND DONALD B. MCCORMICK

VOLUME 124. Hormone Action (Part J: Neuroendocrine Peptides)
Edited by P. MICHAEL CONN

VOLUME 125. Biomembranes (Part M: Transport in Bacteria, Mitochondria, and Chloroplasts: General Approaches and Transport Systems)
Edited by SIDNEY FLEISCHER AND BECCA FLEISCHER

VOLUME 126. Biomembranes (Part N: Transport in Bacteria, Mitochondria, and Chloroplasts: Protonmotive Force)
Edited by SIDNEY FLEISCHER AND BECCA FLEISCHER

VOLUME 127. Biomembranes (Part O: Protons and Water: Structure and Translocation)
Edited by LESTER PACKER

VOLUME 128. Plasma Lipoproteins (Part A: Preparation, Structure, and Molecular Biology)
Edited by JERE P. SEGREST AND JOHN J. ALBERS

VOLUME 129. Plasma Lipoproteins (Part B: Characterization, Cell Biology, and Metabolism)
Edited by JOHN J. ALBERS AND JERE P. SEGREST

VOLUME 130. Enzyme Structure (Part K)
Edited by C. H. W. HIRS AND SERGE N. TIMASHEFF

VOLUME 131. Enzyme Structure (Part L)
Edited by C. H. W. HIRS AND SERGE N. TIMASHEFF

VOLUME 132. Immunochemical Techniques (Part J: Phagocytosis and Cell-Mediated Cytotoxicity)
Edited by GIOVANNI DI SABATO AND JOHANNES EVERSE

VOLUME 133. Bioluminescence and Chemiluminescence (Part B)
Edited by MARLENE DELUCA AND WILLIAM D. MCELROY

VOLUME 134. Structural and Contractile Proteins (Part C: The Contractile Apparatus and the Cytoskeleton)
Edited by RICHARD B. VALLEE

VOLUME 135. Immobilized Enzymes and Cells (Part B)
Edited by KLAUS MOSBACH

VOLUME 136. Immobilized Enzymes and Cells (Part C)
Edited by KLAUS MOSBACH

VOLUME 137. Immobilized Enzymes and Cells (Part D)
Edited by KLAUS MOSBACH

VOLUME 138. Complex Carbohydrates (Part E)
Edited by VICTOR GINSBURG

VOLUME 139. Cellular Regulators (Part A: Calcium- and Calmodulin-Binding Proteins)
Edited by ANTHONY R. MEANS AND P. MICHAEL CONN

VOLUME 140. Cumulative Subject Index Volumes 102–119, 121–134

VOLUME 141. Cellular Regulators (Part B: Calcium and Lipids)
Edited by P. MICHAEL CONN AND ANTHONY R. MEANS

VOLUME 142. Metabolism of Aromatic Amino Acids and Amines
Edited by SEYMOUR KAUFMAN

VOLUME 143. Sulfur and Sulfur Amino Acids
Edited by WILLIAM B. JAKOBY AND OWEN GRIFFITH

VOLUME 144. Structural and Contractile Proteins (Part D: Extracellular Matrix)
Edited by LEON W. CUNNINGHAM

VOLUME 145. Structural and Contractile Proteins (Part E: Extracellular Matrix)
Edited by LEON W. CUNNINGHAM

VOLUME 146. Peptide Growth Factors (Part A)
Edited by DAVID BARNES AND DAVID A. SIRBASKU

VOLUME 147. Peptide Growth Factors (Part B)
Edited by DAVID BARNES AND DAVID A. SIRBASKU

VOLUME 148. Plant Cell Membranes
Edited by LESTER PACKER AND ROLAND DOUCE

VOLUME 149. Drug and Enzyme Targeting (Part B)
Edited by RALPH GREEN AND KENNETH J. WIDDER

VOLUME 150. Immunochemical Techniques (Part K: *In Vitro* Models of B and T Cell Functions and Lymphoid Cell Receptors)
Edited by GIOVANNI DI SABATO

VOLUME 151. Molecular Genetics of Mammalian Cells
Edited by MICHAEL M. GOTTESMAN

VOLUME 152. Guide to Molecular Cloning Techniques
Edited by SHELBY L. BERGER AND ALAN R. KIMMEL

VOLUME 153. Recombinant DNA (Part D)
Edited by RAY WU AND LAWRENCE GROSSMAN

VOLUME 154. Recombinant DNA (Part E)
Edited by RAY WU AND LAWRENCE GROSSMAN

VOLUME 155. Recombinant DNA (Part F)
Edited by RAY WU

VOLUME 156. Biomembranes (Part P: ATP-Driven Pumps and Related Transport: The Na,K-Pump)
Edited by SIDNEY FLEISCHER AND BECCA FLEISCHER

VOLUME 157. Biomembranes (Part Q: ATP-Driven Pumps and Related Transport: Calcium, Proton, and Potassium Pumps)
Edited by SIDNEY FLEISCHER AND BECCA FLEISCHER

VOLUME 158. Metalloproteins (Part A)
Edited by JAMES F. RIORDAN AND BERT L. VALLEE

Volume 159. Initiation and Termination of Cyclic Nucleotide Action
Edited by Jackie D. Corbin and Roger A. Johnson

Volume 160. Biomass (Part A: Cellulose and Hemicellulose)
Edited by Willis A. Wood and Scott T. Kellogg

Volume 161. Biomass (Part B: Lignin, Pectin, and Chitin)
Edited by Willis A. Wood and Scott T. Kellogg

Volume 162. Immunochemical Techniques (Part L: Chemotaxis and Inflammation)
Edited by Giovanni Di Sabato

Volume 163. Immunochemical Techniques (Part M: Chemotaxis and Inflammation)
Edited by Giovanni Di Sabato

Volume 164. Ribosomes
Edited by Harry F. Noller, Jr., and Kivie Moldave

Volume 165. Microbial Toxins: Tools for Enzymology
Edited by Sidney Harshman

Volume 166. Branched-Chain Amino Acids
Edited by Robert Harris and John R. Sokatch

Volume 167. Cyanobacteria
Edited by Lester Packer and Alexander N. Glazer

Volume 168. Hormone Action (Part K: Neuroendocrine Peptides)
Edited by P. Michael Conn

Volume 169. Platelets: Receptors, Adhesion, Secretion (Part A)
Edited by Jacek Hawiger

Volume 170. Nucleosomes
Edited by Paul M. Wassarman and Roger D. Kornberg

Volume 171. Biomembranes (Part R: Transport Theory: Cells and Model Membranes)
Edited by Sidney Fleischer and Becca Fleischer

Volume 172. Biomembranes (Part S: Transport: Membrane Isolation and Characterization)
Edited by Sidney Fleischer and Becca Fleischer

Volume 173. Biomembranes [Part T: Cellular and Subcellular Transport: Eukaryotic (Nonepithelial) Cells]
Edited by Sidney Fleischer and Becca Fleischer

Volume 174. Biomembranes [Part U: Cellular and Subcellular Transport: Eukaryotic (Nonepithelial) Cells]
Edited by Sidney Fleischer and Becca Fleischer

Volume 175. Cumulative Subject Index Volumes 135–139, 141–167

VOLUME 176. Nuclear Magnetic Resonance (Part A: Spectral Techniques and Dynamics)
Edited by NORMAN J. OPPENHEIMER AND THOMAS L. JAMES

VOLUME 177. Nuclear Magnetic Resonance (Part B: Structure and Mechanism)
Edited by NORMAN J. OPPENHEIMER AND THOMAS L. JAMES

VOLUME 178. Antibodies, Antigens, and Molecular Mimicry
Edited by JOHN J. LANGONE

VOLUME 179. Complex Carbohydrates (Part F)
Edited by VICTOR GINSBURG

VOLUME 180. RNA Processing (Part A: General Methods)
Edited by JAMES E. DAHLBERG AND JOHN N. ABELSON

VOLUME 181. RNA Processing (Part B: Specific Methods)
Edited by JAMES E. DAHLBERG AND JOHN N. ABELSON

VOLUME 182. Guide to Protein Purification
Edited by MURRAY P. DEUTSCHER

VOLUME 183. Molecular Evolution: Computer Analysis of Protein and Nucleic Acid Sequences
Edited by RUSSELL F. DOOLITTLE

VOLUME 184. Avidin–Biotin Technology
Edited by MEIR WILCHEK AND EDWARD A. BAYER

VOLUME 185. Gene Expression Technology
Edited by DAVID V. GOEDDEL

VOLUME 186. Oxygen Radicals in Biological Systems (Part B: Oxygen Radicals and Antioxidants)
Edited by LESTER PACKER AND ALEXANDER N. GLAZER

VOLUME 187. Arachidonate Related Lipid Mediators
Edited by ROBERT C. MURPHY AND FRANK A. FITZPATRICK

VOLUME 188. Hydrocarbons and Methylotrophy
Edited by MARY E. LIDSTROM

VOLUME 189. Retinoids (Part A: Molecular and Metabolic Aspects)
Edited by LESTER PACKER

VOLUME 190. Retinoids (Part B: Cell Differentiation and Clinical Applications)
Edited by LESTER PACKER

VOLUME 191. Biomembranes (Part V: Cellular and Subcellular Transport: Epithelial Cells)
Edited by SIDNEY FLEISCHER AND BECCA FLEISCHER

VOLUME 192. Biomembranes (Part W: Cellular and Subcellular Transport: Epithelial Cells)
Edited by SIDNEY FLEISCHER AND BECCA FLEISCHER

VOLUME 193. Mass Spectrometry
Edited by JAMES A. MCCLOSKEY

VOLUME 194. Guide to Yeast Genetics and Molecular Biology
Edited by CHRISTINE GUTHRIE AND GERALD R. FINK

VOLUME 195. Adenylyl Cyclase, G Proteins, and Guanylyl Cyclase
Edited by ROGER A. JOHNSON AND JACKIE D. CORBIN

VOLUME 196. Molecular Motors and the Cytoskeleton
Edited by RICHARD B. VALLEE

VOLUME 197. Phospholipases
Edited by EDWARD A. DENNIS

VOLUME 198. Peptide Growth Factors (Part C)
Edited by DAVID BARNES, J. P. MATHER, AND GORDON H. SATO

VOLUME 199. Cumulative Subject Index Volumes 168–174, 176–194

VOLUME 200. Protein Phosphorylation (Part A: Protein Kinases: Assays, Purification, Antibodies, Functional Analysis, Cloning, and Expression)
Edited by TONY HUNTER AND BARTHOLOMEW M. SEFTON

VOLUME 201. Protein Phosphorylation (Part B: Analysis of Protein Phosphorylation, Protein Kinase Inhibitors, and Protein Phosphatases)
Edited by TONY HUNTER AND BARTHOLOMEW M. SEFTON

VOLUME 202. Molecular Design and Modeling: Concepts and Applications (Part A: Proteins, Peptides, and Enzymes)
Edited by JOHN J. LANGONE

VOLUME 203. Molecular Design and Modeling: Concepts and Applications (Part B: Antibodies and Antigens, Nucleic Acids, Polysaccharides, and Drugs)
Edited by JOHN J. LANGONE

VOLUME 204. Bacterial Genetic Systems
Edited by JEFFREY H. MILLER

VOLUME 205. Metallobiochemistry (Part B: Metallothionein and Related Molecules)
Edited by JAMES F. RIORDAN AND BERT L. VALLEE

VOLUME 206. Cytochrome P450
Edited by MICHAEL R. WATERMAN AND ERIC F. JOHNSON

VOLUME 207. Ion Channels
Edited by BERNARDO RUDY AND LINDA E. IVERSON

VOLUME 208. Protein–DNA Interactions
Edited by ROBERT T. SAUER

VOLUME 209. Phospholipid Biosynthesis
Edited by EDWARD A. DENNIS AND DENNIS E. VANCE

VOLUME 210. Numerical Computer Methods
Edited by LUDWIG BRAND AND MICHAEL L. JOHNSON

VOLUME 211. DNA Structures (Part A: Synthesis and Physical Analysis of DNA)
Edited by DAVID M. J. LILLEY AND JAMES E. DAHLBERG

VOLUME 212. DNA Structures (Part B: Chemical and Electrophoretic Analysis of DNA)
Edited by DAVID M. J. LILLEY AND JAMES E. DAHLBERG

VOLUME 213. Carotenoids (Part A: Chemistry, Separation, Quantitation, and Antioxidation)
Edited by LESTER PACKER

VOLUME 214. Carotenoids (Part B: Metabolism, Genetics, and Biosynthesis)
Edited by LESTER PACKER

VOLUME 215. Platelets: Receptors, Adhesion, Secretion (Part B)
Edited by JACEK J. HAWIGER

VOLUME 216. Recombinant DNA (Part G)
Edited by RAY WU

VOLUME 217. Recombinant DNA (Part H)
Edited by RAY WU

VOLUME 218. Recombinant DNA (Part I)
Edited by RAY WU

VOLUME 219. Reconstitution of Intracellular Transport
Edited by JAMES E. ROTHMAN

VOLUME 220. Membrane Fusion Techniques (Part A)
Edited by NEJAT DÜZGÜNEŞ

VOLUME 221. Membrane Fusion Techniques (Part B)
Edited by NEJAT DÜZGÜNEŞ

VOLUME 222. Proteolytic Enzymes in Coagulation, Fibrinolysis, and Complement Activation (Part A: Mammalian Blood Coagulation Factors and Inhibitors)
Edited by LASZLO LORAND AND KENNETH G. MANN

VOLUME 223. Proteolytic Enzymes in Coagulation, Fibrinolysis, and Complement Activation (Part B: Complement Activation, Fibrinolysis, and Nonmammalian Blood Coagulation Factors)
Edited by LASZLO LORAND AND KENNETH G. MANN

VOLUME 224. Molecular Evolution: Producing the Biochemical Data
Edited by ELIZABETH ANNE ZIMMER, THOMAS J. WHITE, REBECCA L. CANN, AND ALLAN C. WILSON

VOLUME 225. Guide to Techniques in Mouse Development
Edited by PAUL M. WASSARMAN AND MELVIN L. DEPAMPHILIS

VOLUME 226. Metallobiochemistry (Part C: Spectroscopic and Physical Methods for Probing Metal Ion Environments in Metalloenzymes and Metalloproteins)
Edited by JAMES F. RIORDAN AND BERT L. VALLEE

VOLUME 245. Extracellular Matrix Components
Edited by E. RUOSLAHTI AND E. ENGVALL

VOLUME 246. Biochemical Spectroscopy
Edited by KENNETH SAUER

VOLUME 247. Neoglycoconjugates (Part B: Biomedical Applications)
Edited by Y. C. LEE AND REIKO T. LEE

VOLUME 248. Proteolytic Enzymes: Aspartic and Metallo Peptidases
Edited by ALAN J. BARRETT

VOLUME 249. Enzyme Kinetics and Mechanism (Part D: Developments in Enzyme Dynamics)
Edited by DANIEL L. PURICH

VOLUME 250. Lipid Modifications of Proteins
Edited by PATRICK J. CASEY AND JANICE E. BUSS

VOLUME 251. Biothiols (Part A: Monothiols and Dithiols, Protein Thiols, and Thiyl Radicals)
Edited by LESTER PACKER

VOLUME 252. Biothiols (Part B: Glutathione and Thioredoxin; Thiols in Signal Transduction and Gene Regulation)
Edited by LESTER PACKER

VOLUME 253. Adhesion of Microbial Pathogens
Edited by RON J. DOYLE AND ITZHAK OFEK

VOLUME 254. Oncogene Techniques
Edited by PETER K. VOGT AND INDER M. VERMA

VOLUME 255. Small GTPases and Their Regulators (Part A: Ras Family)
Edited by W. E. BALCH, CHANNING J. DER, AND ALAN HALL

VOLUME 256. Small GTPases and Their Regulators (Part B: Rho Family)
Edited by W. E. BALCH, CHANNING J. DER, AND ALAN HALL

VOLUME 257. Small GTPases and Their Regulators (Part C: Proteins Involved in Transport)
Edited by W. E. BALCH, CHANNING J. DER, AND ALAN HALL

VOLUME 258. Redox-Active Amino Acids in Biology
Edited by JUDITH P. KLINMAN

VOLUME 259. Energetics of Biological Macromolecules
Edited by MICHAEL L. JOHNSON AND GARY K. ACKERS

VOLUME 260. Mitochondrial Biogenesis and Genetics (Part A)
Edited by GIUSEPPE M. ATTARDI AND ANNE CHOMYN

VOLUME 261. Nuclear Magnetic Resonance and Nucleic Acids
Edited by THOMAS L. JAMES

VOLUME 262. DNA Replication
Edited by JUDITH L. CAMPBELL

VOLUME 263. Plasma Lipoproteins (Part C: Quantitation)
Edited by WILLIAM A. BRADLEY, SANDRA H. GIANTURCO, AND JERE P. SEGREST

VOLUME 264. Mitochondrial Biogenesis and Genetics (Part B)
Edited by GIUSEPPE M. ATTARDI AND ANNE CHOMYN

VOLUME 265. Cumulative Subject Index Volumes 228, 230–262

VOLUME 266. Computer Methods for Macromolecular Sequence Analysis
Edited by RUSSELL F. DOOLITTLE

VOLUME 267. Combinatorial Chemistry
Edited by JOHN N. ABELSON

VOLUME 268. Nitric Oxide (Part A: Sources and Detection of NO; NO Synthase)
Edited by LESTER PACKER

VOLUME 269. Nitric Oxide (Part B: Physiological and Pathological Processes)
Edited by LESTER PACKER

VOLUME 270. High Resolution Separation and Analysis of Biological Macromolecules (Part A: Fundamentals)
Edited by BARRY L. KARGER AND WILLIAM S. HANCOCK

VOLUME 271. High Resolution Separation and Analysis of Biological Macromolecules (Part B: Applications)
Edited by BARRY L. KARGER AND WILLIAM S. HANCOCK

VOLUME 272. Cytochrome P450 (Part B)
Edited by ERIC F. JOHNSON AND MICHAEL R. WATERMAN

VOLUME 273. RNA Polymerase and Associated Factors (Part A)
Edited by SANKAR ADHYA

VOLUME 274. RNA Polymerase and Associated Factors (Part B)
Edited by SANKAR ADHYA

VOLUME 275. Viral Polymerases and Related Proteins
Edited by LAWRENCE C. KUO, DAVID B. OLSEN, AND STEVEN S. CARROLL

VOLUME 276. Macromolecular Crystallography (Part A)
Edited by CHARLES W. CARTER, JR., AND ROBERT M. SWEET

VOLUME 277. Macromolecular Crystallography (Part B)
Edited by CHARLES W. CARTER, JR., AND ROBERT M. SWEET

VOLUME 278. Fluorescence Spectroscopy
Edited by LUDWIG BRAND AND MICHAEL L. JOHNSON

VOLUME 279. Vitamins and Coenzymes (Part I)
Edited by DONALD B. MCCORMICK, JOHN W. SUTTIE, AND CONRAD WAGNER

VOLUME 280. Vitamins and Coenzymes (Part J)
Edited by DONALD B. MCCORMICK, JOHN W. SUTTIE, AND CONRAD WAGNER

VOLUME 281. Vitamins and Coenzymes (Part K)
Edited by DONALD B. MCCORMICK, JOHN W. SUTTIE, AND CONRAD WAGNER

VOLUME 302. Green Fluorescent Protein
Edited by P. MICHAEL CONN

VOLUME 303. cDNA Preparation and Display
Edited by SHERMAN M. WEISSMAN

VOLUME 304. Chromatin
Edited by PAUL M. WASSARMAN AND ALAN P. WOLFFE

VOLUME 305. Bioluminescence and Chemiluminescence (Part C) (in preparation)
Edited by MIRIAM M. ZIEGLER AND THOMAS O. BALDWIN

VOLUME 306. Expression of Recombinant Genes in Eukaryotic Systems (in preparation)
Edited by JOSEPH C. GLORIOSO AND MARTIN C. SCHMIDT

VOLUME 307. Confocal Microscopy
Edited by P. MICHAEL CONN

VOLUME 308. Enzyme Kinetics and Mechanism (Part E)
Edited by VERN L. SCHRAMM AND DANIEL L. PURICH

VOLUME 309. Amyloid, Prions, and Other Protein Aggregates
Edited by RONALD WETZEL

VOLUME 310. Biofilms
Edited by RON J. DOYLE

VOLUME 311. Sphingolipid Metabolism and Cell Signaling (Part A) (in preparation)
Edited by ALFRED H. MERRILL, JR., AND Y. A. HANNUN

VOLUME 312. Sphingolipid Metabolism and Cell Signaling (Part B) (in preparation)
Edited by ALFRED H. MERRILL, JR., AND Y. A. HANNUN

VOLUME 313. Antisense Technology (Part A: General Methods, Methods of Delivery, and RNA Studies) (in preparation)
Edited by M. IAN PHILLIPS

VOLUME 314. Antisense Technology (Part B: Applications) (in preparation)
Edited by M. IAN PHILLIPS

VOLUME 315. Vertebrate Phototransduction and the Visual Cycle (Part A) (in preparation)
Edited by KRZYSZTOF PALCZEWSKI

VOLUME 316. Vertebrate Phototransduction and the Visual Cycle (Part B) (in preparation)
Edited by KRZYSZTOF PALCZEWSKI

VOLUME 317. RNA-Ligand Interactions (Part A: Structural Biology Methods) (in preparation)
Edited by DANIEL W. CELANDER AND JOHN N. ABELSON

Volume 318. RNA-Ligand Interactions (Part B: Molecular Biology Methods) (in preparation)
Edited by Daniel W. Celander and John N. Abelson

Volume 319. Singlet Oxygen, UV-A, and Ozone (in preparation)
Edited by Lester Packer and Helmut Sies

Volume 320. Cumulative Subject Index Volumes 290–319 (in preparation)

Volume 321. Numerical Computer Methods (Part C) (in preparation)
Edited by Michael L. Johnson and Ludwig Brand

Volume 322. Apoptosis (in preparation)
Edited by John C. Reed

Section I

Molecular Biology of Biofilm Bacteria

[1] Reporter Systems for Microscopic Analysis of Microbial Biofilms

By DARREN R. KORBER, GIDEON M. WOLFAARDT, VOLKER BRÖZEL, RAYNARD MACDONALD, and TANJA NIEPEL

Introduction

Microorganisms flourish in most terrestrial and aquatic ecosystems as biofilms. The biofilm microenvironment is typified by a wide range of biological, chemical, and physical factors, the complexity of which is only now being appreciated. Such complex associations between microbial community members within biofilms have developed over evolutionary time in such a way that nearly all possible environments on earth support the growth of attached microorganisms. The interactions that occur between biofilms and their physical and chemical micro- and macroenvironment determine, to a large extent, the manner and success whereby these biological systems cycle nutrients, degrade toxicants, survive in hostile macroenvironments, and resist antimicrobial agents.

The examination of native biofilm communities was complicated greatly by the difficulty in identifying constituent biofilm members *in situ,* in quantifying physical, chemical, and spatial aspects of biofilms, and in linking particular processes and activity with specific biofilm bacteria. Technological and methodological advances, especially those enabling the nondestructive, direct analysis of native biofilms, have facilitated the task of delineating some of the processes driven by complex biofilm communities. The foundation of these developments has relied to a large extent on the maturation of ribosome-directed oligonucleotide probe methodology, the refinement of fluorescent molecular probes sensitive to either the microbial condition or the microbial microenvironment, as well as advances in microscopy and digital imaging systems. This article describes developments in microscopy-based, fluorescent reporters amenable for the spatial and temporal analysis of microbial phylogeny, antigenicity, physiology, and genetic control within complex biofilm systems.

Methodology

Flow Cell Systems for Study of Microbial Biofilms

Flow cell systems have proven highly valuable for the cultivation and on-line digital analysis of microbial biofilms. In general, flow cells permit

0076-6879/99 $30.00

the careful control of the laminar flow velocity as well as the type, concentration, and duration of exposure to nutrients or test compounds.[1–4] Multichannel flow cells allow convenient experimental replication and control, analysis of a range of treatments (e.g., flow velocity or substrate concentration), as well as the sacrifice of channels during destructive analyses.[5,6] Within our group, flow cells have seen use in a variety of studies, including investigations of microbial colonization behaviors,[7] flagellar motility,[8] bacterial attachment,[9] biofilm formation kinetics,[7,8] antimicrobial efficacy,[3,5] and degradative microbial communities.[1,4,6,10] Irrigation solutions used with flow cells may range from highly defined growth media, to liquids obtained directly from the environment, or water perfused over environmental sediments.

Flow cell designs have evolved from single-channel glass slide devices used during phase- and dark-field computer-enhanced microscopy studies[7,11,12] to multichannel flow cells constructed from 5-mm Lexan stock, 45 × 50 mm No. 1 glass coverslips, silicone tubing, and assembled using pourable silicone adhesive (RTV 110, GE Silicones, Waterford, NY).[1,3,5,10] The flexibility and low cost associated with the use of minimilling machines facilitates the construction of experiment-specific devices greatly when using substrates such as Lexan plastic (aluminum and stainless-steel devices may also be used for flow cell construction, but are more difficult and expensive to mill due to increased material hardness). Our group has designed and manufactured flow cells incorporating single channels, parallel multiple channels, multiple channels with different cross-sectional areas (providing different laminar flow velocities when a constant bulk flow rate is used),

[1] G. M. Wolfaardt, J. R. Lawrence, R. D. Robarts, and D. E. Caldwell, *Appl. Environ. Microbiol.* **60,** 434 (1994).

[2] D. R. Korber, A. Choi, G. M. Wolfaardt, and D. E. Caldwell, *Appl. Environ. Microbiol.* **62,** 3939 (1996).

[3] D. R. Korber, A. Choi, G. M. Wolfaardt, S. C. Ingham, and D. E. Caldwell, *Appl. Environ. Microbiol.* **63,** 3352 (1997).

[4] S. Møller, D. R. Korber, G. M. Wolfaardt, S. Molin, and D. E. Caldwell, *Appl. Environ. Microbiol.* **63,** 2432 (1997).

[5] D. R. Korber, G. A. James, and J. W. Costerton, *Appl. Environ. Microbiol.* **60,** 1663 (1994).

[6] G. M. Wolfaardt, J. R. Lawrence, R. D. Robarts, and D. E. Caldwell, *Microb. Ecol.* **35,** 213 (1998).

[7] D. E. Caldwell and J. R. Lawrence, *Microb. Ecol.* **12,** 299 (1986).

[8] D. R. Korber, J. R. Lawrence, B. Sutton, and D. E. Caldwell, *Microb. Ecol.* **18,** 1 (1989).

[9] D. R. Korber, J. R. Lawrence, L. Zhang, and D. E. Caldwell, *Biofouling* **2,** 335 (1990).

[10] G. M. Wolfaardt, J. R. Lawrence, J. V. Headley, R. D. Robarts, and D. E. Caldwell, *Microb. Ecol.* **27,** 279 (1994).

[11] D. E. Caldwell and J. R. Lawrence, *in* "CRC Handbook of Laboratory Model Systems for Microbial Ecology Research" (J. Wimpenny, ed.), p. 117. CRC Press, Boca Raton, FL, 1988.

[12] J. R. Lawrence, D. R. Korber, and D. E. Caldwell, *J. Microbiol. Methods* **10,** 123 (1989).

connected serpentine multiple channels (allowing sequential channel sacrifice), channels with small void volumes for use with oligonucleotide or immunological probes, channels with large volumes for fill with sediments or sands, channels with microcapillaries installed for pulse or plume delivery of test compounds, cleanable flow cells for the on-line study of plaque growth and regrowth, and devices fitted with porous ceramic plates for diffusion-controlled delivery of test compounds in one or two dimensions. Note that the flow cells should be constructed so that the device can be mounted directly to the microscope stage, thereby keeping the field of observation motionless, parallel with the stage, as well as facilitating temporal studies of the same field or repeated movement to different fields using a computer programmable microscope stage.

Detection of Biofilm Bacteria Using Ribosome-Directed Probes

Ecological studies of mixed-species biofilms ideally require the reliable localization of members of specific bacterial populations in relation to others, as well as information on the activity of individual cells. In the past, such information could only be obtained using light microscopy and was limited to the early stages of biofilm formation as well as constrained in terms of the amount and types of information that could be derived.[13] More detailed information on biofilm population structure could be obtained using disruptive techniques whereby the biofilm cells were scraped off a surface and then plated on selected media.[14] This approach allowed the determination of the relative abundance of different bacteria, but precluded determining any information on spatial relationships of different bacteria within the biofilm community and, more importantly, biased the results against organisms that could not be cultivated on agar plates or outside of the community context.

The application of molecular methods, including DNA base ratios, DNA–DNA hybridization, and 16S and 23S rDNA probes, has revolutionized the routine identification of bacteria from environmental and industrial samples.[15–17] Central to these developments are indicator-tagged (fluorescent, radiolabel, or antigen) rDNA probes[18] for the *in situ* hybridization to

[13] R. I. Amann, W. Ludwig, and K. H. Schleifer, *Microbiol. Rev.* **59,** 143 (1995).

[14] R. I. Amann, J. Stromley, R. Devereux, R. Key, and D. A. Stahl, *Appl. Environ. Microbiol.* **58,** 614 (1992).

[15] C. R. Woese, *Microbiol. Rev.* **51,** 221 (1987).

[16] K. H. Schleifer, W. Ludwig, and R. I. Amann, *Fresenius J. Anal. Chem.* **343,** 47 (1992).

[17] G. J. Olsen and C. R. Woese, *FASEB J.* **7,** 113 (1993).

[18] R. I. Amann, *in* "Molecular Microbial Ecology Manual" (A. D. L. Akkermans, J. D. van Elsas, and F. D. de Bruijn, eds.), p. 1. Kluwer, Dortrecht, 1995.

ribosomes. Phylogenetic fluorescent probes have the advantage of binding target ribosomes independent of the growth conditions (with the exception of severe low ribosome numbers seen in starved or very slowly growing cells) and may, under certain circumstances, be developed without the need for laboratory culture of a pure isolate.[13,18] Such probes are available for an ever-expanding list of 16S and 23S rDNA sequence data for specific bacterial species, genera, subdivisions, or divisions.[14,19–21] Thus, it is now possible to correlate the spatial distribution and abundance of specific biofilm members with their functional roles *in situ* (determined using gene expression vectors[22–24]).

Such *in situ* analyses of biofilm population structure are generally considered nondestructive, or at least minimally destructive, when fluorescence *in situ* hybridization (FISH) protocols are adopted.[25] These approaches have also undergone significant development in terms of sensitivity, specificity, and intrusiveness; FISH was once only suitable for the analysis of heat-fixed cell material on slides, but now is used routinely for the analysis of biofilms, sludge granules, sediments, and other biological materials.[19,23,26–32]

[19] T. J. DiChristina and E. F. DeLong, *Appl. Environ. Microbiol.* **59,** 4152 (1993).

[20] W. Manz, U. Szewzyk, P. Ericsson, R. Amann, K. H. Schleifer, and T. A. Stenström, *Appl. Environ. Microbiol.* **59,** 2293 (1993).

[21] J. B. Risatti, W. C. Capman, and D. A. Stahl, *Proc. Natl. Acad. Sci. U.S.A.* **91,** 10173 (1994).

[22] J. B. Andersen, C. Sternberg, L. K. Poulsen, S. P. Bjørn, M. Givskov, and S. Molin, *Appl. Environ. Microbiol.* **64,** 2240 (1998).

[23] S. Møller, C. Sternberg, J. B. Andersen, B. B. Christensen, J. L. Ramos, M. Givskov, and S. Molin, *Appl. Environ. Microbiol.* **64,** 721 (1998).

[24] S. Stretton, S. Techkarnjanaruk, A. M. McLennan, and A. E. Goodman, *Appl. Environ. Microbiol.* **64,** 2554 (1998).

[25] M. Wagner, P. Hutzler, and R. Amann, *in* "Digital Image Analysis of Microbes: Imaging, Morphometry, Fluorometry and Motility Techniques and Applications" (M. H. F. Wilkinson and F. Schut, eds.), p. 467. Wiley, New York, 1998.

[26] M. Wagner, B. Assmus, A. Hartmann, P. Hutzler, and R. Amann, *J. Microsc.* **176,** 181 (1994).

[27] M. Wagner, B. Assmus, A. Hartmann, P. Hutzler, N. Springer, and K.-H. Schleifer, *J. Microsc.* **176,** 251 (1994).

[28] M. Wagner, G. Rath, H.-P. Koops, J. Flood, and R. Amann, *Wat. Sci. Technol.* **34,** 237 (1996).

[29] W. Hönerlage, D. Hahn, and J. Zeyer, *Arch. Microbiol.* **163,** 235 (1995).

[30] A. Neef, A. Zaglauer, H. Meier, R. Amann, H. Lemmer, and K.-H. Schleifer, *Appl. Environ. Microbiol.* **62,** 4329 (1996).

[31] J. Snaidr, R. Amann, I. Huber, L. Ludwig, and K.-H. Schleifer, *Appl. Environ. Microbiol.* **63,** 2884 (1997).

[32] B. Zarda, D. Hahn, A. Chatzinotas, W. Schoenhuber, A. Neef, R. Amann, and J. Zeyer, *Arch. Microbiol.* **185,** (1997).

rRNA-Based Whole Cell in Situ Hybridization

By definition, whole cell *in situ* hybridization involves the identification of target sequences within specific cells by oligonucleotide probes.[33] Oligonucleotides (~20 nucleotides) are preferred to polynucleotides (~50 nucleotides) as they allow for single mismatch discrimination of target nucleic acids.[33] Ribosomal RNA is a suitable target for phylogenetic probes for a number of reasons: (1) ribosomes are ubiquitous. (2) Ribosomal sequences are functionally conserved molecules and are nontransferable between species. (3) The primary structures of 16S and 23S rRNA are composed of regions of higher and lower evolutionary conservation.[34,35] If microbial species or subspecies need to be distinguished, probe specificity can be adjusted freely to target the most variable regions of the molecule in contrast to highly conserved regions, which may be targeted for universal probes.[13,18,36–38] (4) Extensive rRNA sequence databases are available, especially for 16S rRNA, allowing the computer-assisted design and testing of oligonucleotide probes (i.e., ARB software[39]). (5) Ribosomes usually occur in high copy numbers throughout the cytoplasm (e.g., >1000 copies) so that the entire cell content becomes fluorescent on hybridization to a specific probe.[33] (6) The ribosome count per cell can be used to provide a measure of growth and activity of biofilm cells.[40]

Fluorochromes for Quantitative Fluorescence Microscopy

The principle conjugate fluorochromes are derivatives of fluorescein or rhodamine. Although these substances fluoresce brightly when excited with the appropriate wavelength, they show some serious problems, including pH sensitivity and/or rapid photoinactivation [e.g., fluorescein isothiocyanate (FITC)], even in the presence of antifading agents. This makes quantitative analysis with these fluorophores problematic. Cyanine dyes have been developed that are very photostable, and show narrow emission bands and

[33] R. I. Amann, F. O. Glöckner, and A. Neef, *FEMS Microbiol. Rev.* **20,** 191 (1997).

[34] G. E. Fox, K. R. Pechman, and C. R. Woese, *Int. J. Syst. Bacteriol.* **27,** 44 (1977).

[35] R. I. Amann, W. Ludwig, and K. H. Schleifer, *ASM News* **60,** 360 (1994).

[36] W. Manz, R. Amann, W. Ludwig, M. Wagner, and K. H. Schleifer, *Syst. Appl. Microbiol.* **15,** 593 (1992).

[37] G. Muyzer and N. B. Ramsing, *Wat. Sci. Tech.* **32,** 1 (1995).

[38] R. I. Amann, W. Ludwig, R. Schulze, S. Spring, E. Moore, and K. H. Schleifer, *Syst. Appl. Microbiol.* **19,** 501 (1996).

[39] O. Strunk and W. Ludwig, ARB software program package; htpp://www.biol.chemie.tu-muenchen.de/pub/ARB/ (1997).

[40] S. Møller, C. S. Kristensen, L. K. Poulsen, J. M. Carstensen, and S. Molin, *Appl. Environ. Microbiol.* **61,** 741 (1995).

high extinction coefficients.[41] The characteristics of some of the most common fluorescent conjugates are summarized in Table I.

It is important to note that when performing multilabeling experiments, it is essential to choose fluorophores whose excitation/emission wavelengths overlap minimally. It is also important to note that alternate markers, such as digoxigenin or biotin, can be used for *in situ* hybridization experiments. Such compounds can then be detected using enzyme-labeled antibodies, thereby increasing the overall sensitivity of the hybridization.[42] Using this approach, the probe-linked hapten can be visualized using an antibody–peroxidase conjugate and suitable substrate, which becomes fluorescent and is precipitated within the cell on hydrolysis.[43]

Fixation and Hybridization Procedures

Procedure 1. Analysis of Biofilm Bacteria Using rRNA Probes and Microscopy

1. Fix the biofilm with 4% (w/v) paraformaldeyde solution[44] for 1 hr at 4°.
2. Wash the biofilm carefully with phosphate-buffered saline [PBS; 130 m*M* sodium chloride, 10 m*M* sodium phosphate (pH 7.2)].
3. Dehydrate the biofilm in an ethanol series [50, 80, and 96% (v/v) for 3 min each].
4. Break off small pieces of the coverslip and position on slides with the biofilm side up.

[41] R. P. Haugland, *in* "Handbook of Fluorescent Probes and Research Chemicals." Molecular Probes, Eugene, OR 1996.

[42] B. Zarda, R. I. Amann, G. Wallner, and K.-H. Schleifer, *J. Gen. Microbiol.* **137,** 2823 (1991).

[43] R. I. Amann, B. Zarda, D. A. Stahl, and K. Schleifer, *Appl. Environ. Microbiol.* **58,** 3007 (1992).

[44] Over recent years, various fixation protocols have been evaluated for cell stabilization as well as cell membrane permeabilization for oligonucleotide probe penetration. While fixation with 4% formaldehyde is considered standard for gram-negative bacteria,[18] the *in situ* visualization of some groups of gram-positive species remains problematic. Because the hydrophobic cell walls of these bacteria are much more difficult to permeabilize, targets are not accessible to fluorescently labeled probes without pretreatment. Methods such as enzymatic digestion,[29,45] mild acid hydrolysis,[46] or combinations of enzymes and acids have been applied to increase the permeability of bacterial cells. Application of a 0.1% solution of lysozyme [1 mg (37, 320 U) in 1 ml of 100 m*M* Tris–HCl, pH 7.5, 5 m*M* EDTA at room temperature for 15 min] after fixation has been described for permeabilizing the outer membrane of the cell.[29] Bacteria should be washed with PBS solution, described in step 2, to halt enzymatic activity.

[45] M. Schuppler, M. Wagner, G. Schoen, and U. B. Goebel, *Microbiology* **144,** 249 (1998).

[46] S. J. MacNaughton, A. G. O'Donnel, and T. M. Embley, *Appl. Environ. Microbiol.* **62,** 4632 (1994).

TABLE I
FLUORESCENT DYES COMMONLY USED FOR FISH[a]

Fluorophore	Color of fluorescence	Excitation maximum (nm)	Emission maximum (nm)	Extinction coefficient (M^{-1} cm^{-1})
FITC	Green	494	520	—
Alexa 488	Green	490	520	62,000
Carboxytetramethylrhodamine	Orange	550	576	93,000
Cy3	Orange	550	565	150,000
Tetramethylrhodamine	Orange	555	580	80,000
Alexa 546	Orange	555	570	104,000
Texas Red	Red	596	615	85,000
Cy5	Far-red	649	670	250,000

[a] In combination with confocal laser microscopy listed by increasing excitation/emission wavelength.

5. Apply 9 μl of hybridization buffer [X% formamide (note that formamide concentration varies for different bacterial species[47]), 0.9 M NaCl, 0.01% sodium dodecyl sulfate (SDS), 5 mM EDTA, 20 mM Tris–HCl (pH 7.2)] to the fixed biofilms.
6. Add 1 μl of probe (25–50 ng) and incubate the slides in a moist chamber for 3 hr at 46°. Note that the addition of nonfluorescent competitor DNA material may be required to optimize staining efficacy. Note also that for staining with more than one oligonucleotide probe, both probes may be added to the sample simultaneously, provided that the formamide concentrations for the probes are the same. Multiple-labeled probes and multiple-labeled probe techniques are described in detail by Amann.[18]

[47] Formamide concentration is dependent on the characteristics of the oligonucleotide probe, including the number of bases, and the %GC ratio [Eq. (1)]:

$$T_d = 81.5 + 16.6 \log_{10}(M) + 0.41(\%GC) - 820/n \tag{1}$$

where M is NaCl molarity, %GC is the base composition as a percentage, and n is the probe length, allows the estimation of T_d, the temperature which provides half the maximal signal. T_d is then used to calculate formamide concentration (FA) [Eq. (2)]:

$$37° = T_d - 0.7\%FA \tag{2}$$

Optimizing the specificity and stringency of the probe[13] requires empirical confirmation using a range of FA concentrations (e.g., 0, 10, 20, 30, 40, 50), testing with appropriate nontarget controls, and so on.

7. Remove the probe gently by washing with buffer [30 min at 48°; *X* m*M* NaCl,[48] 20 m*M* Tris–HCl (pH 7.2), 0.01% SDS, 5 m*M* EDTA] prewarmed to the hybridization temperature.
8. Rinse with distilled water and dry at room temperature.
9. Mount the slides in glycerol/PBS (pH 8.5) (Citifluor, Chemistry Laboratory, University of Kent, Canterbury CT2 7NH, UK) in order to minimize photobleaching.

Procedure 2. In situ Analysis of Microbial Biofilms in Flow Cells Using rRNA Probes and Microscopy[49]

1. Grow biofilm bacteria in multichannel flow cells with small channel dimensions (i.e., 0.5 mm deep × 1 mm wide × 10 mm long) to minimize the amount of probe required for hybridization.
2. Cut silicone tubing on each side of the channel 1.27 cm from the edge of the Lexan base. Tubes should be plugged with sterile Teflon connectors to prevent leakage during fixation and hybridization. Introduction of air bubbles and high shear forces must be avoided at all stages of the procedure.
3. Perform steps 1–3 from Procedure 1 using sufficient volumes of fixative,[50] PBS, and ethanol to fill and/or rinse the entire flow cell channel. Solutions should be delivered slowly and carefully using a syringe and 27-gauge hypodermic needle. Note that when solutions are added, the downstream tubing plugs must be removed, after which they are replaced. Note that the flow cell channel should always be filled with liquid.

[48] The stringency of the washing step is also probe dependent. It is generally adjusted by changing the NaCl concentration in the washing buffer. In general, the higher the formamide concentration, the lower the concentration of NaCl (e.g., at 5% *FA*, 636 m*M* NaCl is used; at 30% *FA*, 112 m*M* NaCl is used).

[49] Some publications refer to the stabilization of biofilm material using polyacrylamide gels[23] or other materials. Our work has shown that this is not necessary in all cases. Biofilms imaged before and after FISH showed that the architecture and thickness of biofilms at reference positions (determined using SCLM and a computer-controlled microscope stage) remained unchanged during and after fixation and hydridization without the need for gel stabilization.

[50] Readers should also note that current studies failed to demonstrate a difference in the brightness of probe signal (using GPHGC, Eub 388, γ subclass of the class proteobacteria, and species-specific probes for *Aureobacterium* sp., *Terrabacter* sp., and *Cellulomonas* sp.) of native PCB-degrading biofilms fixed with 4% paraformaldehyde and then ethanol versus those simply stained with ethanol. Furthermore, little or no difference was noted between biofilms stained with the standard 50, 80, and 96% ethanol series and those simply stained with 96% ethanol for 3 min. It is suggested that readers evaluate the efficacy and necessity of excessive fixation, especially during analysis of biofilms where native structure is of importance.

4. Appropriate volumes, according to flow cell dimensions, of hydridization buffer and probe are added, ensuring that the volumes of the silicone tubing and chamber inlet are considered during calculations. For the flow cell described in step 1 of Procedure 2, ~99 μl of hybridization buffer and 1 μl probe (1000 ng/μl DNA for Cy3-labeled probes; 2000 ng/μl DNA for FITC-labeled probes) were sufficient. A 5 μl volume of 4′,6-diamidino-2-phenylindole dihydrochloride (DAPI) may be added directly to the probe-hybridization solution for total cell counterstaining. Once the flow cell channels have been resealed, the flow cell is placed in a 46° water bath for 3 hr.
5. Perform steps 7 and 8 (Procedure 1) using appropriate volumes of solutions, keeping the flow cell chamber fully hydrated at all times. At no time should an air–water liquid interface pass over the biofilm.
6. Thick biofilms are best viewed using scanning confocal laser microscopy (SCLM).

Potential Problems

Some methodological problems that may have an impact on the success of FISH protocols include cell permeability, target site accessibility, target site specificity, and sensitivity. The permeability of fixed cells may also be affected by their state of growth[51]; alterations in the cell wall structure of dormant cells (e.g., spores) increase their resistance to adverse environmental conditions.[52] Cell permeability may be enhanced using fixation methods relying on denaturants such as alcohols, cross-linkers such as formaldehyde or paraformaldehyde, solvents, acids, or cell wall degrading enzymes (see earlier discussion).

When the probes are labeled with high molecular weight enzymes such as horseradish peroxidase (HRP) (molecular mass ca. 44 kDa versus 330 Da for fluorescein), steric hindrance is increased.[53,54] Even once cells have been rendered permeable, there is no guarantee that hybridization will occur within the cell. This could be the result of the inaccessibility of the target sequence.[55] Amann *et al.*[13,18] discuss some solutions to this problem and provide an extensive list of accessible target sites. Another major problem, especially when probing cell material from natural systems, is the

[51] T. Boehnisch, *in* "Handbook of Immunochemical Staining Methods" (S. J. Naish, ed.), p. 2. Dako, Carpinteria, CA, 1989.

[52] D. B. Roszak and R. R. Colwell, *Microbiol. Rev.* **51,** 365 (1987).

[53] W. Schönhuber, B. Fuchs, S. Juretschko, and R. Amann, *Appl. Environ. Microbiol.* **63,** 3268 (1997).

[54] E. Bidnenko, C. Mercierm, J. Tremblay, P. Tailliez, and S. Kulakauskas, *Appl. Environ. Microbiol.* **64,** 3059 (1998).

[55] I. M. Head, J. R. Saunders, and R. W. Pickup, *Microb. Ecol.* **35,** 1 (1998).

signal strength of the probes. Cells in nutrient-limited biofilms often have a low ribosome count and the attainable signal per cell is weak, necessitating signal amplification.[56]

Enhancing Weak Signals. Industrial cooling water system biofilms are predominated by slowly growing bacteria in a low nutrient, saline environment. Low ribosome numbers per cell result in very weak signals following *in situ* hybridization. Treating these biofilms with chloramphenicol, glucose, and yeast extract for 3 hr prior to fixation improved the fluorescent signal-to-noise ratio by increasing the cell volume and ribosome content while concurrently suppressing cellular division.[57] A sterile solution of 5 g/liter glucose and 5 g/liter yeast extract can simply be incubated with the sample for 2 hr at 30°. Twenty microliters per milliliter of chloramphenicol (5 μg/μl) should then be incubated with the sample for an additional hour.

Multiple single-labeled probes, multiple-labeled probes, enzyme-linked probes, or detection systems have been used for signal amplification.[58,59] Lebaron *et al.*[60] used a tyramide signal amplification system with biotinylated oligonucleotide probes and streptavidin–horseradish peroxidase to increase the sensitivity of FISH. A 7- to 12-fold amplification of fluorescent signal was observed, but the signal was heterogeneous for mixed cells. Cell permeabilization efficiency was recognized as being responsible, as different cells have different permeabilization efficiencies. Therefore this would cause a problem if used in diverse natural systems.

Background Fluorescence. A problem often encountered during the study of natural biofilms (e.g., those from the environment, biofilms cultivated using natural materials such as soils, sediments, or sewage sludge) is autofluorescence, predominantly in the green range. This is thought to be the result of organic/inorganic particle autofluorescence, or even due to cellular interference is some cases.[37] Interfering material is often diffuse and displays high-intensity fluorescence, making target probe signal recognition, especially using fluorescein-labeled probes, difficult or impossible. Furthermore, autofluorescent debris is commonly resistant to photobleaching and generally insoluble in a range of detergents and solvents.

This problem has been overcome by hybridizing target and nontarget bacteria with a Cy5-labeled eubacterial probe (i.e., Eub 388), binding to all eubacterial cells and fluorescing in the far-red range. Cohybridized Cy3- or fluorescein-labeled target cells can then be visualized using dual channel

[56] L. K. Poulsen, G. Ballard, and D. A. Stahl, *Appl. Environ. Microbiol.* **59,** 1354 (1993).

[57] C. C. Ouverney and J. A. Fuhrman, *Appl. Environ. Microbiol.* **63,** 2735 (1997).

[58] S. Lee, C. Malone, and P. F. Kemp, *Mar. Ecol. Prog. Ser.* **101,** 193 (1993).

[59] R. Amann, R. Snaidr, M. Wagner, W. Ludwig, and K. H. Schleifer, *J. Bacteriol.* **178,** 3496 (1996).

[60] P. Lebaron, P. Catala, C. Fajon, F. Joux, J. Baudart, and L. Bernard, *Appl. Environ. Microbiol.* **63,** 3274 (1997).

SCLM. Cells detected in the far-red channel may then be assigned blue pseudocolor, whereas target cells fluorescing in the red or green channel are colored red or green. Target cells that hybridize with fluorescein-labeled probes thus appear turquoise, those hybridized to Cy3 appear purple, and all other cells appear green and hence may be disregarded. Alternatively, the total cell population may also be visualized using DAPI if an ultraviolet laser source is available.

Penetration Limitations. Some biofilms interfere with the free diffusion of fluorescent probes, especially where the concentration of glycocalyx-associated cations is high and cell exopolymer is abundant. This has been observed by the authors in industrial cooling system biofilms where high ambient calcium concentrations are typical. In such cases we have opted to scrape the biofilm off the surface, sacrificing spatial information for improved oligonucleotide hybridization, as described next.

1. Biofilms are scraped from the respective surface and diluted in 10 ml 1× PBS. A 5-ml volume of the cell suspension is centrifuged for 5 min (10,000*g*) at 20°. The supernatant is discarded.
2. A 100-μl volume of 0.5% (w/v) glucose and 0.5% (w/v) yeast extract is added to the sample and incubated for 2 hr at 30°. A 20-μl volume of chloramphenicol (5 μg/μl) is then added, followed by an additional 1-hr incubation.
3. The pellet is then washed twice in 1× PBS. Cells are then fixed for 3 hr at room temperature by adding 0.1 volume 37% (w/v) formaldehyde.
4. Cells are harvested and resuspended in 50 μl 1× PBS and 50 μl 96% ethanol and are stored at −20°.
5. A 3-μl volume of fixed cells in suspension is spotted onto gelatin-coated slides (see later) and air dried for 2 hr at 37°. Slides are washed consecutively in 50, 70, and 90% ethanol for 3 min, respectively.
6. Gene frames (1 cm^2 −20 μl) (Advanced Biotechnologies, Ltd., Epsom, Surrey, UK) are pasted onto slides to surround the fixed cell spots.
7. A 10-μl volume of prewarmed (46°) hybridization buffer [5× SET; 0.01% (w/v) SDS] is added to a polyester coverslip [1× SET contains 0.15 *M* NaCl, 1.0 m*M* EDTA, 20 m*M* Tris base (pH 7.8)], followed by the addition of fluorescent probes [3.5 μl in the case of fluorescein- and rhodamine-labeled probes, and 4.5 μl in the case of Cy5-labeled probes (50 ng/μl)].
8. Slides with adhesive frames are pressed onto a polyester cover, placed in a 50-ml polypropylene tube, and incubated at 46° for 18 hr. Unhybridized and nonspecifically bound probes are removed by washing the slides in prewarmed washing buffer [0.03 *M* NaCl, 0.2 m*M* EDTA,

4 m*M* Tris base (pH 7.8)] at 48° for 30 min. Slides are then stored at 4° in the dark.

Gelatin-coated slides are prepared by soaking in 10% (w/v) KOH in ethanol for 1 hr, rinsing with distilled H_2O, washing with ethanol, rinsing with distilled H_2O, and air drying. Slides are then soaked in a solution containing 0.1% (w/v) gelatin and 0.01% (w/v) $KCr(SO_4)_2$ at 70° for 30 min and air dried.

Future Prospects

Although *in situ* hybridization studies provide a number of advantages over culture-based methodologies, they are not without limitations. Further methodological refinements will likely see the development of even more highly specific probes. For example, the recent development of DNA analogs [i.e., peptide nucleic acids (PNAs)] have significant potential. PNAs contain neutral amide backbone linkages, resist degradation by enzymes, and hybridize complementary nucleic acid sequences with higher affinity than analogous DNA oligomers.[61] Although PNA development is still in its infancy, it may soon provide an attractive alternative for those currently using DNA-based hybridization methods.

Reporter Genes for Use in Biofilm Systems

There is increasing evidence that bacteria sense the proximity of a surface and respond by substantially upregulating a number of genes.[62,63] The regulation and expression of other genes of interest, notably within the spatial and temporal context of a complex microbial community, include those involved during the degradation of recalcitrant pollutant compounds (e.g., regulation of catabolic promoters from the TOL and NAH plasmids). Unfortunately, the direct and temporal quantification of specific gene products within biofilm bacteria is frequently problematic or impossible. Also, there is an important difference among oligonucleotide probes based on microbial phylogeny, oligonucleotide probes specific for genes encoding specific enzymes such as those involved in the degradation of aromatic pollutants (functional probes), and fluorescent gene reporters.

Linking the expression of attachment-inducible or biofilm-specific target genes to fluorescent reporter gene constructs provides an indirect method whereby the regulation of these genes may be analyzed quantitatively and, more importantly, over time as fluorescent reporters are nonintrusive and

[61] D. R. Corey, *Trends Biotechnol.* **15,** 226 (1997).
[62] D. G. Davies and G. G. Geesey, *Appl. Environ. Microbiol.* **61,** 860 (1995).
[63] G. A. O'Toole and R. Kolter, *Mol. Microbiol.* **28,** 449 (1998).

nondestructive. Reporter genes include, for example, the *lacZ* gene (encoding β-galactosidase requiring a fluorescent substrate analog such as fluorescein di-β-D-galactopyranoside), the *lux* or *luc* systems (for bacterial bioluminescence or firefly luciferase gene, respectively), or the green fluorescent protein (*gfp*) gene. These gene constructs are commercially available in various cloning vectors from molecular supply companies and may subsequently be placed under the control of a specific host promoter for study. Alternatively, immunodetection has seen application in catabolic gene regulation studies.[64] Studies utilizing these fluorescent reporter genes have been used to examine starvation gene expression[65] and community-level interactions involved in the induction of the meta-pathway promoter (*Pm*) from the TOL plasmid[23] and to monitor chitinase gene expression.[24]

Most reporter constructs produce fluorophores with an extended intracellular half-life (e.g., once formed, GFP is quite stable), consequently real-time measurements are either difficult or impossible as there is limited responsiveness to changes in the transcriptional activity of the target gene to which the reporter gene is fused. A range of GFP mutants has been constructed that display a short half-life in selected proteobacteria (pd1EGFP N1 $t_{1/2} = 1$ hr, pd2EGFP N1 $t_{1/2} = 2$ hr, pd4EGFP N1 $t_{1/2} = 4$ hr; Clontech, Palo Alto, CA). The excitation maximum of this enhanced GFP is at 488 nm and the emission maximum is at 507 nm. Others[22] have independently developed GFP variants with short half-lives by ligating short peptide sequences to the C-terminal end of intact *gfp*mut3 (a mutant GFP approximately 20 times more fluorescent than the wild-type GFP when excited by 488 nm light). These peptide sequences are cleaved by cellular proteases, resulting in the reduced duration of fluorescence output. Construction of *gfp*-based reporter systems is described by Suarez *et al.*[66]

Isolation of Biofilm-Specific Regulatory Elements

Attachment-inducible or biofilm-specific promoters can be isolated using various protocols. O'Toole and Kolter[63] have isolated such elements by screening an insertion mutant library and selecting clones not able to form biofilms at the wall of microtiter plate wells. Whereas this approach is effective, it fails to identify those genes which, although regulated by the biofilm mode of growth, are not essential for the growth as a biofilm. A second approach by Brözel *et al.*[67] is to construct a gene library in a reporter

[64] A. Cebolla, C. Guzmán, and V. de Lorenzo, *Appl. Environ. Microbiol.* **62,** 214 (1996).

[65] G. Wolfaardt, A. Matin, and D. Caldwell, unpublished data (1996).

[66] A. Suarez, A. Güttler, M. Strätz, L. H. Staendner, K. N. Timmis, and C. A. Guzmán, *Gene* **196,** 69 (1997).

[67] V. S. Brözel, G. M. Strydom, and T. E. Cloete, *Biofouling* **8,** 195 (1995).

vector with a promoterless β-galactosidase gene and to determine the expression in each clone while growing in suspension and as a biofilm. Expression can be determined quantitatively by lysing cells, adding a chromogenic substrate for β-galactosidase such as 2-nitrophenyl-β-D-galactopyranoside, and determining the hydrolyzate color spectrophotometrically (405 nm). This will yield all regulatory elements affected by the biofilm mode of growth. Glass wool in microfuge tubes may be used to supply a large surface area in a small volume to facilitate the sensitive detection of upregulated biofilm clones.

Green Fluorescent Protein

Since the heterologous expression of the green fluorescent protein gene was described by Chalfie *et al.*[68] in 1994, GFP has seen wide and varied usage, particularly in those studies utilizing scanning confocal laser microscopy and flow cytometry. Examples of GFP usage in ecological studies other than gene expression (see earlier discussion) include monitoring and quantifying horizontal and vertical plasmid transfer among different species of bacteria,[69–72] as well as to provide a conservative bacterial tracer for quantifying relative species abundance within biofilms.[73] For example, a *Pseudomonas putida* strain carrying a chromosomal *lacI*^{q} gene (repressor) as well as a plasmid encoded *lacp-gfp* construct was used to monitor the transfer of new genetic traits to other members of a mixed biofilm community.[70]

While the methodologies involved in the custom development of *gfp* clones (e.g., *gfp* gene expression vectors) require a degree of competence in molecular biology, random *gfp* mutations may be achieved using relatively simple conjugal matings or transformation protocols. *gfp* clones or vectors for these purposes are available from either research groups or commercial suppliers (Clontech; www.clontech.com). Furthermore, variants (red- or blue-shifted GFPs) with enhanced or modified spectral properties from that of the original GFP protein have also been developed, enhancing the signal-to-noise ratio as well as the wavelengths of excitation and emission. (For protocols on transformation of gram-negative bacteria with *gfp* constructs such as pEGFP from Clontech, see Ref. 73a.)

[68] M. Chalfie, Y. Tu, G. Euskirchen, W. W. Ward, and D.C. Prasher, *Science* **263,** 802 (1994).

[69] B. B. Christensen, C. Sternberg, and S. Molin, *Gene* **173,** 59 (1996).

[70] B. B. Christensen, C. Sternberg, J. B. Andersen, L. Eberl, S. Møller, M. Givskov, and S. Molin, *Appl. Environ. Microbiol.* **64,** 2247 (1998).

[71] C. Dahlberg, M. Bergström, and M. Hermansson, *Appl. Environ. Microbiol.* **64,** 2670 (1998).

[72] B. Normander, B. B. Christensen, S. Molin, and N. Kroer, *Appl. Environ. Microbiol.* **64,** 1902 (1998).

[73] S. Karthikeyan, G. Wolfaardt, D. Korber, and D. Caldwell, unpublished data (1998).

[73a] "Current Protocols in Molecular Biology." Wiley, New York.

Random, chromosomal *gfp* mutations can be obtained using a triparental mating system as described by Christensen *et al.*[69] In our studies, an *Escherichia coli* strain containing the 1.9-kb *Not*I fragment [harboring the RBSII-*gfp*(mut 3)-T_0-T_1-cassette] in the unique site of a pUT-kana vector,[74] an *E. coli* strain with the RK600 helper plasmid containing the rep/mob region of RP4, and a recipient gram-negative target strain was used as follows.

1. Wash 1 ml of overnight culture from each of the three strains two times in 0.9% (w/v) NaCl. Resuspend the cells in 50 μl of 0.9% (w/v) NaCl, combine, and transfer the mixture spotwise to one-fourth strength Luria broth (LB) plates. Incubate overnight.
2. Remove colonies from the plates aseptically and suspend in 100 μl of 0.9% NaCl. Three serial 10× dilutions should then be performed before plating on selective agar plates: minimal medium agar plates with 1 g/liter, citrate (*E. coli* cannot grow on citrate, whereas *Pseudomonas* spp. can[75]) supplemented with 50 μg/ml kanamycin (Km). Approximately 1–5% of Km-resistant clones should be GFP$^+$, with signal intensity varying with the strength of the promoter downstream of which the *gfp* gene has become inserted.
3. Plates are then screened using an epifluorescence microscope equipped with an FITC filter set and UV light source. A low-power lens (e.g., 6–10×) facilitates the screening of GFP$^+$ colonies.
4. Transfer GFP$^+$ bacteria to a new LB plate, aseptically and verify the presence of GFP$^+$ clones after 24–48 hr.

The identification of exconjugants bearing the *gfp* genotype and phenotype is dependent on the random insertion of the Tn*5* construct downstream from a strong chromosomal promoter. A number of studies have demonstrated that the isolation of *gfp* mutants without altered growth rates or characteristics may be obtained readily.

Fluorescent Antibodies

Fluorescent poly- and monoclonal antibodies offer sensitivity and flexibility for both the applied and the ecological study of microbial biofilms. Generally, fluorescent antibody methods are less complicated and less destructive than those involved with oligonucleotide probe hybridizations,

[74] V. de Lorenzo, M. Herrero, U. Jakubzik, and K. N. Timmis, *J. Bacteriol.* **172,** 6568 (1990).

[75] Note that if the recipient strain cannot grow on citrate, spontaneous rifampicin mutants (100 μg/ml^{-1}) need to be obtained. Cells from step 2 should thus be plated spotwise on LB plates supplemented with 50 μg/ml kanamycin as well as 100 μg/ml rifampicin.

the reagents are less expensive (in the case of polyclonals), and the signal-to-noise ratio is as good or better. The disadvantage to this method is that it is most useful as a detection method for known and well-characterized bacteria, and cross-reactivity between serologically related species must be eliminated as a potential source of error. However, immunological methods have and will continue to be a key method for the detection of pathogen and indicator organisms within industrial and medical settings. Within the context of complex microbial communities, antigenic relatedness need not necessarily be viewed as a limitation; class-specific fluorescent antibodies may provide a temporal index of overall change in biofilm community structure in response to stress.

The adaptation of fluorescent antibody methods[76] for the analysis of biofilms, rather than planktonic cells, has proven relatively straightforward. The following generalized procedure outlines the major steps in using polyclonal antibodies for the study of biofilms cultivated in flow cells.

1. Fix the biofilm with 1% (w/v) formaldehyde for 30 min–1 hr.
2. Wash with PBS or other appropriate buffer.
3. Block with 1% (w/v) bovine serum albumin (BSA) for 30 min–1 hr.
4. Incubate with primary antibody, diluted in 0.1% BSA for 30 min–1 hr. The dilution factor depends on the antibody titer. We have successfully used a 1:100 dilution of a 1:128,000 primary antibody titer solution.
5. Wash with PBS or other appropriate buffer.
6. Incubate with secondary (fluor-conjugated; a 1:100 dilution of a 1-mg/ml solution) antibody for 30 min–1 hr (note the same fluor conjugates may be used as described for FISH).
7. View using epifluorescence or scanning confocal microscopy equipped with the appropriate filter set.

Indicators of Bacterial Metabolic Condition

Reliable measurement of the metabolic activity of microbial cells from various environments represents a common goal of many ecological investigations. Somewhat overshadowed by developments in probing methods and the construction of novel reporter genes, the vast majority of industry-related biofilm interests remain focused on the killing, inactivation, or control of biofilm bacteria in various water distribution systems, pipelines, or processing plant equipment. On-line or model flow cell systems, which can be used to demonstrate the relative efficacy of various control strategies, continue to be the major thrust of many industrial research programs. Thus,

[76] J. Winkler, K. N. Timmis, and R. A. Snyder, *Appl. Environ. Microbiol.* **61,** 448 (1995).

the question is not so much what the phylogenetic affiliation of these organisms is or how their genes are regulated, but rather how many living bacteria are left following treatment and how long until they regrow. Generally, for native microbial biofilms consisting of a spectrum of different bacterial species, the variable permeability of the cell membrane of different bacteria (of different size, different growth rates, and differing intrinsic sensitivities) typically results in differential survival following exposure to an antimicrobial.

A number of fluorescent indicators have been examined for their sensitivity to cell parameters such as cell membrane permeability, cell membrane potential, enzymatic activity, and reduction potential and many were used in flow cytometric applications prior to the advent of SCLM.[77] For example, propidium iodide, cyanoditolyl tetrazolium chloride (CTC), rhodamine 123, acridine orange, and fluorescein diacetate have all seen application for determining cellular metabolic condition or viability. While approaches based on each of these parameters have specific value under certain study conditions, most have not proven generally applicable for all types of samples or analyses. However, the integrity of the cell membrane used in conjunction with nucleic acid probes with differential membrane permeabilities has served well as an indicator of cell viability.[2] Commercial preparations based on cell membrane integrity adequately reflect indices of cell viability in pure cultured and indigenous biofilms treated with various antimicrobial agents when compared with viable plate counts. It is also noteworthy that few alternatives are available for the routine measure of bacterial metabolism and viability for direct observation at the single cell level.

Regardless of which indicator of viability is being used, its application to biofilms cultivated in flow cells is generally the same. For example, a solution of the compound, prepared to the appropriate concentration, is simply pulsed into the flow cell, allowed to react for an appropriate period, and then washed out. Observation with either epifluorescence or scanning confocal laser microscopy, used in conjunction with a digital imaging program such as NIH image, may then be used to compare the relative numbers of bacteria that react positively or negatively with the probe. In some instances, a counterstain (e.g., DAPI, Syto 17) may be used to provide a measure of total bacterial numbers or biomass. For example, if CTC were being used to provide an index of actively respiring cells, counterstaining

[77] J. R. Lawrence, D. R. Korber, G. M. Wolfaardt, and D. E. Caldwell, *in* "Manual of Environmental Microbiology" (C. J. Hurst, G. R. Knudsen, M. J. McInerney, L. D. Stetzenback, and M. V. Walter, eds.), p. 79. ASM Press, Washington, DC, 1997.

with DAPI has been shown to be sufficient.[78] The two-component BacLight probe (i.e., BacLight Viability Probe, Molecular Probes, Inc., Eugene, OR) is composed of two nucleic acid stains: a membrane-impermeant, red fluorescing compound (propidium iodide; stains the dead cells) and a green fluorescent membrane-permeant compound (Syto 9; stains the living cells). The following protocol outlines the steps used by our group[2,3] for the analysis of biofilms cultivated in a 3 × 1 × 40-mm flow cell channel.

1. Prepare the two components of the BacLight probe in equal amounts in sterile media to a final concentration of 3 μg/ml.
2. Inject approximately 300 μl of the probe mixture into the flow cell chamber in the absence of flow, allowing approximately 15 min to react in darkness. It is advantageous to perform the staining directly on the stage of the photomicroscope or SCLM system, as this permits before and after comparisons to ensure that no air bubbles or liquid pulse disrupt the biofilm. Note that the reaction times must be kept constant, regardless of which type of probe is being used.
3. The fluor can then be washed from the flow cell by simply turning the pump, which delivers media, on or by pulsing fluor-free media into the flow cell.
4. Analyze microscopic fields using dual-channel SCLM and image analysis.

[78] J. J. Smith, J. P. Howington, and G. A. McFeters, *Appl. Environ. Microbiol.* **60,** 2977 (1994).

[2] Molecular Tools for Study of Biofilm Physiology

By Bjarke Bak Christensen, Claus Sternberg, Jens Bo Andersen, Robert J. Palmer, Jr., Alex Toftgaard Nielsen, Michael Givskov, and Søren Molin

Introduction

Most bacterial activity in nature occurs in microbial communities on substrata (biofilms). The life of such bacteria is, in many respects, quite different from the life of planktonic bacteria in suspension—the traditional laboratory scenario. Important differences between these two life forms are (1) biofilm communities develop internal heterogeneities, (2) structure/function relationships exist in biofilms that are important to biological activity, and (3) individual cells and entire communities respond to environmental changes. A number of molecular methods and tools useful in the

0076-6879/99 $30.00

field of microbial physiology and ecology, especially in the analysis of microbial community activity, have been developed. These methods allow detailed investigations, at the single-cell level, of bacterial physiological activities, specific gene expression, gene transfer, and cell-to-cell communication. Community features such as surface colonization, metabolic interactions, utilization of carbon and energy sources, bacterial motility, and microcolony structure/activity have also been analyzed in biofilm communities of different complexities.

In natural systems, biofilms are often located in places that are difficult to access, which makes direct analysis of the individual organisms very complicated. However, just as laboratory-based batch culture experiments have been used to study the physiological behavior of suspended bacteria, it is possible to obtain significant information about the bacterial behavior of organisms growing in a complex biofilm using simple experimental biofilm model systems. The use of the confocal microscope in combination with model systems provides a good basis for the on-line microscopic examination of biofilms and, combined with new genetic marker systems, such investigations facilitate detailed studies of biofilms at the single-cell level and at the community level.

This article describes methods for the handling and analysis of microbial behavior of organisms in biofilm communities at both microscopic and macroscopic levels. Only methods and reporter systems that can be applied without disturbing the spatial organization of the organisms in the biofilm are presented.

Monitoring Biofilm Development in Flow Cells

Numerous laboratory devices have been developed for studies of biofilm development (the rototorque bioreactor, the Robbins device, flow cells, etc.).[1,2] We have mainly worked with flow cells due to their relatively simple design and their applicability to direct on-line microscope examinations of the biofilms.

Traditional transmission light microscopy or epifluorescence microscopy may be used to follow biofilm development. However, as the biofilm thickness increases, it becomes more difficult to obtain good images due to the contribution from unfocused parts of the viewing field. The scanning

[1] D. E. Caldwell, "Microbial Biofilms" (H. M. Lappin-Scott and J. W. Costerton, eds.), p. 64. Cambridge Univ. Press, Cambridge, UK, 1995.

[2] R. J. Palmer, Jr., *Methods Enzymol.* [12] **310** (1998) (this volume).

confocal laser microscope (SCLM)[3–5] solves this problem by collecting returned fluorescent light from only the thinnest focal plane afforded by the objective lens. By scanning several planes interspersed by short distances, it is possible to reconstruct virtual three-dimensional images of the biofilm. Although most commercial confocal software permits such rendering, the highest quality renderings come from much more powerful Unix-based platforms/software such as IMARIS (Bit-Plane AG, Zurich, Switzerland). Laser microscopy is an epifluorescence in which digital images are acquired by photomultiplier tube detection of fluorescence excited by a laser light source. In some cases, organisms are slightly autofluorescent and can be visualized in an unstained biofilm. However, autofluorescence is often weak, and normally only a small fraction of a total biofilm population is autofluorescent. Fluorescent dyes that stain either the cells in the biofilm or their immediate surroundings (negative stains) may also be used to label bacteria in biofilms. Some stains (such as fluorescein) have a minimal effect on biofilm growth and may be used in a nondestructive way.[6,7] Other dyes often used for the identification of bacteria are acridine orange, 4′,6′-diamidino-2-phenylindole (DAPI), and mitramycin that bind to the DNA or the RNA in the organisms, but most of these seem to have adverse side effects and may inhibit the biofilm growth more or less dependent on the organisms. In general, these dyes seem to work better if the biofilm cells are fixed before the dyes are introduced.

Design of Flow System

The flow cells are, with only a few modifications, the same as those described previously by Wolfaardt *et al.*[8] We have introduced a bubble trap at the influent side of the flow cells to prevent air bubbles from reaching the biofilms (Fig. 1). After overnight sterilization with 0.5% (v/v) hypochlorite and a distilled H_2O rinse, the inoculum (an overnight culture or sample from a bioreactor) is diluted to a final OD_{450} of about 0.05 and injected using a syringe (0.25 ml liquid culture) at the inlet of the flow cell while the pump is turned off and the flow cell is inverted (cover glass turned downward). The influent line is clamped to prevent back-growth during

[3] G. J. Brakenhoff, H. T. v.d. Voort, M. W. Baarslag, B. Mans, J. L. Oud, R. Zwart, and R. v. Driel, *Scan. Microsc.* **2,** 1831 (1988).

[4] D. E. Caldwell, D. R. Korber, and J. R. Lawrence, "Advances in Microbial Ecology" (K. C. Marshall, ed.), Vol. 12, p. 1. Plenum Press, New York, 1992.

[5] D. E. Caldwell, D. R. Korber, and J. R. Lawrence, *J. Appl. Bact. Suppl.* **74,** 52S (1993).

[6] D. E. Caldwell, D. R. Korber, and J. R. Lawrence, *J. Microbiol. Methods* **15,** 249 (1992).

[7] S. Møller, D. R. Korber, G. M. Wolfaardt, S. Molin, and D. E. Caldwell, *Appl. Environ. Microbiol.* **63,** 2432 (1997).

[8] G. M. Wolfaardt, J. R. Lawrence, R. D. Robarts, S. J. Caldwell, and D. E. Caldwell, *Appl. Environ. Microbiol.* **60,** 434 (1994).

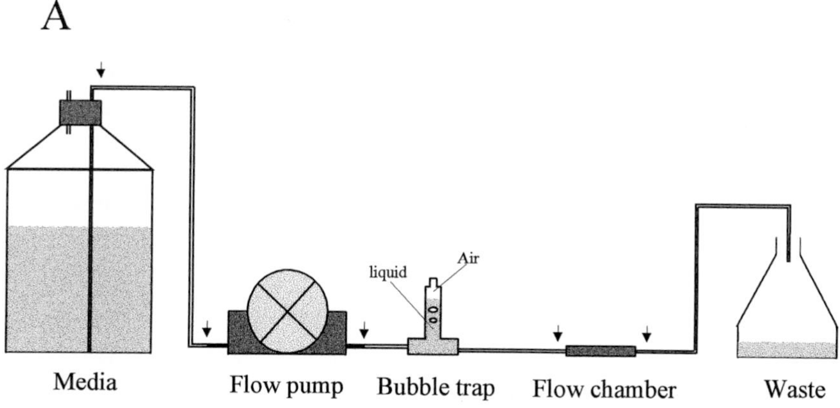

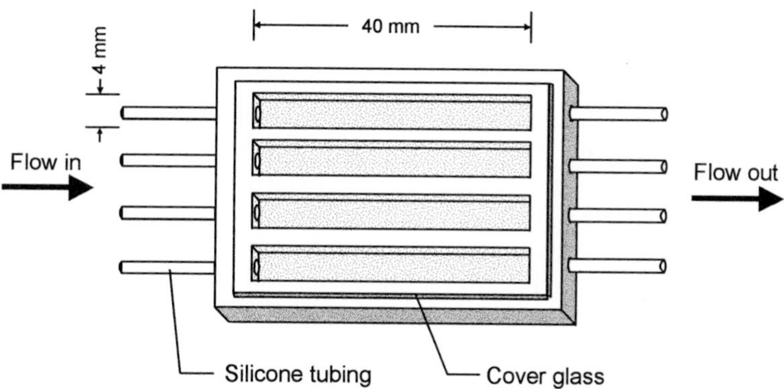

FIG. 1. Schematic presentation of a biofilm flow system. The entire flow system (A) is connected by standard silicone tubing, except for the tubing going into the pump where we use the more sustainable Marprene tubing. The normal tubing dimensions are o.d. 3.0 mm and i.d. 1.0 mm. Care should be taken to make the tubing between the bubble trap and the flow channel long enough for mounting the flow chamber on the microscope. Arrows indicate the positions where tubing is assembled with connectors. A typical flow chamber (B) consists of four parallel flow channels. Each channel is milled 1 mm deep. Holes with a diameter of 3.0 mm for flow inlets and outlets are drilled in each end of the Plexiglas. A coverslip (50 × 40 mm) is mounted on top of the channels using silicone glue (3M, St. Paul, MN). Silicone tubing, (o.d. 3.0 mm and i.d. 1.0 mm) is inserted at each end. Before introducing the tubing, add a small amount of silicone glue on the outside of the tubing that is to be inserted into the hole.

the inoculation. One hour later the flow chamber is turned right-side up and medium flow is started at a flow rate of 0.2 mm/sec (flow channel cross section is 4 mm^2). The biofilms are grown with the coverslip turned upward to prevent the sedimentation of flocks detached from other parts of the biofilm. Optimal microscopic resolution and hydrodynamic considerations within the flow cell set the maximal biofilm thickness at 100–150 μm. [See Palmer *et al.* (this volume)[2] for more details on flow cell design and experimental setup.]

Fixation and Embedding of Hydrated Biofilm Samples

On-line microscopy performed after general staining as indicated briefly yields information about changes in biofilm structure over time. Often it is of interest to relate this information to the composition of organisms in the biofilm community. One of the most widely used techniques for the identification of organisms in microbial samples is that of rRNA probing.[9–12] Targeting of species-specific rRNA sequences in cells using oligonucleotide probes labeled with fluorochromes is a very sensitive method for the *in situ* microscopic detection of individual organisms. However, rRNA hybridization cannot be performed on living cells; fixation is required for the probes to cross the cell membrane. Thus, the entire biofilm must be fixed in order to perform rRNA hybridization.

To prevent a structural collapse of the biofilm and to allow easy handling, the samples are embedded in polyacrylamide, which will maintain the native three-dimensional hydrated biofilm structure. Acrylamide possesses a number of advantageous features for embedding: it solidifies at room temperature, polyacrylamide is transparent, it is nonfluorescent, and the cells remain within the polyacrylamide block after solidification (almost no cells can be detected on the coverslip after removal of the embedded biofilm). The polyacrylamide slabs should be kept in humid environment to prevent drying.

It is possible to use other embedding materials. Agarose has been used,[13] but it has a high melting temperature. Also, rRNA hybridizations seem to be less efficient (i.e., lower signal) in this material. Another method may be to treat the biofilm with polysaccharide-specific antibodies. This will

[9] N. R. Pace, D. A. Stahl, D. L. Lane, and G. J. Olsen, *Adv. Microb. Ecol.* **9,** 1 (1986).

[10] G. J. Olsen, D. J. Lane, S. J. Giovannoni, N. R. Pace, and D. A. Stahl, *Annu. Rev. Microbiol.* **40,** 337 (1986).

[11] C. R. Woese, *Microbiol. Rev.* **51,** 221 (1987).

[12] R. I. Amann, W. Ludwig, and K. H. Schleifer, *Microbiol. Rev.* **59,** 143 (1995).

[13] A. A. Massol-Deyá, J. Whallon, R. F. Hickey, and J. M. Tiedje, *Appl. Environ. Microbiol.* **61,** 769 (1995).

introduce cross-links in the polymeric matrix, thereby preserving the biofilm structure on dehydration.[14,15] However, for each new type of biofilm community, new antibodies are needed, which may indeed be a bottleneck in many studies. Standard histological embedding materials may be used, but these normally result in hard blocks that must be sectioned in a microtome, which is less convenient for this application.

Procedure for Biofilm Fixation

Prepare a fixative (4% paraformaldehyde solution) by first adding 10 μl 10 *M* NaOH and then 2.0 g of paraformaldehyde to 33 ml of doubly distilled H_2O, stirring in a flask at 60°. Stir until all the paraformaldehyde is dissolved (approximately 5 min). The solution should not be heated for longer than needed for dissolving. Add 16.5 ml of 3× PBS buffer (390 m*M* NaCl, 30 m*M* $NaPO_4$, pH 7.2) and cool the solution on ice. Store the solution on ice until use; it is recommended to use the solution within 24 hr. The solution should, for safety reasons, be handled in the fume hood.

When the fixative is ready, stop the flow pump and clamp the effluent tubing. Disconnect the flow chamber from the inlet tubing approximately 3 cm from the flow channel. Attach the effluent tubing to the flow pump and remove the clamps. This should be done gently to prevent pulses in the solution, which may disturb the biofilm structure in the flow chamber. Reversing the pumping direction briefly until there is a small drop at the end of each inlet tube will ensure that the inlet tubing is free of air bubbles. Transfer the fixative to a small beaker and submerge the flow chamber influent tubing into the fixative. Draw the fixative into the channels by a flow rate of 0.2 mm/sec for approximately 15 min. The standard incubation time is 1 hr; however, this time may be varied depending on the later steps in the biofilm treatment.

Biofilm Embedding

Wash the fixed biofilm by pumping 1× PBS through the channel for 20 min (flow rate: 0.2 mm/sec). Make sure that the inlet tubing is just long enough to reach the bottom of a 2-ml reaction tube and make sure that the tubing is free of air bubbles. Prior to embedding, mix 1.0 ml of 20% (w/v) acrylamide monomer (200 : 1, acrylamide–bisacrylamide) and 8 μl *N,N,N′,N′*-tetramethylethylenediamine (TEMED) in 2-ml reaction tubes. Just before embedding add 20 μl of 1% (w/v) freshly prepared ammonium persulfate (APS). Note that this is one-tenth of what is normally

[14] E. B. Mackie, K. N. Brown, and J. W. Costerton, *J.Bacteriol.* **138,** 609 (1979).
[15] M. E. Bayer and H. Thurow, *J. Bacteriol.* **130,** 911 (1998).

used for acrylamide solutions for molding, e.g., acrylamide sequencing gels. *Note:* Always perform a test in a reaction tube before starting the embedding procedure on the flow chamber biofilms. After addition of the activator, mix gently by inverting the reaction tube a few times. Avoid shaking/whirli mixing of the reaction tube, as the polymerization process is inhibited by high concentrations of oxygen. After the addition of APS, the acrylamide mixture should solidify within 3–5 min. The solidifying time varies with the temperature and can be modulated by the amount of APS added to the acrylamide solution.

Because the acrylamide solution solidifies rapidly, it is important that the steps following APS addition are performed relatively fast (prepare a working routine). Put the tubing from the flow channels into the reaction tube with the activated acrylamide solution. [There is enough acrylamide solution in one tube to embed two (4 × 1 × 40 mm) channels simultaneously.] Pump for 1 min 50 sec (use a stop watch) by allowing the solution into the channels at a speed of 0.5 mm/sec. Note the increased pumping rate, which is necessary to avoid solidification of the acrylamide solution before the flow channel is filled. Immediately after, gently clamp off the effluent tubing and remove the flow system from the pump. Let the acrylamide solidify in the channels for at least 1 hr before further treatment.

After solidification, the embedded biofilm can be removed from the flow channel. Cut the glass between each channel with a knife and remove the glass by loosening it from the Plexiglas base with a scalpel. During the process the glass will break into pieces, but will be connected due to the silicone glue. It is important not to touch the acrylamide block. After the glass is completely removed from the acrylamide block, remove the embedded biofilm from the open flow channel. The extracted acrylamide block should be covered with doubly distilled H_2O to ensure that it will not begin to desiccate. This acrylamide-embedded biofilm can be stored until use at 4° in a humidified chamber, e.g., a petri dish sealed with Parafilm.

By monitoring the biofilm structure throughout an embedding routine we have found that this method is nondestructive and preserves the native three-dimensional structure well.

Ribosomal RNA Hybridization of Biofilm Samples

Polyacrylamide-embedded biofilms can be probed directly by rRNA hybridization methods. If more than one species are to be identified, then hybridization stringency must be sufficient for proper discrimination. We normally modulate the formamide concentration rather than change the temperature to maintain a high stringency (mismatch discrimination). A 1% increase in the formamide concentration decreases the melting temperature about 0.7° (for more details, see Stahl and Amann).[16] Also, changing the salt concentration can be used to vary stringency, but this approach may be more difficult to use routinely. Finally, the probes should be designed to have about the same melting temperature, as the probe with the lowest melting temperature will dictate the stringency of the hybridization solution.

Procedure for rRNA Hybridization on Embedded Biofilms

Cut out an embedded biofilm block of approximately 5 mm from a representative region of the biofilm. Transfer the biofilm sample to a glass slide with wells of approximately 8–10 mm diameter (Novakemi AB, Enskede, Sweden) and ascertain that the substratum side (the side facing the coverslip in the flow cell) is facing downward. To ensure that the polyacrylamide block is saturated with formamide (FA) solution, add 50 μl of the solution [*X*% FA, 0.9 *M* NaCl, 0.1 *M* Tris (pH 7.5), 0.1% sodium dodecyl sulfate (SDS)] to the block and incubate at 37° for 30 min. In the meantime, the hybridization solution (formamide solution + fluochrome-labeled 16S rRNA probes) is prepared (30 μl per block). It is possible to use several probes labeled with different fluorochromes simultaneously, e.g., fluorescein isothiocynate (FITC, green), CY3 (red), or CY5 (far-red). Each probe is added to obtain a final concentration of approximately 2.5 μg/ml in the hybridization solution. This is only an empirically derived initial concentration and should be adjusted if the probes are particularly well or poorly labeled with fluorophore. After removal of the prehybridization solution, add the hybridization solution containing the probe and incubate the biofilm block at 37° in a formamide-saturated atmosphere (e.g., in a sealed large test tube with tissue paper wetted with the prehybridization solution). The minimum incubation time required seems to depend on a number of undefined factors, which may vary from one biofilm community to another. In our biofilm systems, an incubation time of at least 2 hr is required for optimal hybridization signals.

To remove unspecifically bound probes, two washing steps are performed. After removal of the hybridization solution, the biofilm block is incubated at 37°, first with 50 μl prewarmed prehybridization solution for 30 min, and subsequently with 50 μl washing solution [0.1 *M* Tris (pH 7.5), 0.9 *M* NaCl] at 37° for 30 min. Finally, to avoid precipitation of salt, the block is rinsed in 50 μl doubly distilled H_2O by pipetting a few times up and down before removal.

During the washing procedure a specimen holder for microscopic inspection should be prepared. Our preferred holder is a microscope slide with

[16] D. A. Stahl and R. I. Amann, "Nucleic Acid Techniques in Bacterial Systematics" (E. Stackebrandt and M. Goodfellow, eds.), p. 205. Wiley, New York, 1991.

a silicone rubber gasket (length 60 mm × width 24 mm × depth 1 mm) mounted on top. The hybridized biofilm block can be fit into a hole cut in the center of the rubber gasket. The biofilm block is placed in the gap with the side that faced the substratum in the flow channel turned upward. One drop of antifade solution [we use the 2× concentrated part of the SlowFade kit (Molecular Probes, Eugene, OR)] is added on top of the biofilm block. This antifade solution has been found to be particularly good for applications that employ long illumination times, such as in confocal microscopy, because it provides a nearly constant emission intensity over time. Other antifade agents, such as *p*-phenylenediamine,[17] provide nearly the same or even better features than SlowFade, but these may have other disadvantages, e.g., some of these chemicals are colored in a way that may interfere with the experiment. A coverslip (24 × 60 mm) is mounted on top of the silicone/biofilm block and the microscope slide/silicone gasket/coverslip sandwich is then taped together. The biofilm sample is then ready for microscopic inspection. Alternatively, the biofilm sample may be stored in a humidified petri dish sealed with Parafilm. An example of a triple rRNA hybridization of a biofilm is shown in Fig. 2. A three-dimensional reconstruction of the biofilm using the IMARIS software package (Bit-Plane AG, Zurich, Switzerland) shows the structural organization of the individuals in the biofilm.

It is possible to combine other labeling methods (such as fluorescent immunolabels) with the rRNA technique, but antibody-labeling procedures should be performed prior to embedding because these molecules penetrate only poorly in the polyacrylamide. Optimal antibody labeling requires a number of washing steps of the unembedded sample and, unless great care is taken, may lead to structural deterioration.

Molecular Reporters for Bacteria

Monitoring occurrence and behavior of a specific strain, or tracing a specific plasmid, may be achieved by molecular tagging of the organism or plasmid with a reporter gene that encodes a fluorescent protein or an enzyme that catalyzes a reaction leading to a fluorescent product. Such reporter genes may not only be used for tagging, but also for monitoring specific gene activities nondestructively by the fusion of specific promoters to the reporter genes. These genetic manipulations must occur before introduction of the tagged organism into the community. In natural systems this is rarely possible, but in laboratory model systems this strategy is excellent. Two detailed examples of widely used reporter genes and their use in biofilm studies are described.

[17] A. Longin, C. Souchier, M. French, and P. Bryon, *J. Histochem. Cytochem.* **41,** 1833 (1993).

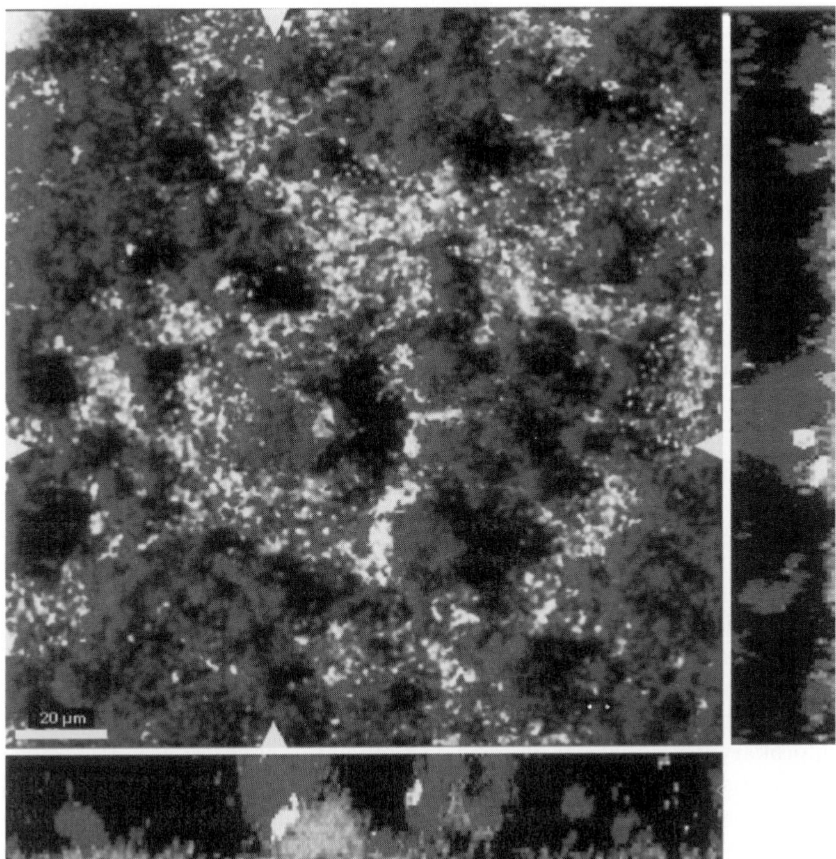

FIG. 2. Illustration of the spatial distribution of three different biofilm community strains—*P. putida* RI cells (dark gray), *Acinetobacter* C6 (light gray), and *Burkholderia cepacia* spp. D8 (white)—analyzed 8 days after initial colonization of the mixed community cells. The three organisms were all identified by fluorescent *in situ* hybridization using the probe D8-647 labeled with FITC targeting *B. cepacia,* ACN449 labeled with CY3 targeting *Acinetobacter* C6, and PP986 labeled with CY5 targeting *P. putida* RI. The large panel shows a horizontal view presented as a simulated fluorescent projection where long shadows (dark areas) indicate large/high microcolonies. Shown to the right and below are vertical cross sections through the biofilm collected at positions indicated by the white triangles.

Green Fluorescent Protein and Its Derivatives

The *gfp* gene [encodes green fluorescent protein (GFP), originally isolated from the jellyfish *Aequorea victoria*] has several characteristics advantageous for studies of gene expression in individual cells. GFP fluoresces without the addition of any external substrate, apart from low levels of oxygen needed for catalyzing an internal rearrangement of the protein into its fluorescent form.[18,19] The versatility of the protein is illustrated by the diversity of its applications. For example, the fusion of specific promoters to the *gfp* gene has been used to monitor intracellular protein localization during sporulation in *Bacillus subtilis*[20] and to analyze mycobacterial interactions with macrophages.[21] GFP has also been used to monitor plasmid transfer between bacteria growing in different environments.[22,23] The *gfp* gene has been transferred and expressed in a variety of bacteria,[18] as well as in higher organisms including yeasts,[24] plants,[25] and mammals.[26]

The original wild-type GFP protein absorbs long-wavelength UV light (absorbance maximum at 395 nm and a minor, but significant, peak at 470 nm) and emits green light at 508 nm. GFP has been mutated extensively in order to alter the properties of the wild-type protein. Heim *et al.*[19] described variants generated by single or few amino acid changes in the 238 amino acid protein. Random or directed mutagenesis of the *gfp* gene sequence has resulted in several spectral variants and mutants with enhanced fluorescence efficiency (see Table I). One family of spectral mutants is the "red-shifted" group, in which the emission spectrum is identical to that of the parent protein, but which has the absorbance maximum shifted toward the red (from long-wavelength UV to 488 nm).

Emission spectrum variants primarily fall into four groups: BFP (blue fluorescent protein), CFP (cyan), GFP (green), and YFP (yellow) (Table I). These are available commercially, but some are available only with a mammalian optimized codon usage and may not be expressed efficiently in prokaryotes. However, in most cases the mutants were originally made on the basis of the jellyfish sequence (which works reasonably well in prokaryotes), and these may be obtained from the laboratories where the variants were created.

Combinations of the various mutant proteins facilitate on-line monitoring of several traits at the same time. It should be noted, however, that some combinations work better than others. For instance, it is possible to use the "mut" mutants of GFP[27] together with uvGFP, especially if the

[18] M. Chalfie, Y. Tu, G. Euskirchen, W. W. Ward, and D. C. Prasher, *Science* **263,** 802 (1994).
[19] R. Heim, D. C. Prasher, and R. Y. Tsien, *Proc. Natl. Acad. Sci. U.S.A.* **91,** 12501 (1994).
[20] C. D. Webb, A. Decatur, A. Teleman, and R. Losick, *J.Bacteriol.* **177,** 5906 (1995).
[21] S. Dhandayuthapani, L. E. Via, C. A. Thomas, P. M. Horowitz, D. Deretic, and V. Deretic, *Mol. Microbiol.* **17,** 901 (1995).
[22] B. B. Christensen, C. Sternberg, and S. Molin, *Gene* **173,** 59 (1996).
[23] B. Normander, B. B. Christensen, S. Molin, and N. Kroer, *Appl. Environ. Microbiol.* **64,** 1902 (1998).
[24] R. K. Neidenthal, L. Riles, M. Johnston, and J. H. Hegemann, *Yeast* **12,** 773 (1996).
[25] S. J. Casper and C. A. Holt, *Gene* **173,** 69 (1996).
[26] B. Ludin, R. Doll, S. Meili, S. Kaech, and A. Matus, *Gene* **173,** 107 (1996).
[27] B. P. Cormack, R. H. Valdivia, and S. Falkow, *Gene* **173,** 33 (1996).

TABLE I
RELEVANT GREEN FLUORESCENT PROTEIN DERIVATIVES

Name	Excitation maximum (nm)[a]	Emission maximum (nm)[a]	Maximal fluorescence intensity relative to wild type when excited with blue light[b]	Mutations (comments)	Refs.
Wild-type GFP	396 (476)	508	1		*d*
GFP					
S65T	489	511	6 (37°)	S65T	*e*
Gfpmut1	488	507	35 (30°)[c]	F64L/S65T	*f*
Gfpmut2	481	507	19 (30°)[c]	S65L/V68L/S72A	*f*
Gfpmut3	501	511	21 (30°)[c]	S65G/S72A	*f*
BFP					
Y66H	382	448	0.6 (22°)	Y66H	*g*
GFPA (Y66H)	384	448	18 (37°), 3 (25°)	Y66H/V163A/S175G	*h*
CFP					
W7	433 (453)	475 (501)	3 (37°)	Y66H/N146I/M153T/V163A/N212K	*i*
W2	432 (453)	480	1.5 (37°)	Y66W/I123V/Y145H/H148R/M153T/V163A/N212K	*i*
YFP					
Gfp-10C	513	527	6 (37°)	S65G/V68L/S72A/T203Y	*j*
UV-GFP					
Gfp-mutant C3	396 (476)	508	42 (37°)	F100S/M154T/V164A	*k*
GFP5	(396) 476	508	111 (37°)	V163A/I167T/S175G (more thermostable)	*h*

[a] Numbers in parentheses refer to a second minor excitation peak.

[b] The fluorescence of wild-type GFP decreases approximately four-fold from 25° to 37°.[h] This should be taken into account when the different GFP derivatives are compared. Numbers in parentheses indicate the temperature at which the different variants were compared. The relative fluorescent intensities are approximate values based on the information given in the literature. Where the investigators compared fluorescence intensity to other mutant GFP's we have converted the data to be the ratio relative to wild-type GFP.

[c] Compared with wild-type GFP at an excitation wavelength of 488 nm. The maximal fluorescence intensity relative to the wild-type GFP is higher, but was not reported by the investigators.

[d] M. Chalfie, Y. Tu, G. Euskirchen, W. W. Ward, and D. C. Prasher, *Science* **263,** 802 (1994).

[e] R. Heim, D. C. Prasher, and R. Y. Tsien, *Proc. Natl. Acad. Sci. U.S.A.* **91,** 12501 (1994).

[f] B. P. Cormack, R. H. Valdivia, and S. Falkow, *Gene* **173,** 33 (1996).

[g] R. Heim, A. B. Cubitt, and R. Y. Tsien, *Nature* **373,** 663 (1995).

[h] K. R. Siemering, R. Golbik, R. Sever, and J. Haseloff, *Curr. Biol.* **6,** 1653 (1996).

[i] R. Heim and H. C. Tsien, *Curr. Biol.* **6,** 178 (1996).

[j] M. Ormö, A. B. Cubitt, K. Kallio, L. A. Gross, R. Y. Tsien, and J. J. Remington, *Science* **273,** 1392 (1996).

[k] A. Crameri, E. A. Whitehorn, E. Tata, and P. C. Stemmer, *Nature Biotechnol.* **14,** 315 (1996).

uvGFP and mut GFP are in two separate organisms because the mut variants are poorly excitable by UV light. In this case the two populations can be differentiated on the basis of UV-induced fluorescence. Nevertheless, it is easier to use variants with entirely different excitation and emission properties. In this context the fluorescence intensities of the different variants should also be taken into consideration.

The rather slow maturation rate (approximately 4 hr at 22° [19] in the case of the wild-type GFP) may be a problem in some gene expression studies. However, the maturation rate increases with increasing temperature. Most of the new GFP variants have a significantly accelerated fluorescence development compared to the wild-type protein. Finally, because the protein needs oxygen to fluoresce, it cannot be used under completely anaerobic conditions. We have found that GFPmut3 is fully activated at an oxygen concentration 300 times below the saturation level in water.[28] We believe this to be sufficiently effective, at least for flow chamber-grown biofilms, where there is constant flow of fresh nutrients (and oxygen) and the thickness normally does not exceed 100–150 μm.

It may be of interest to monitor changes in the gene expression of specific bacteria, and for that purpose the extreme stability of the GFP protein[29] is a disadvantage. The intracellular GFP pool will not immediately change after reduction of the level of gene expression, but only as a result of continued cell proliferation (dilution of the protein). To overcome this problem, we have constructed a series of unstable variants of the GFP protein. The proteins are rendered susceptible to degradation[30] through addition of a target sequence recognized by housekeeping intracellular, tail-specific proteases (one of which is homologous to the *clpXP* from *Escherichia coli*[31–33]). We have added the sequence AANDENYAXXX to the C-terminal end of GFP, where variation of the last three positions yields several different half-lives of the protein in living cells[34] (Table II). Because the tail-specific proteases are found in a wide range of gram-negative and gram-positive bacteria, the unstable GFP variants may be utilized in many different species. We have successfully tested *E. coli* and *Pseudomonas* spp. (Table II). It should be noted that the proteases may be expressed at different levels depending on the chosen organism, the cellular growth phase, or other factors, which may affect the rate of degradation of GFP proteins.

β-Galactosidase Substrates

Another marker protein is β-D-galactosidase (the product of the *E. coli lacZ* gene). Fluorophores and chromophores can be coupled to galactose

[28] A. T. Nielsen, unpublished results (1998).

[29] R. Tombolini, A. Unge, M. E. Davey, F. J. de Bruijn, and J. K. Jansson, *FEMS Microbiol. Ecol.* **22,** 17 (1997).

[30] K. C. Keiler, P. R. H. Waller, and R. T. Sauer, *Science* **271,** 990 (1996).

[31] S. Gottesmam, E. Roche, Y. Zhou, and R. T. Sauer, *Genes Dev.* **12,** 1338 (1998).

[32] C. Herman, D. Thevenet, P. Bouloc, G. C. Walker, and R. D'Ari, *Genes Dev.* **12,** 1348 (1998).

[33] K. C. Keiler and R. T. Sauer, *J. Biol. Chem.* **271,** 2589 (1996).

[34] J. B. Andersen, C. Sternberg, L. K. Poulsen, S. P. Bjørn, M. Givskov, and S. Molin, *Appl. Environ. Microbiol.* **64,** 2240 (1998).

TABLE II
USTABLE GFP VARIANTS AND THEIR HALF-LIVES IN *Escherichia coli* AND *Pseudomonas putida*

Name	$t_{1/2}$	
	E. coli (MT102)[a,b]	*P. putida* KT2442[a]
Gfpmut3	∞	∞
Gfp(LAA)	40	60
Gfp(LVA)	40	190
Gfp(AAV)	60	190
Gfp(ASV)	110	190
Gfp(AGA)[b]	300–500	ND

[a] J. B. Andersen, C. Sternberg, L. K. Poulsen, S. P. Bjørn, M. Givskov, and S. Molin, *Appl. Environ. Microbiol.* **64,** 2240 (1998).
[b] Jens Bo Andersen, unpublished (1998).

(a state in which they are nonfluorescent/chromogenic) and be rendered fluorescent or chromogenic by enzymatic hydrolysis. Chromogenic substrates such as 5-bromo-4-chloro-3-indolyl-β-D-galactopyranoside (X-Gal) are the basis of bioreporter assays for plate-grown microorganisms. UV-excitable fluorophores have also been used for many years; the best example is methylumbelliferylgalactoside (MUG), which has been used in general nonmicroscopic tests of exoenzyme activity in microbial ecology and as a single-cell reporter of alginate synthesis in a microscopic assay of attached bacteria.[35] The substrates have also been applied to flow cytometer studies of bacteria and yeast cells entrapped in agarose beads[36] and of differentiating *Myxococcus* cells.[37]

We have worked with a marine strain, *Oceanospirillum* sp. (carrying a chromosomal copy of the *lacZ* gene inserted by transposon mutagenesis), and with *E. coli* DH5α (carrying the *lacZ* gene on a plasmid). Both bacterial strains are grown at relatively high salt concentrations (sea salts for *Oceanospirillum* and 10 g/liter NaCl for *E. coli*). For the enzyme substrate to enter the cells, a hypotonic shock is required, and providing the substrate in a medium lacking salts established this condition easily. For strains grown at lower salt concentrations, this condition may, however, be difficult to achieve. It has been suggested that a cycling of hypertonic/hypotonic buffers may achieve loading.[37]

[35] D. G. Davies and G. G. Geesey, *Appl. Microbiol. Biotechnol.* **61,** 860 (1995).
[36] R. Nir, Y. Yisraeli, R. Lamed, and E. Sahar, *Appl. Environ. Microbiol.* **56,** 3861 (1990).
[37] F. Russo-Marie, M. Roederer, B. Sager, L. A. Herzenberg, and D. Kaiser, *Proc. Natl. Acad. Sci. U.S.A.* **90,** 8194 (1993).

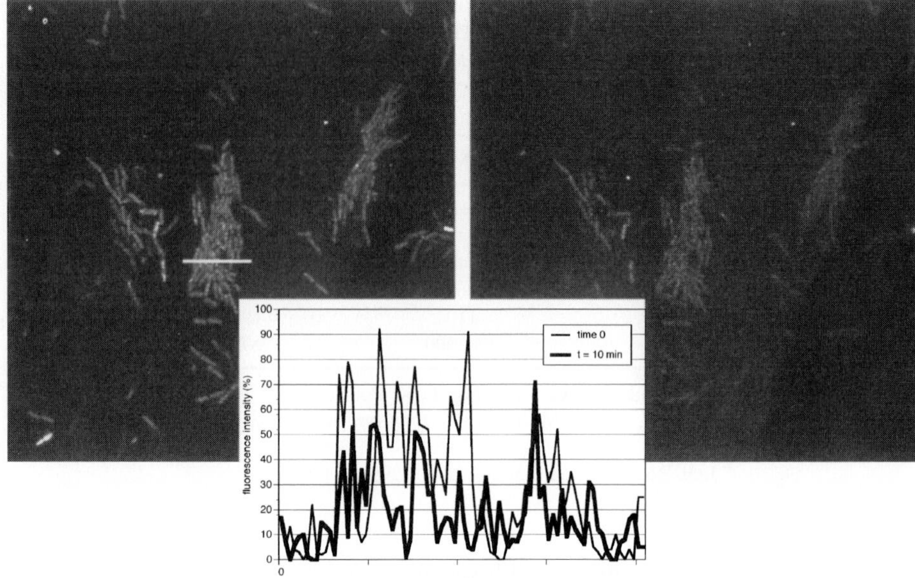

Fig. 3. Monolayer of *E. coli* grown in a flow cell using 1/10 LB broth. The cells harbor a transposon that confers constitutive expression of the *lacZ* gene. Fluorescein digalactoside (50 μM in distilled H_2O) was injected; the left image (time 0) was acquired 1 min later. Ten minutes later, an image was acquired using the same photomultiplier (PMT) settings. The graph shows relative fluorescence intensity in both images along the line drawn in the left image. The images show an area of 100 × 100 μm. The bacterial strain was kindly provided by Tanya Kuritz, Oak Ridge National Laboratory, Oak Ridge, TN.

Fluorescein-based substrates for β-galactosidase often consist of one fluorescein molecule attached to two galactose molecules (e.g., fluorescein digalactoside, carboxymethylfluorescein digalactoside). Therefore, these "disaccharide" molecules require two cleavage steps before fluorescence occurs, a potential problem in cases where enzymatic activity is low. However, the fluorescence intensity of fluorescein is high relative to that of fluorophores attached to a single galactose such as resorufin. Thus, an empirical assessment of the fluorophore–substrate combinations is often required. We have found fluorescein digalactoside and resorufin-galactoside to be interchangeable with respect to the two organisms noted earlier.

Diffusion of the fluorophore out of the cell is a significant problem under conditions of a hypotonic shock. Over a period of 20 min, fluorescence localized within the cells drops and background fluorescence increases to the point where it is difficult to localize cellular fluorescence (Fig. 3). It is therefore recommended that flow be stopped during the assay, that microscopy be initiated immediately after introduction of the substrate, and that all data be collected within 10 min after substrate introduction. The fluorescein digalactoside substrate is available in covalently modified lipophilic forms (C_{12}FDG, C_8FDG) and as a chloromethylfluorescein digalactoside, which are presumably less susceptible to diffusional loss from cells. The lipophilic forms apparently are incorporated into membranes, whereas the chloromethyl derivative reacts with intracellular thiols to create a membrane-impermeable peptide-linked product.

Finally, the effect of hypotonic shock effect on the biofilm state cannot be ignored. In our work with *Oceanospirillum,* we noted disruption that appeared to be related to thickness. Monolayer biofilms were relatively stable, whereas biofilms on the order of 20 μm in thickness suffered loss of cells within 5 min, after shock. However, if timed carefully, the assay could be carried out prior to noticeable deterioration.

Genetic Tools

We have designed a toolbox of general cloning vectors that allow the combination of any promoter of interest with the most convenient marker gene for a given application. The resulting monitor cassette can subsequently be introduced into the chromosome of virtually any gram-negative bacterium (Fig. 4).

The expression system consists of a synthetic ribosomal-binding sequence (RBSII) at optimal distance from a start codon (ATG) placed within a *Sph*I restriction site, followed by *Hin*dIII and two strong transcriptional termination sites T0 (derived from phage λ) and T1 (derived from the *rrnB* operon of *E. coli*), and with translational stop codons in all three reading frames. The RBSII is preceded by a multicloning site with several unique restriction sites.

Two *Not*I restrictions sites for easy excision flank this entire cassette. Two different derivatives of the cloning vector have been constructed, one derived from the high copy number vector pUC18Not[38] and another based on the low copy number plasmid pLOW2.[39] In cases where high promoter activities may disturb the growth of the host cell, the latter may be more feasible due to the lower copy number (about 10–15 per cell in *E. coli*).

To construct a monitor system, the reporter gene of choice is polymerase chain reaction (PCR) amplified using primer sets that produce PCR fragments flanked by the *Sph*I and *Hin*dIII or *Bgl*II restriction sites (neither *gfp* nor *lacZ* reporter genes contain these sites). Subsequently, these fragments are inserted into either of the cloning vectors [pJBA23 (pUC18Not) or pBBC597 (pLOW2)]. For insertion of the reporter gene into pLOW2, the

[38] M. Herrero, V. de Lorenzo, and K. N. Timmis, *J. Bacteriol.* **172,** 6557 (1990).
[39] L. H. Hansen, S. J. Sørensen, and L. B. Jensen, *Gene* **186,** 167 (1998).

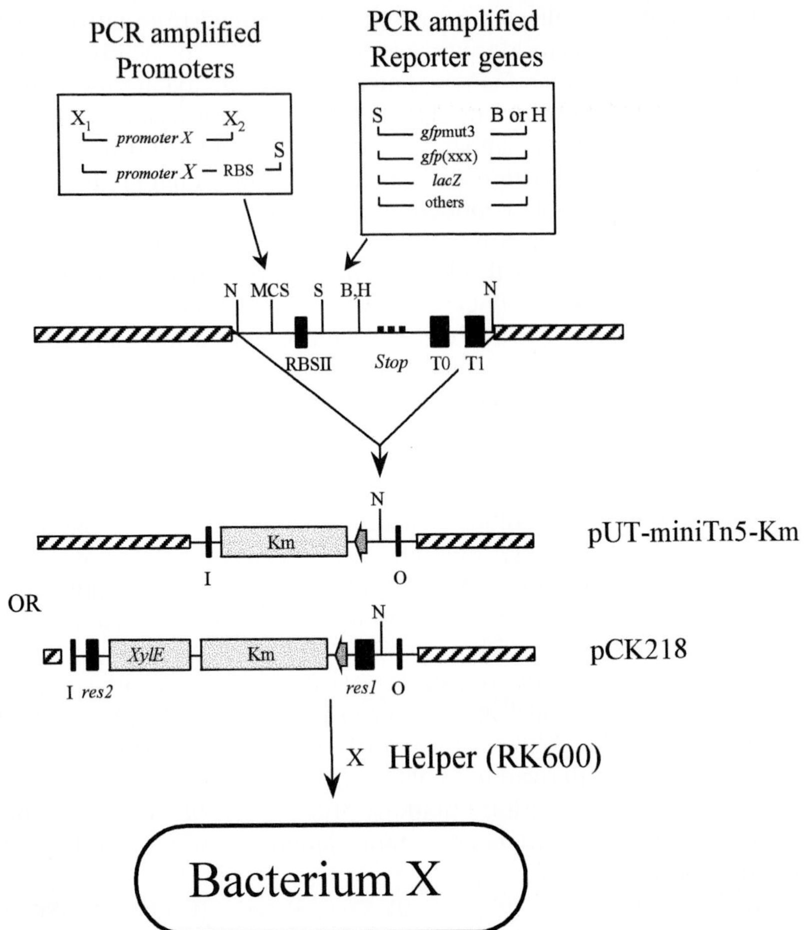

FIG. 4. Schematic presentation of the three-step cloning strategy used for the construction of specific promoters/reporter gene cassettes. The chromosomal organization of the relevant regions in the cloning cassettes and the delivery plasmids are shown. Indicated in the cloning cassette is the synthetic ribosomal-binding site (RBSII), the stop codons in all three reading frames (Stop), and the two termination sites (T0 and T1). Also indicated are the cloning regions for the reporter genes and promoters. The multicloning site (MCS) contains the following unique restriction sites: *Eco*RI, *Sac*I, *Kpn*I, *Sma*I, and *Xba*I. Other restriction sites are N, *Not*I; S, *Sph*I; H, *Hin*dIII; and B, *Bgl*II. X_1 and X_2 should be one of the unique restriction sites present in the MCS. In both delivery vectors the transposon cassette is flanked by the 19-bp transposon inverted repeats, I and O. In plasmid pCK218 the genes *npt* (Km^R) and *XylE* encoding the catechol 2,3-dioxygenase (C2,3O) are encompassed by two tandem core *res* sites, *res1* (305 bp) and *res2* (158 bp), which allows for excision of the *npt* and *XylE* genes when *trans* complemented with the RP4 site-specific resolvase ParA. Boxed arrows indicate promoters and the hatched regions the vector part of the different cloning and delivery plasmids.

*Bgl*II restriction site is preferred because the plasmid contains an additional *Hin*dIII site in the kanamycin (km) gene.

The promoter is inserted in a way similar to that described for reporter genes. The promoter cloning region offers different unique cloning sites, and the promoter of interest can therefore, in most cases, be amplified by PCR using primer sets that yield fragments flanked by specific cloning sites that are not present in the reporter gene (see Fig. 4). It may be desirable to insert a promoter along with its native ribosome-binding site. This can be obtained by replacing the synthetic ribosome-binding site in the cloning cassette with the original ribosome-binding site by designing the downstream end of the amplified promoter/ribosome-binding site region for ligatation to the *Sph*I site in the toolbox vector.

As a delivery system for the reporter gene cassette into gram-negative bacteria, we use the versatile pUT vectors constructed by de Lorenzo and co-workers.[38,40] These are R6K plasmid derivatives, which only replicate when complemented with the gene product of the R6K *pir* gene. Hence, pUT vectors act as suicide vectors in hosts devoid of the *pir* gene (practically all bacteria except for the few specific strains constructed as hosts for these vectors). The pUT vectors harbor the broad host range Tn*5* transposon, which has been modified by moving the transposase gene outside the inverted repeats of the transposon. Thus, when inserted in the chromosome of the target host, the transposon element is immobilized due to the lack of transposase activity.

The *Not*I fragment carrying the reporter gene cassette is inserted into the *Not*I restriction site of the transposon delivery vector pUT-miniTn*5*-km. To integrate the cassette into a suitable host, a triparental mating is performed in which the helper strain *E. coli* HB101(RK600)[41] is used to mobilize the plasmid carrying the reporter cassette from the donor strain, *E. coli* CC118(λ*pir*), into the recipient strain. Plasmid RK600 is a RP4 derivative encoding the *tra* mobilization genes.[42] The transconjugants are selected by kanamycin resistance (from Tn*5*-km) and by any additional selective trait carried by the recipient.

An improvement of the pUT-miniTn*5* delivery plasmid system was obtained by construction of a delivery system in which the kanamycin gene is flanked by two directly repeated multimer resolution sites (*res*) from the broad host range plasmid RP4 multimer resolution system.[43] By *trans* complementation with the RP4 site-specific resolvase, ParA, the DNA

[40] V. de Lorenzo, M. Herrero, U. Jacubzik, and K. N. Timmis, *J. Bacteriol.* **172,** 6568 (1990).
[41] B. Kessler, V. de Lorenzo, and K. N. Timmis, *Mol. Gen. Genet.* **233,** 293 (1992).
[42] D. H. Figurski and D. R. Helinski, *Proc. Natl. Acad. Sci. U.S.A.* **76,** 1648 (1979).
[43] L. Eberl *et al., Mol. Microbiol.* **12,** 131 (1994).

sequence interspersed between the two *res* sites is deleted effectively.[44] This delivery system may, in principle, allow for the introduction of many different bioreporters into the same strain without increasing the number of resistance markers in the bacterium and may also be used to avoid the unnecessary release of resistance genes to the environment.

To ensure that the transposition events do not interfere with any essential parts of the host strain genome, the strains must be tested in batch culture experiments prior to their use in biofilms. The promoter fusions should also be tested to comply with the expected behavior before their use in biofilm experiments.

Finally, in some cases, it may be necessary to insert the reporter gene cassette in a high copy number plasmid to ensure expression levels that are sufficient for detection as a fluorescent signal in single cells. High copy number, broad host range plasmids can be stably maintained in a number of different species.[45,46]

Combining rRNA Hybridization and Fluorescent Reporters

It is possible to combine the *in situ* rRNA hybridization procedure with the detection of GFP in single cells because fixation of the biofilm has little effect on GFP fluorescence. However, the GFP signal does decrease during the hybridization procedure; the hybridization solution apparently influences fluorescence yield or protein stability. Consequently, contact with hybridization solutions should be kept as short as possible when GFP is to be monitored. Preliminary studies on cell smears have indicated that sodium dodecyl sulfate (SDS) in the hybridization solution is the cause of GFP instability. We do not know if it is possible to completely remove SDS from the hybridization solution or to replace it with another detergent without also significantly decreasing the hybridization efficiency in the biofilms.

The unstable variants of the GFP protein can also be monitored in combination with the rRNA hybridization. However, it is important to emphasize that the instability of the protein causes problems, as the tail-specific proteases appear to be slowly inactivated during the fixation procedure. Unstable GFP is consequently degraded during the fixation procedure, although at lower rate than in live cells, until the proteases have been inactivated. To minimize this problem, the flow chamber should be moved

[44] C. S. Kristensen, L. Eberl, J. M. Sanchez-Romero, M. Givskov, S. Molin, and V. d. Lorenzo, *J. Bacteriol.* **177,** 52 (1995).

[45] J. v. d. B. Arjan, L. A. De Weger, W. T. Tucker, and B. J. J. Lutgenberg, *Appl. Environ. Microbiol.* **62,** 1076 (1998).

[46] J. Laville, C. Blumer, C. v. Schrötter, G. Defago, and D. Haas, *J. Bacteriol.* **180,** 3187 (1998).

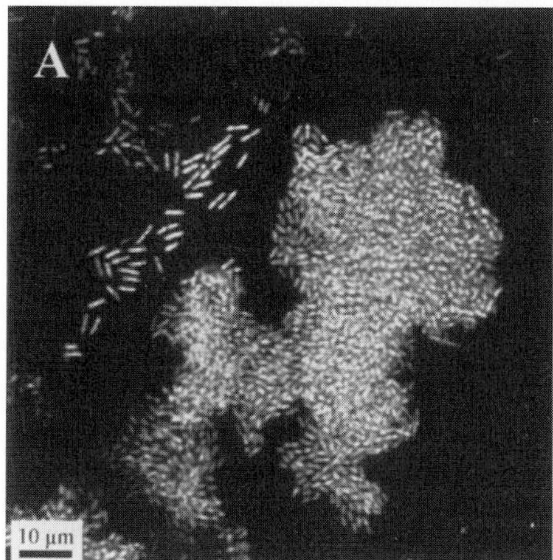

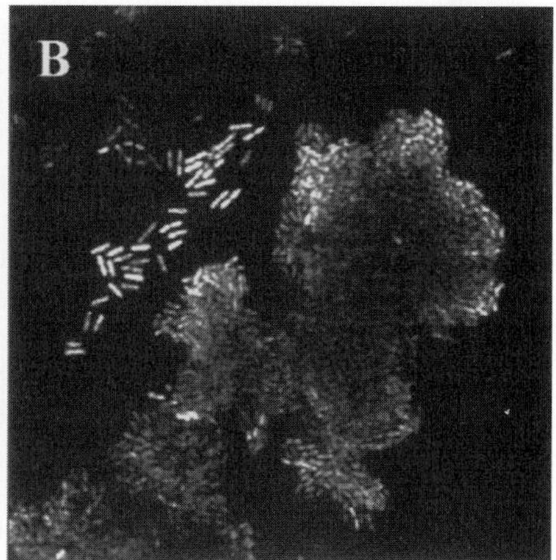

FIG. 5. Simultaneous monitoring of GFP expression and ribosomal probing in a fixed and embedded biofilm. (A) Strain *P. putida* R1 [*rrnB* P1-*gfp*(AAV)] was hybridized with the probe PP986 labeled with CY3. (B) The same part of the biofilm was observed for GFP(AAV) fluorescence.

directly from the flow system and placed on ice immediately after the experiment is stopped, and the fixation procedure must be started immediately thereafter. Keep all solutions on ice and continue the fixation procedure overnight to ensure effective inactivation of the protease (for an example, see Fig. 5).

Single cell detection of *lacZ* expression can also be performed on fixed and embedded biofilm. Fixation does not result in the immediate inactivation of β-galactosidase, and although the standard fixation procedure (1 hr) may cause some protein inactivation, detectable β-galactosidase activity remains. Until now, we have not been able to develop a common solution that permits simultaneous rRNA hybridization and detection of the fluorescent β-galactosidase.

In situ Markers for Gene Transfer

The flow of genetic information in microbial communities can be monitored directly by tagging the mobile DNA molecules with appropriate marker genes. We routinely use transposon mutagenesis as a method for tagging conjugative plasmids.[22,47] First, the actual tagging is performed by a triparental mating procedure as described earlier. Second, it is important

[47] B. B. Christensen, C. Sternberg, J. B. Andersen, L. Eberl, S. Møller, M. Givskov, and S. Molin, *Appl. Environ. Microbiol.* **64,** 2247 (1998).

to test the tagged (plasmid-harboring) strain carefully. A second round of conjugation with the transconjugant as donors is carried out to ensure that only clones with the marker gene inserted in the plasmid are selected. In many cases the insertion may have an effect on important plasmid functions. For instance, the cassette may be inserted in regions affecting transfer efficiency or plasmid maintenance or in genes encoding essential metabolic functions or antibiotic resistance. Even if the insertions may be mapped to regions without known essential genes, they may still influence the general plasmid behavior.

By combining different fluorescent marker genes with different excitation and emission spectra, it may be possible to follow donor, recipient, and transconjugants in live biofilms simultaneously. This can be achieved by tagging the conjugative plasmid with a cassette comprising one constitutively expressed fluorescent marker, and the recipient being chromosomally tagged with another fluorescent marker gene. In this case the donor cells will appear as emitting light in one color, the recipient cells in another, and the transconjugants as a combination of both colors.

Although developed specifically for plasmid conjugation studies, these methods may be modified for studies of transduction or transformation.

Two Marker Gene Strategies for in Situ Investigations of Gene Transfer

We have developed different marker systems for the *in situ* analysis of plasmid transfer in complex environments such as flow chamber biofilms. In one approach, the β-galactosidase promoter, P*lac*, has been fused to the *gfp* gene (Fig. 6A). The rationale was that the P*lac* promoter is not repressed

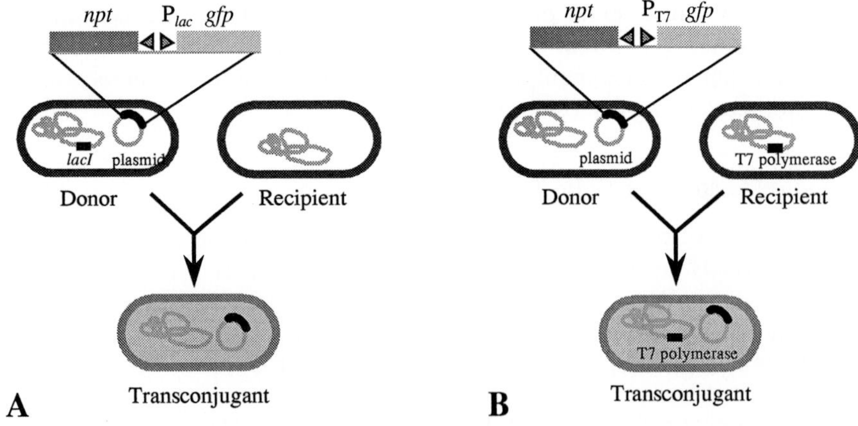

FIG. 6. Two different strategies for the *in situ* monitoring of plasmid transfer in biofilms.

in most bacteria as it is in the original *E. coli* host by the specific repressor protein, LacI. Therefore, if the *lacI* repressor gene is inserted into the chromosome of the donor strain, no P*lac*-induced green fluorescence phenotype is detectable as long as the plasmid resides in this host. However, on conjugation, the plasmid enters a host that does not express the repressor and, provided P*lac* is active in that host, GFP under control of the *lac* promoter will be synthesized (zygotic induction).

An alternative approach has been developed to study the transfer of plasmids to a *specific* recipient strain in the biofilm. The PΦ*10* promoter from the *E. coli* bacteriophage T7 is active only in the presence of the corresponding bacteriophage T7 RNA.[48] Consequently, a construction where the marker gene is under the control of PΦ*10* will not be expressed in any other host than the one expressing the T7 RNA polymerase. If the donor cells harbor a conjugative plasmid comprising the PΦ*10* controlled repressor cassette and if the recipient strain of interest carries a stable insertion of the T7 RNA polymerase gene expressed from a bacterial promoter, a successful mating between the donor and the specific recipient can be detected as cellular expression of the reporter (Fig. 6B).

Concluding Remarks

The *in situ* methods described in this article can be used for more than just identifying or tracing cells or genes in biofilms. By combining promoters that respond to specific environmental signals with appropriate marker genes, it may be possible to tag specific organisms and use these as monitor systems to estimate local chemical composition directly in the biofilms: the tagged bacteria become living biosensors. For instance, it may be possible to construct biosensors that report the limitation of nutrients such as carbon, nitrogen, or phosphorus. The presence or absence of, e.g., oxygen, heavy metals, toxins, certain metabolites, or stress factors, such as high temperature, suboptimal pH, or ionic strength, can also be monitored. An alternative may be to use random insertions by promoter probe cassettes followed by phenotypic screening for the desired traits.

One example of monitoring environmental conditions was described by Møller *et al.*[49] A fusion between the *gfp* gene and the TOL promoter, P*m,* which is induced in the presence of a nutrient (benzoate) above a certain concentration, was inserted into a strain of *P. putida*. This strain fluoresces bright green in the presence of benzoic acid derivatives. Using this biosensor, it was possible to demonstrate metabolic cross-feeding between this

[48] S. Tabor and C. C. Richardson, *Proc. Natl. Acad. Sci. U.S.A.* **82,** 1074 (1985).

[49] S. Møller, C. Sternberg, J. B. Andersen, B. B. Christensen, J. L. Ramos, M. Givskov, and S. Molin, *Appl. Environ. Microbiol.* **64,** 721 (1998).

strain and another strain (which degrades benzyl alcohol to benzoic acid derivatives) when the strains were grown together on benzyl alcohol as the sole carbon source.

Changes in environmental conditions will, however, also have significant effects on the physiological state of the organisms. Such shifting conditions may result in several responses, such as altered growth rates, stress response, starvation, or even cell death. Most of these responses can be visualized directly using specific promoter–reporter fusions.

The cellular growth rate is one essential parameter that is difficult to approach directly, as counting of cells in a developed biofilm is virtually impossible. The ribosome number is a reliable indicator of growth rate in bacteria growing in balanced growth[50] and has been used as a standard for growth rates in biofilm-embedded bacteria as well.[51] However, because ribosomes are normally quite stable, cells that experience sudden changes in growth conditions, i.e., famine or other kinds of growth inhibition, will maintain a relatively high ribosome number for some time, despite a low growth rate. This implies that in environments with fluctuating growth conditions, ribosome counts are not reliable measures of the physiological status of the cells. By monitoring the rate of ribosome synthesis directly, the activity of the cells can be estimated. We have constructed a monitor cassette with the ribosomal *rrnB*P1 promoter fused to a *gfp* derivative, which encodes a GFP protein with reduced stability. By inserting this monitor cassette into the chromosome of selected bacteria, we have been able to monitor the physiological state of the cells.[52] In addition, when grown in biofilms, we could relate growth to the spatial distribution of the cells (Fig. 5).

In conclusion, the combination of ribosomal RNA hybridization with monitor strains tagged with specific promoters fused to marker genes allows determinations of the spatial distribution of organisms and a number of their activities in biofilms (structure–function relationships). Such determinations may be related to many kinds of biofilm phenomena, ranging from gene transfer to microbial community behavior and specific, local chemical compositions of microenvironments in the biofilm.

[50] M. Schaechter, O. Maaløe, and N. O. Kjeldgaard, *J. Gen. Microbiol.* **19,** 592 (1958).

[51] L. K. Poulsen, G. Ballard, and D. A. Stahl, *Appl. Environ. Microbiol.* **59,** 1354 (1993).

[52] C. Sternberg, B. B. Christensen, T. Johansen, A. T. Nielsen, J. B. Andersen, M. Givskov, and S. Molin, submitted.

[3] Quorum Sensing Signals in Development of *Pseudomonas aeruginosa* Biofilms

By MATTHEW R. PARSEK and E. PETER GREENBERG

Introduction

Cell density-dependent regulation of gene expression has been termed quorum sensing.[1] A type of quorum sensing used by many gram-negative bacteria involves the use of freely diffusible acylhomoserine lactone (acyl-HSL) signaling molecules. At high population densities the environmental concentrations of acyl-HSLs can reach a critical level at which the signal can interact with a transcriptional regulatory protein and activate the expression of specific sets of genes.

Quorum sensing has been shown to regulate the expression of a number of virulence factors in human and plant pathogens. For example, the opportunistic pathogen *Pseudomonas aeruginosa* possesses at least two quorum sensing systems: *las* and *rhl*. Each quorum sensing system has its own acyl-HSL synthase (LasI and RhlI), transcriptional regulator (LasR and RhlR), and acyl-HSL signal.[2–7] Signals for *las* and *rhl* systems are *N*-(3-oxododecanoyl)-HSL and *N*-butyryl-HSL, respectively. These systems have been reported to regulate a battery of virulence factors, as well as the gene encoding the stationary phase sigma factor, *rpoS*.[2,8–11] For reviews of quorum sensing in gram-negative bacteria, see Refs. 12–16.

[1] W. C. Fuqua, S. C. Winans, and E. P. Greenberg, *J. Bacteriol.* **176,** 269 (1994).
[2] L. Passador, J. M. Cook, M. J. Gambello, L. Rust, and B. H. Iglewski, *Science* **260,** 1127 (1993).
[3] J. P. Pearson, K. M. Gray, L. Passador, K. D. Tucker, A. Eberhard, B. H. Iglewski, and E. P. Greenberg, *Proc. Natl. Acad. Sci. U.S.A.* **91,** 197 (1994).
[4] J. P. Pearson, L. Passador, B. H. Iglewski, and E. P. Greenberg, *Proc. Natl. Acad. Sci. U.S.A.* **92,** 1490 (1995).
[5] M. J. Gambello and B. H. Iglewski, *J. Bacteriol.* **173,** 3000 (1991).
[6] U. A. Ochsner, A. K. Koch, A. Fiechter, and J. Reiser, *J. Bacteriol.* **176,** 2044 (1994).
[7] A. Latifi, K. M. Winson, M. Foglino, B. W. Bycroft, G. S. A. B. Stewart, A. Lazdunski, and P. Williams, *Mol. Microbiol.* **17,** 333 (1995).
[8] U. A. Ochsner and J. Reiser, *Proc. Natl. Acad. Sci. U.S.A.* **92,** 6424 (1995).
[9] J. M. Brint and D. E. Ohman, *J. Bacteriol.* **177,** 7155 (1995).
[10] E. C. Pesci, J. P. Pearson, P. C. Seed, and B. H. Iglewski, *J. Bacteriol.* **179,** 3127 (1997).
[11] A. Latifi, M. Foglino, K. Tanaka, P. Williams, and A. Lazdunski, *Mol. Microbiol.* **21,** 1137 (1996).
[12] W. C. Fuqua, S. C. Winans, and E. P. Greenberg, *Annu. Rev. Microbiol.* **50,** 727 (1996).
[13] C. Fuqua and E. P. Greenberg, *Curr. Opin. Microbiol.* **1,** 183 (1998).

0076-6879/99 $30.00

Biofilms of *P. aeruginosa* have been characterized extensively. These biofilms can develop into highly differentiated communities of bacteria enmeshed in exopolysaccharide (EPS) separated by water channels.[17–19] The developmental process involved in the conversion of single *P. aeruginosa* cells attached to a surface into highly differentiated communities could be controlled by some type of intercellular communication.[19,20] Therefore, we hypothesized that quorum sensing might be related to biofilm development. We now know that biofilms of *P. aeruginosa* lacking a functional *las* quorum sensing system are abnormal.[21] However, it is unclear which gene(s) regulated by the *las* system is required for normal biofilm development.

This article describes the experimental methods used to examine the role of quorum sensing in *P. aeruginosa* biofilm development. It also describes the manipulations of the strains used in our studies and the continuous culture reactor system used to observe developing biofilms. This article should provide a basic outline for studying the role of quorum sensing in the development of biofilms for organisms other than *P. aeruginosa.*

Overall Strategy and Preparation of Bacterial Strains

Our analysis of the role quorum sensing plays in biofilm development involved the generation of the proper bacterial mutant strains. As mentioned earlier, *P. aeruginosa* possesses two quorum sensing systems. The initial strategy was to study biofilms of strains lacking functional *las* and *rhl* systems and compare these biofilms to a wild-type *P. aeruginosa* biofilm. The mutant strain PAO-JP2 contains deletions in *rhlI* and in *lasI* and is unable to synthesize acyl-HSL signals.[9,22] Strains containing single mutations in the autoinducer synthase genes, PAO-JP1 (*lasI*) and PDO100 (*rhlI*), were also studied. For construction of strains containing inactivated

---

[14] G. P. C. Salmond, B. W. Bycroft, G. S. A. B. Stewart, and P. Williams, *Mol. Microbiol.* **16,** 615 (1995).

[15] D. M. Sitnikov, J. B. Schineller, and T. O. Baldwin, *Mol. Microbiol.* **17,** 801 (1995).

[16] E. C. Pesci and B. H. Iglewski, *Trends Microbiol.* **5,** 132 (1997).

[17] J. R. Lawrence, D. R. Korber, B. D. Hoyle, J. W. Costerton, and D. E. Caldwell, *J. Bacteriol.* **173,** 6558 (1991).

[18] D. DeBeer, P. Stoodley, and Z. Lewandowski, *Biotech. Bioeng.* **44,** 636 (1994).

[19] J. W. Costerton, Z. Lewandowski, D. E. Caldwell, D. R. Korber, and H. M. Lappin-Scott, *Annu. Rev. Microbiol.* **49,** 711 (1995).

[20] D. G. Allison and P. Gilbert, *J. Industr. Microbiol.* **15,** 311 (1995).

[21] D. G. Davies, M. R. Parsek, J. P. Pearson, B. H. Iglewski, J. W. Costerton, and E. P. Greenberg, *Science* **280,** 295 (1998).

[22] J. P. Pearson, E. C. Pesci, and B. H. Iglewski, *J. Bacteriol.* **179,** 5756 (1997).

chromosomal genes we used standard techniques.[23,24] Targeting the autoinducer synthase genes (e.g., *lasI* and *rhlI*) for inactivation is important. This allows phenotypic complementation of an autoinducer synthase mutant by adding the appropriate acyl-HSL to the growth medium. Subsequent study of a strain with a mutation in the respective quorum sensing regulator gene would provide verification of the role of a particular quorum sensing system in biofilm development.

Once Flow-Through Continuous Culture System

There are a number of ways to grow and visualize bacterial biofilms[19,25] (see elsewhere in this volume). We use a once flow-through continuous culture system with a single flow cell.[26] The flow cell is 0.1 mm deep, 1.4 cm wide, and 4.0 cm long and is constructed out of polycarbonate. A glass coverslip (1.5 oz, 22 × 60 mm) is affixed to the top of the flow cell by means of a metal plate, which screws into the polycarbonate and seals an O ring of rubber around the viewing chamber. The glass coverslip is the surface on which biofilm development is monitored. The system (Fig. 1) consists of a reservoir (6-liter flask) from which the medium is drawn into a mixing flask (250-ml flask). An air pump is used to oxygenate the medium in the mixing flask. The medium is then pumped from the mixing flask through the flow cell and then into an effluent flask (4 liters). A septum positioned upstream of the flow cell allows for inoculation of the medium in the flow cell. A bubble trap shunt upstream of the flow cell (see Fig. 1) allows for the redirection of flow and bypass of the flow chamber (required for chamber inoculation and removal of bubbles). Flow in the system is driven by a peristaltic pump, which provides an even, pulseless, laminar (Reynolds number 0.17) flow at a rate of 130 μl/min. The tubing used is made of silicone and is 0.06 mm in diameter (Cole-Parmer, Vernon Hills, IL) except for the tubing threaded around the peristaltic pump, which is made of the more durable Norprene (Cole-Parmer). The tubing is joined using polypropylene connectors (Cole-Parmer). Atmospheric pressure in this closed system is maintained through the use of air vents (0.2-μm pore size; Gelman Sciences, Ann Arbor, MI) attached to the reservoir, mixing, and effluent flasks (see Fig. 1).

[23] F. J. DeBruijn and S. Rossbach, *in* "Methods for General and Molecular Bacteriology" (P. Gerhardt, ed.), p. 387. ASM Press, Washington DC, 1994.

[24] V. L. Miller and J. J. Mekalanos, *J. Bacteriol.* **170,** 2575 (1988).

[25] P. Gilbert and D. G. Allison, *in* "Microbial Biofilms: Formation and Control" (S. P. Denyer, S. P. Gorman, and M. Sussman, eds.), p. 29. Blackwell Press, London, 1993.

[26] D. G. Davies and G. G. Geesey, *Appl. Environ. Microbiol.* **61,** 860 (1995).

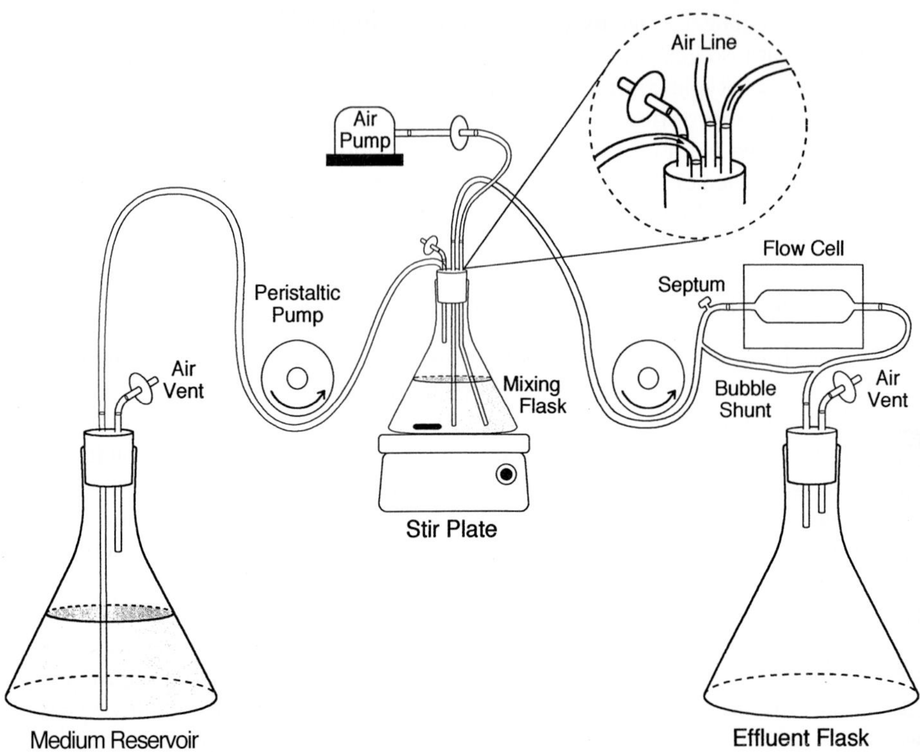

FIG. 1. Once flow continuous culture system. This diagram indicates the different elements of the system. Flow through the system is driven by a peristaltic pump. The medium is aerated in the mixing flask.

This system is assembled prior to autoclaving. Twenty-five milliliters of culture medium is placed in the mixing flask. The tubing running out of the culture medium reservoir and lines into and out of the mixing flask are clamped to prevent overflow of the medium during autoclaving. Aluminum foil is placed over the air vents to prevent water from entering the filters. The air vent lines in the reservoir, mixing, and effluent flasks are not clamped. The metal plate is partially screwed into the polycarbonate viewing chamber of the flow cell with the glass coverslip in place. After 1 hr of autoclaving the apparatus is cooled, the clamps removed, and the metal plate tightened to make a seal between the rubber O ring and the glass coverslip. Flow is initiated and allowed to proceed for 2 hr, at which time the tubing, mixing flask (about 50 ml), and viewing chamber are filled with the medium.

Sample Preparation and Inoculation

The following protocol is used to prepare and inoculate samples.

1. A 10-ml late exponential phase culture grown in EPRI medium [0.005% sodium lactate, 0.005% sodium succinate, 0.005% ammonium nitrate, 0.00019% KH_2PO_4, 0.00063% K_2HPO_4 (pH = 7.0), 0.001% Hutner salts, 0.1% glucose, and 0.001% L-histidine][21] is used to inoculate the system.
2. Inoculation is as follows: Flow in the reactor system is stopped and a clamp is placed upstream of the septum. The 10-ml culture is injected as a single dose into the reactor through the septum.
3. Bacteria are incubated in the chamber for approximately 1 hr. The clamp is then removed and flow reinitiated in order to wash free, nonadherent bacteria out of the system. Bacteria adhering to the glass coverslip are then monitored throughout biofilm development.

Use of Green Fluorescent Protein for Imaging *P. aeruginosa* Biofilms

Biofilm bacteria grown in the reactors just described can be resolved easily with conventional transmitted light microscopy during the initial stages of biofilm formation. However, when a biofilm matures and thickens, resolution of the bacteria within the biofilm becomes more difficult. To facilitate imaging of the three-dimensional composition of a biofilm, fluorescent stains such as acridine orange and fluorescein have been used.[27] Stained bacteria can be resolved using either epifluorescence microscopy or scanning confocal laser microscopy (SCLM). SCLM allows resolution of individual bacteria throughout a biofilm. There are several problems associated with the use of stains. Some stains are toxic to bacterial cells, and penetration of the stain to cells located deep within the biofilm is often diffusion limited. Furthermore, staining is invasive and serves as an end point to the experiment.

The green fluorescent protein (GFP) of *Aequorea victoria* has been used in a number of eukaryotic and prokaryotic systems to tag cells fluorescently or as an indicator of gene expression.[28,29] Green fluorescent protein contains a cyclic tripeptide that serves as a fluorophore, so no cellular cofactors or metabolites are required for fluorescence and it is stable for days inside the cell once synthesized.[30,31] A number of mutants of GFP

[27] D. E. Caldwell, D. R. Korber, and J. R. Lawrence, *J. Microbiol. Methods* **15,** 249 (1992).
[28] M. Chalfie, Y. Tu, G. Euskirchen, W. W. Ward, and D. C. Prasher, *Science* **263,** 802 (1994).
[29] S. Inouye and F. I. Tsuji, *FEBS Lett.* **341,** 277 (1994).
[30] C. W. Cody, D. C. Prasher, W. M. Westler, F. G. Prendergast, and W. W. Ward, *Biochemistry* **32,** 1212 (1993).
[31] W. W. Ward and S. H. Bokman, *Biochemistry* **21,** 4535 (1982).

TABLE I
CONSTRUCTS EXPRESSING GREEN FLUORESCENT PROTEINS THAT REPLICATE IN PSEUDOMONADS

Plasmid	GFP allele	Maxima[a] Excitation	Maxima[a] Emission	Selection[b]	Comments
pEX1.8 series	10C	513	527	Cb	~8.2-kb plasmid has *lacI*^{Q} gene and GFP under control of *Ptac*
	W7	433	475		
	P4-3	381	445		
pBBR1mcs5 series	10C	513	527	Gm	~5-kb plasmid has GFP under control of *Plac*
	W7	433	475		
	P4-3	381	445		
pAMTGFP	mutGFP2	481	507	Cm	Same as pBBR1mcs5 with Cm instead of Gm marker
pMRP9-1	mutGFP2	481	507	Cb	~5.2-kb plasmid has GFP under control of *Ptac*

[a] Maxima in nanometers and as determined previously.[36]

[b] Antibiotic concentrations used in *P. aeruginosa*: carbenicillin (Cb, 300 μg/ml), gentamicin (Gm, 200 μg/ml), and chloramphenicol (Cm, 200 μg/ml).

that show an increase in the quantum yield or a shifted excitation and emission wavelengths are also available.[32–34] Two advantages of using GFP over fluorescent stains are that GFP is nontoxic and it allows the continuous observation of a biofilm as it develops.

Many of the plasmids currently available for GFP expression in bacteria contain *Escherichia coli* origins of replication or are designed for use in eukaryotic cells. Because these plasmids do not replicate in pseudomonads, we and others have constructed a series of broad host range plasmids that contain various GFP alleles, expression systems, and antibiotic resistance markers (for our constructs, see Table I and Fig. 2).[21,35] These vectors also have origins of replication that will allow propagation in a variety of strains. To introduce these vectors into *P. aeruginosa,* we used electroporation.

Electroporation of P. aeruginosa

All centrifugation steps are carried out for 5 min at 6000*g* at 4°, and the 300 m*M* sucrose solution is kept ice cold.

[32] R. Heim, D. C. Prasher, and R. Y. Tsien, *Proc. Natl Acad. Sci. U.S.A.* **91,** 12501 (1994).

[33] S. Delagrave, R. E. Hawtin, C. M. Silva, M. M. Yang, and D. C. Youvan, *Biotechnology* **13,** 151 (1995).

[34] B. P. Cormack, R. H. Valdivia, and S. Falkow, *Gene* **173,** 33 (1996).

[35] G. V. Bloemberg, G. A. O'Toole, B. J. J. Lugtenberg, and R. Kolter, *Appl. Environ. Microbiol.* **63,** 4543 (1997).

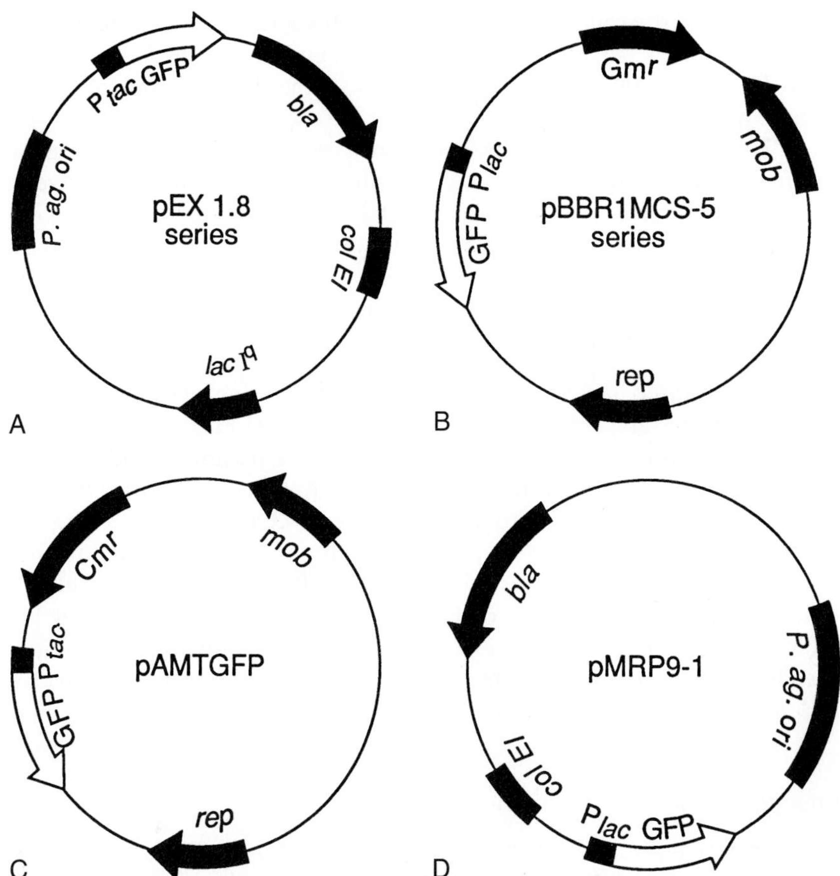

FIG. 2. Different GFP plasmids capable of replication in pseudomonads. (A) The pEX1.8 series harbors 10C, P4-3, or GFPmut2 GFP alleles under control of the *tac* promoter. These plasmids contain a *lacI^q* gene, so expression from the *tac* promoter requires the addition of IPTG. (B) The pBBR1MCS-5 series has the 10C, P4-3, and GFPmut2 GFP alleles under control of the *lac* promoter. pBBR1MCS-5 series plasmids do not have *lacI^q* but they do have mobilization (*mob*) genes for conjugation. (C) The construct pAMTGFP has the GFPmut2 allele under control of the *tac* promoter. (D) The construct pMRP9-1 has the GFPmut2 allele under control of the *lac* promoter.

1. Grow a culture of *P. aeruginosa* overnight at 30° in Luria-Bertani (LB) broth.
2. Subculture this into fresh LB broth 1:100 and grow at 30° until an OD_{600} of 0.400 is reached.
3. Centrifuge the culture.

4. Suspend the cells in a volume of 300 m*M* sucrose equal to the original culture volume. Centrifuge and remove the supernatant fluid.

5. Resuspend the cells in 300 m*M* sucrose at one-half the original culture volume and centrifuge.

6. Resuspend the cells in 300 m*M* sucrose at one-fiftieth the original culture volume. Add 100 μl of this suspension to an electroporation cuvette (1 mm) containing 5–100 ng of plasmid DNA and incubate for 5 min on ice.

7. Electroporate cells using the following parameters: 2.5 V, 200 Ω, 25 μF.

8. Add 700 μl of fresh LB broth and incubate for 3 hr at 30°. Plate out the cells on selective media.

Electroporated bacteria are assayed for GFP by using an epifluorescence microscope. A standard fluorescein isothiocyanate (FITC) filter is used to visualize GFP fluorescence. For strains containing plasmids producing the P4-3 allele of GFP (blue fluorescence), an ultraviolet (UV) filter block is required, although the quantum yield of the P4-3 allele is not as high as the other mutations.[36] Specific filter blocks for the blue variants of GFP are available.[36] Although we did not use the P4-3 allele in our study, it should prove a useful tool in combination with the 10C, GFPmut2, or W7 GFP alleles for dual-labeling experiments. The antibiotic concentrations used for plasmid maintenance in *P. aeruginosa* strains are carbenicillin (300 μg/ml), gentamicin (100 μg/ml), and chloramphenicol (200 μg/ml).

Imaging *Pseudomonas aeruginosa* Biofilms

A variety of techniques can be used to monitor biofilm development. Transmitted light microscopy is effective during the initial stages of biofilm formation. We use an Olympus BH2 microscope (Olympus, Melville, NY) with a 60× S PlanApo long working distance lens. Images are captured using an Optronics charge-coupled device (Optronics Engineering, Goleta, CA) and the imaging program Image-Pro Plus 3.0 for Windows 95 (Media Cybernatics, Silver Spring, MD).

However, as a biofilm begins to mature and EPS is produced, cells within the EPS matrix become difficult to resolve with transmitted light microscopy. To overcome this problem we introduce the GFP expression vector pMRP9-1 into the *P. aeruginosa* strains. This type of GFP strategy has been used previously to visualize cells within a biofilm.[37] Epifluorescence microscopy or SCLM allows increased resolution of individual cells within

[36] R. Heim and R. Y. Tsien, *Curr. Biol.* **6,** 178 (1996).

[37] B. Christensen, C. Sternberg, and S. Molin, *Gene* **173,** 59 (1996).

a more mature biofilm. Epifluorescence is observed by using the Olympus BH2 microscope fitted with a mercury lamp and a FITC filter block. Scanning confocal laser microscopy is performed with a Bio-Rad MRC600 confocal microscope (Hercules, CA). The excitation band is 488 nm with a 514 nm cutoff. A 50X ULWD Olympus lens is used with a Kalman filter pass of 5.

Pseudomonas aeruginosa biofilms are monitored over a period of 14 days. Average biofilm depth and cell density are catalogued over this time period. Cell density is measured using transmitted light images. A minimum of 3000 cells in 10 fields of view are analyzed by measuring the nearest cell centroid to each study cell centroid. The average distance between nearest cells is used as a measurement of average cell density. This analysis was developed by G. Harkin (Montana State University, Bozeman, MT). Average biofilm depth is measured by taking 20 separate fields of view and using SCLM and a focus motor to determine the depth of the biofilm.

Mature biofilms (14 days) of *P. aeruginosa* PAO1 (wild-type) and PAO-JP2 (*lasI rhlI*) have been compared (Fig. 3). The PAO-JP2 biofilms are thinner and more densely packed with cells (Fig. 3). We have also tested strains containing single mutations, PDO100 (*rhlI*) and PAO-JP1 (*lasI*). This analysis reveals that the mutant biofilm phenotype is due to the *lasI* mutation (Fig. 3). To confirm that this mutant phenotype results from the lack of *N*-3-oxododecanoyl-HSL production in the mutant, we grew PAO-JP1 biofilms in the presence of 10 μM 3-oxo-dodecanoyl-HSL. Acyl-HSL restores the wild-type biofilm phenotype (Fig. 3). Many acyl-HSLs are now available commercially. A consideration in designing these experiments is that acyl-HSLs are somewhat unstable at alkaline pH. This is due to the

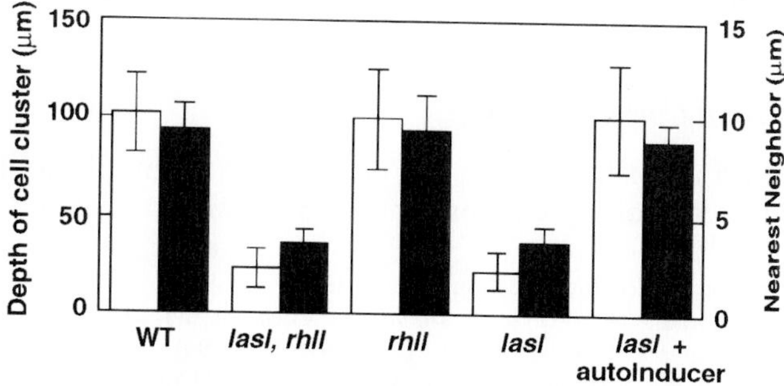

FIG. 3. Statistical representation of biofilm depth and density of tested strains. Open bars indicate the depth of *P. aeruginosa* cell clusters. Filled bars indicate the density of cells within the biofilm or nearest neighbor cell.

hydrolysis of the homoserine lactone ring. A formula approximating the half-life (in days) of acyl-HSL decomposition is $1/(1 \times 10^7 [OH^-])$.[38]

Analysis of Biofilm Exopolysaccharide

Exopolysaccharide is produced by *P. aeruginosa* biofilm cells. The EPS matrix is partially responsible for the spacing of bacteria within the biofilm.[17,19] A difference in the makeup of the EPS vs the wild-type may explain the mutant biofilm phenotype. To study this possibility we performed an initial characterization of the EPS in *P. aeruginosa* PAO1 and PAO-JP1 biofilms. Two tests were conducted: a test measuring total carbohydrates as an indicator of total biofilm EPS levels and an analysis of uronic acids, which is indicative of the relative amounts of the exopolymer alginate. Alginate is composed of $(1 \rightarrow 4)$-linked β-D-mannuronic acid and α-L-guluronic acid and has been implicated as an important virulence factor in chronic cystic fibrosis infections as well as an agent conferring protection of biofilm cells from antibiotics.[39–41]

Biofilms were grown in a once flow-through continuous system configured similarly to the system described earlier. However, the biofilms were grown on size 13 silicone tubing with a flow rate of 130 μl/min or on size 15 silicone tubing at a flow rate of 800 μl/min.

Preparation of Biofilm Samples for EPS Analysis

1. Biofilms are grown for 7 days and then harvested by draining the fluid out of the tubing and washing the biofilm with sterile EPRI medium.
2. The tubing is sliced lengthwise with a razor and biofilm material is scraped from the tubing.
3. Material is centrifuged for 10 min at 13,000 rpm in a microcentrifuge. The sedimented material is then used in the total carbohydrates and uronic acids assay.

Assay for Total Carbohydrates[42]

1. Add 0.5 ml of the sample just prepared to 0.5 ml phenol [5% (v/v) in H_2O] and 2.5 ml of H_2SO_4 (containing 2.5 g/liter hydrazine sulfate). The

[38] A. Eberhard, personal communication.

[39] J. R. W. Govan and J. A. M. Fyfe, *J. Antimicrob. Chemother.* **4,** 233 (1978).

[40] J. R. W. Govan and V. Deretic, *Microbiol. Rev.* **60,** 539 (1996).

[41] W. W. Nichols, S. M. Dorrington, M. P. E. Slack, and H. L. Walmsley, *Antimicrob. Agents Chemother.* **35,** 518 (1988).

[42] P. K. I. Kintner and J. P. V. Buren, *J. Food Sci.* **47,** 756 (1982).

sample is very viscous and pipetting may be difficult. To facilitate pipetting, add an estimated equal volume of water to the material.

2. Incubate the sample in the dark for 60 min.

3. Measure absorbance at 490 nm.

4. Generate a standard curve with glucose and convert results to glucose equivalents.

5. The protein concentration of the initial biofilm material should be determined and the glucose equivalents normalized to protein concentration.

Uronic Acids Assay[43]

1. Stock solutions of *m*-hydroxydiphenyl [0.15% (v/v) in 0.5% NaOH] and sodium borate (0.0125 *M* $Na_2B_4O_7$ in concentrated sulfuric acid) are prepared in chromic acid-washed glassware while cooled in an ice bath.

2. Add 200 μl of sample to the glass test tube.

3. Add 1.2 ml of the sodium borate solution.

4. Vortex the sample for 10 sec and incubate for 5 min in a 100° water bath.

5. Cool the sample in an ice bath for 3 min and then add 20 μl of the *m*-hydroxydiphenyl solution and vortex.

6. Within 5 min of vortexing, measure absorbance at 520 nm using the sample generated in step 7 as a blank.

7. Carbohydrates other than uronic acids are measured by adding 20 μl of 0.5 *M* NaOH to the sample at step 5 instead of *m*-hydroxydiphenyl and then measuring absorbance at 520 nm.

8. A standard curve is generated with glucuronic acid (0–100 μg/liter), and the readings from step 6 are converted to glucuronic acid equivalents.

The sample may have to be diluted in water to generate a measurement that falls on the glucuronic acid standard curve.

We have found that there is little difference in levels of EPS in mutant and wild-type biofilms.[21] These tests are useful for analyzing total EPS and a major component of *P. aeruginosa* EPS, but further investigation is required to understand the relationship between EPS synthesis and the abnormal biofilms of the *lasI* mutant.

[43] M. Dubois, K. A. Giles, J. K. Hamilton, P. A. Rebers, and F. Smith, *Anal. Chem.* **28,** 350 (1956).

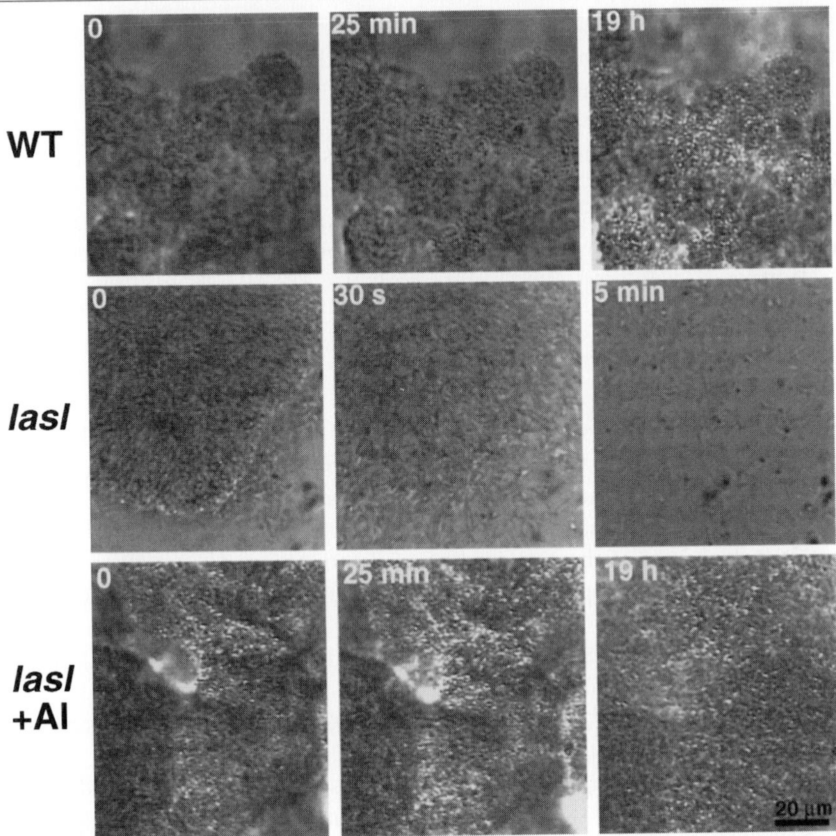

FIG. 4. SDS-induced detachment of biofilm cells. These are phase-contrast photomicrographs of mature biofilms of PAO1 (wild type), PAO-JP1 (*lasI*), and PAO-JP1 with exogenously added 3-oxododecanoyl-HSL to the growth medium. These photomicrographs were taken at times immediately before SDS addition and at the times indicated subsequent to addition.

Sodium Dodecyl Sulfate Treatment for Testing Biofilm Integrity

Biofilms can be extremely resistant to antimicrobial treatments.[44–46] *Pseudomonas aeruginosa* biofilms are resistant to treatment with low levels of the surfactant/biocide, sodium dodecyl sulfate (SDS). As described pre-

[44] I. G. Duguid, E. Evans, M. R. W. Brown, and P. Gilbert, *J. Antimicrob. Chem.* **30,** 803 (1992).

[45] H. Anwar, M. Dasgupta, K. Lam, and J. W. Costerton, *J. Antimicrob. Chemother.* **24,** 647 (1989).

[46] W. W. Nichols, M. J. Evans, M. P. E. Slack, and H. L. Walmsley, *J. Gen. Microbiol.* **135,** 1291 (1989).

viously, we have found that wild-type biofilms are resistant to prolonged exposure to SDS (0.2%).[47] However, the PAO-JP1 biofilm is dispersed completely following a 5-min exposure to SDS (Fig. 4). Resistance of the mutant biofilm to SDS is restored by the addition of 10 μM 3-oxododecanoyl-HSL to the medium (Fig. 4). This analysis provides further evidence that the mutant biofilm phenotype is significantly different than the wild-type.

SDS Treatment of Established Biofilms

1. Mature biofilms (10–14 days) are established in the reactors of the once flow-through continuous culture system as described earlier.
2. Flow through the reactor is stopped, and SDS (0.2%) in 10 ml of the growth medium is injected through the septum and flushed into the flow cell.
3. After 30 min the flow is reinitiated. The process is monitored over time using transmitted light microscopy.

This study demonstrates that quorum sensing in *P. aeruginosa* plays a role in the development of biofilms. However, a number of obvious follow-up studies remain to be performed. Although production of the acyl-HSL signal by LasI is crucial for normal biofilm development, whether this signal is acting through the regulator LasR remains to be determined. If a *lasR* mutant forms abnormal biofilms, then identification of the target gene(s) regulated by LasR involved in biofilm development is the next step. We use only a single flow rate, type of medium, and temperature in our experiments. The effects of these parameters on signaling in biofilms remain to be determined. Additionally, the only biocide tested in our study was SDS. The susceptibility of mutant biofilms to other biocides, such as the clinically relevant antibiotic tobramycin, is not known. Finally, does quorum sensing play a role in the development of biofilms in other gram-negative or gram-positive bacteria? Is intercellular signaling a common means for coordinating gene expression in biofilm communities? These studies will help determine the program of gene expression required for normal biofilm development and how cell-to-cell signaling influences this program.

Acknowledgments

We are grateful for research support from the National Science Foundation (MCB9808308). M.R.P. is a National Research Science Award Postdoctoral fellow supported by Grant GM 18740-01A1. We are grateful to A. L. Schaefer, B. L. Hanzelka, and M. Whiteley for the critical reading of this manuscript.

[47] D. G. Davies, Ph.D. thesis in microbiology, Montana State, Bozeman, MT, 1996.

[4] Monitoring Gene Expression in Biofilms

By CLAIRE PRIGENT-COMBARET and PHILIPPE LEJEUNE

Introduction

Gene Expression in Biofilm

Biofilms are matrix-enclosed collections of microorganisms adherent to each other and/or to biological and abiotic surfaces or interfaces.[1,2] This definition includes cells grown on media solidified by a gelling agent, such as agar, and attachment of pathogens or symbiotes to host eukaryotic cell surfaces and tissues. This article focuses attention on cells grown on abiotic or agar surfaces.

It is generally accepted that immersion of a clean substratum in natural liquids is immediately followed by the rapid adsorption of organic molecules and ions to the surface[3] to form a so-called "conditioning film." The composition of this film may be influenced by the physicochemical properties of the substratum and may affect the attachment of the microorganisms.[4] Once cells have attached to a substratum, growth and division occur in three dimensions. Microorganisms are organized in mushroom-shaped microcolonies embedded in polymer matrix. These cellular islands are initially separated by extensive void regions that progressively tend to coalesce.

Because the adhesion of microorganisms to solid surfaces is a very common phenomenon, biofilms develop on every surface that comes in contact with natural liquids, such as blood, urine, and seawater. Many detrimental biofilms exist, including dental plaque, implant-associated infections, and corrosive biofilms. Given the important medical and economic consequences of biofilm development, there is a strong need to understand the colonization processes in order to design nontoxic coating treatments able to prevent biofilm formation.

Environmental conditions at surfaces and within the biofilm are different from those in the aqueous phase. The differences concern pH, ion concentration, osmolarity, viscosity, nutrient availability, and gas exchange rates.

[1] J. W. Costerton, Z. Lewandowski, D. E. Caldwell, D. R. Korber, and H. M. Lappin-Scott, *Annu. Rev. Microbiol.* **49,** 711 (1995).

[2] R. J. Palmer, Jr., and D. C. White, *Trends Microbiol.* **5,** 435 (1997).

[3] C. E. ZoBell, *J. Bacteriol.* **46,** 39 (1943).

[4] M. C. M. van Loosdrecht, J. Lyklema, W. Norde, and A. J. B. Zehnder, *Microbiol. Rev.* **54,** 75 (1990).

0076-6879/99 $30.00

Abundant evidence indicates that cells grown on solid surfaces or on agar show physiological properties that are different from cells grown in liquid media. A spectacular example is swarmer cell differentiation on agar surfaces of *Vibrio, Proteus, Clostridium, Bacillus,* and *Serratia* species.[5] Numerous studies reported by Goodman and Marshall[6] have also shown differences in growth rate, exopolymer production, fimbriae and flagella synthesis, and susceptibility to antibiotics between biofilm and liquid grown cells.

Increasing attention is now being paid to bacterial surface sensing at the gene expression level. Studies have shown that genes can be activated or repressed when bacteria come into contact with an inert surface and as the biofilm develops.[6,7] Specific targeting of the genes whose expression is modified when cells undergo changes from a planktonic to a sessile state using bioreporter technology [i.e., fusions with genes encoding β-galactosidase, the green fluorescent protein (GFP), or luciferase] may improve our knowledge of biofilm development. The goal of this article is to describe methods for studying the switching off or switching on of genes in biofilms and monitoring their expression.

Reporter Gene Systems

Rapid advances in understanding gene regulation have arisen from the development of reporter gene techniques.[8] As target gene products are usually difficult to assay directly, reporter gene fusions are powerful tools as they allow easy quantification of reporter protein activity.

Genetic fusions can be constructed by recombinant DNA technology. A DNA fragment carrying the promoter of the target gene is inserted into the cloning site of a plasmid vector containing the promoterless reporter gene. Depending on the plasmid vector, transcriptional or translational fusions can be generated.

Numerous transposon-based reporter systems that generate transcriptional and translational fusions are also available. In these tools, the promoterless reporter gene is placed at one extremity of the transposable element. Insertion of the transposon in proper orientation in a target gene places the reporter gene under the control of the promoter. The insertion event can often be selected by means of drug resistance carried by the transposon.

[5] J. Henrichsen, *Bacteriol. Rev.* **36,** 478 (1972).

[6] A. E. Goodman and K. C. Marshall, *in* "Microbial Biofilms" (H. M. Lappin-Scott and J. W. Costerton, eds.), p. 80. Cambridge Univ. Press, Cambridge, 1995.

[7] C. Prigent-Combaret, O. Vidal, C. Dorel, and P. Lejeune, submitted for publication.

[8] T. J. Silhavy, M. L. Berman, and L. W. Enquist, eds., *in* "Experiments with Gene Fusions." Cold Spring Harbor Laboratory, Cold Spring Harbor, NY, 1984.

Transposon delivery vehicles are, in most cases, nonreplicating phages or suicide plasmids. After the transposition event, delivery vehicles are lost in the host cell.

Transposons carrying the transposase gene often provide unstable fusions. Stable insertions that are unable to undergo additional rounds of transposition can be obtained by using a delivery system in which the transposase gene is located outside of the transposon itself (in the ideal case, on the transposon donor molecule, which is lost following transposition). Such transposons are generally referred to as minitransposons.

Random transposon mutagenesis can be obtained with Tn*5*, Tn*3*, Tn*9*, and Mu derivatives as their specificity of insertion is sufficiently low, whereas Tn*10* derivatives exhibit the highest degree of specificity so that insertions into favored "hot spots" can occur.

Many Mu derivatives with *lacZ* or *lux* reporter genes exist.[9,10] These are defective phages that are unable to enter a lytic cycle but can transpose, as Mu *A*, Mu *B* genes and sites required for transposition are present. To mutagenize genes in the chromosome of *Escherichia coli,* lysates are usually prepared by thermoinduction of mini-Mu and Mu*c*ts62 lysogen strain. The *c* repressor controls the transcription of Mu genes and, in the presence of *c*ts allele, the repressor is thermosensitive. The phage lysate is then used to transduce mini-Mu transposons into the recipient strain. Mu has a very broad host range and can be introduced into species that are resistant to Mu infection by transduction with a phage of different host range, conjugation, or transformation.

Compilations of fusion construction and quantification methods in enterobacteria, myxobacteria, cyanobacteria, and other groups are available in Volume 204 of *Methods in Enzymology.*[9]

Reporter Genes Mainly Used for Monitoring Gene Expression in Biofilms

Two main approaches could be used to monitor gene expression in biofilms: *in situ* and real-time measurement of reporter gene activity on cells attached to surfaces or indirect measurement on cells detached from the substratum.

In situ monitoring of gene expression using microscopic methods requires the development of reporter systems that can be applied to the visualization of bacteria living as single cells or in multiple layers in biofilm

[9] J. H. Miller, *Methods Enzymol.* **204** (1991).

[10] N. Symonds, A. Toussaint, P. Van de Putte, and M. M Howe, eds., *in* "Phage Mu." Cold Spring Harbor Laboratory, Cold Spring Harbor, NY, 1987.

structures. Important criteria in selection of the microscopic methods are either spatial resolution (particularly with regard to the thickness of samples and permeability of biofilm cells to substrates) and temporal resolution, i.e., rapid response to both induction and cessation of gene activity.

In many cases, attached bacteria can be removed from the substratum, suspended in cold medium, and reporter gene expression assayed. In this case, spatial resolution is not required and any reporter gene (*lacZ, lux, gfp, pho*A, *gus*) may be used.

Biofilms have become the subject of study by several groups and numerous strategies have been employed to monitor gene expression in biofilms. This article presents the work of several teams even if many approaches are very similar to each other.

We first describe the main reporter gene systems used in biofilm gene expression studies and methods for their quantification.

lacZ Gene of Escherichia coli

The *lacZ* gene, which encodes β-galactosidase, is one of the most commonly used reporter genes in prokaryotic and eukaryotic cells. A large variety of chromogenic and fluorogenic substrates permits the quantification of β-galactosidase activity by spectrophotometric assays (chromogenic substrate) or using fluorimetry, epifluorescence, and confocal laser scanning microscopy (CLSM; fluorogenic substrate).

β-Galactosidase Spectrophotometric Assay with Chromogenic Substrate o-Nitrophenyl-β-D-galactoside. *o*-Nitrophenyl-β-D-galactoside (ONPG) is hydrolyzed by β-galactosidase, yielding *o*-nitrophenol that absorbs at 420 nm. A commonly employed assay using this substrate has been described by Miller.[11] Cells (after $A_{600\,\text{nm}}$ determination) are permeabilized with chloroform, a given volume (V) is suspended in Z buffer, and incubated at 28° with ONPG (final concentration: 0.7 mg/ml). After a given incubation time (t), the reaction is stopped with Na_2CO_3 (final concentration, 0.3 M). Absorption of the mixture is then measured at 420 and 550 nm (for correction of cellular debris absorption) with a spectrophotometer. β-Galactosidase specific activity is expressed in Miller units using the relationship:

$$AS = 1000 \frac{A_{420} - (1.75\,A_{550})}{V\,tA_{600}}$$

β-Galactosidase Fluorogenic Substrate: Fluorescein Di-β-D-galactopyranoside. Two fluorogenic substrates are commonly used to monitor β-ga-

[11] J. H. Miller, *in* "A Short Course in Bacterial Genetics." Cold Spring Harbor Laboratory Press, Cold Spring Harbor, NY, 1992.

lactosidase activity: fluorescein di-β-D-galactopyranoside (FDG) and methylumbelliferyl-β-D-galactoside. FDG is probably the most used and sensitive substrate for detecting β-galactosidase activity. Nonfluorescent FDG is hydrolyzed sequentially by β-galactosidase, first to fluorescein monogalactoside and then to highly fluorescent fluorescein that emits at 530 nm. Fluorescence may then be detected by epifluorescence microscopy using a fluorescein filter block or by CLSM under laser light with an excitation wavelength of 490 nm and an emission wavelength of 530 nm. In confocal laser microscopy, interference signals from out-of-focus objects are eliminated with pinholes near the laser emitter and detectors; the image formed could be assimilated as a thin optical section of the sample.[12] Digitization of a sequential series of x–y images obtained while focusing through the third dimension (z) of the specimen permits its three-dimensional (3D) reconstruction. Moreover, CLSM enables the examination of specimens on opaque supports and is thereby suitable to examine natural biofilms. Despite reports about the need for initial permeabilization of bacterial cells for FDG to penetrate,[13] Alvarez and co-workers[14] reported measurements by flow cytometry of the β-galactosidase activity, on different *E.coli* strains, without any permeabilization or cell lysis: FDG [1 mg/ml in H_2O containing 1% dimethyl sulfoxide (DMSO) and 1% (v/v) ethanol] was incubated at a final concentration of 0.08 mg/ml (i.e., 0.13 m*M*) with 10^7 cells for an ideal incubation time of 20 min at 37° before fluorescence measurements. After longer incubation periods, fluorescein diffuses out of the cells. In other studies, FDG was used at higher concentrations, such as 15 m*M* with gram-positive cells and 0.4 m*M* with gram-negative cells immobilized on coverslips,[15,16] but with a smaller incubation time (5 min) before microscopic examinations. Finally, it is worth noting that more lipophilic derivatives, such as C_{12}-FDG, have been developed. These are able to penetrate live animal cells without previous permeabilization and give good retention of the fluorescent product at 37°. Unfortunately, these substrates are unable to penetrate into gram-negative viable bacterial cells as their outer membrane may function as a barrier for lipophilic and hydrophobic solutes.[17] Depending on the sample (species, thickness, etc.), it seems that FDG labeling methods have to be focused (concentration and incubation time,

[12] D. E. Caldwell, D. R. Korber, and J. R. Lawrence, *Adv. Microbial Ecol.* **12,** 1 (1992).

[13] R. Nir, Y. Yisraeli, R. Lamed, and E. Sahar, *Appl. Environ. Microbiol.* **56,** 3861 (1990).

[14] A. M. Alvarez, M. Ibáñez, and R. Rotger, *Biotechniques* **15,** 974 (1993).

[15] P. J. Lewis, C. E. Nwoguh, M. R. Barer, C. R. Harwood, and J. Errington, *Mol. Microbiol.* **13,** 655 (1994).

[16] C. E. Nwoguh, C. R. Harwood, and M. R. Barer, *Mol. Microbiol.* **17,** 545 (1995).

[17] A. Plovins, A. M. Alvarez, M. Ibañez, M. Molina, and C. Nombela, *Appl. Environ. Microbiol.* **60,** 4638 (1994).

necessity of cell permeabilization). The quantitative measurement of β-galactosidase activity with FDG could be performed using image analysis as described by Nwoguh and co-workers.[16] The principle is described later. Digitized images of the same field obtained with phase-contrast and epifluorescence microscopy were analyzed: fluorescence intensity values were attributed to cells seen by phase microscopy using any image processing software to delineate (with a binary mask) the surface area of each cell. The pixel gray scale intensity obtained for all pixels attributed to each cell was summed to give values of fluorescence per cell in arbitrary units. This method allows the relative levels of β-galactosidase activity to be determined in individual cells. It must be emphasized, however, that the relationship between the observed fluorescence and the actual number of β-galactosidase molecules per cell is complex and is influenced by variations in the permeability of the cell envelope to FDG and depends on the real penetration of substrate into the inner layers of biofilms.

β-Galactosidase Fluorogenic Substrate: Methylumbelliferyl-β-D-galactoside. Methylumbelliferyl-β-D-galactoside is used as a substrate for β-galactosidase to evaluate *in situ algC-lacZ* expression in individual cells of *Pseudomonas aeruginosa* during the initial stages of biofilm development on glass surfaces.[18,19] Methylumbelliferyl-β-D-galactoside is hydrolyzed by β-galactosidase in 7-hydroxy-4-methylcoumarin, a fluorescent compound that can be detected by epifluorescence microscopy, under UV excitation, with a blue filter block or by CLSM under laser light with an excitation wavelength of 488 nm and an emission wavelength of 510 nm. Methylumbelliferyl-β-D-galactoside can be dissolved in *N,N*-dimethylformamide and added to culture medium to a final concentration of 10 mg/liter. Problems of cell permeability were not reported with this substrate. However, fluorescence intensity also depends on cell wall permeability and on access of the substrate to cells of the whole sample.

gfp Gene of Jellyfish Aequorea victoria

The gene encoding green fluorescent protein has become an important visual marker (fluorescence) of gene expression, mainly because GFP requires no special substrate or cofactor for detection.[20] Through an autocatalytic reaction, GFP forms a cyclic peptide that is highly fluorescent and stable. The wild-type green fluorescent protein of the jellyfish *A. victoria* (238 amino acids) absorbs blue light at 396 and 475 nm (minor peak), which

[18] D. G. Davies, A. M. Chakrabarty, and G. G. Geesey, *Appl. Environ. Microbiol.* **59,** 1181 (1993).

[19] D. G. Davies and G. G. Geesey, *Appl. Environ. Microbiol.* **61,** 860 (1995).

[20] M. Chalfie, Y. Tu, G. Euskirchen, W. W. Ward, and D. C. Prasher, *Science* **263,** 802 (1994).

permits good matching to standard fluorescein filter sets, and emits green light at 509 and 540 nm.[21] Various groups have obtained mutants of GFP exhibiting diverse spectral properties that will allow simultaneous analysis of gene expression from a number of different promoters.[22,23] The *gfp* gene has also been mutated to improve the fluorescence intensity of the reporter protein and chromophore formation kinetics.[24,25] These mutations ameliorate a potentially significant limitation in using GFP as a reporter protein for rapid gene inductions. Up until now, when downshifts in gene expression were considered, the *gfp* reporter system could not be used as the wild-type GFP is very stable in its mature fluorescent form. Andersen and collaborators[26] had constructed new variants showing decreased stability. The addition of short destabilizing peptide tails to the C-terminal end of wild-type GFP increases its sensitivity to the action of proteases. Shuttle vectors containing the improved *gfp-mut2*[24] gene expressed from *lac, tac,* or *npt-2* promoters have been constructed to fluorescently tag *E. coli, Pseudomonas* spp.[27] and other gram-negative bacteria.[28] An improved *gfp* cassette is available to create plasmidial transcriptional fusions.[29] Suicide plasmids[30] and several Tn*5*[31]- and Tn*10*[32]-based transposons containing promoterless-improved *gfp* genes have also been constructed to generate genomic *gfp* fusions in diverse bacterial species. For this reporter system, spatial resolution criteria have been largely fulfilled through the development of confocal microscopy (*z* sectioning) and because there is no need of substrate or cofactor for detection. Some new improved GFP variants also show good temporal response.

[21] D. C. Prasher, V. K. Eckenrode, W. W. Ward, F. G. Prendergast, and M. J. Cormier, *Gene* **111,** 229 (1992).

[22] S. Delagrave, R. E. Hawtin, C. M. Silva, M. M. Yang, and D. C. Youvan, *BioTechnology* **13,** 151 (1995).

[23] T. Ehrig, D. J. O'Kane, and F. G. Prendergast, *FEBS Lett.* **367,** 163 (1995).

[24] B. P. Cormack, R. H. Valdivia, and S. Falkow, *Gene* **173,** 33 (1996).

[25] R. Heim, A. B. Cubitt, and R. Y. Tsien, *Nature* **373,** 663 (1995).

[26] J. B. Andersen, C. Sternberg, L. K. Poulsen, S. P. Bjørn, M. Givskov, and S. Molin, *Appl. Environ. Microbiol.* **64,** 2240 (1998).

[27] A. G. Matthysse, S. Stretton, C. Dandie, N. C. McClure, and A. E. Goodman, *FEMS Microbiol. Lett.* **145,** 87 (1996).

[28] G. V. Bloemberg, G. A. O'Toole, B. J. J. Lugtenberg, and R. Kolter, *Appl. Environ. Microbiol.* **63,** 4543 (1997).

[29] W. G. Miller and S. E. Lindow, *Gene* **191,** 149 (1997).

[30] V. S. Kalogeraki and S. C. Winans, *Gene* **188,** 69 (1997).

[31] R. S. Burlage, Z. K. Yang, and T. Mehlhorn, *Gene* **173,** 53 (1996).

[32] S. Stretton, S. Techkarnjanaruk, A. M. McLennan, and A. E. Goodman, *Appl. Environ. Microbiol.* **64,** 2554 (1998).

lux Genes of Vibrio fisheri

Luciferase activity (bioluminescence) is used frequently as a reporter system in eukaryotic and prokaryotic cells. Luciferase is an oxidase that catalyzes the oxidation of reduced flavin ($FMNH_2$) to form an intermediate peroxide, which then reacts with a long-chain aldehyde to give blue-green luminescence emitting at 490 nm and oxidized flavin plus the corresponding long-chain fatty acid as products.[33] The color of the emitted light may differ according to the nature of the substrate, the enzyme structures, and the environment of the emitter (the presence of optical biological filters and accessory lumiphores).

In *V. fisheri,* seven *lux* genes encode the regulatory and biochemical activities necessary for light production and are organized into two operons.[34] The *luxR* and *luxl* genes control the transcription of one operon containing *luxl, C, D, A, B,* and *E.*[35] Light production occurs in dense bacterial cultures and is controlled at the transcriptional level by an extracellular signal molecule termed autoinducer. The α and β subunits of luciferase are encoded by *luxA* and *luxB. lux C, D,* and *E* genes are necessary for producing (recycling and *de novo* synthesis) the long-chain aldehyde (tetradecanal). Therefore, construction of a transposon that would generate *lux* fusions requires the insertion of a fragment containing the promoterless *lux* genes *A, B, C, D,* and *E* into a transposon. Mini-Mu*lux* (Km^R) and Mini-Mu*lux* (Tet^R) transposons have been constructed by Engebrecht and co-workers,[36] replacing the *lac* genes of mini-Mu*lac* (Mu dI1681 or Mu dII1681) by *lux* genes.

Luciferase-based reporters are particularly adapted to monitoring gene expression in biofilms because of their good temporal response. Light can be measured with X-ray or photographic films, by visual or microscopic observations, and using a luminometer or a scintillation counter in chemiluminescence mode. Assays using a scintillation counter are very sensitive and accurate. However, as enzyme activity requires oxygen, $FMNH_2$, and tetradecanal, care must be taken to ensure that these components do not become rate limiting in certain physiological conditions. The stability or activity of luciferase and associated *lux* enzymes and the capacity to exprime *lux* genes may also be influenced by the particular host. *Escherichia coli* and *Vibrio parahaemolyticus* are suitable hosts for light production and mini-Mu*lux* transposition.

[33] J. W. Hastings, *Gene* **173,** 5 (1996).

[34] J. Engebrecht, K. Nealson, and M. Silverman, *Cell* **32,** 773 (1983).

[35] J. Engebrecht and M. Silverman, *Proc. Natl. Acad. Sci. U.S.A.* **81,** 4154 (1984).

[36] J. Engebrecht, M. Simon, and M. Silverman, *Science* **227,** 1345 (1985).

To conclude, *lux* and *gfp* reporter systems are suitable for *in situ* monitoring gene expression in biofilms as problems of wrong spatial resolution are not raised. The sole restriction is that oxygen limitation in the lower layers of biofilms could give some artifactual results. Finally, double labeling using *lux* and *gfp* reporter systems would be desirable to monitor, in parallel, the expression of two genes or to associate the expression of one gene to a given species. It seems, unfortunately, that GFP and luciferase activity could not be detected in a single cell.[37] Both processes may require too much cellular energy.

Genetic Responses on Agar Surfaces

Experiments showing modulated expression of genes in bacteria grown on agar surfaces are interesting to consider because the methods used in these types of studies can be applied easily to biofilms developed on solid surfaces.

Gene Activation Measurement with lux and lacZ Fusions

Vibrio parahaemolyticus occupies a variety of habitats in marine and estuarine environments, where it can be isolated as free-living bacteria or attached to submerged surfaces (sediments, shellfish, chitinaceous plankton, etc.).[38] It is also a human pathogen that causes gastroenteritis transmitted by seafood and seawater, usually in coastal areas during warm months.[39] When grown in liquid media, *V. parahaemolyticus* produces a single polar flagellum (encoded by *fla* genes) that propels the cells through the liquid environment (swimming). When these bacteria are propagated on media solidified with agar, the cells undergo morphological changes[40]: the cells cease septation, begin to elongate, and are surrounded with, in addition to polar flagella, numerous peritrichous lateral flagella (encoded by *laf* genes) that ensure the bacteria to move over surfaces (swarming). This differentiation process is reversible. Cells with lateral flagella adsorb to surfaces and move over the surface, resulting in expansion of the area of colonization.[41,42]

[37] R. J. Palmer, Jr., C. Phiefer, R. Burlage, G. Sayler, and D. C. White, *in* "Bioluminescence and Chemiluminescence: Molecular Reporting with Photons" (J. W. Hastings, L. J. Kricha, and P. E. Stanley, eds.), p. 445. Wiley, New York, 1996.

[38] T. Kaneko and R. R. Colwell, *J. Bacteriol.* **113,** 24 (1973).

[39] P. A. Blake, R. E. Weaver, and D. G. Hollis, *Annu. Rev. Microbiol.* **34,** 341 (1980).

[40] S. Shinoda and K. Okamoto, *J. Bacteriol.* **129,** 1266 (1977).

[41] M. R. Belas and R. R. Colwell, *J. Bacteriol.* **151,** 1568 (1982).

[42] J. R. Lawrence, D. R. Korber, and D. E. Caldwell, *J. Bacteriol.* **174,** 5732 (1992).

Many studies on the surface induction of lateral flagellar genes using different reporter gene fusions are now described. Transcriptional *laf::lacZ*[43] and *laf::lux*[44] fusions using mini-Mu transposons derived from Mu dI1681 (Km^r, *lac, cts*)[45] are constructed to study swarmer cell differentiation (see Table I). Transduction of mini-Mu*lac* (Tet^R) or mini-Mu*lux* (Tet^R) into *V. parahaemolyticus* is performed after the thermoinduction of *E. coli* K12 strains [MC4100 mini-Mu*lac* (Tet^R) or mini-Mu*lux* (Tet^R), P1 *clr-100 CM*]. The P1 repressor and the cts repressor of Mu are inactivated by a temperature shift to 42°. The lytic cycle of phage P1 and mini-Mu replication can then be coinduced. P1 *clr-100 CM* is capable of mini-Mu DNA packaging and infection of recipient cells, but does not replicate in *V. parahaemolyticus.*[43] After infection of the recipient cell, nonswarming (Laf$^-$), tetracycline-resistant, Lac$^+$ or Lux$^+$, transductants are selected.

The expression of the *laf661*::mini-Mu*lac* fusion is measured by quantitative ONPG spectrophotometric assay on cells grown in liquid medium and on cells detached from the surface of the same but agar-solidified medium (indirect measurement).[43] Authors reported a higher β-galactosidase activity in cells grown on agar. This result was later validated by the same team with another reporter gene and by direct quantification. Expression of the *laf-3639*::mini-Mu*lux* fusion was indeed compared between cells grown in liquid marine medium (500 μl) contained within a plastic 1.5-ml centrifuge tube or on the surface of agar cylindrical cores.[44] Luminescence is measured at 30-min intervals using the chemiluminescence mode of the scintillation counter. At the same intervals, samples are removed to determine cell numbers. Authors reported that the luminescence per cell is more than 10-fold higher in cells grown in solid medium than in liquid. Using this method, the authors also demonstrated that light production is not influenced by the composition and osmolarity of the medium, the gelling agent used, or the nature of the surface. Induction of *laf* expression is influenced by the stage of growth of the inoculum and by medium viscosity. Finally, confinement of cells between agar core or in an agar matrix induces *laf.* All these observations led the authors to suggest that the surface induction of lateral flagella expression may involve sensing of forces that obstruct movement of the polar flagellum (as during cultivation on solid media, on liquid media with high viscosity, or after agglutination of polar flagellum with antibodies).[46] Furthermore, McCarter and collaborators have shown

[43] R. Belas, A. Mileham, M. Simon, and M. Silverman, *J. Bacteriol.* **158,** 890 (1984).
[44] R. Belas, M. Simon, and M. Silverman, *J. Bacteriol.* **167,** 210 (1986).
[45] B. A. Castilho, P. Olfson, and M. J. Casadaban, *J. Bacteriol.* **158,** 488 (1984).
[46] L. McCarter, M. Hilmen, and M. Silverman, *Cell* **54,** 345 (1988).

TABLE I

SURFACE-MODULATED REPORTER GENE FUSION

Designation	Type	Reporter gene[a]	Species	Localization	Genetic tool[b]	Marker[c]	Vector	Ref.
Pathogenic strain								
sfaA-lacZ	Protein	*lacZ', Y*	*E. coli* (uropathogenic strain)	Chromosomal	*In vitro* construction followed by site-specific recombination of whole plasmid pTTS703	Ap	pTTS703	48
flaA-lacZ	Operon	*lacZ*	*P. mirabilis*	Plasmidial	*In vitro* construction in *E. coli*	Ap	pCED-Mob	52
flaB-lacZ	Operon	*lacZ*	*P. mirabilis*	Plasmidial	*In vitro* construction in *E. coli*	Ap	pCED-Mob	52
algC-lacZ	Operon	*lacZ*	*P. aeruginosa*	Plasmidial	*In vitro* construction	Cb	pNZ63	18, 19
algD-lacZ	Operon	*lacZ*	*P. aeruginosa*	Plasmidial	*In vitro* construction	Gm	pSDF13,15	61
Marine biofilm communities								
laf-lacZ	Operon	*lacZ, Y, A*	*V. parahaemolyticus*	Chromosomal	mini-Mu*lac* (Tc^R) transposition	Tet	P1 *clr*-1.00 CM	43
X-lacZ	Operon	*lacZ, Y, A*	*Deleya marina* and *Pseudomonas* S9	Chromosomal	mini-Mu*lac* (Tc^R) transposition	Tet	pJO100	54
laf-lux	Operon	*luxC, D, A, B, E*	*V. parahaemolyticus*	Chromosomal	mini-Mu*lux* (Tc^R) transposition	Tet	P1 *clr*-100 CM	44

chi-gfp	Operon	*gfp*	*Pseudoalteromonas* sp. *Vibrio* sp. *Psychobacter* sp.	Chromosomal	mini-Tn*10*-*gfp*-kan transposition	Km	pLOFKm*gfp*	32
Other environmental biofilm communities								
Pu-gfp	Operon	*gfp*	*P. putida*	Chromosomal	Tn*5*-based transposition	Km	pJAB30	65
Pm-gfp	Operon	*gfp*	*P. putida*	Chromosomal	Tn*5*-based transposition	Str	pJAB26	65
Laboratory biofilm model								
X-lacZ (Mu dX)	Operon	*lacZ, Y, A*	*E. coli* K12	Chromosomal	Mu dX transposition	Ap, Cm	SCH135	7
X-lacZ (Mu dl1681)	Operon	*lacZ, Y, A*	*E. coli* K12	Chromosomal	Mu dl1681 transposition	Km	*E. coli* strain POl1681	7
csgA-uidA	Operon	*uidA* (GUS)	*E. coli* K12	Chromosomal	*In vitro* construction before homologous recombination	Km	*E. coli* strain pOV874	7

[a] A "prime" indicates that a particular gene is truncated of a few amino acids at its amino terminus.

[b] The transposons mini-Mu*lac* (Tc^R) and mini-Mu*lux* (Tc^R) are derivatives of Mu dI1681. All of the Mu elements carry the *c*ts62 mutation. Tc^R, tetracycline resistance gene.

[c] Resistance to Ap, ampicillin; Cb, carbenicillin; CM, chloramphenicol; Gm, gentamicin; Km, kanamycin/neomycin; Sm, streptomycin; and Tet, tetracycline.

that swarmer cell differentiation is also controlled at the transcriptional level by a second signal, limitation for iron, using northern blot analyses.[47]

Gene activation in cells grown on agar surfaces has also been reported for a pathogenic *E. coli* strain. S fimbrial adhesins (Sfa) enable this strain to bind to sialic acid-containing eukaryotic receptor molecules. A *sfaA-lacZ* translational fusion has been constructed by Schmoll and co-workers[48] and used to study the influence of environmental conditions (such as contact with a surface) on the production of S fimbriae. This chromosomal fusion is constructed in two steps: a promoterless part of the *sfaA* gene fused to *lacZ* is introduced in pJM703-1 (Ap^R) to give pTTS703. This suicide plasmid carrying the R6K origin is unable to replicate into Pir^- strains, as the Pir protein is required for replication initiation of replicons with R6K origin.[30] This plasmid is introduced in a Pir^-, Lac^- uropathogenic strain and an ampicillin-resistant Sfa^- derivative is selected. Site-specific recombination of the whole plasmid into the *sfa* locus occurs, and the resulting *lacZ* fusion is thereby under the control of the promoter of *sfaA*. β-Galactosidase activity of the *sfaA–lacZ* fusion (see Table I) is measured using the spectrophotometric assay in cells grown in LB liquid medium and in LB agar plate. The authors reported a fourfold increase of β-galactosidase activity in cells grown on solid medium. In this study, an accurate inner control was performed: the authors checked that, in a wild-type Lac^+ strain, no induction of β-galactosidase is observed in cells grown on solid medium versus cells grown in liquid medium.

Indirect Measurement of Gene Activation by Quantification of mRNA Transcripts

When grown on surfaces or in viscous environments, *Proteus mirabillis* also undergoes swarmer cell differentiation. This process involves 40 to 60 genes that are expressed coordinately in response to inhibition of the rotation of the flagella[49] and to the presence of glutamine in the surrounding medium.[50] *Proteus mirabilis* is frequently associated with urinary tract infections, particularly in patients with chronic urinary catheterization.[51] The *fla* locus comprises two tandemly linked, nearly identical copies of flagellin-encoding genes, *flaA* and *flaB*. In order to study surface sensing of these

[47] L. McCarter and M. Silverman, *J. Bacteriol.* **171,** 731 (1989).

[48] T. Schmoll, M. Ott, B. Oudega, and J. Hacker, *J. Bacteriol.* **172,** 5103 (1990).

[49] R. Belas, D. Erskine, and D. Flaherty, *J. Bacteriol.* **173,** 6279 (1991).

[50] C. Allison, H.-C. Lai, D. Gygi, and C. Hughes, *Mol. Microbiol.* **8,** 53 (1993).

[51] R. Belas, *in* "Urinary Tract Infections: Molecular Pathogenesis and Clinical Management" (H. L. T. Mobley and J. W. Warren, eds.), p 271. Am. Soc. for Microbiology, Washington, DC, 1996.

genes, Belas,[52] using several reliable methods, compared the expression of *flaA* and *flaB* between cells grown in liquid medium and on agar surfaces. First, transcriptional fusions with *lacZ* as reporter gene are used. These fusions (see Table I) are constructed using the low copy number plasmid pCED-Mob (Ap^R) carrying a promoterless *lacZ* gene and the *mob* gene to permit plasmid transfer from *E. coli* strain S17-1 (λpir) to *P. mirabilis* by conjugal filter mating.[53] The β-galactosidase activity of the *flaA–lacZ* and *flaB–lacZ* plasmidial fusions are compared between swimmer cells that have grown in L broth versus swarmer cells that have grown (for a same incubation time) on the same medium solidified by the addition of 1.5% agar. Swarmer cells are scraped off the agar surface with an L-shaped glass rod and suspended in ice-cold phosphate-buffered saline (PBS). Swimmer and swarmer cell suspensions are washed in ice-cold PBS, and β-galactosidase is then measured by the quantitative ONPG assay. A 7.5-fold increase of β-galactosidase is observed in swarmer cells for the *flaA–lacZ* fusion. For the *flaB–lacZ* fusion, no β-galactosidase activity is measured in both populations. To validate these results obtained with plasmidial *lacZ* transcriptional fusions, measurements of *flaA* and *flaB* mRNAs in swarmer and swimmer cells are undertaken by northern blot analyses. The protocol used is described next. In a wild-type strain, the cellular RNAs are purified from swimmer and swarmer cell suspensions prepared as described earlier. Equal amounts of total RNAs are applied on a 1.2% agarose–formaldehyde gel. After electrophoresis, RNA molecules are transferred to nitrocellulose filters and revealed by [γ-^{32}P]ATP-specific probes for *flaA* and *flaB*. The *flaA*-specific probe reveals an identical 1.2-kb monocistronic message in both populations. Densitometric analyses of blots on the photographic film are performed and confirmed that *flaA* transcription is increased eightfold in swarmer cells. No *flaB* mRNA is detected with the *flaB*-specific probe. Finally, primer extension experiments were also performed on swarmer and swimmer cell transcripts using [γ-^{32}]ATP *flaA*- and *flaB*-specific primers. After reverse transcription, treatment with RNase, phenol extraction, and ethanol precipitation, DNA samples are loaded onto a polyacrylamide DNA sequencing gel. Primer extension experiments confirm that *flaA* transcription, initiated at the same site in both swarmer and swimmer cells, is increased eightfold in swarmer cells and that *flaB* is a silent gene. This study demonstrates surface sensing of lateral flagella genes using a good approach as three suitable and accurate methods were used and led to the same result.

[52] R. Belas, *J. Bacteriol.* **176,** 7169 (1994).

[53] R. Belas, D. Erskine, and D. Flaherty, *J. Bacteriol.* **173,** 6289 (1991).

Genetic Responses on Solid Surfaces

lacZ Fusions in Marine Bacteria

In 1991, Dagostino *et al.*[54] reported on the use of mini-Mu*lac* transposon mutagenesis in marine bacteria, *Deleya marina* and *Pseudomonas* S9, to target gene switched on at surfaces. This was one of the first studies to focus on screening methods for surface sensing genes. *Escherichia coli* J100 containing the plasmid pOJ100 [pRK2013::mini-Mu (Tet^R, *lacZ*)] was used as the donor strain[55] and mutagenesis was performed using filter matings as described by Belas *et al.*[3] Transposition mutants were selected on agar media containing appropriate antibiotics and X-Gal. Mutants failing to express β-galactosidase on X-Gal agar were tested for their response to a solid inert surface, on agar, and on polystyrene surfaces of microtiter wells. *Deleya marina* transconjugants showed unstable *lacZ* insertions. A *Pseudomonas* S9 mutant (among 192) with pale blue coloration on agar gave a dark blue color in X-Gal culture in the microtiter plate. When a X-Gal-containing colorless liquid culture of this mutant was transferred to a petri dish surface, a dark blue color developed within 3 hr. This screening method allowed the identification of genes switched on the surface. Unfortunately, in this study, β-galactosidase specific activity was not quantified nor the target promoter identified.

lacZ Fusions in Escherichia coli

We have been interested in searching for genes switched on or switched off when *E. coli* cells develop biofilms. In our laboratory, a mutant of *E. coli* K12 able to develop biofilms on hydrophilic and hydrophobic surfaces in less than 24 hr has been isolated.[56] Random insertion mutagenesis of this strain has been performed with Mu dX[57] (*lacZ,* Cm^R, Ap^R) and Mu dI1681 (*lacZ,* Kan^R)[45] following the procedure described by Miller.[11] These two Mu derivatives generate transcriptional *lac* gene fusions (see Table I). Mu dX is a derivative of Mu dI, which contains a Tn*9* insertion in the Mu *B* gene in order to reduce secondary transpositions. However, we have found that Mu dX insertions are difficult to transfer by P1 transduction. Moreover, it is important to ascertain that mutants contain only one Mu dX insertion (by Southern blot using a *lacZ* probe). The same control must

[54] L. Dagostino, A. E. Goodman, and K. C. Marshall, *Biofouling* **4,** 113 (1991).

[55] J. Östling, A. E. Goodman, and S. Kjelleberg, *FEMS Microbiol. Ecol.* **86,** 83 (1991).

[56] O. Vidal, R. Longin, C. Prigent-Combaret, C. Dorel, M. Hooreman, and P. Lejeune, *J. Bacteriol.* **180,** 2442 (1998).

[57] T. A. Baker, M. M. Howe, and C. A. Gross, *J. Bacteriol.* **156,** 970 (1983).

be performed with Mu dI1681 insertions (because of the presence of wild-type Mu*A* and Mu*B* genes in this mini-Mu).

Screening Method of Surface-Modulated Promoters (See Fig. 1).[7] Transposed clones obtained with Mu dX and Mu dI1681 are grown in 200 μl of M63 mannitol (0.2%) medium in the wells of a 96-well microtitration plate. After 24 hr of incubation at 30°, a visible biofilm is present on the wall of each well. The medium containing the free-living bacteria of each well is

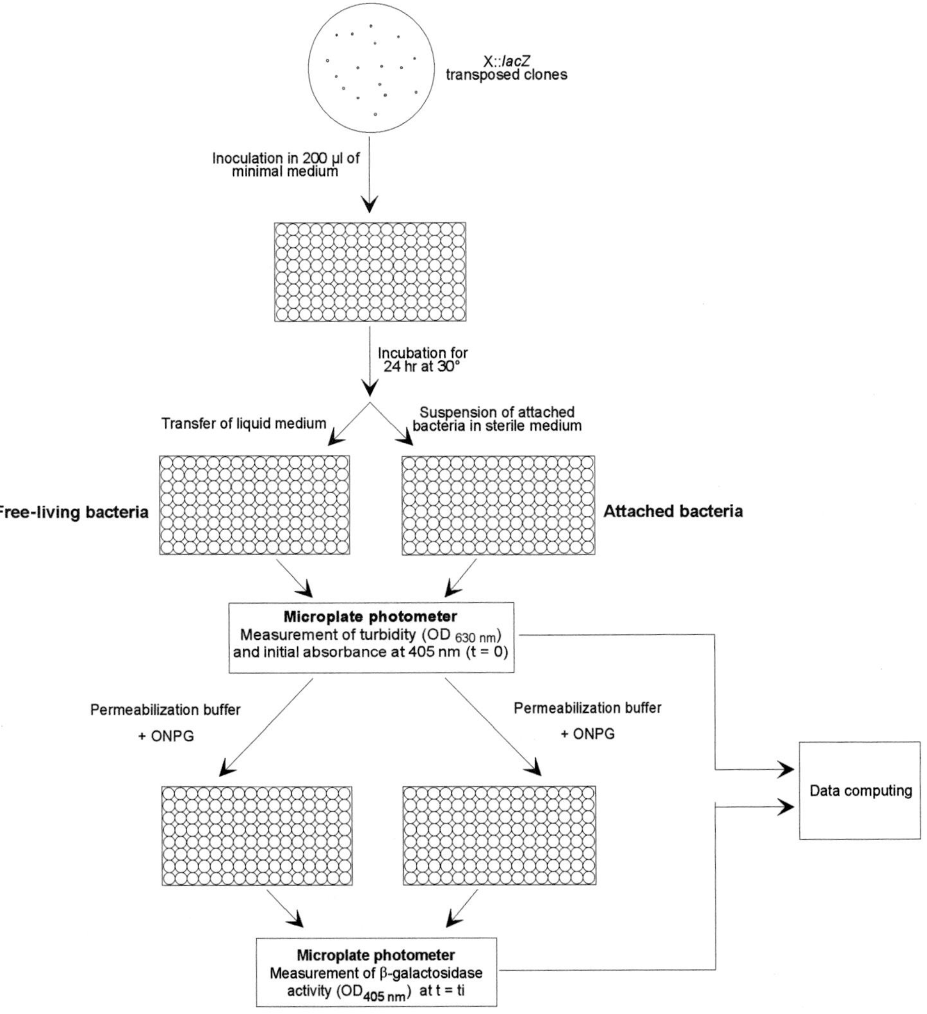

FIG. 1. Screening method of biofilm-modulated *lacZ* fusions.

then removed carefully and is introduced in the corresponding well of a precooled microtitration plate. Two hundred microliters of cold M63 mannitol medium is introduced in each well of the plate containing the biofilms, and bacteria are suspended by vigorous shaking of the plate. The turbidity (optical density at 630 nm) and the initial absorbance of suspensions at 405 nm are estimated using a microplate photometer (Labsystems or BioMérieux). Bacteria are then permeabilized by the addition of 20 μl of the permeabilization reagent (100 mM Tris–HCl, pH 7.8, 32 mM $NaPO_4$, 8 mM dithiothreitol, 8 mM CDTA, 4% Triton X-100, polymyxin B sulfate 200 μg/ml) described by Schupp and co-workers.[58] This buffer is preferred to toluene as it does not damage plastic surfaces. After 3 min of gentle shaking, ONPG is added at a final concentration of 2.7 mM. This substrate is preferred to X-Gal because its product (ONP) is highly soluble whereas the blue compound produced by the hydrolysis of X-Gal is not. Moreover, the quantification of β-galactosidase activity can thus be performed easily at 405 nm with a microplate photometer after various incubation periods at room temperature (before saturation). For each clone, the ratio between the optical density at 405 nm and at 630 nm is calculated at time t_i for free-living bacteria (R_F) and for bacteria from the biofilm (R_B) using the relationship:

$$R_{\mathrm{F\ or\ B}} = \frac{A_{405}|_{t=t_i} - A_{405}|_{t=0}}{A_{630}}$$

The final ratio ($R = R_B/R_F$) is a good and reproducible approximation of the difference in β-galactosidase specific activity between the two bacterial populations of a given clone. This semiquantitative method is used to test 446 Lac^+ fusions. We observed that the transcription of 38% of *E. coli* genes is modified during biofilm development: 22% of fusions are more expressed in the biofilm (ranging from 2- to more than 10-fold) and 16% are less expressed in the biofilm (with ratio R ranging from 0.5 to less than 0.1). Some target promoters are identified by cloning and sequencing the junction between the Mu S end and the target gene using a *lacZ*-targeted oligonucleotide.[7]

Monitoring Gene Expression in Biofilms.[7] Differences in *lacZ* expression in free-living and attached bacteria can be monitored precisely during biofilm development. To confirm the semiquantitative results obtained with the method described earlier, petri plates are filled with 20 ml of liquid M63 mannitol (0.2%) medium and inoculated with 100 μl of an overnight culture of the *lacZ* mutant of interest. After different incubation times at

[58] J. M. Schupp, S. E. Travis, L. B. Price, R. F. Shand, and P. Keim, *Biotechniques* **19,** 18 (1995).

30°, the plates (at least three for each incubation time) are incubated on ice. The liquid medium containing the free-living population is removed carefully and is kept on ice. On the bottom of the plate, the biofilm developed on the polystyrene surface is rinsed with cold sterile medium to remove nonadherent cells. Attached bacteria are then suspended in 2 ml of cold medium by pipetting up and down until there are no more aggregates. Suspensions are kept on ice. The total number of bacteria is estimated by measuring the optical density at 600 nm. We controlled that the same equivalence between one unit of optical density at 600 nm and protein amount is obtained for free-living bacteria and attached bacteria. One $OD_{600\ nm}$ unit contains 0.162 mg of proteins per milliliter (1-cm path length). We observed that a weak difference exists in the equivalence of optical density at 600 nm and the colony-forming unit (cfu) for free-living and attached bacteria: the cfu per $OD_{600\ nm}$ ratio is 1.2 times lower for attached bacteria versus free-living bacteria at different incubation times between 15 and 37 hr. This difference may be due to the fact that suspensions of attached bacteria may still contain small aggregates of bacteria producing a single colony on the plates used for the determination of cfu. Using cfu method, biofilm cells are often underestimated despite vigorous shaking or ultrasonication of detached biofilm cell suspensions.[1,59] The β-galactosidase activity is then measured on toluenized samples by following the hydrolysis of ONPG into ONP at 420 nm. This method allows us to demonstrate that different genes are induced (*pepT, wcaB, cpxA, nikA, ompC,* and *proU*) or repressed (*fliC*) in biofilm cells versus planktonic cells and to reveal some environmental parameters, such as pH, osmolarity, and quorum sensing, involved in gene expression in biofilms.[7] Fusions with the reporter gene *uidA* encoding β-glucuronidase (GUS) are also used. Samples are prepared as described earlier. The β-glucuronidase activity is then measured on toluenized samples by following the hydrolysis of *p*-nitrophenyl-β-D-glucuronide into *p*-nitrophenol at 405 nm and 37°. The β-glucuronidase activity assay is very sensitive. A chromosomal fusion between the uidA reporter gene and csgA, encoding curlin in *E. coli,* is constructed by recombination of a 7.25-kb *Pvu*II fragment of plasmid pOV874[56] containing the *csgA*::*uidA* fusion into the chromosome (O. Vidal, unpublished result). The *csgA* gene is thus inactivated and the adherent phenotype is retrieved after introduction of a high copy number plasmid (pCSG4[60]) carrying the wild-type *csgA* region. We observed a higher expression of the fusion in biofilm versus planktonic cells.

[59] G. D. Christensen, L. Baldassarri, and W. A. Simpson, *Methods Enzymol.* **253,** 477 (1995).

[60] A. Olsén, A. Arnqvist, M. Hammar, S. Sukupolvi, and S. Normark, *Mol. Microbiol.* **7,** 523 (1993).

lacZ Fusions in Pseudomonas aeruginosa

Expression of alginate genes during *P. aeruginosa* biofilm development was studied by different laboratories using *algC*[18,19] and *algD–lacZ*[61] fusions. Alginate is an exopolysaccharide produced by *P. aeruginosa* during infection of the lungs in cystic fibrosis patients and constitutes a major virulence factor. The alginate capsule is believed to allow *P. aeruginosa* cells to adhere to the epithelial cells.

algC and *algD* genes (encoding phosphomannomutase and GDP-mannose dehydrogenase, respectively) represent key regulation points in the alginate biosynthetic pathway. The promoter sequence of *algC* and *algD* are fused to a promotorless *lacZ* reporter gene (see Table I) to form pNZ63 (carbenicillin resistance)[62] and pSDF13,15 (gentamicin resistance).[61] These plasmids are introduced in *P. aeruginosa* strains and used to monitor alginate gene expression during different types of biofilm cultures.

Indirect Measurement of Alginate Gene expression. For the *algC–lacZ* fusion, a Teflon mesh is used as a substratum for biofilm growth. It is introduced aseptically into shaked Erlenmeyer flasks containing sterile medium[18] or into the reactor of a chemostat during a continuous culture experiment.[19]

At different times of culture in the presence of carbenicillin, the Teflon mesh introduced into flasks is removed and rinsed gently with sterile medium to remove nonadherent cells. The remaining attached cells (biofilm population) are removed by scraping in chilled sterile medium. Planktonic cells are cultivated in parallel in flasks containing no mesh and samples are taken. Samples of both biofilm and planktonic populations are washed by centrifugation, suspended in chilled sterile medium, and assayed quantitatively for β-galactosidase specific activity using the ONPG assay. The authors reported that between 12 and 96 hr of incubation, the expression of *algC–lacZ* is higher in biofilm cells than in planktonic cells, especially after 24 hr, when a 5-fold higher expression is observed. This preliminary result was then reproduced by the same authors during a continuous culture experiment over a period of 18 to 22 days where *P. aeruginosa* is allowed to develop biofilm structures on Teflon mesh, here in the absence of antibiotic pressure. During the first 24 hr, the flow is turned off and bacteria are allowed to colonize the mesh. An aliquot of the surrounding growth medium is collected to represent the planktonic population. The flow is than turned on gently to achieve washout of biofilms. After stabilization of effluent cell densities, bacteria shed from the biofilms under flowing conditions are

[61] B. D. Hoyle, L. J. Williams, and J. W. Costerton, *Infect. Immun.* **61,** 777 (1993).

[62] N. A. Zielinski, A. M. Chakrabarty, and A. Berry, *J. Biol. Chem.* **266,** 9754 (1991).

collected each day from the effluent. At the end of the experiment, the Teflon mesh-containing biofilms are removed and the attached bacteria are collected as described earlier. The authors reported that the expression of β-galactosidase is nearly 20-fold higher in biofilm cells and 6-fold higher in shed cells than in planktonic cells. Studies of reporter gene activity on plasmids are valid when the copy number variation from cell to cell does not influence the interpretation of reporter gene activity measurement. With this in mind, the authors measured and compared β-lactamase activity, which is proportional to the copy number of pNZ63, in biofilm and planktonic cells. β-lactamase activity assay is performed on lysed cells mixed with β-lactam substrate (benzylpenicillin, 0.208 mg/ml) for 10 min at 25° using a suitable colorimetric method described by Cohenford *et al.*[63] based on orange color development (absorption maximum at 454.5 nm) of the reaction between the neocuproine–copper reagent with penicilloic acid. The authors observed that pNZ63 is retained in biofilm cells over a period of 22 days in the absence of antibiotic pressure and that β-lactamase specific activity is 1.5-fold higher in biofilm cells than planktonic cells. They suggested that this weak difference could be neglected with regard to the high surface activation of alginate genes. Authors also verified that there is a higher uronic acid (alginate compound) amount in biofilm cells than in planktonic cells, indicating an increase of alginate synthesis.

For the *algD–lacZ* fusion, biofilms are constructed on disks of silicone rubber during a 7-day continuous culture in a modified Robbins device.[61] Using the ONPG assay, a transient higher expression of the plasmidial fusion in attached versus planktonic cells is observed. Here, the authors did not check plasmid copy numbers in biofilm and planktonic bacteria.

In Situ Measurement of Alginate Gene Expression. Continuous culture devices are used to monitor *in situ algC–lacZ* expression in individual cells during biofilm development on surfaces of a glass capillary[18] or on a glass coverslip in a flow cell system.[19] Aerated liquid medium containing carbenicillin and methylumbellilferyl-β-D-galactoside (0.01 g/liter) is pumped from a sterile reservoir through the reactors to a waste reservoir at flow rates that allow only biofilm bacteria to be retained within the reactors. Direct epifluorescence microscopic observations of attached bacteria are performed. Continuous culture devices are placed under an epifluorescence microscope, and fluorescence due to conversion of methylumbelliferyl-β-D-galactoside by β-galactosidase activity is detected under UV excitation. An observation of the same field is also performed using transmitted light with phase contrast to quantify all cells attached to the surfaces. Digitization of microscopic images and computer analysis indicate that cells change the

[63] M. A. Cohenford, J. Abraham, and A. A. Medeiros, *Anal. Biochem.* **168,** 252 (1988).

regulation of *algC* transcription during biofilm development; the majority of attached cells switch from nonfluorescent to fluorescent. This emphasizes the result obtained with indirect ONPG assay.

gfp Fusions

Stretton and co-workers[32] have described the construction of a *gfp* Tn*10*-based transposon and its use to monitor gene expression in marine strain biofilms. This transposon yields stable transcriptional *gfp* fusions, with high frequencies of transposition, in marine strains (*Pseudoalteromonas* sp. S91, *Vibrio* sp. S141, and *Psychrobacter* sp. SW5H). The promoterless *gfp* gene is cloned upstream of a kanamycin resistance gene in pLOFKm to give pLOFKm*gfp*. Mobilization of pLOFKm*gfp* from *E. coli* DH5α to the marine strains is done by plate mating with *E. coli* (pNJ5000) as a helper.[64] The mini-Tn*10*-*gfp*-*kan* transposon gives single random insertions after transposition on the genome of the marine strains, and the delivery vector is lost in 99.5% of the transconjugants. Chitinase activity-negative mutants of *Pseudoalteromonas* sp. S91 are then selected on broth containing 0.1% of colloidal chitin. By epifluorescence microscopy (excitation, 488 nm; emission, 520 nm) on liquid culture, a *chi–gfp* mutant is identified. This mutant is allowed to develop biofilms on 1-cm^2 pieces of natural biodegradable substrata, squid pen, which consists of 40% chitin and 60% protein. Although it is unable to degrade chitin, this mutant is able to grow on squid pen by digesting its proteinaceous portion. After 24 hr or 7 days, small slices are cut aseptically from a piece of squid pen and placed between a glass slide and a coverslip. CLSM microscopy is used to investigate biofilm formation of the *chi–gfp* mutant on squid pen. Excitation at 488 nm is used. At 515-nm emission, both GFP and the autofluorescent squid pen surface are visualized. AT 522- to 535-nm emission, only Gfp$^+$ microcolonies adhering to the surface of squid pen are visualized. Images of *xy* and *xz* sections are collected at both wavelength emissions, digitized, and used to generate composite images or for quantitative examination with a computer image analysis software. This work demonstrates that mini-Tn*10*-*kan*-*gfp* is a useful tool for constructing transcriptional *gfp* fusions and monitoring gene expression in single living cells of biofilm structures.

The *gfp* system was also used by Møller and collaborators[65] to investigate the expression of genes involved in the biodegradation of toluene in a bacterial community. Genes required for toluene degradation are orga-

[64] N. H. Albertson, S. Stretton, S. Pongpattanakitshote, S. Kjelleberg, and A. E. Goodman, *FEMS Microbiol. Lett.* **140,** 287 (1996).

[65] S. Møller, C. Sternberg, J. B. Andersen, B. B. Christensen, J. L. Ramos, M. Givskov, and S. Molin, *Appl. Environ. Microbiol.* **64,** 721 (1998).

nized into two separate operons and transcription is driven from the Pu (upper) and Pm (meta) promoters. The upper pathway operon is induced by toluene, xylenes, and their alcohol derivatives, whereas the meta-pathway operon is induced by benzoates. Fusions between improved *gfp* gene (*mut3a*) and Pu or Pm promoters are inserted together with the genes encoding their regulators (*xylR* or *xylS*) into the chromosome of *Pseudomonas putida,* one of the seven strains of the biofilm community. These fusions are carried by mini-Tn*5* delivery plasmid pJAB26 and pJAB30 and are allowed to transpose into the chromosome. Pure culture biofilms and mixed culture biofilms containing *P. putida* R1 strains with *Pu*::*gfp* or *Pm*::*gfp* fusion are grown in a flow chamber in the presence of benzyl alcohol. Biofilms are then fixed in 3% paraformaldehyde and embedded in 20% acrylamide. By combining GFP expression and hybridization with 16S rRNA targeting probes labeled with different fluorochromes (red or blue), the authors showed by CLSM (equipped with three detectors for green, blue, and red fluorescence) that the Pu promoter is induced continuously in *P. putida* in pure or mixed culture biofilms in the presence of benzyl alcohol, whereas the Pm promoter is only induced in *P. putida* cells growing close to *Acinetobacter* strains. This study demonstrates metabolic interactions between different species of a biofilm community using an elegant method combining direct monitoring of promoter activity and 16S rRNA labeling.

In Situ Polymerase Chain Reaction Methods for Quantification of Transcripts

Methods for prokaryotic *in situ* polymerase chain reaction (PI-PCR) adapted to the visualization of specific gene expression in bacterial communities have been developed by Hodson and co-workers.[66] Two techniques, reverse transcription PI-PCR and RNA probe extension, have been carried out successfully on cells immobilized on glass slides and may be applied to biofilm cells grown on substratum.

A model bacterial community is assembled from eight strains, including *P. putida* AC10 R-7, *P. aeruginosa* 19712, and *E. coli* HB101. The production of *nahA* mRNA transcripts in the *P. putida* strain is monitored in a bacterial community grown in the presence of 0.5% salicylate (for induction of *nahA*). After growth, cells are treated with lysozyme (0.5 mg/ml, 30 min) and proteinase K (0.10 μg/ml, 10 min) at room temperature to permeabilize the cell envelope so that enzymes and reagents diffuse freely in and out of the cell while the amplified PCR products remain inside. The cells are

[66] R. E. Hodson, W. A. Dustman, R. P. Garg, and M. A. Moran, *Appl. Environ. Microbiol.* **61,** 4074 (1995).

then fixed on a 10-mm-diameter well of a printed glass slide by ethanol dehydration using successive 3-min incubations in 50, 80, and 98% ethanol. A plastic 15-mm-diameter Gene Cone chamber is mounted around the periphery of the well to serve as a reservoir for the PCR mixture. A pretreatment of cells with RNase-free DNase is performed to eliminate genomic DNA to which primers might bind nonspecifically.

In case of reverse transcription PI-PCR, the *nahA* cDNA is synthesized with an adequate primer by the avian myeloblastis virus reverse transcriptase. After a wash in PBS, cDNA is amplified by *in situ* PCR with digoxigenin (DIG) or fluorescein-labeled nucleotides (included in the nucleotide mix containing 1 m*M* dATP, dCTP, dGTP; 0.65 m*M* dTTP; and 0.35 m*M* DIG-11-dUTP or fluorescein-12-dUTP for 10^6 to 10^7 cells). Initial denaturation at 94° for 3 min is followed by amplification of the target gene during 45 cycles with denaturing at 94° for 1 min, annealing at 42° for 1 min, and extension at 72° for 1 min. Cells are then washed in PBS buffer.

In case of reverse RNA probe extension, amplification of RNA sequences by extension of a *nahA*-targeted single primer with labeled nucleotides (included in the nucleotide mix described earlier) is carried out using reverse transcriptase (rTth DNA polymerase). Primer extension is carried out for 45 cycles with denaturing at 94°, annealing at 42°, and extension at 72°. Cells are then washed in PBS buffer.

Fluorescein–labeled nucleotide cells are viewed directly via epifluorescence microscopy (fluorescein filter block), whereas DIG-labeled nucleotide cells are detected using antibodies anti-DIG (at a ratio of 1:100, 1 hr at room temperature) conjugated to alkaline phosphatase (color detection) or to fluorescein and rhodamine (fluorescence detection). In the presence of alkaline phosphatase, 5-bromo-4-chloro-3-indolylphosphate toluidinium and nitro blue tetrazolium react to give a blue precipitate, which can be detected by light microscopy.

The authors have shown that, as expected, a signal (blue color or fluorescence) could be detected with the two *in situ* PCR methods in *P. putida* cells induced by 0.5% salicylate, but not in *P. aeruginosa* and *E. coli* strains among the marine bacterial community. Further development of quantitative *in situ* PCR methods would provide interesting insights for monitoring gene expression in biofilm.

Conclusions

Increasing attention has been paid to bacterial surface sensing at the gene expression level. Several techniques based on reporter gene technology and on assays on mRNA transcripts (Northern blots or RT-PCR) have been developed. These can be divided into two main approaches: indirect measurement on biofilm cells detached from the substratum or *in situ* measurement on biofilm cells still attached to the substratum. Further studies of the bacterial structures involved in surface sensing and biofilm formation may provide avenues for preventing biofilm formation and interfering with typical biofilm properties, such as increased resistance to antimicrobial agents.

Acknowledgments

We thank Sue Ferguson-Gottschall, Anne and Christian Tessier, and Olivier Vidal for comments on the manuscript and Christophe Combaret for helping with computer graphics. We also acknowledge C. Dorel, T. T. Le Thi, and the other members of the Laboratoire de Génétique Moléculaire des Microorganismes for their kind interest in our work.

[5] *In Situ* Analysis of Microbial Biofilms by rRNA-Targeted Oligonucleotide Probing

By WERNER MANZ

Introduction

In situ hybridization allows the detection of specific nucleic acid sequences in eukaryotic and prokaryotic cells by binding oligonucleotide probes to their complementary target sequences. The combination of these probes with indirect detection methods using haptens (digoxigenin, biotin) or fluorochromes and enzymes directly coupled to the nucleotides significantly broaden the application of this technique. Comparative sequence analysis of small subunit rRNA[1,2] provided the basis for the development of synthetic oligonucleotide probes that are specific for defined phylogenetic groups of organisms[3] and can be used as molecular tools for the determination of the bacterial phylogeny as well as the identification and *in situ* detection of bacteria. The first applications involved radioactively labeled oligonucleotides for dot-blot hybridizations against immobilized nucleic acids[4] and for *in situ* whole cell hybridizations

[1] G. E. Fox, K. J. Pechman, and C. R. Woese, *Int. J. Syst. Bacteriol.* **27,** 44 (1977).

[2] C. R. Woese, *Microbiol. Rev.* **51,** 221 (1987).

[3] G. J. Olsen, D. J. Lane, S. J. Giovannoni, N. R. Pace, and D. A. Stahl, *Annu. Rev. Microbiol.* **40,** 337 (1986).

[4] D. A. Stahl, B. Flesher, H. R. Mansfield, and L. Montgomery, *Appl. Environ. Microbiol.* **54,** 1079 (1988).

0076-6879/99 $30.00

with glutaraldehyde-fixed organisms.[5] Radiolabeling was replaced by probes linked to fluorochromes[6–8] or haptens such as digoxigenin.[9] Fluorescent dye-labeled oligonucleotide probes combined with epifluorescence microscopy are advantageous because of their superior spatial resolution and their convenient detectability. For that reason, fluorescent *in situ* hybridization (FISH) in combination with advanced microscopic techniques, such as confocal laser scanning microscopy (CLSM) and digital image analysis, became an important part of a polyphasic approach in microbial ecology for the identification and localization of microorganisms within complex environments.[10] In addition to the inherent phylogenetic information of this technique, the correlation among growth rates of bacterial cells, their average ribosome content, and the strength of the probe-conferred fluorescence can be used to determine the metabolic potential of individual cells in biofilms.[11] Defined bacterial populations in biofilms were investigated by FISH in anaerobic multispecies biofilms,[12] in drinking water and associated biofilms,[13–16] in marine photosynthetic biofilms,[17] and in activated sludge, which can be regarded as immobilized biofilm.[18–21] This article provides a brief description of the design, evaluation, and fluorescent labeling of oligonucleotides, the preparation of biofilms for hybridization, and the use of FISH for the identification

[5] S. J. Giovannoni, E. F. DeLong, G. J. Olsen, and N. R. Pace, *J. Bacteriol.* **170,** 720 (1988).

[6] E. F. DeLong, G. S. Wickham, and N. R. Pace, *Science* **243,** 1360 (1989).

[7] R. I. Amann, L. Krumholz, and D. A. Stahl, *J. Bacteriol.* **172,** 762 (1990).

[8] D. A. Stahl, and R. I. Amann, *in* " Sequencing and Hybridization Techniques in Bacterial Systematics," p. 205. Wiley, Chichester, 1991.

[9] B. Zarda, R. Amann, G. Wallner, and K. H. Schleifer, *J. Gen. Microbiol.* **137,** 2823 (1991).

[10] R. Amann, W. Ludwig, and K. H. Schleifer, *Microbiol. Rev.* **59,** 143 (1995).

[11] L. K. Poulsen, G. Ballard, and D. A. Stahl, *Appl. Environ. Microbiol.* **59,** 1354 (1993).

[12] R. I. Amann, J. Stromley, R. Devereux, R. Key, and D. A. Stahl, *Appl. Environ. Microbiol.* **58,** 614 (1992).

[13] W. Manz, U. Szewzyk, P. Eriksson, R. Amann, K. H. Schleifer, and T. A. Stenström, *Appl. Environ. Microbiol.* **59,** 2293 (1993).

[14] U. Szewzyk, W. Manz, R. Amann, K. H. Schleifer, and T. A. Stenström, *FEMS Microbiol. Ecol.* **13,** 169 (1994).

[15] S. Kalmbach, W. Manz, and U. Szewzyk, *FEMS Microbiol. Ecol.* **22,** 265 (1997).

[16] S. Kalmbach, W. Manz, and U. Szewzyk, *Appl. Environ. Microbiol.* **63,** 4164 (1997).

[17] N. B. Ramsing, M. Kühl, and B. B. Jørgensen, *Appl. Environ. Microbiol.* **59,** 3840 (1993).

[18] M. Wagner, R. Amann, H. Lemmer, and K. H. Schleifer, *Appl. Environ. Microbiol.* **59,** 1520 (1993).

[19] W. Manz, M. Wagner, R. Amann, and K. H. Schleifer, *Water Res.* **28,** 1715 (1994).

[20] W. Manz, R. Amann, W. Ludwig, M. Vancanneyt, and K. H. Schleifer, *Microbiology* **142,** 1097 (1996).

[21] W. Manz, M. Eisenbrecher, T. R. Neu, and U. Szewzyk, *FEMS Microbiol. Ecol.* **25,** 43 (1998).

of individual bacterial cells in complex biofilms. It does not deal with oligonucleotide synthesis.

Probe Design and Evaluation

The primary structure of rRNA molecules composed of highly conserved, variable and hypervariable sequence regions enables the design of diagnostic oligonucleotide probes with defined specificities on different phylogenetic levels ranging from domains to bacterial species, which might even be uncultured so far. The specificity and sensitivity of oligonucleotide probes depend on (i) properties of the oligonucleotide itself (length, G + C content, hairpin formation), (ii) hybridization parameters (temperature of hybridization and washing steps; concentration of oligonucleotide; presence and amount of competitor oligonucleotides; presence and concentration of monovalent cations and denaturing agents), and (iii) the amount and accessibility of the target sites within the bacterial ribosomes. In general, most of the gram-negative bacteria are penetrable by oligonucleotides after (para)formaldehyde fixation followed by ethanol dehydration. For gram-positive bacteria with a high G + C content of DNA, fixation with ethanol or treatment with HCl is recommended.[22]

The construction of hybridization probes requires the localization of specific rRNA stretches (signatures) that are unique for different phylogenetic levels. For this purpose, aligned rRNA sequence data have to be screened and evaluated. Data in a plain or processed (aligned) format can be obtained from the GenBank Sequence Database (http://www.ncbi.nlm.nih.gov/Web/Genbank/index.html), the DNA Data Bank of Japan (http://www.ddbj.nig.ac.jp/), and the EMBL Nucleotide Sequence Database (http://www.ebi.ac.uk/ebi_docs/embl_db/ebi/topembl.html), which incorporate DNA sequences from all available public sources. Regular data update and exchange within these databases ensure comprehensive worldwide coverage.

To evaluate the specificity, potential signature sequences for analytical *in situ* probes have to be checked for other organisms that may coincidentally have the identical nucleotide sequences. This check should be repeated in regular time intervals in order to implement new rRNA sequence data. Different software interfaces provide useful tools for data acquisition and computing, such as interfaces for the Check Probe, Sequence Match, Sequence Align, and Chimera Check of the Ribosomal Database Project

[22] C. Roller, M. Wagner, R. Amann, W. Ludwig, and K. H. Schleifer, *Microbiology* **140,** 2849 (1994).

(http://www.cme.msu.edu/RDP/), the sequence similarity search tool BLAST offered through the NCBI (http://www.ncbi.nlm.nih.gov/BLAST/), software packages offered through the ODP oligonucleotide probe database (http://www.cme.msu.edu/OPD), and the ARB software package (http://www.biol.chemie.tu-muenchen.de/pub/ARB).

Oligonucleotides intended for *in situ* hybridization have to fulfill additional requirements. The probes of a typical length ranging from 15 to 25 nucleotides should present a balanced G/C to A/T ratio and a resulting melting point between 50° and 70°. Although one mismatch centrally located in an 18-mer oligonucleotide can be sufficient for discrimination in whole cell hybridization assays,[23] sequences of probe and nontarget rRNA should have at least two mismaches for the reliable identification of individual cells in complex biofilm communities. As empirically determined, mismatches located in the center of the oligonucleotides are discriminated more easily than terminally located mismatches. The target region of the probe should be located in a rRNA region known to be accessible for whole cell hybridizations, as rRNA targeted probes suitable for dot-blot assays with immobilized DNA or RNA do not necessarily result in satisfying signal strength with *in situ* hybridizations. To prevent self-complementarity of the probe, target sites located at both sites of helical structures longer than six nucleotides should be avoided. To minimize cutting loose of terminal A/T base pairs, the oligonucleotides should have G/C base pairs at the ends. Finally, the specificity of the probes should be tested by whole cell hybridizations against multiple reference organisms presenting one, two, and three mismatches within their rRNA sequences. Additional bacterial strains, including closely related nontarget species for which rRNA sequences are not yet available, should be examined thoroughly.

Preparation of Fluorescent Oligonucleotides

Deoxyoligonucleotides may be labeled chemically or enzymatically either directly with fluorochromes and enzymes (alkaline phosphatase, horseradish peroxidase) or indirectly using haptens (biotin, digoxigenin, gold particles). Fluorochromes are visualized directly using fluorescence microscopy. Enzyme-labeled oligonucleotides are detected via the transformation of a suitable chromogenic substrate (e.g., Fast Red-Blue and Nitro-blue tetrazolium chloride-based substrates). Hapten-labeled oligonucleotides additionally need a detection step by antibodies linked to an enzyme able to catalyze a chromogenic reaction, e.g., alkaline phosphatase, prior to

[23] W. Manz, R. Amann, W. Ludwig, M. Wagner, and K. H. Schleifer, *Syst. Appl. Microbiol.* **15,** 593 (1992).

TABLE I
SPECTRAL PROPERTIES OF COMMON FLUOROCHROMES

Fluorochromes	Absorbance wavelength (nm)	Emission wavelength (nm)
AMCA	349	448
Fluorescein	494	518
CY3	550	570
CY5	649	670
Tetramethylrhodamine	555	580
Texas Red	595	615

substrate transformation. The enzymatic approach is described in detail elsewhere.[24] The application of hapten- or enzyme-linked probes can be limited by the larger molecular size of the antihapten–antibody or the oligonucleotide–enzyme conjugate, which might be restricted to penetrate biofilm matrices and cell walls of certain target cells. For chemical labeling, reactive thiol or amine groups are attached to the 5′ end of the oligonucleotides in the final step of synthesis. After purification, the resulting reactive oligonucleotides can be coupled with various activated substrates such as *N*-hydroxysuccinimide esters of commercially available fluorochromes. Alternatively, synthetic oligonucleotides labeled directly at the 5′ end with different fluorochromes can be purchased from custom synthesis laboratories. The most widely distributed fluorochromes include fluorescein isothiocyanate (FITC) and tetramethylrhodamine isothiocyanate (TRITC) (Molecular Probes Inc., Eugene, OR), other rhodamine stains, e.g., Texas Red (Molecular Probes Inc.) the blue fluorescing aminomethylcoumarin (AMCA, Molecular Probes Inc.), and, more recently, cyanine dyes (CyDye, Amersham Pharmacia Biotech Europe, Freiburg, Germany). Generally, shorter excitation wavelengths of the fluorochromes are advantageous for higher resolution, but they also excite more autofluorescence and cause more rapid bleaching. Longer wavelengths are able to penetrate deeper into biofilm structures, but both sensitivity and resolution are decreased (see Table I).

Because emission spectra of the fluorochromes are environment sensitive, they may vary about 10 nm, depending on the mode of coupling to the oligonucleotide. Specific monolabeled fluorescent rRNA probes result in weaker fluorescence signals than general cell stains, such as acridine orange or 4′,6-diamidino-2-phenylindole (DAPI, Sigma, Deisenhofen, Ger-

[24] Anonymous, " Nonradioactive *in Situ* Hybridization Application Manual," 2nd ed. Boehringer Mannheim GmbH, Mannheim, 1996.

many). Thus, the use of antifading reagents in mounting media is necessary. For this purpose, a number of antifading reagents for fluorescence and laser scanning microscopy are available commercially (e.g., Citifluor, Citifluor Ltd., London, England; Slow Fade, Molecular Probes Inc.).

Chemical Fluorescent Dye Labeling of Synthetic Oligonucleotides (Modified after Refs. 7 and 8)

All buffers and solutions should be sterile and DNase free. To this end, they should be prepared with freshly deionized water. For short periods, buffer solutions may be stored at room temperature, whereas longer storage should be at 4°.

Materials and Solutions

Loading buffer: Sucrose (20%, w/v), 0.01% (w/v) bromphenol blue
Electrophoresis buffer: 140 m*M* Tris base, 45 m*M* boric acid, 3 m*M* EDTA
TE buffer: 10 m*M* Tris–HCl, 1 m*M* EDTA, pH 7.4
Doubly distilled water
20% acrylamide solution in electrophoresis buffer
Vertical electrophoresis unit
Two wavelength (254 and 365 nm) hand-held UV lamp (Bachofer, Reutlingen, Germany)
Scalpel or razor blade
UV spectrophotometer

Procedure

1. Synthesized oligonucleotides should be desalted and purified by reversed-phase high-performance liquid chromatography (HPLC) to separate full-length products from truncated sequences. Perform chemical coupling reactions as recommended by the dye manufacturers (e.g., Amersham Pharmacia Biotech; Molecular Probes Inc.).
2. Separate unreacted dyes from the oligonucleotides by passing the reaction mixture through a NAP-5 column (Amersham Pharmacia Biotech). After equilibration of the Sephadex G-25 medium, elute the oligonucleotides in fractions of 200 μl doubly distilled H_2O from the column and monitor the absorbance of the eluates at λ 260 nm. Pool fractions containing dye-coupled oligonucleotides, usually fractions 3 to 6.
3. Reduce the pooled eluates containing the oligonucleotides to a volume of 50 μl in a vacuum concentrator.
4. Add 50 μl of loading buffer to the oligonucleotides and load them

on a 20% nondenaturing acrylamide gel. Separate unlabeled oligonucleotides by electrophoresis at about 300 V (15 V/cm) for 2 to 3 hr.
5. Visualize gel bands by fluorescence shadowing using a fluorescent thin-layer chromatography plate (silica gel 60-F254, Merck, Darmstadt, Germany) and an UV lamp at λ 254 nm and λ 365 nm to visualize unlabeled and labeled oligonucleotides, respectively. Excise the fluorescent gel band exactly with a sterilized scalpel.
6. Place the gel piece into a tube and grind the gel piece with a small sterile pestle. Add 800 μl TE buffer, vortex thoroughly, and incubate in the dark at room temperature overnight. Recover colored TE buffer, add 800 μl TE buffer, and incubate for 3 hr.
7. Collect fractions and concentrate them using a NENsorb 20 column (DuPont, Bad Nauheim, Germany) as recommended by the manufacturer. The labeled oligonucleotide will be eluated with 50% (v/v) methanol (Merck).
8. Dry the conjugate in a vacuum concentrator and dissolve it in 100 μl autoclaved doubly distilled H_2O. To ascertain the labeling efficiency, the ratio of OD_{260} and the peak absorbance of the corresponding fluorescent dye at its absorption maximum ($OD_{fluorochrome}$) can be measured using a UV spectrophotometer. The average ratio of $OD_{260}:OD_{fluorochrome}$ should be around 3:1.
9. Store stock solutions of fluorescent probes at $-20°$.

Fixation of Biofilms for *in Situ* Hybridization

During the hybridization procedure, the biofilms are exposed to elevated temperature, detergents, and osmotic gradients. In order to preserve the morphological integrity of the cells, biofilms have to be fixed prior to hybridization. Cross-linking fixatives such as formaldehyde, paraformaldehyde, and glutaraldehyde preserve both RNA and cell morphology. However, fixation with glutaraldehyde results in high levels of unspecific binding of the probe, which can be minimized by fixation in filtered (para)formaldehyde solutions. The permeability of certain gram-positive and methanogenic (para)formaldehyde fixed cells might be limited. For these organisms, fixation in ethanol,[22] additional lysozyme treatment of fixed cells prior to hybridization, or increasing concentrations of detergents in the hybridization buffer[9] may improve the penetration of probes into the fixed cells. Because prolonged storage in the fixative reduces the permeability of cells for oligonucleotide probes, biofilms should not be stored in the fixative longer than 4 hr. If biofilms were grown on solid supports (glass, polyethylene, polycarbonate), small pieces of biofilms can be cut off from the support with ethanol-sterilized scissors or scalpels and fixed onto standard microscope slides or alternatively in small petri dishes with silicone glue (Dow

Corning, Midland, MI). Generally, it is recommended to work with filter-sterilized solutions.

Materials and Solutions

Phosphate-buffered saline (PBS): 130 m*M* sodium chloride, 10 m*M* sodium phosphate buffer, pH 7.4
It is convenient to prepare 10× PBS and to dilute it to 1 and 3× PBS
Paraformaldehyde (Sigma, Deisenhofen, Germany)
37% (v/v) formaldehyde solution (Merck)
50%, 80% and 98% (v/v) ethanol (Merck)
10 *M* NaOH
5 *M* HCl
0.2-μm pore size filter
Doubly distilled water

Procedure

Preparation of 3% Paraformaldehyde Fixative

1. Heat 70 ml of doubly distilled H_2O to 65° and add 3 g paraformaldehyde.
2. Add one drop of 10 *M* NaOH to stimulate dissolution and stir rapidly until the solution is nearly clear (5–10 min).
3. Remove solution from heat source and add 30 ml of 3× PBS.
4. Adjust pH to 7.0 with 5 *M* HCl and volume to 100 ml.
5. Filter solution through a 0.2-μm filter.
6. Quickly cool to 4° and store in the refrigerator until use.

Use paraformaldehyde solution within 24 hr. Alternatively, a 37% (v/v) formaldehyde solution can be diluted with doubly distilled H_2O to a final concentration of 3% (v/v) and filter sterilized (0.2-μm pore size filter).

Fixation

1. It is convenient to perform fixation, washing, and ethanol dehydration steps in small glass vessels or petri dishes. For this purpose, native biofilms grown on solid substrates (glass, polycarbonate, polyethylene) are cut into small pieces under sterile conditions using ethanol-sterilized scissors, scalpel, or razor blades and mounted with silicon glue on microscope slides or on the bottom of petri dishes. Cross sections of biofilms can be obtained after fixation and embedding in

paraffin[25] or after fixation and subsequent embedding of biofilms in 20% (w/v) acrylamide.[26] Prevent drying of biofilms prior to fixation.

2. Submerse wet biofilms immediately after sampling in 3% (v/v) (para)-formaldehyde solution and incubate them for 1 hr at 4°.
3. For improved hybridization of gram-positive bacteria, fix biofilm samples by using ice-cold 96% (v/v) ethanol instead of (para)formaldehyde solution and incubate them for at least 30 min.
4. Wash biofilms by overlaying them with 1× PBS or submersion in appropriate vessels.
5. After fixation, biofilms are subjected to dehydration. For this purpose, submerse biofilm samples in an ethanol series with 50, 80, and 96% ethanol for 3 min each.
6. Allow dehydrated biofilms to air dry and store at room temperature. Fixed biofilms can be stored for several months without apparent decline in quality as assessed by hybridization.

Hybridization of Biofilms

The hybridization procedure is based on the formation of heteroduplex molecules by deoxyoligonucleotides binding reversibly to their complementary target sites on the 16S or 23S rRNA. As mentioned before, the efficiency of hybridization is determined by the accessibility of the target rRNA in the biofilms and by the length of the probe. Typically used *in situ* probes are 18 to 24 nucleotides in length and the resulting hybridization temperatures range from 40° to 60°. The hybridization conditions should allow the formation of stable heteroduplex molecules between complementary sequences, but should discriminate simultaneously against duplex formation between oligonucleotides and target sequences with one or more mismatches. In order to minimize the hybridization of the probe to related but not completely identical target sequences, hybridizations have to be performed under optimum stringency conditions. The most critically important value of an oligonucleotide is its melting temperature (T_m), which is affected mainly by the salt concentration [Eq. (1)], the presence of denaturants such as formamide [Eq. (2)], and the probe concentration. At a constant hybridization temperature, the hybridization stringency can be adjusted by varying the salt and formamide concentrations [Eq. (3)], resulting in an alteration of T_m (ΔT_m). The higher the temperature and the concentration

[25] C. Rothemund, R. Amann, S. Klugbauer, W. Manz, C. Bieber, K. H. Schleifer, and P. Wilderer, *Syst. Appl. Microbiol.* **19,** 608 (1996).

[26] S. Møller, C. Sternberg, J. B. Andersen, B. B. Christensen, J. L. Ramos, M. Givskov, and S. Molin, *Appl. Environ. Microbiol.* **64,** 721 (1998).

The assays for biofilm formation described in this article, which are based on the attachment to multiwell plates, have proven to be remarkably useful in characterizing biofilm formation. These systems can be used to monitor biofilm formation from initial attachment to development of the complex architecture characteristic of biofilms, to the transition from biofilm back to planktonic growth, and are also amenable to large-scale screening for biofilm-defective mutants. This article focuses on the description of rapid and reliable assays for isolating mutants that are unable to form a biofilm on an abiotic surface. It also discusses additional methods that are useful for efficiently classifying and characterizing these mutants.

Biofilm Assay

The following assay is based on growth and biofilm formation by bacteria in multiwell microtiter dishes (typically with 96 wells), is useful for both small- and large-scale screening, and is a modification of earlier protocols.[10,11] In addition to testing existing mutants for possible defects in attaching to a surface, this approach can be used for the large-scale screening of new mutants defective in biofilm formation. A general screen for mutants serves as a relatively unbiased method for identifying new factors involved in biofilm formation. To begin the molecular genetic analysis of biofilm formation in any microorganism, one must (i) determine growth conditions that promote biofilm formation and (ii) generate mutants defective in biofilm formation.

The microtiter dish-based approach for monitoring biofilm formation is a powerful method for studying not only the early stages of biofilm formation, but later developmental events as well. Variations on this assay method have already been used successfully to identify biofilm formation mutants in *Pseudomonas aeruginosa, Pseudomonas fluorescens, Vibrio cholerae, Staphylococcus epidermidis, Escherichia coli,* and *Shewanella putrefaciens.*[2,4–8,12,13] In *P. aeruginosa,* for example, we have used phase-contrast microscopy to show that biofilm formation in the wells of a microtiter dish encompasses both initial attachment of a monolayer of cells on the polyvinyl chloride (PVC) plastic and the subsequent development of microcolonies. These microcolonies may be the precursors of the complex architecture that is a hallmark of biofilms formed by this organism.[5] Furthermore, we

[10] G. D. Christensen, W. A. Simpson, A. L. Bisno, and E. H. Beachey, *Infect. Immun.* **37,** 318 (1982).

[11] M. Fletcher, *Can. J. Microbiol.* **23,** 1 (1977).

[12] D. Newman and R. Kolter, unpublished data, 1999.

[13] E. Muller, J. Hubner, N. Gutierrez, S. Takeda, D. A. Goldmann, and G. B. Pier, *Infect. Immun.* **61,** 551 (1993).

have observed that after extended incubation (>24 hr) in microtiter dish wells, *P. aeruginosa* in a biofilm begin to detach from the PVC plastic.[14] These observations suggest that this assay may be useful for studying many aspects of biofilm development.

Protocol for Biofilm Formation Assay

Inoculation of Cells into Multiwell Microtiter Dish. The microtiter dishes typically used are made of PVC (such as the Falcon 3911 Microtest III flexible assay plates available from Becton-Dickinson Labware) or polystyrene (PS, such as those available from Costar, Cambridge, MA). However, we and others have also found that microtiter dishes or tubes made of polypropylene and polycarbonate plastic or borosilicate glass also serve as surfaces on which biofilms form.[2,5–8] These microtiter dishes can be inoculated in one of three ways: (i) overnight cultures are diluted 1:100 into fresh medium and then 100–200 μl of the freshly inoculated medium is dispensed into wells, (ii) cells are patched to agar plates in a grid pattern that matches the spacing of the wells on a 96-well dish and then transferred to liquid medium-filled microtiter dishes using a multipronged device (48 and 96 metal-pronged devices are available from Dan-Kar Corp., Wilmington, MA), or (iii) the wells are inoculated individually by picking a colony directly from an agar plate with a toothpick. The organisms discussed in this article all make biofilms in response to a nutrient-replete environment. Because some bacteria form biofilms in response to starvation,[15] growth conditions for each organism must be determined empirically.

Growth Conditions for Bacteria. Table I shows the growth conditions that promote biofilm formation for several model organisms. The microtiter dishes are typically incubated without agitation at the appropriate temperature. For testing biofilm formation by facultative anaerobes in the absence of oxygen, we have found that overlaying the wells with 125 μl of mineral oil allows for a sufficiently anaerobic growth environment.[12]

Detection of Biofilms. Biofilms are detected by staining with crystal violet (CV) or safranin (as indicated in Table I). Both dyes are prepared as 0.1% (w/v) solutions in H_2O. Typically 10 μl of the dye solution is added to each medium-containing microtiter dish well. After the stain is added, the plates are incubated for 10–15 min at room temperature and then rinsed thoroughly and vigorously with water to remove unattached cells and residual dye. After staining, we observe the phenotypes shown in Fig. 1. It is important to note we estimate that an attachment ring composed of $\sim 5 \times 10^6$ total cells (or $\sim$1 bacterium/μm^2) is necessary in order to

[14] G. A. O'Toole and R. Kolter, unpublished data, 1999.

[15] K. C. Marshall, *ASM News* **58,** 202 (1992).

TABLE I
BIOFILM FORMATION CONDITIONS SPECIFIC TO VARIOUS MODEL ORGANISMS

Strain	Media promoting biofilm formation	Temperature (°C)	Time to biofilm[e]	Solvent to remove stained biofilms[f]
Pseudomonas aeruginosa	LB[a] or minimal medium[b] plus any one of the following additions: glucose, citrate, glycerol, lactate, glutamate, CAA, succinate, or gluconate	25–37	8 hr (4–24 hr)	95% ethanol
P. fluorescens	LB or minimal medium[b] plus and one of the following additions: glucose, citrate, glycerol, xylose, CAA, glutamate, malate, or mannitol	25–30	8 hr (4–24 hr)	95% ethanol
Escherichia coli	LB or minimal medium[b] plus any one of the following additions: glucose/CAA, glycerol/CAA, or CAA	25	24 hr (4–48 hr)	80% ethanol 20% acetone
Vibrio cholerae	LB or minimal medium[b] plus any one of the following additions: glucose/CAA or glycerol/CAA	25–30	30 hr (16–60 hr)	100% dimethyl sulfoxide (DMSO)
Shewanella putrefaciens	LB or minimal medium[c] plus lactate	25–30	12 hr (4–48 hr)	100% DMSO
Staphylococcus epidermidis	TSB, TSB plus 0.25% glucose, or minimal medium[d] plus glucose/CAA	37	48 hr	95% ethanol or 100% DMSO
Streptococcus mutans	Minimal medium[d] plus sucrose/CAA	37	48 hr	95% ethanol or 100% DMSO

[a] Luria–Bertani broth.

[b] In all cases, "minimal medium" indicates the base minimal medium for the particular organism supplemented by one of the carbon and energy sources listed. For *E. coli*, *Pseudomonas*, and *V. cholerae*, the minimal medium is M63 (J. H. Miller, "A Short Course in Bacterial Genetics." Cold Spring Harbor Press, Cold Spring Harbor, NY, 1992). CAA, casamino acids added to a final concentration of 0.5%. All carbon sources are added to a final concentration of 0.2% except arginine (0.4%).

[c] An amino acid-supplemented medium that approximates a freshwater environment [C. R. Meyers and K. H. Nealson, *Science* **240,** 1319 (1988)].

[d] Defined minimal medium for streptococci from H. F. Jenkinson, *J. Gen. Microbiol.* **132,** 1575 (1986). TSB, tryptic soy broth.[2]

[e] Top number indicates typical assay time. The time in parentheses indicates range of time when at least low levels of biofilm formation can be detected.

[f] These solubilization conditions are for crystal violet-stained biofilms. *Staphylococcus* strains can also be stained by 0.1% safranin and the absorbance monitored directly (without solubilization) at A_{490}.[2] The use and solubilization of other dyes should be determined empirically.

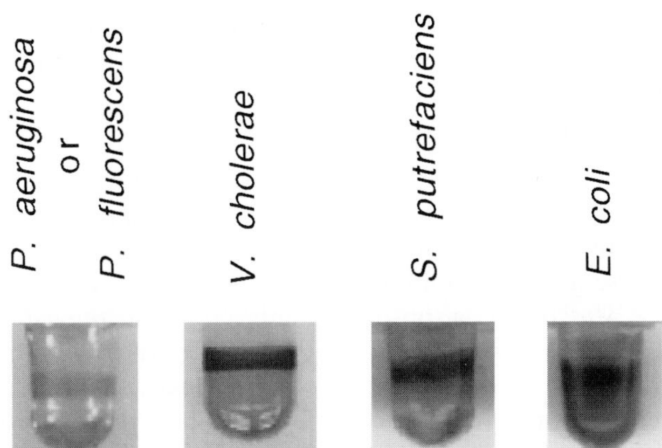

FIG. 1. Representative biofilm formation phenotypes of various organisms on abiotic surfaces. The CV-stained biofilm of *P. aeruginosa* and *P. fluorescens* on polyvinylchloride (PVC) forms at the air–medium interface, which is also shown for *V. cholerae* and *S. putrefaciens* on glass. In contrast, *E. coli* completely coats the bottom of the PVC microtiter dish wells in addition to the dark CV-stained biofilm at the air–medium interface.

visualize attached bacteria by CV staining. Rinsing the wells of the microtiter dishes before adding a dye to detect the biofilm will usually lower background significantly and is necessary when cells are grown in rich medium. For minimal medium-grown cells, especially in large-scale screening experiments, prerinsing is not required. Other investigators report the use of fixative before staining,[7,16] but we find that this is not required.

At this point it is useful to revisit our definition of a biofilm. Given the operational nature of the definition, the incubation and rinsing conditions themselves determine the properties of the biofilm. For example, the rinsing solution can contain a detergent such as sodium dodecyl sulfate (SDS) if the experimentalist wishes to define that particular biofilm as resistant to treatment with that detergent.

Quantitation of Biofilm Formed. In experiments where determination of the extent of biofilm formation is desirable, the dye that has stained the biofilm cells can be solubilized. The absorbance of the solubilized dye ($A_{570-600}$ for CV and A_{490} for safranin) can then be determined with a spectrophotometer or, more conveniently, a microtiter plate reader for large-scale experiments. As a control, uninoculated medium is typically used to determine the background. An example of the increase in CV staining of a *P. aeruginosa* biofilm grown over a period of 8 hr is shown

[16] S. McEldowney and M. Fletcher, *Appl. Environ. Mibrobiol.* **52,** 460 (1986).

in Fig. 2. Table I shows conditions for solubilizing the dyes used to stain the biofilms; conditions for other systems can be determined empirically.

Screening for Biofilm-Defective Mutants

A general screen for mutants defective in biofilm formation is outlined.

Determine Growth Conditions That Promote Biofilm Formation. Table I shows growth conditions that can be used for some model organisms, otherwise these growth conditions must be determined empirically.

Generate Mutants. Transposon mutants are most amenable to rapid molecular genetic analysis, although other mutagens can also be used.

Screen for Mutants Defective in Biofilm Formation. Using the method described earlier, large numbers of mutants can be screened for those with defects in biofilm formation. The wells of the microtiter dishes can be inoculated with mutants using any of the methods described earlier.

Recheck Mutants. All putative mutants should be rechecked for their biofilm formation phenotype to ensure that the mutant phenotype is repeatable and stable.

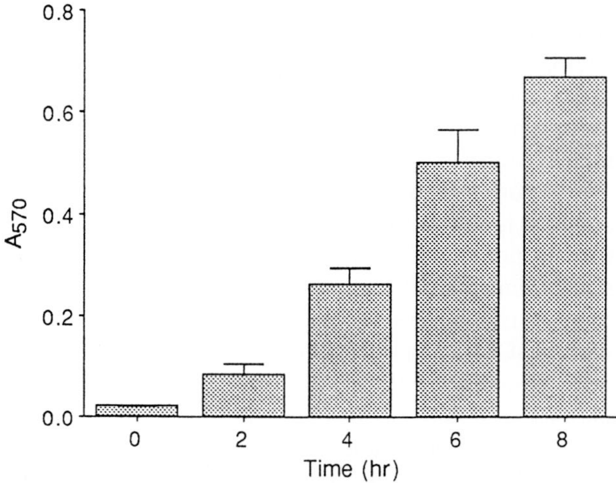

FIG. 2. Quantitation of biofilm formation. The extent of biofilm formation can be determined quantitatively by first staining the biofilm with a dye such as crystal violet (CV), solubilizing the bacterial cell-associated dye, and determining its absorbance at the appropriate wavelength. Shown here is the quantitation of the biofilm formed by *P. aeruginosa* PA14 on PVC over a period of 8 hr in minimal medium supplemented with glucose (0.2%) and casamino acids (0.5%) and grown at 37°. The biofilm was stained with CV, the dye solubilized with 95% ethanol, and the absorbance determined at 570 nm.

Perform Secondary Screens and Microscopy. The tests described later will help classify mutants for subsequent molecular genetic analysis rapidly.

Molecular Genetic Analysis of Mutants. The ultimate goal of these studies is to determine the genes and gene products required for biofilm formation and the role these gene products play in attachment to a surface. For the analysis of transposon mutants we describe a polymerase chain reaction (PCR)-based method for determining the exact site of the transposon insertion (see later).

Secondary Phenotypic Screens

Once mutants defective in biofilm formation have been isolated, a number of simple tests can aid in the initial classification of these mutants. These secondary phenotypic screens may give clues to the function of the genes disrupted in the mutants and may identify candidates for subsequent molecular studies rapidly.

Determination of Growth Rate

The goal of the screen for biofilm-defective mutants is to isolate strains that are specifically impaired in attaching to surfaces. As the screen is designed, however, strains defective in growth have a phenotype identical to biofilm-defective mutants. To identify mutants with growth defects, the growth rate of putative mutants can be determined under conditions that are similar to those used in the original screen for biofilm-defective mutants (see earlier discussion and Table I). To determine growth rates, each mutant strain is subcultured 1:100 into a test tube and grown at the appropriate temperature. The optical density (select the wavelength appropriate for the particular organism, usually 600–650 nm) is monitored over time with a spectrophotometer and the growth rate is compared to the wild-type parent. Alternatively, each mutant strain can be subcultured into a 96-well flat-bottomed microtiter dish and grown at the appropriate temperature, and the optical density is monitored with a microtiter dish plate reader over the course of growth. These data can be used to calculate rates of growth for wild-type and mutants strains. Although the first method is useful, the second method is more efficient and more closely resembles the conditions used in the mutant screen. Any mutants that exhibit growth defects should be studied with caution as the mutations they harbor may not directly affect adhesion to a surface.

Other growth properties of putative mutants can also be tested easily to help classify mutants further. Some of the mutations that affect biofilm development may affect colony morphology or pigmentation. Such mutants are easily identifiable when the strains are grown on nutrient or minimal medium agar plates. Mutants that were isolated in screens using rich media may harbor lesions that confer an auxotrophic phenotype. These mutants can be identified by their inability to grow in minimal media.

Flagellar-Mediated Motility Assay

Genes encoding components required for flagellar-mediated motility have been shown to be important for biofilm development in several organisms.[5,8,17–22] Therefore, it is important to assay the ability of each mutant isolated in the initial screen for its ability to swim. A rapid method for such analysis is a swarm assay. In this test, each strain is stabbed into motility agar (0.3% agar, see Table II for recipe), and the plates are incubated at the temperature appropriate for the organism.[23,24] As nutrients become depleted, cells swim outward toward a more nutrient-rich environment, which results in the formation of increasingly larger circles of cells, or "swarms." Cells defective in flagellar-mediated motility (Mot^-) do not form swarms, but rather remain clustered in the area in which the cells were first stabbed into the agar.

Twitching Motility Assay

Twitching motility is a form of movement mediated by some type IV pili and is independent of flagella.[25–27] In *P. aeruginosa,* type IV pili-mediated twitching motility has been shown to be important for biofilm formation on an abiotic surface.[5] Therefore, if the organism under study can move via twitching motility, it is important to test biofilm-defective mutants for

[17] L. A. de Weger, C. I. M. van der Vlught, A. H. M. Wijfjes, P. A. H. M. Bakker, B. Schippers, and B. Lugtenberg, *J. Bacteriol.* **169,** 2769 (1987).

[18] C. C. R. Grant, M. E. Konkel, J. Cieplak, and L. S. Tompkins, *Infect. Immun.* **61,** 1764 (1993).

[19] D. R. Korber, J. R. Lawrence, and D. E. Caldwell, *Appl. Environ. Microbiol.* **60,** 1421 (1994).

[20] K. M. Ottemann and J. F. Miller, *Mol. Microbiol.* **24,** 1109 (1997).

[21] D. A. Simpson, R. Ramphal, and S. Lory, *Infect. Immun.* **63,** 2950 (1995).

[22] G. Smit, J. W. Kijne, and B. J. J. Lugtenberg, *J. Bacteriol.* **171,** 569 (1989).

[23] A. J. Wolfe and H. C. Berg, *Proc. Natl. Acad. Sci. U.S.A.* **86,** 6973 (1989).

[24] J. Adler, *Science* **153,** 708 (1966).

[25] D. E. Bradley, *Can. J. Microbiol.* **26,** 146 (1980).

[26] C. B. Whitchurch, M. Hobbs, S. P. Livingston, V. Krishnapillai, and J. S. Mattick, *Gene* **101,** 33 (1990).

[27] A. Darzins, *Mol. Microbiol.* **11,** 137 (1994).

TABLE II
MEDIA RECIPES/PROTOCOLS

Motility/swarm agar: This agar is used to assay the function of flagellar mediated motility and is prepared as follows:

Add 1.5 g Difco agar (Detroit, MI) to 400 ml distilled H_2O (a final concentration of 0.3% agar) and autoclave.

When agar has cooled to ~65°, 100 ml of a 5× stock of the minimal medium appropriate for the organism.

Pour agar into petri dishes. When the plates have solidified, stab the agar with the desired strains.

Plates are incubated at 25° for 24–48 hr. Motility positive (Mot^+) strains will have migrated from the point of inoculation to form a zone of growth. Nonmotile strains will form a small colony at the point of inoculation. The agar is very soft so the plates should not be inverted.

Twitching motility assay: This assay is used to assay a surface-based motility that requires functional type IV pili. The medium used standard Luria–Bertani-rich medium supplemented with 1.5% agar as follows:

Per 1 liter distilled H_2O add:
10 g tryptone
5 g NaCl
5 g yeast extract
15 g Difco agar

Autoclave. Plates should be poured thin (~3–5 mm) and allowed to solidify. The desired strain is stabbed into the agar (all the way through the agar to the petri plate) with a toothpick and the plates are incubated for 1–3 days at the appropriate temperature for the organism. Strains proficient for type IV pili-mediated twitching motility ($Twitch^+$) form a hazy zone of growth at the interface between the agar and the petri plate.

their ability to display twitching motility. To test for twitching motility, strains are stabbed into a thin LB plate (see Table II for details of the assay).[25–27] Biofilm-defective mutants that retain the ability to move via twitching motility form a haze of growth at the interface between that plate and the agar. In contrast, strains defective in twitching motility ($Twitch^-$) remain clustered in the area in which they were first inoculated.

Attachment to Other Surfaces

The range of surfaces on which a particular mutant is defective for biofilm formation should be determined by testing its ability to form biofilms on a variety of surfaces. Each mutant should be grown in the original environmental conditions used in the screen (temperature and growth medium), but exposed to different surfaces. These surfaces can include surfaces that are relevant to industrial, medical, and environmental settings (e.g., plastics including polystyrene and polycarbonate, glass, chitin, and metals such as iron, as well as biotic surfaces such as eukaryotic cells). The surfaces chosen to be tested should reflect the nature of the organism being studied.

Rescue of Biofilm-Defective Mutants by Changes in Growth Conditions

We have shown previously that a subset of mutants identified for their inability to form a biofilm under one growth condition can be rescued for their biofilm formation defect by growth of the same strains under different nutritional conditions. For example, about half of the *P. fluorescens* biofilm-defective mutants isolated on a minimal glucose plus casamino acids medium were rescued for biofilm formation by growth in minimal medium with citrate as the sole source of carbon and energy. These data were interpreted to mean that an alternative genetic pathway for biofilm formation was induced by growth on citrate.[6] In order to perform this analysis on newly isolated biofilm-defective mutants, it is first necessary to catalog the growth conditions that stimulate biofilm formation in the particular wild-type strain. These growth conditions may vary widely[2,4–6,8,9] and once they are determined, mutants isolated on one growth medium should be tested for biofilm formation (and growth) in the other growth conditions. This sort of nutritional analysis may define different genetic pathways involved in biofilm formation by a particular organism.

Using Microscopy to Analyze Biofilm Formation

Microscopy is a useful tool for identifying and characterizing biofilm-defective mutants. For example, biofilm-defective mutants may be arrested at a specific stage in biofilm formation, including initial attachment to a surface, cell-to-cell attachment, formation of microcolonies, and development of a multilayered biofilm.[3,5,8,28,29] Using microscopy allows an observable defect in the development of the biofilm to be linked to a specific mutation, possibly providing insight into the role of a particular gene product in biofilm formation. The microscopic method of choice and the method chosen to allow formation of the biofilm will depend on the stage in biofilm development that one wishes to observe and the incubation period required to reach this stage for the particular organism.

[28] D. Mack, W. Fischer, A. Krokotsch, K. Leopold, R. Hartmann, H. Egge, and R. Laufs, *J. Bacteriol.* **178,** 175 (1996).

[29] C. Heilmann, M. Hussain, G. Peters, and F. Gotz, *Mol. Microbiol.* **24,** 1013 (1997).

Phase-Contrast Microscopy and Confocal Scanning Laser Microscopy

The following methods outline various procedures for the formation of biofilms on surfaces such that subsequent microscopic analysis is straightforward. These approaches allow the rapid analysis and classification of biofilm-defective mutants using either phase-contrast microscopy (for biofilms ~1–5 μm in depth) or confocal scanning laser microscopy (CSLM, for biofilms >5 μm in depth). Phase-contrast microscopy does not require the staining of cells and the microscopes are typically widely available. The use of phase-contrast microscopy allows the rapid visual analysis of many mutants in relatively simple model systems for biofilm formation. CSLM requires that the cells to be studied are fluorescently tagged, uses a laser as a light source, and the equipment is very expensive. However, CSLM is the method of choice for analyzing the three-dimensional structure of thick, hydrated biofilms in a nondestructive manner.[1,30] CSLM permits optical sectioning of the biofilm in either the horizontal or the vertical dimension by selective excitation and light collection from a single point in the sample, which may be varied in the *x, y,* or *z* dimensions. Thus, one can obtain both vertical and horizontal cross-sectional images of the biofilm in a noninvasive fashion. In order to detect the sample, it must be labeled fluorescently. For bacteria, it is convenient to express the green fluorescent protein (GFP) from either a plasmid or the chromosome.[31–34] We have found that neither the addition of the GFP plasmid nor the presence of antibiotics to select for maintenance of the plasmid interferes with biofilm formation. Therefore, the biofilm need not be fixed or stained but rather may be observed nondestructively as it is forming. Biofilms may also be visualized by CSLM using a variety of fluorescent dyes, as described in other articles in this volume.[34a,34b]

Methods for Visualization of Biofilms by Phase-Contrast Microscopy

Biofilm Formation on Microscope Slide. Initial bacterial attachment generally occurs within seconds or minutes of exposure to a surface. Because

[30] J. R. Lawrence, D. R. Korber, B. D. Hoyle, J. W. Costerton, and D. E. Caldwell, *J. Bacteriol.* **173,** 6558 (1991).

[31] S. Moller, C. Sternberg, J. B. Anderson, B. B. Christensen, J. L. Ramos, M. Givskov, and S. Molin, *Appl. Environ. Microbiol.* **64,** 721 (1998).

[32] A. G. Matthysse, S. Stretton, C. Dandie, N. C. McClure, and A. Goodman, *FEMS Microbiol. Lett.* **145,** 87 (1996).

[33] A. C. Olofsson, A. Zita, and M. Hermansson, *Microbiology* **144,** 519 (1998).

[34] G. V. Bloemberg, G. A. O'Toole, B. J. J. Lugtenberg, and R. Kolter, *Appl. Environ. Microbiol.* **63,** 4543 (1997).

[34a] D. Korber, G. M. Wolfaardt, V. Brözel, R. MacDonald, and T. Niepel, *Methods Enzymol.* **310,** [1], 1999 (this volume).

[34b] P. S. Stewart and G. Mcfeters, *Methods Enzymol.* **310,** [13], 1999 (this volume).

nutrient depletion, oxygen depletion, and desiccation are not concerns over this period of time, the simplest approach is to put a drop of a dilute bacterial culture in exponential phase on a glass or plastic slide and observe attachment of the bacteria using a phase-contrast microscope (400 to 600× magnification is sufficient for this purpose). A microscope equipped with a CCD camera interfaced with a computer is especially useful for documenting attachment by time-lapse photography. In our laboratory, we use the Scion Image software program (a modification of NIH Image, Scion Corporation, Frederick, MD) to acquire and process images and to make real-time or time-lapse movies.

Biofilm Formation in Multiwell or Petri Plate. An alternative approach for visualizing bacteria attached to a surface is to use sterile, flat-bottomed, multiwell plates (plates should not be tissue culture treated) or petri plates.[5,29] In our laboratory, we use 24-well plates filled with 250–500 μl of growth medium. The microtiter dish well or petri plate is inoculated with a dilution of a saturated liquid culture (e.g., 1:100 dilution). The plate is placed on the stage of an inverted microscope and the bottom of the plate is monitored by phase-contrast microscopy (400 to 1000× magnification). Over the course of the experiment, planktonic cell numbers increase and these cells can potentially interfere with visualization of the biofilm. To eliminate planktonic cells to facilitate visualization of the biofilm, the liquid phase can be removed from the well and replaced with fresh growth medium. Figure 3 shows an example of such an experiment where *P. aeruginosa* has been allowed to form a biofilm over a period of 8 hr. Images from the biofilm at 2 and 8 hr postinoculation are shown. In this case, a monolayer of cells has formed by 2 hr and by 8 hr this monolayer has become very dense and is punctuated by microcolonies. As mentioned earlier, biofilm formation may also be documented by time-lapse photography. An advantage of this system is that a surface can be observed continuously over a period of many hours with a simple experimental design. This is in contrast to the more complicated systems for the observation of biofilms over extended periods, such as the continuous flow systems that have been reported.[30,35–37]

Biofilm Formation on Plastic Tabs. In the assay for biofilm formation in microtiter dishes described earlier, bacteria are incubated in the wells of multiwell plates and a biofilm eventually forms on the walls of the

[35] K. D. Xu, P. S. Stewart, F. Xia, C.-T. Huang, and G. A. McFeters, *Appl. Environ. Microbiol.* **64,** 4035 (1998).

[36] D. G. Davies, M. R. Parsek, J. P. Pearson, B. H. Iglewski, J. W. Costerton, and E. P. Greenberg, *Science* **280,** 295 (1998).

[37] P. Stoodley, D. DeBeer, and Z. Lewandowski, *Appl. Environ. Microbiol.* **60,** 2711 (1994).

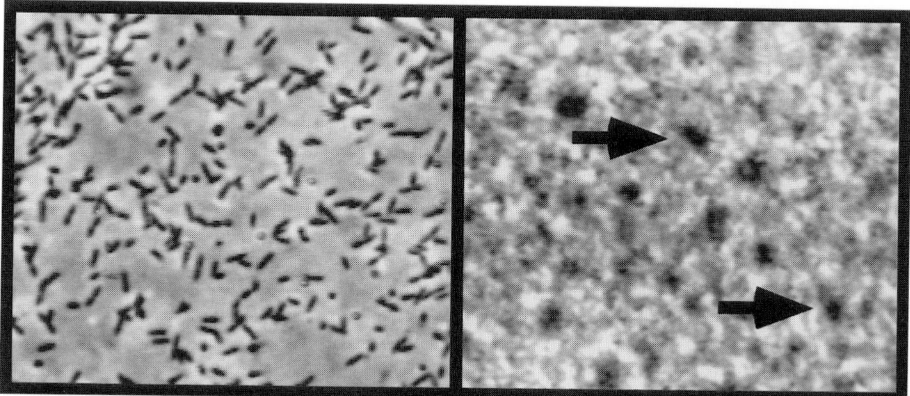

FIG. 3. Direct observation of the biofilm by phase-contrast microscopy. A biofilm of *P. aeruginosa* PA14 allowed to form on polystyrene (PS) plastic was observed 2 (left) and 8 (right) hr postinoculation. This culture was grown at 25° with minimal M63 medium supplemented with glucose (0.2%) and casamino acids (0.5%). These phase-contrast micrographs (400× magnification) show that *P. aeruginosa* PA14 first forms a dispersed monolayer on the PS. By 8 hr there is a dense monolayer of cells on the surface punctuated by microcolonies of cells (indicated by the arrows).

microtiter dish wells. It is also useful to directly visualize the biofilms as they form in this assay system. However, the 96-well microtiter dishes typically used in these assays are difficult to monitor directly under the microscope. In order to bypass this difficulty, a small tab of PVC plastic can be cut from a PVC microtiter dish (~3 × ~6 mm). This plastic tab serves as a surface on which a biofilm is able to form and can be sterilized readily with 100% ethanol or UV light. The plastic tab is placed in a well that contains 100 μl of the freshly inoculated medium appropriate for the bacterium being studied. After incubation for the desired time and at the appropriate temperature, the plastic tab is removed from the well, rinsed with sterile medium to remove planktonic cells (any medium will serve to rinse the tab), blotted gently on a paper towel to remove excess liquid, and mounted on a microscope slide. The plastic tab can then be examined at 400 to 1000× magnification and the attached bacteria observed directly.[34]

Methods for Visualization of Biofilms by CSLM

The following two methods are used to grow bacterial biofilms that can be analyzed by CSLM.

Batch Culture. A batch culture method can be used to allow for the formation of biofilms over a period of up to several days. This approach is convenient for observing biofilms made by many bacterial strains or to test a series of mutants at one time point. Growth medium (6 ml) is placed in a 50-ml Falcon tube (Becton-Dickinson, Chicago, IL), and the medium is inoculated with a 1:100 dilution of an overnight bacterial culture. A sterile glass or plastic coverslip is placed in the medium, and the tube is incubated for up to several days at the appropriate temperature. After incubation, the coverslip is removed and placed on a microscope slide (both sides of the coverslip have a biofilm, so the side placed on the slide does not matter). We use microscope slides that have a single center well that is 1.5 mm in depth (microslide, culture, single depression from VWR Scientific, Boston, MA). The well is filled with sterile medium (the same medium used for growth of the bacteria), and the coverslip is sealed to the slide (e.g., with nail polish or glue) to eliminate desiccation and shifting of the coverslip during observation. By placing the coverslip on a medium-filled well, the biofilm is not pressed directly onto a glass surface and it remains hydrated during the microscopic analysis.

For *V. cholerae,* this technique yields biofilms of approximately 5–15 μm thickness (as compared to ~3 μm after the slide is incubated 24 hr in the 50-ml Falcon tube) with the characteristic three-dimensional structure reported for thick biofilms allowed to form over a period of days or weeks.[1] An example of a CSLM image of a *V. cholerae* biofilm grown in this system is shown in Fig. 4. Although this technique involves a simple experimental setup, disadvantages include the accumulation of metabolites in the culture vessel (which may alter or inhibit biofilm formation) and potential nutrient starvation if the cultures are incubated for extended periods of time.

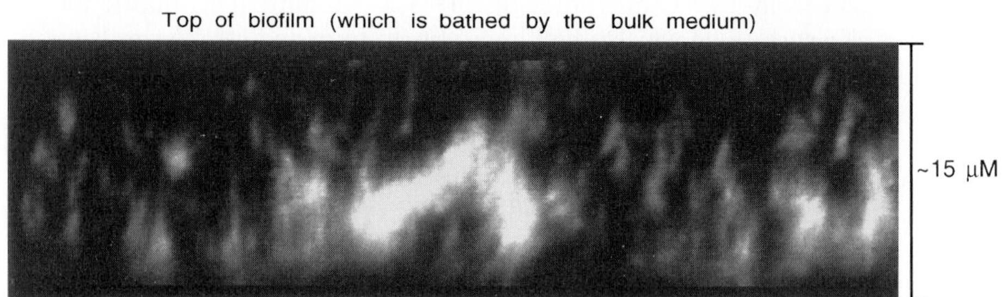

FIG. 4. Confocal scanning laser microscopy. A vertical section (a side view) of a biofilm of wild-type *V. cholerae* (carrying a plasmid constitutively expressing green fluorescent protein) formed on borosilicate glass. Bacteria were grown on Luria–Bertani broth for 3 days at 25°. The biofilm is ~15 μm in depth with the borosilicate glass/biofilm interface at the bottom of the figure and the biofilm/medium interface at the top of the figure. The bright spots represent fluorescent microcolonies of bacteria and the dark regions are those devoid of bacteria. These dark regions may be filled with extracellular polysaccharides and/or may be water channels that allow the flow of liquid from the bulk medium into the deeper portions of the biofilm.

Continuous Flow Systems. The standard method of forming thick biofilms involves construction of a flow cell (i.e., a vessel with inflow and outflow ports through which fresh growth medium is pumped continuously).[30,35–37] A bacterial culture is inoculated into the flow cell and, after several hours, fresh sterile medium is pumped through the vessel and the biofilm is allowed to develop over a period of days to weeks. This method requires an optically transparent, flat, water-tight system, a variable speed peristaltic pump (or some other means to reliably control the flow rate), and a sterilizable inflow tract for the delivery of fresh, sterile growth medium. Continuous flow systems have the following advantages: (i) the system is thought to more closely mimic the conditions under which bacterial biofilms form in some natural environments, (ii) the biofilm may be maintained indefinitely under relatively constant nutrient conditions, (iii) the biofilm may continue to develop over a long period of time, and (iv) the developing biofilm may be observed under the microscope continuously without disturbing the culture system. For a detailed description and discussion of continuous flow systems see chapters elsewhere in this volume.[37a–37c]

Molecular Analysis

Arbitrary-Primed PCR

Once mutants defective in biofilm formation are isolated, the next step is to determine which genes have been mutated. For cases in which the mutations have been generated with a transposon, the DNA sequence (generally 200–400 bp) flanking the transposon insertions can be determined without the need for a cloning step by a simple PCR-based protocol termed arbitrary-primed PCR.[38] In an era where the DNA sequence of entire microbial genomes has been determined, the short DNA sequences flanking the transposon that are generated with this method are often sufficient to identify the gene(s) disrupted in biofilm-defective (or any transposon) mutants rapidly.

The arbitrary-primed PCR method uses primers specific to the ends of the transposon and primers of random sequence that may anneal to chromosomal DNA sequences in close proximity to a transposon insertion.

[37a] D. Bradshaw and P. D. Marsh, *Methods Enzymol.* **310,** [22], 1999 (this volume).

[37b] A. Kharazini, B. Giwercinan, and N. Høiby, *Methods Enzymol.* **310,** [16], 1999 (this volume).

[37c] R. J. C. McClean, M. Whiteley, B. C. Hoskins, P. D. Majors, and M. M. Sharma, *Methods Enzymol.* **310,** [20], 1999 (this volume).

[37d] M. S. Zinn, R. D. Kirkegaard, R. J. Palmer, Jr., and D. C. White, *Methods Enzymol.* **310,** [18], 1999 (this volume).

[38] G. Caetano-Annoles, *PCR Methods Appl.* **3,** 85 (1993).

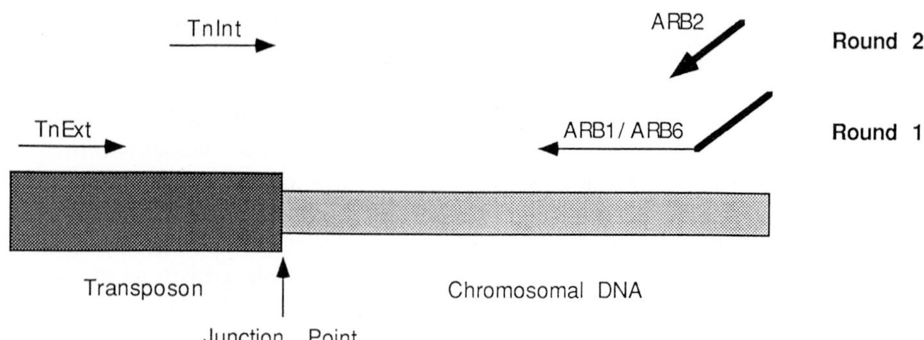

FIG. 5. Schematic of arbitrary-primed PCR. A hypothetical transposon/chromosome junction point is shown along with the relative positions of the two sets of PCR primers used to amplify DNA flanking the transposon insertion. In PCR round 1, the primers ARB1, ARB6, and TnExt (which is complementary to the transposon sequence) are used. Among the many sequences amplified in this round may be the desired PCR product primed from the transposon and flanking DNA sequence. This product is enriched by in the second round of PCR using primers ARB2 (the sequence of this primer is identical to the 5′ end of ARB1 and ARB6) and TnInt (which is also complementary to the transposon sequence and is closer to the transposon/chromosome junction relative to TnExt). The TnInt primer is also used to sequence the purified product(s) of PCR round 2.

Two rounds of amplification are used to specifically amplify and enrich for the DNA flanking the insertion site (the protocol is diagrammed in Fig. 5). The first round of PCR uses a primer unique to one of the end of the transposon and two different arbitrary primers (see Table III for primer sequences and later for experimental details). Among the many possible amplified regions from the first round of PCR are products primed from the transposon and flanking chromosomal DNA. Products flanking the transposon are specifically amplified in the second round of PCR. The following detailed protocol is used for arbitrary-primed PCR.

Components of PCR Reactions. The final concentration of each component is indicated in parentheses.

- DNA polymerase: We routinely use Vent (exo$^+$) DNA polymerase (2 U/100-μl reaction, NEB, Beverly, MA) or Qiagen Tag DNA polymerase (5 U/100-μl reaction, Qiagen, Valencia, CA). The best enzyme should be determined empirically for each organism.
- DNA polymerase buffer (1×, supplied with the enzyme)
- $MgSO_4$ (1 mM)
- dNTPs (0.25 mM of each)
- Oligonucleotide primers (1 μg/100-μl reaction)
- Template DNA (see later)
- PCR reactions can be overlayed with a volume of mineral oil equal to the reaction volume, if necessary.

TABLE III
ARBITRARY PCR PRIMERS

Transposon	Primer	Round[d]	Sequence
Any	ARB1	First	5′-GGCCACGCGTCGACTAGTACNNNNNNNNNNGATAT-3′
Any	ARB6	First	5′-GGCCACGCGTCGACTAGTACNNNNNNNNNNACGCC-3′
Any	ARB2	Second	5′-GGCCACGCGTCGACTAGTAC-3′
Tn*5*[a]	Tn5ext	First	5′-GAACGTTACCATGTTAGGAGGTC-3′
Tn*5*	Tn5int	Second	5′-CGGGAAAGGTTCCGTTCAGGACGC-3′
Tn*10d*(Cm)[b]	CmExt	First	5′-CAGGCTCTCCCCGTGGAGG-3′
Tn*10d*(Cm)	CmInt	Second	5′-CTGCCTCCCAGAGCCTG-3′
Tn*10d*(Kn)[c]	KnExt	First	5′-CCGCGGTGGAGCTCC-3′
Tn*10d*(Kn)	KnInt	Second	5′-ATGACAAGATGTGTATCCACC-3′

[a] These primers can be used with transposons Tn*5B21*(Tc^r, '*lacZ*), Tn*5B30*(Tc^r), and Tn*5B22*(Gm^r) reported in R. Simon, J. Quandt, and W. Klipp, *Gene* **80,** 160 (1989).
[b] This transposon is described in N. Kleckner, J. Bender, and S. Gottesman, *Methods Enzymol.* **204,** 139 (1991).
[c] This transposon is described in M. F. Alexeyev and I. N. Shokolenko, *Gene* **160,** 59 (1995).
[d] Refers to whether the primer is used in PCR round 1 or PCR round 2 (see text for details).

PCR Round 1. The PCR reaction is performed in a final volume of 50 μl. The primers used are the external transposon primer (e.g., for the Tn*5B22* transposon this primer is Tn5Ext, see Table III), ARB1, and ARB6. ARB1 and ARB6 are identical except for 5 bp at the 3′ end of the primer; this difference at the 3′ end of the primer accounts for differences in the GC content of different bacteria. It is important to note that the transposon primers must be complementary to a unique sequence from one end of the transposon to prevent priming from both ends of the transposon. Because the product(s) of the round 2 PCR reaction will be sequenced without purifying individual PCR products (see later), priming from both ends of the transposon would lead to the amplification of two distinct DNA sequences (flanking each side of the transposon insertion) and would result in an unreadable DNA sequence. For most organisms, 5 μl of an overnight culture grown in rich medium will serve as the DNA template. If no PCR products are generated using an aliquot of cell culture (as is the case in our experience for *P. aeruginosa*), 0.5 μg of chromosomal DNA can serve as a template (see Pitcher *et al.*[39] for a rapid chromosomal DNA purification method that we have used successfully to generate a DNA template for arbitrary-primed PCR).

Thermocycler conditions for PCR round 1 are as follows.

1. 95° × 5 min
2. 6 × [94° × 30 sec, 30° × 30 sec, 72° × 1 min]

[39] D. G. Pitcher, N. A. Saunders, and R. J. Owen, *Lett. Appl. Microbiol.* **8,** 151 (1989).

3. 30 × [94° × 30 sec, 45° × 30 sec, 72° × 1 min]
4. 72° × 5 min
5. 4°

PCR Round 2. The second round reaction is performed as just described except that it is performed in 100 μl final volume, the DNA template for the reaction is 5 μl of the PCR reaction from round 1, and the primers used are the internal primer (e.g., for the Tn*5B22* transposon this primer is Tn*5Int,* see Table III) and ARB2. The internal primer is so named because it is closer to the transposon/chromosome junction and should be complementary to the sequence 20–30 bp from the end of the transposon. The internal primer is also used to sequence the final PCR products (see later) and must be located sufficiently far from the transposon–chromosome junction to allow the DNA sequence to be obtained from across this junction. Obtaining the DNA sequence from the transposon–chromosome junction serves as a control to ensure that the PCR product has been primed specifically from the transposon.

Thermocycler conditions for PCR Round 2 are as follows

1. 30 × [94° × 30 sec, 45° × 30 sec, 72° × 1 min]
2. 72° × 5min
3. 4°

After both rounds of PCR are completed, the products are visualized on a 2% agarose gel. As a control, an arbitrary PCR reaction is performed on wild-type DNA template (without a transposon insertion). In our experience, >90% of arbitrary PCR reactions yield a distinguishable product (or products), which usually range from 200 to 400 bp in length. Given the somewhat random nature of the round 1 amplification, if the first attempt at arbitrary-primed PCR does not yield a product, simply repeat both rounds of PCR; this often leads to the successful amplification of a product. Alternatively, the annealing temperature of both rounds of PCR can be changed from the suggested 45°. The PCR products may be purified using the QIAquick Spin PCR purification kit (Qiagen, Valencia, CA) or an equivalent procedure that purifies the PCR products away from primers, nucleotides, etc. These PCR products may then be subsequently subjected to DNA sequence analysis using the internal transposon primer (e.g., for Tn*5* insertions this would be Tn*5Int,* see Table III). The resulting nucleotide sequence can be compared with DNA sequence databases. Reactions that yield multiple PCR products can be subjected to DNA sequence analysis without purifying individual bands because all PCR products primed from the transposon share a common transposon-derived DNA sequence at one end.

Conclusions

The genetic approach described in this article has the potential to expand our understanding of biofilm development profoundly. The methods outlined here are simple and can be applied readily to study many different types of bacteria that inhabit surface-associated communities. Subsequent microscopic and molecular analyses of the mutants uncovered by these screens are also straightforward. The results of these screens for mutants defective in biofilm formation in *P. aeruginosa, P. fluorescens, E. coli, V. cholera, S. putrefaciens,* and *S. epidermidis* have identified genes of both known and unknown function.[2,4–6,8,12,40,41] This suggests that by focusing on surface association, these screens have the potential to provide insight that will assist in assigning roles to genes of previously unknown function. These results also suggest that the study of biofilm formation represents the exploration of an aspect of microbial physiology not yet studied at the molecular genetic level. A molecular understanding of the control of biofilm development may eventually be applied to limit both the virulence and the distribution of pathogens in various environments, serve as a means to regulate microbial attachment to and degradation of solid materials, and improve our control of biofilm formation in industrial settings.

[40] O. Vidal, R. Longin, C. Prigent-Combaret, C. Dorel, M. Hooreman, and P. Lejune, *J. Bacteriol.* **180,** 2442 (1998).

[41] C. Heilmann and C. Gotz, *Zentbl. Bakteriol.* **287,** 69 (1998).

[7] Identifying *in Vivo* Expressed Streptococcal Genes in Endocarditis

By LIN TAO and MARK C. HERZBERG

Viridans streptococci normally reside in the oral cavity to form a special biofilm on the tooth called dental plaque. In the oral cavity, these streptococci are nonpathogenic except for the mutans group, which causes caries.[1] In other anatomic sites, however, these streptococci may be pathogenic.[2] For example, viridans streptococci are the most common organisms causing endocarditis. These bacteria can enter the circulation during dental proce-

[1] S. Hamada and H. D. Slade, *Microbiol. Rev.* **44,** 331 (1980).

[2] R. Bayliss, C. Clarke, C. M. Oakley, W. Somerville, A. G. W. Whitfield, and S. E. J. Young, *Br. Heart J.* **50,** 513 (1983).

0076-6879/99 $30.00

dures and cause transient bacteremia. If predisposing factors such as rheumatic or congenital heart diseases (valve defects) exist, the bacteria can colonize the heart by forming a different biofilm on the endocardium to cause endocarditis.

Once the bacteria colonize the heart, virulence appears to depend in part on the expression of specific virulence genes. To date, the specific conditions that occur *in vivo* in the heart to modify streptococcal gene expression are largely undefined. While it will be of interest to know if virulence depends on streptococcal genes that are induced in the host, *in vivo* animal models[3–5] of infective endocarditis contain a robust set of environmental signals that may not be well simulated by *in vitro* bioassays.[6,7]

To learn about genes that are expressed specifically *in vivo* but not *in vitro,* new approaches are being developed. Mahan *et al.*[8] have reported the detection of host-induced *Salmonella* genes encoding potential virulence factors. This *in vivo* expression technology (IVET) uses an integration plasmid vector carrying two promoterless reporter genes, *purA* and *lacZ,* for both *in vivo* and *in vitro* selection. Using this and similar IVET systems, which target unique metabolic pathways or genetic markers of the seleted pathogens, many host-induced genes have been identified in several different bacterial species, including *Salmonella typhimurium,*[8–10] *Pseudomonas aeruginosa,*[11] *Vibrio cholerae,*[12] and *Staphylococcus aureus.*[13,14] Each IVET system for a specific bacterial genus or species must be designed to exploit certain fastidious, unique growth requirements or other traits that can be used for selection. Therefore, this article describes an IVET system applicable to studying experimental endocarditis and other biofilm problems, permitting selection of *in vivo*-induced (*ivi*) genes in the gram-positive *Streptococcus gordonii.*

[3] H. A. Dewar, M. R. Jones, S. G. Griffin, A. Oxley, and J. Marriner, *J. Comp. Pathol.* **97,** 567 (1987).

[4] D. T. Durack and P. B. Beeson, *Br. J. Exp. Pathol.* **53,** 44 (1972).

[5] M. C. Herzberg, G. D. MacFarlane, K. Gong, N. N. Armstrong, A. R. Witt, P. R. Erickson, and M. W. Meyer, *Infect. Immun.* **60,** 4809 (1992).

[6] M. C. Herzberg, K. Gong, G. D. MacFarlane, P. R. Erickson, A. H. Soberay, P. H. Krebsbach, G. Manjula, K. Schilling, and W. H. Bowen, *Infect. Immun.* **58,** 515 (1990).

[7] C. H. Ramirez-Ronda, *J. Clin. Invest.* **62,** 805 (1978).

[8] M. J. Mahan, J. M. Slauch, and J. J. Mekalanos, *Science* **259,** 686 (1993).

[9] E. M. Heithoff, C. P. Conner, P. C. Hanna, S. M. Julio, U. Hentschel, and M. J. Mahan, *Proc. Natl. Acad. Sci. U.S.A.* **94,** 934 (1997).

[10] R. H. Valdivia and S. Falkow, *Science* **277,** 2007 (1997).

[11] J. Wang, A. Mushegian, S. Lory, and S. Jin, *Proc. Natl. Acad. Sci. U.S.A.* **93,** 10434 (1996).

[12] A. Camilli and J. Mekalanos, *Mol. Microbiol.* **18,** 671 (1995).

[13] A. M. Lowe, D. T. Beattie, and R. L. Deresiewicz, *Mol. Microbiol.* **27,** 967 (1998).

[14] M. A. Lane, K. W. Bayles, and R. E. Yasbin, *Gene* **100,** 225 (1991).

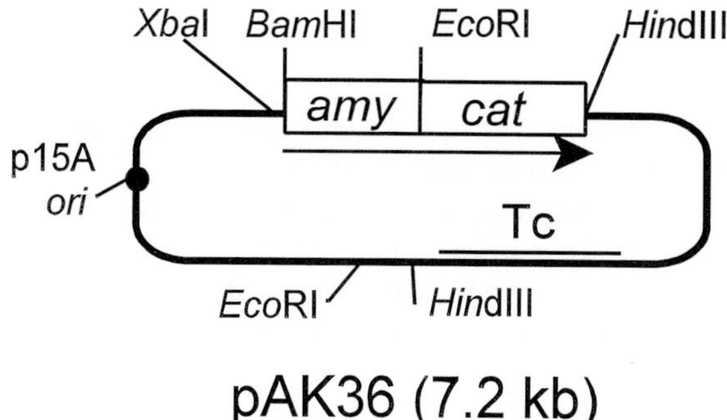

FIG. 1. The streptococcal IVET vector pAK36.

Method

Streptococcal IVET Vector pAK36

In vivo-induced streptococcal genes can be identified using a specially designed IVET vector, pAK36 (Fig. 1). It has two promoterless reporter genes of gram-positive bacterial origin, *amy* and *cat,* from pRQ200[14] and pMH109,[15] respectively. It also has a tetracycline (Tc) resistance marker and replication origin (*ori*) from the streptococcal integration vector pSF143.[16] The plasmid can replicate in *Escherichia coli* with a medium copy number (the p15A *ori*), but it cannot replicate extrachromosomally in streptococci. The plasmid, however, can insert into the streptococcal chromosome by homologous, Campbell-like integration if it carries a streptococcal gene fragment. The *cat* reporter gene provides a positive *in vivo* selection for inducible promoter genes in the reporter gene-fused strains in an animal host to which chloramphenicol (Cm) is given. The *amy* gene confers amylase production, which is suitable for a negative selection *in vitro* and can be detected readily on agar plates containing 0.5% (w/v) starch by flooding with an iodine solution. Unlike constitutively expressed genes, the host-induced genes only express *in vivo,* but not *in vitro*. Therefore, after selection *in vivo, in vitro* selection is also necessary to exclude from consideration constitutively expressed genes. Once a host-induced IVET–fusion clone is identified, the gene flanking the insertion site can be retrieved readily by self-ligating the *XbaI*-digested chromosomal DNA of the IVET–fusion clone and transforming *E. coli* and is characterized by subsequent sequencing or insertion inactivational analysis.

[15] M. C. Hudson and R. Curtiss III, *Infect. Immun.* **58,** 464 (1990).
[16] L. Tao, D. J. LeBlanc, and J. J. Ferretti, *Gene* **120,** 105 (1992).

Construction of Reporter Gene-Fusion Strain Library in Streptococcus gordonii V288

Introduction. To identify streptococcal genes induced *in vivo* in endocarditis, a reporter gene-fusion library of more than 10,000 clones in one streptococcal strain is needed. In these clones, the IVET vector should be inserted randomly in the chromosome. *Streptococcus gordonii* strain V288 is recommended because it is highly competent for natural transformation but is low in background amylase activity.

Procedure

1. Digest 10 μg *S. gordonii* chromosomal DNA with 3 units of *Sau*3A1 overnight at 37°.
2. Digest 10 μg pAK36 DNA with 3 units of *Bam*HI overnight at 37°.
3. Purify the two restriction enzyme-digested DNA preparations with the GeneClean (Bio-101, Inc., Vista, CA) method to remove enzymes and buffering salts.
4. Ligate the two DNA preparations with 3 units of T4 ligase overnight at 14°. Test 1% of the ligation mixture on a 0.7% agarose gel to verify the ligation result and estimate the DNA concentration. The DNA in the remaining ligation mixture should be close to 20 μg.
5. Prepare competent *S. gordonii* V288 cells for natural transformation. [Inoculate a single colony from a blood agar plate into 2 ml Todd–Hewitt broth (THB) supplemented with 10% horse serum (THBS) in a glass test tube. Incubate the culture at 37° for 16 hr. Transfer the 2-ml culture into 80 ml of prewarmed THBS medium (in 10 test tubes). Incubate the 1:40 diluted culture at 37° for exactly 2 hr to achieve maximal competence.]
6. Add one-half of the ligation mixture (about 10 μg DNA) to the competent *S. gordonii* V288 cells and keep the other half at −20°.
7. Continue to incubate the culture at 37° for 1 hr.
8. Concentrate the culture by centrifugation and spread the cells on at least 100 Todd–Hewitt agar plates containing 10 μg/ml Tc.
9. Incubate the plates in candle jars or anaerobically for 24 hr at 37°.
10. Estimate the total number of transformant colonies. (If the procedures are followed correctly, a total of more than 50,000 transformant colonies should be obtained; proceed to step 11. If the colony count is less than 10,000, however, competence may have developed insufficiently. A low transformation rate may occur due to insufficient DNA, incorrect timing of addition of DNA, and impurity of the culture. Repeat steps 5 to 10 precisely using the remaining half of the ligation mixture.)

11. Harvest all colonies with a sterile cotton swab and resuspend the cells in 10 ml THB supplemented with 10 μg/ml Tc.
12. Incubate the culture for 4 hr at 37° to adapt the cells to liquid medium.
13. Store the cells in 10 aliquots with 15% glycerol at −70° as the *S. gordonii* V288 IVET reporter gene-fusion library.

Comments. Normally, the ligation mixture can be amplified in *E. coli* to enrich the DNA to be transformed into streptococci. In the case of *S. gordonii,* however, amplification of the ligation mixture in *E. coli* can cause selective deletions to the cloned streptococcal DNA because some *S. gordonii* genes are toxic to *E. coli* and thus can create a bias in the overall representation within the library. To ensure a thorough representation of the reporter gene-fusion clone library, it is necessary to transform the ligation mixture *directly* into *S. gordonii* as described earlier. Because *S. gordonii* is highly competent for natural transformation, no amplification of ligated DNA in *E. coli* is needed. It is important, however, to start with a larger amount of highly purified DNA. Therefore, the CsCl gradient ultraspeed centrifugation method is highly recommended to prepare the DNA before the restriction enzyme digestion procedure.

Isolation of Streptococcus gordonii in Vivo-Induced Genes in Endocarditis

Introduction. This section includes several major steps: an endocarditis animal model preparation, bacterial inoculation, *in vivo* selection, *in vitro* selection, and gene cloning and sequencing. Although different animals such as mice, rats, pigs, and dogs have been used to study endocarditis, we recommend the rabbit model because it has a larger heart than mice or rats, easily accessible ear veins for intravenous injections, and the anatomy is similar to humans.[17,18]

Procedure

1. Inoculate one sample (about 1 ml) frozen stock culture of *S. gordonii* V288 IVET reporter gene-fusion strain library into 10 ml THB supplemented with 10 μg/ml Tc and incubate for 16 hr at 37°.
2. Dilute the culture 1:4 in the same prewarmed broth and continue to incubate for 3 hr to reach the exponential phase.

[17] T. G. Sarphie, *Am. J. Anat.* **174,** 145 (1985).
[18] H. Masuda, *Int. J. Biochem.* **16,** 99 (1984).

3. Harvest the cells by centrifugation at 4000*g* for 10 min at room temperature and wash in sterile saline once. Resuspend in 2 ml sterile saline (about 10^9 cells/ml).
4. Inject the cell suspension intravenously through an ear vein into a New Zealand White rabbit that has been prepared for experimental endocarditis by placing an indwelling catheter as described previously.[5]
5. Beginning 6 hr after the inoculation of the library, give the rabbit Cm intravenously two times a day for 3 days to a final serum level of 5 μg/ml. (The minimal inhibitory concentration of chloramphenicol for the wild-type *S. gordonii* V288 strain is 0.5 μg/ml.)
6. Euthanize the rabbit and dissect out the aortic valve vegetation. Immerse the valve specimen in sterile saline. Under aseptic conditions, disperse the specimen with a mortar and pestle to recover viable bacterial cells.
7. Spread the bacterial cells onto THB agar plates supplemented with 10 μg/ml Tc and incubate the plates in a candle jar at 37° for 24 hr.
8. Isolate single colonies and replica plate the colonies onto THB agar (master plate), THB agar supplemented with 5 μg/ml Cm, and THB supplemented with 0.5% (w/v) starch. Incubate the plates in candle jars for 24 hr at 37°.
9. Detect amylase activity by flooding the starch plate with an iodine solution [0.2% (w/v) I_2 and 0.2% (w/v) KI]. Identify bacterial colonies that display negative amylase activity and sensitivity to chloramphenicol. Select the same colonies from the master plate as the *ivi* clones.
10. Isolate chromosomal DNA from each of the selected clones. Digest the DNA with *Hin*dIII and hybridize the DNA with labeled pAK36 by the method of Southern.[19] Identify clones that display unique pAK36 insertion patterns to rule out redundant siblings.
11. Digest the chromosomal DNA of the *ivi* clones displaying unique pAK36 insertion patterns with the restriction enzyme *Xba*I or *Sty*I. Remove the enzyme and buffering salt by the GeneClean method. Perform self-ligation with T4 ligase, and transform *E. coli* JM109 to Tc^R.[20] If the transformation is not successful with JM109, use the *E. coli recA*-positive strain C600 instead.
12. Isolate plasmids from the Tc^R transformant colonies with the Qiagen plasmid isolation kit (Qiagen, Inc., Chatsworth, CA) and analyze them by restriction enzyme digestion.

[19] E. Southern, *J. Mol. Biol.* **98,** 503 (1975).
[20] C. T. Chung, S. L. Niemela, and R. H. Miller, *Proc. Natl. Acad. Sci. U.S.A.* **86,** 2172 (1989).

13. Sequence the cloned genes with both forward and reverse primers designed according to the sequence of *amy* and *ori* genes, which flank the cloned streptococcal gene fragment in the plasmid. To sequence DNA fragments upstream to the *amy* gene, an oligonucleotide, 5′-AGC GCA AAT AAC AGC GTC AGC AA-3′, complementary to the 5′ end of the *amy* coding region of *Bacillus licheniformis* (GenBank accession A17930), can be used as the reverse primer. Another oligonucleotide, 5′-CAA GAG ATT ACG CGC AGA CC-3′, derived from the plasmid *ori* sequence of pACYC184 (X06403), can be used as the forward primer. Nucleotide sequences can be determined either by the dideoxynucleotide chain-termina-

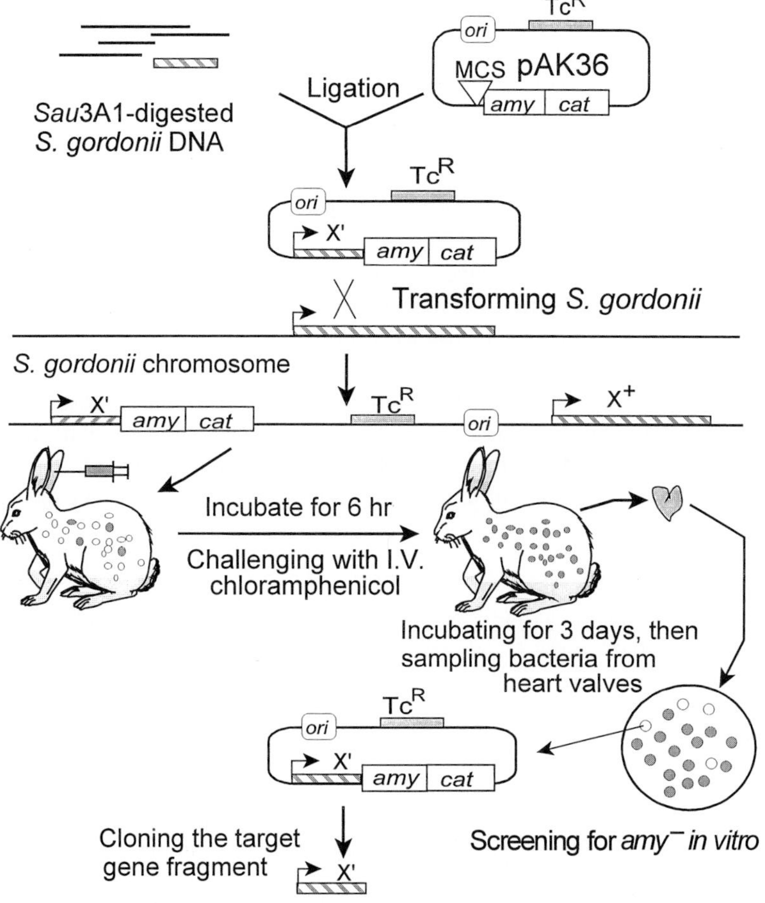

FIG. 2. Isolation of *S. gordonii ivi* genes induced in endocarditis.

tion method[21] or by a DNA sequencing service using an automated DNA sequencer.

14. Analyze the DNA sequence by translation in all six reading frames to search the nonredundant sequence database (National Center for Biotechnology Information, National Institute of Health) by the BLASTX program.[22]

Comments. Certain streptococcal genes may be lethal to *E. coli* and thus cannot be cloned in *E. coli* without deletions or rearrangement. Because these genes may not be cloned with the *E. coli* $recA^-$ strain JM109, an alternative $recA^+$ *E. coli* strain C600 can be used. Deletions of *S. gordonii* genes in *E. coli* C600, however, may occur in about one-half of the plasmids, in which a portion of the *amy* gene may also be lost.[23] Therefore, these plasmids can be sequenced only by the *ori* gene-derived forward primer. In this case, the gene sequenced with the forward primer may not be proximal upstream to the *ivi* gene fused to the *amy–cat* cassette and induced during endocarditis. This can be determined from a comparison of the size of the entire inserted DNA fragment with the size and orientation of the sequenced gene. The *ivi* gene, if deleted, may also be lethal to *E. coli* and not be cloned readily in that host. The gene sequence can be obtained by chromosomal walking with the inverse polymerase chain reaction method[24] once the sequence of its adjacent gene is available.

We have used this method successfully to identify 13 *S. gordonii* V288 *ivi* genes induced in endocarditis with the rabbit model (see Fig. 2).[23] The availability of the new IVET system for *S. gordonii* has expanded the repertoire of genetic tools for the identification of *in vivo* expressed microbial genes. The plasmid pAK36 may also be applicable to the identification of *in vivo*-induced genes from other streptococci and closely related gram-positive bacterial species in endocarditis or other types of infections.

Acknowledgments

This study was supported by NIH Grants DE11336, DE05501, DE08590, and DE00270 and by a Grant-in-Aid (KS-96-GB-56) from the American Heart Association, Kansas Affiliate, Inc.

[21] F. Sanger, S. Nicklen, and A. R. Coulson, *Proc. Natl. Acad. Sci. U.S.A.* **74,** 5463 (1977).

[22] S. F. Altschul, W. Gish, W. Miller, E. W. Myers, and D. J. Lipman, *J. Mol. Biol.* **215,** 403 (1990).

[23] A. O. Kiliç, M. C. Herzberg, M. W. Meyer, X. Zhao, and L. Tao, *Plasmid,* in press (1999).

[24] H. Ochman, A. S. Gerber, and D. L. Hartl, *Genetics* **120,** 621 (1988).

[8] Quorum Sensing, Gene Expression, and *Pseudomonas* Biofilms

By TERESA R. DE KIEVIT and BARBARA H. IGLEWSKI

Introduction

The ability to adhere to solid surfaces and establish microbial communities known as biofilms is an important survival strategy for bacteria in nature. *Pseudomonas aeruginosa* biofilms formed on solid surfaces exposed to a continuous flow of liquid develop into complex mushroom- and stalk-like structures.[1] Water channels separate the clusters of polysaccharide-encased cells, which allow nutrients to flow in and waste products to flow out. How this complex architecture is achieved has only recently begun to be understood at a molecular level.[2]

Quorum sensing is a mechanism used by a large number of gram-negative bacteria to monitor their cell density and, in response, to regulate the expression of specific genes. As an example, *P. aeruginosa* uses quorum sensing to regulate virulence factor production. Comparison of biofilms produced by a wild-type strain of *P. aeruginosa* with those of quorum-sensing mutants revealed that cell-to-cell signaling is required for the formation of a normal *P. aeruginosa* biofilm.[2] Thus *P. aeruginosa* uses quorum sensing to elaborate the complex structure of mature biofilms.

This article contains a brief overview of *P. aeruginosa* quorum sensing and presents two methods for detecting the presence of autoinducer (AI) molecules in biofilms. It also discusses the generation of quorum-sensing mutants and the use of reporter gene fusions for studying gene expression in *P. aeruginosa* biofilms.

Quorum Sensing in *Pseudomonas aeruginosa*

Quorum sensing is a mechanism that enables bacteria to monitor their cell density and, in response, to activate certain genes. Quorum-sensing systems make use of two components; a small diffusible signaling molecule known as the autoinducer and a positive transcriptional activator, or R-protein. At low cell density the AI is produced at a basal level; however,

[1] J. W. Costerton, Z. Lewandowski, D. E. Caldwell, D. R. Korber, and H. M. Lappin-Scott, *Annu. Rev. Microbiol.* **49,** 711 (1995).

[2] D. G. Davies, M. R. Parsek, J. P. Pearson, B. H. Iglewski, J. W. Costerton, and E. P. Greenberg, *Science* **280,** 295 (1998).

0076-6879/99 $30.00

as the cell density increases, so does the concentration of AI. Once a threshold leve! of AI is reached, sufficient AI/R-protein complexes accumulate, enabling the activation of target genes (Fig. 1). Two distinct quorum-sensing systems have been identified in *P. aeruginosa*: the *las* system, which consists of the transcriptional activator LasR and its cognate AI, 3-oxo-C_{12}-HSL [*N*-(3-oxododecanoyl)-L-homoserine lactone); and the *rhl* system, which is composed of the transcriptional activator Rh1R and its cognate AI, C_4-HSL (*N*-butyryl-L-homoserine lactone). Synthesis of the two AIs, 3-oxo-C_{12}-HSL and C_4-HSL, is directed by the products of the *lasI* and *rhlI* genes, respectively. A list of *P. aeruginosa* genes regulated by these quorum-sensing systems is given in Table I. Data indicate that in the development of a normal *P. aeruginosa* biofilm, only the *las* quorum-sensing system is essential.[2] Mutants deficient in the production of 3-oxo-C_{12}-HSL were able to initiate biofilm development; however, the biofilm lacked the complex architecture observed in that of the wild-type strain. The biofilm formed by a *lasI*-mutant was much thinner and it lacked water channels, and the *lasI*-biofilm cells were significantly more sensitive to the detergent sodium dodecyl sulfate (SDS), as compared to the wild-type.[2] Thus biofilm formation can be added to the list of *P. aeruginosa* phenotypes regulated by quorum sensing.

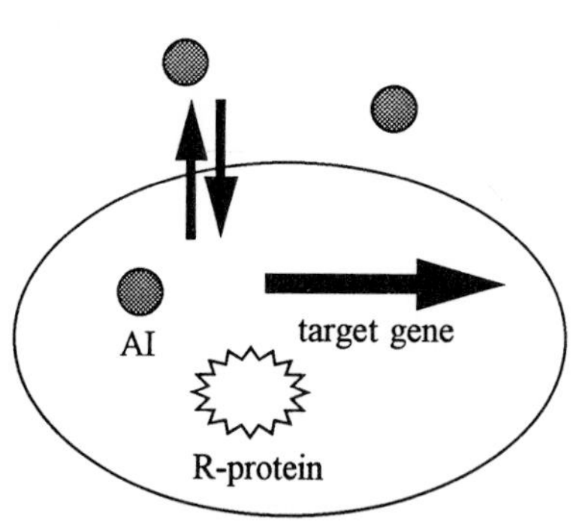

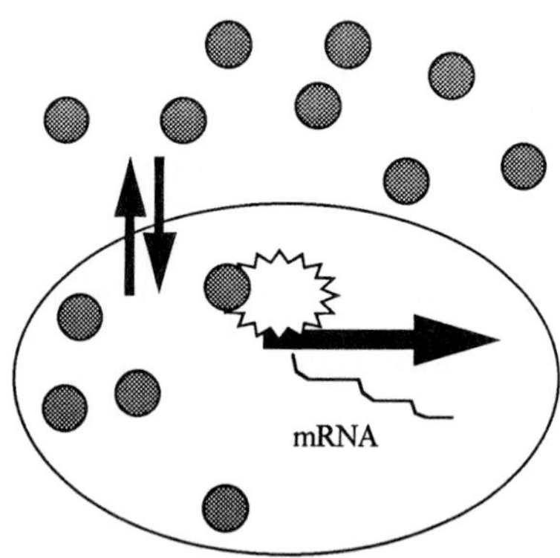

FIG. 1. A general model of quorum sensing is depicted. At low cell density, the diffusible AI is produced at a basal level. With increasing cell density, the level of AI increases until a threshold concentration is reached and the AI is able to bind to its cognate R-protein. The activated R-protein/AI complex can then bind upstream of target genes to induce their expression.

TABLE I
GENES REGULATED BY QUORUM SENSING IN *Pseudomonas aeruginosa*

Gene	Gene product	Quorum-sensing system[a]	Refs.
lasA	lasA elastase	*las*	*b*
lasB	lasB elastase	*las/rhl*	*c, d, e*
aprA	Alkaline protease	*las/rhl*	*f*
toxA	Exotoxin A	*las*	*f*
xcpP	Type II secretion protein	*las/rhl*	*g*
xcpR	Type II secretion protein	*las/rhl*	*g*
rhlAB	Rhamnolipid	*rhl/las*	*d, h, i, j*
rpoS	Stationary-phase sigma factor	*rhl*	*k*
lasR	Transcriptional activator	*las*	*k, l*
lasI	Autoinducer synthase	*las*	*m*
rhlR	Transcriptional activator	*las*	*k, l*
rhlI	Autoinducer synthase	*rhl/las*	*k*

[a] In cases where more than one quorum-sensing system controls gene expression, if the dominant regulatory system is known, it is indicated by appearing first.
[b] D. S. Toder, M. J. Gambello, and B. H. Iglewski, *Mol. Microbiol.* **5,** 2003 (1991).
[c] M. J. Gambello and B. H. Iglewski, *J. Bacteriol.* **173,** 3000 (1991); L. Passador, J. M. Cook, M. J. Gambello, L. Rust, and B. H. Iglewski, *Science* **260,** 1127 (1993).
[d] U. A. Ochsner and J. Reiser, *Proc. Natl. Acad. Sci. U.S.A.* **92,** 6424 (1995).
[e] J. M. Brint and D. E. Ohman, *J. Bacteriol.* **177,** 7155 (1995).
[f] M. J. Gambello, S. Kaye, and B. H. Iglewski, *Infect. Immun.* **61,** 1180 (1993).
[g] V. Chapon-Hervé, M. Akrim, A. Latifi, P. Williams, A. Lazdunski, and M. Bally, *Mol. Microbiol.* **24,** 1169 (1997).
[h] U. A. Ochsner, A. K. Koch, A. Flechter, and J. Reiser, *J. Biol. Chem.* **269,** 19787 (1994).
[i] U. A. Ochsner, A. K. Koch, A. Flechter, and J. Reiser, *J. Bacteriol.* **176,** 2044 (1994).
[j] J. P. Pearson, E. C. Pesci, and B. H. Iglewski, *J. Bacteriol.* **179,** 5756 (1997).
[k] A. Latifi, M. Foglino, K. Tanaka, P. Williams, and A. Lazdunski, *Mol. Microbiol.* **21,** 1137 (1996).
[l] E. C. Pesci, J. P. Pearson, P. C. Seed, and B. H. Iglewski, *J. Bacteriol.* **179,** 3127 (1997).
[m] P. C. Seed, L. Passador, and B. H. Iglewski, *J. Bacteriol.* **177,** 654 (1995).

Detection of Acylhomoserine Lactones or Autoinducers in Biofilms

It is becoming clear that many gram-negative bacteria have components of quorum-sensing systems (see Fuqua *et al.*[3] for a review), and in the systems examined so far, both the R-proteins and the AI molecules are remarkably similar. Despite these similarities, studies have shown that the R-proteins are able to discriminate cognate and noncognate AIs; however,

[3] W. C. Fuqua, S. C. Winans, and E. P. Greenberg, *Annu. Rev. Microbiol.* **50,** 727 (1996).

some infidelity does occur.[4,5] In the gram-negative systems examined thus far, the AI molecules have both common features, such as a lactone ring structure, and unique aspects, including the length and substitution of the fatty acyl side chain. In the case of *P. aeruginosa,* studies have shown that while various acyl chain substitutions influence AI activity, the most important determinant is the length of the acyl side chain.[5,6] In general, compounds differing from the cognate AI by only a few carbons in the acyl chain length induce weak to moderate gene expression, albeit higher concentrations of AI may be required. Conversely, compounds deviating significantly in chain length from the cognate AI are less active.[6] The presence of acylhomoserine lactones, or AI, in naturally occurring biofilms was first reported by McLean and co-workers.[7]

Assay for Detecting Autoinducer Production

A number of gram-negative organisms employ quorum-sensing systems to control gene expression; however, an important first step in determining whether quorum sensing plays a role in biofilm development is to detect AI activity in the biofilm. AI-responsive reporter strains can be employed for this exact purpose. Because the R-proteins can generally be activated by heterologous AI molecules if they differ by only a few carbons in their acyl chain, using quorum-sensing systems that utilize 4, 8, and 12 carbon AI molecules should enable a wide range of AIs to be detected. Potential candidates for these systems include the *P. aeruginosa rhl* quorum-sensing system (4 carbon AI), the *Agrobacterium tumefaciens* system (8 carbon AI), and the *P. aeruginosa las* quorum-sensing system (12 carbon AI). Two methods are presented for detecting the presence of AIs in biofilms. Method 1 involves assaying the crude biofilm sample for AI activity and Method 2 employs an AI purification procedure. For these studies, bacterial strains carrying genes encoding the R-proteins activated by the 4, 8, and 12 carbon AIs, as well as a target gene fused to *lacZ* or another reporter gene, are needed. Using the *P. aeruginosa* LasR/3-oxo-C_{12}-HSL system as an example, *lasI* expression is controlled by LasR/3-oxo-C_{12}-HSL. Only minute quantities of the cognate AI are required to activate LasR and induce expression of the *lasI* gene; therefore, *lasI–lacZ* is an ideal reporter fusion.

[4] K. M. Gray, L. Passador, B. H. Iglewski, and E. P. Greenberg, *J. Bacteriol.* **176,** 3076 (1994).

[5] J. P. Pearson, K. M. Gray, L. Passador, K. D. Tucker, A. Eberhard, B. H. Iglewski, and E. P. Greenberg, *Proc. Natl. Acad. Sci. U.S.A.* **91,** 197 (1994).

[6] L. Passador, K. D. Tucker, K. R. Guertin, M. P. Journet, A. S. Kende, and B. H. Iglewski, *J. Bacteriol.* **178,** 5995 (1996).

[7] R. J. C. McLean, M. Whiteley, D. J. Stickler, and W. C. Fuqua, *FEMS Microbiol. Lett.* **154,** 259 (1997).

For the indicator strain, either an *Escherichia coli* strain harboring *lasR* and *lasI–lacZ* on a multicopy plasmid(s) or a *P. aeruginosa* autoinducer synthase mutant (PAO-JP2; deficient in the production of 3-oxo-C_{12}-HSL and C_4-HSL)[8] harboring a *lasI–lacZ* fusion plasmid can be used. When choosing an appropriate indicator bacterium, it is worth noting a study performed by Zhu and colleagues[9] in which an *A. tumefaciens* strain constitutively overproducing the TraR R-protein was induced by a much wider range of AI analogs as compared to a strain expressing wild-type TraR levels. Thus for detection of HSLs in biofilms, overexpression of the R-proteins in the indicator strain may increase the sensitivity of the bioassay to noncognate AIs.

Method 1

Method 1 is adapted from McLean *et al.*[7] employing components of the *P. aeruginosa las* quorum-sensing system as an example. The general principle of this method is as follows: the indicator strain [e.g., *E. coli* (*lasI–lacZ; lasR*)] and the biofilm sample are plated, side by side, on L-agar containing 5-bromo-4-chloro-3-indolyl-β-D-galactoside (X-Gal). If AI is present, it should diffuse through the agar to the indicator strain and activate LasR, resulting in the production of a blue color in the assay medium.

Reagents

L-agar plates overlaid with 50 μl of X-Gal (20 mg/ml stock solution)
Sterile phosphate-buffered saline (PBS), pH 7.4: 137 mM NaCl, 2.7 mM KCl, 4.3 mM $Na_2HPO_4 \cdot 7\ H_2O$, 1.4 mM KH_2PO_4

Procedure. Streak the indicator strain, i.e., *E. coli* harboring a plasmid with both *lasR* and *lasI–lacZ,* on L-agar containing X-Gal. Remove the biofilm sample from the adhering surface either by scraping the polysaccharide-encased cells into sterile PBS or sonication in sterile PBS. Alternatively, an appropriately sized portion of the biofilm, including the substratum to which it is attached, can be laid directly on the bioassay medium. Place either the culture or the attached biofilm sample approximately 1 inch from the indicator strain and incubate overnight at 37°. Observe for β-galactosidase activity, viewed as a blue coloration of the media. Control test strains should include *E. coli* expressing the 3-oxo-C_{12}-HSL AI synthase

[8] J. P. Pearson, E. C. Pesci, and B. H. Iglewski, *J. Bacteriol.* **179,** 5756 (1997).

[9] J. Zhu, J. W. Beaber, M. I. Moré, C. Fuqua, A. Eberhard, and S. C. Winans, *J. Bacteriol.* **180,** 5398 (1998).

gene (*lasI*) and *lasR* on a plasmid and PAO-JP2 (deficient in the production of both 3-oxo-C_{12}-HSL and C_4-HSL)[8] as positive and negative controls, respectively. Detectable β-galactosidase activity, seen as a blue color, indicates the presence of AI that is capable of activating LasR to induce *lasI* expression.

Method 2

A. Purification of AI

Reagents

M9 minimal medium (other minimal media may be substituted for M9), per liter: 6 g Na_2HPO_4, 3 g KH_2PO_4, 1 g NH_4Cl, 0.5 g NaCl, supplemented with 0.4% (w/v) glucose and 1 m*M* $MgSO_4$
Ethyl acetate acidified with 0.001% glacial acetic acid
Anhydrous sodium sulfate

Procedure. Harvest the largest volume of biofilm sample obtainable from the substratum. Add an equal volume of minimal medium to the sample and suspend, dispersing the cells and polysaccharide material. Separate the cells from the culture media by centrifuging at 10,000*g* for 10 min at 4°. To the culture supernatant, add an equal volume of ethyl acetate acidified with 0.001% glacial acetic acid and mix vigorously for 3–5 min. Centrifuge at 8000*g* for 20 min at 4° and place the ethyl acetate phase (top layer) in a clean tube. After the culture supernatant has been extracted a second time, pool the ethyl acetate phases and remove any water present by mixing vigorously over anhydrous sodium sulfate. Place the ethyl acetate in a round-bottom flask and remove ethyl acetate from the sample using rotary evaporation at 30–37°. Two further extractions of the residue should be performed in 2 ml ethyl acetate each, removing the ethyl acetate after each extraction by rotary evaporation. Dissolve the remaining material in 500 μl ethyl acetate and store at −20°.

B. Detection of AI in Bioassays

For detection of AI activity in the purified sample, the required components are essentially the same as those described in Method 1; i.e., indicator strains carrying genes encoding the R-proteins activated by 4, 8, and 12 carbon cognate AI molecules and a target reporter gene fused to *lacZ*. Features that distinguish this protocol from Method 1 include: (1) the indicator strain is grown in liquid culture instead of on solid media in the presence of the AI sample and (2) AI activity is detected with a standard β-galactosidase assay.[10]

Reagents

Minimal medium, such as M9, or culture media that contain limited amounts of yeast extract (in our hands, yeast extract has been found to decrease β-galactosidase activity).

Procedure. Dilute overnight cultures of the indicator bacteria in M9 medium to an OD_{600} of 0.08 and allow the cultures to grow until an OD_{600} of 0.3 is reached. Meanwhile, add the purified AI samples in ethyl acetate to tubes, and remove ethyl acetate by evaporation under a stream of nitrogen gas. Next, add 1-ml samples of the indicator culture to the tubes and allow to grow for 90 min in both the presence and the absence of AI. β-Galactosidase activity can then be measured according to standard methods.[10] As a positive control, the cognate AI should be included in a range of concentrations so that a dose–response curve can be generated.

Pseudomonas aeruginosa Quorum-Sensing Mutants

To study the contribution of a specific gene to a given trait, such as the ability to form biofilms, gene replacement technology represents a powerful tool. In our laboratory, we have generated a number of quorum-sensing mutants of *P. aeruginosa* using gene-replacement strategies. Null mutants of the R-proteins, the autoinducer synthases, and the products that they regulate have been invaluable for studying gene regulation in *P. aeruginosa*. Furthermore, the availability of these mutants led to the discovery that quorum sensing is involved in biofilm development.[2]

Considerations for Generating Quorum-Sensing Mutants

For reasons discussed later, we have found that AI mutants are the most versatile for studying the role of quorum sensing in *P. aeruginosa*. Following generation of a null mutant, complementation studies are performed to ensure that a given phenotype is directly attributable to a specific gene. With the R-mutants, this can only be accomplished by providing a copy of the wild-type gene *in trans*. In contrast, the AI mutants can be complemented in two ways; either by providing the AI synthase gene *in trans* on a plasmid or by adding purified or synthetic AI to bacterial cultures. By adding back the AI instead of the gene, the requirement for a selective

[10] J. H. Miller, "Experiments in Molecular Genetics." Cold Spring Harbor Laboratory, Cold Spring Harbor, NY, 1972.

pressure, such as antibiotic resistance, to ensure the maintenance of plasmids carrying the complementing allele is eliminated. In many cases, growing cells without antibiotics is desirable as their presence may affect normal cell function. Moreover, if the expression of gene fusions on plasmids is being analyzed, issues such as plasmid incompatibility and distinct antibiotic resistance markers must be taken under consideration when complementing with plasmids. Finally, complementation with the AI instead of the gene allows more precise control of the experimental parameters, as both the AI concentration and the time of AI addition can be manipulated easily.

A number of gene-replacement strategies have been used to successfully generate null mutants of *P. aeruginosa* (see Toder[11] for a review). However, when designing a cloning scheme for gene replacement, a few points should be kept in mind. First, we typically use antibiotic resistance cassettes to insertionally inactivate genes when generating chromosomal knockout mutants. Despite the finding that these mutants are stable, even after passage in the absence of antibiotics, the possibility does exist that the cassette could be excised from the chromosome, resulting in restoration of the wild-type gene. One way to circumvent this problem is to delete a portion of the gene before inserting the cassette. This step ensures that in the unlikely event the cassette is excised, the gene remains nonfunctional. Second, by deleting a piece of DNA approximately the same size as that of the cassette to be inserted, one minimizes the potential for distortion of DNA due to changes in size. Finally, we have observed that the probability of obtaining the double cross-over event in *P. aeruginosa* is increased greatly if there is at least 1 kb of chromosomal DNA flanking each side of the cassette.

Gene Expression in Biofilms

The physiology of cells grown in a biofilm is vastly different from planktonic cells,[12,13] and the physiology varies from cell to cell, depending on where the cells are located within the biofilm.[14] Presumably, cells near the periphery of the biofilm, or adjacent to water channels, are more actively metabolizing than cells embedded deeply within the polysaccharide matrix. How these biofilm cells regulate gene expression is still an enigma. We know that *P. aeruginosa* cells use quorum sensing during the formation of a differentiated biofilm, allowing the exchange of nutrients and waste prod-

[11] D. Toder, *Methods Enzymol.* **235,** 466 (1994).

[12] M. C. M. Van Loosdrecht, J. Lyklema, W. Norde, and A. J. B. Zehnder, *Microbiol. Rev.* **54,** 75 (1990).

[13] H. Anwar, M. K. Dasgupta, and J. W. Costerton, *Antimicrob. Agents Chemother.* **34,** 2043 (1990).

[14] H. Anwar, J. L. Strap, and J. W. Costerton, *Antimicrob. Agents Chemother.* **36,** 1347 (1992).

ucts and thereby preventing overcrowding and starvation.[2] These findings are very provocative and set the stage for many new questions to be addressed, such as (1) which genes regulated by LasR/3-oxo-C_{12}-HSL are required for normal biofilm development; (2) does expression of these genes differ in cells grown planktonically versus cells grown in a biofilm; and (3) how do bacteria regulate gene expression according to their location within the biofilm and according to different stages of biofilm development?

Reporter Systems

The assay for some gene products is relatively simple; therefore, by fusing the target gene with a gene whose product is assayed easily, expression of the target gene can be monitored easily. Table II summarizes reporter systems that have been found useful for gene expression studies in *P. aeruginosa*. A more detailed description of reporter fusions available for analyzing gene expression in biofilms is found elsewhere in this volume.[14a]

Monitoring Gene Expression in Pseudomonas aeruginosa Biofilms

In the case of *P. aeruginosa,* the expression of quorum-sensing genes can be assayed for either directly or indirectly. Because the regulators of the quorum-sensing systems, i.e., the R-proteins and their cognate AIs regulate the expression of a number of genes, fusions of the genes encoding the R-proteins or autoinducer synthases can be generated. Alternatively, expression of the genes controlled by the quorum-sensing systems can be monitored. As an example, the transcriptional activator LasR and its cognate autoinducer, 3-oxo-C_{12}-HSL, regulate the expression of *lasB,* the gene encoding elastase. Therefore, we could assay for the expression of *lasR* or the AI synthase gene, *lasI,* in either one of two ways: we could generate *lasR* and *lasI* gene fusions and monitor their expression directly or we could generate a *lasB* gene fusion to monitor expression of these genes indirectly.

Direct Analysis of lasR Expression Using a lacZ Reporter System

Reagents

Fluorogenic substrate for β-galactosidase, such as fluorescein di-β-D-galactopyranoside (FDG; Molecular Probes, Eugene, OR)

PBS

50% PBS containing 1 m*M* FDG

[14a] D. R. Korber, G. M. Wolfaardt, V. Brözel, R. MacDonald, and T. Niepel, *Methods Enzymol.* **310,** [1], 1999 (this volume).

TABLE II

REPORTER FUSION SYSTEMS FOR ANALYZING GENE EXPRESSION IN BIOFILMS

Fusion system	Gene(s)	Basic principle of method	Equipment required	Advantages	Disadvantages
β-Galactosidase	*lacZ*	5′ end of target gene, including promoter, cloned into *lacZ* fusion vector. Expression of *lacZ* fusion protein monitored using fluorogenic substrate for β-galactosidase [e.g., fluorescein di-β-D-galactopyranoside (Molecular Probes, Eugene, OR)]	Standard microscope and optical filters for detection of fluorescein	*lac* operon well studied. Numerous fusion vectors and commercial products for β-galactosidase detection available. Large DNA fragments encoding up to 500 amino acids can be expressed[a,b]	Requires exogenously added substrate and permeabilization of cell membranes; therefore limited use with living cells
Bacterial luciferase	*luxA, B, C D, E*	5′ end of target gene, including the promoter, cloned into *lux* fusion vector. Expression of *lux* fusion monitored using instrumentation capable of quantifying photon emission	Scintillation counter or luminometer	Allows real-time *in vivo* gene expression analysis. Light production requires a functional biochemistry, thus can be used as reporter for cell viability	
Green fluorescent protein	*gfp*	5′ end of target gene, including the promoter, cloned into *gfp* fusion vector. When excited by near-UV or blue light, GFP emits a strong green fluorescence. Expression of *gfp* fusion monitored using equipment and optical filters capable of detecting fluorescein	Standard microscope and optical filters for detection of fluorescein	Bacterial communities can be examined *in vivo,* at single-cell level	GFP is very stable; thus temporal gene expression analysis is not possible. New GFP variants more susceptible to cellular proteases should be more useful for real-time studies[c]

[a] P. A. Silver, L. P. Keegan, and M. Ptashne, *Proc. Natl. Acad. Sci. U.S.A.* **81,** 5951 (1984).

[b] S. Froshauer and J. Beckwith, *J. Biol. Chem.* **258,** 10896 (1984).

[c] J. B. Anderson, C. Sternberg, L. K. Poulsen, S. P. Bjørn, M. Givshov, and S. Molin, *Appl. Environ. Microbiol.* **64,** 2240 (1998).

Procedure. Clone the 5′ end of the *lasR* gene, including the promoter, into a *lacZ* fusion vector, such as pLP170.[15] Transform the recombinant plasmid into the bacterial strain of interest, e.g., *P. aeruginosa* strain PAO1, and allow the recombinant PAO1 strain to initiate biofilm development. To monitor *lasR* expression, incubate cells in 50% PBS containing 1 m*M* FDG for 1 min at 37° (the 50% hypotonicity of the PBS/FDG solution permeabilizes the cells, allowing entry of FDG). Replace the PBS/FDG solution with ice-cold 100% PBS and visualize the biofilm sample using a standard fluorescence microscope and optical filters for the detection of fluorescein. The sample can be fixed before microscopic analysis; however, many fixatives permeabilize the cell membranes, which may cause leaching of the fluorescein and decreased fluorescence. The level of fluorescence produced by the background strain containing the backbone vector with no insert [e.g., PAO1(pLP170)] should also be determined.

Summary

Quorum sensing has been shown to be important for the development of a normal *P. aeruginosa* biofilm, and it follows that other microorganisms may employ a similar mechanism in the development of mature biofilms. Two methods for detecting the presence of AI activity in biofilms are presented that employ an AI-responsive reporter strain harboring a *lacZ* fusion. Method 1 involves detection of AI activity in crude biofilms, whereas Method 2 employs an AI purification procedure. By using multiple indicator strains activated by AIs of various acyl chain lengths, a wide range of AI molecules can be detected.

Chromosomal knockout mutants are extremely useful for examining the contribution of a given gene to a specific phenotype. For quorum-sensing gene expression studies, mutants deficient in the production of AI offer more versatility than R-protein mutants. The main advantage of the AI mutants is that they can be complemented by either the AI synthase gene or the AI itself. Complementation with the AI circumvents having to grow the cells in the presence of antibiotics and allows experimental parameters such as AI concentration and time of addition to be manipulated easily.

Finally, three reporter systems suitable for monitoring gene expression in *P. aeruginosa* biofilms are summarized in Table II. The choice of reporter fusion depends mainly on whether *in vivo* analysis is required, whether temporal gene expression is to be examined, and the availability of equip-

[15] M. J. Preston, P. C. Seed, D. S. Toder, B. H. Iglewski, D. E. Ohman, J. K. Gustin, J. B. Goldberg, and G. B. Pier, *Infect. Immun.* **65,** 3086 (1997).

ment. In the case of *P. aeruginosa,* expression of quorum-sensing genes can be monitored either directly, by examining fusions of the R genes or AI synthase genes, or indirectly, by analyzing expression of genes controlled by these quorum-sensing systems.

Acknowledgments

Supported by National Institutes of Health Research Grant R01A133713-04 (B.H.I.) and a Canadian Cystic Fibrosis Foundation Postdoctoral Fellowship Award (T.R.K.).

Section II

Microscopic Methods of Biofilm Formation and Physiology

[9] Confocal Laser Scanning Microscopy for Analysis of Microbial Biofilms

By JOHN R. LAWRENCE and THOMAS R. NEU

Confocal Laser Scanning Microscopy: An Introduction

Among the most versatile and effective of the nondestructive approaches for studying biofilms is confocal laser scanning microscopy (CLSM). CLSM reduces greatly the need for pretreatments such as disruption and fixation that reduce or eliminate the evidence for microbial relationships, complex structure, and organization in biofilms. Although confocal imaging has a relatively long history of application in the physical sciences and in medical research, its applications to biofilms began in the early 1990s.[1] There has been increased application of this tool in conjunction with an increasingly wide range of fluorescent probes and other novel techniques for biofilm research. The reason for this increased interest is that CLSM allows optical thin sectioning of intact fully hydrated biofilm material creating images with enhanced resolution, clarity, and information content. As such, it is an ideal tool for studying spatial distribution of a wide range of biofilm properties. This capacity is evident in a series of CLSM-based research publications.[1–15] CLSM has also been used extensively in combination with fluorescent *in situ* hybridization techniques.[16–22]

[1] J. R. Lawrence, D. R. Korber, B. D. Hoyle, J. W. Costerton, and D. E. Caldwell, *J. Bacteriol.* **173,** 6558 (1991).

[2] J. R. Lawrence, G. M. Wolfaardt, and D. R. Korber, *Appl. Environ. Microbiol.* **60,** 1166 (1994).

[3] J. R. Lawrence, Y. T. J. Kwong, and G. D. W. Swerhone, *Can. J. Microbiol.* **43,** 178 (1997).

[4] J. R. Lawrence, T. R. Neu, and G. D. W. Swerhone, *J. Microbiol. Methods* **32,** 253 (1998).

[5] J. R Lawrence, G. D. W. Swerhone, and Y. T. J. Kwong, *Can. J. Microbiol.,* **44,** 825 (1998).

[6] T. L Bott, J. T. Brock, A. Battrup, P. A. Chambers, W. K. Dodds, K. Himbeault, J. R. Lawrence, D. Planas, E. Snyder, and G. M. Wolfaardt, *Can. J. Fish. Aqu. Sci.* **54,** 715 (1997).

[7] T. R. Neu and J. R. Lawrence, *FEMS Microbiol. Ecol.* **24,** 11 (1997).

[8] A. A. Massol-deya, J. Whallon, R. F. Hickey, and J. M. Tiedje, *Appl. Environ. Microbiol.* **61,** 769 (1995).

[9] G. M. Wolfaardt, J. R. Lawrence, R. D. Robarts, and D. E. Caldwell, *Appl. Environ. Microbiol.* **60,** 434 (1994).

[10] G. M. Wolfaardt, J. R. Lawrence, R. D. Robarts, and D. E. Caldwell, *Appl. Environ. Microbiol.* **61,** 152 (1995).

[11] G. M. Wolfaardt, J. R. Lawrence, R. D. Robarts, and D. E. Caldwell, *Microbial. Ecol.* **35,** 213 (1998).

0076-6879/99 $30.00

With the advent of krypton–argon, commercial UV, and two-photon and multiphoton laser microscopy systems, multiple parameter imaging of biofilms is practical, allowing the collection of multiple quantitative data sets at a single location within a biofilm.[5]

CLSM provides a digital database that is amenable to image processing and analysis. Thus the user may obtain quantitative information on a wide variety of parameters, including cell numbers, cell area, object parameters such as minimum/maximum dimensions, orientation, and average gray value. Data sets may also be analyzed to determine diffusion coefficients within biofilms[2,23,24] and growth rates[25] of microorganisms. When combined with the variety of fluorescent antibodies, oligonucleotide probes, or physiological probes (see other chapters in this volume),[25a] additional specific information may be derived.

An additional very significant aspect of confocal laser microscopy and optical sectioning is the creation of sets of images in perfect register allowing the production of three-dimensional (3D) reconstructions, renderings, and animations of data sets.[1,7,16] This provides additional power to scientific visualization of biofilm materials.

The primary goal of this article is to provide a primer for CLSM observation of biofilm materials.

---

[12] S. Møller, C. Sternberg, J. B. Anderson, B. B. Christensen, J. L. Ramos, M. Givskov, and S. Molin, *Appl. Environ. Microbiol.* **64,** 721 (1998).

[13] M. N. Mohamed, J. R. Lawrence, and R. D. Robarts, *Microbiol. Ecol.* **36,** 121 (1998).

[14] B. Assmus, P. Hutzler, G. Kirchhof, R. Amann, J. R. Lawrence, and A. Hartmann, *Appl. Environ. Microbiol.* **61,** 1013 (1995).

[15] M. Wagner, G. Rath, H. P. Koops, J. Flood, and R. Amann, *Wat. Sci. Technol.* **34,** (1/2), 237 (1996).

[16] R. Amann, W. Ludwig, and K. H. Schleifer, *Microbiol. Rev.* **59,** 143 (1995).

[17] R. Amann, R. Snaid, M. Wagner, W. Ludwig, and K. H. Schleifer, *J. Bacteriol.* **178,** 3496 (1996).

[18] S. Miller, A. R. Pedersen, L. K. Poulsen, J. M. Carstensen, and S. Molin, *Appl. Environ. Microbiol.* **62,** 4632 (1996).

[19] A. Neef, A. Zaglauer, A. H. Meier, R. Amann, H. Lemmer, and K. H. Schleifer, *Appl. Environ. Microbiol.* **62,** 4329 (1996).

[20] W. C Ghiorse, D. N. Miller, R. L. Sandoli, and P. L. Siering, *Microsc. Res. Technol.* **33,** 73 (1996).

[21] M. Schuppler, M. Wagner, G. Shon, and U. B. Gobel, *Microbiology* **144,** 249 (1998).

[22] D. de Beer, P. Stoodley, F. Roe, and Z. Lewandowski, *Biotechnol. Bioeng.* **53,** 151 (1997).

[23] J. J. Birmingham, N. P. Hughes, and R. Treloar, *Philos. Trans. Soc. Lond. B Biol. Sci.* **350,** 325 (1995).

[24] L. K. Poulsen, G. Ballard, and D. A. Stahl, *Appl. Environ. Microbiol.* **59,** 1354 (1993).

[25] R. P. Haugland, "Handbook of Fluorescent Probes and Research Chemicals," 4th ed. Molecular Probes, Eugene, OR, 1996.

[25a] *Methods Enzymol.* **310** [1–6] (1999) (this volume).

General Considerations

CLSM setups are available from most of the major microscopy companies, with a wide range of options (software and hardware) and peripheral devices. CLSM is a combination of traditional epifluorescence microscope hardware with a laser light source, specialized scanning equipment, and computerized digital imaging. A general schematic diagram is shown in Fig. 1. Lasers used include argon ion, helium–neon, krypton–argon, helium–cadmium, and UV excimer lasers. The most commonly used CLSM systems are equipped with a helium–neon (543 or 633 nm) and mixed-gas krypton–argon lasers (488-nm blue, 568-nm yellow, and 647-nm red lines). Helium–cadmium lasers are seldom used but can provide a strong 442-nm line. UV/VUV excimer lasers (157–351 nm) may be obtained with commercially available CLSM systems. These lasers provide the advantage of a wide range of usable fluorochromes and simultaneous excitation of up to three fluorochromes with little spectral emission overlap. In most instances the laser is connected to the scanning head via a fiber optic connection. The scan head is a unit with galvanometric mirrors to scan the beam onto the specimen and a system of mirrors and beam splitters that direct the return signal to specific photomultiplier tubes (PMTs). The laser beam is scanned point by point in a raster fashion to build a gray scale image of the specimen under observation. The scanned areas may consist of 512 × 512, 512 × 768, or 1024 × 1024 pixels. The scan rate may usually be varied; however, when the scan rate is increased, resolution is lost and when the scan rate is low photobleaching is increased. Thus these factors must be balanced by the user to obtain optimum results. The presence of a pinhole or pinholes in the light path allows only those fluorescence signals that arise from a focused *XY* plane to be detected by a PMT. These pinholes are said to be confocal and thus prevent fluorescent signals originating from above, below, or beside the point of focus from reaching the photodetector. In addition, sets of wavelength specific filters are used to supply specific excitation and emission wavelengths for the fluorescent probes used. For example, when using a Kr–Ar laser, the following combinations of excitation and emission wavelengths are commonly available: 488 nm excitation, 522/32 nm emission (green); 568 nm excitation, 605/32 emission (red); and 640 nm excitation 680/32 nm emission (far-red). Most CLSM systems also incorporate filter sets for reflection imaging and a separate system for imaging nonconfocal-transmitted laser images using, for example, differential interference contrast (DIC) or phase-contrast optics.

Multiphoton excitation fluorescence imaging systems are a relatively recent development and offer a number of features that may be very valuable for biofilm studies, including longer observation times with living

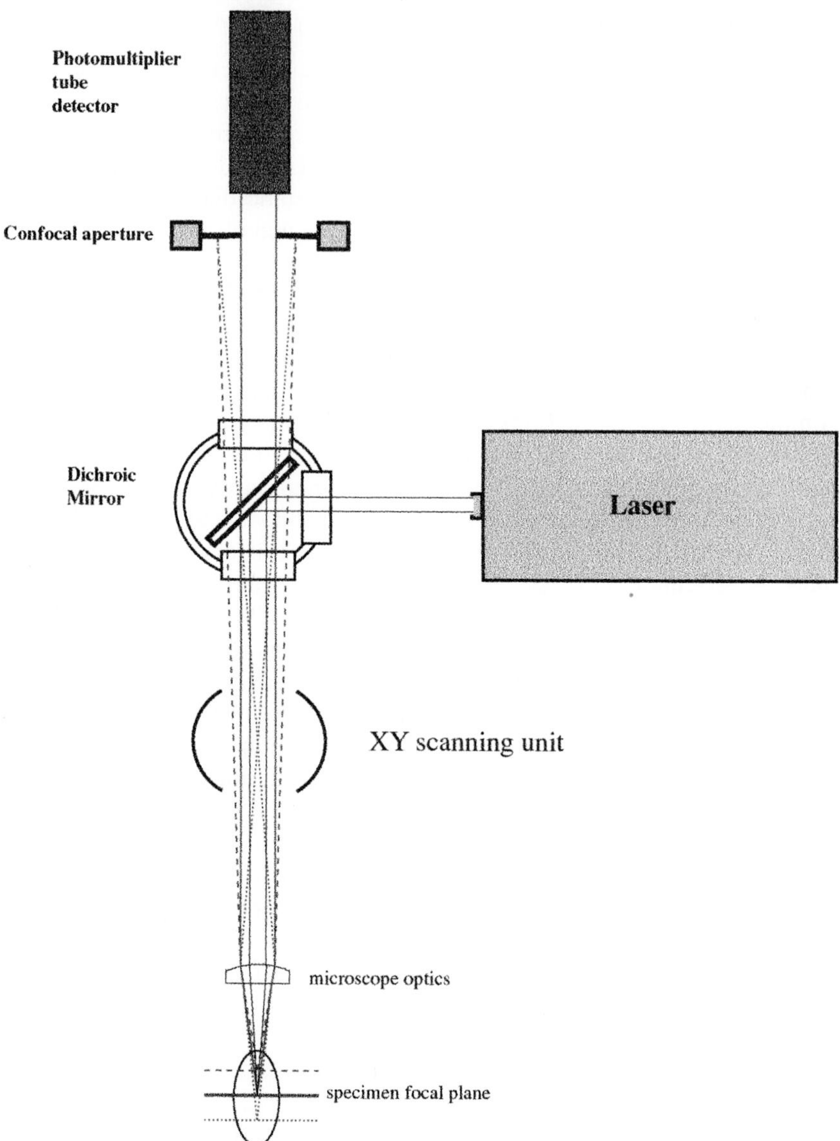

FIG. 1. Schematic diagram of a confocal laser scanning microscope showing generalized construction and components.

specimens, increased fluorescence emission, increased depth for optical sectioning, and reductions in interference and photobleaching. These systems utilize a pulsed laser source such as titanium–sapphire with a major emission at 1047 nm. However, at the time of writing there have not been published applications of this technology in microbial ecology or biofilm studies (see further discussion later). All systems are operated using a combination host computer or workstation equipped with proprietary software specific to the manufacturer. These systems usually have two monitors, one displaying the control software and the other for the images. Linking the computer to additional workstations to separate image collection from image analysis and display is an important consideration. In addition, because image data require an enormous amount of storage, a hard drive with a minimum of 5 GB capacity and equipment for production of CDs is essential (see later).

Getting Started

Turning on the system and checking alignment is usually done in advance of preparation of materials for observation, this allows the laser to warm up for 30 min prior to initiating imaging. This is important with the Kr–Ar laser where production of the far-red line (640 nm) will vary during the warmup period. With Kr–Ar lasers the working lifetime may be rather short, in the range of 1000–1400 hr; problems that develop with age are instability and loss of far-red excitation wavelengths. With most of the current commercial CLSM systems there are few critical adjustments that can be made by the user. However, prisms and test samples may be provided to check alignment of mirrors, laser beam, and image quality. The user can make up slides consisting of fluorescent beads or Focal Check beads (Molecular Probes, Eugene, OR), which will allow them to routinely check whether image brightness and alignment are remaining within desired specifications. The use of Focal Check beads is particularly useful in ensuring that images obtained from the same location but with different excitation emission combinations (multiparameter imaging or colocalization studies) are in perfect register. A frequently overlooked adjustment is that of the gray scale and alignment of the computer screens themselves, which is vital to viewing the images collected correctly.

Sample Preparation

Biofilm sample materials range from prepared fixed materials, i.e., fluorescent *in situ* hybridization, which are covered by a glass coverslip, to biofilms grown on prepared substrata, to observations on natural irregular surfaces recovered from various environments. All of these may be used for CLSM studies. The use of biofilm incubation systems such as the rotating annular bioreactor, continuous flow slide culture, Robbins devices, and others provides convenient surfaces for observations that are made for microscopic study. However, natural biofilms are usually not so conveniently located or growing on optically perfect surfaces. Usually biofilm samples are associated with a solid interface. This so-called substratum covered with the biofilm must be mounted for CLSM observation. To examine the fully hydrated living features of a biofilm, the sample is preferably used directly without any fixation or embedding procedures. In general, the preparation of the biofilm sample is dependent on the geometry of the substratum and the type of microscope available.

Upright versus Inverted Microscope

Dependent on the microscopic setup, upright or inverted, the following considerations are necessary. For the normal microscope the biofilm sample covered with the original liquid phase can be fixed to the bottom of a small petri dish (diameter 5 cm) with acid-free silicone glue, placed in a well slide, or a dam created with plasticine, wax, or silicone glue. All staining techniques are then applied in the small volume of the liquid layer covering the biofilm or within the larger volume of the whole container.

If an inverted microscope is employed, the sample has to be mounted upside down in a chamber having a coverslip bottom (Nunc, Roskilde, Denmark). Depending on the expected thickness of the biofilm and the working distance of the objective lens used, spacers with a thickness of 50–500 μm may be glued into the chamber to avoid damage to the biofilm. The staining is performed in the space in between the coverslip bottom of the chamber and the biofilm sample mounted upside down.

Flat/Irregular Substratum

Biofilm samples from flow-through devices, the Robbins device, rotating annular biofilm reactors, or other sampling ports are usually flat. They can be easily mounted and stained. In addition, this volume contains detailed descriptions of several methods for biofilm cultivation that are suitable or may be adapted for microscopic study.

If the biofilm sample to be examined is located on an irregular surface, e.g., a piece of rock, some points need to be considered. With an upright microscope, the sample should be placed in a small petri dish and kept covered with the original liquid phase. We have found that mounting samples in petri dishes using wax, plasticine, or neutral chemistry silicone glues or the creation of reservoirs on rock or wood surfaces using silicone dams provides easy access to the observation surface using water-immersible lenses (see later) and upright microscopes. If an inverted microscope is used, the biofilm is immersed in the original liquid phase, but access to the sample is limited and the working distance of the objective lens may further limit examination of the biofilm sample.

Although care must be taken in handling biofilm samples, in real-world environments, biofilms are exposed to physical stress and are usually resilient enough to be manipulated and mounted for staining and observation as described.

Staining Options

After determination of sample type, one must select the type of staining or fluorescent or reflective probe to be used; this is based on the nature of the information desired. Important considerations when using combinations of fluors are relative intensity, narrowness of the emission band, photostability, potential for interference, additive effects, and quenching effects. The user has a wide range of stains and probes that may be used in conjunction with CLSM imaging to obtain information on cell position, identity, diffusion rates, chemistry, etc.[2,14,24,25,26] Fundamentally, staining in CLSM may be either positive or negative in nature. Negative staining of biofilms through the flooding of the sample with a fluor such as fluorescein was described in detail by Caldwell *et al.*[27] Positive staining encompasses the entire range of nucleic acid stains, i.e., the SYTO series (Molecular Probes, Eugene, OR), acridine orange (AO), 4′,6-diamidino-2-phenylindole (DAPI), protein stains, fluorescein isothiocyanate (FITC), 5-(4,6-dichloroz-triazin-2-yl)aminofluorescein(DTAF), tetramethylrhodamine isothiocyanate (TRITC), other rhodamine stains, Texas Red, and lipophilic stains such as Nile Red. Other probes may be labeled using fluorescein (FLUOS, FITC), rhodamine (TRITC), cyanins (CY2, CY3, CY5), aminomethylcoumarin (AMCA), or phycoerythrin (PE). A new series of stains, the Alexa dyes, are also now available from Molecular Probes. In addition, autofluorescence can be a useful source of information, i.e., detecting and imaging algae and some bacteria.[4]

It should also be noted that unstained controls should be imaged with all samples using the same settings as for the stained materials to ensure that autofluorescence artifacts are not present in the resulting images.

When staining for observation using water-immersible lenses, the sample of the substratum with the attached biofilm should be covered with

[26] T. R. Neu and J. R. Lawrence, *Methods Enzymol.* **310** [10] (1999) (this volume).

[27] D. E. Caldwell, D. R. Korber, and J. R. Lawrence, *J. Microbiol. Methods* **15,** 249 (1992b).

original water. Alternately, the biofilm sample may be kept in a moist chamber such as a petri dish with a wet tissue. In general, many staining procedures for CLSM study do not require fixation of the biofilm sample. The complete staining procedure may be carried out in the liquid droplet covering the biofilm while it is still attached to the substratum. An example of a specific staining procedure is given by Neu and Lawrence[26] for use with fluor-conjugated lectins.

Consideration should also be given to application of fade retardants[28] such as Citifluor (UKC Chemlab, Canterbury, UK) or Slow-Fade preparations (Molecular Probes). Although these are generally used with fixed stained samples, their use should not be restricted to this type of sample preparation.

Objective Lenses

Establishment of the sample type leads to clear decisions regarding the primary imaging tool, the objective lens. The major limitation of all objective lenses, particularly when applied for CLSM imaging, is that the axial or *Z* dimension resolution is poor relative to the lateral resolution. Additional concerns arise from the fact that some objective lenses are not corrected for imaging in the far-red.[29] Similarly, when excitation wavelengths are extended into the ultraviolet there may be loss of transmission and serious image aberration. However, in general, the use of high numerical aperture (NA) oil or water-immersion lenses (i.e., NA 1.2–1.4) is recommended. Some of the water-immersion lenses are designed with confocal microscopy applications in mind and may allow imaging through up to 200 μm of biological materials. This, however, presupposes the use of fixed stained materials, optically appropriate mounting media, and high-quality coverslips. However, the real power of CLSM in biofilm studies comes from the enhanced ability to observe and analyze fresh, undisturbed materials in real time. Thus for many studies of biofilms the ideal lenses for the examination of fluorescently stained biofilm samples are water immersible. These include relatively high NA water-immersible objectives supplied by Leica, Nikon, and Zeiss, such as the Zeiss 0.90 NA 63× water-immersible lens. Advantages for biofilm research are long working distance, high numerical aperture, and superior brightness. Furthermore, they can be employed for direct observation without the need to use a coverslip. We have found that we can effectively section through several hundred microns with 63× 0.9 NA water-immersible lenses and up to 1 mm and more when using

[28] R. J. Florijn, J. Slats, H. J. Tanke, and A. K. Raap, *Cytometry* **19,** 177 (1995).
[29] C. Cullander, *J. Microsc.* **176,** 281 (1994).

extra long working distance lenses 20× ELWD or 40× ELWD or 40× 0.55 NA water-immersible lenses. Studies such as those by Neu and Lawrence,[7] Bott *et al.*,[6] and Lawrence *et al.*[3–5] illustrate the application of water-immersible lenses to a variety of biofilms and substrata.

Sampling Considerations

The essential question in any analysis is how many or how often is enough. Few authors have considered this question in detail for biofilm studies. However, it is essential to the advancement of biofilm research that each study considers how to achieve a statistically valid impression of samples or treatment effects. For example, the study of Korber *et al.*[30] used the combination of a computer-controlled microscope stage and CLSM imaging to create large-scale montages of biofilm materials and used a representative elements analysis to determine that analysis areas exceeding 10^5 μm^2 were required for statistically valid comparisons of the biofilms examined in their study. We have adopted a procedure of using five replicate microscope fields per treatment replicate, allowing application of analysis of variance to determine significant effects at $p < 0.05$.[4]

Collecting Images

After the sample is secure (and unlikely to leak fluid on the microscope), it is customary to use phase-contrast or epi-fluorescence microscopy to examine the sample and find suitable microscope fields for further observation using CLSM. With experience, or by necessity due to the nature of the sample, this preliminary step may be omitted. The microscope should be set up with the correct excitation and emission filters in place for the fluor that was selected by the user. Then, with the gain (white level) set to a low sensitivity, the laser intensity at its lowest level, and the pinhole at its smallest aperture, the operator scans the sample and adjusts the pinhole, the gain, and laser intensity to produce a well-defined image of the biofilm material under observation. These optimum settings should correspond to the smallest pinhole aperture, lowest laser intensity, and lowest PMT sensitivity. The image should be illuminated evenly and contain relatively few saturated pixels (i.e., white with a value of 255). Optimal settings must also be established with reference to any autofluorescence signals emitted by the sample being scanned. This is particularly important for interpretation of the distribution of fluorescence or reflective probes in the sample and later use of images for quantitative analyses. After the

[30] D. R. Korber, J. R. Lawrence, M. J. Hendry, and D. E. Caldwell, *Biofouling* **7,** 339 (1993).

settings are optimized (a procedure that may have to be repeated many times) for the sample material under observation, the user must then determine which imaging options will be useful. The choices include a single *XY* optical thin section, a series of *XY* optical thin sections through the material, or a single or a series of *XZ* sections through the specimen. CLSM systems may also be programmed to collect images through time allowing the user to capture 4D data sets. If additional magnification is desired, various zoom functions that reduce the area scanned and thereby increase magnification may also be selected. Images may also be collected using mathematical filters such as Kalman or running average filters to reduce noise in the primary image. For low signal samples, options such as photon counting, summation, or cumulative collection of the image may be applied.

The optical sections may be collected at the same location using the full range of excitation and emission options provided by the particular system, thereby collecting information on several variables within a single microscope field. Lawrence *et al.*[4] demonstrated the application of multiple parameter imaging to determine the abundance of algae, bacteria, and exopolymers in river biofilms. Although some systems allow the simultaneous collection of images, sequential collection in general results in less photobleaching and optimal image quality. For example, in the Kr–Ar laser the user may collect images in the green, red, and far-red with the option of also collecting a reflection image of the biofilm materials.[5] If the system is provided with a UV laser, an additional channel is available for staining and observation. The introduction of two photon and multiphoton systems may expand the range of imaging options; however, as noted earlier, these have not been used in practice on biofilm materials at this writing.

Image Processing and Presentation

The crisp high-quality CLSM image may be improved for presentation purposes through the application of basic image processing techniques. Readers should consult Russ[31] and Gonzalez and Wintz[32] for extensive detail on image processing. Although every effort is made to obtain the highest quality primary image, some processing or enhancement may be required before analysis. Common processing steps include histogram analysis, gray level transformation, normalization, contrast enhancement, application of median, lowpass, Gaussian, Laplacian filters, image subtraction, addition, multiplication, and erosion and/or dilation of objects to be mea-

[31] J. C. Russ, "The Image Processing Handboook," 2nd ed. CRC Press, Boca Raton, FL, 1995.

[32] R. C. Gonzalez and P. Wintz, "Digital Image Processing." Addison-Wesley, Reading, MA, 1977.

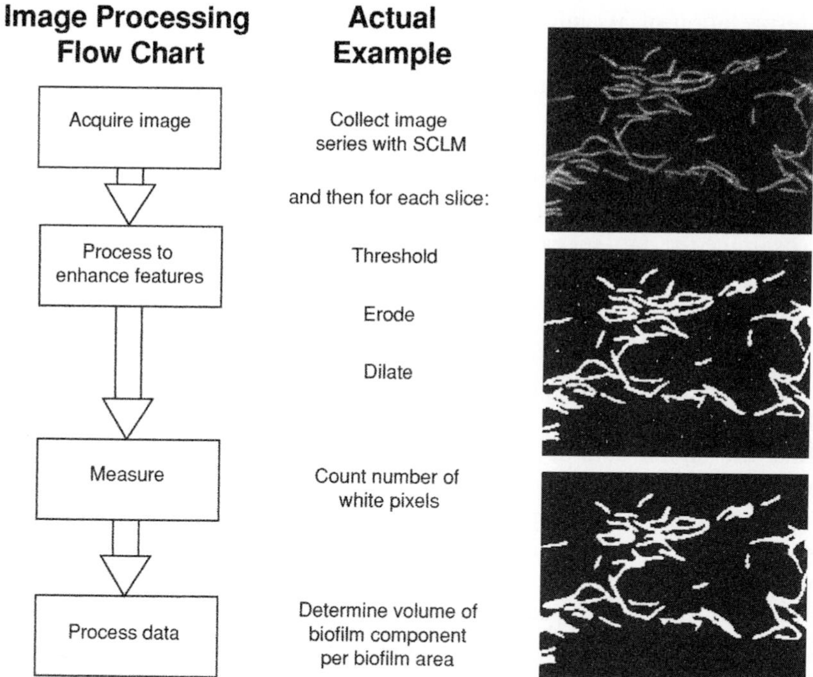

FIG. 2. A flow chart showing a sequence of image processing and analysis steps carried out on CLSM images or image stacks to define objects for measurement of cell area, including application of erode and dilate functions to reduce noise.

sured. A typical series of steps is shown in the flow chart in Fig. 2. Deconvolution may also be applied to CLSM images to sharpen the image through mathematical removal of out-of-focus information.[33,34] All of these functions are applied to smooth the image, reduce noise, and thus more accurately define the objects to be measured. Manual editing of digital images may also be performed.

There are many options for the visual presentation of CLSM images, gallery display showing each section, stereo pairs,[1] red–green anaglyph projections,[7,16,35] three-color stereo pairs.[4,5] Figure 3 (see color insert) shows a (3D) red–green anaglyph projection of a river biofilm. Stereo projections may also be color coded by depth so that materials present at the same depth appear the same color. This approach can be very useful for the

[33] D. A. Agard, *Biophys. Bioeng.* **13,** 191 (1984).

[34] G. L. Gorby, *J. Histochem. Cytochem.* **42,** 297 (1994).

[35] D. E. Caldwell, D. R. Korber, and J. R. Lawrence, *Adv. Microb. Ecol.* **12,** 1 (1992).

interpretation of 3D information. Another option for display of 3D data sets is "simulated fluorescence," whereby the material is viewed as though it were illuminated from an oblique angle and the surface layer was fluorescent. Projections may be made as a solid body or surface projection and animated to show the entire data set. The application of 3D rendering through ray tracing or surface contour-based programs may also provide a useful presentation of three-dimensional data sets, allowing the reader to examine the data set from various perspectives (Fig. 3B, see color insert).

Image Analysis Options

Having obtained a high-quality primary image from the CLSM, the user then has various options for extracting as much information as possible from the digital data set. Although the images are striking and visually pleasing, it is through the application of digital image analysis that the user can extract and present quantitative data. Image analysis is a critical tool for use in conjunction with the 2D, 3D, and even 4D data sets that can be created by CLSM imaging. The tools available range from relatively straightforward image analysis, including object recognition, counting, and gray level measurements, to increasingly sophisticated dedicated programs allowing 2D and 3D image analysis. The essential tools may be found in a number of analytical packages such as the Quantimet system[36]; Möller *et al.*[37] used Cellstat, which is available for UNIX workstations (see http://www.lm.dtu.dk/cellstat/index.html). NIH image, a versatile analysis package for a Macintosh platform, is available as freeware over the internet at http://rsb.info.nih.gov/nih-image/ and is compiled for Windows 95 or NT-based systems (Scion ImagePC at www.scioncorp.com). Neural network systems have been proposed for the analysis of fluorescence images.[38] Silicon Graphics-based software is also offered by Molecular Dynamics; this is a versatile package that also allows three-dimensional image analysis of CLSM *XY* image series. To date, studies have been limited to the analysis of serial 2D images rather than the application of true 3D analyses. Although all of the CLSM manufacturers offer supplementary analytical packages that work with their operating systems and image formats, none offer all the required options for image processing.

[36] J. Bloem, M. Veninga, and J. Sheperd, *Appl. Environ. Microbiol.* **61,** 926 (1995).

[37] S. Miller, C. S. Kristensen, L. K. Poulsen, J. M. Carstensen, and S. Molin, *Appl. Environ. Microbiol.* **61,** 741 (1995).

[38] N. Blackburn, A. Hagstrom, J. Wikner, R. Cuadros-Hansson, and R. K. Bjornsen, *Appl. Environ. Microbiol.* **64,** 3246 (1998).

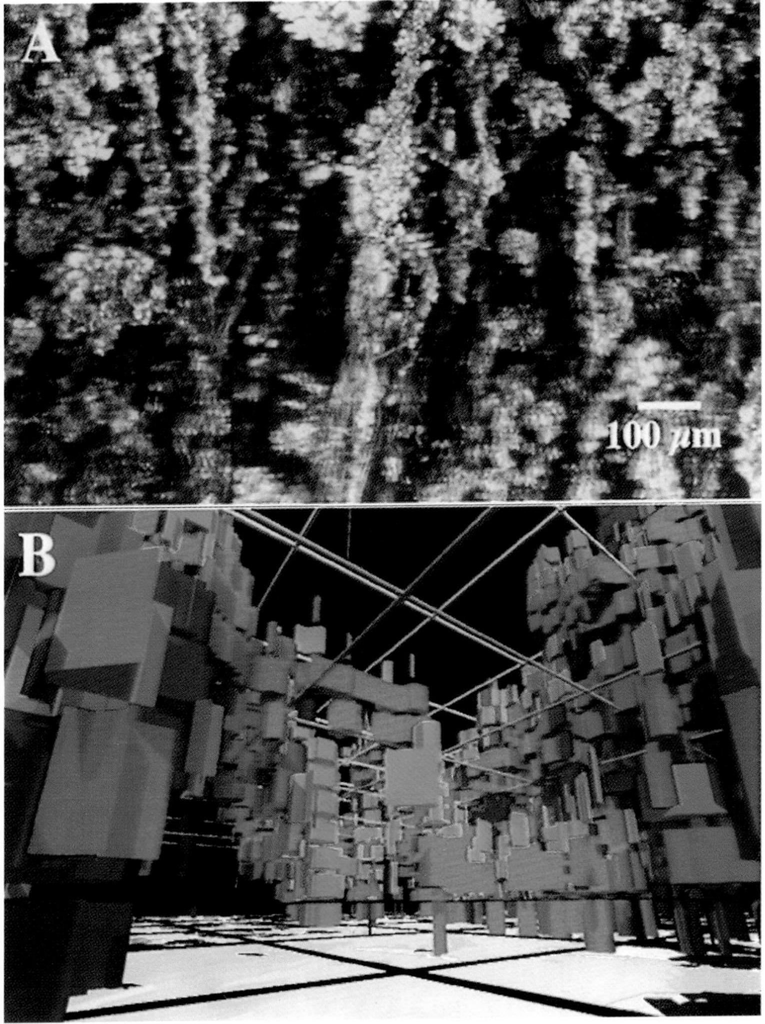

FIG. 3. (A) A series of confocal laser images of a river biofilm stained with the nucleic acid probe SYTO 9 showing the distribution of bacterial cells and general biofilm structure. (B) A three-color rendering of a confocal image series using a stacked height fields approach and the rendering package POV Ray. The image shows the distribution of exopolymeric substances, with *Limulus polyphemus*–FITC lectin (green), *Ulex europeaus*–TRITC lectin (red), and *Arachis hypogaea* CY5 (blue) in a river biofilm. The gridlines are 25 μm apart. The application of rendering allows the viewer to observe the data set from a variety of perspectives, including this one, which places the observer within the biofilm looking up.

TABLE I
CONVENTIONAL CONFOCAL LASER SCANNING MICROSCOPY (CLSM) VERSUS TWO-PHOTON LASER SCANNING MICROSCOPY (2-PLSM)

Feature	CLSM	2-PLSM
Laser	Ar–Kr and UV	TiSph
Excitation volume	Whole sample	Extremely small (femtoliter)
Out of focus bleaching	Yes	No
Out of focus	Yes	No
Optics	Chromatic aberration due to UV laser	No UV optic necessary
Pinhole	Yes, pinhole throughput loss	Not necessary
Light penetration	Small (50–100 μm)	High (200–1000 μm)

Image Archiving/Printing

Application of CLSM techniques results in the creation of vast image and data sets; our facilities can produce several gigabytes per day. Thus, the final consideration is how one archives all this information. First, it is critical to have the largest hard drive available for the operating computer. Second, many options exist for long-term storage, including optical drives (write once and rewriteable formats), Bernoulli drives, Syquest, ZIP, and CDs. For cost-effective, secure, portable, relatively universal storage media, CDs remain the best recommendation. However, it is likely that digital video disks may replace CD technology. Images may be stored in a variety of formats, such as tagged image formats (TIFF), GIFF, RAW, PICT, EPS, JPEG, and BioRadTIFF. Each of these has advantages and disadvantages, such as degrees of image fidelity and their ability to compress images (i.e., JPEG). In general, TIFF are used the most universally and will be opened by most software such as NIH Image or Adobe. Archiving represents another major hurdle that should be considered early in the process of developing a CLSM-based research program.

Images may be printed for publication using a variety of means, including video printers, dye sublimation printers and slide printers.

Perspectives

Since the first application of CLSM for studying biofilms in 1991, it has become the key technique for the microscopic study of interfacial microbial communities. CLSM offers the only means for real-time, in-depth analysis of undisturbed biofilms. However, rapid advancement in the field is both occurring and required in some areas. Fluor creation is extensive and the commercially available selection increases monthly; this is an area in which the user must take particular care to stay current. Future software needs lie in the areas of 3D image processing and analysis. On the hardware side, considerable work is required to improve the axial resolution of objective lenses. Current research in this field is investigating so-called 4Pi and Theta microscopy to improve axial resolution. There are several combinations possible to set up a hybrid microscope with elements from confocal, 4Pi, and Theta microscopes. With this approach, the axial resolution may be enhanced by a factor of 7.6 if, for example, a two-photon/4Pi-confocal Theta microscope is employed.[39] Additional rethinking of standard corrections for objective lenses is also required.

In the meantime, however, new developments have created significant potential advantages over conventional CLSM. One of the new techniques is called two-photon laser scanning microscopy (2-PLSM).[40,41] Several CLSM companies already offer this option within their product line. The major advantage of 2-PLSM over normal CLSM is an extremely small excitation volume and thus dramatically reduced photodamage to the sample. Furthermore, there is no need to use a UV laser, thus reducing chromatic aberration and cell damage. A summary of CLSM versus 2-PLSM is given in Table I. More recently, even three-photon excitation has been reported for UV fluorochromes.[42] Thus multiphoton laser scanning microscopy will become part of a new generation of laser scanning microscopes for three-dimensional imaging of interfacial microbial communities. In conclusion, future progress in three-dimensional imaging will further reduce the observation volume in all three dimensions, thereby leading to the ultimate resolution possible in light/laser microscopy. As a consequence for biofilm research, the freedom of imaging in four dimensions without significant disadvantages will become a reality.

Acknowledgments

The authors acknowledge the financial support of the Canada–Germany Agreement on Science and Technology and Environment Canada. The technical support of George D. W. Swerhone and Ute Kuhlicke is gratefully acknowledged.

[39] S. Lindek, E. H. K. Stelzer, and S. Hell, in "Handbook of Confocal Microscopy" (J. B. Pawley, ed.), p. 417. Plenum Press, New York, 1995.

[40] W. Denk, J. H. Strickler, and W. W. Webb, *Science* **248,** 73 (1990).

[41] W. Denk, D. W. Piston, and W. W. Webb, *in* "Handbook of Confocal Microscopy" (J. B. Pawley, ed.), p. 445. Plenum Press, New York, 1995.

[42] C. Xu, W. Zipfel, J. B. Shear, R. M. Williams, and W. W. Webb, *Proc. Natl. Acad. Sci. U.S.A.* **93,** 10763 (1996).

[10] Lectin-Binding Analysis in Biofilm Systems

By THOMAS R. NEU and JOHN R. LAWRENCE

Lectins: An Introduction

Proteins with lectin characteristics have been known for more than a century.[1] They were described as agglutinins that were carbohydrate inhibitable. In 1954 the term lectin was suggested and accepted.[2,3] Initially, lectins were regarded as bi- or multivalent carbohydrate-binding proteins. However, new findings suggest that some lectins may have another binding site that is specific for a noncarbohydrate ligand. Now lectins are discussed as proteins with a lectin–carbohydrate and a lectin–protein-binding site. Consequently lectins were redefined as nonenzymatic and nonimmunogenic proteins with a bifunctional binding pattern. In general, lectins are characterized on the basis of the best interaction with a certain monosaccharide. However, there is increasing evidence that complex glycoconjugates represent far better ligands.[4] Lectins are produced by many organisms, including plants, vertebrates, protists, slime molds, and bacteria. The biological function of lectins may be best expressed as cell/surface recognition molecules.[5,5a] The specificity of the recognition process is determined by stereochemical principles. In addition, the carbohydrates interact with lectins via hydrogen bonds, metal coordination, van der Waals, and hydrophobic interactions.[6] Despite the hydrophilic character of most carbohydrates, forces such as hydrophobic interactions have been recognized to be important in lectin interaction processes.[7] The structural basis of lectin–carbohydrate interactions is a topic of several review articles.[8,9] Detailed information on lectins can be gathered in several comprehensive

[1] N. Sharon and H. Lis, *Trends Biochem. Sci.* **12,** 227 (1987).

[2] W. C. Boyd and E. Shapleigh, *Science* **119,** 419 (1954).

[3] J. Kocourek, *in* "The Lectins: Properties, Functions and Applications in Biology and Medicine" (I. E. Liener, N. Sharon, and I. J. Goldstein, eds.), p. 1. Academic Press, New York, 1986.

[4] S. H. Barondes, *Trends Biochem. Sci.* **13,** 480 (1988).

[5] N. Sharon and H. Lis, *Science* **246,** 227 (1989).

[5a] N. Sharon and H. Lis, *Science* **246,** 227 (1989).

[6] S. Elgavish and B. Shaanan, *Trends Biochem. Sci.* **22,** 462 (1997).

[7] J.-L. Ochoa, A. Sierra, and F. Cordoba, *in* "Lectins: Biology, Biochemistry, Clinical Biochemistry" (T. C. Bog-Hansen, ed.), Vol. 1, p. 73. de Gruyter, Berlin, 1981.

[8] G. N. Reeke and J. W. Becker, *Curr. Top. Microbiol. Immun.* **139,** 35 (1988).

[9] W. I. Weis and K. Drickamer, *Annu. Rev. Biochem.* **65,** 441 (1996).

0076-6879/99 $30.00

books.[10–12] Further information may be found in a practical microscopy handbook on lectin histochemistry.[13]

Using Lectins to Collect Information

The challenge to analyze biofilm polysaccharides has already been discussed.[14] The difficulty of isolating a single polymer type from a complex biofilm matrix may be comparable to the situation at the cellular level.[15] As a consequence, there is a need to establish an *in situ* technique for the assessment of glycoconjugate distribution in biofilm systems. At present, the most promising approach to achieve this is the application of lectin-binding analyses in combination with confocal laser scanning microscopy (CLSM).

The applicability of lectins to probe living and fully hydrated biofilm systems is based on the definition of biofilms. A general definition for biofilms is the collection of microorganisms and their extracellular polymeric substances (EPS) associated with an interface.[16] Extracellular polymeric substances are defined as organic polymers of biological origin, which in biofilm systems are responsible for the interaction with interfaces.[17] The EPS may include polysaccharides, proteins, nucleic acids, and amphiphilic polymeric compounds.[18,19] Thus, in biofilm systems, one can expect two types of polymeric carbohydrate structures: (1) those located on cell surfaces and (2) those located extracellularly throughout the biofilm matrix. The significance of EPS is basically twofold; EPS represent a major structural component of biofilms and are responsible for sorption processes.[20,21]

[10] T. C. Bog-Hansen, "Lectins: Biology, Biochemistry, Clinical Biochemistry," Vols. 1, 2, and 3. de Gruyter, Berlin, 1981.

[11] R. J. Doyle and M. Slifkin, "Lectin–Microorganism Interactions." Dekker, New York, 1994.

[12] I. E. Liener, N. Sharon, and I. J. Goldstein, "The Lectins." Academic Press, Orlando, FL, 1986.

[13] S. A. Brooks, A. J. C. Leathem, and U. Schumacher, "Lectin Histochemistry." Bios Scientific Publishers, Oxford, 1997.

[14] T. R. Neu, *in* "Microbial Mats, Structure, Development and Environmental Significance" (L. J. Stal and P. Caumette, eds.), p. 221. Springer-Verlag, Berlin, 1994.

[15] R. I. Amann, W. Ludwig, and K. Schleifer, *Microbiol. Rev.* **59,** 143 (1995).

[16] K. C. Marshall, "Microbial Adhesion and Aggregation." Life Sciences Research Report No. 31. Springer, Berlin, 1984.

[17] K. E. Cooksey, *in* "Biofilms: Science and Technology" (L. F. Melo, T. R. Bott, M. Fletcher, and B. Capdeville, eds.), p. 137. Kluwer Academic, Dordrecht, 1992.

[18] T. R. Neu, *Microbiol. Rev.* **60,** 151 (1996).

[19] T. R. Neu and J. R. Lawrence, *in* "Microbial Extracellular Polymeric Substances" (J. Wingender, T. R. Neu, and H.-C. Flemming, eds.). Springer, Heidelberg, in press.

[20] H.-C. Flemming, J. Schmitt, and K. C. Marshall, *in* "Sediment and Toxic Substances" (W. Calmano and U. Förstner, eds.), p. 115. Springer, Berlin, 1996.

[21] T. R. Neu and K. C. Marshall, *J. Biomat. Appl.* **5,** 107 (1990).

Why not use antibodies to probe the biofilm for carbohydrate distribution? The production of antibodies against carbohydrates is, in general, difficult if compared with proteins. In addition, it requires the isolation of pure polysaccharide material from the complex polysaccharide matrix of a complex microbial biofilm community. This is another obstacle that makes wet chemistry of biofilm polysaccharides a nearly impossible task. If the antibody could be produced, its specificity would allow only the detection of a small and well-characterized fraction of the carbohydrates present in a complex biofilm community. In contrast, the many lectins available offer a huge and diverse group of carbohydrate-specific binding molecules waiting to be employed for an *in situ* approach.[22]

Application of Lectins

Lectins are available commercially as purified proteins or even fluorescently labeled. They are usually supplied freeze-dried in lots of 1–5 mg previously diluted in buffer. The fluorescence labels include mostly fluorescein isothiocyanate (FITC) and tetramethylrhodamine isothiocyanate (TRITC). However, some other fluorochromes are available. In addition, there are commercial kits by which lectins may be labeled with, for example, the cyanin dyes, such as CY5 or other fluorescent markers. Furthermore, lectins with reflective labels such as colloidal gold are available. However, their signal may be difficult to distinguish in environmental biofilm samples from other reflective material of similar size (e.g., mineral compounds).

Most lectins are hazardous and some of them are extremely toxic. Many are supplied and delivered in triple-sealed containers! Thus the following precautions are necessary if they are employed for staining biofilms: (1) Wear gloves. (2) Never touch lectins or breathe in the powder. (3) Avoid manipulation of freeze-dried lectins on a balance. (4) If necessary, handle lectins in a fume hood. (5) Do not open crimp-sealed bottle, prepare stock solution in the vial in which the lectins are supplied. (6) Transfer lectins only in solution. (7) Collect waste according to laboratory safety rules for hazardous compounds (lectin and fluorescence label).

Prepare a stock solution of 1 mg protein per milliliter buffer by adding filter-sterilized (0.2 μm) water or buffer with a syringe to the vial in which the lectin is supplied. From this stock solution it may be convenient to prepare subsamples of 100 μl in Eppendorf tubes to be kept at $-20°$ for subsequent use. These subsamples are finally diluted with 900 μl water to have a working solution at a concentration of 100 μg/ml. The solution

[22] R. D. Cummings, *Methods Enzymol.* **230,** 66 (1994).

can be used for about 5–10 biofilm samples that are each approximately 100 mm^2.

Single Lectin Staining

Specific steps for the removal of added lectin will vary with the nature of the biofilm substratum or culture method. The following outline assumes that the biofilm is attached to an unenclosed smooth surface.

1. Take the fresh biofilm sample covered with original water and mount it according to the microscope (upright or inverted) available (for details, see Lawrence and Neu,[22a] this volume).
2. Use a filter paper or similar absorbent material to carefully draw off the excess water from the biofilm surface.
3. Add lectin working solution and leave for 20 min at room temperature.
4. The sample then has to be "washed" carefully to remove unbound lectin. This is done again with a filter paper followed by adding filter-sterilized original water to cover the biofilm.
5. Repeat washing step four times.
6. The sample may then be examined with an upright CLSM equipped with a water-immersible lens.[22a]

The single and direct lectin staining is the fastest and most straightforward technique to demonstrate glycoconjugate distribution in complex biofilms. By selecting an appropriate general lectin stain, this approach may also be used to monitor the distribution and abundance of EPS in biofilms. Mohamed *et al.*[23] used lectin staining to monitor the change in quantity of EPS as a result of exposure of biofilms to nutrients and pulp mill effluents. Lawrence *et al.*[24] demonstrated the use of *Triticum vulgaris* lectin as a general EPS stain to quantify the EPS component in river biofilms. In addition, the technique may be combined with other fluorescent stains, e.g., nucleic acid stains such as the SYTO series (Molecular Probes, Eugene, OR). See, for example, Neu and Lawrence[25] and Lawrence *et al.*[24,26]

[22a] J. R. Lawrence and T. R. Neu, *Methods Enzymol.* **310** [9] (1999) (this volume).
[23] M. N. Mohamed, J. R. Lawrence, and R. D. Robarts, *Microb. Ecol.* **36,** 121 (1998).
[24] J. R. L. Lawrence, T. R. Neu, and G. D. W. Swerhone, *J. Microbiol. Methods* **32,** 253 (1998b).
[25] T. R. Neu and J. R. Lawrence, *FEMS Microbiol. Ecol.* **24,** 11 (1997).
[26] J. R. Lawrence, G. M. Wolfaardt, and T. R. Neu, *in* "Digital Analysis of Microbes: Imaging, Morphometry, Fluorometry and Motility Techniques and Applications" (M. H. F. Wilkinson and F. Schut, eds.), p. 431. Wiley, Sussex, 1998.

Multiple Staining

With the standard CLSM equipped with an Ar–Kr laser (lines 488 nm, 568 nm, 647 nm) and three channels for detection, a maximum of three stains can be applied and recorded sequentially or simultaneously. If photosynthetic organisms are present, their autofluorescence signal may be collected in the far-red region of the spectrum.[24] With the expensive UV laser option, a maximum of four different stains may be applied to the biofilm sample. With respect to fluorochrome exitation and emission, this is presently the utmost limit in the combination of fluorescent stains.

Thus, depending on the autofluorescence of the sample, the nucleic acid stain used, and the type of lectin labeling, the following combinations are possible with a Ar–Kr laser and two or three channel recording: (1) lectin + autofluorescence, (2) nucleic acid stain + lectin, (3) nucleic acid stain + lectin + autofluorescence (Figs. 1A–1C, Fig. 2A, see color insert), (4) nucleic acid stain + lectin + lectin (Figs. 1D–1F, Fig. 2B and Fig. 1J–1L, Fig. 2D), (5) lectin + lectin, (6) lectin + lectin + autofluorescence (Fig. 3, see color insert), and (7) lectin + lectin + lectin (Figs. 1G–1I, Fig. 2C).

However, the actual application of lectin combinations is limited, due to specific characteristics of lectins. Some lectins have a very similar specificity and thereby may give a very similar signal.[26a] Other lectins may bind to each other, resulting in precipitates that appear in the case of FITC–lectin (green) and TRITC–lectin (red), a color overlay of the precipitate in yellow (T. R. Neu, unpublished observation, 1998). Furthermore, factors such as the order of addition and the nature of the fluor conjugated to the lectin may also influence the effectiveness and interpretation of multiple staining (T. R. Neu *et al.,* in preparation).

Indirect Methods

In standard immunology handbooks, several indirect or sandwich techniques are described. Some of these work via primary or secondary antibodies, whereas others are by means of alkaline phosphatase or peroxidase in combination with antibodies. More recently, avidin–biotin techniques have been developed. These indirect methods may be adapted to lectin-binding analysis.[13] Their advantages are that native, unlabeled lectins can be used and that indirect methods have generally a higher specificity. The following indirect technique combinations may be suggested for lectin-binding analysis: (1) carbohydrate: lectin and lectin antibody with fluorescence label; (2) carbohydrate: lectin, lectin antibody, and antiantibody with fluorescence

[26a] T. R. Neu, G. D. W. Swerhone, and J. R. Lawrence, submitted for publication.

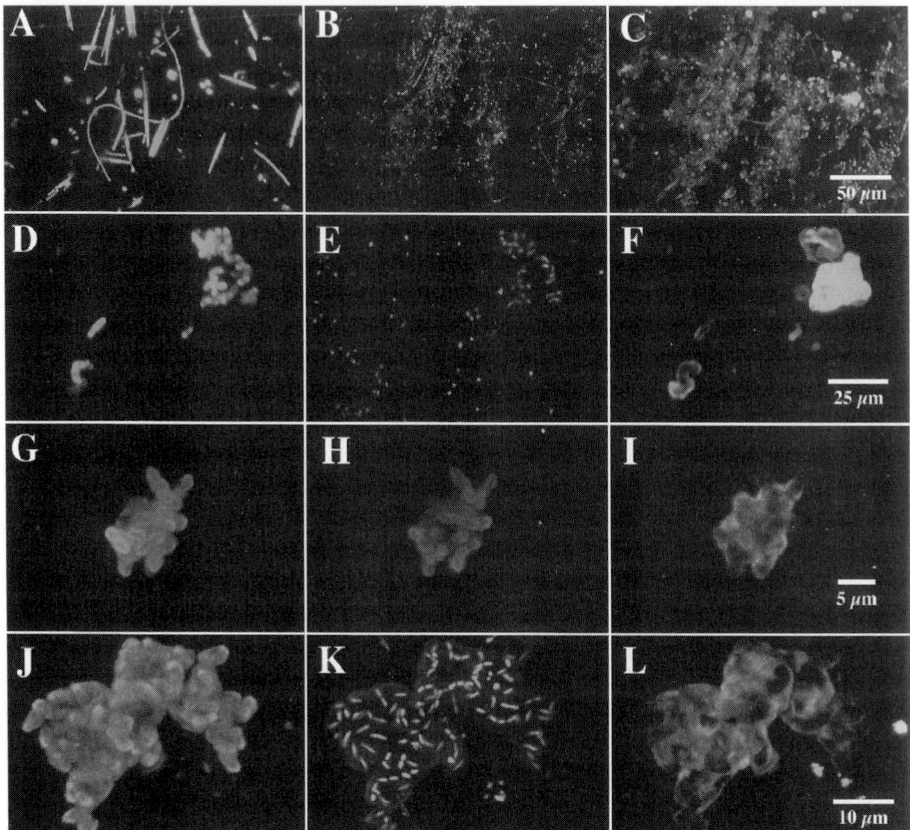

FIG. 1. Single-channel information as detected by the photomultiplier of the CLSM. The images are related to the false color images in Fig. 2. (A–C) River biofilm showing signals of chlorophyll autofluorescence (A), SYTO 9 staining of bacteria (B), and *Glycine max* lectin staining (C). (D–F) Microcolony of a river biofilm community showing signals of *Triticum vulgaris* lectin (D), nucleic acid-specific SYTO 9 stain (E), and *Tetragonolobus purpurea* lectin (F). (G–I) River biofilm microcolony with three lectin signals of *Canavalia ensiformis* lectin (G), *T. purpurea* lectin (H), and *Arachis hypogaea* lectin (I). (J–L) Large lotic microcolony with signals of *C. ensiformis* lectin (J), bacteria stained with SYTO 9 (K), and *A. hypogaea* lectin (L).

label; (3) carbohydrate: lectin, lectin antibody with biotin label, and avidin with fluorescence label; and (4) carbohydrate: lectin, lectin antibody, antiantibody with biotin label, and avidin with fluorescence label.

Control Experiments

Several positive or negative controls should be set up in parallel to evaluate lectin-binding analysis. A positive control, e.g., with pure cultures

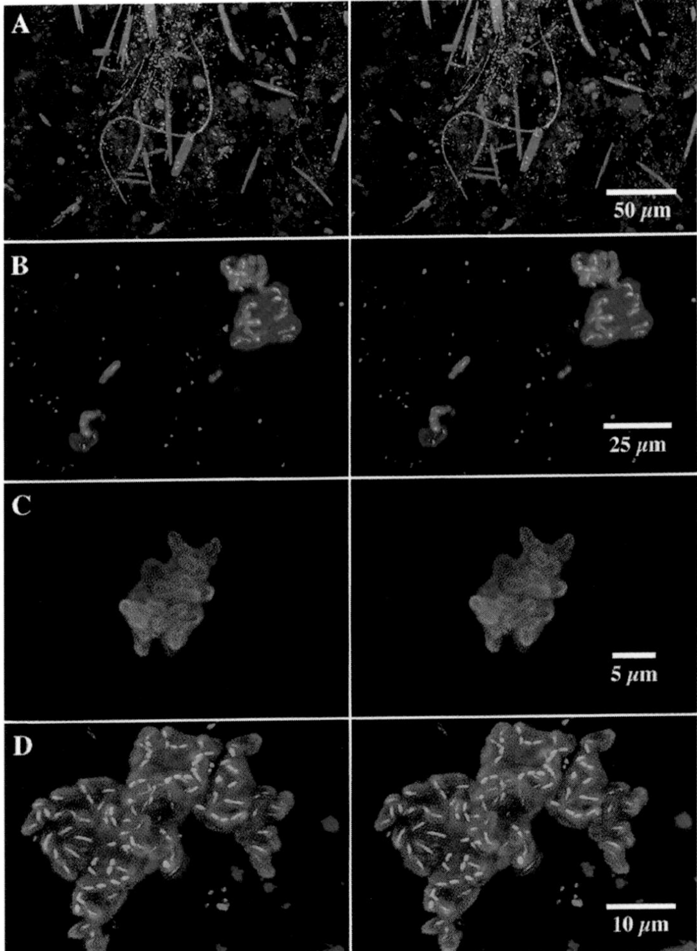

FIG. 2. False color presentation of images from Fig. 1. The images have been shifted to allow examination with a stereo viewer to get a three-dimensional impression in color. (A) Three-channel CSLM images showing a river biofilm stained with SYTO 9 nucleic acid stain (green), *Glycine max* lectin (blue), and far-red imaging of the autofluorescence of algae (red). (B) Projection of a series of CSLM optical thin sections showing the distribution of two types of lectin-binding sites [*Triticum vulgaris* lectin (red) and *Tetragonolobus purpurea* lectin (blue)] and the position of bacterial cells (SYTO 9, green) in a microcolony growing as part of a river biofilm community. Note that there are cells and microcolonies that are stained with SYTO 9 but not with either lectin. (C) Combined projection of three-channel CSLM images showing the distribution of three lectin-binding sites [*Canavalia ensiformis* lectin (red), *T. purpurea* lectin (green), and *Arachis hypogaea* lectin (blue)] within a single microcolony growing in a river biofilm community. (D) Extended focus view of a lotic microcolony showing the distribution of two lectin-binding sites [*C. ensiformis* lectin (red), *A. hypogaea* lectin (blue)] and bacteria stained with SYTO 9 (green).

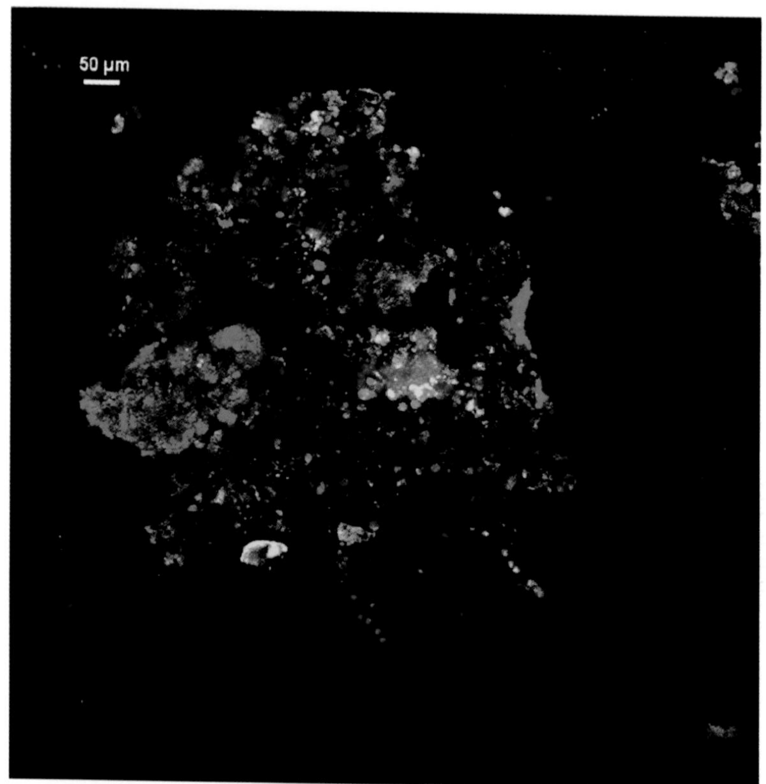

Fig. 3. Extended focus image of a fully hydrated, unfixed lotic microbial aggregate. The sample was stained with *Limulus polyphemus* (green) and *Triticum vulgaris* (red) lectins. The signals of the two lectins and autofluorescence of photosynthetic organisms (blue) were recorded in three channels simultaneously. Note the chemical heterogeneity of extracellular polymeric substances throughout the aggregate.

showing a clear lectin binding, would be ideal. The obvious negative control can be done by just leaving out the lectin from the staining protocol. This control is compulsory and is identical with recording the autofluorescence of the sample. This has to be measured in each of the channels used for signal detection. Other controls to perform are inhibition experiments with the lectin-specific carbohydrates. In general, most authors have assumed that the specificity indicated by the supplier or in the literature adequately defines the lectin. However, this is not the case for application of lectins in biofilm systems, therefore inhibition experiments are critical for interpretation. For these tests, several monosaccharide concentrations have to be measured. Starting at the minimal inhibition concentration, carbohydrates should be tested at three different carbohydrate molarities[27] (T. R. Neu *et al.*, in preparation). If an indirect staining method is employed, other controls may be performed.

Perspectives

Lectin binding has been used fairly extensively in microbiology to characterize cell surfaces of bacteria grown in pure culture. In the meantime, several reports have been published describing their applicability to complex biofilm communities.[25,27,28] However, the application of lectin-binding analysis is not restricted to certain types of biofilms on solid surfaces. The same technique can be used for motile biofilms in the form of aquatic aggregates. This has been demonstrated for fixed marine snow[29] and for nonfixed river snow (Fig. 3, see color insert).[26a] Furthermore, lectins are powerful probes for the localization of biofilm subdomains such as conditioning films, adhesive "footprints,"[30] cell surfaces, microcolonies (Figs. 1 and 2), EPS "clouds" (Figs. 1C, 1I, and 1L, blue signal in Fig. 2, green signal in Fig. 3),[26a] and biofilm ridges.[25]

What is needed to further fine-tune the lectin-binding analysis? The main necessity is a more detailed characterization of the various lectins and a greater understanding of the nature of the lectin-binding sites for protein and carbohydrates.[31] The natural binding specificity and the properties of multiple lectin-binding sites must be known. These binding specificities should be determined using sites that are relevant to microbial and biofilm applications. In addition, new lectins specific for bacterial and algal

[27] T. Michael and C. M. Smith, *Mar. Ecol. Progr. Ser.* **119,** 229 (1995).

[28] G. M. Wolfaardt, J. R. Lawrence, R. D. Robarts, and D. E. Caldwell, *Microb. Ecol.* **35,** 213 (1998).

[29] C. F. Holloway and J. P. Cowan, *Limnol. Oceanogr.* **42,** 1340 (1997).

[30] T. R. Neu and K. C. Marshall, *Biofouling* **3,** 101 (1991).

[31] A. Gohier, J. F. Espinosa, J. Jimenez-Barbero, P.-A. Carrupt, S. Perez, and A. Imberty, *J. Mol. Graph.* **14,** 322 (1996).

carbohydrate sequences should be isolated. These highly specific lectins would be ideal as molecular probes for the assessment of biofilm sorption properties as well as for the structural examination of microbial biofilm communities.

Acknowledgments

The study was supported by the Canada–Germany Agreement on Scientific Research (Grant ENV46/2). Excellent technical assistance was provided by Ute Kuhlicke (Germany) and George Swerhone (Canada).

[11] Spatially Resolved, Quantitative Determination of Luciferase Activity by Photon-Counting Microscopy

By ROBERT J. PALMER, JR. and DAVID C. WHITE

Luciferase activity offers some distinct advantages as a bioreporter. It can be selected for using relatively simple equipment (bioluminescent bacterial colonies can be seen in a dark room, whereas early experiments used photographic film as a detection/recording method), and signal (light output) ceases after the translation of target gene/luciferase gene fusion stops[1] (in contrast to fluorescent molecules such as unmodified Green Fluorescent Protein, which remains fluorescent for extended periods). One major advantage for biofilm researchers is that with the type of equipment described in this article, the origin of luciferase activity can be spatially resolved and quantitated in two dimensions, thereby allowing quantitative discrimination at the level of the single bacterial cell.[2] Several quantitative detection techniques are used to determine luciferase-mediated light production (scintillation counters, luminometers, photomultipliers) but none of these offers the spatial resolution provided by photon-counting camera setups. Hamamatsu Photonics (Bridgewater, NJ) and Science Wares (East Falmouth, MA) market turnkey photon-counting camera/microscope systems, and at least one laboratory has developed a cruder, yet effective, system "in house."[3] Such systems are designed to operate at photon fluxes

[1] R. J. Palmer, Jr., B. Applegate, R. Burlage, G. Sayler, and D. C. White, *in* "Bioluminescence and Chemiluminescence: Perspectives for the 21st Century" (A. Roda, M. Pazzagli, L. J. Kricka, and P. E. Stanley, eds.), p. 609. Wiley, Chichester, 1999.

[2] R. J. Palmer, Jr., C. Phiefer, R. Burlage, G. S. Sayler, and D. C. White, *in* "Bioluminescence and Chemiluminescence: Molecular Reporting with Photons" (J. W. Hastings, L. J. Kricka, and P. E. Stanley, eds.), p. 445. Wiley, Chichester, 1997.

[3] J. Elhai and C. P. Wolk, *EMBO J.* **9,** 3379 (1990).

0076-6879/99 $30.00

of <10^{-5} lux and are at least 1 order of magnitude more sensitive than "intensified" or cooled charged-coupled devices (CCD), ISIT cameras, etc. Although much of the methodology presented in this article is hardware specific, the principles are those which must be applied to any quantitative, spatially resolved method.

Hardware

The Hamamatsu camera consists of a CCD camera attached to an image intensifier (Fig. 1). The image intensifier contains a photocathode, anodes, microchannel plate, and a phosphor plate, all located within an evacuated

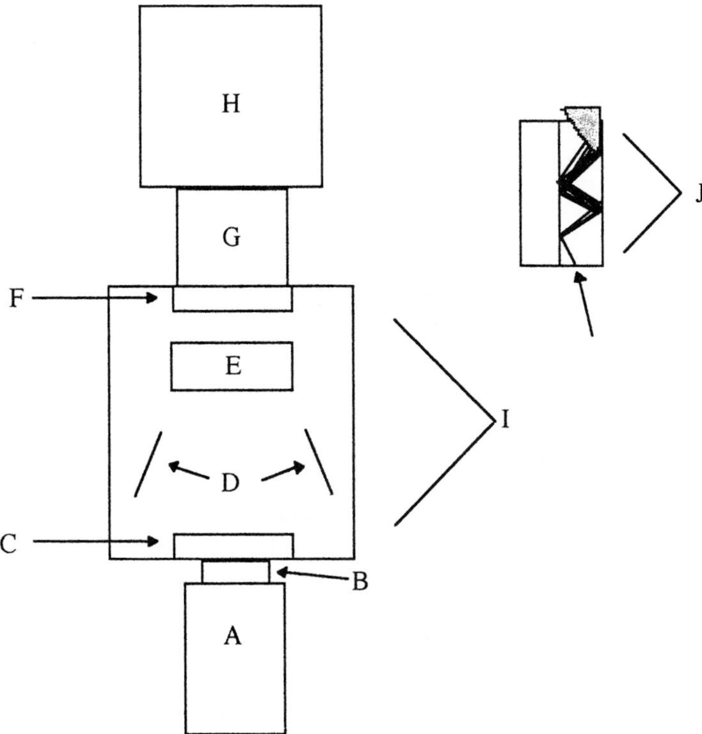

FIG. 1. Diagram of a turnkey photon-counting device (Hamamatsu Photonics). (A) Microscope phototube or adapter to C-mount, (B) C-mount, (C) photocathode, (D) anodes, (E) microchannel plate, (F) phosphor plate, (G) relay lens, (H) CCD camera, and (I) evacuated chamber. (J) Representation of photon multiplication within the microchannel plate. Two microchannels are shown. The photon enters one channel (arrow) and strikes the walls of the chamber, thereby generating additional electrons that exit the top of the channel in a burst (gray).

chamber. Photons strike the photocathode and are converted to photoelectrons and accelerated past the anodes into the microchannel plate. Within the microchannel plate, the photoelectrons strike the walls of the channel, resulting in the generation of additional electrons up to a multiplication of approximately 10^6 for each photoelectron that initially enters (see inset J in Fig. 1). A burst of electrons thus exits the microchannel plate, strikes the phosphor plate, and is converted back to photons that then strike the CCD camera. Data are collected as a 16-bit image that can be pseudocolored (blue, lowest intensity; green, yellow, red, white, highest intensity) and displayed on a video monitor. The voltage applied across the microchannel plate is controlled in two places: a button is used to turn the high voltage on and off and the voltage is increased/decreased using the sensitivity knob. Both of these controls are located on an interface box between the camera and a computer used for data storage. The camera/intensifier unit is connected to a microscope via a C-mount. For use as a macro camera (e.g., to record bioluminescence of bacterial colonies on plates), a photographic lens can be attached to the C-mount and the camera/intensifier/lens unit held by a tripod or other support.

Software

The burst of photons that illuminate the CCD exposes more than one pixel. In order to obtain higher spatial resolution, a software/hardware combination called the "center of gravity" board remaps the likely origin of the original photon to one pixel based on the size and shape of the multipixel illuminated spot. This feature is required for quantitative work. In addition, several algorithms and macros are included that allow the determination of the number of exposed pixels within user-defined areas, automatic collection of time-resolved data, and graphs of pixel intensity across user-defined lines within an image. The software package includes rudimentary image processing hardware and software (Argus 50) that are used to operate many different types of cameras. This software can be used to acquire transmitted light images with the camera (see later).

Dark Box

For true photon-counting applications, the camera must be isolated from background light. Clearly, a windowless room with a well-sealed door is a prerequisite. Even with this precaution, a dark box that encloses the microscope or the entire microscope/camera assembly is ordinarily required. When designing a dark box, careful consideration should be given to the types of experiments to be conducted. For example, flow-through (perfusion) chambers are routinely employed in this laboratory (see Chapter 12 of this volume); therefore, ports for tubing are required. If ambient room light is kept to a minimum (e.g., computer monitors only), it is usually sufficient to have these ports constructed out of 5 to 8 cm of black-painted 1-cm-diameter polyvinyl chloride pipe. Clearance for the operator's hands (for focusing the microscope) and any external light sources or cameras (such as might be required for correlative work) must be incorporated into the design, otherwise frequent (and cumbersome) removal of the box will be necessary. In this laboratory, because upright and inverted microscopes are used in conjunction with the photon-counting system, two different dark box designs were required. In both designs, consideration was given to mounting a laser confocal scan head to allow sequential photon-counting/confocal microscopy on the same field of view (correlative microscopy, see later). The front of the boxes (the front of the microscopes) is open; during photon counting, the opening is covered with photographer's dark cloth that seals to the edges of the box with Velcro strips. All microscopy light sources (transmitted, UV) must be turned off completely (not just turned down) prior to photon counting. Many modern microscopes, particularly those with electronic focus or lens switching, have lighted information panels built into the body of the microscope that are operational whenever the microscope is turned on. These panels are a light source within the dark box; mask them with a piece of aluminum foil.

Example: Photon Counting of Attached *Vibrio fischeri* Cells

As a preface, it should be emphasized that the high voltage on the photon-counting camera should be turned off (or, in dimly lit rooms, the sensitivity dial should be set at its lowest setting) before sending light to the camera. This avoids the sudden delivery of a high photon flux to the microchannel plate. The voltage should always be increased slowly while watching the monitor to assure that the system does not saturate. If saturation is reached, the system shuts off automatically, but these accidents can reduce the effective life span of the instrument.

Test for Light Leaks

Before an experiment is begun, it is useful to check for light leaks in the system, particularly if the dark box is removed frequently. All microscope light sources are switched off, the dark box is sealed, and the photon-counting camera is turned on. The voltage is turned up slowly until the display shows saturation (indicative of a large light leak) or maximum voltage is reached (photon-counting mode). If photon-counting voltage is reached, a count is made over 1 min accumulation time. Our system has very low counts when no light leaks are present (80–100 counts per field per minute) and we attempt to control leakage if counts of >200 per field per minute are reached.

Background Light Levels

Background light levels should be established under exactly the same conditions as those for the experiment. Changes in the optics (numerical aperature of lens, magnification, oil immersion vs dry lens) will affect the background; the procedure just described for light leak detection should be repeated with a "blank" (nonluminescent sample preparation similar to that to be examined in the experiment) prior to recording photon-counting data. The blank sample should be brought into focus using the transmitted light illumination, the light turned off, and the photon-counting procedure initiated. Once the investigator becomes sufficiently experienced with the system, the leak detection step will be replaced by the background determination step. Our background levels are very close (100–120 counts per field per minute) to the levels recorded during the procedure described earlier.

Collection of Data

It is almost always necessary to collect two types of data. First, a transmitted light image should be acquired. This image is useful for the correlation of light output (the .IMA image) with cellular location. We usually acquire these data using the photon-counting camera as the detector in the following manner.

a. bring the specimen into focus in the oculars using transmitted light, turn off the light, switch the light path from oculars to camera, check that the photon-counting sensitivity (voltage) dial is set to the lowest (counterclockwise) position, and depress (turn on) the high-voltage button.

b. within the MONITOR window, set DISPLAY (image source) to RAW (= live) and uncheck the "color" box to produce a pseudocolored image.

c. turn on the transmitted light illumination at its lowest voltage

d. the transmitted light image may be visible at this point or the transmitted light voltage may have to be increased to make the image visible. If no color is visible after reaching the maximum voltage of the transmitted light source, slowly turn up the camera sensitivity (voltage) dial. Readout on the voltage is from 1 to 10. If no image is visible on reaching 3, it is likely that problems with the light path (i.e., no light is reaching the camera) or camera electronics exist. It is not advisable to turn the voltage higher. Once a color image is visible, the pseudocolor should be turned off (uncheck the "color" box to obtain a gray scale image). Some refocusing may be required to compensate for differences in light paths between the binocular (view) port and the camera port of the microscope; this can be accomplished without saturating the camera if the room is dark and the sensitivity knob is set relatively low. The focused image can be acquired using the Argus software commands "freeze" or "integrate" located under the IMAGING menu. "Freeze" stores a single video frame in the memory, whereas "integrate" stores the sum of several (number is user defined) frames. The latter function is generally preferable because lower transmitted light and sensitivity levels can be used. Once the transmitted light image is acquired to memory, it must then be saved to disk using the Argus software "save as TIFF" command under the FILE menu. Figure 2a shows a transmitted light image of bacterial cells obtained in this manner.

After the transmitted light image is obtained and saved, the photon-counting image is acquired in the following manner.

a. Turn off the transmitted light source.

b. Reset DISPLAY to RAW (display defaults to PROCESSED during acquisition of the transmitted light image) and check (select) the "color" box to produce the pseudocolor image.

c. Slowly turn up the sensitivity knob until the maximum (10) is reached. If the system saturates prior to reaching this setting, the photon flux is too high to count with this setup (it is possible to count with the voltage set below 10; however, these voltage levels are difficult to reproduce accurately and the quantitative data are comparable only with those obtained at exactly the same voltage setting). At this point, centers of high light production should be visible. If no light is being produced, only scattered photon events (noise) will be seen. If no heterogeneity of light production exists, then high (relative to background) but noisy activity will be seen. If very low levels of light are being produced, it may be necessary to do an actual count (light accumulation) to see the light production. To count, select the command "photon counting" under the IMAGING menu. A dialog box appears in which the accumulation time for the image file (.IMA) can be defined (frames, seconds, minutes, hours). In our experience, accumulation times of 30 sec to 20 min are sufficient for bacterial luciferase bioluminescence using 100× oil-immersion optics. "Mode" should be set as "slice/gravity"; this results in storage in memory of both the "slice" image (the image prior to center of gravity calculation) and the "gravity" image (after center of gravity calculation). The slice image is frequently more useful as a visual data presentation than is the gravity image; however, only the gravity image is useful for quantitative purposes. After the acquisition

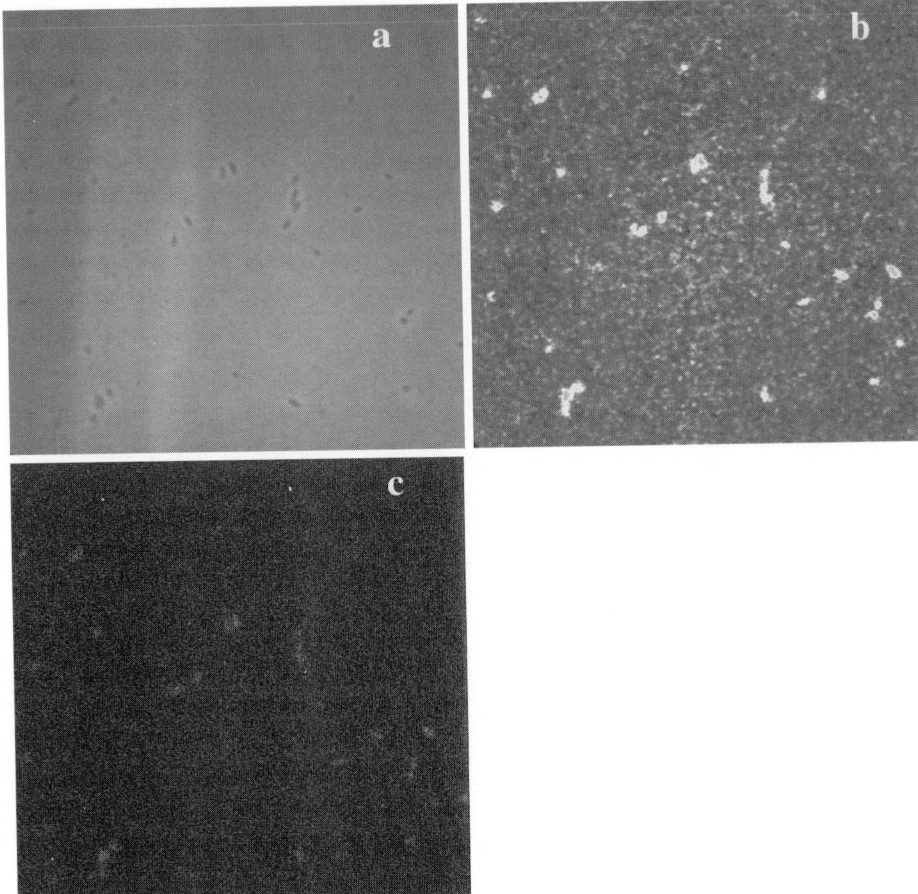

FIG. 2. Images acquired by a photon-counting camera. (a) Transmitted light image of *Vibrio fischeri* cells grown in a microscopy flowcell (perfusion chamber; see Palmer and White, this volume). (b) "Slice" mode image of same field shown in (a). (c) "Gravity" mode image of field shown in (a). Accumulation time for (b) and (c) was 5 min.

parameters have been defined, "start" is selected and the accumulation begins. The process occurs in real time on the video monitor, the remaining accumulation time is displayed on a counter within the dialog box, and, after the user-defined end point is reached, the resulting image(s) is stored in memory. The images must now be saved (as .IMA file) to disk using the "save image" command from the FILE menu. If desired, the acquisition process can be continued (effectively extending the accumulation period) by selecting "continue"; the RAM images are modified and must be saved as noted earlier. It should also be noted that any images currently in memory are overwritten when a new acquisition process is started. Some warnings appear to inform users of this potential problem; however, images can be overwritten easily. An .IMA file can be reloaded into memory and the "slice" image or the "gravity" image can be saved to disk as a TIFF image for export. Figures 2b and 2c show, respectively, "slice" and "gravity" images of the field in Fig. 2a.

Quantitation and Normalization

Quantitative measurements can be done in many ways; the method employed routinely in this laboratory is to use the "area analysis" command from the ANALYSIS menu. This function permits the user to define a box or circle around the region of light production and query the computer as to how many photons were accumulated within that region. The box is then moved to a region that contains no cells (correlate with transmitted light image) and the computer is queried as to the number of counts within this region (the background noise). It is important to perform the background count on every image as it will vary (generally parallels increase or decrease in overall light production). Background counts are subtracted from the microbially produced counts to give the final result. To obtain numbers comparable from different sized colonies, the final counts can be normalized to colony area. This procedure cannot be performed with the Argus software; a third-party image analysis package is required. The TIFF format-transmitted light image is exported to the image analysis software, spatially calibrated, thresholded to create a binary image encoding the colony only, and the area determined by particle analysis procedures. Difference imagery can be used to determine the area of separate colonies within the same image. A review by Caldwell *et al.*[4] provides examples of these image analysis methods as applied to biofilms, and a book by Russ[5] provides an in-depth theoretical treatment.

Correlative Microscopic Techniques

It is often useful to acquire additional data on colonies from which light production data have been obtained. For example, multiple centers of light production can occur within a single colony. In this case it is useful to know if these centers represent areas in which the thickness of the colony, or the relative activity of cells, within those regions differs from that outside the

[4] D. E. Caldwell, D. R. Korber, and J. R. Lawrence, *Adv. Microb. Ecol.* **12,** 1 (1992)

[5] J. C. Russ, "The Image Processing Handbook." CRC Press, Boca Raton, FL, 1995.

regions. This laboratory has employed confocal microscopy together with photon-counting microscopy to assess just these issues in a study on a *Pludomonas putida* strain with a *tod/lux* fusion (grows on toluene as the sole carbon source and produces light in response to *tod* promoter activity).[1]

Absolute Photon Flux Values

A drawback of this methodology is that the numbers are relative. Data are not absolute photon flux values because the loss of photons occurs from light scattering, from optical transfer, and from conversions within the intensifier. However, if a calibration method based on objects approximately the size of bacterial cells were to be developed, then an estimate of the efficiency of the camera/microscope combination could be arrived at and used to calculate absolute photon flux values.

[12] Microscopy Flowcells: Perfusion Chambers for Real-Time Study of Biofilms

By ROBERT J. PALMER, JR.

If flowcells are defined as perfusion chambers for the observation of growth/physiology of stationary cells, then they have been used by eukaryotic cell biologists for about a century; more elaborate systems were designed in the 1950s[1,2] that were the forerunners of those used today.[3–6]

Flowcells offer many advantages. A defined, constant environment is provided by laminar flow. They can be made inexpensively and therefore can be disposable (thereby eliminating the necessity of cleaning between uses). The design can be versatile in the selection of material for the substratum. Perhaps most importantly, they are the only sample preparation device by which time-resolved, nondestructive measurements at the level of the single bacterial cell can be made. As with all methodology, flowcells do have some disadvantages. Sterilization can be difficult because many of

[1] G. S. Christiansen, L. Belty-Danes, L. Allen, and P. J. Leinfelder, *Exp. Cell Res.* **5,** 10 (1953).
[2] W. Schwöbel, *Exp. Cell Res.* **6,** 79 (1954).
[3] H. E. Berg and S. M. Block, *J. Gen. Microbiol.* **130,** 2915 (1984).
[4] J. Sjollema, H. C. van der Mei, H. M. Uyen, and H. J. Busscher, *J. Adhesion Sci.* **4,** 765 (1990).
[5] G. M. Wolfaardt, J. R. Lawrence, R. D. Robarts, S. J. Caldwell, and D. E. Caldwell, *Appl. Environ. Microbiol.* **60,** 434 (1994).
[6] R. J. Palmer, Jr., and D. E. Caldwell, *J. Microbiol. Methods* **24,** 171 (1995).

0076-6879/99 $30.00

the common materials used for flowcell construction do not respond well to autoclaving, especially to repeated autoclaving. Other sterilization methods require equipment not routinely accessible to most researchers (e.g., ethylene oxide chambers) or are, in fact, disinfection methods (e.g., hypochlorite). Liquid must be delivered to the flowcell; peristaltic pumps are commonly used for this purpose. However, even the most expensive peristaltic pumps produce some pulsation in liquid delivery. High-quality syringe pumps are pulseless but are more expensive than peristaltic pumps and require reloading for long-term experiments. One laboratory has used gravity feed for liquid delivery[4]; that this method has not become widespread attests to its difficulty, although the advantages (low cost, pulseless flow) are clear. Any enclosed system is subject to wall effects: aberrations in flow in the corners of the chamber. Finally, in cases of high biomass within the flowcell, it is conceivable that a gradient in nutrients could be established over the length of the flowcell. However, the disadvantages listed here are not severe and are frequently present in other systems; flowcells seem to be the best approach for most studies of bacterial biofilms.

Experimental Considerations for All Flowcells

Organic matter at the substratum can play a role in the physiology of attached cells, particularly in initial attachment and monolayer development. Therefore, a disposable flowcell design has the attraction of eliminating carryover from one experiment to the next: only a design that could survive heating to 450° would be strictly comparable. However, some designs simply cannot be realized in a disposable form, e.g., large units such as those described elsewhere in this volume[7] were designed for use with a fiber optic probe. In such designs, easily cleanable (e.g., acid-washable) materials should be employed (Teflon, polyvinyl carbonate) and the units should be thoroughly cleaned immediately after an experiment is terminated.

The choice of substratum is regulated primarily by the manner in which data are to be collected and by the type of microbial system to be investigated. Flowcells that use microscopy coverslips or other transparent windows as the substratum are perhaps the most well known. Opaque materials can be used as the base of a flowcell that incorporates an overlying viewing window. Universal equations for the calculation of hydrodynamic parameters within flowcells are provided by Zinn *et al.* in this volume.[7]

[7] M. S. Zinn, R. D. Kirkegaard, R. J. Palmer, Jr., and D. C. White, *Methods Enzymol.* **310,** [18] (1999) (this volume).

Design Considerations for Microscopy Flowcells

Flowcells for biofilm microscopy have been built around two designs (Fig. 1). The most commonly used design[5] is one in which parallel grooves (approximately 4 mm^2) are milled into a plexiglass base—the grooves stop a few millimeters before the ends of the plexiglass. Microscopy cover glass is used to cover the open side of the grooves, thereby forming a closed channel, the upper side of which is cover glass. Inlet and outlets are made by boring holes in the end of the plexiglass through to the channel; tubing

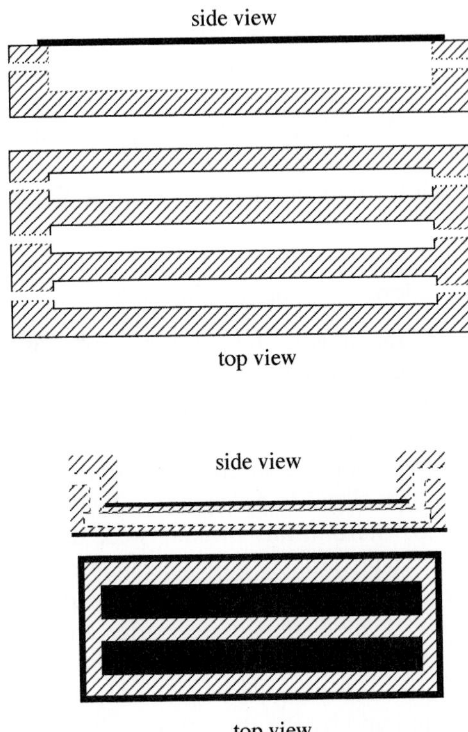

FIG. 1. Two basic flowcell designs. (Top) Plexiglass flowcell (after Wolfaardt *et al.*[5]) in a cutaway view through a flow channel (upper diagram) and in top view (lower diagram). Hatched regions are plexiglass. The thick dark line represents the cover glass. The size of the unit can vary, but is generally about 50 mm on a side and 8–10 mm in total thickness. The flow channels measure 4 mm on a side. (Bottom) Glass and silicone flowcell (after Palmer and Caldwell[6]) in a cutaway view through a flow channel (upper diagram) and in top view (lower diagram; inlet/outlet ports are not shown for clarity). The black area represents the cover glass, whereas the hatched area is silicone. The size of the unit is identical to that of a microscope slide.

is cemented into the holes. The design is boundary-free in that the size of the flowcell is limited only by the limits of attachment to the microscope. A second design[6] is one in which the coverslip is used for the top and the bottom of a channel that is formed by a molded silicone rubber gasket.[6] Elbow-shaped inlet and outlet ports are molded out of silicone. The design has two channels and has the same dimensions as a standard microscope slide, thereby making it compatible with most microscope stage hardware. Experimental considerations will contribute to decisions on which basic design is to be used and on the final size of the flowcell. Frequently, the biofilms are examined by confocal microscopy and therefore accommodation of the phase-contrast optics normally used to examine bacterial cells is irrelevant. In this case, the flowcell with a plexiglass (or other opaque) base is adequate because transmitted light is not required. However, if one wishes to investigate cell monolayers using high-resolution, transmitted-light phase-contrast optics (e.g., in photon-counting microscopy; see Palmer and White, this volume[8]), then the working distance of the lens and the travel of the substage condenser are important in setting the total thickness of the flowcell, and the flowcell must have glass as the base and as the top. For example, for a flowcell constructed of two coverslips held apart by a spacer, the maximum thickness to observe colonization hanging from the underside of the upper coverslip (or on the upper side of the lower coverslip for inverted microscopes) using 100× Koehler-illuminated, oil-immersion optics is on the order of 1 mm for most commercial microscopes. Frequently, the substage condenser cannot be moved close enough to the bottom of the flowcell to achieve Koehler illumination unless a release button is pushed (this is a safety feature to prevent novices from pushing the preparation against the lens and scratching the condenser optics). Observation of colonization on the opposite coverslip (that most distant from the objective lens) is impossible with high-magnification, oil-immersion optics because the working distances of such lenses are too small. Use of epifluorescence techniques to observe the cells eliminate the restrictions imposed by phase-contrast Koehler illumination, but not that of working distance. Long working distance lenses are available from most vendors; these are especially useful for examining cells on opaque substrata through a window up to about 3 mm from the substratum. However, these lenses are of lower magnification and lower numerical aperture (NA: less suitable for observation of single bacterial cells) than oil-immersion optics. Finally, water-immersible optics (dipping lenses) can be used to examine substrata in flowcells that have a removable cover; these lenses also suffer from having a lower NA and a lower magnification than oil-immersion objectives; however

[8] R. J. Palmer, Jr., and D. C. White, *Methods Enzymol.* **310** [11] (1999) (this volume).

(just as with long working distance lenses), they can play a vital role in obtaining data from samples that otherwise could not be imaged.

Typical Experimental Setup

A basic setup consists of a medium reservoir, tubing connecting the reservoir to the flowcell, tubing connecting the flowcell to the pump or other liquid delivery device, and a waste collection vessel. In our laboratory, the pump is placed on the downstream (outflow) side of the flowcell and medium is drawn through at a slightly negative pressure. This approach reduces the occurrence of air bubbles in the flowcell (frequently generated on the downstream side of the pump) and creates an enhanced physical barrier to contamination from the open outlet against the direction of flow and through the pressure of pump head on the tubing; sterile filters on the outlet to the waste are generally not required for experiments lasting up to a week. The medium reservoir does require a sterile "air in" vent and a down tube through which the medium is pumped out into the flowcell. These can be fashioned by attaching silicone tubing to glass or metal tubing inserted through a stopper on an Erlenmeyer flask. The free ends of the tubing (to which the flowcell will be connected) are simply wrapped prior to autoclaving; presterilized filters can be attached and tubing connections to the flowcell (with autoclavable barbed connectors) can be made after sterilization. For media that cannot be autoclaved, the stopper insert containing the tubing connections should be autoclaved separately from the Erlenmeyer flask (sterilization is often poor when the unit is in place without liquid in the flask); after autoclaving of the hardware, the medium can be decanted into the flask and the stopper inserted. It is helpful to weight the end of the flexible tubing on the down tube by inserting a piece of glass or metal tubing so that it remains positioned at the bottom of the flask. For anaerobic experiments (Fig. 2), a sparging inlet to the medium reservoir is required; this can be constructed in the same manner as the down tube and a sterile filter is attached on the gas inlet (tank) side. The entire unit of the reservoir/flowcell/pump can have a footprint of less than 0.5×0.5 m; it can be located on a laboratory cart and moved to and from the microscope with relative ease. If the gas tank during anaerobic experiments cannot be moved easily (normally the case), then the reservoir/flowcell/pump unit can be disconnected at the "gas in" side of the filter to maintain sterility during microscopic observation.

Inoculation Procedures

The most common inoculation method is the injection of culture directly into the flowcell through the silicone tubing. This is performed exactly like

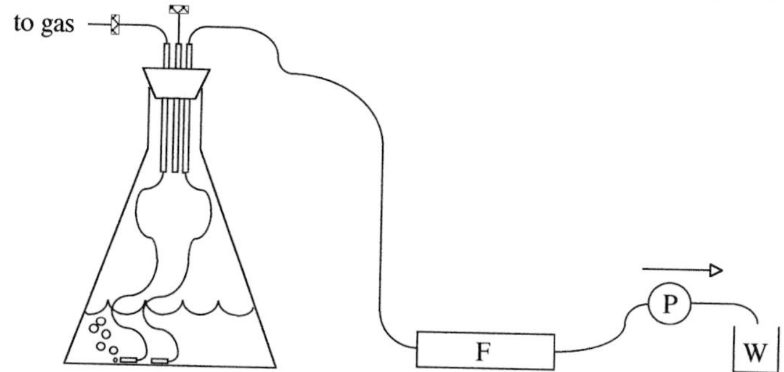

Fig. 2. Components of an anaerobic flowcell system. (Left) Medium reservoir (sparged flask) with an overpressure outlet. (Center) Flowcell (F) is connected with silicone rubber tubing to the medium reservoir and pump (P; arrow shows direction of flow). (Right) Waste container (W) is immediately downstream of the pump. Small hatched rectangles show the location of sterile filters. The setup for the aerobic system is identical except that the sparging line is not required. Tubing within the medium reservoir is weighted to maintain position.

a venipuncture: wipe with 70% isopropanol, be certain that the tubing is blocked at an appropriate site to cause the bolus to move in the proper direction, remove air bubbles from the syringe, and inject slowly and steadily. In some cases, repeated injections may be necessary. Pick new sites on the tubing for subsequent injections. If the tubing should begin to leak from an injection site, a small drop of silicon can be used to plug the hole.

Continuous or interrupted flow can also be used to introduce cells. A T junction is inserted in the tubing immediately prior to the flowcell and culture is pumped through the flow channel. This can be done aseptically by setting up the culture reservoir in a manner similar to that of the medium reservoir. Anaerobic cultures can be inoculated by direct pumping from Bellco pressure tubes after insertion of a needle for venting.

Suppliers of Flowcell Construction Materials

Microscopy cover glass can be obtained in standard thicknesses/dimensions from most scientific supply houses. Custom sizes can be ordered from Corning Glass Work [OEM Department, P.O. Box 5000, Corning, NY 14830, (www.corninglabware.com)].

Several types of silicone rubber can be purchased from Newark Electronics (4801 Ravenswood Avenue, Chicago, IL 60640), which has an extensive international distribution network. Medical-grade silicon rubber tubing can be purchased from most scientific supply houses, as can barbed tubing connectors.

Frequently it is necessary to have parts custom-made (generally by a milling process). Most universities provide shop services that can perform these operations or can refer the customer to an outside machine-shop company. For those interested in milling parts for smaller flowcells "in house," Sherline Products Inc. (2350 Oak Ridge Way, Vista, CA 92083; www.sherline.com) makes bench-top milling machines of the sophistication necessary for most work.

Note Added in Proof

The anaerobic system described herein has been used for biofilm culture of capnophilic organisms (*Actinobacillus actinomycetemcomitans* and *Capnocytophaga* spp.).[9] Recent experiments using a transformed streptococcal strain demonstrated that residual O_2 levels were sufficient to result in oxidation (fluorescence) of Green Fluorescent Protein. Modifications have resulted in a setup in which an O_2 concentration of ≤ 0.025 mg/l is maintained.[10]

[9] R. J. Palmer, Jr., unpublished.
[10] M. C. Hansen, R. J. Palmer, Jr., and D. C. White, *J. Microbiol. Meth.*, submitted.

[13] Fluorescent Probes Applied to Physiological Characterization of Bacterial Biofilms

By JOHN T. LISLE, PHILIP S. STEWART, and GORDON A. MCFETERS

Introduction

Efforts to describe the physiological activities of microbial populations have been hindered by the inability of culture-based techniques to recover a significant proportion of bacteria present in a sample[1] and by our traditional dependence on methods that yield community averages of physiological indices, such as oxygen consumption, evolution of carbon dioxide, and radiolabeled substrate incorporation into macromolecules. Progress in the development of *in situ,* intracellular fluorescent labels and probes has facilitated the rapid assessment of microbial physiology without cultivation. In the context of this article, labels are defined as fluorescent dyes that bind or interact with specific targets within the cell [e.g., 4′6-diamidino-2-phenylindole (DAPI)]. Labels most often have inherent fluorescence

[1] R. Amann, W. Ludwig, and K. Schleifer, *Microbiol. Rev.* **59,** 143 (1995).

METHODS IN ENZYMOLOGY, VOL. 310
0076-6879/99 $30.00

regardless of their being free, nonspecifically bound, or bound to their cellular target. Probes are fluorescent dyes that change their emission spectra in response to an intracellular activity (e.g., enzyme) or parameter (e.g., pH). The probes most applicable to biofilm systems are those that remain nonfluorescent until acted on by the cellular target.

The advantages of using these types of labels and probes include the direct, single cell assessment of (1) physiological status, (2) specific metabolic activities, (3) gene expression, and (4) total cell densities.[2] Additionally, when used in conjunction with fluorescent antibodies, these fluorescent markers can assess physiological activity at the species or serotype level.[3] Other nonculture-based techniques, for example, fluorescent *in situ* hybridization (FISH) using 16S rRNA probes, provide only indirect evidence pertaining to the physiological activity of the sampled community or individual cells.[4–6]

Biofilms are highly heterogeneous in their structure and composition. The traditional methods used to investigate biofilms have relied on the removal of biofilms from the substrata, followed by disaggregation. These approaches provide nominal spatial information regarding the structure of and physiological activity of individual cells within intact biofilms. Biofilm composition heterogeneity is influenced directly by constituents from the overlying bulk phase adsorbing or precipitating onto and into the biofilm matrix and intrabiofilm deposition of microbially secreted products. Individually or collectively, these constituents may influence the performance of a label or probe. Table I lists characteristics that should be considered when choosing a label or probe. Table I does not list the types of chemical modifications that have been used to stabilize and optimize a label or the permeability and fluorescence of a probe. More detailed discussions of these factors and those listed in Table I have been published elsewhere.[7–9]

Table I provides some insight into the complexities of selecting an appropriate label or probe for investigating the microbial physiology of biofilm systems. We have found that Molecular Probes, Inc. (Eugene, OR,

[2] G. McFeters, F. Yu, B. Pyle, and P. Stewart, *J. Microbiol. Methods* **21,** 1 (1995).

[3] B. Pyle, S. Broadaway, and G. McFeters, *Appl. Environ. Microbiol.* **61,** 2614 (1995).

[4] L. Poulsen, G. Ballard, and D. Stahl, *Appl. Environ. Microbiol.* **59,** 1354 (1993).

[5] S. Williams, Y. Hong, D. Danavall, M. Howard-Jones, D. Gibson, M. Frischer, and P. Verity, *J. Microbiol. Methods* **32,** 225 (1998).

[6] C. Buswell, Y. Herlihy, L. Lawrence, J. McGuiggan, P. Marsh, C. Keevil, and S. Leach, *Appl. Environ. Microbiol.* **64,** 733 (1998).

[7] R. Haugland, "Handbook of Fluorescent Probes and Research Chemicals." Molecular Probes, Eugene, OR, 1996.

[8] J. Slavik, "Fluorescent Probes in Cellular and Molecular Biology." CRC Press, Boca Raton, FL, 1994.

[9] P. Stewart, *Biotech. Bioeng.* **59,** 261 (1998).

TABLE I
CHARACTERISTICS TO CONSIDER WHEN SELECTING LABELS OR PROBES

Label or probe characteristic	Contributing factors	Effect on label or probe[a]
Preparation solution	pH, water, ethanol, methanol, Tris–EDTA, dimethyl sulfoxide, growth medium	±
Excitation and emission	Large Stokes shift	+
	Quenching	−
	pH	±
Permeability and retention in biofilms	Thick biofilms (> monolayers)	−
	Thin biofilms (monolayers)	+
	Cationic, anionic or neutral charge	±
	Hydrophilic or hydrophobic	±
	pH	±
	Molecular weight (effective diffusion coefficient)	±
Permeability and retention within cells	Active or facilitated diffusion	+
	Uni- or bidirectional active transport	±
	Acetylated modifications	+
	Cationic, anionic, or neutral charge	±
	Hydrophilic or hydrophobic	±
	pH	±
	Molecular weight	±

[a] +, increases label or probe efficiency; −, decreases label or probe efficiency; and ±, may increase or decrease a label or the efficiency of a probe dependent on local conditions.

http://www.probes.com) offers an extensive selection of labels, probes, and technical information for these types of applications.

Biofilm Recovery and Microscopic Visualization

Cryosectioning and Fluorescent Microscopic Visualization

The visualization of monolayer biofilms is accomplished easily by either light or fluorescence microscopy. However, due to the resolution limits of optical microscopy, studies on thicker biofilms require their removal from the substratum prior to visualization. Cryosectioning effectively removes biofilms from substrata, while retaining their complex structures.[10,11] This technique allows the visualization of biofilm cross sections using fluorescent microscopy. A similar technique has been developed for sectioning sludge granules.[12]

1. Biofilms should be grown on a substratum that will be able to withstand exposure to dry ice for 20 min. A commonly used substratum is 316L stainless steel (17 × 75 mm). A larger size can be used, but one must be able to manipulate the substratum easily without physically disrupting the biofilm or inhibiting the freezing of the specimen.
2. Transfer the biofilm sample slide to a dry ice slab, biofilm side up.
3. Immediately dispense a thick layer of Tissue-Tek OCT compound (Miles, Inc., Diagnostics Division, Elkhart, IN) on top of the biofilm. Allow the sample to freeze until the entire sample turns opaque white.
4. Gently bend the substratum to pop off the frozen sample and immediately replace the embedded biofilm sample on the dry ice with the surface that was attached to the substratum pointed up.
5. Dispense a thick layer of OCT compound on this surface of the biofilm sample and allow to freeze until the sample turns opaque white. Label the substratum side of the sample with a permanent marker.
6. Wrap the embedded biofilm sample in aluminum foil and store at −70°.
7. Mount the frozen biofilm sample on a precooled specimen chuck and then slice sections (5 μm thick) using a cryostat operated at −20°. Some trimming may be required to obtain a smooth surface, prior to collecting sections for visualization. Collect the biofilm sections onto polylysine-coated slides, such as Superfrost Plus slide (Fisher Scientific, Pittsburgh, PA).
8. The biofilm sections can then be visualized directly using fluorescence microscopy. If the addition of a coverslip is required, the mounting medium should be evaluated as to its ability to quench, dissolve, or release the label or probe.

[10] F. Yu, G. Callis, P. Stewart, T. Griebe, and G. McFeters, *Biofouling* **8,** 85 (1994).

[11] C. Huang, P. Stewart, and G. McFeters, The study of microbial biofilms by classical fluorescence microscopy, pp. 411–429. *in* "Digital Image Analysis of Microbes: Imaging, Morphometry, Fluorometry and Motility Techniques and Applications" (M. Wilkins and F. Schut, eds.), p. 411. Wiley, Chichester, UK, 1998.

[12] H. Harmsen, H. Kengen, A. Akkermans, A. Stams, and W. de Vos, *Appl. Environ. Microbiol.* **62,** 1656 (1996).

Confocal Laser Scanning Microscopy

Confocal laser scanning microscopy (CLSM) allows direct, nondestructive, and three-dimensional visualization of biofilm structure. When cells residing within these biofilms have been labeled or probed with fluorescent dyes, information on physiological activity at the intracellular and microniche level can also be ascertained. Several of the techniques described in the following sections have used CLSM to visualize the respective biofilm systems. Fluorescent microscopy can be used with these systems as well, but the biofilms will have to be embedded and cryosectioned.

It is not the objective of this article to discuss the application of CLSM to the visualization of biofilms. Lawrence *et al.*[13] have published an excellent introduction and review on using CLSM to study biofilms (see chapter 9 of this volume). Also, a compilation of other methods to visualize microbes has been recently published.[14] Questions and information about the use of CLSM and image analysis for the study of biofilms can be submitted to the Center for Biofilm Engineering at Montana State University at http://www.erc.montana.edu.

Autofluorescence Quenching

Autofluorescence from bacteria, algae, and abiotic materials (e.g., clay, soil) has emission spectra very similar to one or more of the commercially available labels and probes. Therefore, differentiation between labeled or probed bacterial cells and autofluorescing cells or debris may be practically impossible. Selecting the correct combination of microscope filters can partially reduce autofluorescence, especially if the autofluorescence occurs within a narrow wavelength. As an alternative, quenching solutions can be applied to the biofilms. Slavik[8] lists commonly used quenchers for types of labels and probes that have been used with planktonic cells, but does not address autofluorescence. However, some of the quenchers may be applicable to biofilm systems.

Huang *et al.*[15] evaluated sodium borohydride (50 m*M* after ethanol dehydration), Evans blue [0.5% (w/v) in phosphate-buffered saline and 0.1 *M* sodium azide], and crystal violet (2 mg/ml) as autofluorescence quenchers in industrial biofilm systems. These quenchers shifted the autofluorescence

[13] J. Lawrence, G. Wolfaardt, and T. Neu, The study of biofilms using confocal laser scanning microscopy, pp. 431–465. *In* "Digital Image Analysis of Microbes: Imaging, Morphometry, Fluorometry and Motility Techniques and Applications" (M. Wilkins and F. Schut, eds.), p. 431. Wiley, Chichester, UK, 1998.

[14] M. Wilkinson and F. Schut, eds., "Digital Image Analysis of Microbes: Imaging, Morphometry, Fluorometry and Motility Techniques and Applications." Wiley, Chichester, UK, 1998.

[15] C. Huang, G. McFeters, and P. Stewart, *Biofouling* **9,** 269 (1996).

emission spectra to green, bright red, and dark red, respectively. The selection of an appropriate quencher is dependent on emission spectra of the label or probe and quencher so as to optimize contrast. The use of quenchers in conjunction with optimized microscope filter combinations may minimize autofluorescence effectively.

Labeling and Probing Bacterial Biofilms

Total Cell Counts Using 4′,6-Diamidino-2-phenylindole

DAPI is a DNA-specific probe that forms a fluorescent complex when bound in the minor groove of A-T-rich sequences.[16] The following version of this technique describes the staining of a bacterial biofilm with DAPI for total cell counts within the biofilm and a two-dimensional view of its structure.

A binary biofilm of *Klebsiella pneumoniae* and *Pseudomonas aeruginosa* was grown in a continuous flow annular reactor containing 316L stainless-steel slides. The biofilm growth medium was minimal salts medium supplemented with 20 mg/liter glucose. Biofilm thickness ranged from 50 to 100 μm. A similar approach has been used to label bacteria in cooling tower water biofilms that were monolayers to several cells thick.[17]

1. Remove the biofilm sample from the growth medium and place it in a staining container, biofilm side up, containing formalin [5% (v/v), final concentration]. Incubate at room temperature for 5 min to fix the cells.
2. Immediately add DAPI (1 μg/ml, final concentration) and incubate at room temperature for 5 min.
3. Gently remove the solution from the staining container and transfer the biofilm sample to a dry ice slab, biofilm side up, and process for cryosectioning as described earlier.
4. Cells that have retained DAPI will fluoresce blue.

The fixation step may not be compatible with all labels and probes, as some fixatives may quench, dissolve, or release the intracellular label or probe. Controls to assess the compatibility of labels, probes, and fixatives should be performed on planktonic cells prior to their use in biofilm systems.

DAPI is used routinely for total cell counts in planktonic and biofilm systems. Additionally, it has also been used to indirectly determine viability based on the presence of an intact genome or nucleoid. Studies have shown

[16] J. Kapuscinski, *Biotech. Histochem.* **70,** 220 (1995).

[17] G. Wolfaardt, R. Archibald, and T. Cloete, *Biofouling* **4,** 265 (1991).

that labeling cells with DAPI may overestimate the number of nucleoid-containing bacteria due to nonspecific binding; this can be removed by gently rinsing.[18–21]

Physiological Activity in Biofilms

To date, there has been limited use of fluorescent labels and probes for the assessment of bacterial physiology within biofilms. The techniques in this section are not necessarily restricted in their application to the described biofilm systems. They should be used as general guidelines for the respective application of the labels and probes to any biofilm system. As discussed previously, the characteristics described in Table I should always be considered.

The following techniques represent a collection of methods for labeling and probing different types of biofilms. Each technique is introduced by a brief description of the conditions under which the biofilms were grown and their thicknesses. Also, references to using fluorescent or confocal laser scanning microscopy are included as described in the referenced literature. However, depending on the desired result, either technique may be used with any of the following labeling and probing techniques.

Assessing General Bacterial Activity.[11,22] *Klebsiella pneumoniae* biofilms were grown in a continuous flow annular reactor using minimal salts medium supplemented with 40 mg/liter glucose as the growth medium. Biofilms were grown on 316L stainless-steel slides at room temperature for 7 days. The biofilm thickness ranged from 50 to 150 μm.

1. Embed and cryosection the biofilm samples as described earlier.
2. Fix cryosections of biofilm in an acidic fixative [5 ml formalin, 2.5 ml glacial acetic acid, 42.5 ml 95% (v/v) ethanol] for 10 min at 4°.
3. Rinse the sections twice with 85% ethanol at 4° and allow to air dry.
4. Add 5 μl of 4 μg/ml acridine orange solution on each section and incubate for 1 min at room temperature in the dark.
5. Gently remove the excess acridine orange with a paper towel or pipe.
6. View using fluorescence microscopy. Cells within the biofilm that fluoresce orange are presumed to have an elevated RNA content and high activity. Cells that fluoresce green are presumed to have a reduced RNA content and low activity.

[18] U. Zweifel and A. Hagstrom, *Appl. Environ. Microbiol.* **61,** 2180 (1995).
[19] J. Choi, E. Sherr, and B. Sherr, *Limnol. Oceanogr.* **41,** 1161 (1996).
[20] M. Suzuki, E. Sherr, and B. Sherr, *Limnol. Oceanogr.* **38,** 1566 (1993).
[21] J. Vosjan and G. van Noort, *Aquat. Microb. Ecol.* **14,** 149 (1998).
[22] E. Wentland, P. Stewart, C. Huang, and G. McFeters, *Biotechnol. Prog.* **12,** 316 (1996).

Assessing Membrane Integrity Using LIVE/DEAD BacLight Viability Kit in Laboratory Biofilms.[7,23] The LIVE/DEAD *Bac*Light Viability kit (Molecular Probes, Inc.) uses the principle of dye exclusion to determine the integrity status of the bacterial membrane. The first label, SYTO 9, is a membrane-permeant, DNA-labeling dye that labels all cells and fluoresces green. The second label is propidium iodide, which also labels DNA, but is excluded from cells with intact membranes. Propidium iodide is able to enter cells with compromised membranes and competes with and quenches SYTO 9, making such cells fluoresce red.

Salmonella enteritidis was grown as a batch culture to mid exponential phase in tryptic soy broth (10%, v/v) at 21° and then used to inoculate modified flow cells. The dimensions of the chamber were 1.3 × 5 × 80 mm with a volume of approximately 0.5 ml. Biofilms were grown within the modified flow cells on glass coverslips for 72 hr at 21°. The biofilm thickness ranged from 3 to 9 μm.

1. Prepare a solution of the LIVE/DEAD *Bac*Light viability assay as described by the manufacturer.
2. Turn the flow to the flow cell off and add approximately 0.3 ml of the LIVE/DEAD *Bac*Light solution to the flow chamber.
3. Incubate for 15 min at 21° in the dark.
4. Reinitiate the flow to the flow cell to wash the unbound label from the system.
5. View the labeled biofilms using CLSM. Cells that fluoresce green have intact cell membranes, whereas cells that fluoresce red have compromised cell membranes.

Use of the LIVE/DEAD *Bac*Light viability assay has been used by numerous researchers to assess the status of membrane integrity in bacterial systems. There has been some question as to its ability to penetrate bacterial exopolymers.[24] However, Korber *et al.*[25] have shown that penetration of the labels is not inhibited in biofilms of *S. enteritidis.* Additionally, some cells may demonstrate both green and red fluorescence.[26,27] This dual fluo-

[23] D. Korber, A. Choi, G. Wolfaardt, S. Ingham, and D. Caldwell, *Appl. Environ. Microbiol.* **63,** 3352 (1997).

[24] D. Caldwell, G. Wolfaardt, D. Korber, and J. Lawrence, *Adv. Microb. Ecol.* **15,** 1 (1997).

[25] D. Korber, A. Choi, G. Wolfaardt, and D. Caldwell, *Appl. Environ. Microbiol.* **62,** 3939 (1996).

[26] S. Terzieva, J. Donnelly, V. Ulevicius, S. Grinshpun, K. Willeke, G. Stelma, and K. Brenner, *Appl. Environ. Microbiol.* **62,** 2264 (1996).

[27] J. Lisle, B. Pyle, and G. McFeters, *Let. Appl. Microbiol.* (1999).

rescence has not been addressed in the technical information that is supplied by the manufacturer, but has been discussed elsewhere.[27,28]

Assessing Respiratory Activity Using 5-Cyano-2,3-ditolyltetrazolium chloride (CTC) in Dechlorinated Drinking Water Biofilms.[29] CTC is a soluble nonfluorescent tetrazolium salt that forms intracellular, insoluble, red fluorescent formazan crystals on reduction.[30] The presence of intracellular CTC–formazan crystals indicates that the cell has an active respiratory system, as succinate dehydrogenase and NADPH dehydrogenase have been shown to be responsible for its reduction.

Treated drinking water was fed into a polymethylene methacrylate (PMMA) flow cell that contained removable coupons. The coupons were made of stainless-steel, copper, high-density polyethylene and PMMA. The dimensions of the chamber were 217 × 100 × 65 mm or approximately 1.4 liters. Biofilms of the uncharacterized natural population water were grown for 7 days at 12°. Biofilms were heterogeneous in their spatial distribution and were one to several cells thick.

1. Remove the coupons from the flow cell, submerge in a mixture of R2A medium (50%, v/v) and CTC (final concentration, 5.0 m*M*), and incubate for 1 hr at room temperature.
2. Gently rinse the coupons and air dry.
3. Counterstain the cells by submerging the coupon in DAPI (2.0 μg/ml) for 20 min at room temperature.
4. Gently remove the excess DAPI with a paper towel or pipette.
5. Using fluorescence microscopy, cells in the CTC/DAPI-probed biofilm will fluoresce red if they have reduced CTC, within an intracellular background blue fluorescence.

CTC has been the most frequently used fluorescent probe to assess physiological activity in biofilms. However, several factors should be considered prior to its use and interpretation of the results. Several studies have addressed the effects of nutrient addition or removal on the efficiency of CTC reduction.[31–36] Additionally, inorganic constituents (i.e., phosphates)

[28] P. Millard and B. Roth, *Biotech. Int.* **1,** 291 (1997).

[29] G. Schaule, H. Flemming, and H. Ridgeway, *Appl. Environ. Microbiol.* **59,** 3850 (1993).

[30] J. Smith and G. McFeters, *J. Microbiol. Methods* **29,** 161 (1997).

[31] G. Rodriguez, D. Phipps, K. Ishiguro, and H. Ridgway, *Appl. Environ. Microbiol.* **58,** 1801 (1992).

[32] L. Gribbon and M. Barer, *Appl. Environ. Microbiol.* **61,** 3379 (1995).

[33] A. Braux, J. Minet, Z. Tamanai-Shacoori, G. Riou, and M. Cormier, *J. Microbiol. Methods* **31,** 1 (1997).

[34] F. Joux, P. Lebaron, and M. Troussellier, *FEMS Microbiol. Ecol.* **22,** 65 (1997).

[35] J. Coallier, M. Prévost, and A. Rompré, *Can. J. Microbiol.* **40,** 830 (1994).

[36] M. Prévost, A. Rompré, J. Coallier, P. Servais, P. Laurent, B. Clément, and P. LaFrance, *Water Res.* **32,** 1393 (1998).

and pH values >6.8 have been shown to affect CTC reduction negatively.[30,37,38] CTC has also been found to be toxic at concentrations of ≥5.0 μM.[39] All of these studies assessed CTC reduction using planktonic cells.

Assessing Respiratory Activity Using CTC in Industrial Biofilms.[15] Biofilms were grown under the following conditions: (a) simulated cooling tower water (mixed culture of bacteria, algae, and 5 g of fresh soil in dechlorinated tap water), bentonite/kaolinite (50 mg/liter) was added as a slurry after 1–2 months of biofilm growth, and polyvinyl chloride (PVC) was used as the biofilm growth substratum. The biofilms were grown for approximately 13 months at 30° and pH 8.0. Biofilm thickness ranged from 2 to 5 mm. (b) Artificial seawater was fed continuously into an outdoor recirculating cooling tower. Biofilms were grown for 2 months on silicone tubing placed in the effluent line at 25–30°and pH 8.0. Biofilm thickness was less than 1 mm. (c) Biofilms from paper mill water (wood fibers, alum, rosin size, starch, and a polymer retention aid) were grown on the clear leg wall of a flotation device in an acid fine paper machine. Biofilms were grown for 3–4 weeks at 43–49° and pH 4.9. Biofilms were removed with a razor blade by gently scraping and lifting the biofilms from the substratum and immediately placing them on stainless-steel slides. Biofilm thickness ranged from 3 to 5 mm. (d) Polycarbonate coupons were placed in a treated wastewater effluent ditch. Biofilms were grown for 6 months at 5–25° and pH 7.0. Biofilm thickness ranged from 1 to 3 mm.

1. Collect biofilm coupons, slides, or samples and submerge in CTC (final concentration, 1.3 mM) for 2 hr at room temperature.
2. Fix the biofilms by submerging the coupon, slide, or sample in formaldehyde (5% v/v) for 5 min at room temperature.
3. Embed and cryosection the fixed biofilm as described earlier and view under fluorescence microscopy.

Thicker biofilms may be difficult to section due to abiotic materials adsorbed to and embedded within the biofilm matrices. Also, the presence of autofluorescing bacteria and algae may make the assessment of intracellular CTC reduction difficult. Autofluorescence will increase as the thickness of the biofilm sections increases. Methods to reduce autofluorescence in biofilms have been described previously in this article.

Assessing Alkaline Phosphatase Activity in Phosphate-Starved Biofilms.[40] Biofilms of *K. pneumoniae or P. aeruginosa* were grown in a defined minimal medium with 1 g/liter of Na_2HPO_4 and 0.1 g/liter of glucose for

[37] B. Pyle, S. Broadaway, and G. McFeters, *Appl. Environ. Microbiol.* **61,** 4304 (1995).

[38] J. Smith and G. McFeters, *J. Appl. Bacteriol.* **80,** 209 (1996).

[39] S. Ullrich, B. Karrasch, H. Hoppe, K. Jeskulke, and M. Mehrens, *Appl. Environ. Microbiol.* **62,** 4587 (1996).

[40] C. Huang, K. Xu, G. McFeters, and P. Stewart, *Appl. Environ. Microbiol.* **64,** 1526 (1998).

96 hr at room temperature on 316L stainless-steel slides. Prior to the alkaline phosphatase assay, the phosphate concentration was lowered to 0.01 g/liter of Na_2HPO_4. Samples were collected prior to and after 8, 12, 24, and 36 hr of exposure to the low phosphate medium. The biofilm thicknesses ranged from 60 to 110 μm. A similar approach has been used to assess the effect of oxygen availability on the spatial heterogeneity in *P. aeruginosa* biofilms.[41]

1. Prepare the ELF-97 phosphatase substrate solution (Molecular Probes, Inc.) as described by the manufacturer.
2. Submerge the biofilm sample slide in the ELF-97 solution for 30 min at 37° in the dark.
3. To counterstain the sample, submerge the slide in propidium iodide (10 mg/liter) (Molecular Probes, Inc.) for 5 min at room temperature.
4. Fix the labeled biofilm sample with formaldehyde (1% v/v) for 5 min at room temperature.
5. Embed and cryosection the fixed biofilm sample as described earlier and view under fluorescence microscopy.
6. Areas of the biofilm that fluoresce yellow–green have active alkaline phosphatase activity. If the sample has been counterstained, cells in areas with no alkaline phosphatase activity will fluoresce red.

Assessing Respiratory Activity Using CTC and Membrane Potential Using Rhodamine 123 (Rh123) in Disinfected Biofilms.[42] Rh123 is an anionic, lipophilic, fluorescent probe that is distributed across membranes having a membrane potential.[7,8] Cells that have lost their membrane potential are unable to distribute and accumulate Rh123 and, as a result, are nonfluorescent.

Klebsiella pneumoniae cultures were grown in tryptic soy broth (TSB) (10% v/v) for 24 hr at 35°. A subsample of this culture (1%) was used to inoculate another TSB (10% v/v) culture and was incubated for 24 hr at 35° before being used as the inoculum for the biofilm reactor. The biofilms were grown in batch reactors on 316L stainless-steel slides using TSB (10% v/v) as the growth medium. Biofilms were grown for 36 hr at 25°. The thickness of the biofilm was not given. However, the probed samples were assessed directly using fluorescence microscopy, suggesting that the biofilms were monolayers or only a few cells thick prior to disinfection. This probing technique requires at least three biofilm sample slides for probing with CTC and Rh123 and total cell counts (DAPI).

[41] K. Xu, P. Stewart, F. Xia, C. Huang, and G. McFeters, *Appl. Environ. Microbiol.* **64,** 4035 (1998).

[42] F. Yu and G. McFeters, *Appl. Environ. Microbiol.* **60,** 2462 (1994).

A similar approach has been used with a *K. pneumoniae* and *P. aeruginosa* biofilm system.[43] In this study, biofilms were grown in a continuous flow annular reactor using a defined minimal medium supplemented with 20 mg/liter glucose. Biofilms were grown for 7–10 days at room temperature and ranged in thickness from 60 to 125 μm prior to disinfection. These biofilms were cryosectioned.

1. Gently rinse biofilm sample slides in sterile distilled water and then transfer to an acid-washed batch reactor containing chlorine demand-free phosphate-buffered water and a magnetic stirring bar.
2. Add the disinfectants, chlorine (0.25 mg/liter, pH 7.2) or monochloramine (1.0 mg/liter, pH 9.0), to the stirred reactors. Record the temperature of the reaction system.
3. After a predetermined exposure time, remove the biofilm samples and submerge them in sodium thiosulfate [final concentration, 0.01% (w/v)] to neutralize the disinfectant.
4. For CTC probing, gently rinse the neutralized biofilm sample and then submerge it in CTC (1.3 m*M*) for 2 hr at 35°.
5. For Rh123 probing, gently rinse the neutralized biofilm sample and submerge in Rh123 [final concentration, 5 μg/ml in permeabilizing solution (50 m*M* Tris–HCl, 5 m*M* disodium EDTA, pH 8.0)] for 2 hr at 35°.
6. For total cell counts, gently rinse the neutralized biofilm sample and submerge it in DAPI and incubate as described earlier. DAPI can be used as a counterstain with CTC-probed biofilms. However, DAPI is not compatible with Rh123 and cannot be used as a counterstain. A separate biofilm sample must be labeled with DAPI, counted, and then compared to Rh123 data.
7. Using fluorescence microscopy, cells will fluoresce red within an intracellular background blue fluorescence that has reduced CTC. Cells that have not reduced CTC will fluoresce blue. Cells that have accumulated Rh123 will fluoresce green.

Interpretation of Results

The term "viability" is commonly used to describe a wide range of physiological states and activities, as measured by a variety of methods in planktonic and biofilm systems. It is not surprising, therefore, that microbiologists frequently fail to agree on a unified definition for bacterial viability

[43] C. Huang, F. Yu, G. McFeters, and P. Stewart, *Appl. Environ. Microbiol.* **61,** 2252 (1995).

other than demonstrable growth.[44–46] Hence, the interpretation and comparison of results from studies assessing viability is problematic. A contributing factor to this problem is the incorrect assumption that data from methods assessing different aspects of bacterial physiology (e.g., culturability and membrane potential) are equivalent. Often a single aspect of physiology is measured and conclusions are drawn as to the overall viability of the cell. A more sensible approach is to assess multiple indices of physiological activity, such as respiratory enzyme activity, membrane integrity, and membrane potential. This approach allows a more comprehensive assessment of which aspects of cellular physiology are affected by a given environmental stressor or treatment.

The application of fluorescent labels and probes to the assessment of physiological activity makes the multiple indices approach practical because of their sensitivity and ability to provide rapid results. Also, fluorescent labels and probes have the potential to provide information pertaining to the intracellular site and degree of lethal and sublethal injury,[27,47] unlike culture-based techniques.

[44] D. Roszak and R. Colwell, *Microbiol. Rev.* **51,** 365 (1987).
[45] D. Kell, A. Kaprelyants, D. Weichart, C. Harwood, and M. Barer, *Antonie van Leeuwenhoek* **73,** 169 (1998).
[46] D. Lloyd and A. Hayes, *FEMS Microbiol. Lett.* **133,** 1 (1995).
[47] J. Lisle, B. Pyle, and G. McFeters, *Appl. Environ. Microbiol.* **64,** 4658 (1998).

[14] Deconvolution Fluorescence Microscopy for Observation and Analysis of Membrane Biofilm Architecture

By DON PHIPPS, GRISEL RODRIGUEZ, and HARRY RIDGWAY

Deconvolution Fluorescence Microscopy for Observation and Analysis of Membrane Biofilms

The ability of nutrients, biocidal agents, cleaning compounds, and other substances to penetrate sensitive targets within the biofilm matrix is controlled by biofilm composition and architecture. Consequently, detailed knowledge of the microscale morphology of membrane biofilms is essential for understanding the mechanisms of water and solute transport phenomena in separation membranes. Using scanning laser confocal microscopy,

0076-6879/99 $30.00

Stoodley *et al.*[1] and DeBeer *et al.*[2] demonstrated fluid channels in biofilms that facilitate the diffusional transport of solutes and colloids within the biofilm. This article provides a detailed description of a unique microscope system that permits both two-dimensional (2D) and three-dimensional (3D) viewing of membrane biofilms from any desired perspective. This system, which is outlined schematically in Fig. 1 and depicted in Fig. 2, combines advanced optical microscope imaging and deconvolution techniques with robotic sampling and digital image processing methods to create virtual models of microbial biofilms on water processing membranes. The theory and application of the deconvolution techniques employed have been reviewed by Shaw.[3] Several 3D image stacks may be acquired under different imaging (filtration) conditions and combined to produce a color-coded volume that simultaneously maps multiple biofilm features (e.g., dual or triple dye staining). The resulting models may be used to investigate specific biofilm features such as the distribution of specific microorganisms and extracellular polysaccharides, the nature of biofilm fluid channels, or the penetration of biocides or other molecules into the biofilm matrix (e.g., Fig. 3, see color insert).

Epifluorescence Light Microscopes and Objectives

Biofilm specimens are imaged using either an Olympus IX-70 inverted microscope or an Olympus AX-70 upright microscope. These microscopes feature infinity corrected optics and are each equipped with three epifluorescent filter cubes: a wide UV (U-MWU, excitation 330–385 nm, dichroic cutoff 400 nm, emission >420 nm), a wide blue (U-MWB, excitation 450–480 nm, dichroic cutoff 500 nm, emission >515 nm), and a wide green (U-MWG, excitation 510–550 nm, dichroic cutoff 570 nm, emission >590 nm). Microscopes and filter cubes are stock configurations and are readily available from Olympus (Olympus America, Inc., Melville, NY). Both microscopes are also equipped with a UPlanApo 60× 1.2 numerical aperature (NA)W PSF objective (Olympus). This special water-immersion objective is corrected for cover glass thickness and is designed to produce nearly aberration-free images over its entire working distance through aqueous media. It is suitable for work with thin window flow cells (see later). In addition, planapochromatic 40× air-and 100× oil-immersion UV-transmission objectives are also employed occasionally (Olympus).

[1] P. Stoodley, D. DeBeer, and Z. Lewandowski, *Appl. Environ. Microbiol.* **60,** 2711 (1994).

[2] D. DeBeer, P. Stoodley, and Z. Lewandowski, *Biotechnol. Bioeng.* **53,** 151 (1997).

[3] P. J. Shaw, *in* "Handbook of Biological Confocal Microscopy" (J. B. Pawley, ed.), p. 373. Plenum Press, New York, 1995.

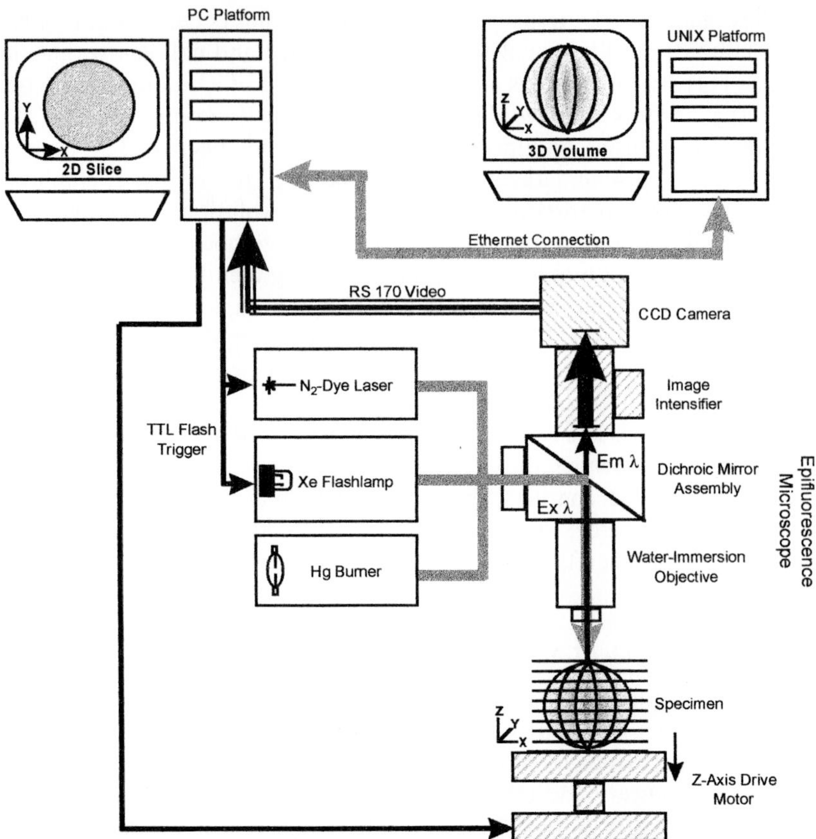

FIG. 1. Schematic of a digital deconvolution microscope and 3D imaging system for investigating membrane biofilm structure. The PC platform coordinates the cooled camera, xenon flash/N-laser systems, and automated (z-axis) stage mechanism to collect 2D image stacks. Basic 2D image processing operations include background correction, deconvolution, and contrast/brightness adjustments. The UNIX platform combines 2D image stacks into 3D volumes. Volume processing operations include 3D viewing of one to three overlaid volumes, volume color assignments and false color rendering, morphometric measurements and volume dissections, and movie loop productions. See text for details.

Fluorescence Excitation Sources

One of two discontinuous (flash) sources are typically utilized for imaging. The first source is a 2600 series EG&G flashlamp kit modified to fit an Olympus microscope condenser lamp housing and equipped with a FX-249-U short-arc xenon flashlamp (EG&G Electro-Optics, Salem, MA). This system provides a broadband source of white light when driven with

FIG. 2. Main components of a digital deconvolution microscope system: inverted microscope (a), mercury burner (b), xenon flash units, (c) nitrogen laser-pumped pulsed dye laser with optical fiber couplings (d), cooled CCD camera (e), image intensifier (f), Ludl z-axis stage drive mechanism (g), live image monitor (h), video intensity analyzer (i), upright microscope system (j), vibration dampening tables (k), gain control for intensifier (l), PC platform for 2D image capture, image manipulation, and z-axis sectioning control (m), and UNIX platform for 3D volume reconstruction, volume morphometry, and dissections (n). See text for details.

1 J of electrical energy per pulse. Manufacturer's specifications indicate that the overall flash duration from this source is 100 μsec; however, peak illumination intensity occurs at 15 μsec and the majority of the energy is emitted between 10 and 30 μsec. Overall light energy output is rated at 1000–2000 μJ/pulse emitted over a spectral range from 280 to 1100 nm. Because high-pressure xenon arcs emit almost continuous spectra from 300 to 800 nm, the excitation properties of this source are essentially that of the particular filter cube being used. The flash energy is collimated and directed into the microscope using a quartz condenser assembly (Opti Quip, Highland Mills, NY).

The second excitation system consists of a nitrogen laser-pumped dye laser (Oriel Model 79111 low-pressure N_2 laser pumping an Oriel Model 79120 dye laser module; Oriel Instruments, Stratford, CT). This system may be used to provide a source of monochromatic pulsed light for fluorochrome excitation. Either the 337-nm nitrogen laser line or the tunable dye laser output (350–750 $\pm$ 0.4 nm) is directed into the microscope via a 600-μm-diameter single quartz fiber optic and quartz microscope coupler (Laser Sciences, Inc., Franklin, MA). Laser transmission through such large single fibers is characterized by uneven light output as a result of constructive and destructive interference between the multiple transmission modes of the fiber. Transmission mode "scrambling" is achieved by providing two sharp bends in the fiber using a small pair of clamps set at right angles to each other near the microscope adapter. This treatment produces an evenly illuminated circular output beam of monochromatic light from the fiber. The nitrogen laser pump produces >300 μJ/pulse with a flash duration of

5 nsec. Dye laser energy output varies, depending on the coupling efficiency of the dye, but according to the manufacturer, outputs should range from about 25 to 63 μJ/pulse at 4 nsec/pulse. Dyes for laser operation are obtained premixed from Laser Science, Inc. (Franklin, MA). Laser excitation with the UV filter cube is achieved by either using the nitrogen laser output directly or using the dye laser with BBQ (4,4‴-bis[(2-butyloctyl)oxy]-1,1′:4′,1″:4″,1‴-quaterphenyl, 1 m*M* in *p*-dioxane; lasing range: 380 to 395 nm). Excitation with the blue and green filter cubes is achieved using Coumarin 500 [7-(ethylamino)-4-(trifluoromethyl)-2*H*-1-benzopyran-2-one, 9.33 m*M* in methanol; lasing range: 485 to 570 nm]. Dyes are stored in the dark at room temperature in completely filled and sealed individual 1 × 1 cm quartz cuvettes. A magnetic "flea" stir bar in the cuvettes driven during laser operation by a magnetic stirrer mounted below the cuvette holder provides sufficient dye circulation for a stable laser output. Attenuation of laser excitation energy is accomplished when desired by tuning the dye laser toward the upper or lower pass band limits of the particular filter cube in use.

This discontinuous excitation source allows extremely rapid micro- to nanosecond image acquisition. Thus the specimen is exposed to excitation radiation for the minimum time required to obtain an image. Therefore, fluorochrome fading is minimized without resorting to adulteration of the biofilm with antioxidant compounds. In addition, the extreme speeds of exposure render image acquisition nearly insensitive to vibration error so that vibration isolation of the microscope is unnecessary. It is indeed possible to obtain 2D images of highly mobile specimens using these flash sources, such as bacteria carried with the bulk flow in a flow cell or fluorescent beads suspended in thin detergent films.

The stock 100-W mercury lamp provided with the Olympus microscopes produces continuous wave (CW) excitation, and this source is used when fluorochrome fading is not significant or when it is desirable to make visual observations of the specimen to quickly check results of a labeling protocol or to aid in locating an area of interest (AOI). A three-way mirror housing (Opti Quip) fitted on each microscope allows selection of any one of the three excitation sources prior to imaging.

Image Acquisition

Microscope adjustments are made to properly accommodate the particular fluorochrome being imaged (choice of filter cube, light source, etc.). The biofilm specimen, either contained in a flow cell (see later) or prepared on a cover glass or coupon, is placed under the microscope, and the fluores-

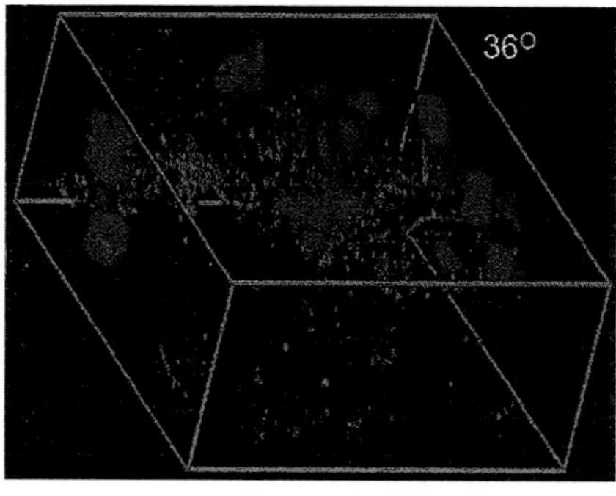

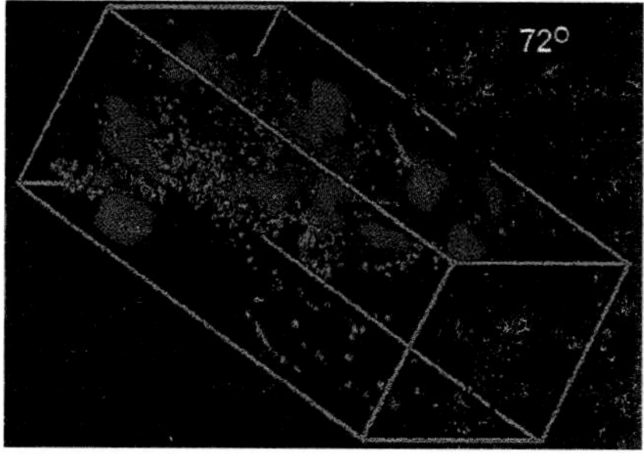

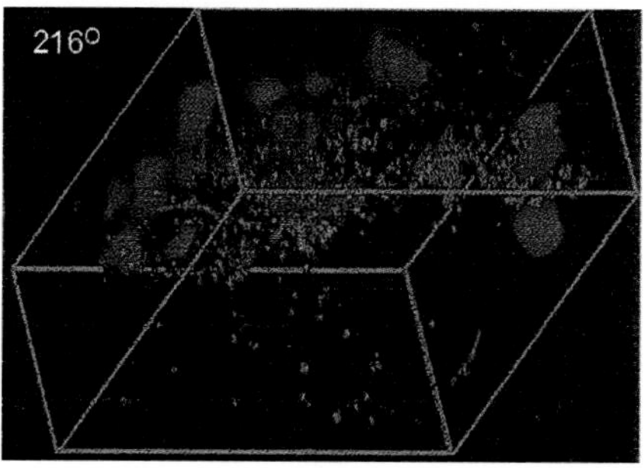

FIG. 3. Three perspective views of a 48-hr-old native multispecies membrane biofilm reconstructed in 3D from 150 2D image sections. Red objects: individual bacteria stained with propidium iodide. Blue material: extracellular polymeric substances (EPS) stained differentially with Calcofluor white. Images were acquired and processed using the digital deconvolution microscope system described in the text. The biofilm was developed on a CA-coated cover glass coupon in a continuous-flow microscope flow cell fed with secondary municipal effluent at Water Factory 21, an advanced water reclamation facility located in southern California.

cence emission image from the specimen is detected and amplified using a modified (manual gain control added) third generation microchannel plate-type image intensifier providing 46 line pairs/mm spatial resolution and a maximum of 30,000× luminous gain (B. E. Meyers, Redmond, WA). The resulting intensified image is recorded with a cooled (−30° below ambient), optically coupled HyperHAD $\frac{1}{2}$-inch interline transfer charge-coupled device (CCD) video camera (Dage IFG-300, Dage MTI, Inc., Michigan City, IN) acquiring at a standard video frame rate (30 frames/sec). Resulting images are inspected in real time with an in-line video monitor (Sony Trinitron PVM-1342Q, Sony Corp.). Real-time image intensity histogram data are visualized during image optimization procedures using an on-screen display analyzer (RasterScope, Dage MTI, Inc.).

Optimization of the camera and intensifier is accomplished by first setting the camera gain to maximum (+12-dB gain) and, with the intensifier off, increasing the camera black level to eliminate dark noise. Following this, the image intensifier is activated and the intensifier gain is increased to achieve video signal saturation of the brightest image highlights. This method produces a well-contrasted image providing the full video signal range; however, adjustment of the camera black level and the intensifier gain beyond this point is sometimes required for badly contrasted specimens. In addition, when the laser excitation source is used, the laser excitation wavelength is also adjusted in coordination with the intensifier gain to provide the highest image contrast.

Image Digitization and Flash Control

The resultant analog (RS 170) video signal from the camera is digitized to 640 × 480 rectangular pixels at 8 bits (256 gray levels) using a 266 Mhz PC clone platform equipped with a FlashPoint 128 frame grabber (Integral Technologies, Indianapolis, IN). This board includes an output trigger and hardware/firmware to allow the use of a flash source to illuminate a video image. The board achieves flash synchronization using the video vertical sync portion of the RS-170 signal, and a user-selectable field delay allows time for the camera to process its illuminated fields before frame grabbing occurs (typically a one field delay was sufficient). Both flash sources require a TTL trigger signal (5-μsec square wave pulse, 3 to 5 V), which is provided by connecting a 3-V battery source in series with the closure switch output trigger of the frame grabber board. Programming allows the continuous display of images for composition at two to three frames/sec; in addition, the laser provides an integral 30-Hz pulse generator that is used for real-time image visualization. Captured images are saved in standard tagged image file format (TIFF).

Calibration of Two-Dimensional Image X–Y Axes

For a given optical configuration, pixel dimensions (*dx* and *dy*) of the 2D images are determined by imaging an optical micrometer consisting of precise parallel rulings spaced at 10-μm intervals (Bausch and Lomb, Rochester, NY). Measuring the number of pixels corresponding to the distance between the maximum number of lines across the field of view allows accurate calculation of the micrometer/pixel ratio. The micrometer is mounted in water with a 0.17-μm-thick (No. 1) cover glass placed over it and is imaged using bright-field (transmission) microscopy with water in the path between the objective and the cover glass to simulate specimen-mounting conditions. This procedure is performed with the micrometer rulings in both vertical (to determine *dx*) and horizontal (to determine *dy*) orientations.

Volume Sampling

A 3D "volume" of the biofilm specimen is sampled by obtaining a stack of 2D *x–y* plane "optical slices." Typically, 10 images are averaged with the PC platform into each optical slice using Image-Pro Plus image processing software (Media Cybernetics, Silver Spring, MD) to compensate for intensity variations encountered with the flash excitation sources. The sequential optical slices (typically 50 to 170) are collected along the *Z* axis by controlled movements of the microscope stage using a geared stepper motor system connected to the microscope fine focus control (Ludl MAC 2000 system, Ludl Electronic Products, Ltd., Hawthorne, NY) under the direction of software designed to coordinate image acquisition with stage movements (VolumeScan, VayTek, Inc., Fairfield, IA). The distance between slices (*dz*) chosen is typically a multiple of the optical slice pixel dimensions (*dx, dy*), usually 1 : 1 or 2 : 1. A *dz* of 0.3 to 0.6 μm is used during slice acquisition if the *dx, dy* of the slice is 0.3 μm, for example. Specimens are sampled from the bottom up due to requirements of the 3D reconstruction software. The use of custom macro programs simplifies volume acquisition and processing greatly. Oversampling (obtaining slices spaced less than the depth of field of the objective and thus duplicating some of the information in the images) is desirable because it ensures full representation of data along the *Z* axis during 3D reconstruction and increased *Z*-axis resolution. The time required for acquisition of a volume varies as a function of the number of optical slices in the volume and the number of images per slice. Collection of a volume composed of 100 slices with 10 images per slice requires approximately 5 min.

Volumes are inspected along the *Z* axis using a "top view" program that overlays all the optical sections of the volume and projects their brightest pixels (maximum point projection) into the resultant composite image (VayTek, Inc.). This method is useful as a quality control step before more extensive image processing is attempted. Dimensions of sampled volumes vary depending on microscope optics used, but with the Olympus 60× PSF water-immersion objective they typically range from 97–145 μm in the X axis, 72–109 μm in the Y axis, and 16–45 μm in the Z axis (for 150 slices).

Volume Preprocessing

All 2D image processing is carried out using Image-Pro Plus image processing software (Media Cybernetics, Silver Spring, MD). This package possesses a complete set of image processing tools and, in addition, uses a rich macro language that allows the user to combine and automate all the tools into a broad spectrum of image processing filters. The background of uniformly recurring noise (defects in the image intensifier or camera, dust flecks on the optics, etc.) is eliminated from the volume by a user-defined macro that performs a running two-slice average over the entire volume to detect and then subtract the noise image from each slice in the volume. The processed volume thus contains only nonuniformly repeating information (e.g., the specimen). The image quality of all volume slices is enhanced when required by adjustment of the brightness, contrast, and gamma settings (BCG) using functions provided by the image processing package combined in another user-defined macro. Results are checked over the entire volume by examining a top view.

Digital Deconvolution of Volume

Defocused information (haze) is eliminated from each slice in the volume using a nearest-neighbor deconvolution algorithm in accordance with methods suggested by the manufacturer (MicroTome AT, VayTek, Inc.) based on the theoretical point-spread function (PSF) of the objective. Haze removal is typically set at >90% and the dz value used for determination of the PSF is sometimes varied from the actual dz to obtain best results. In most cases, it is possible to render focused objects in the fluorescent images (typically consisting of bright bacteria scattered in a dark background) with confocal quality so that only objects in the depth of field of the objective remain in the slice (see Fig. 4). Thus, following deconvolution, the "focused" volume contains little or no out-of-focus haze, just the focused information remains. The quality of the deconvolution is monitored both by checking selected individual slices and by rendering the deconvolved volume with a top view as described previously. Processing time on a 266-MHz PC averaged about 10 sec/slice.

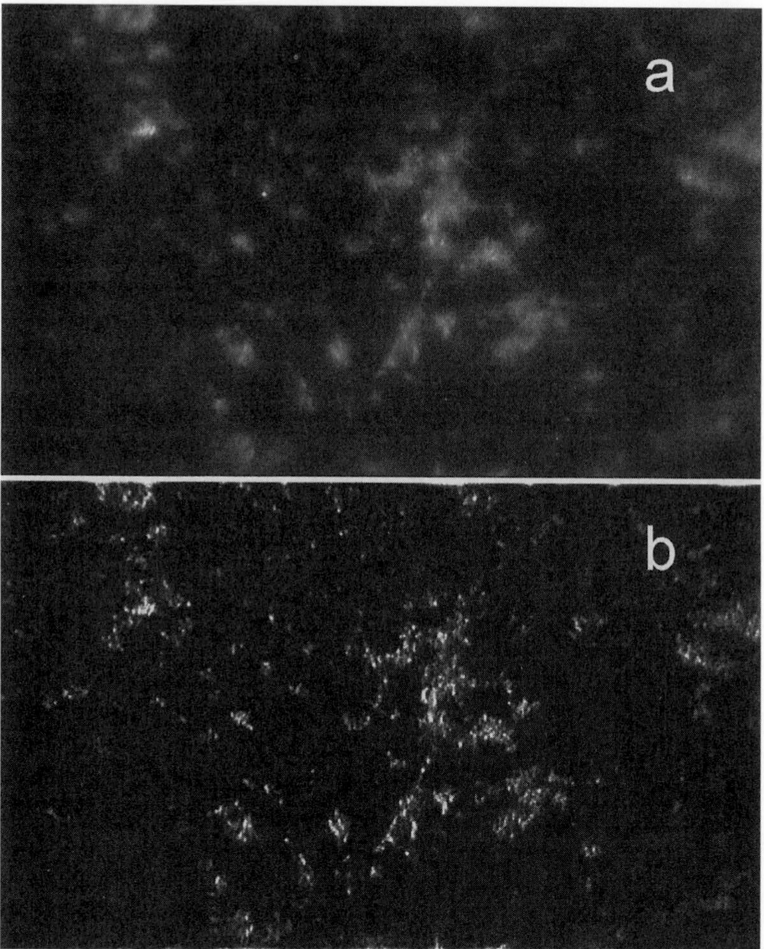

FIG. 4. A raw (a) and digitally deconvolved image (b) of one optical section (thickness 0.3 μm) of a wastewater biofilm on a CA-coated cover glass coupon. The 48-hr-old biofilm was formed in a special microscope flow cell (see text). Biofilm cells were stained with 0.1 μg/ml propidium iodide.

Volume Postprocessing

BCG adjustment is occasionally required to enhance the appearance of slices in the focused volume. In addition, the deconvolution algorithm occasionally enhances very minute differences in the intensity of pixels comprising the odd and even video fields when using flash illumination, resulting in subtle one-pixel-wide horizontal "stripes" over the entire image.

In this case, use of a macro that first applies a fast Fourier transform (FFT) to each slice, excises the frequency component corresponding to the horizontal noise stripes, and finally inverts the FFT to reconstitute the remaining image data effectively eliminates this artifact. In addition to these filters, volume slices are treated with other image filters (convolution algorithms such as sharpening, edge detection or enhancement algorithms, histogram optimization algorithms, etc.) as required to enhance the features of interest prior to 3D rendering.

Three-Dimensional Rendering of Volume

Volume slices are exported via an Ethernet link to a UNIX workstation (SGI Indigo Elan 4000, Silicon Graphics, Inc., Irvine, CA). A 3D rendering package (VoxBlast, VayTek, Inc.) uses the 2D X–Y axis information (pixels) present in the optical slices and, based on the spacing between the slices and an interpolation algorithm, constructs a virtual model of the biofilm specimen made up of regular image cubes (voxels). Thus, all of the primary intensity and position data of objects in the original biofilm are represented in the rendered volume. The volume is rendered in low resolution (one out of three voxels rendered) to speed image processing while setting the 3D view angle, but palette and transparency (alpha channel) adjustments are made from highest resolution (every voxel) renderings to provide complete data for analysis and recording.

The volume requires trimming to remove thin "dead" regions at its extreme X and Y faces. False coloring of gray values is achieved using a user-defined palette map, and colors are chosen to enhance features of interest in the rendered volume. Where required, two palettes are applied to halves of the volume to help reveal internal details masked by structures of low transparency. Lighting models are also occasionally applied to enhance the 3D effect and emphasize opaque features (such as bacteria). A polygon rendering feature allows multiple, color-coded polygon overlays to be applied to the volume. This overlay feature is typically used to outline the outer dimensions of the volume to aid visual orientation during volume rotation and display.

Calibration of the Three-Dimensional Volume Z-Axis

Accurate spatial rendering of the volume along the Z axis is achieved by calibration of the 3D rendering software with targets of known Z-axis geometry. Fluorescent latex spheres with diameters ranging from 3.0 to 20 μm (Fluoresbrite Calibration Grade Size Range Kit, Polysciences, Inc., Warrington, PA) embedded in a gel matrix of distilled water solidified with 0.7% purified agar (Becton-Dickinson Microbiological Systems,

Cockeysville, MD) are used for this purpose. The bead diameter used should be significantly larger than the objective's depth of field. Stocks produced by mixing beads of specific diameters with the melted agar in 24-ml scintillation vials are prepared in advance and stored at 4° until use. Bead specimens are prepared by melting a bead/agar stock in a microwave oven, placing a drop of the melted mixture on a glass slide at room temperature and immediately dropping a cover glass on top of the drop. The agar solidifies quickly and immobilizes the beads, creating a sample over 100 μm thick containing evenly distributed fluorescent targets of defined geometry.

The bead target slide is sampled as described earlier and reconstructed in 3D. The shape of the beads is examined in a top view of the volume to confirm dx/dy accuracy and is then examined along the X–Y plane (perpendicular to the optical slice plane), where the rendering software synthesizes the view entirely. The magnitude of deviation from a spherical appearance of the beads is determined throughout the length of the Z axis. The software interslice distance is adjusted until the interpolation algorithm properly renders the beads as spheres. The degree of this adjustment is typically very minor for the 60× water-immersion objective, which is specifically engineered not to introduce significant spherical aberration with deep specimens. However, volumes sampled using the more common (and far less expensive) oil-immersion or air objectives produce significant compression or elongation artifacts when viewed along the X–Y plane. Alteration of the interslice distance using the bead targets as a guide allows limited compensation of these distortions, and accurate 3D reproduction is, in fact, possible with these more common microscope objectives as long as the biofilm specimen thickness does not exceed 5 μm or so and is not viewed through an excessive depth of media (it is possible to accurately render a thin biofilm on the inner surface of a flow cell window, for instance).

Processing of Rendered Volume

It is possible to examine the rendered image volume from any desired orientation. In addition, two-dimensional slice planes may be passed through the volume either parallel to the original slice planes or at random orientations within the volume, and the resulting view is extracted to a 2D image for examination. An independent color palette is assigned to this extracted image to emphasize elements of interest not obvious within the original 3D volume. In addition, resampling along an axis perpendicular to the random slice plane allows extraction of all or part of the original volume into a volume with a user-defined orientation, such as down the axis of a biofilm channel.

Measurements performed on the rendered volume include morphometric determinations in 3D and 2D (including biofilm thickness, distances between biofilm bacteria or microcolonies, and dimensions of biofilm channels). Analysis of gray level distributions allows relative determination of fluorochrome concentrations in both 2D and 3D.

Views of the rendered volume are extracted and saved as 24-bit RGB TIFF images. A movie loop generator is used to produce rotating views of the volume, providing up to 360° of rotation in either altitude or azimuth (individually or simultaneously). This technique provides a smooth rotation of the rendered biofilm specimen and presents a view from all sides that often aids the visualization of biofilm details hidden behind the specimen. In addition, by producing a full rotation using 120 frames, 60 stereo pairs (3° tilt) result that represent the biofilm from all visual aspects of its rotation.

Combination of Multiple Volumes for Multifluorochrome Imaging

Biofilms stained with multiple fluorochromes (up to three) are sampled sequentially under conditions designed to detect each of the fluorochromes in turn. The resulting volumes are saved separately (a macro has been designed for this purpose) and processed independently using the methods described earlier. Each volume (up to a total of three) is then imported into the rendering software as the red, green, or blue channel and a composite renderer is used to produce a colored volume in which all the features of the component volumes are displayed simultaneously. This allows the spatial distribution of bacteria (imaged using generic nuclear stains) in the biofilms to be compared directly with the location of polysaccharides (e.g., detected using calcofluor white) or with the location of specific organisms identified by fluorescence *in situ* hybridization or specific immunostains. When performing this type of imaging, special considerations must be given to the intensity histograms of each of the volumes; the gray level distributions should be matched as well as possible to maximize the final image dynamic range (intensifier gain for each volume must be adjusted independently due to differences in the sensitivity of the intensifier at different wavelengths and possibly the effects of different excitation sources if more than one is used).

Real-Time Visualization and Quantification of Bacterial Adhesion Using Microscope Flow Cells

Microscope flow cells (MFCs) of various design configurations have been developed to allow direct microscopic visualization and quantification of attached biofilm cells in real time under dynamic flow conditions.[1,2,4] Microscope flow cells may be used exclusively in the laboratory to grow and observe axenic or native mixed species biofilms under controlled physicochemical and hydrodynamic conditions or they may be placed on line in an actual membrane facility to monitor biofilm formation on designated feed waters. In membrane biofouling studies, MFCs may be used to explore and identify (at the microscopic scale) physical or chemical conditions that disrupt biofilms (e.g., candidate chemical cleaning agents) or that interfere with or otherwise influence biofilm growth kinetics (e.g., biocides). The MFC should be sufficiently flexible in design to accommodate different flow channel geometries (e.g., channel shape and dimensions). Poor MFC design or machining invariably results in flow cells that leak or that are too cumbersome or time-consuming to use on a routine basis. Well-designed, thermo-regulated MFCs meeting these criteria are available commercially (e.g., Bioptecs, Inc., Butler, PA).

A custom MFC design that has worked well with the fluorescence deconvolution microscope system described earlier consists of round upper (top) and lower (base) stainless-steel plates, each possessing optically transparent glass or sapphire windows for viewing by transmitted or epi-illumination (Fig. 5). The window of the lower plate consists of a round 47-mm-diameter cover glass of standard thickness ($\sim$170 μm) that seats into a precision-machined circular recess (of the same depth) in the base plate. The dimensions of the MFC are such that it mates directly with the stage of an Olympus inverted (IX-70) or upright (AX-70) microscope (see Fig. 5b). When using an inverted microscope, the objective lens is moved into position directly beneath the base plate (cover glass) window. The cover glass may be used without any modification, but for membrane studies it is coated with a layer of CA, PA, or other polymer to simulate as closely as possible an actual membrane surface.

The hydrodynamic properties of the MFC may be calculated using standard fluid mechanic formulas (Fig. 6).[5] The shape and dimensions of the flow channel (and hence the flow characteristics) of the MFC are defined solely by the silicon rubber spacer used to seal and separate the upper and lower stainless-steel plates. Typically, a 1.46 cm $\times$ 2.22 cm $\times$ 100 μm thick channel spacer is employed at flow rates of 0.05 to 1.0 ml/min. At these loading rates and spacer thicknesses, the calculated Reynolds number is

[4] R. F. Mueller, W. G. Characklis, W. L. Jones, and J. T. Sears, *Biotechnol. Bioeng.* **39,** 1161 (1992).

[5] J. W. Murdock, *in* "Mark's Standard Handbook for Mechanical Engineers" (T. Baumeister, E. A. Avallone, and T. Baumeister III, eds.), 8th ed., p. 3–33. McGraw-Hill, New York, 1978.

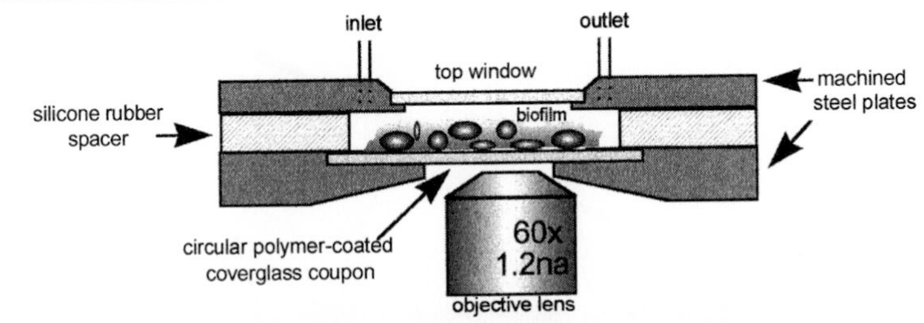

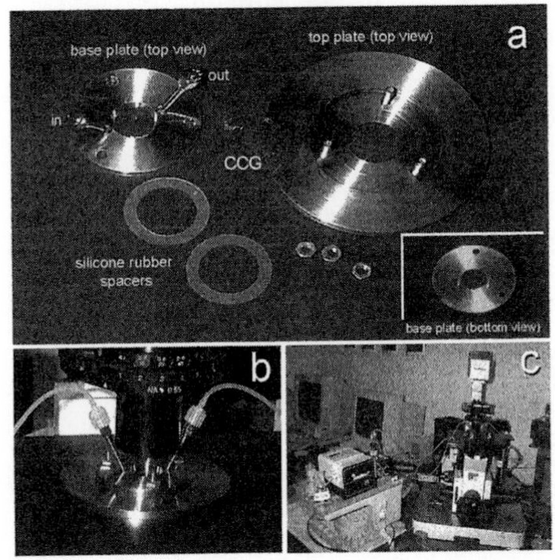

Fig. 5. (Top) Schematic of basic MFC construction and flow pattern. (Bottom) MFC components disassembled (a), MFC mated to inverted microscope stage (b), and MFC system integrated with microscope and peristaltic pump (c), CCG, circular polymer coated cover glass window.

$\ll$1.0; thus flow is considered to be laminar at the polymer film surface. Dye injection experiments conducted using the MFC have confirmed the laminar nature of the flow over the range of spacer thickness and loading rates employed (data not shown).

The calculated flow velocities that attached bacteria might experience in the MFC are generally lower than those in an actual RO membrane module. However, computer modeling studies have revealed that flow in an actual membrane module is highly variable, depending on the position of the attached cells relative to the module feed channel spacer material (R. L. Riley, Separation Systems, Inc., San Diego, CA, personal communication). Indeed, large areas of the membrane surface (especially in the vicinity of the flow spacer) experience flows comparable to those in the MFC. Thus, cell attachment and biofilm growth rate data obtained using the MFC are

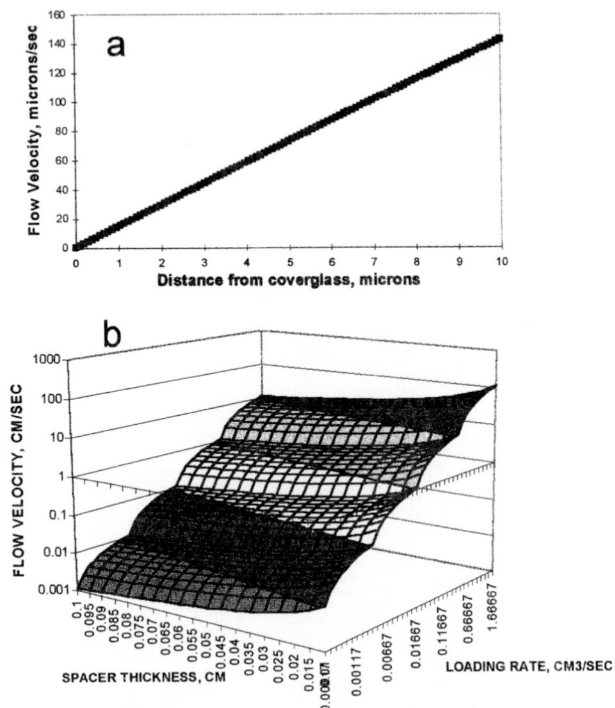

FIG. 6. Calculated hydrodynamics of the MFC described in the text and depicted in Fig. 5. Note that, the flow velocity is linear over the range of 0–10 μm from the coupon surface for a rectangular flow channel with dimensions of 1.46 cm $\times$ 2.22 cm $\times$ 100 μm (a). Early membrane biofilms typically fall well within this thickness range. MFC flow velocity is a function of spacer thickness and loading rate as indicated (b).

valid for those areas of actual membrane systems where flow velocities are low (e.g., proximal to the Vexar spacer). Furthermore, use of generally lower flow velocities and shear forces in the MFC provide a more conservative experimental approach, as cell detachment or biofilm disruption observed at the lower flow rates of the MFC could then be expected to also occur at the considerably higher flow rates of actual membrane modules.

An important distinction between the MFC and an actual separation membrane system is that there is no pressure-driven water or solute transport in the MFC. Thus, biofilm formation results exclusively from (i) initial microbial adhesion to the membrane surface as determined by the inherent affinity of bacterial cells for the synthetic polymer membrane surface and (ii) growth of attached cells at the expense of feed water nutrients. There

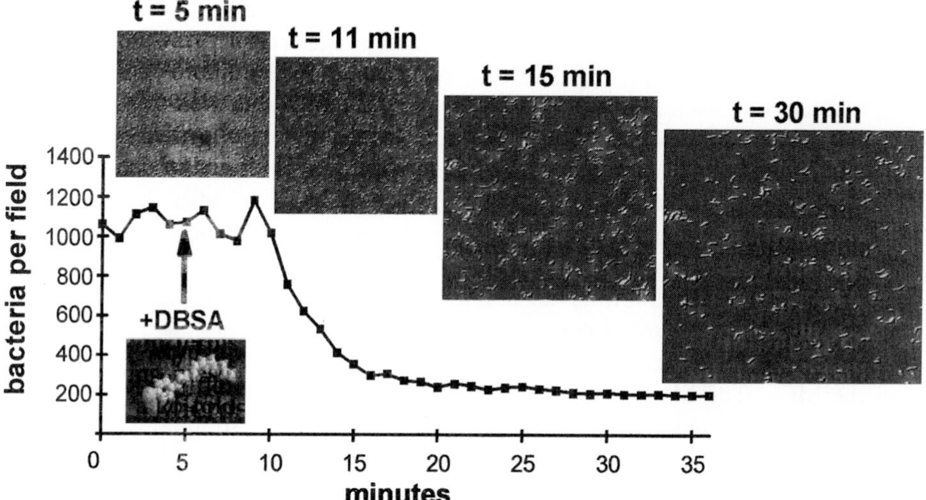

FIG. 7. Results of an MFC experiment designed to determine how early biofilm cells respond to 0.1 wt% of the anionic surfactant/biocide dodecylbenzenesulfonic acid (DBSA) in MS buffer. After 5 min of MS buffer rinse, DBSA was introduced into the MFC containing a 48-hr-old biofilm of *Mycobacterium chelonae* strain BT12-100 (arrow). The substratum was a CA-coated cover glass (see text) and the MFC was mounted on the stage of an Olympus IX-70 inverted microscope equipped with a 60× objective lens. Following a 5-min lag corresponding to the time needed for DBSA to reach the interior of the MFC, the mycobacteria began to detach. Image analysis software was used to quantify the number of bacteria per field in the Nomarski DIC images.

is no opportunity for hydraulically mediated "filtration" of bacteria onto the membrane surface in the MFC. Thus, biofilm formation under MFC operating conditions might be expected to be slower than that in an actual RO membrane system. Despite this limitation, the MFC has proven extremely useful for rapidly screening conditions that impede initial bacterial adhesion to selected polymer membrane materials or that disrupt existing biofilms.

For operation of the MFC, the polymer film on the cover glass surface must first be completely hydrated and preconditioned by introducing a suitable buffer solution (e.g., sterile-filtered MS buffer) via syringe or peristaltic pump. The buffer solution is typically pumped through the MFC for a period of no less than 30 min (at room temperature; ~23°), although the exact time depends on the specific swelling characteristics of the polymer film under investigation. Following complete hydration of the membrane substratum, bacterial adhesion kinetics may be conveniently and directly determined by introducing a cell suspension in MS buffer at time 0. Epifluorescent or Nomarski differential interference contrast (DIC) images from single or multiple regions of interest are captured digitally at intervals to document bacterial adhesion kinetics. Alternatively, it is possible to first establish a monospecies or native mixed species biofilm and then introduce experimental cleaning agents, biocides, or physiological probes (e.g., 5-cyano-2,3-ditolyltetrazolium chloride).[6] To establish an MFC biofilm, a bacterial suspension (about 10^8 cells/ml) is pumped through the prehydrated MFC for about 1 hr or until initial bacterial adhesion (of a desired density) to the membrane surface has occurred (determined microscopically). Following initial cell adhesion to the membrane surface, a sterile nutrient solution (e.g., MS growth medium; see Ref. 5a) is introduced into the MFC. The MFC is then operated in a continuous flow mode (e.g., at 50–100 μl/min) until confluent cell coverage has been attained (typically 24–48 hr at room temperature). Exogenous nutrients and loosely adsorbed bacteria are subsequently flushed away (by sterile MS buffer) from the biofilm, and test chemicals or physiological stains may be introduced in a continuous flow mode to determine how biofilm cells respond. A region of interest is selected and images are captured digitally at prescribed intervals to document the effects of the chemical over time, as illustrated in Fig. 7.

[6] G. G. Rodriguez, D. Phipps, K. Ishiguro, and H. F. Ridgway, *Appl. Environ. Microbiol.* **58,** 1801, (1992).

[15] Bacterial Biofilms: Strategies for Preparing Glycocalyx for Electron Microscopy

By THERESA A. FASSEL and CHARLES E. EDMISTON, JR.

Introduction and Background

Critical to the study of the bacterial biofilm by electron microscopy is successful preservation of the glycocalyx. Conventional fixation techniques sequentially employing the aldehydes, namely glutaraldehyde and paraformaldehyde, followed by osmium tetroxide, fail to preserve the bacterial glycocalyx adequately. This is in part due to an inadequate stabilization of highly substituted and variable polysaccharide moieties that constitute the glycocalyx. In addition, delicate ultrastructural features of the glycocalyx are altered further during the dehydration stages; the highly hydrated and

METHODS IN ENZYMOLOGY, VOL. 310
0076-6879/99 $30.00

fibrous anionic arrays of polymers are potentially condensed, collapsed, or distorted.[1–4] The use of positive contrasting agents such as uranyl acetate and lead citrate often fails to adequately stain any remaining remnants of structures in thin sections. Additionally, these features have no electron density differences from the background plastic resin. Because structures with more density appear "darker" on the phosphorescent screen of the electron microscope and those that are "lighter" in density appear brighter, these exopolysaccharides, which both lack innate electron density and do not gain density from uranyl acetate or lead citrate,[3] appear indistinguishable from the background embedding medium.

To ensure that there is adequate glycocalyx stabilization to withstand the rigors of dehydration with the addition of contrast requires reagents that react with highly polymerized and highly charged anionic mucopolysaccharides and glycoconjugates. Ruthenium red[5,6] and alcian blue,[5,7,8] which have been used traditionally to improve the visualization of acidic polysaccharides in light microscopy, have also been adapted for enhancing ultrastructural features in electron microscopy. The molecule of ruthenium red is a highly charged sphere (+6) of 1.1 nm.[5,9] The more planar copper phthalocyanin, alcian blue, has a +4 charge distributed on its quaternary amino group side chains.[10]

By the methods of Luft,[5,6] ruthenium red is included in all stages of fixation, buffer washes, and dehydration through 70% (v/v) ethanol. Improved visualization of the bacterial glycocalyx has been observed, although the abundance of material was sometimes limited as for the clinically relevant staphylococci[11]; or glycocalyx was collapsed or condensed, most notably where attachment surfaces or sufficient protein sites were lacking that would have provided additional stability through processing.[1] When alcian blue was similarly used, agreement with ruthenium red observations of extended fibrous matrices was observed for some bacterial species,[12] but

[1] J. W. Costerton, R. T. Irwin, and K. J. Cheng, *Annu. Rev. Microbiol.* **35,** 299 (1981).

[2] A. Progulske and S. C. Holt, *J. Bacteriol.* **143,** 1003 (1980).

[3] I. L. Roth, *in* "Surface Carbohydrates of the Prokaryotic Cell" (I. W. Sutherland, ed.), p. 5. Academic Press, New York, 1977.

[4] I. W. Sutherland, *Adv. Microbial Physiol.* **8,** 143 (1972).

[5] J. H. Luft, *Anat. Rec.* **171,** 347 (1971).

[6] J. H. Luft, *Anat. Rec.* **171,** 369 (1971).

[7] O. Behnke and T. Zelander, *J. Ultrastruc. Res.* **31,** 424 (1970).

[8] J. E. Scott and J. Dorling, *Histochemie* **5,** 221 (1965).

[9] D. E. Hanke and D. H. Northcote, *Biopolymers* **14,** 1 (1975).

[10] J. E. Scott, *Histochemie* **30,** 215 (1972).

[11] T. A. Fassel, J. E. VanOver, C. C. Hauser, L. E. Buchholz, C. E. Edmiston, J. R. Sanger, and C. C. Remsen, *Cells Mater.* **2,** 37 (1992).

[12] T. A. Fassel, M. J. Schaller, and C. C. Remsen, *Micros. Res. Tech.* **20,** 87 (1992).

not for staphylococci.[11] Another approach to allow better visualization of the staphylococcal exopolysaccharide associated with biomaterial-associated infections was needed to improve application to clinically relevant specimens.

The diamine lysine with glutaraldehyde forms large polymers with cross-linkages of various lengths and directions with the anionic glycocalyx of mammalian tissues.[13] This glutaraldehyde–lysine fixative with ruthenium red visualizes extended glycocalyces for several bacterial species, notably where ruthenium red alone is insufficient.[14] For the staphylococcal glycocalyx, considerable enhancement of abundant fibrous material was obtained that often surrounded cells in a complete layer and extended between cells[11] as well as between cells and biomaterials.[15]

For glutaraldehyde–lysine fixative with alcian blue, condensed structures seen previously in the absence of lysine[11] were eliminated and elaborate glycocalyces were observed.[16] Unfortunately, if the prefix incubation exceeded 20 min, samples were potentially lost due to a solidification of the sample in the prefixative solution. This hampered the application of this approach to biomaterial-associated infection studies in the clinical realm. An improvement was shown where the prefix time barrier could be extended to several hours or occasionally overnight if paraformaldehyde was included in the fixative solution with glutaraldehyde–lysine and ruthenium red.[17] However, some samples would still be lost to solidification in an unpredictable and variable manner. To prevent loss of valuable and potentially irreplaceable clinical samples, a further modification to this approach was developed. The lysine used previously in these fixatives was the free amino form. Replacement of lysine free amino with lysine monohydrochloride or lysine acetate eliminated the solidification artifact in studies carried out to 24 hr.[18] For paraformaldehyde–glutaraldehyde–lysine fixative with alcian blue, similar observations have been made. However, for fixative solutions with the lysine monohydrochloride or lysine–acetate forms at short time intervals of 20 min, some condensed structures occur occasionally coexisting with fibrous features. These are avoided by extending the time

[13] J. K. Boyles, *in* "Science of Biological Specimen Preparation for Microscopy and Microanalysis" (O. Johari, ed.), p. 7. SEM, AMR O'Hare, Chicago, 1984.

[14] M. Jacques and L. Graham, *J. Electron Micros. Tech.* **11,** 167 (1989).

[15] T. A. Fassel, J. E. VanOver, C. C. Hauser, C. E. Edmiston, and J. R. Sanger, *Cells Mater.* **1,** 199 (1991).

[16] T. A. Fassel, J. R. Sanger, and C. E. Edmiston, *Cells Mater.* **3,** 327 (1993).

[17] T. A. Fassel, P. E. Mozdziak, J. R. Sanger, and C. E. Edmiston, *Micros. Res. Tech.* **36,** 422 (1997).

[18] T. A. Fassel, P. E. Mozdziak, J. R. Sanger, and C. E. Edmiston, *Micro. Res. Tech.* **41,** 291 (1998).

of this fixation to 2 hr when elaborate and abundant staphylococcal glycocalyx are observed consistently.[19]

Details of the following methods have been worked out on three species of gram-positive, coagulase-negative staphylococci: *Staphylococcus epidermidis* RP62, *Staphylococcus aureus* ATCC 25923, and *Staphylococcus hominis* SP2. These strains, which include a laboratory reference strain (ATCC 25923) and two clinical isolates (RP62 and SP2), have been used frequently to study biomedical device infections in both laboratory and animal models of infection.[20]

Methods

Cell Culture

Cells, recovered from $-70°$ frozen storage, are plated on blood agar plates to check for viability. After 24 hr, five colonies are inoculated to Trypticase soy broth (TSB) and incubated for an additional 18 hr at 35°. Initially, cells are ultracentrifuged (4000*g*, 10 min at 4°) and the pellets are fully suspended in the first fixative solution. During each further stage of the processing before dehydration, the cells are centrifuged (microcentrifuge between 12,000 and 16,000*g*, 2 min at 4°) and the pellets are suspended carefully in the next reagent. In the last aqueous buffer wash before dehydration, the cells are enrobed in 4% agar and handled as 1-mm^3 blocks thereafter.

Electron Microscopy

The general electron microscopy procedure is detailed in steps A through K. Steps B through K are identical for all fixation method variations of step A. All methods were developed to be used at room temperature, allowing broad application whether in the clinical or in the laboratory environment.

The first fixation step (A) is critical for glycocalyx preservation/*en bloc* staining. The diamine lysine is available as alternative salts or forms. The monohydrochloride and acetate forms have both been demonstrated to be useful in acquiring high-quality ultrastructural preservation and are adaptable to a highly flexible fixation time.[18] The latter advantage is of particular importance when there is an attempt to demonstrate biofilm development on bioprosthetic devices obtained in the clinical environment.

[19] T. A. Fassel, P. E. Mozdziak, J. R. Sanger, and C. E. Edmiston, submitted for publication.

[20] C. E. Edmiston, D. D. Schmitt, and G. R. Seabrook, *Infect. Control Hosp. Epidemiol.* **10,** 111 (1989).

Under these conditions it is often difficult to approximate ideal fixation intervals.

Regardless of the method, the monohydrochloride (Sigma Chemical Company, St. Louis, MO) or acetate (Sigma)[18] forms of lysine (L-2,6-diaminohexanoic) may be used.

The free amino form of lysine (Sigma) provides good ultrastructural features, but its use should be limited to Methods 1 and 2. As the fixation time increases beyond 20 min (Methods 3 through 6), some samples processed with the free amino form of lysine may undergo a solidification that prevents further processing of the sample.[17]

The buffer for all aqueous solutions is 0.1 *M* sodium cacodylate, pH 7.2. Ruthenium red is obtained from Sigma containing 40% dye content. Alcian blue 8GX is obtained from Aldrich (Milwaukee, WI) containing 25% dye content.

Procedure for Electron Microscopy Preparation of Staphylococcal Glycocalyx

A. Fixation methods for the preparation of staphylococcal glycocalyx.

Method 1: Paraformaldehyde–glutaraldehyde–lysine. Prefixation is in buffered 75 m*M* lysine in 2% (w/v) paraformaldehyde, 2.5% (w/v) glutaraldehyde for 20 min. Fixation in buffered 2% paraformaldehyde, 2.5% glutaraldehyde follows for 2 hr.

Method 2: Paraformaldehyde–glutaraldehyde–ruthenium red–lysine. Prefixation is in buffered 75 m*M* lysine in 0.075% (w/v) ruthenium red, 2% (w/v) paraformaldehyde, 2.5% (w/v) glutaraldehyde for 20 min. Fixation in buffered 0.075% ruthenium red, 2% paraformaldehyde, 2.5% glutaraldehyde follows for 2 hr.

Method 3: Paraformaldehyde–glutaraldehyde–alcian blue–lysine. For lysine monohydrochloride or lysine acetate, fixation is recommended in buffered 75 m*M* lysine in 0.075% (w/v) alcian blue, 2% (w/v) paraformaldehyde, 2.5% (w/v) glutaraldehyde for 2 hr. Note this difference in time from Methods 1 and 2. This avoids a condensed artifact of the glycocalyx for these lysine forms with alcian blue at short times.[19] For lysine free amino, processing times follow Methods 1 and 2.

Method 4: Paraformaldehyde–glutaraldehyde–lysine/extended time. Fixation is in buffered 75 m*M* lysine in 2% (w/v) paraformaldehyde, 2.5% (w/v) glutaraldehyde for extended times (24 hr).

Method 5: Paraformaldehyde–glutaraldehyde–ruthenium red–lysine extended time. Fixation is in 75 m*M* lysine in buffered 0.075% (w/v) ruthenium red, 2% (w/v) paraformaldehyde, 2.5% (w/v) glutaraldehyde for extended times (24 hr).

Method 6: Paraformaldehyde–glutaraldehyde–alcian blue–lysine/extended time. Fixation is in buffered 75 m*M* lysine in 0.075% (w/v) alcian blue, 2% (w/v) paraformaldehyde, 2.5% (w/v) glutaraldehyde for extended times (24 hr).

B. Wash three times for 10 min in 0.1 *M* sodium cacodylate buffer (pH 7.2).
C. Postfix in 1% (w/v) OsO_4, 0.1 *M* sodium cacodylate buffer (pH 7.2) for 2 hr.
D. Wash three times for 10 min in 0.1 *M* sodium cacodylate buffer (pH 7.2).
E. Dehydrate through a graded ethanol series of 10, 25, and 50% for 10 min each, 70% overnight, 95%, and two anhydrous 100% stages for 1 hr each.
F. Infiltrate in LR White resin with three changes in fresh resin; one overnight change is followed by two 3-hr stages.
G. Embedment uses well-filled and tightly capped gelatin capsules.
H. Polymerize at 60° in an oven with consistent temperature. Use good LR White polymerization to draw a weak vacuum, as in a vacuum oven, if possible.
I. Cut thin sections with a diamond knife.
J. Postfix thin sections on grids with 2% uranyl acetate and Reynolds lead citrate.
K. Observe thin sections on grids in a transmission electron microscope.

Observations and Discussion

Conventional fixation employing glutaraldehyde and paraformaldehyde in the initial prefixation/fixation is inadequate for observation of the bacterial glycocalyx as illustrated for *S. aureus* in Fig. 1A. The presence of secreted exopolysaccharide material is completely lacking. When ruthenium red or alcian blue is included but is lysine deficient, glycocalyx appears sparse or condensed with some collapse of structures (Fig. 1, arrowheads). This is shown in Fig. 1B for *S. epidermidis* RP62 with paraformaldehyde–glutaraldehyde–ruthenium red fixation (Method 2).

Use of Lysine in Fixation

When lysine is included in the fixative as in Method 1 (paraformaldehyde–glutaraldehyde–lysine) or Method 4 (paraformaldehyde–glutaraldehyde–lysine/extended time), the staphylococcal glycocalyx appears more abundant (Fig. 1, arrowheads). This is illustrated for lysine monohydrochloride-

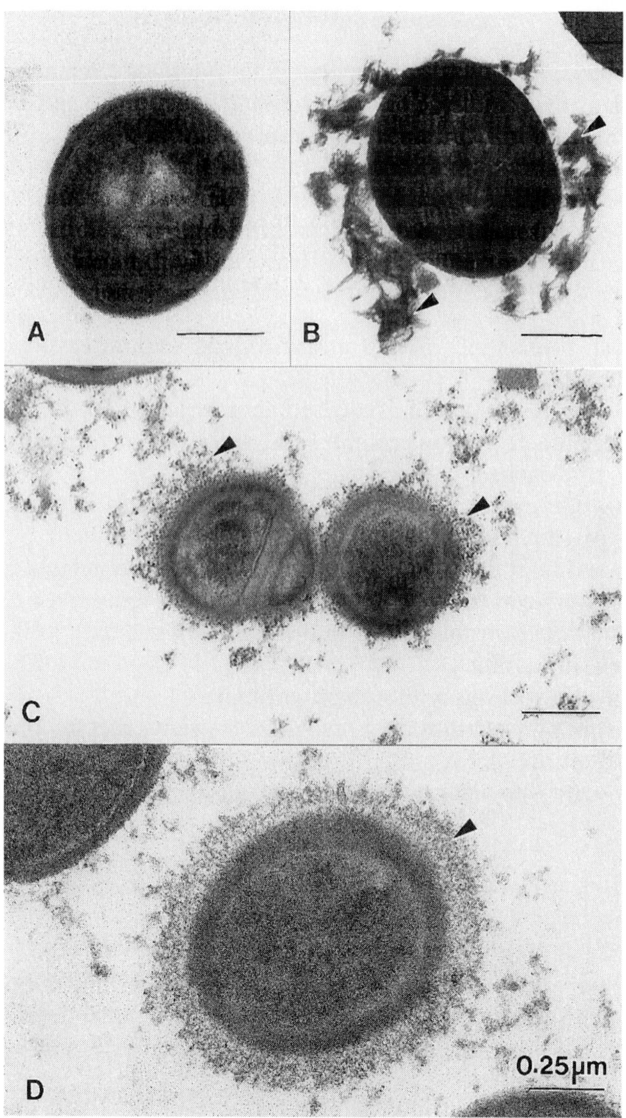

FIG. 1. (A) Inadequate control fixation of glutaraldehyde and paraformaldehyde shows an absence of fibrous bacterial glycocalyx as illustrated for *S. aureus.* (B) Control fixation of glutaraldehyde and paraformaldehyde with the addition of ruthenium red shows sparse glycocalyx with some collapse of structures (arrowheads) as shown for *S. epidermidis* RP62. (C) By Method 4 (paraformaldehyde–glutaraldehyde–lysine/extended time) with lysine monohydrochloride fixation of *S. aureus* cells, glycocalyx material appears abundant (arrowheads). (D) By Method 4 (paraformaldehyde–glutaraldehyde–lysine/extended time) with lysine free amino fixation of *S. hominis* SP2 cells, fibrous glycocalyx (arrowhead) is also more abundant. Bar: 0.25 μm.

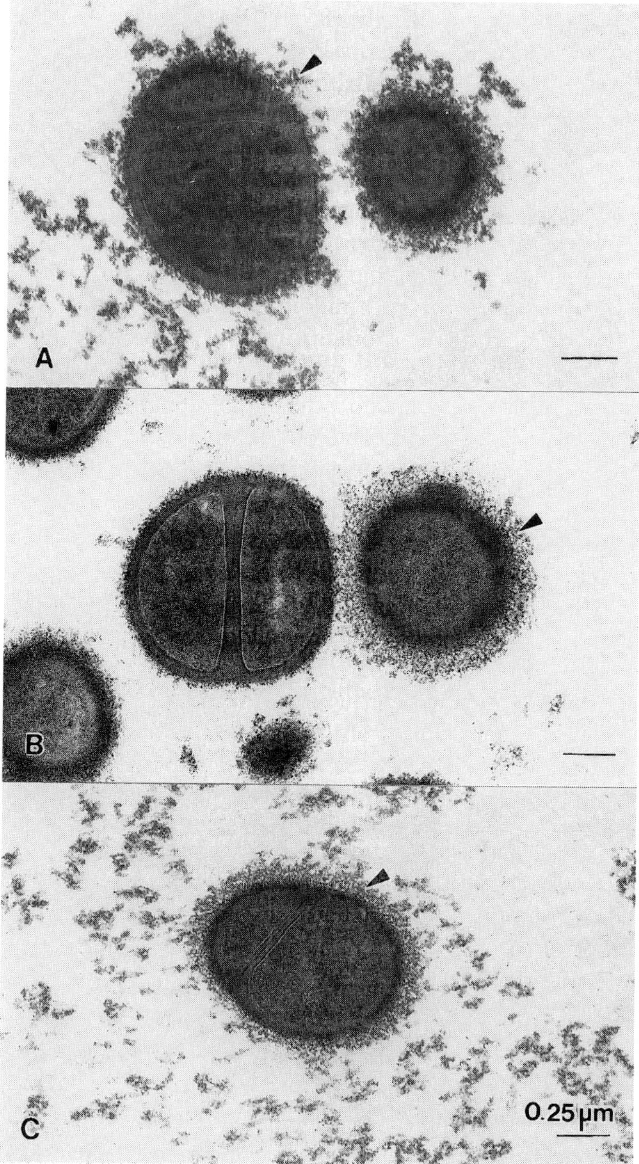

FIG. 2. For fixation of *S. hominis* SP2 cells (A) by Method 5 (paraformaldehyde–glutaraldehyde–ruthenium red–lysine/extended time) with lysine monohydrochloride, (B) by Method 2 (paraformaldehyde–glutaraldehyde–ruthenium red–lysine) with lysine free amino, or (C) by Method 5 (paraformaldehyde–glutaraldehyde–ruthenium red–lysine/extended time) with lysine free amino, abundant fibrous glycocalyx often surrounds cells (arrowheads). Bar: 0.25 μm.

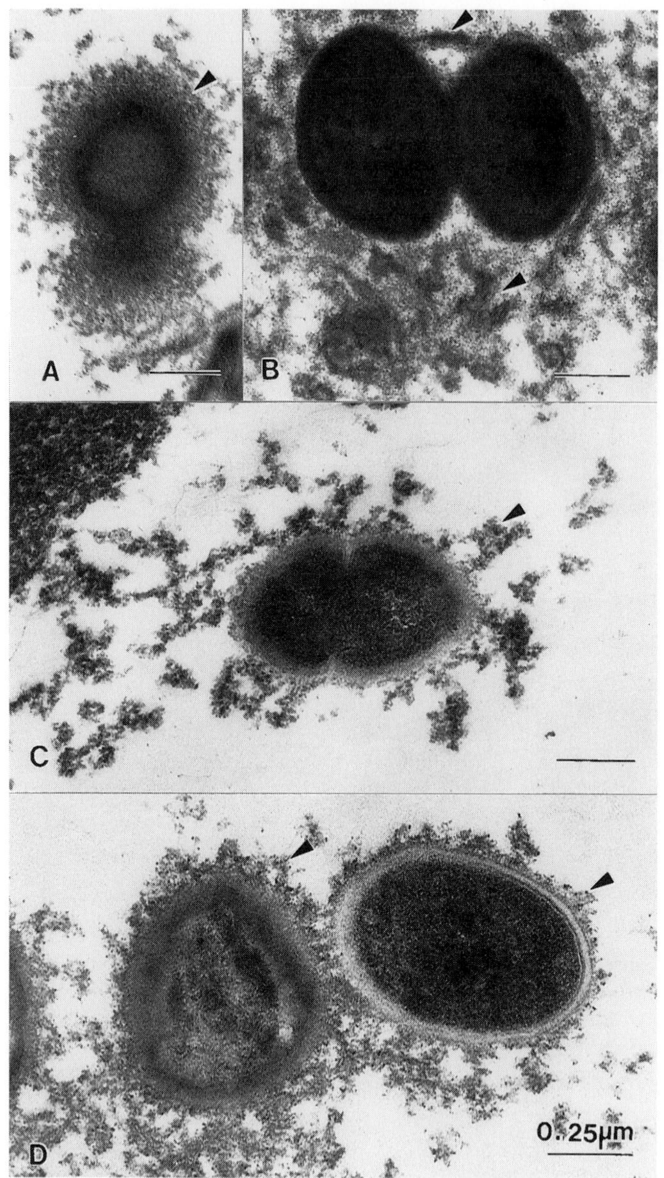

FIG. 3. Elaborate fibrous glycocalyces (arrowheads) are preserved for cells of (A) *S. hominis* SP2 by Method 6 (paraformaldehyde–glutaraldehyde–alcian blue–lysine/extended time) with lysine acetate, (B) *S. epidermidis* RP62 by Method 6 (paraformaldehyde–glutaraldehyde–alcian blue–lysine/extended time) with lysine monohydrochloride, (C) *S. aureus* by Method 6 (paraformaldehyde–glutaraldehyde–alcian blue–lysine/extended time) with lysine monohydrochloride, and (D) *S. aureus* by Method 3 (paraformaldehyde–glutaraldehyde–alcian blue–lysine) with lysine acetate. Bar: 0.25 μm.

treated *S. aureus* cells (Fig. 1C) and for lysine free-amino-treated *S. hominis* SP2 (Fig. 1D) by Method 4 (paraformaldehyde–glutaraldehyde–lysine/extended time).

Use of Ruthenium Red–Lysine in Fixation

With inclusion of ruthenium red and lysine in the fixative as for Method 2 (paraformaldeyde–glutaraldehyde–ruthenium red–lysine) or Method 5 (paraformaldeyde–glutaraldehyde–ruthenium red–lysine/extended time), the glycocalyx material is seen surrounding the staphylococcal cells as a fibrous layer extending well into the space between the bacterial cells. In the example of *S. hominis* SP2 cells, the glycocalyx is observed as abundant (Fig. 1, arrowheads) following the use of Method 5 (paraformaldehyde–glutaraldehyde–ruthenium red–lysine/extended time) with lysine monohydrochloride (Fig. 2A), by Method 2 (paraformaldehyde–glutaraldehyde–ruthenium red–lysine) with lysine free amino (Fig. 2B), and by Method 5 (paraformaldehyde–glutaraldehyde–ruthenium red–lysine/extended time) with lysine free amino (Fig. 2C).

Use of Alcian Blue–Lysine in Fixation

When alcian blue is used in conjunction with lysine, extensive fibrous material is often observed surrounding and extending beyond individual staphylococcal cells (Method 3: paraformaldehyde–glutaraldehyde–alcian blue–lysine and Method 6: paraformaldehyde–glutaraldehyde–alcian blue–lysine/extended time). An elaborate layer (arrowhead) surrounds *S. hominis* SP2 as shown in Fig. 3A by Method 6 (paraformaldehyde–glutaraldehyde–alcian blue–lysine/extended time) with lysine acetate. Considerable fibrous material (arrowheads) extending between cells of *S. epidermidis* RP62 is shown in Fig. 3B when Method 6 is used: paraformaldehyde–glutaraldehyde–alcian blue–lysine/extended time with lysine monohydrochloride. An abundant glycocalyx material (arrowheads) is shown radiating from cells of *S. aureus* by Method 6 (paraform–aldehyde–glutaraldehyde–alcian blue–lysine/extended time) with lysine monohydrochloride (Fig. 3C) and by Method 3 (paraformaldehyde–glutaraldehyde–alcian blue–lysine) with lysine acetate (Fig. 3D).

Thus, adequate stabilization and visualization of the fine detailed ultrastructure of delicate staphylococcal glycocalyx are possible when the cationic reagents ruthenium red and alcian blue are used with selected forms of lysine in the fixation protocols.

Section III

Flow and Steady-State Methods

[16] Robbins Device in Biofilm Research

By ARSALAN KHARAZMI, BIRGIT GIWERCMAN, and NIELS HØIBY

Introduction

Microbial biofilms have been observed on surfaces in many aquatic ecosystems of medical and industrial importance. Until recently, most investigations have been performed on planktonic microorganisms. However, in nature, in some chronic infections, and in industrial situations, these microorganisms present themselves as biofilms. With the discovery of biofilm mode of growth of microorganisms and awareness of the importance of biofilm, attention has been drawn to the study of microorganisms, particularly bacteria in the biofilm mode of growth.[1] Bacteria grown as biofilms differ markedly from free-floating, planktonic cells in their growth kinetics, cellular metabolism, and outer membrane properties.[2,3] The changes of target molecules for biocides, antibiotics, antibodies, and phagocytes make eradication of biofilms very difficult.[4,5] In the case of infections, the time course, pathogenesis, interaction with the immune system of the host, and sensitivity to antibiotics are profoundly different in biofilm than in planktonic bacteria. The mechanism and the extent of damage caused by biofilm in industry are also different from that of planktonic bacteria. The ultimate goal of microorganisms by growing as biofilm is to protect themselves from their hostile environments.

Robbins Device

Several systems have been developed to study the biofilm mode of growth of microorganisms. The Robbins device is the most widely used and well-known system. The device was later modified by McCoy *et al.* in 1981 and was called the modified Robbins device.[6] The Robbins device provides quantifiable samples of biofilms growing on submerged surfaces in aqueous systems.[7] Various adaptations of the device allow low-pressure flow applications, such as biofilm sampling for bacteria, i.e., *Pseudomonas*

[1] J. W. Costerton, K-J. Cheng, and G. G. Geesey, *Annu. Rev. Microbiol.* **41,** 435 (1987).
[2] M. R. W. Brown, D. G. Allison, and P. Gilbert, *J. Antimicrobial Chemother.* **22,** 777 (1988).
[3] M. R. W. Brown and P. Williams, *Annu. Rev. Microbiol.* **39,** 527 (1985).
[4] J. R. W. Govan, *J. Med. Microbiol.* **8,** 513 (1975).
[5] R. S. Baltimore and M. Mitchell, *J. Infect. Dis.* **141,** 238 (1980).
[6] W. F. McCoy, J. D. Bryers, J. Robbins, and J. W. Costerton, *Can. J. Microbiol.* **27,** 910 (1981).
[7] J. W. Costerton and H. M. Lappin-Scott, *ASM News* **55,** 650 (1989).

0076-6879/99 $30.00

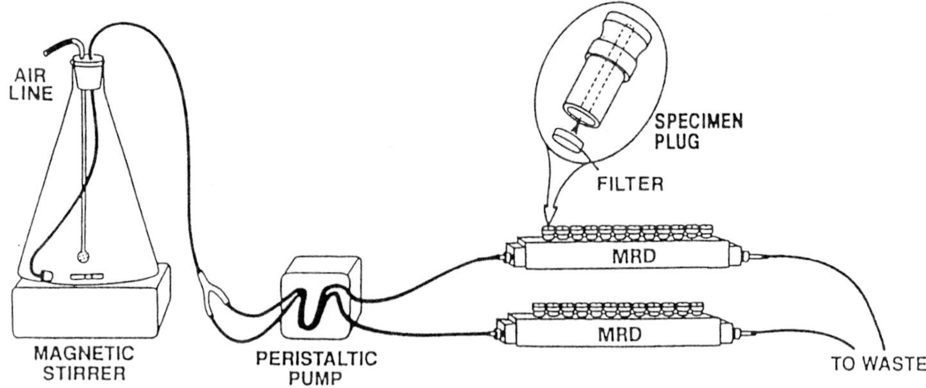

FIG. 1. Diagrammatic representation of two parallel modified Robbins devices (MRD) and the apparatus used to provide a continuous flow of bacteria and the culture medium. From G. Domininque *et al., J. Clin. Microbiol.* **32,** 2564 (1994).

aeruginosa in the studies of bacterial pathogenesis in pulmonary infections of cystic fibrosis patients and osteomyelitis caused by *Staphylococcus aureus* or *Legionella* in hospital water and air conditioning systems. The Robbins device has also been used in industrial studies for biofilm sampling in very high-pressure oil transmission pipelines. It can also be used for the investigations of bacterial corrosion on water and oil pipelines.[1]

The modified Robbins device is an artificial multiport sampling catheter. It is constructed of an acrylic block (41.5 cm long) with a lumen of 2 by 10 mm. Twenty-five evenly spaced sampling ports are devised so that silicone disks (0.5 cm^2) attached to sampling plugs will lie on the inner surface, without disturbing flow characteristics. The sampling plugs can be removed and replaced aseptically (Fig. 1). Inoculation is made by passing an exponential phase culture of the test organism through the reactor for several hours followed by sterile medium. Alternatively, the device is coupled to the outflow of a continuous fermentation vessel. Biofilms develop over time and may be sampled by removal of the pistons or disks. In the Robbins device the medium flows through materials that provide large surface areas for bacterial adhesion. The moving liquid phase imposes shear stresses on the developing biofilm, causing cells to be shed.[8] Planktonic cells shed from the biofilm can also be investigated and compared to the adherent bacteria.

The Robbins device sessile sampler can be used to determine the concentrations of antibiotics and biocides that kill planktonic bacteria in the bulk

[8] V. Deretic, R. Dikshit, M. W. Konyecsin, A. M. Chakarabarty, and T. K. Mishra, *J. Bacteriol.* **171,** 1278 (1989).

fluids. Such information helps in preventing biofilm formation on clean surfaces, such as surgically implanted medical devices. The sessile sampler can help to determine the minimal inhibitory concentrations (MICs) of antibiotics necessary to kill biofilm bacteria.

Typical studies compare the properties of adherent bacteria with those of the equivalent planktonic population of cells passing through the device. The Robbins device has been used extensively to investigate the immune response to biofilm-grown bacteria[9,10] and antibiotic resistance.[11–14] The technique has provided much valuable information on the physiology and metabolism of biofilm bacteria, regulation of various bacterial genes and their products, and resistance to antimicrobial compounds and their interactions with various components of the immune system. This article discusses the use of the Robbins device in infection-related biofilm research.

Biofilm Establishment

Biofilms are established on surfaces exposed to flow in the Robbins device. A typical modified Robbins device consists of silicone disks of 2 mm thickness and 7 mm diameter glued to studs in a row of 25 on the flow channel. The device, the studs, and the disks are washed for 3 days in distilled water at 4°, after which they are sterilized by either autoclaving or exposure to ethylene oxide. The exponential phase broth culture of a given bacterium such as *P. aeruginosa* is pumped through the device at 40–60 ml per hour at room temperature. After 1 to 5 days the studs protruding into the flow channel are taken out and the disks are removed aseptically and rinsed with physiological saline. Bacterial numbers are made by scraping off the disks into 2 ml of saline, mildly sonicating for 10 minutes, and counting by microscopy or estimation of colony-forming units (cfu). Biofilms produced by this procedure usually consist of bacteria enmeshed in a glycocalyx matrix containing lipopolysaccharide (LPS), alginate, and proteins.[15] The LPS of biofilm-grown *P. aeruginosa* appears predominantly

[9] E. T. Jensen, A. Kharazmi, K. Lam, W. J. Costerton, and N. Høiby, *Infect. Immun.* **58,** 2383 (1990).

[10] E. T. Jensen, A. Kharazmi, N. Høiby, and W. J. Costerton, *APMIS* **100,** 727 (1992).

[11] J. C. Nickel, I. Ruseska, J. B. Wright, and J. W. Costerton, *Antimicrobial Agents Chemother.* **27,** 619 (1985).

[12] R. C. Evans and C. J. Holmes, *Antimicrobial Agents Chemother.* **31,** 889 (1987).

[13] A. G. Gristina, R. A. Jennings, P. T. Naylor, Q. N. Myrvik, E. Bart, and L. X. Webb, *Antimicrobial Agents Chemother.* **33,** 813 (1989).

[14] B. Giwercman, E. T. Jensen, N. Høiby, A. Kharazmi, and W. J. Costerton, *Antimicrobial Agents Chemother.* **35,** 1008 (1991).

[15] E. T. Jensen, A. Kharazmi, P. Garred, G. Kronborg, A. Fomsgaard, T. E. Mollnes, and N. Høiby, *Microbial Pathogen.* **15,** 377 (1993).

rough as compared to the smooth planktonically grown bacteria in suspension or on agar plates.[16]

Robbins Device and Immune Response to Infections

Chronic Pulmonary Pseudomonas aeruginosa in Cystic Fibrosis Patients

The modified Robbins device has been used extensively to study the immune response to *P. aeruginosa* grown as biofilm. *Pseudomonas aeruginosa* is the most prevalent infection in patients with cystic fibrosis (CF) and is the most predominant cause of death in these patients. The infection has been shown to be a typical biofilm mode of growth in which the bacteria colonize the smaller airways without invading the deeper tissues.[17] The chronic *P. aeruginosa* lung infection in CF patients is regarded as a model for other chronic infections caused by biofilm-forming bacteria. The persistence of bacteria in the lungs has been attributed to the concept of bacterial biofilms.[1,18] The ability of bacteria to establish themselves as biofilm, where they are enmeshed in glycocalyx and escape elimination by the host defense and antibiotics, has been documented in both clinical cases and under experimental conditions.[19–22] The understanding of the pathogenesis of this infection has been facilitated by *in vitro* experiments of the interaction between *P. aeruginosa* biofilms and the components of the immune system using the modified Robbins device.

1. Leukocyte Response to Biofilm. Polymorphonuclear leukocytes (PMNs) are the most predominant cell type found in the lungs of CF patients infected chronically with *P. aeruginosa*.[20,23,24] Inflammation due to leukocytes has also been observed in other types of bacterial infections

[16] B. Giwercman, A. Fomsgaard, B. Mansa, and N. Høiby, *FEMS Microbiol. Immunol.* **89,** 225 (1992).

[17] R. S. Baltimore, C. D. C. Christie, and G. J. W. Smith, *Am. Rev. Resp. Dis.* **140,** 1650 (1989).

[18] M. R. W. Brown, P. J. Collier, and P. Gilbert, *Antimicrobial Agents Chemother* **34,** 1623 (1990).

[19] T. M. Bergamini, J. C. Peyton, and W. G. Cheadle, *J. Surg. Res.* **52,** 101 (1991).

[20] A. Kharazmi, P. O. Schiøtz, N. Høiby, L. Baek, and G. Döring, *Eur. J. Clin. Invest.* **16,** 143 (1986).

[21] D. A. Lee, J. R. Hoidal, C. C. Clawson, P. G. Quie, and P. K. Peterson, *J. Immunol. Methods* **63,** 103 (1983).

[22] T. J. Marrie, J. Y. Sung, and W. J. Costerton, *J. Gastroenterol. Hepatol.* **5,** 403 (1990).

[23] N. Høiby and C. Koch, *Thorax* **45,** 881 (1990).

[24] S. Suter, *in* "*Pseudomonas aeruginosa* Infection" (N. Høiby, S. S. Pedersen, G. H. Shand, G. Döring, and I. A. Holder, eds.), p. 158, 1989.

where the bacteria grow as biofilms.[25–27] In the studies carried out by Jensen *et al.* in 1992, the interactions of PMNs with *P. aeruginosa* biofilm were investigated. *Pseudomonos aeruginosa* biofilm is produced on silicone disks using a modified Robbins device as decribed earlier. The biofilms are exposed to purified peripheral blood PMNs from healthy subjects. The oxidative burst response of the cells to biofilm bacteria is determined by a chemiluminescence assay and by production of superoxide anion by the activated cells. The PMN response to biofilm is compared with the response to planktonic bacteria from the same culture setup. The chemiluminescence assay is a modification of a previously published method,[28] which is described elsewhere.[10] Briefly, disks with biofilm or the equivalent number of planktonic bacteria with or without sterile disks are mixed with buffer. One milliliter of PMN suspension (1×10^6 cells) is added to the disks. Various cfu : PMN ratios are used in the study. The chemiluminescence response of PMNs to bacteria is determined by a liquid scintillation counter (Beckman L 8000/LS 5000) in an out-of-coincidence mode up to 150 minutes. Superoxide anion production is determined by a cytochrome *c* reduction assay as described perviously.[29] Briefly, biofilms or planktonic bacteria are mixed with cytochrome *c* and added to PMN cell suspension at various ratios of cfu : PMN. After 30 minutes of incubation at 37°, the reaction is stopped with cold buffer, and the reduced cytochrome *c* is measured at an optical density of 550 nm in a spectrophotometer. In these studies, the PMN oxidative burst response to biofilm is reduced significantly to about 30%–80% of the response to an equivalent number of planktonic bacteria. Mechanical disruption of the biofilms elicits a significantly increased response. These findings indicate that the biofilm mode of growth provides an escape mechanism for bacteria from the hostile action of phagocytic cells.[9,10]

2. Complement Activation by Biofilm. The complement consists of a series of inactive proteins, in which on activation by microorganisms such as bacteria are converted to biologically active products. It is known that both gram-positive and gram-negative bacteria can activate complement.[30] Complement activation by bacteria could have positive as well as negative implications for the chronically infected patient. A positive outcome would

[25] M. Jacques, T. J. Marrie, and J. W. Costerton, *Microbial Ecol.* **13,** 173 (1987).
[26] J. Robin, W. A. Roger, H. M. Taylor, D. Everett, B. F. Prowant, L. V. Fruto, and K. D. Nolph, *Ann. Intern. Med.* **92,** 7 (1980).
[27] G. Peters, R. Locci, and G. Pulverer, *J. Infect. Dis.* **146,** 479 (1982).
[28] A. Kharazmi, N. Høiby, G. Döring, and N. H. Valerius, *Infect. Immun.* **44,** 587 (1984).
[29] A. Kharazmi, G. Döring, N. Høiby, and N. H. Valerius, *Infect. Immun.* **43,** 161 (1984).
[30] K. Inada, *Jap. J. Exp. Med.* **50,** 197 (1980).

be eradication of typically serum-sensitive bacteria.[31] However, complement activation by bacteria can enhance the inflammatory response, resulting in tissue damage.[20,32–34] Most studies on complement activation by bacteria have been carried out using planktonic bacteria.[35] However, some studies have also demonstrated complement activation by biofilm-grown bacteria.[36,37] These studies have shown that biofilm-grown *P. aeruginosa* resisted and survived the lytic action of normal human serum in contrast to planktonic bacteria. Studies by our group on interactions between biofilm bacteria and the complement system have involved the use of the modified Robbins device. In these studies the role of biofilm in the activation of complement, a major contributor to the inflammatory process, was investigated.[15] Complement activation by *P. aeruginosa* grown as a biofilm in a modified Robbins device was studied in a complement consumption assay, production of C3 and factor B conversion products assessed by crossed immunoelectrophoresis, C5a generation tested by a chemotaxis assay, and terminal complement complex formation measured by enzyme-linked immunosorbent assay (ELISA). In these studies it was shown that *P. aeruginosa* grown in biofilm activated the complement system less than planktonic bacteria.[15] Other studies have shown that some fragments of activated complement are deposited on the biofilm.[36]

Robbins Device and Antibiotic Resistance

Bacterial infections caused by colonization with sessile bacteria are a major cause of morbidity in patients with cystic fibrosis and in patients who have received medical implants.[38,39] Biofilm bacteria are a major concern for the clinicians engaged in the treatment of infectious diseases, and eradication of biofilm bacteria is often found to be very difficult even when an apparently susceptible strain is cultured from the patient prior to and during treatment.

In biofilms, bacterial cells grow embedded in a thick, highly hydrated anionic matrix that conditions the environment of the individual cells and

[31] N. Høiby and S. Olling, *Acta Pathol. Microbiol. Scand. C* **85,** 107 (1977).

[32] G. Döring, W. Goldstein, P. O. Schiøtz, N. Høiby, M. Dasgupta, and K. Botzenhart, *Clin. Exp. Immunol.* **64,** 597 (1986).

[33] N. Høiby, G. Döring, and P. O. Schiøtz, *Annu. Rev. Microbiol.* **40,** 29 (1986).

[34] M. Berger, *Clin. Rev. Allergy* **9,** 119 (1991).

[35] N. L. Schiller, *Infect. Immun.* **56,** 632 (1988).

[36] G. Pier, M. Grout, and D. DesJardins *J. Immunol.* **147,** 1369 (1991).

[37] H. Anwar, J. L. Strap, and J. W. Costerton, *FEMS Microbiol. Lett.* **92,** 235 (1992).

[38] T. M. Bergamini, D. F. Bandyk, D. Govostis, H. W. Kaebnick, and J. B. Towne, *J. Vasc. Surg.* **7,** 21 (1988).

[39] R. E. Wood, T. F. Boat, and C. F. Doershuk, *Am. Rev. Resp. Dis.* **113,** 833 (1976).

constitutes a different solute phase from the bulk fluids of the system. The matrix may protect the cells against aminoglycoside and peptide antibiotics by binding the antibiotics to the lipopolysaccharide and may retard piperacillin diffusion.[1,40] In addition, high levels of β-lactamase activity in the matrix can afford a very large measure of protection from β-lactam antibiotics.[14] Thus, bacterial biofilms are highly resistant to antibiotics showing MIC values far from realistic *in vivo* concentrations.[1,14] MIC and MBC studies of biofilms can be performed by the use of a Robbins device.

For antibiotic resistance studies, we use a modified Robbins device to prepare a *P. aeruginosa* biofilm.[14] In brief, the device is connected to a 2-liter reservoir at 37°. The reservoir is inoculated with a 2% inoculum of an exponential phase culture so that the reservoir delivers dividing cells to the device. Medium is pumped from the reservoir through the Robbins device by a peristaltic pump set to deliver 60 ml/hr. Within a few hours *P. aeruginosa* will form a monolayer biofilm and a fully developed biofilm within 20–24 hr, depending on the strains used. Bacterial growth within the reservoir is monitored by using standardized turbidity as a growth parameter. Disks containing the biofilms are removed aseptically for analysis. After use, the Robbins device is sterilized with ethylene oxide.[11]

Enumeration of Biofilm Cells

Bacteria in the biofilms can be enumerated by determining viable counts and microscopy. Disks containing the biofilms are removed aseptically and rinsed five times with 1 ml of saline in order to remove planktonic bacteria. The biofilms are scraped off, mixed in 2 ml of saline, and treated ultrasonically for 10 min at 60 W. Dilution series are made up to 10^{-5} and spread on nutrient agar from which quantitative plate counts are made. A modified epifluorescense technique shows all cells and is usually 1 log higher than viable counts due to the clumping artifacts.[11] However, microscopy techniques do not distinguish between live or dead bacteria.

Determination of MIC and MBC Values of Biofilms

Method 1

Biofilms are established and enumerated as described earlier. The reservoir and connection tubes are replaced with sterile ones and medium containing antibiotic is pumped through the device. Bacteria in the biofilms

[40] B. D. Hoyle, J. Alcantatara, and J. W. Costerton, *Antimicrob. Agents Chemother.* **36,** 2054 (1992).

before and after exposure to antibiotics can be enumerated by determining viable counts and microscopy. Biofilm cells of *P. aeruginosa* often survive 1000 times higher MIC values when compared to planktonic cells.[14]

Method 2

Adherent bacteria are established on membrane filters of 0.22 μm pore size instead of silicone disks.[41] The reservoir is inoculated with a 2% inoculum of an exponential phase culture so that the reservoir delivers dividing cells to the device. Medium is pumped from the reservoir through the Robbins device by a peristaltic pump set to deliver 60 ml/hr for 60 min depending on the strain used. Membrane filters are then colonized with approximately 10^4 cfu of exponential phase bacteria per filter. Colonized filters are placed on agar dilution plates and incubated at 37° overnight for MIC determinations. Subsequently, filters are transferred to control plates (without antibiotic) and plates (with antibiotic) and incubated to obtain MBC values. Using this method, Domingue *et al.*[41] demonstrated a higher MBC of such very young biofilm bacteria when compared to planktonic cells. This method incorporates guidelines of the National Committee for Clinical Laboratory Standards.

β-Lactamase and Biofilms

A major resistance factor against β-lactam antibiotics is the production of β-lactamase.[42,43] The chromosomally encoded enzyme is usually expressed at a very low basal level in the absence of antibiotics, but can be induced to much higher levels in their presence. Strains exhibiting stable derepression permanently produce elevated amounts of the enzyme. In a previous study we showed that imipenem induced *P. aeruginosa* biofilms to high levels of β-lactamase production. Piperacillin also induced *P. aeruginosa* biofilms but to a lesser degree.[14] Thus exposure of biofilm cells of *P. aeruginosa* to β-lactamase inducers triggers the production of the enzyme and would be expected to afford these sessile cells a large measure of protection from β-lactam antibiotics. Using a Robbins device, β-lactamase activity can be studied in biofilms.

[41] G. Domingue, B. Ellis, M. Dasgupta, and W. J. Costerton, *J. Clin. Microbiol.* **32,** 2564 (1994).

[42] D. M. Livermore, *J. Clin. Microbiol.* **6,** 439 (1987).

[43] C. C. Sanders and W. E. Sanders, *J. Clin. Microbiol.* **6,** 435 (1987).

Induction of β-Lactamase Production in *P. aeruginosa* Biofilms

Biofilms are established and enumerated as described earlier. After 72 hr, inducer is added, e.g., piperacillin or imipenem, to the modified Robbins device for 6–16 hr using the MIC value of planktonic bacteria. Disks bearing induced biofilms are removed, rinsed with saline to remove free bacteria, and used for β-lactamase quantitation, protein determination, and bacterial enumeration.

β-Lactamase Quantitation

β-Lactamase is released by freeze-thawing of the disks bearing induced or noninduced biofilms, and the activity of the suspended homogenized biofilms is quantitated by a direct spectrophotometric method using nitrocefin as the substrate.[44] For protein investigation, the freeze-thawed biofilm suspension is sonicated on ice for six periods of 30 sec. in the assay buffer, phosphate buffer, pH 6.9. The protein content is determined using the Bio-Rad (Richmond, CA) protein assay.[14]

Conclusions

In summary, studies using the Robbins device indicate that bacteria persisting in biofilm mediate a constant low-grade activation of various parameters of the immune system, thereby protecting themselves from the harmful action of the immune system and, at the same time, contributing to the chronic inflammation and tissue damage in the lungs of cystic fibrosis patients infected with *P. aeruginosa.* Furthermore, bacteria grown as biofilm are highly resistant to antibiotics showing MIC values far from realistic *in vivo* concentrations. In conclusion, the use of the Robbins device has contributed a great deal to the understanding of the mechanisms by which microorganisms protect themselves from the hostile environment in which they live.

Acknowledgment

We acknowledge the Danish Medical Research Council for financial support.

[44] C. H. O'Callaghan, A. Morris, S. M. Kirkby, and A. H. Shingler, *Antimicrob. Agents Chemother.* **1,** 283 (1972).

[17] Controlled Environment Model for Accumulation of Biofilms of Oral Bacteria

By GEORGE H. W. BOWDEN

Introduction

The study of dental plaque has been a significant topic in dental research for decades. As might be expected, this research has included *in vivo* and *in vitro* study of the human oral bacterial microbiota, and bacterial population shifts in plaque *in vivo* have been modeled in suspension in continuous culture *in vitro*. However, since the earliest studies of dental bacteria, the significance of the tooth surface biofilm community in oral health and disease has been recognized. Consequently, several *in vitro* model systems or "artificial mouths," which mimic the oral environment to some degree, have been used to explore the biology of biofilms of pure and mixed cultures of oral bacteria. None of these *in vitro* systems can accurately mimic the oral cavity, and those wishing to carry out studies of biofilms of oral bacteria have to select the system that best suits their requirements.[1–3] The model described in this article does not attempt to mimic the mouth. Essentially, it is a modified chemostat system[4–6] and provides environmental control together with the facility to remove and replace biofilms accumulated on different substrata for different periods of time.

Construction of Model System

The basic fermenter vessel (Fig. 1A) is constructed from 316 grade stainless steel and includes a stainless-steel ring in the base to retain the magnetic stirrer, a medium overflow port, and a port to sample the planktonic phase. The lid is formed from 12-mm stainless steel to add weight to stabilize the apparatus during use and carries ports for inoculation, medium input, pH/eH electrodes (autoclavable pH electrode, Russell pH Ltd.,

[1] G. H. Bowden, *Adv. Dent. Res.* **9,** 255 (1995).

[2] C. H. Sissons, *Adv. Dent. Res.* **11,** 110 (1997).

[3] J. W. T. Wimpenny, *Adv. Dent. Res.* **11,** 150 (1997).

[4] C. W. Keevil, D. J. Bradshaw, A. B. Dowsett, and T. W. Feary, *J. Appl. Bacteriol.* **62,** 129 (1978).

[5] Y.-H. Li and G. H. Bowden, *Oral Microbiol. Immunol.* **9,** 1 (1994a).

[6] D. J. Bradshaw, P. D. Marsh, K. M. Schilling, and D. Cummins, *J. Appl. Bacteriol.* **80,** 124 (1996).

METHODS IN ENZYMOLOGY, VOL. 310
0076-6879/99 $30.00

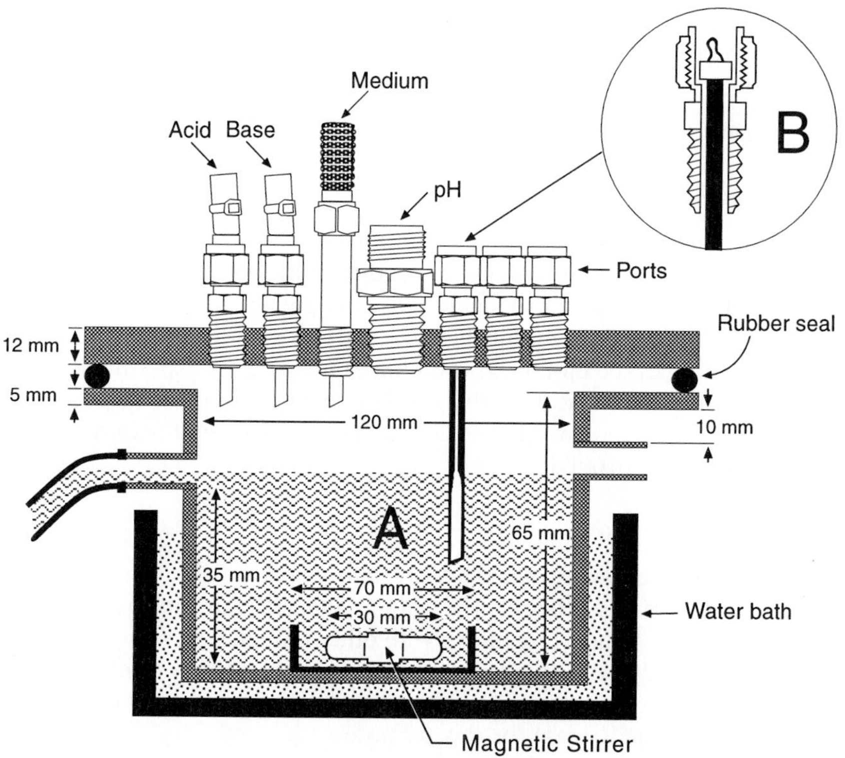

FIG. 1. Diagram of controlled environment biofilm model. (A) Basic fermenter vessel. (B) Detail of substratum port and support of rods for biofilm accumulation.

Station Road, Auchtermuchty, Fife, Scotland KY14 7DP, UK), thermometer, gases, and 10 ports to support rods or flat strips of substrata for the accumulation of biofilms. The ports are constructed from standard Swagelock pipe fittings of sufficient diameter to accept the electrodes, thermometer, and the rods. The lid is secured by three U-shaped screw clamps placed equidistant around the rim, and the junction is sealed by a silicone/neoprene gasket of the type supplied for vacuum desiccators. The fermenter is placed in a Plexiglas bath heated by the flow of water from a heated water circulator. The bath is supported on a nonheated magnetic stirrer, which serves as a base for the system.

The dilution rate in the fermenter vessel is controlled by the rate of delivery of medium by a peristaltic pump (Piper Pump, Dungey Inc., 2429 Kennedy Road, Agincourt, Ontario, Canada M1T 3H2) and the medium pH is maintained by the addition of 0.1 *M* potassium hydroxide or 0.1 *M* lactic acid, monitored by a pH controller (LH Fermentation-Inceltech, 3942 Trust Way, Hayward, CA 94545-3716).

Accumulation of Biofilms

Planktonic Environment

In this model, the planktonic environment is controlled and set to selected parameters. These parameters and the time of accumulation determine the biomass, number of viable cells, and the structure of the biofilms. Given that the physical and biochemical features of the environment influence the nature of the biofilm,[7–9] it should be possible, by experiment in this modified continuous culture system, to define the parameters to produce a biofilm of specific biomass and structure. The environment within the biofilm cannot be controlled.

Nutrient

The nature and availability of a nutrient play a significant role in both the accumulation and probably the cellular and acellular (matrix) structure of the biofilm.[1,7–9] Therefore, along with other parameters, such as gaseous environment and shear, the control of nutrients is an important consideration in deciding on strategies for the accumulation of biofilms. The continuous culture system is ideal for this purpose.

When oral bacteria have been grown, some workers have attempted to reproduce saliva, to mimic the oral environment,[10,11] whereas others have used complex media.[12] Given the significance of nutrients, an essential consideration for the *in vitro* study of biofilms is the reproducibility of composition and production of the medium and a fully defined medium would be optimum for this purpose. This does not rule out complex media, but for these media single batches of the complex components should be purchased in sufficient quantity for a complete set of experiments. Calibration of new materials against previously used batches of peptones and so on using a standard organism is possible with the continuous culture system

[7] M. C. M. van Loosdrecht, D. Eikelboom, A. Gjaltema, A. Mulder, L. Tijhuis, and J. J. Heijnen, *Wat. Sci. Tech.* **32,** 35 (1995).

[8] G. H. Bowden and Y.-H. Li, *Adv. Dent. Res.* **11,** 81 (1997).

[9] J. W. T. Wimpenny and R. Colasanti, *FEMS Microbiol. Ecol.* **22,** 1 (1997).

[10] R. P. Shellis, *Arch. Oral Biol.* **23,** 485 (1978).

[11] H. D. Donoghue, G. H. Dibdin, R. P. Shellis, G. Rapson, and C. M. Wilson, *J. Appl. Bacteriol.* **49,** 295 (1980).

[12] A. S. McKee, A. S. McDermid, D. C. Ellwood, and P. D. Marsh, *J. Appl. Bacteriol.* **59,** 263 (1985).

described. However, throughout our studies we have used a semidefined medium (Table I) that will support the growth of oral streptococci (*Streptococcus mutans, S. oralis, S. mitis* biovar 1 and *S. sanguis*), *Actinomyces* (*Actinomyces israelii, A. viscosus, A. naeslundii,* and *A. odontolyticus*), and *Lactobacillus* species. This medium can be prepared fully defined,[13] but the semidefined medium[14] is convenient with the large volumes that are often required for continuous culture. An important feature of this medium for those who wish to study biofilm or planktonic phase polymers or antigens is that with the exception of the small amount of peptone and mucin, all of the components are dialyzable and the risks of contamination of polymers from components such as yeast extract are reduced.

In the continuous culture system the number of cells in the planktonic phase can be manipulated by adjusting the levels of the limiting nutrient and the dilution rate. In this model, biofilms accumulate in a relatively large volume of fluid and one may wish to stress or reduce the numbers of viable cells within the planktonic phase to provide an advantage to biofilm cells. As an example, under carbon (glucose) limited conditions, the number of cells at a given dilution rate can be controlled by the level of glucose. We have also found that dilution of the medium reduces the number of viable cells in the planktonic phase under a given set of conditions.[5] The defined medium described diluted four times, with 1.25 m*M* of glucose at a dilution rate of $D = 0.1\ \text{hr}^{-1}$ supporting 47.0 and 38.6×10^6 viable cells/ml of *S. mutans* and *A. naeslundii* genospecies 2, respectively.

The medium includes 0.025% hog gastric mucin (Sigma type III), which forms part of the conditioning film on the surface.[5] The inclusion of mucin is also significant because it provides a source of nutrient[15] that can be used by *A. naeslundii* and some streptococci and in bicultures of *Actinomyces* with *Streptococcus*.[8] Apparently mucin contributes to the survival of *Actinomyces* in the planktonic phase.[8]

The medium is prepared from a base to which stock solutions 1, 2, and 3 are added and sterilized by autoclaving. The phosphate buffer solution and the carbohydrate solution are prepared and sterilized separately and are then added aseptically to the basal medium; this procedure avoids caramelizing the carbohydrate during autoclaving. Sterile 2% sodium carbonate (0.5 ml in a liter) is added to the complete medium to supply carbonate. Carbonate is particularly important for the growth of oral *Actinomyces*.

[13] A. O. Christie and J. W. Porteus, *J. Gen. Microbiol.* **28,** 443 (1962).

[14] G. H. W. Bowden, J. M. Hardie, and E. D. Fillery, *J. Dent. Res.* **55,** Special Issue A, 192 (1976).

[15] D. J. Bradshaw, K. A. Homer, P. D. Marsh, and D. Beighton, *Microbiology* **140,** 3407 (1994).

TABLE I

COMPOSITION OF SEMIDEFINED MEDIUM FOR CULTIVATION OF ORAL *Actinomyces* AND *Streptococcus* SPECIES

Composition	Amount for 1 liter of medium
Buffer solution	
Potassium phosphate, monobasic	6 g
Potassium phosphate, dibasic	9 g
Distilled water	to 200 ml
Sterilized separately, autoclave	
Basic solution (make up to 500 ml in distilled water with gentle heat)	
Calcium chloride dihydrate	0.02 g
Magnesium sulfate	0.2 g
Sodium acetate	0.3 g
L-Glutamic acid	0.5g
L-Cysteine hydrochloride	0.2 g
L-Asparagine	0.1 g
Glutathione	0.05 g
L-Tryptophan	0.04 g
Tryptone (Oxoid, Ltd., UK) (other enzymatic digests of casein are suitable)	2.0 g
Hog gastric mucin type III (Sigma, Oakville, Ontario, Canada)	0.25 g
Add the following solutions:	
Solution 1	10 ml
Solution 2	1 ml
Solution 3	1 ml
Composition of solutions 1, 2, and 3	
Solution 1 (amounts in 1 liter of distilled water)	
p-Aminobenzoic acid	0.2 g
Thiamine	0.2 g
Riboflavin	0.2 g
Nicotinic acid	0.2 g
Pyridoxal hydrochloride	0.2 g
Calcium pantothenate	0.2 g
Inositol	0.2 g
Solution 2 (amounts in 100 ml of distilled water)	
DL-Thioctic acid	10 mg
Biotin	10 mg
Hemin (dissolve hemin in 1 drop of water plus 1 drop of 880 ammonium hydroxide)	10 mg
Folic acid	20 mg
Solution 3 (amounts in 100 ml of distilled water)	
Ferrous sulfate heptahydrate	0.4 g
Manganous sulfate hydrate	15 mg
Sodium molybdate dihydrate	15 mg
Sodium carbonate (amount in 100 ml of distilled water)	
Sodium carbonate	2 g
Carbohydrate (amount in 100 ml of distilled water)	
The amount of carbohydrate, e.g., glucose, required for 1 liter should be added to 100 ml of distilled water and sterilized separately from the basal medium. When heat-sensitive carbohydrates are used, the solution should be filter sterilized	

Dilution Rate

As mentioned previously, the number of cells in the planktonic phase can be influenced by the dilution rate. In this system the dilution rate controls the growth rate of the cells in the planktonic phase, as it does in a chemostat. Oral *Streptococcus* and *Actinomyces* spp. will grow in the semidefined medium at pH 7.0 under glucose limitation at dilution rates[5] from $D = 0.05\ \text{hr}^{-1}$ to $D = 0.5\ \text{hr}^{-1}$.

Substrata

Any rod or flat surface that will pass through the ports can be used as a substratum for biofilm accumulation. Glass rods are most convenient and have been used to characterize biofilm accumulation by selected oral bacteria in this system.[5] Sections from glass slides provide surfaces that can be used for microscopic examination. In specific cases, measurement of the impact of different substrata or substrata with different conditioning films on biofilm accumulation may be necessary. In one study on the influence of fluoride liberated from the substratum on biofilm accumulation we found that suspensions of hydroxylapatite powder in Epon, cast as rods, produced a suitable substratum.[16] There seems to be no reason why other test agents should not be included into Epon in a similar manner to test their effects on biofilm accumulation.

The rods are supported in the port by a ring of silicone rubber cut from tubing with a wire loop on the end of the rod (Fig. 1B), which makes them easy to transfer aseptically by use of a small hook. Normally, the screw cover on the port is loosened to the last thread and then the port and cover are flooded with absolute ethanol, which is ignited. As soon as the ethanol is extinguished, the cover is removed and held with sterile forceps, and the rod is removed with a flamed stainless-steel wire with a hook at one end. The cover is then replaced and screwed down. Using this method we have had no problems with contamination of the system during removal and replacement of rods for periods of up to 8 weeks.

Time of Accumulation

Time is a significant factor to be considered in the study of the accumulation of biofilms. In this model, biofilm accumulation follows a sequence of adhesion of cells, growth of adherent cells, and a plateau of accumulation when biofilm cell number doubling times reach a maximum.[5] Usually, in semidefined medium at $D = 0.1\ \text{hr}^{-1}$ and pH 7.0 the plateau is reached

[16] Y.-H. Li and G. H. Bowden, *J. Dent. Res.* **73,** 1615 (1994).

after 24 hr of accumulation and at this time the biofilm is composed of several layers of cells. Allowing biofilms to accumulate for longer time periods gives increased biomass but it seems likely that as the accumulation of the cells increases they become physiologically heterogeneous, depending on their position in the biofilm structure. It is for this reason that an argument could be made for limiting the time of accumulation if cells are required in a particular physiological state. Cells adhere to the substratum during the first 4 hr with little or no growth and then the cells grow rapidly, often exceeding the cell division time of the planktonic phase cells. Calculation has shown that increases in cell numbers are exponential during this phase, suggesting that the division of cells on the surface contributes significantly to accumulation.[5] Subsequently, the plateau is reached. The time of accumulation must be selected based on the scientific question to be asked, the organism, medium, and environment selected. As an example, biofilms accumulated for 72 hr are generally more resistant to acid pH than 24-hr biofilms.[8] When relatively long-term, e.g., several days' accumulations of biofilm are contemplated, it must be remembered that although biofilm forms on the rods, it also forms on other surfaces in the vessel. Depending on the organism and the environment, biofilm accumulations longer than 7 days can begin to interfere with the effectiveness of the pH electrode and cause wider fluctuations in environmental pH.

Shear Forces

Shear forces at the surfaces of the substrata in this model are complex due to the eddies formed behind the rod, relative to the surface that is facing the flow. No attempt has been made to define or calculate such shear forces. However, two factors influencing shear are controlled: the length of the stir bar and the rotations per minute of the stirrer. Data given here are for the model with a 3-cm stir bar and 120 revolutions/min. One potential problem with a stainless-steel vessel is that the stirrer cannot be seen during the run. We normally set up the system and then run the stirrer at a high rate (so that the bar can be heard stirring within the retaining wall). Then the rate is reduced to a set level. Once the run is ended the system is dismantled without switching of the stirrer to confirm that it was running. To date, we have had no problem with the stirring system, which can be tested for reliability in a nonsterile vessel without the lid, filled with water.

Advantages and Limitations of the Model

The model described in this article provides an environment for biofilm accumulation that can be controlled within fine limits. Significantly, it can be used for comparisons of equivalent biomasses of planktonic and biofilm cells grown under identical environmental conditions. The reproducibility of numbers of viable adherent bacteria and biofilm cells is reasonable for up to 20 hr for a single organism in a given environment. Typically, means and standard deviations (SD) of two samples for *S. mutans* viable cells is 0.62×10^6 cm^2 (SD 0.15) at 0.5 hr and 5.12×10^6 cm^2 (SD 1.67) at 20 hr. Longer times of accumulation may give more variation, although viable cell numbers for 5-day biofilms of *S. mutans* and *A. naeslundii* genospecies 2 in basal medium under sucrose limitation were 29.6×10^6 cells/cm^2 (SD 3.6) and 27.0×10^6/cm^2 (SD 4.5), respectively. Also, cell numbers of *A. naeslundii* and *S. mutans* in biculture gave similar results. The accumulation of *S. mutans* and *A. naeslundii* in biculture for up to 24 hr appeared independent of each other, with the number of cells accumulating on the surface being equivalent to monoculture biofilms.[8] These results suggest that compared to monoculture biofilms, biofilm communities of different bacterial populations may accumulate larger amounts of biomass in a given time.

Comparison of the accumulation of oral bacteria in this model, compared to that *in vivo* in experimental animals, reveals some similarities. The phenomenon mentioned earlier of the accumulation of two organisms independent of the other has been reported to occur in gnotobiotic rats inoculated with *S. mutans* and *A. viscosus.*[17] Also, in some cases the biofilm cell number doubling times and the number of biofilm cells accumulated on glass rods in this model resemble those found in humans and experimental animals, although there are exceptions.[5]

The model allows testing of different substrata, and rods or slides can be removed for examination of the biofilm or testing the effects of antibacterials or other agents on biofilm cells *in situ.* Moreover, a rod with a preformed biofilm can be transferred to a second continuous culture system with a different environment to test growth, survival, or succession of biofilm cell populations.

The method has several limitations or aspects that should be considered before it is selected for use. The first is the relatively vast volume of the planktonic phase versus the biofilm phase. This volume is quite unlike the thin layer of saliva that covers the teeth,[18] and very often the planktonic phase will support a large total biomass of cells, which may compete with biofilm cells. Control of the numbers of planktonic cells (see earlier) may be important in this model. It is not possible to view the accumulation and distribution of adherent cells during accumulation, as is done in some flow cell models, without disturbing the system. Unlike flow models the shear

[17] H. J. A. Beckers and J. S. van der Hoeven, *Arch. Oral Biol.* **29,** 231 (1984).

[18] C. Dawes, S. Watanabe, P. Biglow-Lecompte, and G. H. Dibdin, *J. Dent. Res.* **68,** 1479 (1989).

forces operating on the surfaces are not known and may be difficult to calculate given the positioning of the rods. Also, biofilm accumulates on all surfaces in the model, unlike flow cell models. This can be a disadvantage when long-term accumulation biofilms may interfere with the operation of electrodes.

[18] Laminar Flow Chamber for Continuous Monitoring of Biofilm Formation and Succession

By MANFRED S. ZINN, ROBIN D. KIRKEGAARD, ROBERT J. PALMER, JR., and DAVID C. WHITE

Introduction

Microbial biofilms are often found on interfaces between liquids and materials. These biofilms may cause reduction of heat transfer in cooling systems, as well as increased flow resistance and potentiation of microbially influenced corrosion of drinking water systems.[1,2] One means of biofilm control is to alter the chemistry of the substratum to decrease biofilm formation and persistence. Flow cells provide a convenient on-line, nondestructive model for the assessment of biofilm control coating efficacy. This article describes a flow cell that enables quantitative analysis of the biomass under conditions where the flow and chemical properties of the bulk fluid and the substratum composition and topology are controlled. This laminar flow cell system provides valuable comparative data of the efficiency of antifouling (AF) and fouling release (FR) coatings.[3–7] Advantages of the system are (1) laminar flow conditions and controllable defined shear stress over the test surfaces, (2) nondestructive on-line monitoring of the biomass

[1] H. M. Lappin-Scott and J. M. Costerton, *Biofouling* **2,** 323 (1993).

[2] J. C. Block, K. Houdidier, J. L. Paquin, J. Miazga, and Y. Levi, *Biofouling* **6,** 333 (1993).

[3] A. A. Arrage and D. C. White, *in* "Monitoring Biofilm-Induced Persistence of *Mycobacterium* in Drinking Water Systems Using GFP Fluorescence" (J. W. Hastins, L. J. Kricka, and P. E. Stanley, eds.), p. 383. Wiley, New York, 1997.

[4] A. A. Arrage, N. Vasishtha, D. Sundberg, G. Bausch, H. L. Vincent, and D. C. White, *J. Indust. Microbiol.* **15,** 277 (1995).

[5] P. Angell, A. A. Arrage, M. W. Mittelman, and D. C. White, *J. Microbiol. Methods* **18,** 317 (1993).

[6] M. W. Mittelman, J. M. H. King, G. S. Sayler, and D. C. White, *J. Microbiol. Methods* **15,** 53 (1992).

[7] M. W. Mittelman, J. Packard, A. A. Arrage, S. L. Bean, P. Angell, and D. C. White, *J. Microbiol. Methods* **18,** 51 (1993).

0076-6879/99 $30.00

on the test surface, (3) pulsed or continuous inoculation of cells from continuous culture allows testing of antifouling agents in the bulk medium and on the treated test surfaces, and (4) a large test surface enables the harvesting of biofilm cells for quantification using signature biomarker biofilm analyses.

Flow Conditions

A flow is characterized as either laminar or turbulent depending on the strength of its lateral mixing. In laminar flows, excited instabilities are damped, whereas in a turbulent flow they grow to form self-sustained, interacting vortices on many scales. The division between laminar and turbulent flow is characterized for flows in channels (without rotation and thermal effects) by the Reynolds number (Re)[8]:

$$\mathrm{Re} = \frac{\nu d_{\mathrm{H}} \eta_0}{\rho} \tag{1}$$

where ν is a characteristic velocity (cm $\sec^{-1}$), d_{H} is a characteristic length (cm), η_0 is the viscosity in poise (g cm^{-1} $\sec^{-1}$), and ρ is the specific density (g cm^{-3}) of the medium.

For the flow in circular pipes (with d_{H} the diameter of the pipe and ν the center velocity profile) experiments show the onset of turbulent flow at Reynolds numbers of about 1000–2000.[9] The Reynolds number also serves as a basis to compare flows: flow characteristics (e.g., the velocity profile) are identical despite different media, flow velocities, or size of channel as long as the Reynolds numbers are the same.

For flows with more complicated cross sections, the characteristic length is given by the hydraulic diameter:

$$d_{\mathrm{H}} = \frac{4A_{\mathrm{C}}}{s} \tag{2}$$

where A_c is the cross-sectional area (cm^2) and s is the perimeter of the same area (cm).

For our experiments, we use a rectangular flow channel with an aspect ratio (width to height) of 75/3. The large aspect ratio results in uniform flow conditions at the center of the channel. Usually, the experiments will be run with Reynolds numbers between 1 and 10. This is well within

[8] B. R. Munson, D. F. Young, and T. H. Okiishi, "Viscous Flow in Pipes." Wiley, New York, 1990.

[9] S. J. Davies and C. M. White, *Proc. Roy. Soc. Lond. A* **119,** 92 (1928).

the laminar flow range, which results in a better reproducibility of the experiments and still represents a large range of real flow situations.

The test surfaces (coupons) are located sufficiently far away from the entry region of the channel where the flow develops its profile to equilibrium conditions over a distance of about 30 hydraulic diameters.

Shear Stress

The effect of shear stress on the growth of biofilm was investigated by Peyton and Characklis.[10] They found more rigid and homogeneous biofilms with higher shear stresses. The shear stress τ (dynes cm^{-2}) is highest at the wall and is smaller away from the surface depending on the viscosity (η_0), flow velocity (ν), and distance (d_s) from the surface (with $d_s > 0$ cm):

$$\tau = \frac{3\nu\eta_0}{d_s} \tag{3}$$

Thus, additional information on the strength of adhesion can be gained by increasing the shear stress.

Figure 1 summarizes the characteristic parameters of the flow cell that we have in use (see later).

Verification of Laminar Flow

A well-established method to verify laminar flow conditions is to examine the distribution of a soluble stain that is injected upstream of the flow cell. Several stains can be applied, such as bromphenol blue, blue dextran 2000, food stain, or commercial ink. The flow is documented by photography at time intervals. The typical flow profile of the laminar flow should be observed after at least 30 times the hydraulic diameter. At this point the distribution of the flow velocity over the channel bed remains constant as shown in the photographs as a constant hyperbolic front. The inlet into the flow cell is of great importance. A convex inlet decreases the no-flow region to a minimum (Fig. 2).

Nondestructive and On-Line Monitoring of Biofilms

The quantification of biomass on the substratum surface is a general problem in the study of biofilm growth. This quantification must be online and nondestructive for long-term experiments. Fluorescence or biolu-

[10] B. M. Peyton and W. G. Characklis, *Biotechnol. Bioeng.* **41,** 728 (1993).

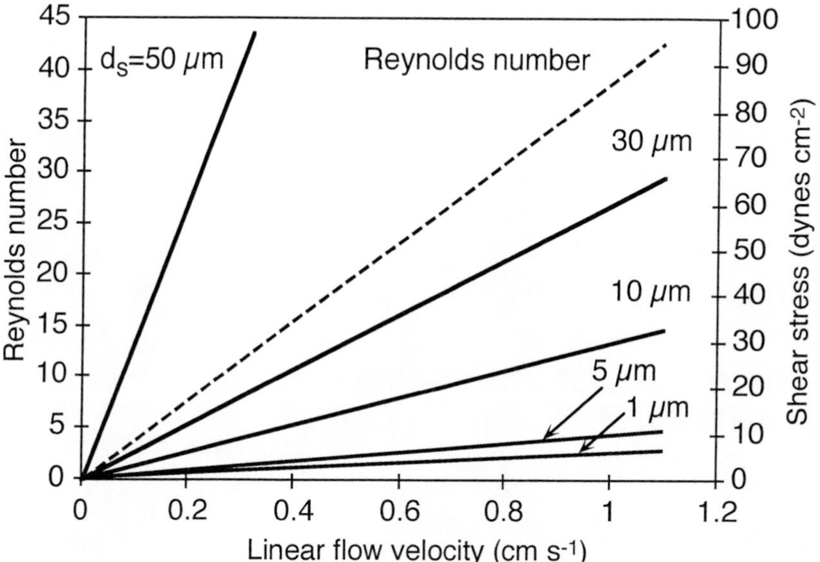

FIG. 1. Flow parameters of the flow channel (hydraulic diameter = 1.52 cm). The linear velocity of the liquid stream determines the fluid dynamics in the flow cell, which can be described by the Reynolds number. The shear stress in the liquid stream increases with the distance (d_s) to the channel surface.

minescence of natural and genetically engineered microorganisms is a nondestructive method to assess AF and FR coating effectiveness. Duysens and Amesz[11] and other authors[12,13] reported that the fluorescence of biomarkers may be used as biomass indicator (see Table I). By this detection method the cells in the biofilm can be followed repeatedly without destruction of the biofilm architecture.

Cleaning of Quartz Windows

For the measurement of fluorescence in the UV range, normal glass windows should not be used. Best results are obtained with quartz windows. Quartz windows need a special cleaning treatment in order to enable reproducible readings (e.g., fingerprints or biofilm formed on the window interfere with tryptophan detection). The procedure consists of several cleaning

[11] L. N. M. Duysens and J. Amesz, *Biochim. Biophys. Acta* **24,** 19 (1957).

[12] J.-K. Li and A. E. Humphrey, *Biotechnol. Bioeng.* **37,** 1043 (1991).

[13] B. Tartakovsky, M. Sheintuch, J. M. Hilmer, and T. Scheper, *Biotechnol. Progr.* **12,** 126 (1996).

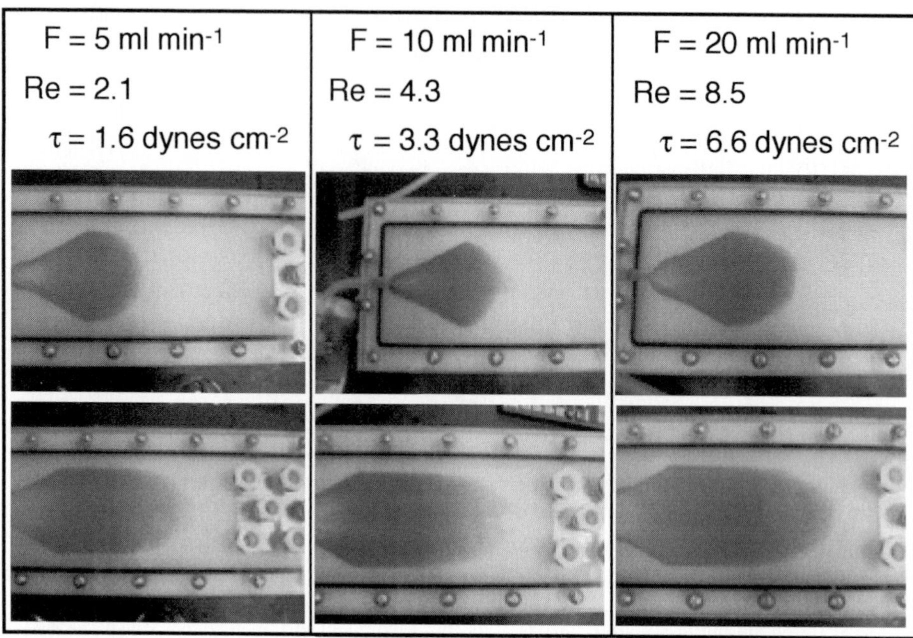

FIG. 2. Verification of laminar flow in a flow cell at three different flow rates. A convex inlet to the flow cell favors the development of the typical laminar flow pattern. After 30 hydraulic diameters, the laminar flow profile is theoretically established and has to remain constant from thereon. (Middle) Inlet of the flow channel, (Bottom) Flow pattern of laminar flow. Figures provided by K. Whitaker.

steps. First, the windows are soaked in a commercial solution of chromic acid/sulfuric acid (Manostat Chromerge, Manostat, New York, NY) overnight, subsequently washed first with distilled water and then with pure methanol. The quartz windows are transferred with forceps into 0.04 *M* KOH in methanol and incubated for about 8 hr. They are washed with this solution as many times as needed until no cloudiness is observed. Finally, the windows are rinsed with distilled water and stored in 65% (w/w) nitric acid until used.

Shortly before use, the windows are washed with distilled water and then methanol. Wearing finger cots, the windows are polished dry with a clean Kimwipe. The windows are then glued with silicone into a hollow polypropylene screw detection port (see later).

Fluorescence and Bioluminescence Measurement

On-line fluorometric measurements are performed with a Fluorolog II spectrofluorometer (Spex Instruments, Edison, NJ) equipped with a

TABLE I
BIOMARKERS TO DESCRIBE PHYSIOLOGY OF BIOFILMS

Biomarker	Detection[a]	Function	Remark	Reference
Tryptophan	Fluorometer (ex 295 nm, em 342 nm)	Biomass	Good correlation to biomass	5
NAD(P)H	Fluorometer (ex 340 nm, em 460 nm)	Cellular activity	Good correlation to biomass during exponential growth	12
ATP	Fluorometer (ex 272 nm, em 380 nm)	Cellular activity		12
Green fluorescent protein	Fluorometer (ex 400 and 480 nm, em 509 nm)	Biomass		3
Lux cassette	Light sensor	Cellular activity and biomass		6

[a] Excitation (ex); emission (em).

bifurcated fiberoptic bundle. Light from a double-grating excitation spectrometer (slit width 2.5 mm) is directed through the fiber-optic cable to the detection window of the flow cell. The emitted light of the biofilm is transported through the same cable to a double-grating emission spectrometer (slit width 2.5 mm). It is important that the spectrometer has a double-grating feature because the reflection of the excitation light can heavily obstruct measurements of emitted fluorescence light.

Bioluminescence of cells containing the lux genes is detected with an Oriel (Stratford, CT) liquid light pipe photomultiplier tube ammeter-monitoring system. All measurements are carried out in a dimmed room in bags made of light-tight fabric.

For long-term studies (>1 day), the detection windows must be cleaned under aseptic conditions with an alcohol pad before each reading.

Description of Flow Cell

The flow cell comprises two separate blocks (Fig. 3) enabling cleaning and efficient sterilization of the flow chamber. The upper block consists of translucent laminated Lexan, whereas the lower block is made of ultrahigh molecular weight polyethylene. The upper block, with a thickness of 2 cm, was designed to be translucent to permit a visual check of the flow cell, e.g., the presence of air bubbles. The chamber, which is cut into the upper

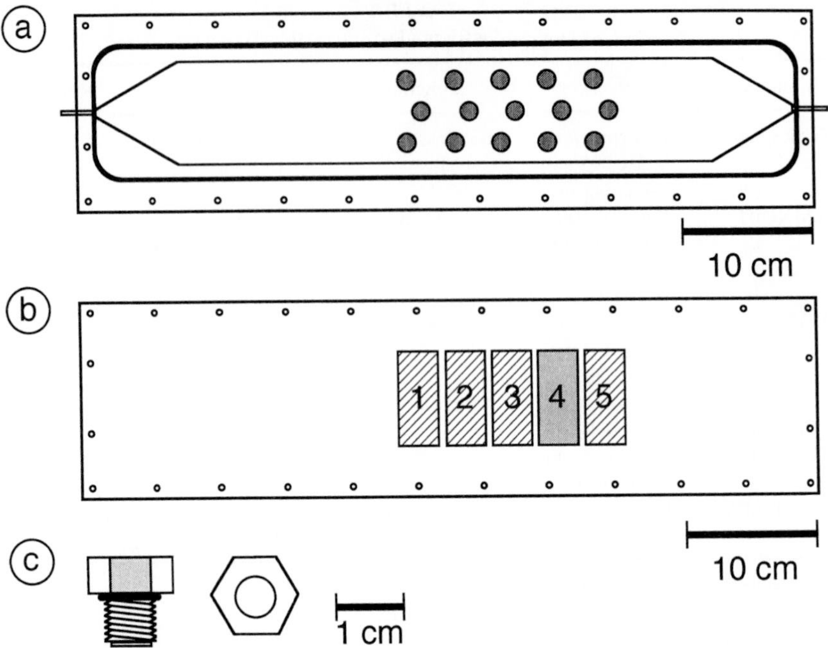

FIG. 3. Schematic view of the flow cell. (a) The upper block of the flow cell is made of translucent Lexan and holds the flow chamber. It also contains the threaded holes with a countersink for the detection windows and the groove for the Viton® O ring that seals the chamber when the flow cell is assembled. (b) The lower block is made of polyethylene and holds five test and reference coupons (numbered in flow direction, number 4 is the test surface). (c) Hollow polypropylene screws are the sample ports for fluorescence and bioluminescence measurements. They are equipped with a quartz window at the lower bottom and a Viton® O ring as a seal.

block, has a depth of 3 mm and holds 75 ml. The chamber is sealed by a Viton® rubber O ring with a cross-cut diameter of 0.5 cm. Metal screws ensure that the O ring is sufficiently compressed to form a seal.

In the bottom half of the channel (1.5 cm thick), five polished stainless-steel coupons (35 × 70 × 3 mm, Metal Samples, Munford, AL) are inserted flush with the flow channel. These coupons are also the carriers for silicone and paint coatings. The hollow polypropylene screws that carry the quartz glass windows (1.2 cm diameter) are placed in the upper half of the flow channel. Thus, fluorescence and bioluminescence measurements can be performed at three locations per coupon.

Sterilization of Flow Cell

Heat sterilization is not possible because the plastic materials warp. The flow cells are assembled with a 3-mm gap between the upper and the lower block and wrapped in a gas-permeable fabric. Sterilization is performed by exposure to ethylene oxide for over 4 hr in a gas chamber of a local hospital. The two blocks are then completely assembled in a laminar flow bench.

Experimental Procedures

The physiology of the bacterial culture used as inoculum should be well defined: the attachment of cells is affected by the growth phase.[14] We generally use a continuous culture at steady state. The dilution rate of the bioreactor is set to about 20% of the maximum growth rate. A continuous sample stream from the bioreactor is then pumped into the flow cell at a rate of 5 ml min^{-1} (Re = 2.1) over a period of 2 hr. Typically, we obtain 10^7 cells cm^{-2} from an inoculum density of 10^6 cells ml^{-1}.

To study bacterial biofouling on test surfaces, sterile medium is provided continuously via peristaltic pumps to the flow cell with sterile medium from a 40-liter reservoir through silicone tubing. The flow is calibrated with a graduated 10-ml pipette inserted into the feed line upstream of the pump through a T connector. The pipette is filled by drawing medium from the reservoir with a 60-ml syringe attached to the pipette through a sterile air filter. Then the tubing to the medium reservoir is clamped while the pump is working; the flow rate is determined by the time required to empty the pipette.

Unfortunately, peristaltic pumps cause pulses in the liquid flow. However, these fluctuations can be damped to some extent by mounting a sterile test tube upside down to the tubing downstream of the pump.

Destructive Biofilm Analysis

At the end of an experiment, a pump is connected to the outlet tubing and the flow cell is tilted 45° with the inlet up. The feed tubing is removed and the pump is started to drain the flow chamber at a rate of approximately 100 ml min^{-1}. The upper block is immediately unscrewed and the test coupons are collected in sterile petri dishes containing wet paper towels to reduce desiccation of the biofilm. The cells are quantitatively removed from the coupons by probe sonication (three times for 1 sec at 20 W) in

[14] S. McEldowney and M. Fletcher, *J. Gen. Microbiol.* **132,** 513 (1986).

10 mM phosphate buffer (pH 7.0) using 1.131-cm^2 glass O-ring extractors (Kontes Glass, Vineland, NJ). The cells can then be processed further for acridine orange direct counts,[15] lipid extractions,[4] or protein analysis.[16]

Data Analysis

From each coupon three data points are collected and are analyzed statistically (t test at a 95% probability level). Generally, three flow cells are run at the same time in order to improve the statistical analysis. For long-term experiments, data of an untreated metal coupon in position 3 (Fig. 3) are taken as a benchmark.[4]

[15] G. A. McFeters, A. Singh, S. Byun, P. E. Callis, and S. Williams, *J. Microbiol. Methods* **13,** 87 (1991).

[16] M. W. Mittelman, D. E. Nivens, C. Low, and D. C. White, *Microb. Ecol.* **19,** 269 (1990).

[19] Perfused Biofilm Fermenters

By DAVID ALLISON, TOMAS MAIRA-LITRAN, and PETER GILBERT

Biofilms may be simply described as functional consortia of microbial cells that are enveloped within extracellular polymer matrices and associated with surfaces.[1] Their physiology, metabolism, and organization are greatly dependent on the nature of those surfaces and also on the prevailing physicochemical environment.[2,3] As such, although biofilms associated with various ecological niches differ significantly from one another, they nevertheless share many common properties. These include increased levels of resistance toward antimicrobial substances, together with increased exopolymer (glycocalyx) and extracellular enzyme synthesis. Indeed, observations made both *in vivo* and *in situ* repeatedly demonstrate the recalcitrance of bacterial biofilms toward inimical agents that show demonstrable efficacy toward planktonic populations.[4,5] Because nutrient and gaseous gradients are strongly evident across mature biofilms, such physiological attributes

[1] J. W. Costerton, K. J. Cheng, G. G. Geesey, T. I. Ladd, J. C. Nickel, M. Dasgupta, and T. J. Marrie, *Annu. Rev. Microbiol.* **41,** 435 (1987).

[2] D. G. Allison and P. Gilbert, *Sci. Prog.* **76,** 305 (1994).

[3] M. Fletcher, *in* "Biofilms: Science and Technology" (L. F. Melo, T. R. Bott, M. Fletcher, and B. Capdeville, eds.), p. 113. Kluwer Academic, Dordrecht, 1992.

[4] D. G. Allison and P. Gilbert, *J. Ind. Microbiol.* **15,** 311 (1995).

[5] P. Gilbert, J. Das, and I. Foley, *Adv. Dent. Res.* **11,** 160 (1997).

0076-6879/99 $30.00

might be related to changes in growth rate and the imposition of specific nutrient limitations within the depth of an established biofilm rather than to the adoption of an attachment specific phenotype.[4] In order to fully evaluate such differences, *in vitro* models are required that differentiate between adhesion and the associated influences of the glycocalyx, nutrient status, and growth rate.[6] Given the ubiquity of biofilms, the now widespread recognition of their unique properties, and the need to develop new methods for their control, numerous laboratory models have been developed in order to reproduce them. These models vary in complexity from growth on solidified nutrient media through to complex biofermenters.[7] Nutrient and oxygen gradients that are present within thick, mature biofilms will change with time in closed, batch growth systems and will result in the starvation of the entire population, dispersal of the attached cells, and the induction of a whole gambit of stringent response genes.[8] Moreover, the concentration of antimicrobial agent applied to a closed growth system will gradually become depleted in terms through reaction with the bulk of cells leaving the residual survivors to develop resistance to that agent. In open, continuous growth systems, however, a pseudo steady state will develop, offering the possibility of growth rate control of the attached population. The choice of an appropriate model system must therefore be tempered not only by technical constraints and an understanding of the *in vivo/in situ* organization of the biofilm, but also by consideration of all the physiological processes that may influence the end result.

Perfused biofilm fermenters allow for the control of growth rate within adherent populations and, in doing so, can often be used to differentiate between the effects of gradients of growth rate and the unique properties associated with adhesion. In perfused fermenters, cells dispersed from the biofilm are washed away continuously whereas fresh nutrients are in constant supply, the result of which is the development of a pseudo steady-state biofilm. Under such conditions the biofilm is accessible to biocide/antibiotic perfusion *in situ* or for susceptibility testing of either intact biofilms or suspended biofilm cells. This technique has generally been used in association with chemostat cultures, maintained at similar growth rates,[9] in order to provide appropriate planktonic controls. Although such open,

[6] M. R. W. Brown, D. G. Allison, and P. Gilbert, *J. Antimicrob. Chemother.* **22,** 777 (1988).

[7] P. Gilbert and D. G. Allison, *in* "Microbial Biofilms: Formation and Control" (S. P. Denyer, S. P. Gorman, and M. Sussman, eds.), p. 29. Blackwell, Oxford, 1993.

[8] M. M Zambrano and R. Kolter, *in* "Microbiological Quality Assurance: A Guide Towards Relevance and Reproducibility of Inocula" (M. R. W. Brown and P. Gilbert, eds.), p. 23. CRC Press, Boca Raton, FL, 1995.

[9] P. Gilbert, D. G. Allison, D. J. Evans, P. S. Handley, and M. R. W. Brown, *Appl. Environ. Microbiol.* **55,** 1308 (1989).

perfused growth systems are atypical of biofilms *in situ* on inanimate surfaces, they are representative of bacterial surface infections with soft tissues and of other situations where the microbes derive the bulk of their nutrients from the substratum. Additionally, perfused biofilm models do not generally allow for the accumulation of high cell density planktonic populations because the detaching cells are washed away before they can divide. Because this is the case then the rate of appearance of cells in the perfusate gives an indication of the growth rate of the sessile population. The former is an important consideration for *in situ* biocide and antibiotic testing where, with most other types of biofilm fermenters, the detached cells will remain to exert a quenching effect on the antimicrobial agent.

This article describes the methodologies and uses of three different but related procedures for the laboratory culture of perfused biofilms, namely (i) the perfused biofilm fermenter (PBF),[9] (ii) the Swinnex biofilm fermenter,[10] and (iii) the Sorbarod biofilm fermenter.[11] Each of these methods provides pseudo steady-state biofilms and varying degrees of growth rate control.

Procedures

Perfused Biofilm Fermenter

The perfused biofilm fermenter[9] is based on an original method used for the selection of populations of bacteria that are synchronized with respect to cell division.[12] In this approach, bacterial cells are attached to one side of a bacteria-proof filter membrane by pressure filtration and then perfused with fresh growth medium from the sterile side. The initial perfusion is rapid and removes loosely attached cells. The majority of cells, however, remain attached to the filter matrix and grow with a proportion of the progeny being lost to the perfusing medium. The originators of this approach noted that cells collected in the perfusate were at a common point in their division cycle and divided synchronously when transferred to batch cultures. An alternate view of the method is, however, that it serves to generate an attached biofilm population and that the continuous flow of fresh medium offers the potential, as in a chemostat, to control the growth rate. This potential of the Helmstetter and Cummings technique has been optimized in the perfused biofilm fermenter.[9] Biofilms produced

[10] S. Gander and P. Gilbert, *J. Antimicrob. Chemother.* **40,** 329 (1997).

[11] A. E. Hodgson, S. M. Nelson, M. R. W. Brown, and P. Gilbert, *J. Appl. Bacteriol.* **79,** 87 (1995).

[12] C. E. Helmstetter and D. J. Cummings, *Biochim. Biophys. Acta* **82,** 608 (1963).

in this device generally have a thickness of 10–20 μm, but the majority of this is composed of exopolymers with cell coverage averaging that of a cell monolayer. Cell densities in the device ($6 \times 10^6/cm^2$) are therefore relatively small and insufficient for biochemical analyses. Despite the low biomass generated in the PBF, some studies have been able to demonstrate that the increased rates of exopolymer, siderophore, and protease synthesis in biofilm communities are growth rate dependent using this approach.[13] As such, the perfused biofilm fermenter provides a useful model of soft tissue infections such as *Pseudomonas aeruginosa* colonization of the cystic fibrotic lung. The most valuable application of the PBF remains, however, its ability to monitor the survival/growth of biofilm communities during their *in situ* exposure to inimical agents.

Establishment of Biofilm Cultures. Midexponential phase batch cultures (60 ml, ca. 10^8 cfu/ml) are pressure filtered (35 kNm^{-2}) through a prewashed (50-ml sterile medium), 47-mm-diameter cellulose acetate membrane filter (0.22 μm, Oxoid, Basingstoke, UK) within a Millipore pressure filtration unit (Millipore Corporation, Watford, UK). The cell-impregnated filter membrane is removed and inverted carefully into the base of a modified glass, jacketed continuous fermentation apparatus (Fig. 1). When fresh medium is passed into the fermentation chamber via the peristaltic pump (1 ml/min), a hydrostatic head develops above the membrane filter. At steady state this forces the perfusion of the filter, from its sterile side, at a rate that is equal to the rate of medium addition to the vessel. When viable counts are performed on the collected perfusates, these then show the number of eluted cells to decrease rapidly over 100–150 min to achieve a steady-state value (Fig. 2) that can be maintained for 5–14 days according to the organism type and medium flow rates. After steady state has been achieved, then the sacrifice of replicate filter units and viable count determination show the adherent population to increase only slightly (less than twofold) over 14 days. The initial rapid decrease in the numbers of shed viable cells has been shown to correspond to the removal of loosely attached cells from the filter bed. Cells eluted at steady state correspond to newly formed daughter cells and divide synchronously when transferred to fresh medium[9] (a 5-ml eluate collected on ice is added to 45 ml growth media in a 250-ml Erlenmeyer flask; synchronization index of >0.9). It should be noted that in order to subsequently control the growth rate of the attached population, it is necessary that the provided growth medium restricts growth, in batch culture, to a stationary phase cell density of ca. 5×10^8 per ml through the availability of a single substrate (i.e., C, N, Fe^{3+}).[14]

[13] E. Evans, M. R. W. Brown, and P. Gilbert, *Microbiology* **140,** 153 (1994).

[14] S. J. Pirt, "Principles of Microbe and Cell Cultivation." Blackwell, Oxford, 1975.

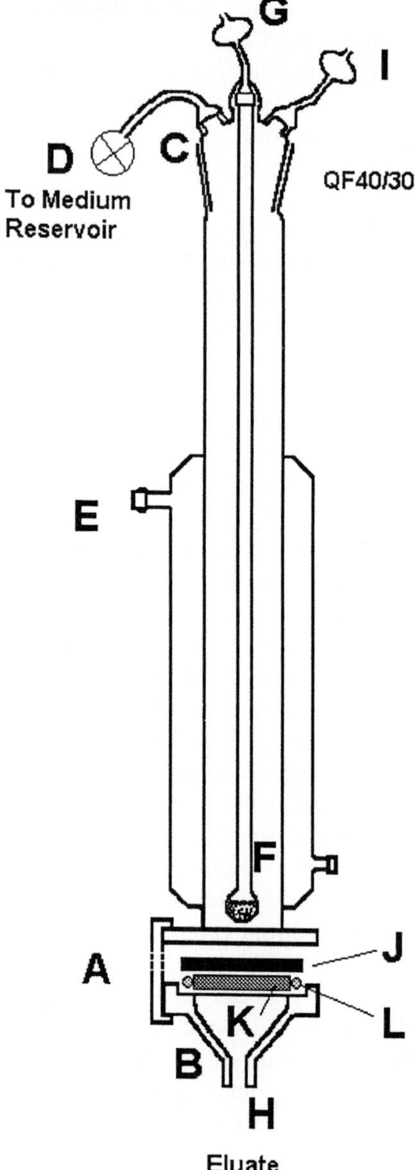
G
I
D
C
To Medium
Reservoir
QF40/30
E
F
A
J
K
L
B
H
Eluate

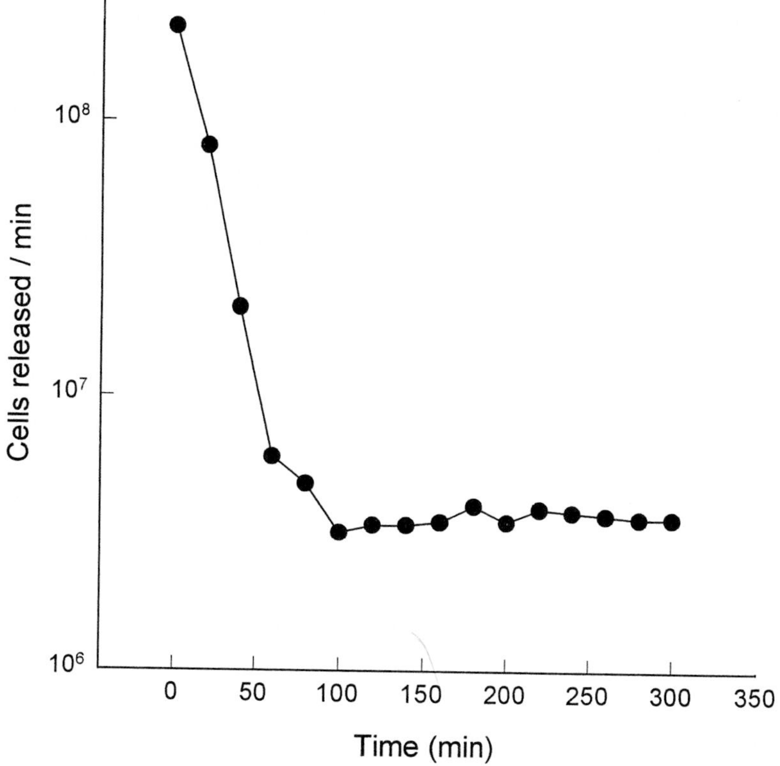

FIG. 2. Elution of *E. coli* from a cellulose acetate filter perfused with medium in the biofilm fermenter at a rate of 1 ml/min. Organisms eluted during the initial 100 min correspond to loosely attached cells. Thereafter, newly formed daughter cells are eluted under steady-state conditions. After P. Gilbert, D. G. Allison, D. J. Evans, P. S. Handley, and M. R. W. Brown, *Appl. Environ. Microbiol.* **55,** 1308 (1989).

FIG. 1. Perfused biofilm fermenter. The biofilm device consists of a glass fermenter (A) and a clamped Teflon base (B). Fresh medium is delivered to the fermenter (C) via a peristaltic pump (D), and the whole device is maintained at incubation temperatures by a water jacket (E). Aeration occurs through a glass sinter (F) connected to an air filter (G) and pump. The medium and air outlets are at H and I, respectively. The filter base assembly shows the positioning of the 0.22-μm pore size filter membrane (J) impregnated with cells on the hydrophobic surface adjacent to the stainless-steel sintered support (K) and O-ring seal (L). After P. Gilbert, D. G. Allison, D. J. Evans, P. S. Handley, and M. R. W. Brown, *Appl. Environ. Microbiol.* **55,** 1308 (1989).

After achieving an initial steady state (ca. 100 min), viable counts made on the perfusate collected at different medium flow rates show that the rate of elution of cells from the filter bed is directly related to the medium flow rate up to a critical flow rate (Fig. 3). Thereafter, the attached biomass and the rate of elution of cells from the filter bed remain constant. The results bear a strong similarity to those of chemostat culture in which the critical flow rate corresponds to achievement of μ_{max} and where the adherent population is under a form of growth rate control brought about by the rate of perfusion of the membrane filter.

As the system has been demonstrated to be at a pseudo steady state,

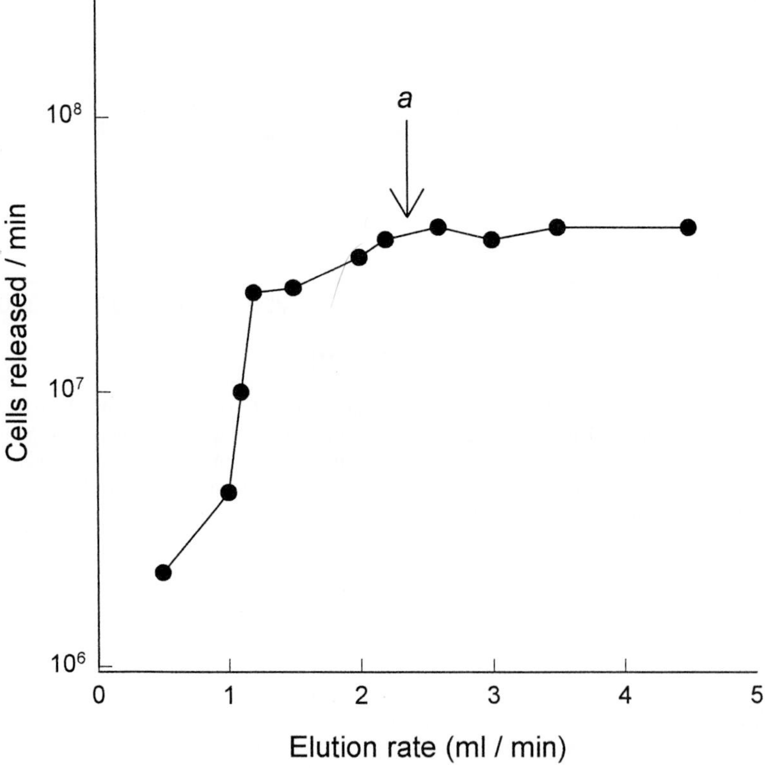

FIG. 3. Relationship between the rate of elution of the filter membrane and the rate of organisms released from the membrane per minute. From the graph it is possible to calculate the critical medium flow rate *a*, which at steady state is equivalent to μ_{max}. Growth rate control is exerted in the period up to *a*. The rate of cell devision in the biofilm at slow elution rates is regulated by the rate of fresh medium flow.

then the doubling time (t_d) of the cells on the membrane can be calculated from the following equation:

$$t_d = \frac{\text{membrane population}}{\text{rate of elution of cells}}$$

The specific growth rate (μ, per hour) of the biofilm cells may then be calculated using

$$\mu = \frac{\log_e 2}{t_d}$$

Harvesting of Cells from Steady-State Biofilms. At various intervals after achieving steady state, membranes are removed carefully from the fermenter using flamed forceps and placed into Universal bottles containing sterile saline (10 ml). The adherent cells are removed by vortexing (2 min). This was shown to remove greater than 99% of the attached cells. Susceptibility testing can be performed either on the suspended biofilm population or on samples of intact biofilm. For the latter, sections of biofilm, together with its membrane support, are immersed in antibiotic solutions for fixed exposure times, removed into sterile saline, vortexed (2 min) to remove and suspend the treated biofilm cells, and viable counts made.[15]

Applications. Sample biofilms, prepared for scanning electron microscopy by low-temperature stage-freeze techniques and/or embedded and sectioned for transmission electron microscopy (TEM), show discrete cells attached to the filter matrix through an extended extracellular polymer matrix (20–30 μm thick), which strongly resembles *in vivo*.[9,16] The technique has been used successfully for the culture of biofilms of a number of bacterial species, which include *Escherichia coli*[17] and clinical isolates of *P. aeruginosa*[18] and gram-positive organisms such as *Staphylococcus epidermidis*[19] and the yeast *Candida albicans*,[20] to investigate the effects of growth rate on cell surface properties such as hydrophobicity and charge,[17–19] production of

[15] D. J. Evans, D. G. Allison, M. R. W. Brown, and P. Gilbert, *J. Antimicrob. Chemother.* **27,** 17 (1991).

[16] I. G. Duguid, E. Evans, M. R. W. Brown, and P. Gilbert, *J. Antimicrob. Chemother.* **30,** 791 (1992).

[17] D. G. Allison, M. R. W. Brown, D. J. Evans, and P. Gilbert, *FEMS Microbiol. Lett.* **71,** 101 (1990).

[18] D. G. Allison, D. J. Evans, M. R. W. Brown, and P. Gilbert, *J. Bacteriol.* **172,** 1667 (1990).

[19] P. Gilbert, D. J. Evans, I. G. Duguid, E. Evans, and M. R. W. Brown, *in* "Microbial Cell Surface Analysis: Structural and Physicochemical Methods" (N. Mozes, P. Handley, H. J. Busscher, and P. G. Rouxhet, eds.), p. 341. VCH Publishers, New York, 1991.

[20] G. S. Baillie and L. J. Douglas, *Antimicrob. Agents Chemother.* **42,** 1900 (1998).

virulence factors,[13] and biocide/antibiotic susceptibility of microbial biofilms.[15,16,20–23]

Because the cells that were shed spontaneously from the perfused membrane filter were at a common point in their division cycle (*S. epidermidis, E. coli,* and *P. aeruginosa*), research using the approach suggested that there might be some involvement of the cell cycle in the programmed dispersal of cells from biofilm communities. The approach offers the potential to make collections of the spontaneously dispersed cells and to study their characteristics. These properties commonly include a much reduced ability to colonize virgin surfaces, or indeed new filter membranes, and a lack of surface-associated materials such as pili and exopolymer. In addition, the perfused biofilm fermenter offers the possibility of exposure of intact biofilm communities to controlled levels of antibiotic that may be included within the medium supply. As such, the effects of antibiotic exposure can be monitored on the overall population dynamics of the system, during and posttreatment. A major drawback of the technique in this respect is the difficulty in operating multiple sets of the apparatus simultaneously. A modification of the PBF was therefore made using Swinnex filter units (later) in order to enable multiple sets to be operated at once. This is particularly important when undertaking screens, with a high throughput of compounds, or when pharmacodynamic data are required.

Swinnex Biofilm Fermenter

The basic principles of the PBF were incorporated into a scaled-down version through the use of modified Swinnex filter units (Millipore Corporation, Bedford, MA).[10] Although these enable, as with the PBF, some degree of growth rate control of the sessile population, they are simple to assemble, inexpensive and less prone to contamination than the PBF.

The developed model consists of Swinnex filter units (Millipore Corporation; 22 mm diameter, 0.22-μm filter) modified such that air may be supplied to the exposed side of the incorporated filter. The modified apparatus is illustrated in Fig. 4. In addition to the provision of an aeration port, the ridge of the base unit was machined so as to give a flat surface, and Teflon spray (Orme Scientific, Manchester, UK) was applied to the inside of the filter unit prior to sterilization. These procedures were found neces-

[21] D. J. Evans, M. R. W. Brown, D. G. Allison, and P. Gilbert, *J. Antimicrob. Chemother.* **25,** 585 (1990).

[22] D. J. Evans, D. G. Allison, M. R. W. Brown, and P. Gilbert, *J. Antimicrob. Chemother.* **26,** 473 (1990).

[23] I. G. Duguid, E. Evans, M. R. W. Brown, and P. Gilbert, P, *J. Antimicrob. Chemother.* **30,** 803 (1992).

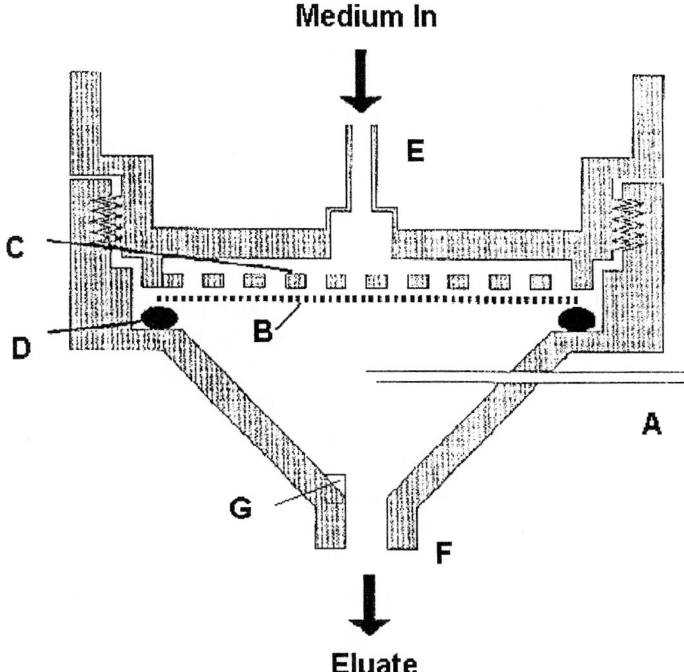

FIG. 4. Schematic representation of the Swinnex biofilm fermenter. Cells are loaded onto a 0.22-μm pore size filter membrane (B) *in situ* through port F. The membrane is secured by an O-ring seal (D) and supported by the sintered surface (C). The whole unit is inverted and medium and air inlet tubing attached. Sterile medium is delivered by means of a peristaltic pump through port E, with F now acting as an outlet. Aeration is achieved using a needle inserted in the side of the Swinnex unit (A). A ridge (G) present in the base of the unit was removed in the modified device. Adapted from S. Gander and P. Gilbert, *J. Antimicrob. Chemother.* **40,** 329 (1997).

sary in order to prevent excessive attachment of the dispersed cells to the exposed surfaces of the filter unit.

Establishment of Biofilm Culture. Swinnex filters are purpose made small-scale filtration devices that may be fitted in line to intravenous delivery lines or used for sterile filtration of fluids directly from a syringe. One side of the filter unit has a Luer-lock connector for attachment to a sterile disposable syringe. In the present method, an adapted, presterilized Swinnex unit containing a 0.45 μm porosity cellulose acetate filter (22 mm diameter) is connected directly (via port F) to a prefilled syringe (20 ml) with a clamp placed on the air-inlet tubing. Initially the membrane is washed through with 20 ml of prewarmed, sterile growth medium. This is followed by 10 ml of a midexponential phase batch culture of the test organism.

Once the membrane has been inoculated with cells, the syringe is removed and the unit is inverted, clamped, using a suitable stand, and held within a static incubator. The medium inlet and output tubing, together with the air inlet tubing, are attached at ports E, F, and A, respectively, from/to remote pumps and reservoirs. The unit is subsequently perfused continuously with fresh growth medium by means of a peristaltic pump (15 ml/hr). The behavior characteristics of the Swinnex filter units are identical to that of the PBF in that after an initial rapid loss of loosely attached cells, the attached biofilm and the numbers of dispersed cells reach a steady state at which the rate of growth of the attached population can be regulated through the rate of supply of fresh medium. The two major advantages of the Swinnex system are that inversion of the filter holder, rather than inversion of an inoculated membrane, significantly reduces the likelihood of contaminants and that through the use of multitubing peristaltic pumps, many devices can be set up and operated simultaneously.

Biofilms of *E. coli* generated using the modified Swinnex filter units can be maintained at steady-state conditions and produce synchronously dividing cells within the eluate, as with the PBF, for 5 to 6 days. Steady-state values are achieved more quickly in the Swinnex filter units (ca. 60 min). When mucoid, clinical isolates of *P. aeruginosa* and clinical isolates of *Staphylococcus aureus* were employed, however, the magnitude of both the atached biofilm and of the eluted population was found to increase steadily. This was related to an increase in the levels of exopolysaccharide, leading to blockage of the membrane at around 60 hr.[10] Blockage did not occur when antibiotics or biocide was incorporated within the perfusate ca. 24 hr after initiation of the fermenters enabling the device to be used, even with these organisms for antibiotic screening. The differences in performance between the Swinnex units and the PBF are likely to relate to the much higher hydrostatic pressures generated with a hand-operated syringe during inoculation than with the controlled, low-pressure filtration of the PBF.

Harvesting of Cells from Steady-State Biofilms. Biofilm populations are harvested from the Swinnex filter units by disassembly and removal of the filter with sterile forceps. The harvested biofilms, together with their membrane supports, can then be used directly or the biofilm cells may be suspended, as before.

Applications. The major application of the Swinnex biofilm fermenter is the *in situ* evaluation of the effects of the controlled exposure of biofilms to both biocides and antibiotics. In such approaches, the antibiotic is incorporated into the medium supply vessel after the unit has achieved steady state. The effectiveness of the agent can be monitored through the numbers of viable cells dispersed from the biofilm with the perfusate. A reduction in colony-forming units indicates either a growth inhibitory effect on the biofilm community or a direct bactericidal effect. These two effects may be differentiated from one another by continual monitoring of the perfusate for viable cells for a posttreatment period. Alternatively, individual units may be sacrificed during an experiment in order to assess the viability and size of the residual attached population of cells. Defined pharmacokinetic profiles of antibiotic concentration with time can be duplicated by the application of an appropriate dilution regime to the antibiotic-containing medium reservoir. In this fashion, Gander and Gilbert[10] utilized the system to differentiate the effects of bolus injection of antibiotic with that of continuous infusion. Gander perfused steady-state biofilms of *E. coli* with medium containing various concentrations of ciprofloxacin for 120 min (60 ml/hr) with posttreatment perfusion with antibiotic-free medium for a further 24 hr (Fig. 5). Such experiments facilitated not only a comparison of the bactericidal effects of the agent, but also allowed study of the postantibiotic effect in biofilm communities.

Sorbarod Biofilm Fermenter

Neither the PBF nor the Swinnex biofilm fermenter is suitable for the long-term culture and growth rate control of heavily mucoid organisms such as *P. aeruginosa.* In addition, whereas these two techniques are suitable for application in studies of antimicrobial susceptibility, where the major analytical tool is the performance of viable cell counts, they do not provide sufficient biomass as to permit biochemical analyses of the biofilm cells and associated exopolymers. Sorbarod filter units were first developed as supports for the micropropagation of plant tissue[24] and had not been used previously in bacteriology. They consist of cylindrical paper sleeves (10-mm diameter by 20 mm) encasing compacted concertinas of cellulose fibers arranged longitudinally within the case. Resembling a cigarette filter, the packed cellulose filling provides a large surface area for bacterial adhesion, whereas the extensive interfiber spaces avert system blockage when perfused. The use of these filters, housed within a rigid perfusable support (Fig. 6), was evaluated as a means of generating perfused biofilm models.[11] Such filter matrices were found to be significantly less prone to blockage than the 0.45-μm porosity membranes used previously, achieved pseudo steady state when perfused at low rates, and generated a relatively large biomass suitable for biochemical analyses. A major difference between these systems and the PBF is that the biofilm is perfused from its outermost surface and a degree of heterogeneity is introduced into the community through its longitudinal arrangement.

[24] D. Conkie, *Orchid Rev.* **6,** 390 (1988).

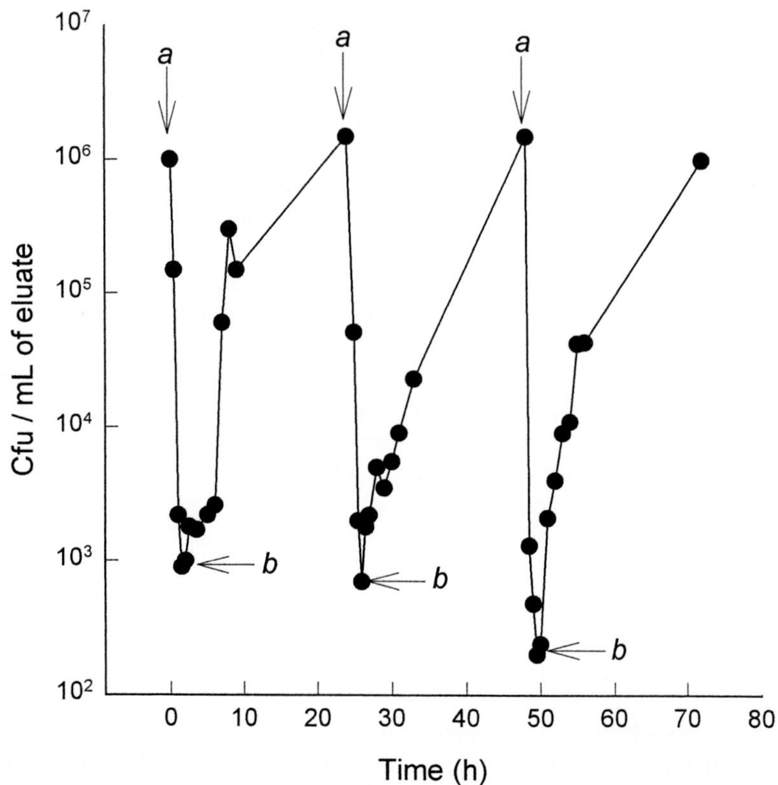

FIG. 5. Perfusion of steady-state *E. coli* biofilms with ciprofloxacin (2 mg/liter) for a 120-min period every 24 hr. Changes in the number of viable cells eluted from the biofilms with the perfusate are shown. Ciprofloxacin was included (indicated as *a*) in the perfusing medium (60 ml/hr) at time 0 and removed (indicated as *b*) after 120 min for each exposure to the antibiotic. From S. Gander and P. Gilbert, *J. Antimicrob. Chemother.* **40,** 329 (1997).

Establishment of Biofilm Cultures. Lengths of clear polyvinyl chloride (PVC) tubing (40 mm long, 10 mm diameter) containing single Sorbarods (Ilacon, Kent, UK) were sterilized by autoclaving. After drying, each Sorbarod was prewetted with 5 ml of sterile saline and inoculated with midexponential phase cultures (1×10^9 cfu in 10 ml media) from a syringe, dropwise onto the Sorbarod. The plunger is withdrawn from a sterile, disposable 2-ml syringe with a central outlet, leaving only the rubber seal within the syringe lumen (Fig. 6B). Figure 6 shows that the syringe can then be introduced into the PVC tubing (E) containing the Sorbarod (F), and a sterile, disposable needle (D; 0.8×40 mm) is inserted through the rubber seal. When assembled, multiples of these units can be clamped upright and

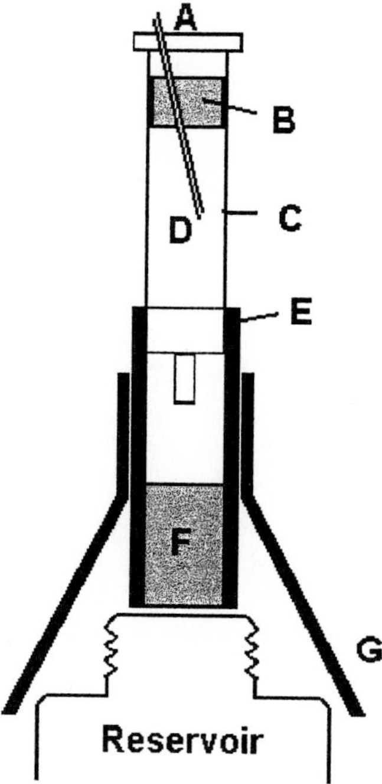

FIG. 6. Schematic diagram of the Sorbarod biofilm fermenter showing silicone rubber tubing connecting to peristaltic pump and media vessel (A), rubber plunger seal (B), 2-ml disposable sterile syringe (C), syringe needle (D), PVC tubing (E), and Sorbarod (F), together with an aseptic collection hood (G) and a collection reservoir. Modified design from that of A. E. Hodgson, S. M. Nelson, M. R. W. Brown, and P. Gilbert, *J. Appl. Bacteriol.* **79,** 87 (1995).

placed in an incubator. Media inlet tubing is attached via the inlet needles (Fig. 6A), and sterile medium may be delivered to each unit via a peristaltic pump at controlled rates.

After inoculation and start of perfusion the dispersed cells may be collected and viable counts made (Fig. 7). At various times after setup, individual Sorbarods may be sacrificed, the biofilm cells suspended, and estimates of the bioburden made (see later). Such results are reminiscent of the PBF in that over the first few hours of perfusion the number of dispersed cells decreases significantly, presumably corresponding to the detachment of loosely bound cells. Unlike with the PBF, however, this is followed by a period where both the magnitude of the biofilm community and the rate of appearance of dispersed cells in the perfusate increase significantly to achieve a new steady state after ca. 30–40 hr (Fig. 7). Preliminary experiments had determined that during this time period cells detaching from the upper regions of the filter were reattaching lower down,

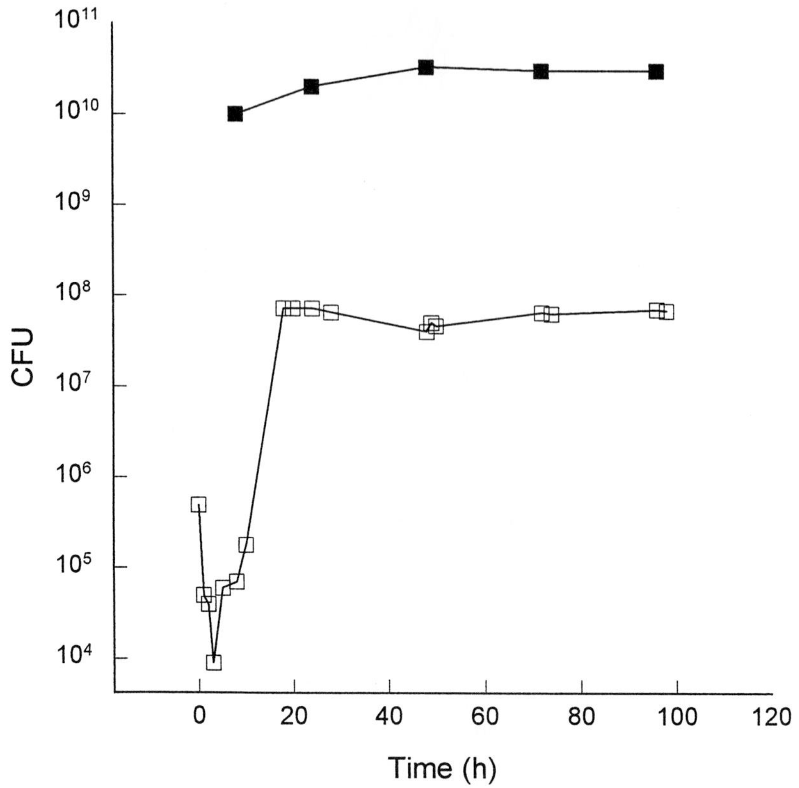

FIG. 7. Number of *Enterococcus faecalis* cells associated, as biofilms, within Sorbarod filters and eluted with the perfusing medium over a 96-hr period after initiation of medium perfusion at a flow rate of 0.1 ml min^{-1}. Data are illustrated for numbers of viable cells suspended from Sorbarods (■) and released into the eluate (□).

gradually saturating the available surface of the filter with attached bacteria. Population growth rates can be determined from the doubling time for the adherent biomass and the rate of release of cells from it. It is important to note that inoculated Sorbarod filters release cells from attached biofilms, which then become entrapped further down the filter matrix. The calculated growth rates are therefore the average of the filter community, which will presumably be heterogeneous in this respect, as with other biofilm devices[25] from its proximal to distal end. The net growth rates obtained, however, are highly reproducible and significantly lower than those obtained for the

[25] H. M. Lappin-Scott, J. Jass, and J. W. Costerton, *in* "Microbial Biofilms: Formation and Control" (S. P. Denyer, S. P. Gorman, and M. Sussman, eds.), p. 1. Blackwell, Oxford, 1993.

same cells grown planktonically in the same media. The yield of biofilm cells from the Sorbarod filters is relatively high (10^{10-11} cfu/filter) and sufficient as to allow biochemical analyses of the community.[26]

Harvesting of Cells from Steady-State Biofilms. Cellular biomass may be recovered relatively simply from the Sorbarod filter units. The filter is removed from its PVC sleeve using sterile forceps and is placed in a glass boiling tube (24 × 146 mm). The outer sheath is cut with a scalpel and the fibers are allowed to unravel as (10 ml) sterile saline is added and the mixture is vortexed for 1 min. The fibers are then physically separated from the fluid using a 24-mm-diameter silicone disk attached to a piston. The latter operates as a cafetiere and can be plunged into the boiling tube, forcing the fibers to the base. The supernatant fluids are decanted and the whole washing procedure is repeated four times before a viable count is made on the accumulated washings. If appropriate, then the residual cellulose fibers may be digested enzymatically with cellulase in order to release "footprint" polymers. The number of cells harvested per filter unit is between 10^{10} and 10^{11} cfu, depending on species and medium employed.

Applications. Sorbarod filters, inoculated and perfused with media, provide slow-growing biofilms for which the rate of growth is reproducible and measurable. In conjunction with continuous culture, this will allow comparisons to be made between physiologies and structures of attached and planktonic cells without the added complication of differences in growth rate. This method of culture has a number of advantages over the perfused biofilm fermenter, not the least of which is its cost effectiveness and ease of use. The simplicity of the device makes it easy to set up and operate in large numbers without the need for any specialist equipment. More significantly, the number of bacterial cells that may be harvested on sacrifice from Sorbarod filters is substantially greater (ca. 10^{11}) compared to those collected from perfused membranes (ca. 10^{8}). This gives the approach great potential for use in screens for biofilm activity,[11] for designing and testing of biocides and antibiotics,[27,28] and for conducting biochemical analysis.[26] Sorbarod filters can also be used for the study of population dynamics occurring within mature biofilms. Using appropriate chromosomally marked strains in conjunction with epifluorescence microscopy, the Sorbarod fermenter is an ideal vehicle for measuring the transfer of genetic information between cells in a biofilm. In addition, it has also been used

[26] H. S. Montgomery, S. Gaskell, J. Andrew, A. E. Hodgson, and P. Gilbert, *Microbiology,* submitted.

[27] R. K. Budhani and J. K. Struthers, *J. Antimicrob. Chemother.* **40,** 601 (1997).

[28] I. Foley and P. Gilbert, *J. Antimicrob. Chemother.* **40,** 667 (1997).

to study the immigration, attachment, and genetic interaction of planktonic cells with the biofilm community.[29]

Comments

It is acknowledged that cells within natural biofilms grow very slowly, and evidence suggests that many of the physiological properties ascribed to the biofilm mode of growth may actually be attributable to slow rates of growth. It is essential, therefore, that care be taken in choosing appropriate planktonic controls if the conclusions made from experiments are to be relevent to adhesion phenotypes per se. Many of these potential problems are overcome by the use of these perfused biofilm models where the growth rate control of the adherent population can be achieved at rates similar to those predominating *in situ/in vivo*. In this respect, not only do perfusion biofilm models provide excellent means of assessing antibiotic susceptibilities but they also enable the comparison of the physiologies of biofilm cells relative to planktonic populations and facilitate the study of population dynamics within mature biofilms.

[29] P. Gilbert, D. G. Allison, A. E. Jacob, D. Korber, G. Wolfaardt, and I. Foley, *in* "Biofilms: Community Interactions and Control" (J. Wimpenny, P. Handley, P. Gilbert, H. Lappin-Scott, and M. Jones, eds.), p. 133. Bioline Press, Cardiff, 1997.

[20] Laboratory Techniques for Studying Biofilm Growth, Physiology, and Gene Expression in Flowing Systems and Porous Media

By ROBERT J. C. MCLEAN, MARVIN WHITELEY, BRIAN C. HOSKINS, PAUL D. MAJORS, and MUKUL M. SHARMA

Introduction

A number of techniques are employed for culturing planktonic (i.e., unattached) bacteria and studying their physiology and patterns of gene expression. These approaches include batch and continuous culture techniques and the development of a variety of defined and complex liquid media. Many studies (reviewed in Ref. 1) have shown that bacteria in most

[1] J. W. Costerton, Z. Lewandowski, D. E. Caldwell, D. R. Korber, and H. M. Lappin-Scott, *Annu. Rev. Microbiol.* **49,** 711 (1995).

0076-6879/99 $30.00

environments grow predominately as surface-adherent biofilm communities. In contrast to their planktonic counterparts, sessile organisms within biofilms are quite resistant to antimicrobial agents.[2,3] This has been attributed to the slow growth of biofilm organisms as well as diffusion gradients mediated by biofilm matrix polymers.[4] Consequently, biofilms are implicated in a number of chronic medical and industrial problems (reviewed in Ref. 1). The increasing interest in sessile (i.e., biofilm) bacteria has spurred the development of laboratory techniques to culture them. Biofilm investigators have often employed culturing techniques for planktonic bacteria and modified them for growing sessile bacteria. The rationale for this approach is that *de novo* biofilm formation occurs on most surfaces as a consequence of planktonic bacterial colonization.[5] There are some instances (e.g., urinary catheters) where biofilm formation begins in one region of a surface. The remaining surface is then colonized as the biofilm creeps along the surface.[6] Many different types of substrata are colonized. Surfaces such as polyethylene are relatively inert in that the colonizing bacteria do not gain any nutrition from the plastic itself, although they may gain nutrients from adsorbed molecules.[7] Some substrata, such as plant fiber, represent sources of nutrition, so the rationale for adhesion is readily apparent.[8] Still, there are a number of instances where biofilm formation occurs on surfaces such as copper that are toxic to planktonic bacteria.[9] The biofilm mode of growth must provide microorganisms a strategy whereby they can survive a normally unfavorable environment.

Physiological studies of biofilm bacteria are still in their infancy. Such investigations in planktonic bacteria have typically relied on growing large quantities of organisms for experimentation. Experimental values of protein concentrations or enzyme activities reflect average values of planktonic populations. We have found that experimental variation between replicates of batch-grown organisms can be minimized by subculturing organisms prior to experimentation. A minimum of two, 18- to 24-hr subcultures are generally sufficient in the case of fast-growing bacteria such as *Escherichia*

[2] J. C. Nickel, I. Ruseska, J. B. Wright, and J. W. Costerton, *Antimicrob. Agents Chemother.* **27,** 619 (1985).

[3] M. W. LeChevallier, C. D. Cawthon, and R. G. Lee, *Appl. Environ. Microbiol.* **54,** 2492 (1988).

[4] P. Gilbert, J. Das, and I. Foley, *Adv. Dent. Res.* **11,** 160 (1997).

[5] J. R. Lawrence, P. J. Delaquis, D. R. Korber, and D. E. Caldwell, *Microb. Ecol.* **14,** 1 (1987).

[6] J. C. Nickel, J. A. Downey, and J. W. Costerton, *Urology* **39,** 93 (1992).

[7] G. Reid, C. Tieszer, R. Foerch, H. J. Busscher, A. E. Khoury, and H. C. Van der Mei, *Cells Mater.* **2,** 253 (1992).

[8] T. A. McAllister, H. D. Bae, G. A. Jones, and K.-J. Cheng, *J. Anim. Sci.* **72,** 3004, (1994).

[9] R. J. C. McLean, A. A. Hussain, M. Sayer, P. J. Vincent, D. J. Hughes, and T. J. N. Smith, *Can. J. Microbiol.* **39,** 895 (1993).

coli, Pseudomonas aeruginosa, or *Selenomonas ruminantium.*[9a] Genetic manipulation is typically done with liquid-grown or agar-grown cultures. Identification of mutants or recombinants is often done by techniques such as positive selection for antibiotic resistance or enzyme activity, e.g., β-galactosidase expression on X-Gal plates or negative selection used for the identification of auxotrophic mutants.[10] Although mutants or recombinants arise in single cells, they are usually identified from colonies that are populations typically containing $>10^6$ cells.

Studies of biofilms are complicated by the heterogeneity of the chemical microenvironments and cell distribution throughout the biofilms.[11] As a consequence, a given chemical environment and associated microbial physiological activity may exist in a very small region of a biofilm (often a few μm^3). Microbial physiology and gene expression within biofilms are therefore best studied at the level of the individual cells and not in large populations. Although biofilm investigations are a relatively new aspect of microbiology, there are a number of different techniques for culturing them (summarized in Ref. 12). The large diversity in biofilm culture techniques increases the difficulty of comparing results between laboratories. This article documents several commonly used strategies for growing and studying biofilms in laboratory conditions. We shall highlight approaches familiar to the authors and commonly used in biofilm investigations. For transparent or translucent biofilms, these approaches include the Robbins' device (RD)[2,13] (see elsewhere in this volume[13a]) and the flow cell used in microscopy[14,15] (see elsewhere in this volume[13b]). For opaque biofilms, the approaches include imaging nuclear magnetic resonance (NMR)[16] and environmental scanning electron microscopy.[14]

[9a] R. J. C. McLean and K. J. Cheng, Unpublished observations, 1984.

[10] E. Eisenstadt, B. C. Carlton, and B. J. Brown, *in* "Methods for General and Molecular Bacteriology" (P. Gerhardt, R. G. E. Murray, W. A. Wood, and N. R. Krieg, eds.), p. 297, ASM Press, Washington, DC, 1994.

[11] J.R. Lawrence, D. R. Korber, B. D. Hoyle, J. W. Costerton, and D. E. Caldwell, *J. Bacteriol.* **173,** 6558 (1991).

[12] J. W. T. Wimpenny and S. L. Kinniment, *in* "Microbial Biofilms" H. M. Lappin-Scott and J. W. Costerton, eds.), p. 99, Cambridge Univ. Press, Cambridge, 1995.

[13] W. F. McCoy, J. D. Bryers, J. Robbins, and J. W. Costerton, *Can. J. Microbiol.* **27,** 910 (1981).

[13a] A. Kharazmi, B. Giwercman, and N. Høiby, *Methods Enzymol.* **310,** [16], 1999 (this volume).

[13b] R. J. Palmer, Jr., *Methods Enzymol.* **310,** [2], 1999 (this volume).

[14] S. B. Surman, J. T. Walker, D. T. Goddard, L. H. G. Morton, C. W. Keevil, W. Weaver, A. Skinner, K. Hanson, D. E. Caldwell, and J. Kurtz, *J. Microbiol. Methods* **25,** 57 (1996).

[15] D. G. Davies, M. R. Parsek, J. P. Pearson, B. H. Iglewski, J. W. Costerton, and E. P. Greenberg, *Science* **280,** 295 (1998).

[16] K. Potter, R. L. Kleinberg, F. J. Brockman, and E. W. McFarland, *J. Magn. Reson. B* **113,** 9 (1996).

Biofilm Growth in Flowing Systems

A number of experimental devices have been used for biofilm colonization.[12] These devices range from simple glass microscope slides suspended in liquid to rather complex laminar flow and turbulent flow systems coupled to a fermentor and computer. Although bacteria attach readily to most surfaces, the diversity in biofilm culturing systems makes it difficult to compare results from one laboratory to another. This section addresses some of the issues involved in culturing bacterial biofilms.

Culture Selection

Culture selection is a key aspect of biofilm experimentation. Here, investigators can employ either one or more laboratory cultures or else natural isolates. Well-defined laboratory cultures are essential during investigations of physiology in that the relevant genes and enzymes may be well characterized. When necessary, recombinant strains can be engineered so that expression levels of a particular gene (and its influence on biofilm physiology) can be controlled. Several strategies can be used for controlling expression levels. One strategy is to insert the gene of interest on a range of plasmids having different copy numbers. A second strategy would be to insert the gene of interest into a well-defined promoter. One tightly controlled promoter that can be employed is the arabinose operon in *E. coli.*[17] To monitor patterns of gene expression, gene fusions can be constructed with reporter genes such as *gfp* (described in other articles in this volume[17a]). While genetic engineering offers many important tools for the study of microbial physiology within biofilms, one must be aware that the physiology of recombinant or mutant organisms can be altered greatly. For example, the synthesis of additional gene products such as reporter proteins (e.g., luciferase, Gfp, or β-galactosidase) requires an allocation of resources from other cellular functions such as the synthesis of capsule polymers (CPS). Should resources be diverted away from key aspects of biofilm physiology, such as CPS synthesis,[18] biofilm formation may be hindered.

One major drawback of using laboratory cultures is that one or more physiological traits (e.g., CPS biosynthesis) may be unstable and lost easily on laboratory subculture. Here, natural isolates have an advantage over laboratory strains in that they are under constant environmental selection to form biofilms.[1] Enrichment culture provides an easy way to isolate biofilm-

[17] D. A. Siegele and J. C. Hu, *Proc. Natl. Acad. Sci. U.S.A.* **94,** 8168 (1997).

[17a] C. Prigent-Combaret and P. Lejeune, *Methods Enzymol.* **310,** [4], 1999 (this volume).

[18] D. G. Davies and G. G. Geesey, *Appl. Environ. Microbiol.* **61,** 860 (1995).

forming organisms. In one example, we isolated biofilm-forming bacteria from a karst aquifer by suspending sterile limestone rocks, encased in a nylon mesh, under the water table. After the rocks had been submerged for several days, the rocks were removed aseptically. Adherent bacteria were then removed by suspending the rocks in sterile water, sonicating the rock–water mixture for 10 min in a bath sonicator, and isolating the organisms by dilution plating.[19]

Shear forces, which are produced during sonication, are employed to dislodge bacteria from surfaces and to separate individual organisms from aggregates. Shearing can also cause cell death due to disruption of the bacterial cell wall or through the generation of heat. We recommend that investigators establish conditions for biofilm enumeration. This can be done by sonicating a colonized surface for varying times. Samples of the bacterial suspension are removed periodically and bacterial densities are measured by dilution plating. The sonication time, which gives the highest bacterial count [expressed as colony-forming units], is then chosen for future experimentation. In addition, samples from the bacterial suspension can be examined by phase-contrast microscopy to determine if microbial clumps have been separated adequately. Once isolated, the accurate identification of environmental isolates can be challenging. There are a number of techniques commonly used, including the Biolog (Biolog, Inc., Hayward, CA) identification system, which employs metabolic tests,[20] membrane lipid analysis,[21] and 16S rRNA typing.[22]

Once isolated, cultures must be catalogued and stored in the laboratory. Here, the investigator has several options, which include freeze-drying, freezing, or frequent subculturing. While freeze-drying and storage in liquid nitrogen has been shown to be the most effective technique for maintaining culture viability, the equipment required is costly. A suitable approach for most laboratories is to grow the cultures overnight in liquid media (e.g., LB broth), add sterile glycerol to a final concentration of 10% (v/v), dispense the cultures into labeled vials and then store the vials in an −80° freezer. Several commercially available presterilized culture storage supplies are available from manufacturers such as Nalge Nunc International (Rochester,

[19] M. Whiteley, E. Brown, and R. J. C. McLean, *J. Microbiol. Methods* **30,** 125 (1997).

[20] P. S. Amy, D. L. Haldeman, D. Ringelberg, D. H. Hall, and C. Russell, *Appl. Environ. Microbiol.* **58,** 3367 (1992).

[21] D. C. White, H. C. Pinkart, and D. B. Ringelberg, *in* "Manual of Environmental Microbiology" (C. J. Hurst, G. R. Knudsen, M. J. McInerney, L. D. Stetzenbach, and M. V. Walter, eds.), p. 91, ASM Press, Washington, DC, 1997.

[22] D. A. Stahl, *in* "Manual of Environmental Microbiology" (C. J. Hurst, G. R. Knudsen, M. J. McInerney, L. D. Stetzenbach, and M. V. Walter, eds.), p. 102, ASM Press, Washington, DC, 1997.

NY). While storage of mixed populations has been proposed by some investigators,[23] it has been our experience that the population composition of a frozen mixed culture will shift over time due to differences of survival in the individual members of the population. Further details of culture storage are described by Gherna.[24]

Community Composition

Most biofilm investigations to date have looked either at natural populations *in situ* or at monocultures *in vitro.*[1] The most common gram-negative organisms employed in monoculture biofilm experiments are *Pseudomonas* sp., primarily *P. aeruginosa.* Gram-positive organisms, primarily *Staphylococcus aureus* and *Staphylococcus epidermidis,* are also used in some studies. These organisms form biofilms readily *in vitro* and are also implicated in medically important biofilm infections.[1] While monoculture biofilms are grown easily in the laboratory, they are not representative of biofilm formation in nature. As a result, there is a rekindling of interest in mixed culture biofilm studies. Detailed studies of mixed culture biofilms must necessarily explore interactions between component members of a community. With binary cultures, one can compare the impact of binary culture on biofilm formation and physiology to those effects seen when the component cultures are grown in monoculture. In the case of tricultures, possible interactions include those among all three components, between any two components, and those seen in monoculture. One can easily see the great complexity involved in studying more complex communities.

Mixed culture biofilms can be established in one of two ways. In the first instance, a mixed population (e.g., a naturally colonized rock from a stream or a naturally colonized catheter from a patient) can be placed directly into a biofilm-growing apparatus (examples described later). In the second instance, the investigator can selectively add individual cultures to generate a mixed population of desired composition. In the first case (inoculation from a natural population), some variability may be encountered between experimental replicates because of the uneven distribution of microbial populations. In the second case, allowing the mixed population to achieve equilibrium in chemostat culture before commencing biofilm growth can minimize variability. In chemostat culture experiments, we typically assess a population equilibrium of mixed cultures by sampling the culture liquid (planktonic population) and measuring bacterial numbers by

[23] D. E. Caldwell, G. M. Wolfaardt, D. R. Korber, and J. R. Lawrence, *Adv. Microb. Ecol.* **15,** 105 (1997).

[24] R. L. Gherna, *in* "Methods for General and Molecular Bacteriology" (P. Gerhardt, R. G. E. Murray, W. A. Wood, and N. R. Krieg, eds.), p. 278, ASM Press, Washington, DC, 1994.

dilution plating and estimating diversity by counting colonies of varying colony morphology. It has been our experience that mixed aquifer and stream populations take 5–7 days to "equilibrate" in a chemostat. Once a steady state has been achieved, biofilm growth is allowed to commence (described later). The great advantage of this approach is that variability between experimental replicates is minimized greatly.

Batch Culture and Continuous Culture

Batch cultures have been widely used in many aspects of microbiology,[25] including biofilm culture.[26] Here, the organisms are placed into sterile media, allowed to grow, and produce biofilms. Many aspects of biofilm physiology, notably antibiotic resistance, have been related to the growth rate.[27] Consequently, chemostat culture has become the dominant approach for biofilm experimentation.[19,27] It has been our experience that the reproducibility of experimental data between experimental runs (a notorious problem with batch-grown biofilms) is enhanced with continuous culture biofilms grown on defined media. Many commercially available chemostats and fermentors are available. The expense of these commercially available systems can be quite formidable. We designed a simple, inexpensive chemostat from a laboratory flask that can be constructed inexpensively with appropriate glass-blowing facilities (Fig. 1).[19]

As with other factors, the choice of growth media will impact biofilms. Many of our studies investigate the role(s) of global gene regulation involved in such aspects as starvation survival, slow growth, and quorum sensing. Here, defined media and growth rates are essential to control physiology. Most of our work has involved biofilms grown under carbon limitation. To develop the appropriate defined media, one inoculates the organism to be investigated into a series of tubes with varying dilutions of a carbon source such as glucose. After overnight incubation, growth of the organism is measured with a spectrophotometer and graphed as a function of glucose concentration. Based on the results, one can then design a medium where bacterial growth is solely limited by glucose concentration.

Biofilms can be formed from batch-grown or chemostat-grown cultures by a number of techniques. The simplest techniques involve suspending sterile surfaces, such as glass microscope slides, into these cultures. While

[25] P. Gerhardt and S. W. Drew, *in* "Methods for General and Molecular Bacteriology" (P. Gerhardt, R. G. E. Murray, W. A. Wood, and N. R. Krieg, eds.), p. 224, ASM Press, Washington, DC, 1994.

[26] J. W. Costerton, K.-J. Cheng, G. G. Geesey, T. I. Ladd, J. C. Nickel, M. Dasgupta, and T. J. Marrie, *Annu. Rev. Microbiol.* **41,** 435 (1987).

[27] S. Gander and P. Gilbert, *J. Antimicrob. Chemother.* **40,** 329 (1997).

FIG. 1. A simple chemostat for biofilm experimentation can be constructed from a glass flask.[19] An aeration tube (A) with a 0.22-μm filter to prevent contamination and a waste outlet (W), are illustrated. For biofilm growth, the culture is pumped from an outlet (O) at the bottom of the flask and returned through the top as illustrated in Fig. 3. Loss of media due to evaporation can be minimized by bubbling air through a water trap (not shown).[19]

biofilms will certainly form on these surfaces, they are highly variable in structure and so are ill suited for any quantitative measurements. The Robbins' device, (Chapter 16) constructed originally of brass,[13] has been used to monitor biofilm formation in a number of industrial settings. A modified RD (MRD) (shown in Fig. 2) constructed of acrylic[2] allowed biofilm formation to be investigated on a number of different substrata. RD and MRD can be purchased from Tyler Research Corporation (Edmonton, AB, Canada). The RD consists of a tube, either circular or rectangular in cross section, through which bacteria-containing liquid is pumped. It contains a number of sample plugs (typically 25) along the length of the device. These plugs are removable sections of the tube wall on which biofilms form. A number of different materials can be inserted into the plugs of the MRD so that biofilm formation and physiology can be studied on a number of different materials. In theory, all sample plugs on a RD or MRD should give equivalent results. We have found that while chemostat-grown cultures produce reproducible results in the 25 sample plugs of a typical MRD, this reproducibility does not hold true for batch-culture biofilms. Prior to conducting any biofilm experiments with RD, MRD, or any other device, the investigator should perform some background experiments to ascertain the reproducibility of biofilm formation within a single experimental run and between experimental runs. We typically assess

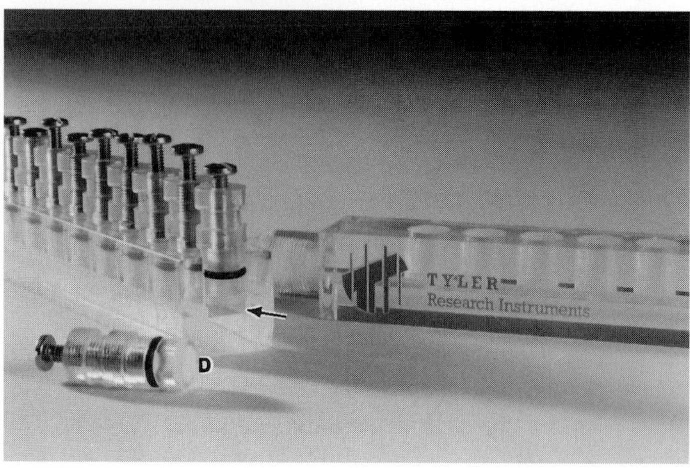

FIG. 2. A cross section of a modified Robbin's device (MRD) showing the path through which liquid flows (arrow).[2] The plug design of the MRD permits bacterial adhesion to be measured on disks (D) of different materials.

biofilm reproducibility by growing our organism of interest in a MRD and measuring biofilm reproducibility based on colony-forming units per sample area (cfu/cm^2). Contamination and leakage are also concerns with RD and MRD experiments. Leakage can be contained by using Teflon tape on all screws and placing the RD or MRD in a tray. The acrylic MRD can be disinfected with 2% (v/v) NaOCl and rinsed with sterile H_2O. Ensuring that all containers and tubing are either sterilized or disinfected can control contamination.

Once background experiments in media development, identifying culture stability (for mixed culture experiments), and biofilm reproducibility are completed, experimentation can commence. The chemostat is prepared and inoculated with the appropriate culture(s) and allowed to grow in batch mode overnight to establish growth before the chemostat pump is started. The planktonic culture is grown for a minimum of one to three dilutions (i.e., 100–300 hr at a dilution rate of 0.01 hr^{-1}) to allow the culture to become stabilized, after which the chemostat is connected to the RD or MRD. The duration of biofilm growth is left up to the investigator. Bacterial colonization and biofilm formation can occur quickly, whereas physiological changes occur within aged biofilms.[28]

[28] H. Anwar, J. L. Strap, and J. W. Costerton, *Antimicrob. Agents Chemother.* **36,** 1347 (1992).

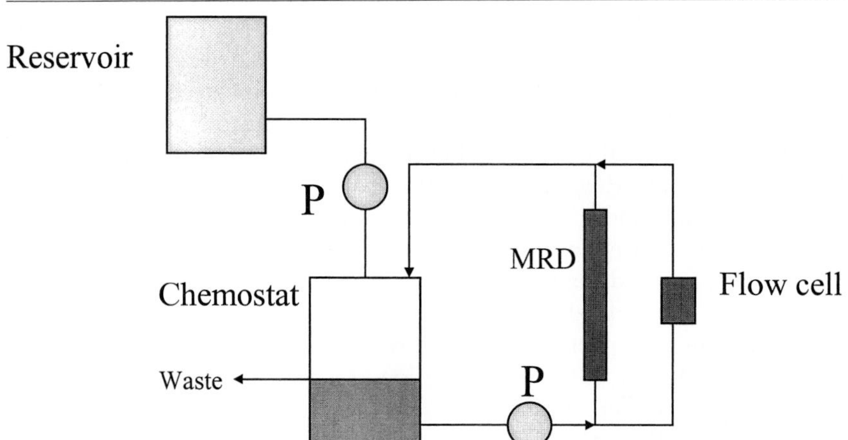

FIG. 3. Schematic diagram, illustrating how a chemostat can be connected to a biofilm-growing device such as a MRD or flow cell (F).[19] This design permits the investigator to use two pumps (P) to control growth rate and flow rate independently.

Imaging Biofilms

There are a number of methods for observing biofilms. Biofilms formed on RD or MRD can be observed by scanning electron microscopy or transmission electron microscopy. Although detailed ultrastructural information can be obtained in this fashion, biofilm samples must be processed and examined under vacuum.[26] Biofilms represent highly hydrated structures. Consequently, the dehydration and staining protocols for electron microscopy cause cell death and may introduce a number of artifacts.[29] Scanning confocal laser microscopy (SCLM) is widely used in biofilm studies as it permits the investigation of living organisms. Details of SCLM sample preparation and fluorescent staining protocols are presented in other articles in this volume. This section addresses some of the concepts involved in growing biofilms for SCLM.

Optical examination of biofilms by SCLM requires the cells, substratum, and liquid to be transparent or translucent. Light scattering, the SCLM staining protocol, and the quality of the optics will influence the success of this protocol. As a result, many studies of biofilm structure are performed on glass using a device such as flow cell.[30] Flow cells can be connected easily to a chemostat using a schematic as presented in Fig. 3. SCLM can

[29] L. L. Graham, R. Harris, W. Villiger, and T. J. Beveridge, *J. Bacteriol.* **173,** 1623 (1991).

[30] D. E. Caldwell and J. R. Lawrence, *in* "CRC Handbook of Laboratory Model System for Microbial Ecosystems" (J. W. T. Wimpenny, ed.), Vol. 1, p. 117, CRC Press, Boca Raton, FL, 1988.

be performed on biofilms adherent to opaque substrata. In this case, the biofilms are grown on the substratum, covered with a coverslip, stained (if necessary), and examined by SCLM. Some compression artifacts may result, but in general this approach can yield much useful information. There are many commercially available fluorescent stains from suppliers such as Molecular Probes (Eugene, OR) that can be employed in SCLM. Image analysis software, including that available with many SCLM systems, permits one to obtain quantitative data from SCLM studies. In this fashion one can acquire and map data such as growth rate, gene expression,[31,32] and viability [33] in individual cells. Wherever possible, one should calibrate fluorescent stains, especially those that will be used to acquire quantitative information. During a SCLM observation of biofilms formed under microgravity,[33a] we observed that bacteria, detected as nonviable (by a fluorescent staining protocol), were actively swimming. Fluorescent labels can also be placed onto antibodies or gene probes in order to identify individual cells within a biofilm. Although a small molecule such as fluorescein penetrates biofilms readily, [34] the introduction of larger molecules may necessitate fixation.[22]

Physiology and Gene Expression

Physiological studies of biofilm organisms can be performed on either the population as a whole or at the individual cell level. Individual cells are best studied using fluorescent probes and SCLM (covered previously). When the physiology of biofilm populations is to be studied, biofilms are grown and then subjected to the appropriate biochemical test. Radiolabeled substrates are often employed and the investigator measures the disappearance of a substrate and the appearance of a product or an enzyme cofactor. For practical purposes, biofilms of a desired characteristic can be grown using a MRD. The plugs, containing the adherent biofilm, are then removed from the MRD, dipped into sterile H_2O (to remove unattached cells), and placed into a scintillation vial containing an appropriate substrate. It is important that the biofilm cells not be dislodged during this process, as one will end up investigating planktonic cells. Physiological data should be

[31] S. Møller, A. R. Pedersen, L. K. Poulsen, E. Arvin, and S. Molin, *Appl. Environ. Microbiol.* **62,** 4632, 1996.

[32] S. Møller, C. Sternberg, J. B. Andersen, B. B. Christensen, J. L. Ramos, M. Givskov, and S. Molin, *Appl. Environ. Microbiol.* **64,** 721, 1998.

[33] M. Virta, S. Lineri, P. Kankaanpää, M. Karp, K. Peltonen, J. Nuutila, and E.-M. Lilius, *Appl. Environ. Microbiol.* **64,** 515, 1998.

[33a] R. J. C. McLean, J. M. Cassanto, M. B. Barnes, and J. Koo, unpublished results, 1998.

[34] J. R. Lawrence, D. R. Korber, and D. E. Caldwell, *J. Microbiol. Methods* **10,** 123, 1989.

normalized on the basis of cell density, total cellular protein, or some other appropriate factor. An example of a protocol for studying the heterotrophic potential of urinary catheter biofilms is presented in Ladd *et al.*[35] In this fashion, decontamination of radiolabeled equipment is minimized.

Biofilm Growth in Opaque Porous Media

Biofilms often form in opaque media. This phenomenon can occur in situations where microbial corrosion products obscure biofilms[36] or when biofilms occur within porous rock formations.[37] Opaque biofilms on corroded metal surfaces can be studied by environmental scanning electron microscopy,[36] although this technique is not amenable to biofilms that form deep within porous media. This section describes a technique based on nuclear magnetic resonance for studying biofilms in opaque environments wherein conventional microscopic techniques do not work. Biofilm formation can be produced in porous media by pumping a bacterial suspension into a substrate such as a core of sandstone or sintered glass beads.[37]

Lewandowski *et al.*[38–41] used NMR flow measurement and imaging techniques to study biofilm growth in various two-dimensional (2D) flow fields. These measurements were used in conjunction with confocal microscopy and microelectrode measurements to study how hydrodynamics and metabolite transport are affected by the biofilm colonies. Stepp *et al.*[42] used high-resolution 3D NMR microscopy to study the growth of various biofilms in a small Bentheim sandstone plug at pore scale resolution. Images were acquired both before and after growth and compared. Differences were attributed to microbial activity, i.e., the exclusion of pore fluid due to gas and biofilm polymer formation. Imaging was used to assess permeability modification and fluid diversion.

Nuclear magnetic resonance techniques have been developed to selectively detect and measure biofilms in the presence of the (often much

[35] T. I. Ladd, D. Schmiel, J. C. Nickel, and J. W. Costerton, *J. Urol.* **138,** 1451, 1987.

[36] B. Little and P. Wagner, *Can. J. Microbiol.* **42,** 367, 1996.

[37] F. A. MacLeod, H. M. Lappin-Scott, and J. W. Costerton, *Appl. Environ. Microbiol.* **54,** 1365, 1988.

[38] Z. Lewandowski, P. Stoodley, S. Altobelli, and E. Fukushima, *Water Sci. Technol.* **29,** 223, 1994.

[39] Z. Lewandowski, S. A. Altobelli, and E. Fukushima, *Biotechnol. Prog.* **9,** 40, 1993.

[40] Z. Lewandowski, S. A. Altobelli, P. D. Majors, and E. Fukushima, *Water Sci. Technol.* **26,** 577, 1992.

[41] Z. Lewandowski, P. Stoodley, and S. Altobelli, *Water Sci. Technol.* **31,** 153, 1995.

[42] Stepp, A. K., R. S. Bryant, F. M. Llave, and R. P. Lindsey, *in* "Proceedings from the Fifth International Conference on Microbiol Enhanced Oil, Recovery and Related Biotechnology for Solving Environmental Problems," p. 389, 1995.

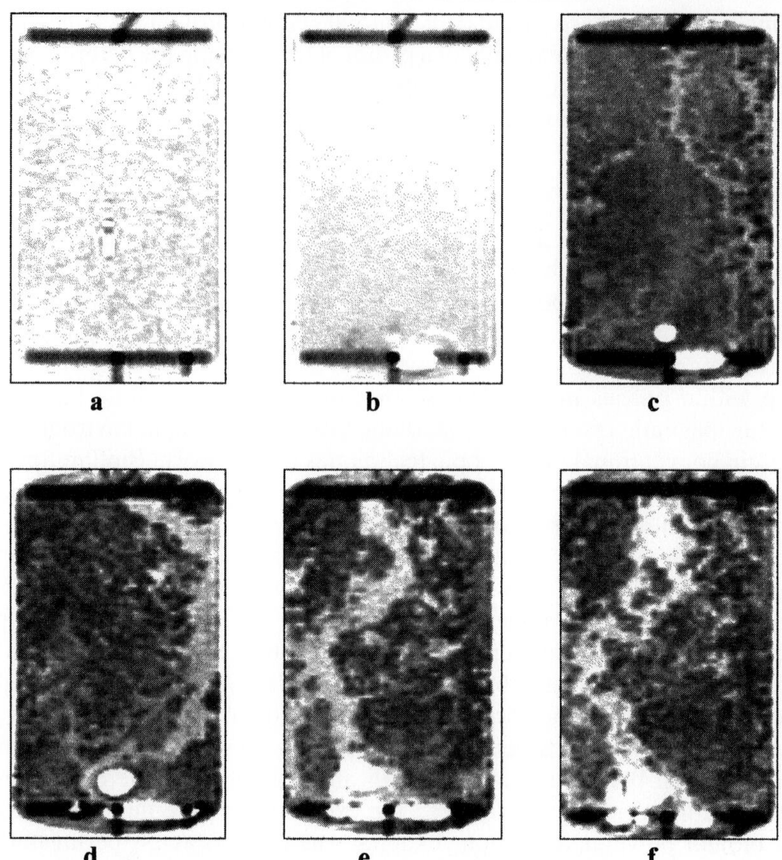

FIG. 4. Two-dimensional T_1 relaxation-weighted NMR images of a parallel-plate flow model containing *E. coli* bacteria at 0 (a), 3 (b), 5 (c), 6 (d), 7 (e), and 8 (f) days growth time. The bright line features observed in all images are flow model connections and troughs, and the round, low-intensity features are air bubbles.

larger) bulk water signal. These techniques typically exploit the effect of biofilms on the NMR relaxation times or apparent molecular diffusion rates of the proximal (intracellular or extracellular) water hydrogen atoms. For example, longitudinal (T_1) or transverse (T_2) relaxation times for intracellular water are shorter than for bulk water.[43] Bacterial cells and their exopolymer by-products lead to changes in the chemical and physical (translational and rotational) molecular environment.

[43] J. W. Pettegrew, "NMR, Principles and Applications to Biomedical Research." Springer-Verlag, Berlin, 1989.

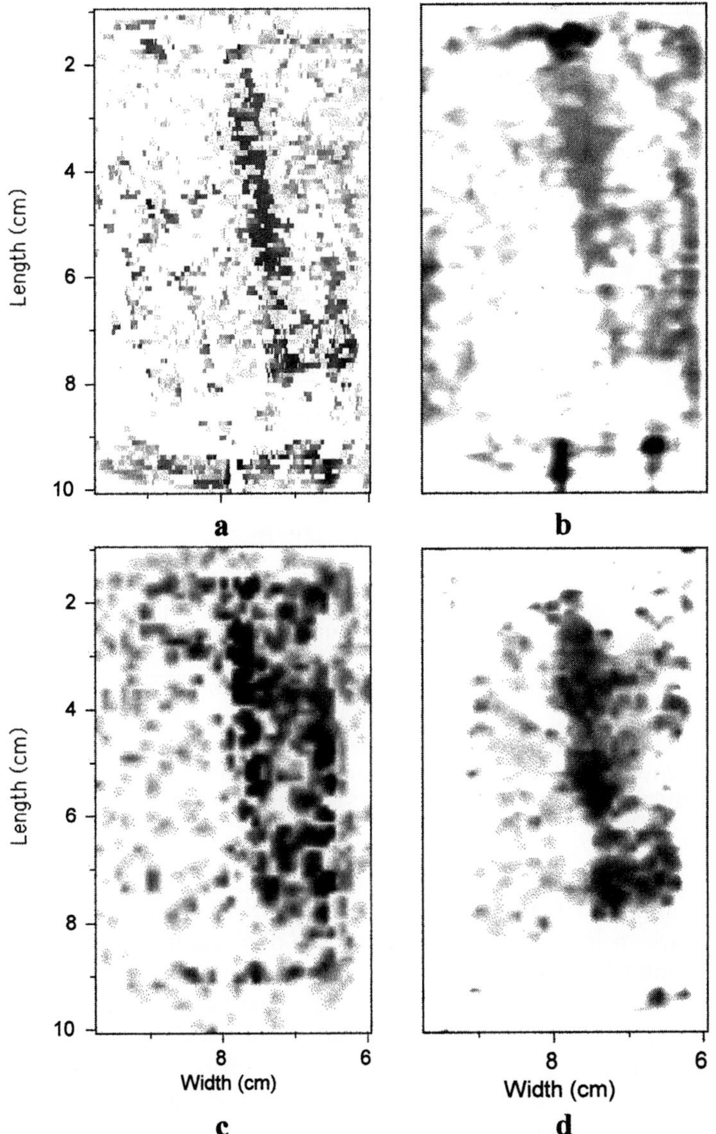

FIG. 5. Comparison of imaging techniques used to identify biofilm in a glass bead-packed bioreactor. (a) T_1-resolved, (b) T_1-weighted, (c) T_2-resolved, and (d) optical image. The resolved NMR images (a and c) distinguish the biofilm from the bulk water more clearly.

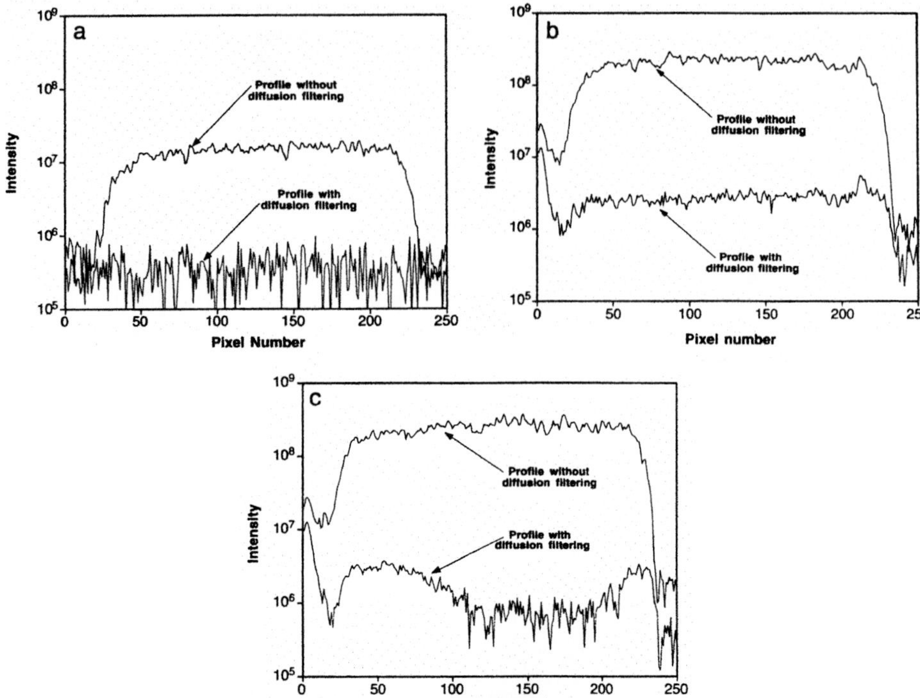

FIG. 6. One-dimensional diffusion-weighted longitudinal NMR profiles of a cylindrical quartz sand pack containing, water only (a) uniformly distributed bacteria (b) and nonuniformly distributed bacteria (c). [Reproduced with permission from Potter *et al.* (1996)].

Fevang[44] used T_1 relaxation-weighted NMR to selectively image *E. coli* biofilm grown in nutrient solution in two-dimensional flow models. (An experimental NMR imaging repetition time was chosen to suppress the slowly relaxing bulk water signal while allowing for nearly complete recovery of the rapidly relaxing biofilm signal.) Figure 4 shows the results of an 8-day biofilm growth experiment that was monitored by NMR imaging. A flow channel was observed to develop and change location with time. NMR images are in excellent agreement with optical images (not shown). Fevang also demonstrated that the biofilm colonies were resolvable within the limits of the image pixel resolution.

[44] L. Fevang, "Nuclear Magnetic Resonance Imaging of Biofilms and Bacterial Adhesion Studies." Thesis, University of Texas at Austin, 1997.

Hoskins[45] introduced relaxation-resolved T_1 and T_2 NMR imaging methods designed to simultaneously correct and resolve biofilm and bulk signals. (Several relaxation-weighted images are collected and a relaxation analysis is performed; the resulting corrected intensities are then sorted according to their corresponding relaxation times.) The T_2-resolved method is much faster (several minutes of image acquisition time versus several hours for the T_1-resolved method), making the real-time monitoring of biofilm growth feasible. Figure 5 compares T_1 and T_2 relaxation-resolved images with T_1-weighted and optical images for an *E. coli* biofilm in a porous medium (a transparent parallel plate bioreactor packed with glass beads). The separability of biofilm and bulk signals by their relaxation times is somewhat diminished in porous media due to the reduced relaxation times for pore fluids.

Potter *et al.*[16] employed a diffusion filtering method to obtain spatial profiles of *Pseudomonas* (*Burkholdia*) *cepacia* cell populations in a small quartz sand pack. This method is based on the selective detection of intracellular water hydrogens in a strong magnetic field gradient due to their restricted diffusion path within the cells. Figure 6 shows longitudinal NMR profiles (1D images) of the cylindrical sand pack with different distributions of bacteria, obtained both with and without diffusion filtering.

Future Developments.

Microbiologists now realize that biofilms are a dominant feature of microbial growth in nature. Studies are in progress to address the physiological and genetic mechanisms required for this mode of growth. A large number of devices are currently employed to grow biofilms in the laboratory. In order to be able to compare data between laboratories, it will be necessary to adopt universal protocols for biofilm production. The RD and MRD are widely used for biofilm research and thus permit a comparison of data between many laboratories. They suffer in that they do not permit the control of laminar and turbulent flow patterns, which are an important controlling feature of biofilm formation. Future developments in biofilm equipment design need to address this issue. NMR techniques provide an exciting approach in studying biofilms in opaque media. Future developments in NMR technology will enable investigators to measure and map biofilm metabolism. The future of biofilm research is indeed quite exciting.

[45] B. C. Hoskins, "Nuclear Magnetic Resonance Imaging of Biofilms in an Open-Flow and Glass Bead-Packed Bioreactor." Thesis, University of Texas at Austin, 1998.

Acknowledgments

Research in the authors' laboratories was funded by the Advanced Research Program of the Texas Higher Education Coordinating Board (RJCM), Research Enhancement Grants from Southwest Texas State University (RJCM), the Environmental Protection Agency (RJCM and MMS), and the National Science Foundation (MMS).

[21] Use of Constant Depth Film Fermentor in Studies of Biofilms of Oral Bacteria

By MICHAEL WILSON

Introduction

The commonest diseases of the oral cavity, dental caries and the inflammatory periodontal diseases, result from the accumulation of dental plaque above (supragingival) or below (subgingival) the gingival (gums) margin. Dental plaque is a classic example of a biofilm, and current preventative and therapeutic regimens employed by the general public and by dental practitioners for these diseases are based on a recognition of this fact as they entail mechanical removal (by toothbrushing and root scaling) of these biofilms. Increasing interest in chemically based preventative and therapeutic measures to supplement these mechanical procedures has created a need for a laboratory-based model for assessing suitable compounds. Unfortunately, most laboratory assessments have employed methods (e.g., determination of the minimum inhibitory concentration of the agent) that are appropriate for systemic infections and have taken no account of the biofilmic nature of oral diseases.[1,2] However, it has been well established that the susceptibility of many bacteria to antimicrobials is much lower when they comprise a biofilm than when they are in aqueous suspension.[2,3] A biofilm-based model is far more likely to be able to predict the effectiveness of a compound for treating or preventing plaque-related diseases *in vivo.* Other areas of applied oral microbiology research that would benefit from and, indeed, necessitate a biofilm-based approach include the determination of the effectiveness of plaque removal devices and the assessment of the corrosion susceptibility of dental materials.

[1] M. Wilson, *Microb. Ecol. Health Dis.* **6,** 143 (1993).

[2] M. Wilson, *J. Med. Microbiol.* **44,** 79 (1996).

[3] P. Gilbert, J. Das, and I. Foley, *Adv. Dent. Res.* **11,** 160 (1997).

0076-6879/99 $30.00

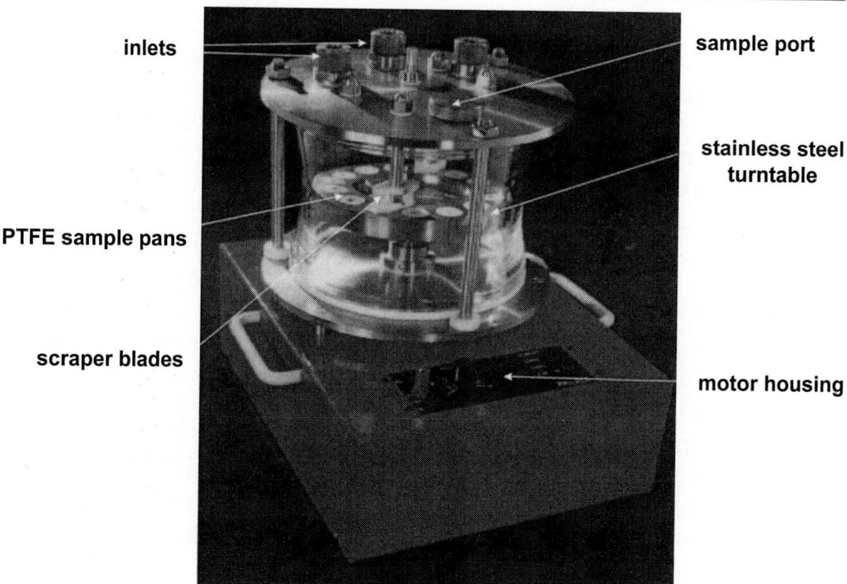

FIG. 1. The constant depth film fermentor.

Over the years a number of biofilm-generating devices have been developed and these have involved the use of disks (or wires) immersed in a microbial suspension, hollow tubes through which a microbial suspension is pumped (e.g., the "Robbins device"), rotating drums immersed in a microbial suspension, packed columns through which the bacterial suspension is pumped, and teeth (or enamel slices or hydroxylapatite) over which saliva is continually dripped. This article describes the use of one of these devices, the constant depth film fermentor[4] (CDFF; Fig 1), to investigate the ability of various agents (i) to prevent biofilm formation, (ii) to kill bacteria in biofilms, and (iii) to evaluate the corrosion potential of oral biofilms.

Constant Depth Film Fermentor

General Description

The CDFF consists of a glass vessel (18.0 cm diameter, 15.0 cm depth) with stainless-steel end plates, the top one of which has ports for the entry

[4] J. W. T. Wimpenny, A. Peters, and M. A. Scourfield, *in* "Structure and Function of Biofilms" (W. G. Characklis and P. A. Wilderer, eds.), p. 111. Wiley, Chichester, 1989.

of medium and gas and for sampling, whereas the bottom end plate has a medium outlet (Fig. 2). The vessel houses a stainless-steel disk (15.0 cm diameter) containing 15 polytetrafluoroethylene (PTFE) sample pans (2.0 cm diameter), which rotates under PTFE scraper bars that smear the incoming medium over the 15 pans and maintain the biofilms, once formed, at a constant predetermined depth. For delivering the medium to the CDFF at a constant rate, we have found the 101 U/R peristaltic pump (Watson Marlow Ltd., Falmouth, UK) to be suitable and reliable. Excess medium is swept away by the scraper bars and flows out through the medium effluent port in the lower end plate. Each sampling pan has up to six cylindrical holes containing PTFE plugs (generally 4.7 or 5.0 mm in diameter) on which the biofilms form, with their upper surfaces recessed to a predetermined depth below the surface of the steel disk. The PTFE plugs may also be used to support disks of any material of interest (or the plugs themselves may be fabricated from the material), enabling the formation of biofilms on a range of substrata. The sampling pans can be removed aseptically during the course of an experimental run and the biofilms subjected to analysis, exposed to antimicrobial agents, etc.

The CDFF is small enough to be housed in an incubator, thus enabling accurate temperature control. It may be purchased from Professor J. Wim-

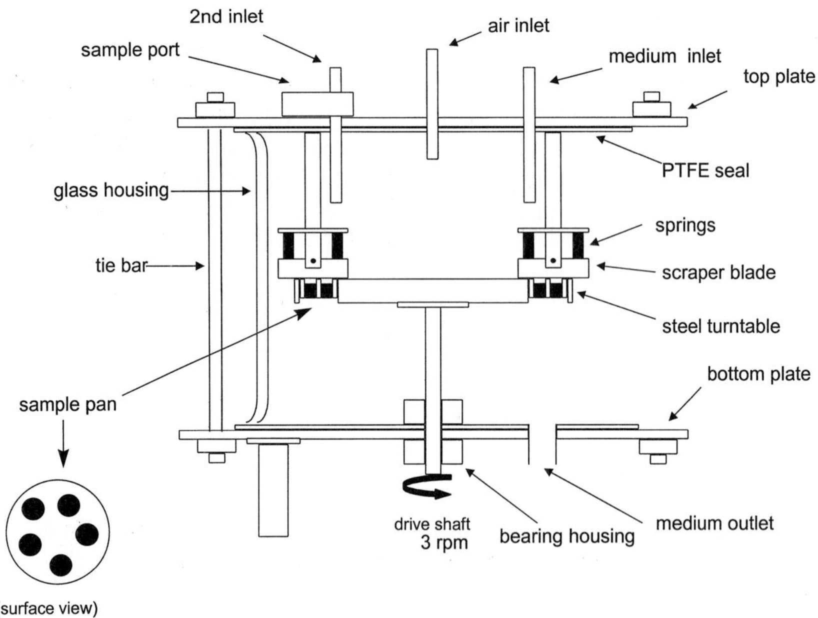

FIG. 2. Diagram illustrating the main components of the constant depth film fermentor.

penny, School of Pure and Applied Biology, University of Wales, Cardiff CF1 3TL, United Kingdom.

Depth of Biofilms

One of the advantages of the CDFF is that, unlike many other biofilm-generating devices, it is possible to regulate the thickness of the biofilms obtained. Hence, biofilm thickness may be included as a variable in investigations. Specially machined tools are available for depressing the PTFE plugs (together with the test substratum disks) to a depth of between 50 and 500 μm. This facility is useful, for example, for mimicking the thin plaques found on the smooth surfaces of teeth and the much thicker plaques found between teeth.

Substratum

Biofilms may be formed directly on the surfaces of the PTFE plugs of the sample pan. The number of plugs per pan will depend on their diameter; this is usually, but not necessarily, 4.7 mm (giving six disks per pan) or 5 mm (giving five disks per pan). Alternatively, the plugs may be used to support disks of any other material of interest or may be fabricated from such materials, as long as the material can be machined to form disks of appropriate diameter and tolerance. In our studies we have successfully used the following substrata: hydroxylapatite, bovine enamel, glass, denture acrylic, orthodontic magnets, and dental amalgam. Hydroxylapatite disks may be purchased from Clarkson Chromatography Products Inc., 213 Main Street, South Williamsport, Pennsylvania 17702. Alternatively, they may be made as follows. Captal R sintering grade hydroxylapatite powder (Plasma Biotal Ltd., Derbyshire, UK) is pressed into 5-mm-diameter disks (3 mm in depth) in a 5-mm pellet die (Graseby Specac Ltd., Orpington, Kent, UK) under a pressure of 250 kg. The disks are then sintered at 1000° for 1 hr in a Carbolite RHF 1500 furnace (Carbolite Ltd., Sheffield, UK).

Inoculation

The CDFF can be used to produce biofilms consisting of a single species[5] or several species[6] as well as to generate microcosm dental plaques.[7] A variety of inoculation procedures may be used depending mainly on the type of biofilm being grown. For monospecies biofilms, an overnight culture of the organism may simply be pumped through the CDFF for an appro-

[5] M. Wilson, H. Patel, and J. Fletcher, *Oral Microbiol. Immunol.* **11,** 188 (1996).

[6] J. Pratten, P. Barnett, and M. Wilson, *Appl. Environ. Microbiol.* **64,** 3515 (1998).

[7] J. Pratten, A. W. Smith, and M. Wilson, *J. Antimicrob. Chemother.* **42,** 453 (1998).

priate period of time. For biofilms of oral streptococci, we have found that inoculation times as short as 4 hr at a flow rate of 30 ml/hr (see later) are adequate to enable biofilm formation.

For multispecies biofilms, a stock of identical inocula needs to be prepared to ensure reproducibility between runs. Each member of the biofilm is grown to confluence on an agar plate of an appropriate medium and harvested in nutrient broth (Oxoid Ltd., Basingstoke, UK). These suspensions are centrifuged and resuspended in nutrient broth containing 10% (v/v) glycerol, and 1.0-ml aliquots are stored at −70°. To provide the inoculum for a run, one of these is then thawed, diluted with 5.0 ml of the medium (see later) to be used in the CDFF, and then pipetted onto the rotating pans through the sampling port. This approach has been used to produce biofilms with a composition similar to that of supragingival plaque,[6] with the constituent organisms being *Streptococcus sanguis* NCTC 10904, *Streptococcus mutans* NCTC 10449, *Streptococcus oralis* NCTC 11427, *Actinomyces naeslundii* NCTC 10951, *Neisseria subflava* ATCC A1078, and *Veillonella dispar* NCTC 11831. Alternatively, the CDFF may be inoculated with the effluent from a chemostat in which a defined community of bacteria is being grown as described by Kinniment *et al.*[8] In this study, a nine-membered community of oral bacteria (*N. subflava, Actinomyces viscosus, Lactobacillus casei, S. oralis, Streptococcus gordonii, S. mutans, V. dispar, Porphyromonas gingivalis,* and *Fusobacterum nucleatum*) was established in a chemostat and the effluent from this was pumped into the CDFF for 8 hr. The chemostat was then disconnected and a mucin-containing medium was pumped into the CDFF at a rate of 50 ml/hr.

A suitable inoculum for producing microcosm plaques is provided by pooled human saliva. Saliva is obtained from a large number of healthy individuals (at least 10) and this is pooled, divided into 1.0-ml samples, and stored at −70°. One or, preferably, two of the samples are thawed in 500 ml of artificial saliva (see later) and this is pumped through the CDFF for 8 hr at a flow rate of 30 ml/hr.[9] The proportions of the major groups of bacteria (streptococci, *Actinomyces* spp., *Veillonella* spp., etc.) in the inoculum should be checked for each run using selective media as described later.

Nutrient Source

Clearly, any of the standard culture media may be used in the CDFF. However, in order to grow oral biofilms under conditions that mimic those

[8] S. L. Kinniment, J. W. Wimpenny, D. Adams, and P. D. Marsh, *Microbiology* **142,** 631 (1996).

[9] J. Pratten, P. Barnett, and M. Wilson, *in* "Biofilms: Community Interactions and Control" (J. W. Wimpenny, P. Handley, P. Gilbert, H. Lappin-Scott, and M. Jones, eds.), p. 245. Bioline, Cardiff, 1997.

found in the oral cavity, careful attention must be paid to choosing an appropriate source of nutrients. For those biofilms growing on tooth surfaces above the gingival margin, the main source of nutrients is saliva. There are obvious practical problems, as well as health and safety issues, with obtaining the large volumes of human saliva that would be needed to operate (possibly for weeks) the CDFF. Consequently, using an artificial saliva then becomes necessary. Unfortunately, the composition of saliva varies from person to person and, of course, shows dramatic variations in an individual depending on the time of sampling (before, during, or after a meal), what the individual chooses to eat that day, and whether the individual indulges in between-meal snacking or drinking.[10] It is, therefore, very difficult to define a "typical" composition for saliva. Nevertheless, analysis of the composition of saliva taken in the morning from healthy, fasting adults ("resting" saliva) provides a starting point for formulating a suitable artificial saliva for experimental purposes. The composition of human saliva is extremely complex and consists mainly of water (>99%), protein (approximately 0.2%; mainly mucins, enzymes, and immunoglobulins), a range of amino acids, urea, and many inorganic ions (mainly Na^+, Ca^{2+}, K^+, Cl^-, Po_4^{3-}, HCO_3^-).

In our studies, we have used an artificial saliva based on the formulations of Russell and Coulter[11] and Shellis,[12] which consists of mucin type III, partially purified from porcine stomach (Sigma Ltd., Poole, UK), 2.5 g/liter; yeast extract (Oxoid), 2 g/liter; proteose peptone (Oxoid), 5 g/liter; lab-lemco (Oxoid), 1 g/liter; NaCl (Sigma), 0.2 g/liter; KCl (Sigma), 0.2 g/liter; $CaCl_2$ (Sigma), 0.3 g/liter; and urea (Sigma), 0.5 g/liter.[13] The pH of the medium is 6.9. A key feature of this formulation is the inclusion of hog gastric mucin, which functions not only as a major source of nutrients for oral bacterial communities, but also covers the surfaces of the substratum to form a "conditioning" film. This mimics the conditioning film (termed the "salivary pellicle"), which is always found on the surfaces of teeth (and prosthetic devices) *in vivo*. This salivary pellicle plays a major role in the adhesion of bacteria to teeth and in the retention of topically applied antimicrobial agents in the oral cavity.

The next problem to resolve is the rate at which the artificial saliva should be added to the CDFF. If the objective is to mimic the rate of saliva flow *in vivo,* then once again there are problems. Hence, there is tremendous variation between individuals and, in a particular individual, the flow rate will depend, among other factors, on the time of day, whether food is being

[10] W. M. Edgar, *Br. Dent. J.* **172,** 305 (1992).

[11] C. Russell and W. A. Coulter, *Appl. Microbiol.* **29,** 141 (1975).

[12] R. P. Shellis, *Arch. Oral Biol.* **23,** 485 (1978).

[13] J. Pratten, K. Wills, P. Barnett, and M. Wilson, *J. Appl. Microbiol.* **84,** 1149 (1998).

consumed, the presence or absence of light, and physical and mental stress. Nevertheless, it has been estimated that the mean salivary flow rate in humans is 0.72 liter per day[14–16] and we have used this flow rate in our studies.

In some experiments it may be necessary to supplement this artificial saliva with other nutrients, which can be added via one of the ports in the upper end plate. For example, in sucrose solutions we have pulsed on a thrice-daily basis to mimic the intake of this nutrient at meal times in order to determine the effect of this on biofilm composition (see later).[17,18]

In the case of biofilms growing below the gingival margin (subgingival plaques), the main source of nutrients is gingival crevicular fluid (GCF), which has a composition similar to that of serum except that the glucose and NaCl contents are higher.[19] We have used a formulation (derived from a knowledge of the protein content of human GCF) consisting of RPMI (Sigma, St. Louis, MO) supplemented with 40% (v/v) horse serum (Oxoid Ltd.) as a convenient, safe, and economically feasible approximation of the normal nutrient source for bacteria growing in subgingival plaques.[20] We have used this medium to grow, from a saliva inoculum, organisms (including gram-negative anaerobes) that would be found in subgingival plaque. An anaerobic gas mixture [95% N_2/5% Co_2 (v/v)] is connected to the CDFF, and the gas mixture is allowed to flow through the system at a rate of 100 ml/hr at a pressure of 1 bar. As the amount of GCF produced *in vivo* per day is of the order of only a few milliliters, which would be impracticable to reproduce *in vitro,* the RPMI/horse serum should be delivered at the lowest possible flow rate that prevents the biofilms drying out. Flow rates of 6 ml/hr are adequate and enable the formation of biofilms containing species characteristic of subgingival plaques.

General Operating Conditions

Temperature control is conveniently achieved by housing the CDFF in an incubator. The medium reservoir is usually maintained at room tem-

[14] G. Bell, D. Emslie-Smith, and C. Patterson, *in* "Textbook of Physiology," p. 36. Churchhill-Livingstone, London, 1980.

[15] J. F. Lamb, *in* "Essentials of Physiology," p. 93. Blackwell Scientific, Oxford, 1991.

[16] A. C. Guyton, *in* "Human Physiology and Mechanisms of Disease," p. 486. Saunders, Pennsylvania, 1992.

[17] M. Wilson, H. Patel, H. Kpendema, J. H. Noar, N. P. Hunt, and N. J. Mordan, *Biomaterials* **18,** 53 (1997).

[18] M. Wilson, H. Patel, and J. H. Noar, *Curr. Microbiol.* **36,** 13 (1998).

[19] G. Cimasoni, *Monogr. Oral Sci.* **12,** 1 (1983).

[20] M. A. Curtis, G. S. Griffiths, S. J. Price, S. K. Coulthurst, and N. W. Johnson, *J. Clin. Periodontol.* **15,** 628 (1988).

perature but may, of course, be kept at 4°. The turntable is set to rotate at 3 rpm.

For modeling supragingival plaques, an aerobic atmosphere is maintained by connecting one of the ports in the upper end plate to a 0.22-μm membrane filter (Hepa Vent filters, Whatman, Maidstone, UK). When modeling subgingival plaques, which grow in an anaerobic environment, the gas port is connected via the membrane filter to a cylinder of 90% N_2/ 10% CO_2 (BOC Gases, Guildford, UK). In this case, it is also important to flush the medium reservoir with this anaerobic gas mixture.

Use of CDFF for Studying Inhibition of Biofilm Formation

Disks of an appropriate substratum (e.g., enamel, hydroxylapatite, denture acrylic) are placed in 5 ml of artificial saliva for at least 1 min to enable the formation of a glycoprotein-containing conditioning film. They are then immersed in 5 ml of the test solution, or in an appropriate placebo, for 1 min. Longer incubation times with the test agents may also be used, although it must be remembered that most agents (especially those delivered in the form of mouthwashes or toothpastes, as is the case with many oral hygiene products) are eliminated rapidly from the oral cavity. It may be possible to test a number of agents in a single run of the CDFF, provided that there is no danger of carryover of the agents from one pan to another. This can be minimized by including pans containing untreated disks to separate those pans containing disks treated with different agents. If a series of concentrations of one agent is being tested in a single run of the CDFF, then disks treated with the lowest concentration of the agent must be placed in the leading pan, with pans containing disks treated with increasing concentrations of the agent following sequentially behind. The CDFF is then autoclaved at 121° for 15 min. For heat-labile antimicrobials, the enamel or hydroxylapatite disks are autoclaved in the PTFE pans and are then treated with the filter-sterilized test agents in a laminar flow cabinet. The pans are then inserted into the autoclaved CDFF in a laminar flow cabinet. Alternatively, the CDFF containing the disks and pans can be autoclaved and then the filter-sterilized agent added aseptically to the surfaces of the disks by means of a pipette through the sampling port.

For monospecies biofilms, the inoculum is prepared as follows. Ten milliliters of nutrient broth (Oxoid) is inoculated with colonies of *S. sanguis* growing on tryptone soya agar (Oxoid) and incubated anaerobically at 37° overnight. This is used to inoculate 2 liter of artificial saliva, which is pumped through the CDFF at a rate of 30 ml/hr.

Sample pans are removed at the desired test intervals (4, 8, and 16 hr are appropriate), and each PTFE plug (together with its associated disk and biofilm) is ejected carefully from the sample pan by gently pushing the base of the plug with a sterile glass rod. Each plug (+disk + biofilm) is ejected into 10 ml of neutralizing broth (Difco Laboratories, East Molesey, UK) to inactivate any residual antimicrobial agent and is then vortex mixed for 60 sec to dislodge and disrupt the biofilm. The number of viable bacteria is then determined by viable counting on tryptone soya agar.

Effect of Antimicrobial or Antiplaque Agents on Viability of Bacteria in Monospecies Biofilms, Multispecies Biofilms, and Microcosm Dental Plaques

Monospecies Biofilms

The inoculum is prepared as described earlier but the *S. sanguis* in artificial saliva is pumped through the CDFF for 24 hr, after which time the inoculating flask is disconnected and the CDFF is connected to a reservoir of sterile artificial saliva that is delivered at a rate of 30 ml/hr.[5,7,13] Growth of the biofilms is monitored by removing sample pans daily and determining the biofilm mass as follows. Each PTFE plug (together with the associated substratum and biofilm) is ejected from the sampling pan as described earlier except that it is ejected into 10 ml of tryptone soya broth instead of neutralizing broth. The number of viable bacteria is determined by viable counting on tryptone soya agar following disruption of the biofilm by vortex mixing. Once the biomass reaches a constant value (approximately 4 days after inoculation; total viable cell count/biofilm is 10^7 cfu, viable cell density is 5×10^5 cfu/mm^2), the sampling pans are removed from the CDFF and the susceptibility of the biofilms to antimicrobial agents is determined as follows. Each pan, containing six biofilms, is placed in a sterile container and 5.0 ml of antimicrobial-containing solution [e.g., 0.2% (w/v) chlorhexidine gluconate or 0.05% (w/v) cetylpyridinium chloride], or sterile water as a control, is added to cover the biofilms. The solutions are prewarmed to 37° and added carefully down the side of the container so as not to disturb the biofilms. After incubation for various periods of time (1, 5, and 60 min are appropriate), the solutions are removed and each PTFE plug and its associated biofilm is transferred to 10 ml of neutralizing broth to inactivate any antiseptic present and is vortex mixed for 60 sec. Serial dilutions are then prepared and the number of surviving organisms is determined by viable counting as described earlier.

Multispecies Biofilms

Biofilms composed of those bacteria representative of a particular habitat within the oral cavity can also be grown in the CDFF. For example, a six-membered community with a composition similar to that of supragingival plaque can be obtained as follows.[6] A 1.0-ml frozen sample of a mixture of the desired species (prepared as described previously under "inoculum") is thawed in 5.0 ml of artificial saliva and this is pipetted through the sampling port of the CDFF directly onto the substratum-containing pans and left for 1 hr. The CDFF is then connected to a reservoir of artificial saliva (flow rate 30 ml/hr). Growth of biofilms is monitored on a daily basis by removing sample pans and analyzing the biofilms as follows. Each biofilm is ejected from the sample pan (as described earlier) into 10 ml of artificial saliva and vortex mixed for 60 sec. Serial dilutions are prepared in artificial saliva and viable counting is performed using a range of media: Wilkins–Chalgren (WC) agar (Oxoid) containing 5% unlysed horse blood (incubated aerobically and anaerobically for the total aerobic and anaerobic counts, respectively), mitis-salivarius agar (Difco) for streptococci, *Veillonella* agar (Difco) for *V. dispar,* cadmium fluoride-acriflavine-tellurite agar[21] for *A. naeslundii,* and Thayer–Martin agar (Oxoid) for *N. subflava.* Plates are incubated anaerobically for 4 days at 37°, except for those containing Thayer–Martin agar (aerobic, 4 days, 37°). Colonies growing on each of the plates are gram stained to confirm their identities and counted; the different streptococcal species are distinguished on the basis of colonial morphology. Once the relative proportions of the various species in the biofilms have stabilized (approximately 7 days), sample pans are removed and the susceptibility of the biofilms to exposure to antimicrobial agents for various periods of time (e.g., 1, 5, and 60 min) can be determined as described previously. Two hundred and sixteen hours after inoculation, biofilms grown in this way contain approximately 10^8 cfu/mm^2, and the mean percentage proportions of the different species are as follows: *N. subflava* (29%), *S. mutans* (23%), *S. sanguis* (20%), *A. naeslundii* (14%), *V. dispar* (7%), and *S. oralis* (6%). The relative proportions of the organisms in these biofilms are similar to those found in supragingival plaque samples.

Microcosm Dental Plaques

These are grown from pooled human saliva as described earlier (under "inoculum").[7,9,18] Growth is monitored by analyzing the biofilms on a daily basis as described in the previous section except that colonies growing on selective media are subjected to the following identification tests. Colonies growing on the mitis-salivarius plates are enumerated, gram stained, and tested for catalase production. Those colonies consisting of catalase-negative gram-positive cocci are considered to be streptococci. Colonies on

[21] L. J. Zylber and H. V. Jordan, *J. Clin. Microbiol.* **15,** 253 (1982).

Veillonella agar are gram stained and those containing gram-negative cocci are subcultured onto WC blood agar and are incubated aerobically and anaerobically. Colonies consisting of gram-negative cocci that grow on WC plates incubated anaerobically but not aerobically are designated *Veillonella* spp. Colonies on the CFAT agar are gram stained and those consisting of gram-positive rods are considered to be *Actinomyces* spp. fifteen days after inoculation, the total anaerobic count of the biofilms is approximately 10^9 cfu with a density of approximately 10^8 cfu/mm^2 and their composition approximates to the following: streptococci (22%), *Actinomyces* spp (32%), and *Veillonella* spp. (14%). The susceptibility of these biofilms to antimicrobial agents may be determined as described previously.

Effect of Sucrose Supplementation on Antimicrobial Susceptibility of Monospecies Biofilms and Microcosm Dental Plaques

The medium used in the studies described so far has had a composition resembling "resting saliva," i.e., saliva produced in the absence of any exogenous stimulation, and this is the main source of nutrients for oral bacteria when their host is neither eating nor drinking. Obviously, when food or drink is present in the oral cavity, the nutrient content of saliva will change dramatically, which can have profound effects on the bacterial composition of plaque. The CDFF has a number of ports in its upper end plate that can be used for adding, either periodically or continuously, additional nutrients (or antimicrobials). From the point of view of mimicking caries-inducing conditions, the most important additional nutrient is, of course, sucrose.

Monospecies biofilms of *S. sanguis* NCTC 10904 are grown in the CDFF as described previously except that, as well as providing a continuous supply of artificial saliva, 333 ml of a 10% (w/v) aqueous solution of sucrose is pulsed into the CDFF three times daily (at 9 AM, 12 PM, and 5 PM) over a 30-min period.[22] The total daily amount of sucrose added to the CDFF (100 g) represents the mean daily intake of sucrose by an adult in the United Kingdom. Growth of the biofilms is monitored daily as described previously until the biomass becomes constant (approximately 5 days). The susceptibility of the sucrose-grown biofilms to antimicrobial agents is determined as described previously.

The effect of sucrose supplementation on both the bacterial composition and the antimicrobial susceptibility of microcosm dental plaques can be determined in a similar manner.[18]

[22] J. V. Embleton, H. N. Newman, and M. Wilson, *Appl. Environ. Microbiol.* **64,** 3503 (1998).

Effect of Repeated Pulsing of Antimicrobial Agent

Manufacturers of oral hygiene products (toothpastes, mouthwashes, etc.) are increasingly incorporating antimicrobial agents into their products. However, little is known about the possible consequences of repeated exposure to these agents on either the susceptibility of oral bacteria to these antimicrobials or the bacterial composition of dental plaques. The CDFF may be used to investigate such effects.[7,9]

Monospecies biofilms of oral bacteria or microcosm dental plaques are grown as described previously. Once the biofilms reach the steady state they can be pulsed with small quantities (e.g., 10 ml) for short periods of time (e.g., 60 sec) of the test agent [e.g., 0.05 or 0.2% (w/v) chlorhexidine gluconate] to reflect the manner in which it would be used *in vivo*. Ideally, the pulsing should also be carried out twice daily to mimic morning and evening use of the agent. Sample pans are removed on a daily basis and the number of survivors (as well as the relative proportions of different species in the case of the microcosm plaques) is determined by viable counting as described previously.

Effect of Light-Activated Antimicrobial Agents on Viability of Bacteria in Oral Biofilms

During the last decade increasing interest has been shown in the use of antimicrobial agents for the treatment of plaque-related diseases. However, the widespread use of such agents to treat so prevalent a group of chronic diseases is of concern as it may result in resistance development in the target organisms. Photodynamic therapy (PDT), which involves the use of light-activated drugs, may offer an alternative approach to the use of traditional antimicrobial agents in the treatment of such diseases.[23] PDT of an infectious disease involves treating the causative organism with a light-activatable compound (a "photosensitizer"), which, on irradiation with light of a suitable wavelength, generates singlet oxygen and free radicals that can kill the organism. We have shown that lethal photosensitization of the causative organisms of both caries and periodontitis is possible when they are in aqueous suspensions.[24] The CDFF can be used to generate biofilms for assessing their susceptibility to light-activated antimicrobial agents.[25–27]

[23] M. Wilson, *Int. Dent. J.* **44,** 181 (1994).
[24] M. Wilson, *J. Appl. Bact.* **75,** 299 (1993).
[25] M. Wilson, *J. Proc. S.P.I.E.* **3191,** 68 (1997).
[26] M. Wilson, T. Burns, and J. Pratten, *J. Antimicrob. Chemother.* **37,** 377 (1996).
[27] M. Wilson, *in* "The Life and Death of Biofilm" (J. W. Wimpenny, P. Handley, P. Gilbert, and H. Lappin-Scott, eds.), p. 143. Bioline, Cardiff, 1995.

Biofilms of *S. sanguis* are grown in the CDFF as described previously, and their susceptibility to the light-activated drugs aluminum disulfonated phthalocyanine ($AlPcS_2$) or toluidine blue O (TBO) is determined as follows. In the case of $AlPcS_2$ (absorption maximum 675 nm), an appropriate light source is a gallium aluminum arsenide (GaAlAs) diode laser (Omega Universal Technologies, London) with a power output of 11 mW. This emits light with a beam diameter of 9 mm at a wavelength of 660 nm in a pulsed mode with a frequency of 20 kHz. Eight microliters of a solution of $AlPcS_2$ (100 μg/ml) in 0.85% (w/v) NaCl is added to each of two biofilms, which are then exposed to 0.8 J of light (energy density of 4.1 J/cm^2) from the GaAlAs laser.[26] The disks are then vortexed to remove the biofilms and the survivors are enumerated by viable counting. A number of controls must be included to determine the effect of the laser light alone and the photosensitizer alone. Additional biofilms are therefore treated with 8.0 μl of $AlPcS_2$ but are not exposed to laser light or are treated with 8.0 μl of 0.85% (w/v) NaCl and exposed to light from the GaAlAs laser.

In the case of the TBO (absorption maximum 633 nm), a helium/neon (HeNe) gas laser (NEC Corporation, Japan) is used.[27] This has a power output of 7.3 mW, emitting light with a beam diameter of 1.3 mm at a wavelength of 632.8 nm. A lens is used to increase the diameter of the light beam so that the whole biofilm is irradiated. Eight microliters of an aqueous solution of TBO (100 μg/ml) is added to each to two biofilms, which are then exposed to 1.1 J of laser light (energy density of 5.5 J/cm^2) and the survivors are enumerated. Additional biofilms are treated in a similar manner to that described earlier to serve as controls.

Corrosion of Dental Materials by Oral Biofilms

A number of materials are used in the oral cavity for repairing (filling materials), replacing (crowns, dentures, implants), or altering the position (orthodontic devices) of teeth. The materials used include steel, acrylic, mercury amalgam, glass ionomers, porcelain, precious metals, and high-density magnets containing iron, neodymium, and boron. One of the most important of the criteria used in the selection of suitable materials for these purposes is their biocompatibility. Corrosion of materials in the oral cavity is influenced by chemical, mechanical, and biological factors and is a cause for concern because of its effects on the mechanical and other properties of the materials and the possible toxicity of the resulting products.[28] Although the chemical and mechanical aspects of the corrosion of such materials have received considerable attention, few investigations have been con-

[28] G. Palaghias, *Swed. Dent. J.* **30,** 1 (1985).

cerned with the contribution of oral bacteria, especially when in the form of biofilms, to the corrosion process.[29] The CDFF is a useful device for evaluating the extent of corrosion of materials by oral biofilms.[17,30]

The CDFF is prepared for each run as described previously except that the sampling pans are fitted with preweighed plugs of the material under test (e.g., orthodontic magnets, dental amalgam). On each sampling occasion, following removal of the biofilm as described previously, the plugs are dried in a desiccator to constant weight to determine the weight loss incurred, which is a useful indicator of the extent of corrosion that has taken place, and may be examined for surface alterations by scanning electron microscopy. The effluent from the CDFF may be collected and tested for its cytotoxic effects on epithelial cells, etc., and its content of elements is derived from the substratum (e.g., Hg, Ca, Fe).

Additional Analytical Techniques

Additional data may be obtained from studies described by the application of three other techniques: vital staining, electron microscopy, and cryosectioning.

Vital Staining of Bacteria

Vital staining of the biofilms (or sections derived from them; see later) is carried out using a BacLight live/dead viability kit (Molecular Probes Inc., Eugene, OR). In the case of whole biofilms, the disks on which the biofilms form are vortexed for 1 min and transferred into vials containing 1.0 ml of phosphate-buffered saline (Oxoid) prior to the addition of the reagents, as per the manufacturer's specifications. After 15 min, 5.0 μl of the suspension is added to a counting chamber. Green fluorescence indicates live bacteria, whereas dead bacteria appear red when viewed with light at a wavelength of 300 nm. We have found that this technique can distinguish between viable and heat-killed cells of the following oral species: *S. sanguis* NCTC 10904, *S. mutans* NCTC 10449, *S. oralis* NCTC 11427, *A. naeslundii* NCTC 10951, *N. subflava* ATCC A1078, and *V. dispar* NCTC11831.

Electron Microscopy

The biofilms are fixed in 3% glutaraldehyde in 0.1 *M* sodium cacodylate buffer at 4° overnight. The specimens are then postfixed in 1% (w/v) osmium

[29] M. M. Vrijhoef, P. R. Mezger, J. M. Van der Zel, and E. H. Greener, *J. Dent. Res.* **66,** 1456 (1987).

[30] M. Wilson, H. Kpendema, J. Noar, and N. Morden, *Biomaterials* **16,** 721 (1995).

tetroxide at 4° for 2 hr before being dehydrated in a graded series of alcohol [20–100% (v/v); 15-min application time]. The biofilms are embedded in fresh Araldite CY212 (Agar Scientific, Stanstead, UK), and 90- to 100-nm sections are cut. The sections are stained sequentially for 5 min each with 1.5% (w/v) lead citrate and 1.0% (w/v) uranyl acetate and are viewed with a JEOL 100 CX transmission electron microscope.

For scanning electron microscopy, the biofilms are fixed in 3% (w/v) glutaraldehyde in 0.1 *M* sodium cacodylate buffer at 4° overnight. These are then postfixed in 1% (w/v) osmium tetroxide at 4° for 2 hr before being dehydrated in a graded series of ethanol (20–100%, 15-min application time). Alcohol is then exchanged with 100% acetone for three 10-min rinses. The sample is then transferred into hexadimethylsilane (TAAB Ltd, Aldermaston, UK) for 1–2 min and left to dry in a desiccator. After 2 days it is removed and mounted onto an aluminum stub using two-tubed Araldite (TAAB). The specimen is then sputter coated with gold/palladium in a Polaron E5000 sputter coater (Bio-Rad, Cambridge, UK) and is then viewed on a Cambridge Stereoscan 90B electron microscope operating at 15 kV.

Cryosectioning

The physical sectioning of frozen biofilms followed by analysis (e.g., viable counting, vital staining) of the sections obtained can be used to obtain information concerning (i) the distribution of bacterial species within multispecies biofilms, (ii) the distribution of viable bacteria throughout mono- or multispecies biofilms, and (iii) the effects of antimicrobial agents on the survival of bacteria at different depths within the biofilm. The cryosectioning methodology is based on the work of Kinniment.[31] The sampling pan containing the biofilms is placed in 5 ml of a 25% (w/v) dextran (Sigma) solution, a cryoprotectant,[32] in artificial saliva for 2 hr at 37° before being removed and placed in 5 ml of 8% formaldehyde for 20 min. Disks are removed and placed onto a thin layer of OCT embedding compound (Raymond A. Lamb, London, UK) on a cryostat chuck. This is then placed in a −70° freezer for 20 min until the OCT compound and the biofilm are frozen. Additional OCT compound is then placed over the top of the biofilm so that the entire sample is embedded. Horizontal sections (30 μm thick) are then cut from the biofilm using a Bright 5030 microtome (Bright Instrument Co. Ltd., Huntington, UK), and each sample is placed in phosphate-buffered saline with cooled forceps. The suspensions obtained

[31] S. L. Kinniment, *in* "Bacterial Biofilms and Their Control in Medicine and Industry" (J. W. Wimpenny, W. Nichols, D. Stickler, and H. Lappin-Scott, eds.), p. 53. Bioline, Cardiff, 1995.

[32] M. J. Ashwood-Smith and C. Warby, *Cryobiology* **8,** 453 (1971).

may then be subjected to viable counting and/or vital staining as described previously.[6]

Conclusions

The CDFF is particularly suited to studies of biofilms of oral bacteria in that it provides an environment similar to that found in the oral cavity, i.e., a biofilm growing on a solid substratum with nutrients being provided in a thin film of liquid, continually replenished, trickling over the surface of the biofilm. Furthermore, removal of the surfaces of the biofilms by the scraper blade simulates the continuous removal of the outermost layers of supragingival plaque due to mastication and tongue movements. The advantages of the system are that (1) it provides many replicates (up to 90) in a single run, hence permitting good statistical analysis; (2) it allows intermittent pulsing of nutrients and/or antimicrobials; (3) it allows sampling of the biofilm (with large numbers of replicates) at various intervals during the course of a run; (4) pure or mixed cultures, saliva, or homogenized plaque samples can be used as an inoculum, which can be added continuously or intermittently; (5) it can be used to model supragingival or subgingival biofilms; (6) the system is autoclavable and temperature controlled; (7) a number of substrata can be investigated, e.g., hydroxylapatite, enamel, dental materials; (8) the depth of the biofilms obtained can be varied; and (9) it can be used to study the effects of pretreatment of substrata with antimicrobial agents on subsequent biofilm formation.

[22] Use of Continuous Flow Techniques in Modeling Dental Plaque Biofilms

By DAVID J. BRADSHAW and PHILIP D. MARSH

Introduction

Dental plaque is the biofilm that is found naturally on teeth where it forms part of the host defenses by preventing colonization by exogenous (and often pathogenic) microorganisms.[1] The microbial composition of dental plaque is diverse and contains many species with conflicting atmo-

[1] P. D. Marsh, *Proc. Finn. Dent. Soc.* **87,** 515 (1991).

0076-6879/99 $30.00

TABLE I
BACTERIAL GENERA FOUND IN THE ORAL CAVITY

Gram positive	Gram negative
Cocci	
Abiotrophia	*Moraxella*
Enterococcus	*Neisseria*
Peptostreptococcus	*Veillonella*
Streptococcus	
Staphylococcus	
Stomatococcus	
Rods	
Actinomyces	*Actinobacillus*
Bifidobacterium	(*Bacteroides*)[a]
Corynebacterium	*Campylobacter*
Eubacterium	*Cantonella*
Lactobacillus	*Capnocytophaga*
Propionibacterium	*Centipeda*
Pseudoramibacter	*Desulfovibrio*
Rothia	*Desulfobacter*
	Eikenella
	Fusobacterium
	Haemophilus
	Johnsonii
	Leptotrichia
	Porphyromonas
	Prevotella
	Selenomonas
	Simonsiella
	Treponema
	Wolinella

[a] The genus *Bacteroides* has been redefined. In time, the remaining oral bacteria still placed in this genus will be reclassified. *Mycoplasma* is also isolated from the mouth.

spheric and nutritional requirements (Table I[2]). This is a consequence of the development of gradients in many key environmental variables within these deep biofilms, producing a mosaic of microenvironments.

The mouth is an ideal habitat for microorganisms. The oral cavity is kept warm (ca. 36°), although the temperature around the gums can rise by several degrees during inflammation.[3] The pH of microbial communities

[2] P. D. Marsh and M. V. Martin, "Oral Microbiology," 3rd ed. Chapman & Hall, London/New York, 1992.

[3] P. F. J. Fedi and W. J. Killoy, *J. Periodontol.* **63,** 24 (1992).

on surfaces in the mouth is kept around neutrality due to the buffering action of saliva. Following the intake of fermentable sugars in the diet, the pH can fall rapidly (e.g., <pH 5.0),[4] and this pH fall produces shifts in the proportions of organisms from bacteria associated with dental health (e.g., *Streptococcus sanguis*) to bacteria associated with dental caries (e.g., *Streptococcus mutans*).[5–7] The mouth is an overtly aerobic habitat, and yet facultative and obligate anaerobes comprise the majority of the organisms recovered from most surfaces, particularly in dental plaque.[2] The redox falls markedly during plaque formation due to the metabolism of the colonizing pioneer species,[8] making conditions suitable for the growth of fastidious groups such as spirochetes and sulfate-reducing bacteria.

Plaque preferentially forms at stagnant or retentive sites and, unless removed by diligent oral hygiene, can predispose a site to either caries or periodontal diseases. In disease, there is a shift in the balance of the composition of the plaque microbiota.[9] In caries, there is an increase in acidogenic and acid-tolerating (aciduric) species (e.g., mutans streptococci and lactobacilli) at the expense of many acid-sensitive species (e.g., *Streptococcus sanguis, S. mitis*).[6,10–12] In contrast, in periodontal diseases there is an increase in plaque mass and eventually in the proportions of obligately anaerobic and often asaccharolytic genera, including *Prevotella, Porphyromonas, Eubacterium,* and *Treponema.*[13]

Because of the difficulties, both practical and ethical, in carrying out experiments on patients or volunteers, a number of laboratory models have been developed in order to study: (1) factors involved in the formation of dental plaque biofilms (e.g., the role of conditioning films and endogenous versus exogenous nutrients); (2) interactions among members of the resident microbiota (e.g., in food webs); (3) factors involved in the transition of the biofilm microbiota from having a commensal to a pathogenic relationship with the host; and (4) mode of action of antimicrobial and antibiofilm (antiplaque) agents.

[4] M. E. Jensen, P. J. Polansky, and C. F. Schachtele, *Arch. Oral Biol.* **27,** 21 (1982).

[5] J. D. de Stoppelaar, J. van Houte, and O. Backer-Dirks, *Caries Res.* **3,** 190 (1969).

[6] D. A. Dennis, T. H. Gawronski, S. Z. Sudo, R. S. Harris, and L. E. Folke, *J. Dent. Res.* **54,** 716 (1975).

[7] G. E. Minah, E. S. Solomon, and K. Chu, *Arch. Oral Biol.* **30,** 397 (1985).

[8] E. B. Kenney and M. Ash, *J. Periodontol.* **40,** 630 (1969).

[9] P. D. Marsh, *Adv. Dent. Res.* **8,** 263 (1994).

[10] R. H. Staat, T. H. Gawronski, D. E. Cressey, R. S. Harris, and L. E. A. Folke, *J. Dent. Res.* **54,** 872 (1975).

[11] G. E. Minah and W. J. Loesche, *Infect. Immun.* **17,** 43 (1977).

[12] G. E. Minah and W. J. Loesche, *Infect. Immun.* **17,** 55 (1977).

[13] W. J. Loesche, F. Gusberti, G. Mettraux, T. Higgins, and S. Syed, *Infect. Immun.* **42,** 659 (1983).

The models that have been developed differ in their complexity and utility; they range from various types of "artificial mouth",[14] the rototorque,[15] the constant depth film fermenter,[16] and flow cells.[17–19] In addition, the chemostat has been of particular value in studies of the human oral microbiota and of other microbiota in humans (reviewed in Ref. 20). Chemostats provide control of environmental conditions and allow variation of single parameters at a time so that cause-and-effect relationships can be established unequivocally. This article concentrates mainly on the use of chemostat models that incorporate surfaces for biofilm formation. As mentioned earlier, a number of other systems, originally not including surfaces and gradually incorporating greater complexity and heterogeneity, have been described for investigating the physiology and metabolism of dental plaque using chemostat systems and other continuous flow techniques. The evolution and applications of some of these models are listed in Table II.

The formation of oral microbial biofilms in chemostat systems[21–24] has permitted greater insights into the ecology of dental plaque in health and disease. The continuous culture approach does not attempt to reproduce, or act as a true model of, all of the physical properties of the habitat; rather, the chemostat permits the modeling, under highly controlled and reproducible conditions, of specific events that occur *in vivo*. Some of the key practical issues for modeling dental plaque using this approach will now be described.

Experimental System

The simplest chemostat systems consist of a single stage reactor. Such systems can be purchased from a number of manufacturers. For biofilm studies, however, the head plate may have to be modified to create extra

[14] C. H. Sissons, T. W. Cutress, M. P. Hoffman, and J. S. Wakefield, *J. Dent. Res.* **70,** 1409 (1991).

[15] R. A. Burne, Y.-Y. M. Chan, and J. E. C. Penders, *Adv. Dent. Res.* **11,** 100 (1997).

[16] M. Wilson, *Methods Enzymol.* **310** [21] 1999 (this volume).

[17] J. Sjollema, H. J. Busscher, and A. H. Weerkamp, *J. Microbiol. Methods* **9,** 73 (1989).

[18] H. J. Busscher and H. C. van der Mei, *Adv. Dent. Res.* **11,** 24 (1997).

[19] S. Singleton, R. Treloar, P. Warren, G. K. Watson, R. Hodgson, and C. Allison, *Adv. Dent. Res.* **11,** 133 (1997).

[20] P. D. Marsh, *J. Chem. Tech. Biotechnol.* **64,** 1 (1995).

[21] C. W. Keevil, D. J. Bradshaw, T. W. Feary, and A. B. Dowsett, *J. Appl. Bacteriol.* **62,** 129 (1987).

[22] D. A. Glenister, K. E. Salamon, K. Smith, D. Beighton, and C. W. Keevil, *Microb. Ecol. Hlth Dis.* **1,** 31 (1988).

[23] Y. H. Li and G. H. Bowden, *Oral Microbiol. Immunol.* **9,** 1 (1994).

[24] D. J. Bradshaw, P. D. Marsh, K. M. Schilling, and D. Cummins, *J. Appl. Bacteriol.* **80,** 124 (1996).

ports through which surfaces for biofilm formation can be inserted or removed aseptically.[21,23,24] Alternatively, it is possible to manufacture in a workshop a custom-designed vessel from, for example a Quickfit glass vessel sandwiched between two steel or titanium plates. The dimensions can be configured to have a high surface-to-volume ratio so as to give a greater surface area on the head plate for ports for electrodes and for surfaces.

Surfaces that have been inserted into continuous cultures include acrylic,[21] glass,[23,25] or hydroxylapatite[24,26]; more than one surface can be attached to titanium wire or a glass rod and immersed below the surface of the culture medium. Hydroxyapatite surfaces reflect the chemical composition of enamel surfaces and can be machined to create retentive areas.[24] The effect of fluoride leached from fluorhydroxyapatite (FHA) has also been examined in a similar way.[26]

In order to facilitate the aseptic removal of surfaces, the gas flow can be increased to create a positive pressure within the culture vessel or, alternatively, the vessel can be placed in a class 2 microbiological safety cabinet in which the fermenter is enclosed in particulate-filtered air and the operator is protected by a "curtain" of air from exposure to airborne microorganisms when the surfaces are removed from the fermenter vessel.

In general, chemostat studies have examined the responses of steady-state cultures to stresses applied by the experimenter. A steady state has been defined as the time taken for a continuous culture to reach "balanced growth" in which the levels of medium components, bacterial enzymes, and so on are constant.[27,28] In the case of mixed cultures, this term also implies a stable balance of the component microbial species. In our studies, we have found empirically that a steady state is achieved after approximately 10× mean generation time (MGT).[29,30] The MGT in the chemostat is related to the dilution rate (D) by the formula:

$$\mathrm{MGT} = \log_e 2/D$$

thus, a steady state is established at about 3–4 days at $D = 0.1\ \mathrm{hr}^{-1}$ (MGT = 6.9 hr) or after 6–8 days at $D = 0.05\ \mathrm{hr}^{-1}$ (MGT = 13.8 hr). The biofilm surfaces are inserted once this state is reached, and key environmental or biochemical parameters can then be either maintained or varied,

[25] T. Takehara, M. Itoh, N. Hanada, and E. Saeki, *J. Dent. Res.* **64,** 447 (1985).

[26] Y. H. Li and G. H. W. Bowden, *J. Dent. Res.* **73,** 1615 (1994).

[27] D. W. Tempest, *Adv. Microbial Physiol.* **4,** 223 (1970).

[28] S. J. Pirt, "Principles of Microbe and Cell Cultivation." Blackwell Scientific, Oxford, 1975.

[29] A. S. McKee, A. S. McDermid, D. C. Ellwood, and P. D. Marsh, *J. Appl. Bacteriol.* **59,** 263 (1985).

[30] D. J. Bradshaw, A. S. McKee, and P. D. Marsh, *J. Dent. Res.* **68,** 1298 (1989).

TABLE II

EVOLUTION OF CONTINUOUS FLOW METHODS FOR STUDYING DENTAL PLAQUE[a]

System	Type	Growth medium	Inoculum	Gas phase	Surface	Application	Refs.
Chemostat	Single stage	Complex	Monoculture	CO_2/N_2	No	Physiology, ecology, metabolism, sugar transport, fluoride, pathogenicity	45, *b–j*
		Single stage complex	Binary	CO_2/N_2	No	Physiology, ecology, interactions, chemostat theory, competition, glucose metabolism, fluoride, mucin degradation	*k–q*
		Single stage complex + mucin	9/10 species	CO_2/N_2	No	Physiology, ecology, metabolism, community development, fluoride, sugars, pH, antimicrobials, mucin degradation	29, 30, 44, *r–w*
		Single stage complex	Plaque	CO_2/N_2	No	Physiology, ecology, sugar metabolism	33
	Single stage	Complex + mucin	Monoculture	Air	Glass, HA, FHA	Biofilm formation, physiology, nutrition, interactions, pH	23, 26
		Complex + mucin, artificial saliva	Pooled plaque	CO_2/N_2	Acrylic, HA/acrylic	Ecology, physiology, sugar metabolism, biofilm formation	22, *x*
	Single, multistage	Complex + mucin	9/10 species	CO_2/N_2, air	HA	Ecology, physiology, biofilm formation, oxygen, coaggregation, antimicrobials	24, 32, 38, 52

CDFF	Single inoculum	Complex + mucin	Single, several	CO_2/N_2	PTFE, HA	Sugar metabolism, biofilm formation, biofilm structure	16
	Chemostat inoculum	Complex + mucin	9/10 species	CO_2/N_2, air	PTFE, HA	Sugar metabolism, biofilm formation, biofilm structure, antimicrobials	*y*, *z*
Artificial mouth	Batch grown	Artificial saliva	Mono/binary culture	Air	Enamel, HA	Sugar metabolism, biofilm formation, biofilm structure	*aa–ee*
	Saliva inoculated	Complex ± mucin ± urea	Plaque	Air	HA, enamel	Sugar metabolism, biofilm formation, biofilm structure, urea metabolism, transport in biofilms	14, *ff–ii*
Flow cell	Batch grown	Complex	Mono/binary culture	Air	Various	Initial adhesion, detachment, coadhesion	17, 18
	Chemostat fed	Complex + mucin	5 species; 8–10 species	CO_2/N_2, air	HA, germanium	Sugar metabolism, biofilm formation, biofilm structure, transport in biofilms	19, 50
Rototorque	Batch grown	Complex + mucin	Monoculture	Air	Polystyrene	Gene expression, gene regulation	15

[a] HA, Hydroxyapatite; FHA, fluorhydroxyapatite; PTFE, polytetrafluoroethylene.
[b] I. R. Hamilton, R. M. Boyar, and G. H. Bowden, *Infect. Immun.* **48,** 664 (1985).
[c] I. R. Hamilton and D. C. Ellwood, *Infect. Immun.* **19,** 434 (1978).
[d] I. R. Hamilton, P. J. Phipps, and D. C. Ellwood, *Infect. Immun.* **26,** 861 (1979).
[e] D. W. S. Harty and P. S. Handley, *J. Appl. Bacteriol.* **65,** 143 (1988).
[f] P. D. Marsh, A. S. McDermid, C. W. Keevil, and D. C. Ellwood, *J. Dent. Res.* **64,** 85 (1985).
[g] P. D. Marsh, M. I. Williamson, C. W. Keevil, A. S. McDermid, and D. C. Ellwood, *Infect. Immun.* **36,** 476 (1982).
[h] A. S. McKee, A. S. McDermid, R. Wait, A. Baskerville, and P. D. Marsh, *J. Med. Microbiol.* **27,** 59 (1988).
[i] C. Vadeboncoeur, L. Thibault, S. Neron, H. Halvorson, and I. R. Hamilton, *J. Bacteriol.* **169,** 5686 (1987).
[j] D. C. Ellwood, J. R. Hunter, I. R. Hamilton, P. D. Marsh, and G. H. Bowden, *in* "Continuous Cultivation of Microorganisms" (B. Sikyta, Z. Fencl, and V. Polacek, eds.), p. 333. Czechoslovak Academy of Sciences, Prague, 1980.

(*continued*)

TABLE II (*continued*)

[k] F. H. M. Mikx and J. S. van der Hoeven, *Arch. Oral Biol.* **20,** 407 (1975).
[l] J. S. van der Hoeven, M. H. de Jong, A. H. Rogers, and P. J. Camp. *J. Dent. Res.* **63,** 389 (1984).
[m] G. H. Bowden and I. R. Hamilton, *Can. J. Microbiol.* **33,** 824 (1987).
[n] G. H. Bowden and I. R. Hamilton, *J. Dent. Res.* **76,** 203 (1988).
[o] I. R. Hamilton and G. H. Bowden, *Infect. Immun.* **36,** 255 (1982).
[p] J. S. van der Hoeven, M. H. de Jong, P. J. M. Camp, and C. W. A. van den Kieboom, *FEMS Microbiol. Ecol.* **31,** 373 (1985).
[q] J. S. van der Hoeven and P. J. M. Camp, *J. Dent. Res.* **68,** 1041 (1991).
[r] A. S. McDermid, A. S. McKee, D. C. Ellwood, and P. D. Marsh, *J. Gen. Microbiol.* **132,** 1205 (1986).
[s] A. S. McDermid, A. S. McKee, and P. D. Marsh, *J. Dent. Res.* **66,** 1315 (1987).
[t] D. J. Bradshaw, A. S. McKee, and P. D. Marsh, *J. Dent. Res.* **69,** 436 (1990).
[u] D. J. Bradshaw and P. D. Marsh, *Caries Res.* **28,** 251 (1994).
[v] D. J. Bradshaw, K. A. Homer, P. D. Marsh, and D. Beighton, *Microbiology* **140,** 3407 (1994).
[w] D. J. Bradshaw and P. D. Marsh, *Caries Res.* **32,** 456 (1998).
[x] C. W. Keevil, D. J. Bradshaw, A. B. Dowsett, and T. W. Feary, *J. Appl. Bacteriol.* **62,** 129 (1987).
[y] S. L. Kinniment, J. W. T. Wimpenny, D. Adams, and P. D. Marsh, *Microbiology* **142,** 631 (1996).
[z] S. L. Kinniment, J. W. T. Wimpenny, D. Adams, and P. D. Marsh, *J. Appl. Bacteriol.* **81,** 120 (1996).
[aa] F. Lagerlof, R. Dawes, and C. Dawes, *J. Dent. Res.* **64,** 405 (1985).
[bb] W. D. Noorda, A. M. A. P. van Montfort, D. J. Purdell-Lewis, and A. H. Weerkamp, *Caries Res.* **20,** 300 (1986).
[cc] W. D. Noorda, D. J. Purdell-Lewis, A. M. van Montfort, and A. H. Weerkamp, *Caries Res.* **22,** 342 (1988).
[dd] H. D. Donoghue, G. H. Dibdin, R. P. Shellis, G. Rapson, and C. M. Wilson, *J. Appl. Bacteriol.* **49,** 295 (1980).
[ee] H. D. Donoghue and C. J. Perrons, *Microb. Ecol. Hlth Dis.* **1,** 193 (1988).
[ff] C. H. Sissons, L. Wong, E. M. Hancock, and T. W. Cutress, *Arch. Oral Biol.* **39,** 497 (1994).
[gg] C. H. Sissons, T. W. Cutress, and E. I. Pearce, *Arch. Oral Biol.* **30,** 781 (1985).
[hh] C. H. Sissons, L. Wong, E. M. Hancock, and T. W. Cutress, *Arch. Oral Biol.* **39,** 507 (1994).
[ii] C. H. Sissons, T. W. Cutress, G. Faulds, and L. Wong, *Arch. Oral Biol.* **37,** 913 (1992).

as required, in the culture vessel. Effects on the planktonic and biofilm communities of the various parameters of interest can then be determined, either by viable counts on a range of selective and nonselective agar plates or by the use of molecular or immunological probes. In a single stage system, however, once the primary vessel is perturbed, the experiment is often effectively terminated, and a fresh chemostat has to be set up for subsequent experiments. Multistage systems offer the benefit that the first stage can be used as a continuous (or intermittent, if required) seed or inoculum vessel for downstream vessels. Biofilms can be generated, and various treatments or experiments can be performed in these downstream vessels, which can be quite destructive on the microorganisms; because the first stage is unaffected, fresh steady states can be established quickly and easily without the need to establish a totally new system. This approach has proved particularly effective in the study of antimicrobial agents.[31] Death rates and mean survival times of the bacteria in the planktonic phase of the second-stage chemostat can then be determined by a calculation of the "viable biomass balance" from viable count data of the two vessels, using the following formulas:

$$\text{Death rate } (r) = \log_e[\text{rate of addition}^*] - \log_e[\text{cfu ml}^{-1} \times \text{flow rate}]^{\ddagger}$$

(* of culture from first stage. ‡ in second "test" stage).

The mean survival time (t_{50}) in the second stage vessel is then calculated from

$$t_{50} = \log_e 2/r$$

This method can be used to determine bacterial survival following antimicrobial treatments and was also used to determine the survival times of anaerobes in aerated cultures.[32]

Inoculum

Because plaque is a mixed culture biofilm, attempts have been made to reproduce some of the diversity of the microbiota in chemostat models by using inocula containing mixtures of bacteria (see Table II). These studies have included the study of the interactions of two or three species, where it is relatively straightforward to interpret the various interactions or responses to environmental stimuli. It may still be difficult, however, to extrapolate from these relatively reductionist situations to the natural habitat. At the other extreme, complex inocula, such as saliva or dental plaque,

[31] P. D. Marsh and D. J. Bradshaw, *Int. Dent. J.* **43**(Suppl. 1), 399 (1993).

[32] D. J. Bradshaw, P. D. Marsh, C. Allison, and K. M. Schilling, *Microbiology* **142,** 623 (1996).

have been used.[21,22,33] Such inocula have the benefit that they possess the full diversity of the natural microbiota of a site (including nonculturable species), but suffer from a number of inevitable drawbacks. For example, such inocula (a) are extremely diverse and, therefore, difficult to analyze and quantify, (b) are ill-defined for replicate experiments both within and outside the particular laboratory, (c) may contain undesirable species or even lack critical microbial populations (although this latter problem could be remedied by the addition of a known species), (d) cannot have their composition manipulated for experimental purposes, and (e) can never be adequately reproduced once a stock of, for example, frozen, pooled-plaque inocula[21] has been exhausted.

An alternative, intermediate approach has been the construction of a defined, but complex inoculum consisting of only those species relevant to the particular study. An advantage of this method is that the organisms can be selected with characteristics that make them easy to identify (e.g., colonial pigmentation, growth on particular selective media, catalase production)[29] (Table III). Such inocula are constructed by growing each organism separately, pooling the components, and using this to inoculate the chemostat. This inoculation procedure may have to be performed on more than one occasion to ensure the establishment of slowly growing, nutritionally fastidious, or obligately anaerobic species.[29] These defined inocula can be quality controlled before being dispensed into numerous vials for use in replicate experiments and then stored in the gas phase over liquid nitrogen until required.[34] An example of such an inoculum, together with criteria used for their discrimination, is shown in Table III.

When this approach was applied to an inoculum of nine plaque bacteria (see Table III), the counts of only one species (*Fusobacterium nucleatum*) were reduced significantly after freezing for 15 weeks. However, the counts of susceptible species can be boosted deliberately during the preparation of the inoculum to counter such losses, although it was found that the initial low counts of *F. nucleatum* did not affect its final steady-state levels in the microbial community in the chemostat.[34]

As mentioned earlier, the species composition of defined inocula can be manipulated for experimental purposes. Thus, in studies of the role of oxygen-consuming species on the establishment of obligate anaerobes in a biofilm, inocula were constructed that contained, or deliberately excluded in sequential experiments, the oxygen-utilizing species *Neisseria subflava*.[32] Later, all of the facultative species were omitted, leaving just four obligate

[33] P. D. Marsh, J. R. Hunter, G. H. Bowden, I. R. Hamilton, A. S. McKee, J. M. Hardie, and D. C. Ellwood, *J. Gen. Microbiol.* **129,** 755 (1983).

[34] D. J. Bradshaw, A. S. McKee, and P. D. Marsh, *J. Microbiol. Methods* **9,** 123 (1989).

TABLE III
DEFINED INOCULUM USED FOR MODELING STUDIES

Species	Distinguishing characteristics[a]
S. mutans	TYC/TYCSB agar. Small, crumbly, "bread crumb" white colonies on TYC. G+ve cocci. Facultatively anaerobic
S. oralis	TYC. Small, soft white/off white colonies on TYC. G+ve cocci. Facultatively anaerobic
S. sanguis	TYC. Hard, rubbery colonies on TYC. G+ve cocci. Facultatively anaerobic
L. rhamnosus	TYC, VBA, TYCSB. Large white colonies on all media. G+ve short rods. Facultatively anaerobic
A. naeslundii	TYC, CFAT. Brick red, domed colonies on all media. G+ve rods/filaments. Facultatively anaerobic
N. subflava	VBA. Mucoid orange/brown colonies on CBA or VBA. G−ve cocci. Aerobic
V. dispar	VBA. Colorless/gray colonies, very small (c. 1 mm after 5-day incubation). G−ve cocci. Obligately anaerobic
P. nigrescens	VBA. Dark brown/black colonies after only 2–3 days of incubation. G−ve short rods/cocci. Obligately anaerobic
F. nucleatum	VBA. Flecked colonies on CBA/VBA, in incident light have appearance of "oil droplet on water in sunlight." G−ve long, pointed ended rods (fusiforms). Obligately anaerobic
P. gingivalis	VBA. Beige to brown colonies after a 3-day incubation, eventually going darker. More rounded appearance, cf., *P. nigrescens*. G−ve short rods/cocci. Obligately anaerobic

[a] Distinguishing characteristics shown are as follows: Media on which organism grows (all bacteria grow on Columbia-blood agar). TYC medium.[5] VBA.[a] CFAT.[b] TYCSB.[c] Colonial appearance (on named medium). Gram reaction and cell morphology. Aerobic, obligately anaerobic, or facultatively anaerobic growth.

N.B. Individual species can be added or deleted for experimental purposes (see text for details).

[a] J. M. Hardie and G. H. Bowden, *in* "Proceedings: Microbial Aspects of Dental Caries. Special Supplements to Microbiology Abstracts" (H. M. Stiles, W. J. Loesche, and T. C. O'Brien, eds.), p. 63. Information Retrieval Inc., Washington, DC, 1976.

[b] L. J. Zylber and H. V. Jordan, *J. Clin. Microbiol.* **15,** 253 (1982).

[c] W. H. van Palenstein Helderman, M. Ijsseldijk, and J. H. J. Huis in't Veld, *Arch. Oral Biol.* **28,** 599 (1983).

anaerobes[35] (see later). Furthermore, this approach can be extended to determine the role of a particular gene product. Identical inocula can be formulated containing wild-type strains and isogenic mutants lacking a known function.

[35] D. J. Bradshaw, P. D. Marsh, G. K. Watson, and C. Allison, *Lett. Appl. Microbiol.* **25,** 385 (1997).

Surfaces and Conditioning Films

Although a number of different types of surfaces have been used in various chemostat systems, e.g., glass, acrylic, hydroxylapatite, or restorative materials,[21,23,26,36] there have not been any systematic comparisons of the pattern of colonization on these different surfaces. In any event, any surface immersed in proteinaceous growth medium, whether of peptone mixtures or artificial saliva, would be likely to become coated rapidly with a layer of proteins, glycoproteins, and even bacterial products and surface material (i.e., a conditioning film analogous to the acquired pellicle on teeth). Thus, bacteria interact with a layer of reactive molecules, irrespective of the nature of the surface below. Some surfaces can have extreme properties (i.e., highly hydrophobic or hydrophilic), and these might be expected to alter the properties of the conditioning film. Similarly, some surfaces may leach out ions with antimicrobial effects on the colonizing organisms.[26]

One study using a chemostat mixed culture biofilm system did attempt to determine the influence of the nature of the conditioning film on the final composition of a biofilm. Hydroxylapatite surfaces were precoated with either human parotid saliva or crude bacterial supernatants enriched for glucosyltransferases and glucans (the latter can act as receptors for glucan-binding proteins on bacterial surfaces).[37] After 4 days, surfaces that had been pretreated with the crude bacterial extracts had higher cell counts than those treated with saliva. Subsequent studies suggested that saliva may have had more of an influence on the earlier stages of biofilm formation.[38] Such studies demonstrate the potential for this type of system to study the influence of individual molecules, when adsorbed to a surface, on biofilm formation.

A study has compared the physiology and gene expression of plaque bacteria when growing in pure culture in a biofilm or planktonic culture[15] (Table II). Gene expression was upregulated when cells were associated with a surface, although the effect was more pronounced in older rather than young biofilms.[15]

Growth Medium

Microorganisms found in the mouth satisfy their nutritional requirements most often from the catabolism of host-derived molecules, particularly proteins and glycoproteins. Dietary components have only a qualitative influence on the composition of the plaque microbiota and do not appear

[36] J. T. Walker, D. J. Bradshaw, P. D. Marsh, and B. Gangnus, *J. Dent. Res.* **77,** 948 (1998).
[37] K. M. Schilling, M. H. Blitzer, and W. H. Bowen, *J. Dent. Res.* **68,** 1678 (1989).
[38] D. J. Bradshaw, P. D. Marsh, G. K. Watson, and C. Allison, *Biofouling* **11,** 217 (1997).

to alter the overall growth rate of dental plaque *in vivo*.[39,40] However, fermentable sugars do favor the growth of aciduric species,[6,7,10] whereas the regular consumption of snack foods containing alternative sweeteners such as xylitol can suppress the growth of mutans streptococci.[41]

The choice of growth medium can be critical to the development of a relevant model system. Early studies tended to use standard bacteriological media, with variations in the concentration of a readily fermentable carbon source.[29,33] In order to mimic more closely the nutritional condition in the mouth, more recent studies have used modified media, with proteins and glycoproteins as the main carbon and energy sources (habitat-simulating media). In particular, hog gastric mucin has been used to replace simple sugars. When dental plaque was used as an inoculum in a chemostat model system, the supplementation of a conventional laboratory medium with mucin resulted in an increase in biomass of the resultant community and a rise in the number of obligate anaerobes and *Actinomyces* spp.; spirochetes were also detected.[22] Replacement of the laboratory medium with an "artificial saliva" supplemented with mucin resulted in a lower cell yield, although levels of many of the fastidious Gram-negative anaerobes were enhanced.

The composition of the growth medium can be modified to reflect other habitats in the mouth. Thus, human serum has been used to mimic the increased flow of gingival crevicular fluid during an inflammatory host response associated with periodontal disease. In this way, a microbiota typical of periodontal pockets was enriched successfully *in vitro*.[42,43]

Atmosphere

As stated earlier, the mouth is essentially an aerobic habitat and yet there are very few truly aerobic organisms within the plaque microbial community (Table I). Plaque is comprised predominantly of obligately anaerobic species or of species, such as streptococci, whose growth is greater under CO_2-enriched conditions. Model systems are therefore usually maintained at a low redox potential by the sparging of the medium reservoir and/or the head space in the culture vessel with an appropriate low oxygen

[39] H. J. A. Beckers and J. S. van der Hoeven, *Arch. Oral Biol.* **29,** 231 (1984).

[40] D. Beighton and H. Hayday, *Arch. Oral Biol.* **31,** 449 (1986).

[41] K. K. Makinen, *in* "Progress in Sweeteners" (T. H. Grenby, ed.), p. 331. Elsevier, London, 1989.

[42] P. F. ter Steeg, J. S. van der Hoeven, M. H. de Jong, P. J. J. van Munster, and M. J. H. Jansen, *Antonie Van Leeuwenhoek* **53,** 261 (1987).

[43] P. F. ter Steeg, J. S. van der Hoeven, M. H. de Jong, P. J. J. van Munster, and M. J. H. Jansen, *Microb. Ecol. Hlth Dis.* **1,** 73 (1988).

or oxygen-free gas. Because some of the predominant oral bacteria are capnophilic (CO_2-requiring), a gas mix of 5% (v/v) CO_2 in nitrogen has often been used to sparge the culture vessel. In addition, conventional silicone rubber tubing can be replaced by oxygen-impermeable butyl rubber, although this was not found to be essential when growing many oral anaerobes in mixed culture[29,44] or even when growing pure cultures of the strict anaerobe *Porphyromonas gingivalis*.[45] Care has to be taken, however, not to use butyl tubing on sections that pass through the rollers on the peristaltic pumps that drive the medium in-flow, as this tubing is less elastic and can be susceptible to splitting.

Applications

Continuous flow techniques have been used in a number of applications, some of which have been referred to in the previous sections describing issues of relevance to modeling studies and many of which are listed in Table II. Some additional examples that illustrate the utility of this approach will now be described.

Ecology of Dental Plaque in Health and Disease

A mixed culture chemostat system, inoculated with a defined inoculum of nine plaque bacteria and using a habitat-simulating medium, was used to prove conclusively for the first time that low pH generated from sugar metabolism rather than freely available sugar per se was responsible for the breakdown of homeostasis in plaque and in the selection of potentially cariogenic species.[30] A steady-state community was established in a mucin-based medium at a constant pH 7.0 in two identical chemostats. Both were pulsed daily, for 10 days, with glucose (as a model fermentable substrate), but in one chemostat the pH was kept constant at pH 7.0 throughout the pulsing regime (by the automatic addition of alkali) to study the impact on the composition of the model oral microbiota of carbohydrate alone; in the other, the pH was allowed to fall for 6 hr after each pulse (driven by bacterial metabolism) before the pH was returned to neutrality for 18 hr prior to the next pulse.[24,30] There was little change in the balance of the community when the pH was kept constant at pH 7.0, but the microbiota in both biofilm and planktonic culture was perturbed severely when the pH fell after each glucose pulse. After the 10th pulse, the communities

[44] D. J. Bradshaw, P. D. Marsh, G. K. Watson, and D. Cummins, *J. Dent. Res.* **72,** 25 (1993).

[45] A. S. McKee, A. S. McDermid, A. Baskerville, A. B. Dowsett, D. C. Ellwood, and P. D. Marsh, *Infect. Immun.* **52,** 349 (1986).

were dominated by acidogenic and aciduric species, including *S. mutans* and *Lactobacillus casei.* Two further key findings emerged from this study. First, acid-sensitive species survived under conditions of low pH when part of a microbial community, which would not have been tolerated in pure cultures. Second, the reaction of bacteria was moderated in the biofilm phase of growth compared to the planktonic phase. Thus, acidogenic species (*S. mutans,* lactobacilli) were enriched to a lesser extent in biofilms grown during glucose/low pH episodes, whereas acid-sensitive species (*S. gordonii, F. nucleatum*) remained in greater numbers in the biofilm than in the planktonic phase during these low pH exposures. Among the critical features of these models are (i) stability and reproducibility, enabling parallel chemostats to be compared directly, (ii) the ability to repeatedly pulse the medium with additional substrates (the continuous flow of medium through the systems prevents buildup of substrate), and (iii) the amplification of inhibitory effects over time.

Evaluation of Antimicrobial Agents

The effects of antimicrobial agents can be studied in chemostat systems by two complementary approaches.[46] First, dosing strategies allow the agent to be added continuously, at a constant rate from a separate reservoir. In this case, the concentration of the inhibitor in the culture increases with time, reaching a steady state that is related inversely to the dilution rate of the culture, according to the formula:

$$I_t = I_f - I_f e^{-Dt}$$

where I_t is the concentration of the inhibitor at time t, I_f is intended (final) concentration of inhibitor; D is the dilution rate; and t is time. At $D = 0.1$ hr^{-1} the concentration of the dosed inhibitor reaches 50% of its maximum at 6.9 hr and 99% of maximum after approximately 48 hr.

The second approach is to pulse inhibitors into the culture vessel by adding a small volume of a relatively high concentration at a particular time point. The concentration of inhibitor then declines with time in a way related directly to the dilution rate, according to the formula

$$I_t = I_o e^{-Dt}$$

where I_o is the original concentration of the inhibitor at the start of pulsing. At $D = 0.1$ hr^{-1}, the concentration of the pulsed dosed inhibitor declines to 50% of its initial concentration at 6.9 hr and to 1% of the initial concentration after approximately 48 hr. In both cases, the formulas assume that the

[46] W. Borzani and M. L. R. Vairo, *Biotechnol. Bioeng.* **XV,** 299 (1973).

inhibitor is not metabolized. These approaches allow the inhibitory effects observed in planktonic or biofilm cultures at a certain time to be related to the precise concentration of the agent at that time. The dosed approach simulates the effect of a constantly applied antimicrobial, analogous to, for example, the relatively low level of fluoride in (some) potable water supplies. In contrast, the pulsed approach simulates an antimicrobial applied at a relatively high concentration at a particular time point, analogous to, for example, the application of an antimicrobial contained in a mouthwash. Chemostat responses to additions of inhibitors have been reviewed elsewhere.[46]

Biofilms can also be removed aseptically from the chemostat and immersed in a known concentration of an antimicrobial agent for a fixed length of time.[31] The microbial composition of the biofilms can then be determined by viable counts on selective and nonselective agar plates before, during, and after treatments.

Antimicrobial agents are being incorporated into a wide range of dental health care products to be used as an adjunct to toothbrushing in the quest to maintain oral health.[47–49] Some of these agents were compared for their efficacy against planktonic and biofilm communities of 10 species using a single-stage chemostat containing removal hydroxylapatite (HA) disks. Four-day-old biofilms and planktonic culture were removed from the chemostat and exposed to physiologically relevant concentrations of the antimicrobials (Triclosan, zinc citrate, pyrophosphate; alone and in various combinations) outside of the chemostat for 1 hr. When used alone, zinc citrate or pyrophosphate had little antimicrobial effects, whereas Triclosan had moderate effects, especially on planktonic cultures. Additive and synergistic effects were found using combinations of Triclosan with zinc citrate or pyrophosphate, particularly against gram-negative species, even when associated with biofilm.[31] This study reinforced the general finding that biofilms were more resistant than planktonic cultures to antimicrobial agents. The important features of this type of experimental approach are the ability (i) to compare simultaneously the effects of an agent on both planktonic and biofilm communities, (ii) the reproducible development of communities for comparative studies, and (iii) the ability to use antimicrobial *ex situ* so as not to perturb the integrity of the seed vessel, thereby facilitating replicate experiments. Chemostats have also been used as seed vessels to inoculate hydroxylapatite disks or germanium surfaces in flow cells downstream of the chemostat. The effect of antimicrobial agents on the biofilms formed

[47] P. S. Hull, *J. Clin. Periodontol.* **7,** 431 (1980).

[48] M. Addy, *in* "Chemical Plaque Control" (J. B. Kieser, ed.). Wright, 1990.

[49] P. D. Marsh, *J. Dent. Res.* **71,** 1431 (1993).

was studied using conventional bacteriological and novel spectrophotometric techniques.[50]

Oxygen, Survival, and Longitudinal Development of Biofilms

The effect of aeration on the development of a defined oral biofilm consortium (Table III) was investigated using a two-stage chemostat system. The inoculum, comprising 10 species, including both facultatively anaerobic and obligately anaerobic bacteria, was inoculated into a "conventional" anaerobic (gas phase, CO_2 in N_2) first-stage chemostat vessel. The effluent from this chemostat was fed into an aerated [200 ml/min of 5% (v/v) CO_2 in air] second-stage vessel in which biofilms were allowed to develop on HA disks. Planktonic and early biofilm communities in the aerated vessel were dominated by the oxygen-consuming species *Neisseria subflava.* Obligate anaerobes persisted in the planktonic culture and they predominated in 7-day-old biofilms. In a subsequent experiment, *N. subflava* was omitted from the inoculum to produce a nine-species culture. Streptococcal species predominated in planktonic cultures instead of *N. subflava.* Biofilms again underwent successional change, with anaerobes such as *Prevotella nigrescens* and *Fusobacterium nucleatum* increasing in proportion with time.[32] Knowledge of the kinetics of the chemostat was utilized to demonstrate that anaerobes were growing rather than merely persisting in the aerated culture. The study showed that the presence of oxygen-consuming, or oxygen-tolerant, species can protect obligate anaerobes from the toxic effects of oxygen, particularly in the biofilm phase. Thus, the consortium was behaving as a true microbial community, as defined by Wimpenny.[51] In the absence of these species, it was demonstrated that anaerobes were killed (at differential rates) in the aerated second-stage culture (using the formulas described earlier), and growth as a biofilm did not mitigate this effect.[35] Subsequent experiments showed that three-species coaggregations among the component species, mediated by *F. nucleatum,* facilitated the survival of anaerobes in these (hostile) aerated conditions.[52] The mechanism probably involved close physical cell contact between oxygen-tolerant and oxygen-sensitive species, thereby forming metabolically functional aggregates. This series of studies illustrates further key features of this type of experimental system, for example, (i) manipulation of the composition of cultures, to

[50] S. Herles, S. Olsen, J. Afflitto, and A. Gaffar, *J. Dent. Res.* **73,** 1748 (1994).

[51] J. W. T. Wimpenny, *in* "Bacterial Biofilms and Their Control in Medicine and Industry" (J. Wimpenny, W. Nichols, D. Stickler, and H. Lappin-Scott, eds.), p. 1. Bioline, School of Pure & Applied Biology, University of Wales College of Cardiff, Cardiff, 1994.

[52] D. J. Bradshaw, P. D. Marsh, G. K. Watson, and C. Allison, *Infect. Immun.* **66,** 4729 (1998).

allow the dissection of the role of individual species in community-level environmental stress responses, (ii) the advantages of two-stage systems in allowing stresses to be applied over time to a community in the second-stage vessel, and (iii) the quantitative, kinetic nature of the chemostat system, allowing the unequivocal demonstration of anaerobe growth in aerated conditions in a complex community and their rate of death in the absence of facultative species.

Concluding Remarks

Continuous flow techniques allow the modeling of key factors that influence the development of *in vivo* dental plaque communities (and microbial communities from other habitats) under controlled conditions in the laboratory. Table II illustrates the evolution and development of model systems, from monocultures grown in simple, chemically defined media to the use of complex mixtures of bacteria grown in peptone-based media, habitat-simulating media, and synthetic saliva. The introduction of surfaces has allowed spatial heterogeneity to be modeled in these systems so that now the experimenter can manipulate the composition of the microbial communities (and hence their metabolic repertoire) and can vary single environmental variables independently. These approaches allow the identification of key determinants of oral microbial biofilm formation and development.

Acknowledgments

The work described has been supported by Unilever Dental Research, the Medical Research Council of Great Britain, and the Public Health Laboratory Service.

[23] Steady-State Biofilm: Practical and Theoretical Models

By GEORGE DIBDIN and JULIAN WIMPENNY

Introduction

Steady-state systems in microbiology are powerful tools for investigating biological and physiological phenomena. This is because, once achieved, the steady state can be easily perturbed to obtain unequivocal data on the effect of any perturbation. The most common steady-state tool in microbiol-

0076-6879/99 $30.00

ogy is the chemostat, in which a homogeneous planktonic culture in a fermenter is fed with nutrient solutions, one of which is limited to an amount that restricts growth. Culture is removed at the same rate as nutrient is added. Over a wide range of dilution rates, the culture will enter a steady state, which can be maintained for prolonged periods. Homogeneous systems are easy to investigate in the chemostat. However, most natural ecosystems, including biofilm, are spatially heterogeneous, and here the concept of a steady state is less easy to define.

Biofilm has a complex structure, which can be broadly divided into three classes: (1) microbial stacks or columns of microcolonies plus estracellular polysaccharides (EPS), which are well separated from their neighbors; (2) penetrated biofilm consisting of interconnected "mushroom" or "tulip" shapes containing pores or channels; and (3) dense biofilm with few or no pores. This last type of biofilm is found at high substrate concentrations, e.g., in the oral environment as dental plaque. Differences in these structures have been interpreted predominantly in terms of substrate concentration.[1–3]

The first section discusses experimental systems for generating steady-state biofilm, particularly the constant depth film fermenter (CDFF). This is followed by a discussion of methods for the analysis of solute movement and reaction, within such a biofilm, using mathematical modeling.

A. Experimental Systems for Generating Steady-State Heterogeneous Biofilm Communities

Chemostat-Based Systems

The chemostat has been used by several workers[4–6] to generate cultures of bacteria that are then allowed to adhere to surfaces suspended in the growth vessel or more often in one or more vessels connected to it. However, while the biofilm is bathed constantly in a steady-state planktonic culture, the film itself passes through the various stages of attachment, growth, and ultimate detachment, characteristic of most natural biofilms.

[1] M. C. M. van Loosdrecht, D. Eikelboom, A. Gjaltema, A. Mulder, L. Tijhuis, and J. J. Heijnen, *Water Sci. Technol.* **32,** 35 (1995).

[2] J. W. T. Wimpenny and R. Colasanti, *FEMS Microbiol. Ecology* **22,** 1 (1997).

[3] C. Picioreanu, M. C. M. van Loosdrecht, and J. J. Heijnen, *Biotechnol. Bioeng.* **57,** 718 (1998).

[4] W. Keevil, D. J. Bradshaw, A. B. Dowsett, and T. W. Feary, *J. Appl. Bacteriol* **62,** 129 (1987).

[5] W. Keevil, *in* "Recent Advances in Microbial Ecology" (T. Hattori, ed.), p. 151. Japan Scientific Society Press, Tokyo, 1989.

[6] P. D. Marsh, *in* "Microbial Biofilms" (H. M. Lappin-Scott and J. W. Costerton, eds.), p. 282. Cambridge University Press, Cambridge, 1995.

Robbins Device

This device consists of a tubular section made of glass, metal, or clear plastic in which are located removable studs on which biofilm can grow. Medium flows through the tube at a predetermined rate generating shear forces.[7,8] The test section is part of a circulation system that may be "straight through" or closed so that medium is constantly recirculating. Flow rates and hence shear forces can be varied so that it is theoretically possible to achieve a quasi-steady-state community in which detachment is approximately equal to growth.

Rototorque

This device consists of two concentric cylinders the inner of which rotates at a set speed and generates a set shear rate. Culture medium flows through the system, giving a dilution rate faster than the μmax of the cell population. A torque converter mounted between the drive unit and the inner cylinder monitors drag forces, which, together with rotational speed, enables fluid frictional forces to be determined. The walls of the outer cylinder are fitted with from 4 to 12 removable slides on which biofilm grows. Nutrient medium is fed to the device, which is equipped with draft tubes that help provide good mixing. In this system, liquid residence time is independent of shear stress.[9,10] As with the Robbins device, a shear rate-dependent steady state may occur when growth is balanced by biomass removal, however, see Gjaltama *et al.*,[11] who noted considerable heterogeneity in rototorque biofilm.

Fluidized Bed Reactors

Fluidized bed reactors were developed especially for wastewater treatment. A biofilm forms on the insoluble support, e.g., sand. The system is fluidized in a vertical tube section by passing culture liquid and gases upward through the column. Attrition between neighboring beads removes excess biofilm from their surfaces, causing growth to approach steady-state conditions. Coelhoso *et al.*[12] investigated wastewater denitrification in a fluidized

[7] F. McCoy, J. D. Bryers, J. Robbins, and J. W. Costerton, *Can. J. Microbiol.* **29,** 910 (1981).
[8] L. Ruseska, J. Robbins, J. W. Costerton, and E. Lashen, *Oil Gas J.,* March, 153 (1982).
[9] G. Trulear and W. G. Characklis, *J. Water Pollut. Control Fed.* **54,** 1288 (1982).
[10] R. Bakke, M. G. Trulear, J. A. Robinson, and W. G. Characklis, *Biotechnol. Bioeng.* **26,** 1418 (1984)
[11] A. Gjaltema, P. A. M. Arts, M. C. M. Vanloosdrecht, J. G. Kuensen, and J. J. Heijnen, *Biotechnol. Bioeng.* **44,** 194 (1994).
[12] I. M. Coelhoso, R. Bonaventura, and A. Rodrigues, *Biotechnol. Bioeng.* **40,** 625 (1992).

bed reactor in which the substratum was 1.69-mm-diameter-activated carbon beads. The system was grown using molasses as nutrient and a thick biofilm (800 μm) developed after 1 week. Other configurations include biofilm growth on suspended particles in the related airlift fermenters.[13]

Constant Depth Film Fermenter

The constant depth film fermenter[14–17] developed by Wimpenny and colleagues is based on an earlier idea of Atkinson and Fowler.[18] It consists of a flat plate sometimes covered by a template with cutouts. Medium flows over the plate, and a wiper blade is passed regularly over the surface, removing any growth above a maximum thickness. We now consider the modern CDFF design in more detail.

Design and Operation of Constant Depth Film Fermenter

Design Criteria for CDFF

A good biofilm model must be sterilizable and allow the aseptic removal of sample. It must allow investigations of natural or defined microbial communities or pure cultures of a single species. Samples must be discrete, representative, and as reproducible as possible. To achieve this the system should generate large numbers of samples, enabling the statistical reproducibility of experiments to be investigated. Samples, together with their substrata, should be removable so that the biofilm is disturbed as little as possible. This allows the deployment of a wide range of analytical techniques, including horizontal and vertical cryosectioning, scanning and transmission electron microscopy, microelectrode techniques, and so on.

The most recent version of the CDFF manufactured and marketed by Cardiff Consultants Ltd. (Fig. 1) is described here; its use is considered further in the article by Wilson in this volume.[18a]

[13] L. Tijhuis, E. Rekswinkel, M. C. M. Vanloosdrecht, and J. J. Heijnen, *Water Sci. Technol.* **29,** 377 (1994).

[14] A. Coombe, A. Tatevossian, and J. W. T. Wimpenny, *in* "Surface and Colloid Phenomena in the Oral Cavity. Methodological Aspects" (R. M. Frank and S. A. Leach, eds.), p. 239. London Information Retrieval, 1981.

[15] A. Coombe, A. Tatevossian, and J. W. T. Wimpenny, *in* "Bacterial Adhesion and Preventative Dentistry" (J. M. ten Cate, S. A. Leach, and J. Arends, eds.), p. 193. IRL, Oxford, 1984.

[16] A. C. Peters and J. W. T. Wimpenny, *Biotechnol. Bioeng.* **32,** 263 (1988).

[17] A. C. Peters and J. W. T. Wimpenny, *in* "Handbook of Laboratory Model Systems for Microbial Ecosystems." CRC Press, Boca Raton, FL, 1988.

[18] B. Atkinson and H. W. Fowler, *Adv. Biochem. Eng.* **3,** 224 (1974).

[18a] M. Wilson, *Methods Enzymol.* **310** [21] 1999 (this volume).

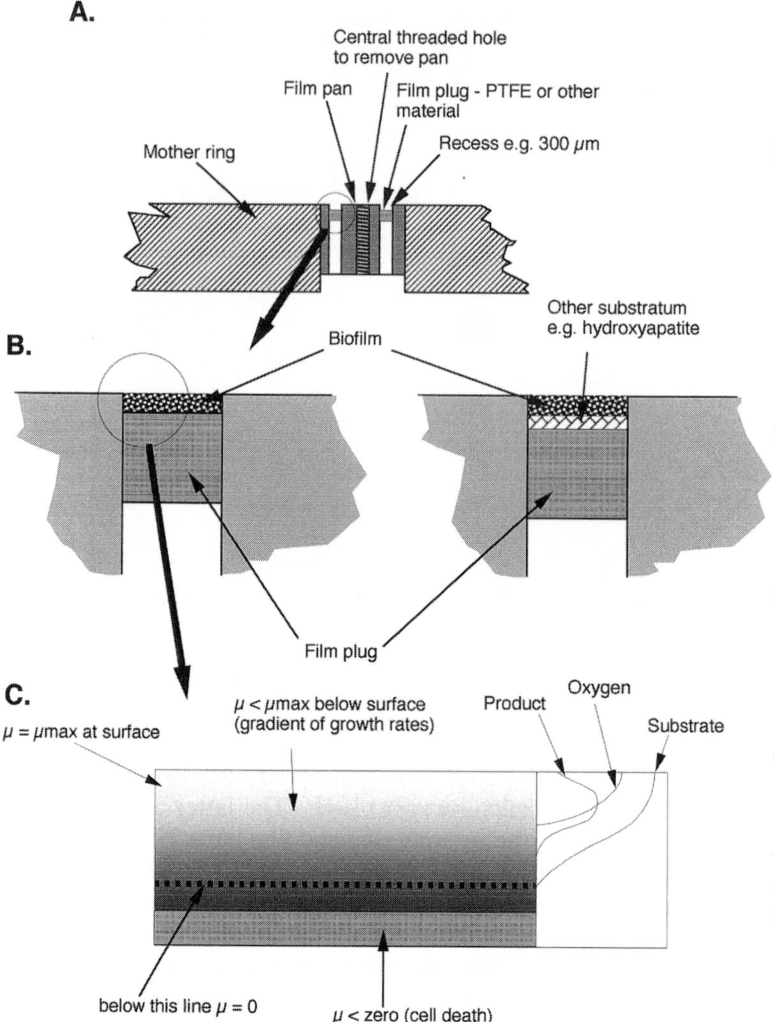

FIG. 3. Details of film assembly. (A) Film pan in mother ring. (B) Film pan and biofilm. (C) Possible structure of a dense biofilm.

vessel or flask containing both medium and inoculum and then circulate this through the CDFF for a time to encourage attachment before switching to the main reservoir and straight through operation; or (iii) grow the inoculum in a chemostat and feed the effluent from the latter directly into the CDFF (can be advantageous).

After the start-up period the CDFF is run for an appropriate time.

Temperature is maintained by running the system in a temperature-controlled laboratory or in an incubator. One version of the CDFF is fitted with a glass water jacket so that it can be run on an ordinary laboratory bench with a circulating constant temperature water bath.

Applications of CDFF

We have applied the CDFF in four separate research areas. These are: (1) to develop steady-state biofilm models with oral bacteria in an attempt to generate reproducible models of dental plaque.[19,20] Some aspects of this have been discussed in detail by Wilson, who has been a strong advocate of the CDFF and has made many useful advances in its use[18a]; (2) as a model to investigate biodeterioration by biofilms, particularly those which often contaminate metal working fluids[21–24]; (3) to investigate natural stream or riverine biofilms forming at low substrate concentrations[16]; and (4) to investigate the formation of biofilms associated with indwelling bladder catheters.[25]

Some of the cultural methods used are described next.

Steady-State Oral Community[19,20]

The nine strains used in this study are *Neisseria subflava* A1078, *Veillonella dispar* ATCC 17745, *Lactobacillus casei* AC 413, *Fusobacterium nucleatum* ATCC 10953, *Actinomyces viscosus* WVU 627, *Porphyromonas gingivalis* W50, *Streptococcus oralis* EF 186, *Streptococcus gordonii* NCTC 7865, and *Streptococcus mutans* R9.

The growth medium is a complex medium in which a model glycoprotein (hog gastric mucin) is used as the major carbon source.[26] The composition (g/liter) is porcine gastric mucin (Sigma) 2.5; KCl 2.5; proteose peptone (Lab-M) 2; yeast extract (Difco) 1; Trypticase peptone (BBL) 1; cysteine hydrochloride 0.1; and hemin 0.001.

[19] S. L. Kinniment, J. W. T. Wimpeny, D. Adams, and P. D. Marsh, *Microbiology* **142,** 631 (1996).

[20] S. L. Kinniment, J. W. T. Wimpenny, D. Adams, and P. D. Marsh, *J. Appl. Bacteriol.* **81,** 120 (1996).

[21] S. L. Kinniment and J. W. T. Wimpenny, *Int. Biodeterior.* **26,** 181 (1990).

[22] S. L. Kinniment and J. W. T. Wimpenny, *Appl. Environ. Microbiol.* **58,** 1629 (1992).

[23] S. L. Kinniment, Ph.D. Thesis, University of Wales, Cardiff, 1992.

[24] J. W. T. Wimpenny S. L. Kinniment, and M. A. Scourfield, *Soc. Appl. Bacteriol. Techn. Ser.* **30,** 51 (1993).

[25] L. Ganderton, Ph.D. Thesis, University of Wales, Cardiff, 1994.

[26] D. J. Bradshaw, A. S. McKee, and P. D. Marsh, *J. Dent. Res.* **68,** 1298 (1989).

The CDFF is inoculated from the output of a chemostat using the same medium under anaerobic conditions. The dilution rate for the chemostat is 0.1 hr^{-1}. The CDFF is run aerobically at 37°. The turntable rotates at 3 rpm, the medium flow rate is 50 ml min^{-1}, and the aeration rate is 250 cm^3 min^{-1}.

Metal Working Biofilms

Metal working fluids often become contaminated, and a well-developed biofilm forms as wall growth.[21,23] A strain of *Pseudomonas aeruginosa* is isolated as the dominant species in a cutting fluid. This is grown on a medium aiming to reproduce some of the properties of a cutting fluid. The amine : carboxylate medium (ACM) contains 0.085% (w/v) of cutting fluid components (60% triethanolamine plus 40% fatty acids of which 66% was oleic acid). In addition the medium contains in g/liter distilled water: $FeSO_4 \cdot 7H_2O$, 1.5×10^{-2}; $MgSO_4 \cdot 7H_2O$, 0.2; $ZnSO_4 \cdot 7H_2O$, 1.75×10^{-3}; $MnSO_4 \cdot 4H_2O$, 1×10^{-4}; $CuSO_4 \cdot 5H_2O$, 1×10^{-4}; NH_4Cl, 1.0; $K_2HPO_4 \cdot 3H_2O$, 1.2; $CaCl_2 \cdot 6H_2O$, 1×10^{-2}; calcium pantothenate, 2×10^{-3}; and biotin, 1.0×10^{-5}. Cutting fluid components are supplied by Castrol Research Laboratories (Pangbourne, Reading).

Growth in the CDFF[21,23] is carried out at 30° under a flow of 250 ml/min sterile water-saturated air with a medium flow rate of 1 ml/min. The turntable rotates at 3 rpm.

Natural River Biofilm Communities

Mixed microbial communities[16,17] are isolated from fresh stream water on glass slides in a 1-liter fermenter vessel. Two consortia, the first of 6 organisms and the second of 24, are isolated and used as inocula. The growth medium containes (in mg/liter): distilled water: KH_2PO_4, 100; $MgSO_4$, 20; $CaCl_2$, 1; NH_4Cl, 10; and glucose, 5. In some experiments this medium is supplemented with 0.1 mg/liter trace elements as follows: $FeSO_4 \cdot 7H_2O$, $ZnSO_4 \cdot 7H_2O$, $CoSO_4 \cdot 7H_2O$, $MnCl_2$, and $NaMoO_4 \cdot 2H_2O$. After inoculation, the CDFF is run aerobically either at ambient temperature or at 20°. Rotation speed is 2.5 rpm.

Indwelling Catheter Model Biofilm

The CDFF is used as a model of catheter biofilm formation by Ganderton,[25] but also see Stickler *et al.*[26a] The following media are used:

[26a] D. J. Stictter, N. S. Morris, and C. Winters, *Methods Enzymol.* **310** [35] 1999 (this volume).

1. Pooled urine from healthy male volunteers.
2. Artificial urine. This is the medium developed by Cox *et al.*[27] from a recipe used by Griffith *et al.*[28] It contains, in g/liter distilled water: $CaCl_2 \cdot 2H_2O$, 0.65; $MgCl_2 \cdot 6H_2O$, 0.65; NaCl, 4.6; Na_2SO_4, 2.3; trisodium citrate, 0.65; sodium oxalate, 0.02; KH_2PO_4, 2.8; KCl, 1.6; NH_4Cl, 1.0; and urea, 25.0. Gelatin [0.5% (w/v)] is added to the medium at 45° and the pH is adjusted to 6.1 before filter sterilization using a Sartorius cartridge filter of 0.22 μm pore size.

The CDFF is set up as usual, but in this case the inoculum enters the vessel via a glass "bladder" containing 10 ml of medium plus inoculum.

Analytical Methods Used with CDFF Studies

Protein Determination

This is currently determined for biofilm on each plug using the Lowry method after digesting the sample for 5 min in 1 *M* NaOH.

Viable Count Determinations

Considerable effort has been made to develop a method for reliably dispersing bacteria in a biofilm. For studies of metal-working biofilms, the use of hand-held homogenizers, a Waring blender, sonication, and vortexing with or without dispersing agents have been compared.[23] The most reliable method is to disperse biofilm in 10 ml of sterile deflocculant solution containing 0.01% (w/v) Cirrasol (ICI Uniquema, Everberg, Belgium) in 0.01% (w/v) sodium pyrophosphate. The suspension is shaken for 90 sec with about 250 2.5- to 3.5-mm-diameter glass beads (BDH/Merck, Lutterworth, Leic, UK) in a universal container. Viable counts are made as usual using plates containing amine : carboxylate medium solidified with agar. The model cutting fluid gives variable results, which depend on the concentration of the amine : carboxylates, the most reproducible biofilm formed using 0.085% of the latter in the growth medium.[23]

Cryosectioning Biofilm

A number of different cryosectioning techniques have been compared,[23] and the following method was finally selected. Sample plugs are removed carefully from film pans and held, biofilm side down, by forceps held in the jaws of a micromanipulator. The cryostat sample holder has disks of

[27] A. J. Cox, J. E. Harries, D. W. L. Hukins, A. P. Kennedy, and T. M. Sutton, *Br. J. Urol.* **59,** 159 (1987).

[28] D. P. Griffiths, D. M. Musher, and C. Itin, *Invest. Urol.* **13,** 346 (1976).

moist tissue paper and agar on its surface. One drop of Cryo-M bed is placed on the agar surface, and the biofilm is carefully lowered onto it so that it just breaks the liquid surface. Liquid nitrogen is then placed into the polystyrene container up to the depth of the agar plug, and the agar plus biofilm is allowed to freeze. Finally, the forceps, together with the Teflon plug, are removed, leaving the biofilm sample bottom side up on the agar.

Cryosectioning biofilm as described earlier is unsatisfactory when measuring viable counts, as a very significant loss in viability (of the order of 4 orders of magnitude) results from the freeze-thawing process. Use of a cryoprotectant largely solves this problem, with dextrans (molecular weight 60,000–90,000 large enough not to enter cells) used at a final concentration of 25% (w/v). When sample pans are removed from the CDFF and immersed in medium plus cryoprotectant for 2 hr at 30°, the effect of dextran on frozen and unfrozen film shows that the loss of viability due to freezing has been reduced to a factor of less than two,[23] a value regarded as acceptable.

In either case, sectioning is performed using the Bright Starlet 2212 cryostat. This is a compact bench-top cryostat with its own refrigeration unit. The frozen sample on its holder is fixed to the microtome in the cryostat, and the knife blade is set to cut 12-μm-thick sections. Adjustments are made to the blade angle to ensure that sections are cut parallel to the base of the biofilm. Sections in duplicate or triplicate are removed using a clean sterile microspatula.

Adenylate Measurements

Adenylates are determined either in whole biofilm or in horizontal frozen sections.[22] The latter are frozen as described earlier. Sections are removed from the knife in pairs or threes and placed into 0.5-ml aliquots of 2.3 *M* perchloric acid plus 6.7 m*M* EDTA extraction agent. These are shaken on ice for 15 min and are centrifuged for 3 min in an Eppendorf bench-top centrifuge at 13,000*g* at room temperature, and the supernatant is neutralized with 2 *M* KOH–0.5 *M* triethanolamine buffer. After leaving for 5 min for the potassium perchlorate to completely precipitate, the mixture is centrifuged for an additional 2 min and the samples are frozen at −70° until assayed. Adenylates are measured using the luciferin–luciferase bioluminescence method.[29]

[29] A. Lundin and A. Thore, *Appl. Microbiol.* **30,** 713 (1975).

Microelectrode Techniques

Measurement of pO_2 Gradients across Biofilm. Oxygen microelectrodes are constructed according to the method described by Coombs and Peters.[30] The electrodes are conditioned for 24 hr before applying a polarizing voltage of −0.75 V. The sample pan removed from the CDFF is located in a Perspex chamber set at an angle of about 7°, which allows a shallow laminar flow of medium across the surface of the biofilm.[31] The chamber is placed in a Faraday cage to isolate the high-impedence electrode from extraneous fields, and the medium inlet is connected to a reservoir containing 250 ml of growth medium. The electrode is attached to a micromanipulator, and a plate-viewing microscope is arranged so that the tip of the electrode could be located accurately at the surface of the biofilm. A silver–silver chloride reference electrode is located in the flow chamber at a distance from the measuring electrode. The latter is calibrated immediately before use in air or in nitrogen-saturated water. Readings are made using a chemical microsensor (Transidyne General Corporation, Model 1201, Ann Arbor, Michigan).

Measurements of pH Gradients across Biofilm. Liquid ion-exchange glass pH microelectrodes are constructed according to Robinson *et al.*[32] and are calibrated in pH 7.0 and 9.0 standard buffers. Film pans are removed from the CDFF and placed in empty petri dishes. Pans are just covered with sterile medium. pH gradients are measured across the medium and biofilm and across the biofilm after removing the medium, using a pH meter (Dulas Engineering, Aberystwyth) enclosed within the Faraday cage.

X-Ray Microanalysis

X-ray microanalysis is used,[23,25] in conjunction with scanning electron microscopy, to examine biofilm surfaces both from the *P. aeruginosa* biofilm isolated from metal working fluids (which showed evidence of pigmented regions across the surface) and from encrusting biofilms generated by catheter communities.

Biofilm from the metal working pseudomonad is fixed[23] in 0.5% (w/v) glutaraldehyde in phosphate buffer (pH 7.4) at 4°. After rinsing in phosphate buffer for 1–2 hr, the samples are dehydrated and subjected to critical point drying. Specimens are mounted on carbon stubs.

[30] J. P. Coombs and A. C. Peters, *J. Microbiol. Methods* **3,** 199 (1985).

[31] A. C. Peters, Ph.D. Thesis, University of Wales, Cardiff, 1988.

[32] T. Robinson, J. W. T. Wimpenny, and R. G. Earnshaw, *J. Gen. Microbiol.* **137,** 2285 (1996).

Biofilm from a mixed culture of *P. aeruginosa* CS 8 and *Proteus mirabilis* isolated from indwelling bladder catheters are freeze dried directly onto carbon stubs.[25]

Samples are coated with aluminum and then examined using the Jeol scanning electron microscope with the X-ray detector and microanalysis system (Link Systems 860 Series II energy dispersive microanalysis system and solid-state energy dispersive detector).

Atomic Absorption Spectrophotometry: Catheter System

Catheter biofilms can show encrustation made up of a mixture of crystals of struvite plus microbial biomass. The crystal components are analyzed by atomic absorption spectrophotometry.[25] Plugs plus biofilm are washed in double distilled water and transferred to 10-ml volumes of 1% (w/v) lanthanum chloride in 1 *N* HNO_3 (Spectrosol or Analar grade, BDH). The solutions are sonicated to disrupt the biofilm and diluted as necessary in water to the required concentrations (0.2–3.0 ppm Ca and 0.05–0.4 ppm Mg). It has been confirmed[25] that where crystals were present, levels of both metals are high.

Testing Antimicrobial Action in Steady-State Biofilm

It is well known that antimicrobials of most types are much less effective on organisms located within a biofilm than planktonic cells. This naturally makes a device capable of generating a steady-state biofilm a valuable tool for investigating antimicrobial activity under well-controlled conditions.

The CDFF has been used (i) to test the effects of a range of biocides against a pseudomonad isolated from contaminated metal working fluids[21,23]; (ii) to test the effect of chlorhexidine against a community of oral bacteria[19]; and (iii) to investigate the effects of six antibiotics against a pseudomonad isolated from an indwelling bladder catheter.[25]

Bladder Catheter System. Biofilms of *P. aeruginosa* CS 8 are generated in the CDFF and removed after different time periods. Plugs are placed in 1 ml synthetic urine plus the antibiotic and either disrupted in an ultrasonic bath followed by vortex mixing for 2 min to disrupt the biofilm or left intact. After 6 hr of incubation, disrupted films are made up to 10 ml with nutrient broth. The test solution is decanted from the test film samples and the latter is washed in 10 ml PBS solution. They are resuspended in nutrient broth and the cells are dispersed by sonication and vortexing. Viable count determinations are carried out as usual.

In one set of experiments,[25] six antibiotics (gentamycin, tobramycin, ciprofloxacin, imipenem, and two developmental compounds obtained from

Smith Kline and Beecham. (Brockham Park, Betchworth) were used in concentrations ranging from 8 to 128 μg/ml. Resistance to the antibiotics increased with the age of the biofilm. In all cases the intact biofilm was more resistant to the antibiotics than the disrupted cells.

Metal Working System. Biocides including formaldehyde, 3-methyl-4-chlorophenol (chlorocresol), and the isothiazolone, Kathon-W, were investigated using the CDFF.[21,23] Biofilms of *P. aeruginosa* are grown to steady state and at 50 hr the antimicrobial is added to fresh medium. Responses to the antimicrobial are determined by measuring total protein and viable counts of dispersed organism from film plugs. Control experiments are performed in planktonic cultures grown to a similar total biomass in conical flasks. In each case the biofilm is (often dramatically) more resistant to the biocide than the planktonic culture.

Achieving a Steady State

Using the different systems just described we can determine how well we have succeeded in the generation of steady-state biofilms.

The stream communities were the slowest growers.[31] This is not surprising considering the low nutrient concentrations used and the low ambient temperatures at which the experiments were run. This system generally took from 10 to 12 days to reach a steady state as ascertained by protein accumulation. The steady state was maintained for a further 14 to 18 days. Interestingly, in all CDFF experiments, viable counts reached their steady-state values earlier.

The *P. aeruginosa* biofilm growing on artificial metal-working medium at 30° showed excellent reproducibility and the biofilm reached steady state in terms of protein level after about 48 hr. This was maintained for at least another 100 hr. Viable counts appeared to enter a steady state a few hours earlier than the protein content. As noted earlier, reproducibility was dependent on the concentration of amine carboxylates in the medium.[23]

Growth of *P. aeruginosa* CS 8 on synthetic urine gave somewhat variable results, with the shortest time to steady state in terms of viable count at about 10 hr. Other experiments revealed rapid growth at first with a slower increase to steady state from 40 to 80 hr. These results were confirmed by protein assays. Scanning electron microscopy across the cut face of the biofilm and support indicated that the biofilm had completely filled the 300 μm-deep film pan.[25]

Other experiments[25] using natural urine and pure or mixed cultures of other bacteria indicated that although the cultures reached steady state as gauged by protein and viable count determinations, they did not completely fill the pans (Tables I and II).

TABLE I
MAXIMUM VALUES FOR EACH ORGANISM GROWN IN CDFF[a]

Organism	Viable count (cfu/plug)	Protein (μg/plug)
Pseudomonas aeruginosa	1.2×10^8 (1.0×108)	110 (65)
Proteus mirabilis	1.8×10^7 (1.1×10^7)	128 (125)
Enterobacter cloacae	2.2×10^6 (2.2×10^6)	64 (62)

[a] On natural urine as growth medium, steady-state values given in parenthesis.[25]

Using the oral community in the CDFF, steady states were reached for total viable counts after about 100 hr, and in terms of protein by 250 hr (Fig. 4). Individual species behaved differently. The aerobic *N. subflava,* already present at high levels, reached a steady state at 200 hr whereas *A. viscosus* and *L. casei* reached peak values at 280 hr but fell toward a lower steady state by about 400 hr. All streptococci plateaued after about 150 hr. Of the three, *S. mutans* was present in the lowest numbers. *S. gordonii* and *S. oralis* were detected at approximately equal levels at steady state. The anerobes showed the most interesting behavior during biofilm development. As might be expected when exposed to the aerobic conditions in the CDFF, all of these fell to very low numbers. *F. nucleatum* and *V. dispar* were detectable at about 10^4/ml. Their numbers then rose to reach a peak at 70–90 hr. *F. nucleatum* remained high thereafter whereas *V. dispar* fell to a steady state 2 logs lower than its peak value, after about 500 hr. Even more dramatic was the behavior of *P. gingivalis.* Its numbers were only just detectable ($<10^3$ ml^{-1}) for the first 120 hr. They then rose rapidly, reaching a steady state value of about 6×10^7 cfu/plug by 250 hr.[19]

TABLE II
APPROXIMATE STEADY-STATE VALUES FOR MIXED CULTURES[a]

Organism		Viable count (cfu/plug) A	Viable count (cfu/plug) B	Protein (μg/plug) approx. steady state
Pseudomonas	(A)	9.6×10^7	1.1×10^8	46
+ group D streptococcus	(B)			
Pseudomonas aeruginosa	(A) plus	1.6×10^8	8×10^6	140
Morganella morganii	(B)			
M. morganii	(A)	5.3×10^6	1.2×10^6	63
+ group D streptococcus	(B)			

[a] Grown in the CDFF on natural urine.[25]

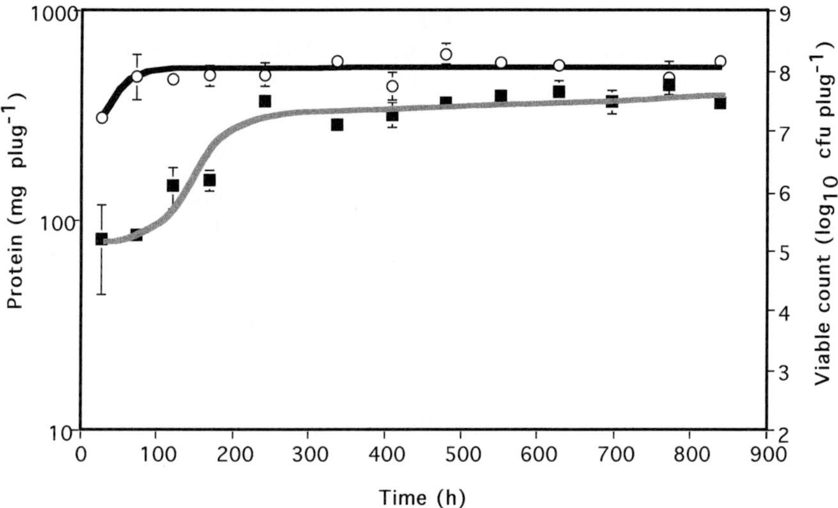

FIG. 4. Protein and viable counts for a nine-membered community of oral bacteria. Redrawn from Kinniment *et al.*[19]

We conclude that in all the systems described here, steady-state conditions in terms of total protein and viable counts could be generated and maintained for long periods.

Practical Problems with CDFF

The CDFF was designed to generate a biofilm that *should* fill the recessed area completely so that shear produced by the scraper bar will only remove growth that appears above the surface of the film pan. Although it is true that some biofilms fulfill this condition, others do not. Those that do include films of *P. aeruginosa,* films of natural and artificial dental plaque and of some river and stream communities, and biofilms of some isolates from indwelling bladder catheters. Other communities and pure cultures do not fill the pan. This might be expected, as each biofilm will behave differently, thus the composition of the EPS matrix will vary, and the degree and strength of adhesion to surfaces and to other community members will also alter. In many cases incomplete films will enter a steady state of sorts as the low shear forces present in the CDFF will remove cells at a rate similar to their regrowth. This steady state although paradoxically more natural than the process of physically scraping the surface to remove excess growth, is less reproducible than the former.

Unfortunately, it is not easy to predict the degree of adhesion or the likelihood of pans filling completely, and this must be a subject for a practical experiment by the investigator.

Transport of Solutes Inside and Outside Biofilm Matrix

Access of substrate into the film itself is more complex than described earlier. The biofilm consists of a mixture of cells and suspending matrix composed of water and EPS. In low nutrient biofilm there may be pores reaching from the surface to the base of the film. If cells are sparse within the EPS matrix, then the mean nutrient consumption rate per volume of biofilm will be reduced and more nutrient will penetrate deeper into the film. We have not yet emphasized the complex nature of solute distribution across a biofilm profile. Even if the film were uniform and lacked any form of water channels, macromolecules present interact with low molecular weight molecules in different ways. This will be dealt with in more detail in the next section, which considers the dynamics of solute transfer from a mathematical point of view.

B. Mathematical Modeling of a Steady-State Biofilm

The intention of this section is (i) to consider the definition of, and requirements for, a steady-state mathematical biofilm model; (ii) to discuss, in a simple way, some of the mathematical techniques for implementing such a model; (iii) to consider ways of obtaining the input data needed; and (iv) to suggest, in the light of recent biofilm research, some desirable future directions in mathematical modeling of steady-state biofilms and strategies for their implementation.

Every equation (e.g., the Michaelis–Menten equation) is a mathematical model, but the use of complex numerical models containing many disparate parts is required to solve problems of the sort discussed here. A common finding in such cases is that the very act of setting up and developing such a model is in itself as important as the predictive power of the model, with new questions being continually posed in a "spiral learning process"[33] as the model develops.[34]

Requirements for Steady-State Homogeneous Biofilm Model

Steady state in the present context is taken to mean that film thickness, bacterial density, and bacterial distribution (plus all intrinsic biochemical/

[33] G. H. Dibdin, *Adv. Dent. Res.* **11,** 127 (1997).

[34] W. G. Charaklis, *in* "Biofilms" (W. G. Charaklis and K. C. Marshall, eds.), p. 20. Wiley, New York, 1990.

physiological characteristics) are invariant with time in the absence of external effects.

It will also be assumed that intrinsic factors are invariant with position, i.e., the biofilm is macroscopically homogeneous. These simplifications are offset by the capacity of the model to encompass a wide range of reactions (reversible and irreversible, charge-coupled, physicochemical, metabolic/enzymatic) together with movement of all the reactants according to the laws of diffusion. Possibilities for extending the model to allow for steady-state growth as discussed in the previous section and of macroinhomogeneity as exists in some biofilms are considered briefly at the end of the section.

Mathematical Methods

Mixed reaction/diffusion processes are not soluble analytically, and numerical methods are necessary. Using the finite difference (FD) method, an imaginary film in the one-dimensional case considered here (Fig. 5) is split up into a series of slices parallel to its surface, from $j = 1$ to $j = j_1$,

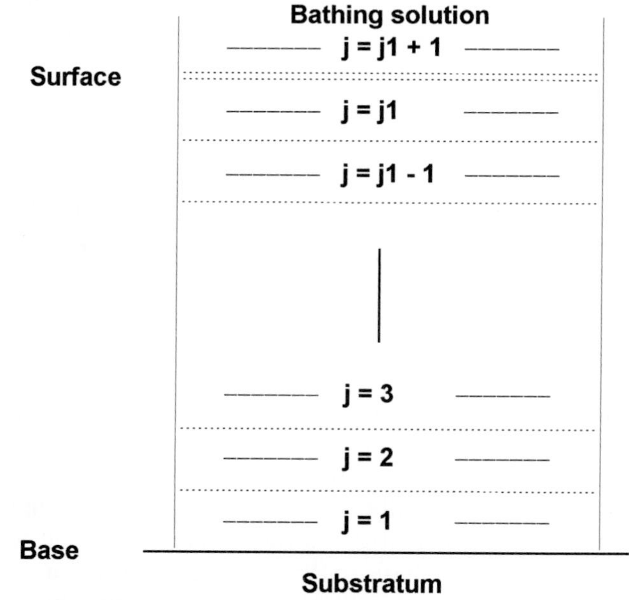

FIG. 5. One-dimensional finite difference scheme for a biofilm. The film is divided into j_1 slices by dashed lines, with a double dashed line marking the interface with the bathing solution, which can be made any chosen volume per unit area. Labeled horizontal lines mark the nodal planes of each slice where all chemical concentrations and reactions are assumed located and between which diffusion is calculated.

with no lateral variations. Variation with time is modeled in a series of discrete time steps. All species concentrations and their reactions are then assumed located at the midplane (node) of each slice. Transport between neighboring nodes then operates according to the laws of diffusion over the period of the time step, during which time the driving concentration gradients are assumed constant. Clearly as slice thicknesses and time steps become smaller and smaller, the model approaches ever closer to the continuous nature of a real system. A more complete mathematical analysis of the discussion that follows is given elsewhere.[35]

Modeling Diffusion in the System

The basic equation for one-dimensional diffusion is

$$\frac{\partial C}{\partial t} = \frac{\partial}{\partial x}\left(D\frac{\partial C}{\partial x}\right) \tag{1}$$

where C is the concentration, D is the diffusion coefficient, x is the distance, and t is the time. Converting to finite differentials Δ and δ and substituting a permeability: $K = D/\delta x$ where Δt is the time step, δx is the distance between neighboring nodes with Δx the width of a node, the relation becomes

$$\frac{\Delta C}{\Delta t} = \frac{1}{\Delta x}\Delta\left(K\,\delta x\frac{\delta C}{\delta x}\right) \tag{2}$$

Adding the suffix i to define each species and substituting w_j for the width of the jth slice, one form of the simple explicit finite-difference equation becomes

$$\frac{n_{i,j} - o_{i,j}}{\Delta t} = \frac{1}{w_j}\{K^{+}_{i,j}[o_{i,j+1} - o_{i,j}] + K^{-}_{i,j}[o_{i,j-1} - o_{i,j}]\} + q_{i,j} \tag{3}$$

The term $q_{i,j}$ is added to represent the rate of reaction. The calculation is swept through the whole calculation zone for each i in turn, starting at $j = 1$ and ending with $j = j_1$, to estimate all the new $(n_{i,j})$ concentrations after the time step from the old $(o_{i,j})$ concentrations before the time step. The term "explicit" means that the diffusion fluxes that lead to the new values are calculated on the basis of the known concentration gradients at the start of a time step. $K^{+}_{i,j}$ and $K^{-}_{i,j}$ define a total permeability between nodes $j + 1$ and j and nodes j and $j - 1$, respectively. When necessary, permeabilities may conveniently be combined[35] by adding their reciprocals

[35] G. H. Dibdin, *Comput. Appl. Biosci.* **8,** 489 (1992).

as "diffusion resistances." However, this explicit form of the FD diffusion equation is only stable if the size of the time steps taken, in relation to the distance between nodes and the rate of diffusion of the fastest diffusing species, is less than a critical value. The stability criterion (cf. Smith[36]; Dibdin[35]) requires that the stability factor (r) where

$$r = \frac{D\,\Delta t}{(\Delta x)^2} \tag{4}$$

must be less than 0.5 for the fastest diffusing species.

The simple explicit calculation of Eq. (3) works well for chemical species that do not undergo reaction; it also applies for species where reaction ($q_{i,j}$) is either instantaneous, as with normal acid/base equilibria,[35] or linear (rate independent of concentration), as in the case of lactic acid production due to carbohydrate metabolism. This excludes diffusion of any substrates of enzyme reactions, whose rates of reaction can vary over a great range and depend on substrate concentration. Here the so-called semi-implicit Crank–Nicolson form[37] is preferred.[35] In this the driving forces for diffusion are assumed to depend on averages between known concentrations ($o_{i,j}$) at the start and unknown concentrations ($n_{i,j}$) at the end of a time step:

$$\begin{aligned}\frac{n_{i,j} - o_{i,j}}{\Delta t} = \frac{1}{2w_j}\{&K^{+}_{i,j}[o_{i,j+1} + n_{i,j+1} - o_{i,j} - n_{i,j}] \\ &+ K^{-}_{i,j}[o_{i,j-1} + n_{i,j-1} - o_{i,j} - n_{i,j}]\} + q_{i,j}\end{aligned} \tag{5}$$

It is more complicated to solve this equation in comparison with Eq. (3) because the $n_{i,j}$ are unknown. Because of the nonlinear reactions represented here in the $q_{i,j}$, iteration is the preferred method.

Biochemical Reactions

The advantage of the iterative method is that the diffusion-controlled concentration changes in Eq. (5) and the reaction-controlled concentration changes implied by Eq. (6) can be calculated and converged simultaneously. For a single reaction, $q_{i,j}$ and $V_{i,j}$ in the two equations become identical:

$$V_{i,j} = f_i(\text{pH}, A, B, \ldots)\left[\frac{C_{i,j}V'_{\text{max i}}}{C_{i,j} + K'_{\text{Mi}}}\right] \qquad (i = 1 \text{ to } i_1; j = 1 \text{ to } j_1) \tag{6}$$

[36] G. D. Smith, "Numerical Solutions of Partial Differential Equations: Finite Difference Methods," 3rd ed. Clarendon Press, Oxford, 1985.

[37] J. Crank, "The Mathematics of Diffusion," 2nd ed., p. 326. Clarendon Press, Oxford, 1975.

where $C_{i,j}$ is the concentration at node j of the ith species (substrate), $V_{i,j}$ is its reaction velocity, and $V'_{\max i}$ and K'_{Mi} are its apparent Michaelis–Menten constants. The f(pH, A, B, ...) represents dependence of the reaction velocity of factors such as pH end product concentration. As the latter vary comparatively slowly, this function is calculated explicitly from values at each node before the previous time step. All that is then required at each time step is to sweep through from $j = 1$ to j_1, solving the appropriate equations, and repeating the whole procedure for increasing i from 1 to i_1. Here $q_{i,j}$ takes a negative value, as the substrate is lost during the reaction, whereas in Eq. (3), if i indicates a product of this same reaction, $q_{i,j}$ is always positive and equal to mr times the substrate $q_{i,j}$, with mr being the product/substrate molar ratio.

Charge Coupling

Charge-charge interactions between different diffusing ions and between these and fixed charges on the bacterial surfaces or extracellular matrix must also be considered. When ions diffuse there is a very strong requirement for local charge neutrality at the scale under consideration (between 10^{-6} and 10^{-3} m, say). Charge separation exists, as at interfaces, but only over atomic dimensions. With just two ions (H^+ and Cl^- say), the more mobile H^+ "pulls" the slower Cl^- along, whereas the Cl^- slows the H^+ down until they are diffusing at the same speed. In the general case of more than two diffusing ions, progress is at different speeds, but charge coupling still alters the diffusion velocity of each according to valency, ionic mobility, and concentration (the ionic "transport number").

Whereas charge coupling can be ignored for uncharged molecules, the effects on charged molecules are usually too important to ignore. There are two conventional ways of dealing with this problem. The first is to minimize the gradient of diffusion potential (see, e.g., McGregor[38]). The second involves replacing all the individual ionic species with their neutral component equivalents and then calculating concentration-dependent diffusion coefficients by matrix inversion methods. Both are very heavy on computational time. There is an alternative method, which, although less rigorous, is fast, accurate, and easy to implement. In this procedure, a short add-on routine[39] is applied separately, at each time step, after calculating the diffusion with charge coupling neglected. It uses the ionic transport number information (concentration, valency, ionic mobility) to redistribute

[38] R. McGregor, "Diffusion and Sorption of Fibers and Films," p. 120. Academic Press, London, 1974.

[39] G. H. Dibdin, *Comput. Appl. Biosci.* **5,** 19 (1989).

all the ions to the charge-coupled positions they would have occupied if charge coupling had been included in the diffusion calculation for that time step. This procedure has been found to work well and has been tested in a number of applications. It has the advantage of being added very easily to a program, but has not yet been tested in a two- or three-dimensional implementation.

Fast Chemical Reactions

Acid/base reactions of buffers such as bicarbonate and phosphate are effectively instantaneous and are essential in any model where pH is important. Such reactions include acid production from glycolysis in dental plaque leading to tooth demineralization[35,40] or in studies of metal corrosion under biofilms. It may also be necessary to define pH simply in order to model pH-dependent enzyme systems. Washed biofilm bacteria possess a large residual-fixed buffer capacity that is associated with the high concentration of charged groups (mainly acidic) on the bacterial cell walls and matrix,[41] which must be included with the other buffers where any acid or base changes are being modeled. It is possible to treat all these buffer reactions in a separate subroutine after each time step once diffusion and charge coupling have been treated. The analysis is detailed elsewhere,[35] but in principle the total concentration of hydrogen (H_{Total}) is written in terms of all the compounds containing it in dissociable form:

$$H_{Total} = [H^+] + [HR] + \sum_{species} x[H_xB^{y-}] \tag{7}$$

Square brackets denote concentrations. In the summation term, y defines the valency of each species and x the number of dissociable hydrogen atoms it contains. Each of the terms in the summation can be substituted by one involving $[H^+]$, the total amount of the acid (B), and its practical dissociation constant (K). In the case of lactate (La), the substitution is

$$[HLa] = \frac{La_{total}}{K_{La} + [H^+]}[H^+] \tag{8}$$

Equation (7) can then be rearranged to give $[H^+]$ (and therefore pH) in terms of total concentrations of all chemical species and their dissociation constants. This equation is then iterated to convergence at each node after a time step when the relative ionized and unionized concentrations of all species are calculated from equations similar to Eq. (8). The term [HR] in

[40] G. H. Dibdin, *J. Dent. Res.* **69,** 1324 (1990).

[41] R. P. Shellis and G. H. Dibdin, *J. Dent. Res.* **67,** 217 (1988).

Eq. (6) represents the fixed ionic groups, already mentioned, to which hydrogen ions are bound, dissociating to $[H^+] + [R^-]$ over a range of site energies. An easy way of approximating this dissociation is via a suitable curve based on practical titration measurements[41] made at low dilution on the biofilm material concerned. Binding of other cations (e.g., calcium) can also be included in the calculation.[35,40]

Diffusion/Reaction Input Data for Use in Model

Ideally, one must obtain values for diffusion coefficients without any interfering effects of reaction, and reaction rates with the effects of diffusion eliminated or minimized.

Diffusion

There are several ways of obtaining diffusion coefficients. If published diffusion coefficients obtained for open solution are used, allowances must be made for tortuosity and excluded volume. In the case of ions, their diffusion coefficients are conveniently obtained from their ionic conductances at infinite dilution.[42]

Diffusion coefficients can be measured in model plaques. In one method,[43] a diaphragm of the biofilm material acts as a separator in a two-chamber diffusion cell, and diffusion fluxes of radiolabeled molecules between the two are obtained by sampling the chambers periodically. In another method, model biofilms of known thickness are first equilibrated with tracer, and then the rate of clearance into a bathing solution is analyzed.[44,45] In a third method,[45,46] tracer molecules are allowed to diffuse into the end of a tube containing biofilm material, and diffusion coefficients are calculated from the tracer distribution along the tube after a known time has elapsed.

It is important to bear in mind the possibility of reversible or irreversible reactions affecting the measurements, especially in method 3, but to a lesser extent in method 2. A detailed discussion of some of these matters is given elsewhere.[33]

[42] R. C. Weast, "Handbook of Chemistry and Physics," 52nd ed. CRC Press, Ohio, 1979.

[43] G. H. Dibdin, *Arch. Oral. Biol.* **26,** 515 (1981).

[44] S. G. McNee, D. A. M. Geddes, C. Main, and F. C. Gillespie, *Arch. Oral. Biol.* **24,** 359 (1980).

[45] G. H. Dibdin, *in* "Cariology Today" (B. Guggenheim, ed.), p. 191. Karger, Basel, 1984.

[46] A. Tatevossian and E. Newbrun, *Arch. Oral. Biol.* **28,** 109 (1983).

Biochemical Reaction Rates

Rates of biochemical reactions unrestricted by diffusion limitations can be determined easily in well mixed reaction vessels. Thus Michaelis constants, rates of glycolysis, and their variations with pH and end product concentration can be measured in a pH stat for inclusion in the model. However, it is important to try to match the biochemical conditions of measurement with those in the biofilm. For example, a biofilm usually contains dead bacteria and bacterial debris, which frequently provide important nutritional supplement for live organisms (Dibdin and Dawes, 1998, unpublished data).

Other enzyme systems that have been included in the biofilm model, and for which velocity data have been obtained, are the phosphatases of dental plaque that hydrolyze monofluorophosphate ions into F^- and orthophosphate as it diffuses into the plaque[47]; the ureases of dental plaque that hydrolyze urea into ammonia and CO_2, giving a net pH rise[48]; and lactamases that degrade β-lactam antibiotics penetrating into biofilms of *P. aeruginosa.*[49]

It should also be remembered that the buildup of metabolic products in the close confines of the extracellular compartment of a biofilm may cause end product inhibition as, for example, in lactate inhibition of glycolysis[50] or orthophosphate inhibition of phosphatase activity.[47] The model has been designed to include such effects, with additional experiments being needed to obtain the required data input.

Acid/Base Dissociation Constants

Except for the fixed buffers on bacterial cell walls already discussed, these may simply be taken from the literature. Because concentrations, not activities, are used in Eq. (8), practical (apparent) values of the dissociation constants K_a apply, but these vary with ionic strength. It may often be adequate to use these directly, assuming an approximately isotonic ionic strength. More properly,[35] the thermodynamic values defined at infinite dilution[42] are used. In this case, the model must calculate the ionic strength at every node and time step before applying a suitable extension of the Debye–Hückel equation to compute activity coefficients, and hence the

[47] E. I. F. Pearce and G. H. Dibdin, *J. Dent. Res.* **74,** 691 (1995).

[48] G. H. Dibdin and C. Dawes, *Caries Res.* **32,** 70 (1998).

[49] G. H. Dibdin, S. J. Assinder, W. R. Nichols, and P. A. Lambert, *J. Antimicrob. Chemother.* **38,** 757 (1996).

[50] S. J. Assinder, L. V. J. Eynstone, and G. H. Dibdin, *FEMS Microbiol. Lett.* **134,** 287 (1995).

apparent dissociation constants. The Davies equation[51] is the most convenient compromise,[35] as it depends solely on ionic strength and valency.

Ideal Requirements for a New Biofilm Computer Model

The computer model discussed earlier is able to model many factors in the behavior of fairly homogeneous biofilms, such as dental plaque, and has been valuable in a number of ways. However, practical developments in confocal laser scanning microscopy (see, e.g., Stoodly *et al.*[52]) have highlighted the inhomogeneous and discontinuous nature of many other types of biofilm. It therefore seems appropriate (see also Dibdin[33]) to consider desirable extensions of the model to deal with factors such as inhomogeneity, and to suggest possible strategies for their implementation by future workers.

1. First, and most importantly, the new program should allow at least limited modeling of the inhomogeneities of many biofilms; limited, because it is not helpful to compromise analyzability with unnecessary complexity or diversity. A sensible move would be to extend the model to two dimensions. As discussed elsewhere,[33,53] this would allow useful analyses of many inhomogeneities and discontinuities, including the presence of bacteria-depleted channels and dense bacterial clumps.
2. If the model is to include such local density variations, then to be useful it must contain a reasonable number of nodes, probably a minimum of five over the diameter/width of one channel or clump. This implies the need for a nodal net with internodal distances of no more than 2–3 μm.
3. The future model should if possible include (with its current ability to model short-term details of enzyme/substrate and acid/base reaction/diffusion) the additional ability to model steady-state growth, as discussed earlier in this article. This will not be too difficult to implement. It will require inclusion of a Monod-type relationship defining a rate of increase of bulk at each node of the computational net in terms of local substrate concentration. The slow incremental outward transfer of matter that then results will allow calculation of the upward expansion rate at the biofilm surface by integrating the rate

[51] C. M. Davies, "Ion Association," 2nd ed. Butterworths, London, 1962.

[52] P. Stoodley, D. de Beer, and Z. Lewandowski, *Appl. Environ. Microbiol.* **60,** 2711 (1994).

[53] W. Gujer and O. Wanner, *in* "Biofilms" (W. G. Characklis and K. C. Marshall, eds.), p. 397. Wiley, New York, 1990.

of increase of bulk at each node between substratum and surface. Stewart[54] has considered this outward expansion mathematically in his "biofilm accumulation model."

4. Any new modeling scheme should be capable, in order to be cost effective, of development into a reasonably robust application that can be handled by a number of end users without detailed knowledge of its structure and code.

Requirements 1 and 2 will be difficult to fulfill, not just because of the greater mathematical complexity involved, but, more fundamentally, as a result of the requirements highlighted by Eq. (4). This shows the heavy penalty, in terms of computing time, imposed by the need to refine the computing mesh. Thus, a 5-fold reduction in internodal distance would require a 25-fold reduction in the size of each time step, and so similar increase in computing time, in order to maintain adequate stability [Eq. (3)] or accuracy [Eq. (5)]. Add to this the fact that the number of nodes for which calculations must be made has been squared by the move to two dimensions (even without the need for a finer computing mesh) and it can be seen that fast computing hardware and an efficient code will be necessary in order to obtain useful results in a reasonable time. It is likely that some improvement will be possible from efficiency-raising methods in the two-dimensional diffusion calculations (see, e.g., Peaceman and Rachford[55]). There are also fully developed applications that could be applied, ready-made, for some parts of any newly designed model.

Conclusion

Section A described some of the methods used for the practical modeling of biofilms at the steady state, especially those using the CDFF, and also described some methods of data measurement and analysis. Section B described a steady-state (strictly short-term, homogeneous) mathematical model of biofilm interactions and indicated how it could be developed by others to model a degree of inhomogeneity, as well as long-term growth toward a steady state as discussed in the first section. We are conscious, however, that the definition of steady state, based purely on the constancy of bacterial count and protein density, is difficult in as complex a real system as a biofilm. For example, slow genetic changes may be occurring all the time, growth rates, varying with depth, may lead to sufficient nutrient depletion at the base of the biofilm as to cause cell death; and complex

[54] P. S. Stewart, *Antimicrob. Agents Chemother.* **38,** 1052 (1994).

[55] W. Peaceman and H. H. Rachford, *J. Soc. Industr. Appl. Math.* **3,** 28 (1955).

local interactions between many organisms of the biofilm community may introduce instability. We nevertheless believe that practical and theoretical steady-state models of the sort we have discussed will continue to be important tools in biofilm research.

[24] Spatial Organization of Oral Bacteria in Biofilms

By PAUL E. KOLENBRANDER, ROXANNA N. ANDERSEN, KAREN KAZMERZAK, ROSEMARY WU, and ROBERT J. PALMER, JR.

Introduction

Biofilms formed by communities of mixed species of human oral bacteria on enamel surfaces *in vivo* are called dental plaque. Modeling dental plaque communities *in vitro* can be accomplished by employing a flowcell that has been described elsewhere in this volume[1] and is based on the original description.[2] The flowcell (Fig. 1C7 for top view and Fig. 1C8 for side view) is the size of a microscope slide (25 × 75 mm), contains two channels in a molded silicone rubber gasket, and has microscopy cover glass surfaces on top and bottom. Inlet and outlet ports are molded into towers of silicone rubber at each end of the gasket and are fused with 3-inch pieces of medical grade silicone rubber tubing. Barbed plastic connectors are used to couple the pieces of tubing of the flowcell with longer tubing connected to media reservoirs and pumps.

Construction of Flowcell

Teflon molds (Fig. 1A2, tower mold; Fig. 1A3, gasket mold) are fabricated in a local machine shop and are of dimensions 0.91 cm (h) × 3.56 cm (w) × 10.16 cm (l) [see Appendix for engineer's drawings of tower mold (Fig. A.1) and gasket mold (Fig. A.2)]. The silicone adhesive (Fig. 1A1) is squeezed onto the mold cavities, which are slightly overfilled. Self-leveling (pourable) silicone adhesive must be used. The molds are covered with a piece of plastic-coated freezer paper that is dampened with water on the paper side. The plastic surface is placed against the adhesive, which in turn is covered with a sturdy wooden block (Fig. 1A4). All pieces are clamped with two standard laboratory/secretarial 1-inch capacity binder

[1] R. J. Palmer, Jr., *Methods Enzymol.* **310** [12] 1999 (this volume).
[2] R. J. Palmer, Jr. and D. E. Caldwell, *J. Microbiol. Methods* **24,** 171 (1995).

0076-6879/99 $30.00

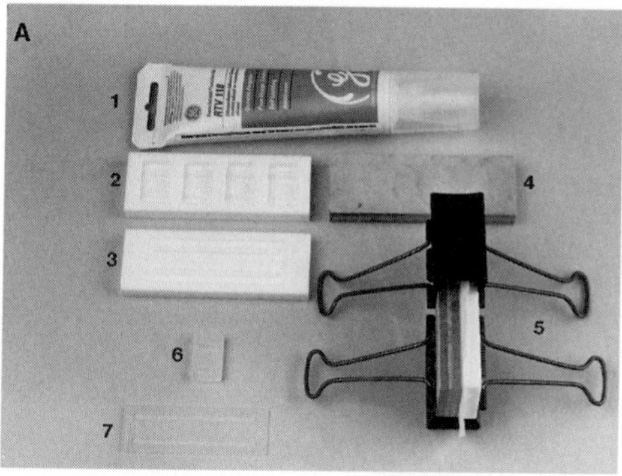

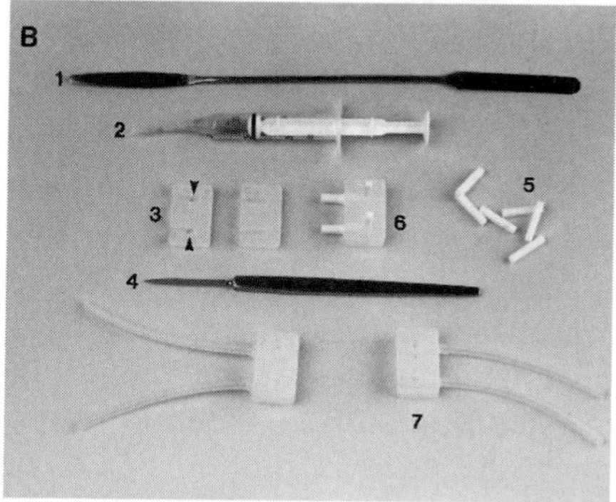

FIG. 1. (A) Materials needed to fabricate the silicone towers and gaskets. Silicone rubber sealant (1), Teflon mold for towers (2), Teflon mold for gaskets (3), wooden block (4), clamped sandwich of wooden block, freezer paper with plastic-coated surface facing Teflon block (5), one-half of a silicone tower (6), and silicone gasket (7). (B) Assembly of completed silicone tower. Thin-bladed metal spatula (1), 3-ml plastic syringe with pulled tip (2), one-half of tower showing cavity and square hole (3) cut with microtechnique scalpel (4) at the interior end of the cavity, Teflon square rods (5) inserted into holes in assembled tower consisting of two halves bonded with silicone adhesive (6), and silicone medical grade tubing inserted into tower channels (7) with excess space in tower cavity filled with silicone adhesive. (C) Assembly of complete flowcell. Diamond glass scriber (1) used to cut full-length cover glass (2) to shorter cover glass (3), silicone gasket (4) mounted on full-length cover glass (5) with shorter cover glass centered on top of gasket exposing four openings (arrows) at each end of the two flowcell channels to complete the biofilm chamber (6), completed flowcell top view (7) and side view (8), piece of 0.41 cm ID silicone tubing with male Luer-lock × 0.41 cm hose barb (9), set of male/female Luer-lock fittings with hose barbs (10), and plastic clamp (11) for stopping flow through silicone tubing.

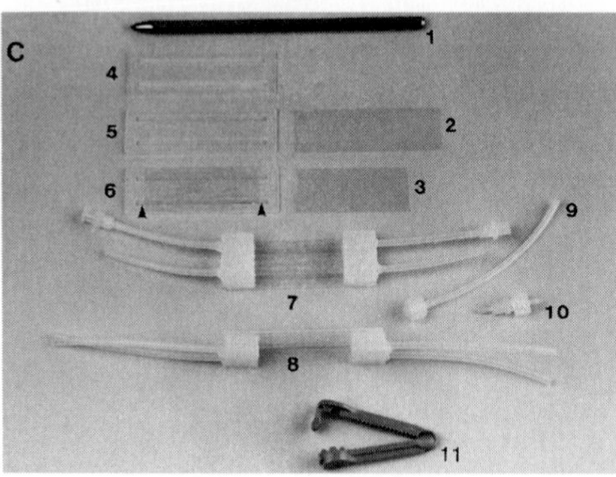

FIG. 1. (*continued*)

clips (Fig. 1A5). Curing is enhanced with wet paper, but it still requires at least 24 hr at ambient temperature to obtain a fully cured silicone gasket (Fig. 1A7). The tower halves (Fig. 1A6) usually require an additional 12 to 24 hr to cure. The silicone items are released from the mold by gently edging the item with a thin-bladed metal spatula (Fig. 1B1), taking care not to score the mold. The removal of the items is made easier by first immersing the empty mold in a 50:50 solution of ordinary dishwashing liquid soap and water; the mold is allowed to dry before applying the silicone adhesive as described previously. Excess cured silicone is removed from the edges of the mold by rubbing with a gloved hand. After the cured pieces are removed from the molds, the soap residue should be rinsed off thoroughly with distilled water.

Additional bonding during construction of the flowcell is accomplished with a 3-ml capacity plastic syringe whose tip has been drawn out by gently heating the syringe barrel in the flame of a bunsen burner, allowing the softened tip to fall toward the floor. The elongated tip is cut at the desired bore size for the application of silicone adhesive (Fig. 1B2). Adhesive is squeezed from the tube (Fig. 1A1) into the barrel of the syringe and the plunger forces the adhesive through the small-bore tip. Adhesive remains pliable in the syringe for 2 hr. Two small square-sided holes (about 2 mm on each side) are cut into the channels of one of the tower pieces (Fig. 1B3) using a microtechnique scalpel (Fig. 1B4). These holes will later be positioned over the channels in the gasket. The tower is constructed by placing the channel cavities of two halves facing each other and bonding the halves with a narrow bead of adhesive applied from the syringe; the tower is allowed to dry overnight. A square-sided Teflon rod (Fig. 1B5) ($2 \times 2 \times 16$ mm) is first inserted into each of the cut holes until it contacts the top inside wall of the tower. A second Teflon rod is inserted perpendicular to the first rod along the top wall of the tower until it abuts the other rod. The unfilled space in the holes of the tower piece is filled with silicone adhesive, which is allowed to dry 3–4 hr (Fig. 1B6). The Teflon rods are removed, and the channels inside the tower are checked for remnants of silicone adhesive, which is removed with a scalpel. This creates an L-shaped flow channel inside the tower that is the conduit between the channel in the gasket and the silicone tubing on either end of the flowcell. The 8-cm pieces of medical grade silicone tubing are inserted into the tower and secured with adhesive (Fig. 1B7).

The biofilm chamber of the flowcell is assembled using a diamond glass scriber (Fig. 1C1) to cut a shorter upper cover glass of 58 mm in length (Fig. 1C3) from the lower cover glass (Fig. 1C2). A narrow adhesive bead is also applied to the gasket (Fig. 1C4), which is then bonded to the lower cover glass surface (Fig. 1C5). A procedure found to be useful is to blot briefly the adhesive-beaded gasket onto a piece of paper, which spreads the adhesive evenly and prevents excess adhesive from oozing into the channel area. The gasket is then placed onto the bottom cover glass and is allowed to cure for 3 to 4 hr. A very narrow bead (1 mm) of adhesive is placed on top of the gasket along the outer long edges, and two 1-mm beads are drawn along the center region of the gasket. The shorter piece of cover glass is centered over the channels in the gasket and is pressed gently to seal it to the gasket, leaving a 5- to 8-mm portion of the channel exposed at each end of the gasket (Fig. 1C6, arrows). We use the eraser end of a pencil to flatten both cover glasses onto the gasket. Any excess silicone that squeezes out from the ends is removed by scraping it off with a flat-sided toothpick. Do not get any adhesive in the channels. It is important to avoid getting silicone adhesive on the exterior of the top and bottom cover glasses, as the flowcell may not sit flat on the microscope stage. A razor blade is used to clean the bottom cover glass. Use care in removing excess adhesive. Do not apply too much pressure as the cover glass can crack. The biofilm chamber is allowed to bond overnight.

Two tower assemblies (Fig. 1B7) are placed onto the biofilm chamber such that the two scalpel-cut holes of a tower assembly are directly over each exposed end section (Fig. 1C6, arrows) of the biofilm chamber, and the assembly is bonded with silicone adhesive. Spread adhesive over the entire bottom surface of the tower, taking care not to get adhesive near the holes. The completed flowcell is shown in a view from above (Fig. 1C7) as used on the stage of a microscope and in a side view (Fig. 1C8). Barbed Luer connectors (Fig. 1C10) are used to facilitate the easy connection of new flowcells to existing tubing. We usually place the female Luer connector on the flowcell (Fig. 1C7) and the male Luer connector on the end of the tubing (Fig. 1C9) from the the reservoir or the pump. Clamps (Fig. 1C11) may be used to stop flow or prevent spillage when the flowcell is disconnected and transported.

The entire process occurs over a period of 4 days. The completed flowcell is tested for leaks by forcing water through the channels. Leaks can usually be repaired by applying silicone adhesive and letting it dry overnight. The flowcell may be autoclaved, but we routinely use ethylene oxide for gas sterilization.

Streptococcal Biofilm Formation in Flowcell

The primary colonizers of the human tooth surface after professional cleaning are members of the genera *Actinomyces, Haemophilus, Streptococcus,* and *Veillonella,* but the vast majority (60–90%) are streptococci.[3–6] Because most of the early colonizing streptococci are *S mitis, S. sanguis, S. oralis,* and *S. gordonii,* we have focused our attention on forming a multistrain, multispecies streptococcal biofilm to model the early colonization process. This section describes the formation of a monospecies oral streptococcal biofilm that is established with only saliva as both the conditioning film and the growth medium.

Paraffin-stimulated saliva from 6 to 10 individuals was collected on ice and pooled before further processing by a slight modification of the protocol of de Jong and van der Hoeven.[7] Dithiothreitol was added to a final concentration of 2.5 m*M* and the saliva was stirred slowly for 10 min at 4°. The saliva was centrifuged at 25,000*g* at 4° for 20 min, and the supernatant was filter sterilized using a large-diameter filter (e.g., 47 mm) (polyether sulfone, low protein binding, 0.22-μm pore size) while keeping the filter on ice. The pooled saliva was stored at −20° in 50-ml portions until use. Some precipitate was observed on thawing the saliva and was removed by centrifugation at 3000*g* for 10 min at 4°.

The components and arrangment of the flowcell system are the same as shown in another article in this volume.[1] We use a Watson-Marlow multichannel pump assembly (Watson-Marlow, Inc., Wilmington, MA) adjusted to give a flow of 0.2 ml/min, which mimics the human salivary flow rate. The reservoir and flowcell are incubated at 37°.

[3] B. Nyvad and M. Kilian, *Scand. J. Dent. Res.* **95,** 369 (1987).

[4] W. F. Liljemark, C. G. Bloomquist, C. L. Bandt, B. L. Pihlstrom, J. E. Hinrichs, L. F. Wolff, *Oral Microbiol. Immunol.* **8,** 5 (1993).

[5] W. F. Liljemark, L. J. Fenner, and C. G. Bloomquist, *Caries Res.* **20,** 481 (1986).

[6] B. Nyvad and M. Kilian, *Caries Res.* **24,** 267 (1990).

[7] M. H. de Jong and J. S. van der Hoeven, *J. Dent. Res.* **66,** 498 (1987).

S. gordonii DL1 (Challis strain) was grown overnight in CAMG medium[8] containing tryptone, 5 g/liter; yeast extract, 5 g/liter; glucose, 2 g/liter; dibasic potassium phosphate, 5 g/liter; and Tween 80, 0.5 ml/liter; and adjusted to pH 7.4. Cells were grown overnight at 37° under anaerobic atmosphere (80% N_2, 10% CO_2, 10% H_2) in Gas Pak systems (BBL Microbiology Systems, Cockeysville, MD) diluted 1:20 with prereduced, fresh media and incubated for 4 hr as described earlier. Cells were harvested by centrifugation at room temperature at a setting of 6 for 10 min in a tabletop centrifuge (IEC clinical centrifuge, Damon/IEC Division, Needham Heights, MA). Cells were washed twice by centrifugation with saliva diluted 1:4 with sterile-distilled water. The cell density was adjusted to 1×10^9 cells/ml [Klett value of 100 (660 nm, red filter), determined with a Klett-Summerson colorimeter] (Klett Manufacturing Co., Inc., New York, NY) and was held at room temperature (for no longer than 5 min) prior to injection into the conditioned flowcell. Conditioning was accomplished by pumping 1:4 diluted saliva (approximately 400 μg salivary protein/ml) into the flowcell, which was then incubated at 37° for at least 15 min to allow salivary molecules to coat the cover glass surfaces in the biofilm chamber. The concentration of protein in the saliva after the initial 15-min cover glass surface conditioning was reduced by 10 to 20%, indicating a significant contribution of salivary proteins to the conditioning film. Bacteria were introduced from a syringe by inserting a 30-gauge 1.27-cm needle into the silicone tubing directly adjacent to the downstream tower. This method is used because the upstream side of the flowcell is open to the atmosphere by virtue of connection to the reservoir, which allows more precise inoculation of the biofilm chamber. Attachment of bacteria is allowed for 15 min and is followed by starting the pump. Approximately 20% of input streptococci bind to the conditioning film. At various intervals, the biofilm in the flowcell channel (biofilm chamber) is visualized by treatment with primary antibody followed by fluorescein-tagged secondary antibody. Usually, several flowcells are running simultaneously, so in order to isolate one channel of a flowcell for the injection of antibody, it is necessary to clamp the silicone tubing downstream of the biofilm chamber to prevent the disturbance of growth in the other flowcell channels.

Staining of Biofilm

1. Clamp the silicone tubing downstream of the flowcell.
2. Disconnect the Luer connector (upstream) attached to the tubing from the saliva reservoir.

[8] J. H. Maryanski and C. L. Wittenberger, *J. Bacteriol.* **124,** 1475 (1975).

3. Reconnect with a Luer connector (upstream) attached to tubing from the 1% bovine serum albumin (BSA) in TBS reservoir [1% BSA in Tris-buffered saline: 3 g Tris base (25 m*M* final concentration), 8 g NaCl, 0.2 g KCl in 1 liter distilled water, adjust to pH 7.4, can be autoclaved]. (If bubbles should arise in the silicone tubing, they can be removed by clamping the upstream tubing on the upstream side of the Luer connector, disconnecting the Luer connector, changing the pump direction, and pumping until the bubbles are passed out of the upstream tubing.)
4. Release the clamp.
5. Continue pumping for 15 min.
6. Clamp the silicone tubing downstream of the flowcell.
7. Inject polyclonal antibody (diluted 1:100) against *S. gordonii* DL1 into the silicone tubing at the downstream tower.
8. Incubate for 15 min.
9. Disconnect Luer connector (upstream) attached to 1% BSA in TBS and reconnect with Luer connector attached to 0.2% BSA in TBS reservoir.
10. Release the clamp.
11. Continue pumping for 15 min.
12. Clamp the silicone tubing downstream of the flowcell.
13. Inject secondary antibody 1:40 as in step 7.
14. Incubate for 15 min.
15. Release the clamp.
16. Continue pumping for 15 min.
17. Clamp silicone tubing at both ends of the flowcell and disconnect Luer connectors.
18. View the biofilm by confocal laser scanning microscopy.

Confocal Laser Scanning Microscopic Images Obtained of Streptococcal Biofilms Formed in Flowcells

We used rabbit antiserum against whole cells of *S. gordonii* DL1.[9] The secondary antibody was fluorescein isothiocyanate (FITC)-conjugated $F(ab')_2$ fragment goat antirabbit IgG (Jackson ImmunoResearch Laboratories, Inc., West Grove, PA). Several flowcells were inoculated with *S. gordonii* DL1. The flowcell system is set up to give unidirectional flow; thus the 1:4 diluted saliva used as the growth substrate is not recycled through the flowcell. At 30 min, 3 hr, and 4 hr, the staining procedure

[9] Y. Takahashi, A. L. Sandberg, S. Ruhl, J. Muller, and J. O. Cisar, *Infect. Immun.* **65,** 5042 (1997).

described earlier was done, and the biofilms were examined using a Leica TCS 4D system equipped with a Kr/Ar laser using a 100×/1.4 objective. Figure 2 shows selected images taken from three different flowcells at the three time intervals. At 30 min the streptococci were spaced evenly (Fig. 2A, *xy* plane) and the biofilm was only one or two cells deep (Fig. 2B, *xz* plane). Between 30 min and 3 hr it appeared that a significant portion of the initial adherent cells detached (Fig. 2C, *xy* plane), but the remaining adherent cells exhibited a biofilm of greater depth (8 μm) (Fig. 2D, *xz* plane). At 4 hr (Fig. 2E, *xy* plane) the biofilm exhibited extensive growth as evidenced by the 25-μm-deep biofilm (Fig. 2F, *xz* plane) and clustering of cells in apparent microcolony formations (Fig. 2E, *xy* plane).

Characteristics of Bacterial Colonization

The flowcell described here is especially useful for studying bacterial biofilms because it is easy to manipulate, is camera ready, and is disposable. It allows real time examination of biofilms by placing the flowcell directly on the microscope stage without interrupting the flow dynamics. Because the flow rate is slow, maintenance of the temperature in the flowcell while on the microscope stage is problematic but can be accomplished by passing the fluid through a heating/cooling device just prior to entering the flow cell. We are using this flowcell to study the characteristics of colonization of several streptococci. Incorporating the results of our earlier studies of coaggregations among oral bacteria, we will investigate the accretion of various other oral bacteria onto the initial colonizing streptococci.[10] In addition to using antibody to stain bacteria, many fluorescent probes are available from commercial sources that will stain bacteria suitable for epifluorescence imaging or confocal laser scanning microscopy. Some probes preferentially stain gram-positive bacteria. The combination of fluorescent probes and the flowcell provides an excellent approach for investigating the spatial organization of oral bacterial biofilms.

Appendix

Silicone Rubber Adhesive Sealant

GE RTV 118 Silicone rubber adhesive sealant
The silicone adhesive is pourable and self-leveling as well as translucent
Supplier: Newark Electronics, Hanover, MD

[10] C. J. Whittaker, C. M. Klier, and P. E. Kolenbrander, *Annu. Rev. Microbiol.* **50,** 513 (1996).

Microtechnique Scalpel

Supplier: Carolina Biological Supply, Burlington, NC

Coverslips

1 1/2 thickness, 24 × 75-mm glass coverslips (special order)
Supplier: Corning Glass, Danville, VA (OEM Department)

Diamond Glass Scriber

Supplier: Electron Microscopic Sciences, Ft. Washington, PA

Drive Unit. 12-Channel Cassette Pump Head, Marprene Manifold Tubing

Watson-Marlow multichannel cartridge pump
Supplier: Waston-Marlow, Inc., Wilmington, MA (also available from Fisher Scientific Co.)
0.63-mm inside diameter Marprene manifold tubing (orange/white)
Supplier: Watson-Marlow, Inc., Wilmington, MA

Tower Mold

Figure A.1 shows the engineer's drawing of the tower mold (James V. Sullivan, Scientific Equipment and Instrumentation Branch, Office of Research Services, NIH). Dimensions are in inches.

Gasket Mold

Figure A.2 shows the engineer's drawing of the gasket mold (James V. Sullivan, Scientific Equipment and Instrumentation Branch, Office of Research Services, NIH). Dimensions are in inches.

Section IV

Biofilms of Archaea

[25] Biofilm Formation in Hyperthermophilic Archaea

By PATRICIA L. HARTZELL, JACK MILLSTEIN, and CHRISTOPHER LAPAGLIA

Members of the domain Archaea represent a diverse group of prokaryotes that thrive in extreme environments: hypersaline, anaerobic, alkaline, acidic, and/or hyperthermophilic. They may be the most ancient of organisms because their habitats, especially hyperthermophilic and anaerobic environments, probably represent the earliest conditions from which life arose on this planet. Although they have many features that distinguish them from the domains Bacteria and Eucarya,[1] discussion about the relationship between the Archaea and other organisms continues to flourish.[2,3]

The Euryarchaeota is a subgroup of organisms within the Archaea[1] and includes organisms that can grow optimally at about 85°, such as members of the genus *Methanococcus* and *Archaeoglobus,* and as high as 110°, such as *Methanopyrus. Archaeoglobus fulgidus* obtains energy by dissimilatory sulfate reduction using H_2, lactate, or pyruvate as electron donors.[4,5] *Archaeoglobus fulgidus* has been isolated from the deep sea vents off the coast of Italy, marine waters in the North Sea,[6,7] terrestrial oil wells in California,[8] and hot springs in Yellowstone National Park in Wyoming.[9]

The abilities of *A. fulgidus* to colonize diverse thermal habitats and to remain viable in cold oxygenated ocean waters suggest that it has evolved mechanisms for surviving fluctuations in temperature, concentrations of nutrients, and potentially toxic compounds. In these "extreme" environments, the formation of biofilm may enable cells to survive exposure to oxygen, lower temperatures, and other suboptimal conditions.[10,11] Consistent with this idea, *A. fulgidus* produces a biofilm material in response to

[1] C. R. Woese, O. Kandler, and M. L. Wheelis, *Proc. Natl. Acad. Sci. U.S.A.* **87,** 4576 (1990).

[2] R. S. Gupta, *Theor. Popul. Biol.* **54,** 91 (1998).

[3] W. F. Doolittle and J. M. Logsdon, Jr., *Curr. Biol.* **8,** 209 (1998).

[4] K. O. Stetter, G. Lauerer, M. Thomm, and A. Neuner, *Science* **236,** 822 (1987).

[5] L. Achenbach-Richter, K. O. Stetter, and C. R. Woese, *Nature* **327,** 348 (1987).

[6] J. Beeder, R. K. Nilsen, J. T. Rosnes, T. Torsvik, and T. Lien, *Appl. Environ. Microbiol.* **60,** 1227 (1994).

[7] K. O. Stetter, R. Huber, E. Blochl *et al., Nature* **365,** 743 (1993).

[8] K. Kashefi and P. L. Hartzell, unpublished (1994).

[9] R. Huber, *Nature* **376,** 57 (1995).

[10] J. W. Costerton, Z. Lewandowski, D. E. Caldwell, D.R. Korber, and H. M. Lappin-Scott, *Annu. Rev. Microbiol.* **49,** 711 (1995).

[11] E. J. Quintero and R. M. Weiner, *J. Ind. Microbiol.* **15,** 347 (1995).

0076-6879/99 $30.00

changes in environmental conditions.[12] *A. fulgidus* can be stimulated to form biofilm by increasing the pH of the medium to >pH 7.5, gradual cooling to 10°, gradual heating to 95°, addition of NaCl, halogenated compounds or heavy metals, exposure to oxygen or ultraviolet light, or addition of antibiotic. Production of biofilm does not appear to be stimulated by acidification of the medium or by nutrient starvation.

Biofilm is produced by other members of the domain Archaea, such as *Haloferax, Thermococcus, Methanosarcina,* and *Methanobacterium,*[10,13–15] and by eubacterial sulfate reducers such as *Desulfovibrio.*[16] Material resembling biofilm has been seen in association with organisms from volcanic vents,[17–19] suggesting that biofilms play an important role in hyperthermophilic environments.

Genetics of Biofilm Formation

The complete sequence of the *A. fulgidus* genome has been released[20] and is available on line [http:// www.tigr.org/tdb/mdb/afdb/afdb_seq_search/]. The wealth of information in this data base may provide clues about the regulation and synthesis of biofilm. Biofilm characterized from mesophilic eubacteria is an organized exopolymer matrix comprised mainly of hydrated polysaccharides.[10] Biofilm from hyperthermophiles, particularly hyperthermophiles in the domain Archaea, is less well characterized. Surprisingly, although biofilm of *A. fulgidus* contains carbohydrate,[12] the genes involved in the synthesis of polysaccharides in eubacteria have not been identified by sequence analysis in *A. fulgidus.*[20] A more detailed analysis of the carbohydrate composition of *A. fulgidus* biofilm is needed to begin to identify the biosynthetic pathways that are required for biofilm production.

The wide variety of conditions that invoke biofilm formation in *A. fulgidus* suggest that the synthesis of biofilm may be regulated through a general stress response pathway. In *Escherichia coli,* disruption of the *surE* gene, which encodes a stationary-phase survival protein, affects the ability of cells to tolerate high temperatures or high salt concentrations and to

[12] C. LaPaglia and P. L. Hartzell, *Appl. Environ. Microbiol.* **63,** 3158 (1997).
[13] J. Anton, I. Meseguer, and F. Rodriguez-Valera, *Appl. Environ. Microbiol.* **54,** 2381 (1988).
[14] K. R. Sowers and R. P. Gunsalus, *J. Bacteriol.* **170,** 998 (1988).
[15] K. D. Rinker and R. M. Kelly, *Appl. Environ. Microbiol.* **62,** 4478 (1996).
[16] S. L. Geiger, T. J. Ross, and L. L. Barton, *Microsc. Res. Tech.* **25,** 429 (1993).
[17] G. Raguenes, P. Pignet, G. Gauthier *et al., Appl. Environ. Microbiol.* **62,** 67 (1996).
[18] P. Vincent, P. Pignet, F. Talmont *et al., Appl. Environ. Microbiol.* **60,** 4134 (1994).
[19] K. O. Stetter, *Nature* **300,** 258 (1982).
[20] H. Klenk, R. A. Clayton, J. F. Tomb *et al., Nature* **390,** 364 (1997).

survive during the stationary phase of growth.[21] These conditions are similar to the conditions that induce biofilm in *A. fulgidus.* The *A. fulgidus* gene AF0942 is predicted to encode a polypeptide that has 36% identity over 261 residues with SurE. If AF0942 has a function like SurE, it may regulate biofilm production.

In *Pseudomonas aeruginosa,* biofilm is regulated by genes involved in quorum sensing such as *lasR* and *lasT.*[22] Homologs of these genes are not found in *A. fulgidus,* consistent with the fact that *A. fulgidus* produces biofilm under conditions that are different from *Pseudomonas.* Flagellar genes have been found to be critical for biofilm formation in bacteria,[23,24] and their homologs in *A. fulgidus* may also encode products needed for attachment to surfaces and to one another. Differences between the type of biofilm formed by *Archaeoglobus* and that of eubacteria suggest that new genes will be identified as we learn more about biofilm formation in this thermophilic archaean. Although efforts are underway to develop genetic tools for *A. fulgidus,* it may be some time before the components required for biofilm formation can be identified using a genetic approach. In the meantime, molecular probes can be used to determine if specific genes, such as *surE,* are regulated in response to conditions that induce biofilm formation.

Protocol 1. Preparation of Media

1. The type strain *Archaeoglobus fulgidus* VC-16 is maintained on sulfate–thiosulfate–lactate (STL) broth medium.[25] Because *A. fulgidus* is a strict anaerobe, perform all manipulations of this organism with the goal of excluding O_2. (Cells can survive exposure to oxygen, but will not grow in the presence of O_2.) Make solutions anoxic by boiling and sparging with N_2 gas. Trace amounts of O_2 must be removed from the N_2 gas by passing it through a copper canister filled with copper metal turnings (Allied Chemical, Morristown, NJ) and heating to 300°. Reduce the oxidized copper in the canister every 3–4 days by flushing with H_2.
2. The STL base medium contains (per liter) 100 ml of 10× salt solution, 20 ml of 1 *M* piperazine-*N,N*-bis(2-ethanesulforic acid) (PIPES)

[21] C. Li, J. K. Ichikawa, J. J. Ravetto, H. C. Kuo, J. C. Fu, and S. Clarke, *J. Bacteriol.* **176,** 6015 (1994).

[22] D. G. Davies, M. R. Parsek, J. P. Pearson, B. H. Iglewski, J. W. Costerton, and E. P. Greenberg, *Science* **280,** 295 (1998).

[23] G. A. O'Toole and R. Kolter, *Mol. Microbiol.* **28,** 449 (1998).

[24] H. M. Dalton and P. E. March, *Curr. Opin. Biotechnol.* **9,** 252 (1998).

[25] K. O. Stetter, "The Genus *Archaeglobus.*" Springer, Berlin, 1992.

buffer at pH 7.0, 1 ml of 1000× trace element stock solution, 1 ml of 0.2% (w/v) $(NH_4)_2Fe(SO_4) \cdot 2H_2O$, 0.5 g yeast extract, 7 ml of 30% (w/v) sodium lactate, and 1050 ml of nanopure (np) water. (The volume of water has been increased by about 18% (179 ml per liter) to compensate for water lost during boiling.) Adjust the base medium to pH 6.7 with 1 *M* NaOH or KOH.

3. To prepare the 10× salt solution, mix 800 ml of salt I, 100 ml of salt II, and 100 ml of salt III. Salt I contains the following materials dissolved in 800 ml of np H_2O in the order of 74 g $MgSO_4$, 3.4 g KCl, 27.5 g $MgCl_2 \cdot 6H_2O$, 2.5 g NH_4Cl, 178 g NaCl, and 5 ml of 0.2% (w/v) resazurin (the redox indicator). Salt II contains 1.4 g $CaCl_2 \cdot 2H_2O$ in 100 ml of np H_2O. Salt III contains 1.37 g K_2HPO_4 in 100 ml of np H_2O. To prepare the 1000× trace element stock, dissolve 29 g of $Na_2EDTA \cdot 2H_2O$ in 800 ml of np H_2O and adjust to pH 8 with KOH. When the EDTA is completely dissolved, add 5 g $MnSO_4 \cdot 6H_2O$, 1.8 g $CoCl_2 \cdot 6H_2O$, 1 g $ZnSO_4 \cdot 7H_2O$, 0.11 g $NiSO_4 \cdot 6H_2O$, 0.1 g $CuSO_4 \cdot 5H_2O$, 0.1 g H_3BO_3, 0.1 g $KAl(SO_4)_2 \cdot 12H_2O$, 0.1 g Na_2 $MoO_4 \cdot 2H_2O$, 0.1 g $Na_2WO_4 \cdot 2H_2O$, 0.05 g Na_2SeO_4, and 0.01 g VaO_5. Adjust the final volume to 1000 ml with np H_2O.
4. To prepare the reducing agent, make a stock of 100 m*M* Na_2S and 200 m*M* $Na_2S_2O_3$. Heat 1100 ml np H_2O to a boil and sparge with N_2 gas for about 5 min. Remove from heat and add 13.4 g Na_2S (60% fused flake Na_2S, Fisher, Pittsburg, PA) and 31.6 g $Na_2S_2O_3$ (Sigma, St. Louis, MO). Stopper, cover with an aluminum seal, and crimp the seal. Autoclave.
5. Boil the base medium under a stream of N_2 gas until the resazurin indicator turns from the original purple color to a pink color (about 10 min). Allow the medium to cool slightly, transfer into serum bottles or tubes that have been gassed with O_2-free N_2, stopper, crimp-seal, and autoclave.
6. After autoclaving, reduce the medium by adding 0.5% volume of sterile reducing agent. Flush a sterile syringe having a 22-gauge needle with sterile N_2 gas. Lightly swab the stopper on the bottle containing the reduction agent with ethanol and flame briefly. Inject the needle through the rubber stopper, invert the bottle, and withdraw the necessary volume of reducing agent. Quickly inject the reducing agent through the ethanol-sterilized stopper of the bottle containing the sterile base medium. For a 50-ml volume of medium, add 0.25 ml of reducing agent. The medium becomes clear on reduction.

Comments. For defined medium, 10 ml of Wolfe's vitamin mixture can be added in place of yeast extract. Wolfe's vitamin mixture contains (per liter) 10 mg pyridoxine, 5 mg each of thiamin hydrochloride, riboflavin, nicotinic acid, calcium DL pantothenate, *p*-aminobenzoic acid, and lipoic acid, 2 mg each of biotin and folic acid, and 0.1 mg cyanocobalamin.[26] The absence of any one of these vitamins, except for nicotinamide, does not affect the growth rate significantly.

Protocol 2. Culture of Archaeoglobus fulgidus

1. Fresh cultures of *A. fulgidus* can be prepared using frozen cell pellets, liquid stocks, or a colony from a roll tube (described later). For routine manipulations, *A. fulgidus* is grown in 50 ml of STL medium for 2–3 days, then stored at 4°. Liquid stocks should be transferred to fresh medium every 2–3 weeks. Cells are transferred to fresh medium using a sterile syringe as described earlier. Maintain cultures of *A. fulgidus* in 10- and 100-ml volumes in stoppered anaerobic culture tubes (Bellco, Vineland, NJ) and serum bottles (120- or 160-ml size, Wheaton, Millville, NJ). Typically, a 5% (vol) inoculum of cells is used for small-scale cultures. Modified culture bottles (1- and 2-liter size, Pyrex, Fig. 1, right-hand side) are used for 250- and 500-ml cultures. Butyl rubber stoppers and aluminum seals can be purchased from Bellco).
2. To obtain a larger amount of biofilm, it is necessary to grow *A. fulgidus* in 19- or 40-liter volumes in glass carboys heated with a drum heating belt (Barnstead Thermolyne, Dubuque, IA) and a hot plate as shown in Fig. 1 (left-hand side). Glass carboys are used instead of steel because the production of H_2S during growth causes pitting (surface oxidation) of metal fermentors.
3. Grow cells for inoculation of 19- and 40-liter glass carboys in 1-liter culture bottles. Prepare 200 ml of fresh (2-day-old) cells to inoculate the 19-liter carboy. A 400-ml inoculum is used for the 40-liter culture. To maintain a uniform temperature, stir the medium and sparge it with N_2 at a rate of 0.07 liter/min throughout growth. Good growth is obtained when the internal temperature is between 75 and 82°. A thermometer mounted on the outside of the carboy with some insulation and duct tape will read about 10° less than the internal temperature.
4. Remove samples (about 5 ml) at 4- to 8-hr intervals to monitor pH and absorbance. To remove a sample, close the valve gas exit and allow pressure to the carboy to increase *slightly*. The increase in pressure will force some of the culture out through the sampling port. After removing a 5-ml sample, be sure to open the exhaust

[26] W. E. Balch, G. E. Fox, L. J. Magrum, C. R. Woese, and R. S. Wolfe, *Microbiol. Rev.* **43,** 260 (1996).

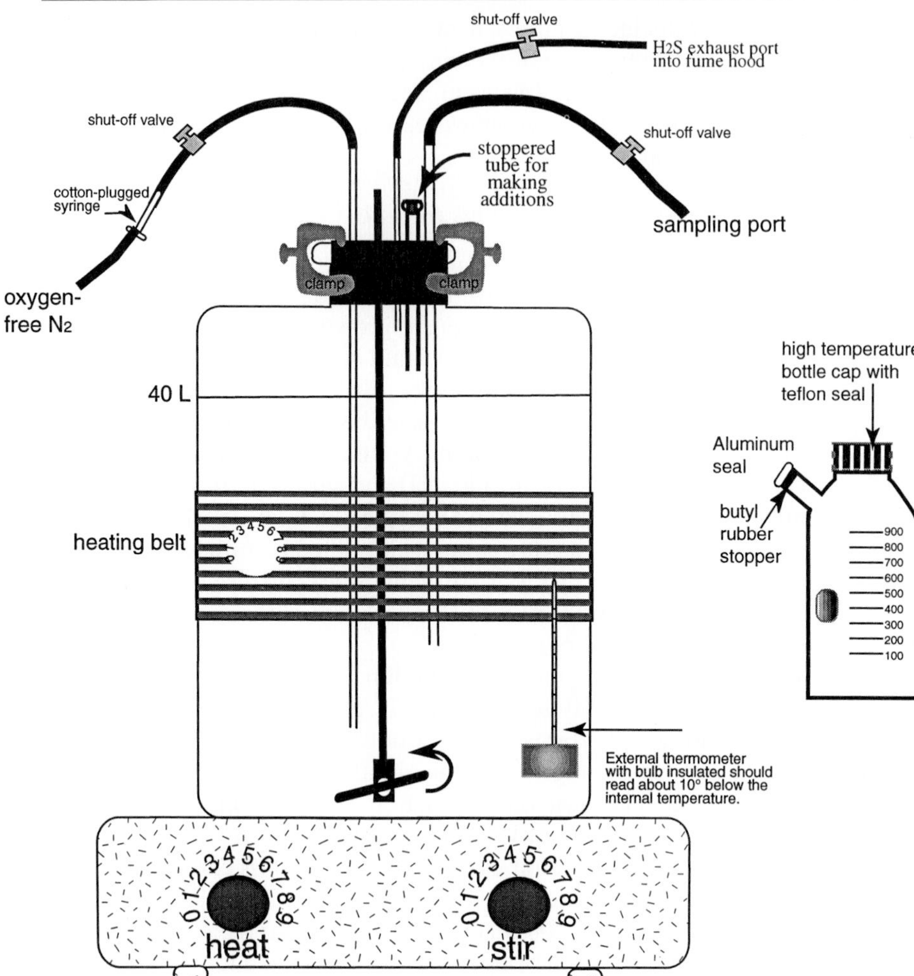

FIG. 1. Equipment for culturing *A. fulgidus*. (Left) Forty-liter cultures are grown in glass carboys maintained at 83° with a heating belt and plate. The stir bar rotates freely on a stationary metal rod suspended from the stopper. Gas and reagents are added and waste and samples are removed through glass tubes inserted through holes in the stopper. Flexible tubing connected to the glass tubes can be purchased at an auto supply store. Cover the ends of the tubing with aluminum foil during autoclaving. (Right) *A. fulgidus* cultures for inoculating the 40-liter carboy are grown in 1- or 2-liter culture bottles. Have the glass blower modify the culture bottles by fusing the top half of an anaerobic culture tube to the upper portion of the bottle. This provides a convenient way to make additions and remove samples. The high-temperature screw-capped lids that fit the culture bottles have Teflon seals that keep the contents anoxic.

valve. Glacial acetic acid (about 10 ml per hour) must be added periodically to maintain a pH near 7–7.2. The rate of increase in pH also can be reduced by decreasing the flow rate of N_2. Growth is measured as a change in absorbance at 600 nm. There is usually a lag period of about 15–20 hr, but growth increases after this time to a density of about 6–8 $\times$ 10^8 cells/ml, which corresponds to an OD_{600} of about 0.6 to 0.7. Additional descriptions of anaerobic techniques are available.[26,27]

Protocol 3. Production of Biofilm

1. To induce formation of biofilm in the carboy, allow the culture to reach an OD_{600} of about 0.3 and then begin to increase the temperature of the carboy. Gradually increase the temperature of the 40-liter carboy by about 3° per hour until the internal temperature is about 90°. If the pH of the medium is allowed to increase to >pH 7.5, the combined effect of high pH and temperature will result in the formation of biofilm within the next 6–12 hr. The yield of biofilm is about 2–3 g wet weight (0.6–0.9 g dry weight) from a 40-liter carboy, whereas the yield of planktonic cells is about 25 g wet weight.
2. The most common forms of biofilm produced are ropes, sheets, or clumps that attach to the walls of the glass carboy at the interface of the medium and the head space. The rope-like biofilm can be spooled from the medium prior to harvest by dropping a long hook into the medium through the wide-mouth opening of the vessel. Biofilm attached to the walls of the carboy can be removed with a long-handled scraper after harvesting the liquid culture. Harvest the remaining planktonic cells by centrifugation in sealed tubes (stainless steel or plastic) that have been flushed with N_2 or by ultrafiltration using an Amicon tangential filtration system.

Comments. *A. fulgidus* begins to form a visible biofilm as cells enter the stationary phase of growth, particularly as the pH of the medium begins to increase. The form of biofilm produced depends on the type of inducing agent and the rate of change.[12] If the pH of the medium is increased gradually to pH 8 over a 12- to-20-hr period, the cells will form a rope-like type of biofilm, some of which sticks to the walls of the glass carboy. In contrast, if the pH is increased rapidly to pH 8 within a 1- to 2-hr period by the addition of a concentrated amount of base, the cells form a clumpy

[27] K. R. Sowers and K. M. Noll, *in* "Archaea: A Laboratory Manual: Methanogens" (K. R. Sowers and H. J. Schreier, eds.), p. 15. Cold Spring Harbor Press, Plainview, NY, 1995.

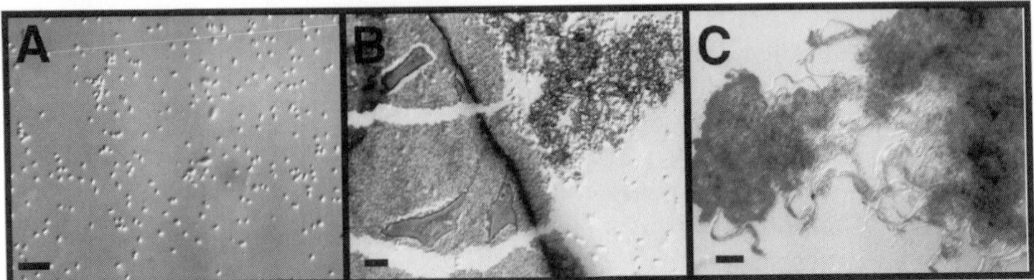

FIG. 2. Microscopic appearance of planktonic cells and biofilm. (A) *A. fulgidus* planktonic cells grown in STL medium. Bar: 5 μm. (B) *A. fulgidus* biofilm stained with Sudan black. Sheet-like biofilm (left), clumps of biofilm (right), and planktonic cells (lower right) can be seen. Bar: 2 μm. (C) *A. fulgidus* rope-like biofilm stained with Congo red. Bar: 5 μm. Photographs were taken with a Nikon FXA microscope.

mass of floating biofilm. Biofilm formation can be suppressed somewhat if the pH is maintained at about 7.2–7.5.

Protocol 4. Microscopic Analysis of Biofilm

1. For routine microscopic analysis, wash the biofilm with a stream of water, immerse in 20 m*M* Tris buffer at pH 7.2 (buffer 1), 10 m*M* Congo red in buffer 1, or 2 m*M* Sudan black in buffer 1, incubate at room temperature for 1 hr, and rinse twice in buffer 1. A sample of planktonic cells should be harvested by centrifugation and treated in parallel (Fig. 2A). Fresh Sudan black is prepared from a 20 m*M* solution of Sudan black in absolute ethanol. Sudan black is a lysochrome used for identifying triglycerides in sections[28] and produces an irregular staining pattern with biofilm (Fig. 2B). As shown in Fig. 2C, the biofilm stains intensely with Congo red (dark areas), which characteristically binds to acidic polysaccharides.[29,30]
2. *A. fulgidus* cells can be distinguished from the surrounding biofilm using a fluorescent microscope because cells contain the deazaflavin F_{420}, which fluoresces when excited with a 390–420 UV–violet excitation filter (excitation 420, emission 468 nm.)[4]
3. A more detailed view of the structure of biofilm can best be obtained through transmission electron microscopy of thin sections of biofilm. To minimize structural damage, spool the biofilm from the growth medium, immerse it in water, wash with a stream of buffer, and fix

[28] W. M. Fredericks, *Histochemistry* **54,** 27 (1977).
[29] J. R. Colvin and D. E. Witter, *Protoplasma* **116,** 34 (1983).
[30] J. W. Arnold and L. J. Shimkets, *J. Bacteriol.* **170,** 5765 (1988).

for 1 hr in a solution containing 2% (w/v) paraformaldehyde and 2% (w/v) glutaraldehyde in 0.1 *M* phosphate buffer (buffer 2) at pH 7. Rinse the sample three times in buffer 2, incubate for 1 hr in 2% (w/v) osmium tetroxide in buffer 2, rinse three times in buffer 2, and dehydrate through successive immersion in 30, 50, 60, 70, 80, 95, and 100% ethanol baths. After two additional 100% ethanol baths, the biofilm is incubated for 45 min in a 1 : 1 mixture of ethanol and LR White (Electron Microscopy Sciences, Fort Washington, PA), transferred to a 100% solution of LR White, and placed in a gelatin capsule. After overnight polymerization at 60°, thin sections can be cut using an ultramicrotome, placed on Formvar-treated grids, and examined by electron microscopy. Osmium tetroxide typically reacts with lipid material and proteins. Cells in the biofilm sample can be seen as dark irregular shapes connected with one another by a stringy material that also stains with OsO_4. Polysaccharides often stain poorly, but the detection of polysaccharide in biofilm may be enhanced by staining with ruthenium red.[31]

Protocol 5. Effect of Metals on Biofilm Formation

1. To stimulate biofilm formation with metals or halogenated compounds, grow *A. fulgidus* in STL medium at 80° and maintain the pH at about 7. Prepare an anoxic 100× stock solution of the inducing agent in a serum bottle and autoclave. Inducing agents, such as metals Cu(II) (as $CuCl_2 \cdot 2H_2O$, ≥3 ppm) and chromium(VI) (as $K_2Cr_2O_7$; 25 ppm), and halogenated compounds, such as pentachlorophenol (PCP, ≥50 ppm) and 5-fluorouracil (≥100 μg/ml), can be used for biofilm production. For a 40-liter culture, add inducer in 10-ml volumes (400 ml total) at about 10-min intervals to allow mixing to occur. Alternatively, to add the inducer slowly over 5 hr, pressurize the serum bottle and invert it on top of the stopper on the carboy. Insert the shorter end of a vacutainer (double barrel) needle into, but not through, the stopper on the serum bottle. Insert the other end of the needle through the carboy stopper. To add the inducer, push the shorter end of the needle into the liquid in the serum bottle. It will be necessary to repressurize the serum bottle periodically.
2. These compounds are toxic to planktonic cells, and formation of the biofilm presumably plays a role in protecting the cells from damage. To determine the level of protection, measure the number of viable cells in the biofilm after it has been transferred to fresh STL medium

[31] P. R. Lewis and D. P. Knight, *in* "Practical Methods in Electron Microscopy" (A. M. Glauert, ed.). North-Holland, Amsterdam, 1977.

lacking the toxic agent. Harvest the biofilm from cultures exposed to the toxic agent, rinse under a stream of buffer 1 or water,[32] weigh the sample, and transfer it to 100 ml of fresh STL medium lacking the toxic compound. Harvest planktonic cells from the same culture by centrifugation, wash in buffer, weigh, and suspend in 100 ml of STL medium. A sample of biofilm and cells from a heat- or pH-shocked culture should be assayed in parallel for comparison.

3. Incubate the cultures at 83° and monitor the change in OD_{600} for 3 to 4 days. This provides a qualitative estimate of cell viability. A lag period of 1 day is typical for untreated cells, and each additional day of lag growth corresponds to about a 20% reduction in cell viability.
4. For a more quantitative measure of viable cells, samples should be plated for colony formation. Weigh out samples of biofilm and cell samples, vortex or sonicate *briefly* (4–5 sec) to break up clumps, and dilute the samples serially in STL medium. Mix a 500-μl amount from each dilution with STL medium Gel-rite (a gellan gum added as a solidifying agent; Merck, Rahway, NJ), stopper, and spin in an ice bath using a tube roller (Bellco), which produces a thin layer of agar in the tube.[33] Incubate upright at 83°. Colonies form efficiently in agar in roll tubes and are easy to see after about 7–10 days because cells produce H_2S, which forms a black FeS precipitate in the colonies (Fig. 3). Count the number of colonies in a roll tube containing about 50–200 colonies and then factor in dilutions to estimate the number of viable *A. fulgidus* cells per milligram of wet weight in the original sample. An estimate of total cell numbers in the original sample is obtained by quantifying the amount of DNA as described in protocol 6 later. The formation of colonies on the surface of STL Gel-rite medium is less efficient, particularly in petri dishes where the loss of moisture due to evaporation may be difficult to regulate. Petri plates can be incubated in an 80° oven in an anaerobic chamber where colony formation can be monitored easily. Under these conditions, cells produce colonies with a film instead of the discrete, compact colonies observed in roll tubes. Although it is much easier to manipulate colonies on petri plates in the anaerobic chamber, the evolved H_2S can poison the catalyst that removes O_2 in the chamber. Therefore, it is necessary to periodically flush the incubator in the anaerobic chamber with N_2 to vent H_2S into a fume hood or to seal

[32] Make buffer or nanopure water anoxic by boiling under a stream of N_2 gas for 10 min.

[33] A. Rodabough, M. S. Foster, and E. C. Neiderhoffer, *in* "Archaea: A Laboratory Manual: Methanogens" (K. R. Sowers and H. J. Schreier, eds.), p. 57. Cold Spring Harbor Press, Plainview, NY, 1995.

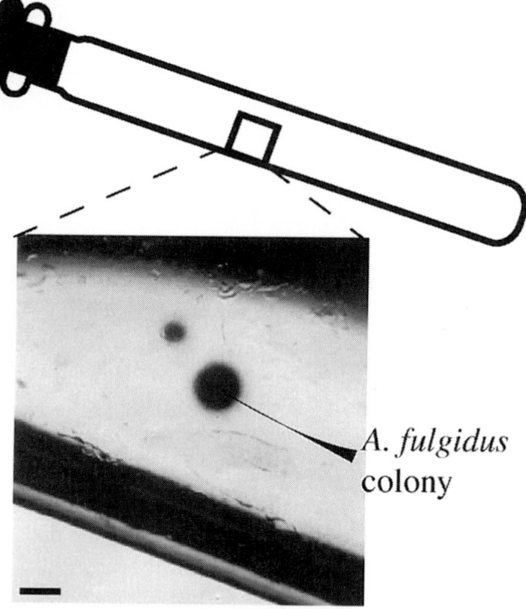

FIG. 3. Colony formation of *A. fulgidus* in a roll tube. Discrete colonies form in STL Gelrite on the walls of the tube and can be picked using a bent wire or glass pipette. Bar: 1.5 mm.

petri plates in canisters.[33] Other descriptions of plate techniques are available.[34]

5. To determine the type of metal(s) in the biofilm and the metal concentration, samples of biofilm and planktonic cells are washed in water, 80% ethanol, or extracted with phenol and dried. Ten micrograms of each dried sample is dissolved in 100 μl of concentrated HNO_3, diluted to 5 ml with np H_2O, and passed through a 0.2-μm filter. Filtered material is then analyzed by inductively coupled plasma (ICP) or atomic absorption against known standards, such as $FeNO_3$, $NiCl_2 \cdot 6H_2O$, or $CoCl_2 \cdot 6H_2O$, at a concentration of 5 or 10 ppm.

Comments. A significant fraction of the metal that induces biofilm formation becomes embedded within the biofilm matrix. It is unclear how metals are held within the biofilm. Proteins in the biofilm may chelate the metal(s), insoluble forms of the metal may become trapped as metal is

[34] G. Erauso, D. Prieur, A. Godfroy, and G. Raguenes, *in* "Archaea: Thermophiles" (F. T. Robb and A. R. Place, eds.), p. 25. Cold Spring Harbor Laboratory Press, Plainview, NY, 1995.

oxidized or reduced, or ionic interactions may retain metals. The identity of metals trapped within the biofilm and their position relative to each other and to the cells can be determined using a transmission electron microscope equipped with an energy dispersive X-ray spectroscope for detecting elements.[35]

Protocol 6. Analysis of Biofilm Composition

1. To quantify the amount of polysaccharide, lipid, and protein in biofilm, it is necessary to account for the contributions from these components due to *A. fulgidus* cells embedded within the biofilm. Wash samples of planktonic cells and biofilm in buffer 1, weigh each sample, and freeze-dry several preparations of each sample to obtain dry weight measurements. Use the remaining wet, weighed samples to measure total polysaccharide, fatty acids and lipids, total protein, and nucleic acids.
2. To determine the average amount of DNA per cell of *A. fulgidus,* remove about 300 μl of planktonic cells from the carboy when the OD_{600} is 0.6. First, calculate the number of cells per milliliter by physically counting cells under the microscope using a Petroff–Hauser chamber (or similar device). Then, using the procedure described by Daniels *et al.,*[36] quantify the DNA. In triplicate, harvest 100 ml of carboy cells by centrifugation and suspend each pellet in 0.25 *N* $HClO_4$ (perchloric acid). Incubate on ice for 30 min and then centrifuge at 12,000*g* at 4° for 15 min. Suspend the pellet in 4 ml of 0.5 *N* $HClO_4$ and incubate at 70° for 15 min, stirring every 5 min. Daniels *et al.*[36] recommend a second $HClO_4$ extraction of the pellet. Transfer 0.5-, 1-, and 2-ml samples from each extraction to fresh tubes and add enough 0.5 *N* $HClO_4$ to bring each volume to 2 ml. Add 4 ml of diphenylamine–acetaldehyde reagent prepared fresh by mixing 40 ml of a 1.5% stock of diphenylamine with 0.2 ml of a 2% aqueous solution of acetaldehyde. Prepare the diphenylamine stock by dissolving 1.5 g diphenylamine in 100 ml of acetic acid and then adding 1.5 ml of sulfuric acid. Prepare a 1-mg/ml stock of DNA in 1 *N* $HClO_4$ and incubate at 70° for 15 min. For standards ranging from 10 to 250 μg/ml, dilute the DNA stock in 2 ml of 0.5 *N* $HClO_4$ and add 4 ml of the diphenylamine–acetaldehyde reagent. Incubate

[35] F. G. Ferris, W. S. Fyfe, and T. J. Beveridge, *Chem. Geol.* **63,** 225 (1987).

[36] L. Daniels, R. S. Hanson, and J. A. Phillips, *in* "Methods for General and Molecular Bacteriology" (P. Gerhardt, R. G. E. Murray, W. A. Wood, and N. R. Krieg, eds.), p. 512. American Society for Microbiology, Washington, DC, 1994.

standards and unknowns at 30° overnight. Read the absorbance at 600 nm. Calculate the ratio of DNA per planktonic cell.

3. Determine the nucleic acid content of a 1-g (wet weight) sample of biofilm using the diphenylamine assay described earlier. Use this value to estimate the number of cells embedded in a 1-g biofilm sample. Use this estimate to determine the percentage of polysaccharide, lipid, and protein contributed by cells within the biofilm.
4. The production of polysaccharide in cultures producing biofilm can be monitored using Congo red.[12,30] To quantify Congo red binding, harvest planktonic cells and biofilm, wash, weigh out equal amounts of each, and incubate with 10 μg/ml Congo red in 75% (v/v) ethanol for 2 hr. Adjust the volume of each sample to 1 ml with np H_2O and centrifuge at 16,000*g* for 15 min at 25° to remove cells, biofilm, and bound Congo red. Dilute the supernatants 1:1 (vol) with np H_2O and measure the %Transmittance (or Abs) at 500 nm spectrophotometrically to determine the amount of dye remaining in each supernatant sample. The amount of Congo red bound by biofilm is calculated by subtracting the amount of Congo red remaining in samples containing cells or biofilm from the control containing only dye.
5. The anthrone reaction[36] provides a simple assay to estimate the percentage of total carbohydrate in biofilm. Wash a sample of cells and biofilm with buffer 1 (Tris), weigh each sample, and freeze-dry. Weigh the dried samples, suspend in 5 ml aqueous saline (0.85% NaCl), add 25 ml of 2 mg/ml anthrone reagent,[37] swirl to mix, and incubate on ice for 30 min. Prepare a set of glucose standards. Boil the samples for 10 min and then return to the ice bath. Centrifuge for 2 min at 5000*g* at 4° if the samples are cloudy. Read the absorbance of standards and samples at 625 nm. Compare the planktonic cells and biofilm samples with the glucose standards to estimate the concentration of pentose and hexose sugars. The volumes of sample and anthrone reagent have been increased to help dissolve the biofilm material. The sample will not dissolve completely until the boiling step, a factor that may reduce the efficiency of the assay. Although this reaction is simple and not subject to interference, it does not detect C_3, C_4, or amino sugars.
6. *A. fulgidus* biofilm stains with the lipophilic dye Sudan black[12] and is enriched with hydrophobic material that can be extracted with methanol and chloroform. To determine the amount of lipid material, wash samples of biofilm and planktonic cells weighing about 2 g (wet

[37] The anthrone reagent is prepared by dissolving 100 mg anthrone in 2.5 ml absolute ethanol and then adding 47.5 ml of 75% H_2SO_4.

weight) twice in water and once in methanol to remove salts. Mix the samples with 10 ml of chloroform, methanol, and water (10:5:4, v:v), vortex, and incubate at room temperature for 3 hr with occasional mixing. Add 7 ml of chloroform and water (1:1, v:v) and continue incubating for another 30 min, with shaking, at room temperature. Filter through Whatman (Clifton, NJ) #1 paper and collect the filtrate in a 30-ml Corex tube. Allow the phases to separate and then remove the water:methanol phase. Transfer the chloroform phase to a tared tube. Place the tube in a 40° water bath in a fume hood and flush gently with N_2 gas until all of the chloroform is evaporated. Weigh the product and the tared tube to determine the amount of crude lipid in cells or biofilm. Additional extraction procedures of polar and neutral lipids have been described for *Archaeoglobus*[38] and other Archaea.[36,39,40]

7. To determine the total protein in cells and biofilm, incubate 1 g (wet weight) with 5 ml of Tris-equilibrated phenol (pH 7.5) at 65° for 48 hr. Centrifuge the samples at 16,000*g* for 15 min at 4° and decant the phenol. Wash the pellet three times in absolute ethanol and air dry overnight. Add 1 ml of 1% (w/v) sodium dodecyl sulfate (SDS) to the pellet and incubate at 70° for 8 hr to solubilize. Nonproteinaceous material embedded in the biofilm may cause cloudiness and can be removed by centrifugation. Protein concentrations can be determined using the Bradford assay[41] or the BCA (Pierce Chemical Co., Rockford, IL). Bradford and BCA assays are rapid and simple but are subject to interference from materials such as Na_2S, which is present in the biofilm. SDS should be added to one set of protein standards to determine the degree of interference from SDS. If possible, these assays should be performed in parallel with a Kjeldahl digestion for organic nitrogen yielding NH_4^+.[42]

Comments. Archaea contain hydrophobic polyisoprenoid ether-linked lipids instead of typical alkane ester-linked lipids found in eubacteria and eukaryotes.[39] A variety of modifications can be found among the Archaea, particularly in the hyperthermophiles, whose modified lipids likely enhance

[38] S. Burggraf, H. W. Jannasch, B. Nicolaus, and K. O. Stetter, *Syst. Appl. Microbiol.* **13,** 24 (1990).

[39] T. A. Langworthy, *in* "The Bacteria" (C. R. Woese and R. S. Wolfe, eds.), p. 459. Academic Press, Orlando, FL, 1985.

[40] D. B. Hedrick and D. C. White, *in* "Archaea: Thermophiles" (F. T. Robb and A. R. Place, eds.), p. 73. Cold Spring Harbor Press, Plainview, NY, 1995.

[41] M. M. Bradford, *Anal. Biochem.* **72,** 248 (1977).

[42] G. H. Grant, F. R. C. Path, and J. F. Kachmer, *in* "Fundamentals of Clinical Chemistry" (N. Tietz, ed.), Vol. 2, p. 300. Saunders, Philadelphia, 1976.

thermal stability. Lipids in biofilm may contain additional modifications that increase resistance to heat and chemicals.

Acknowledgments

These procedures were developed with support from the NSF (MCB-9321893) and EPSCoR (OSR-9350539).

Section V

Physical Methods

[26] Use of Conductance Measurements for Determination of Enzymatic Degradation of Microbial Biofilm

By CHARLOTTE JOHANSEN, BERIT K. BREDTVED, and SØREN MØLLER

Introduction

Electrical effects, especially the impedance, conductance, or capacitance change associated with microbial metabolism, is reported to have significant advantages over other methods currently used for the quantitative detection of microorganisms in complex substances such as biofilms.[1-3] Electrochemical changes in media due to microbial growth have been recognized as being useful in the evaluation of antimicrobial activity.[3-7] Advantages of these techniques are that overall activity of the biofilm can be measured without disrupting the biofilm structure and that the measurements are fast and automatic compared to most other quantitative methods.

Biofilm degrading enzymes are, in general, nonbactericidal and the enzymatic activity against biofilm will be an attack on the extracellular polymers, often by hydrolysis, and not a direct attack on the microbial cell. Conventional quantitative methods for the evaluation of antimicrobial activity against biofilm often involve biofilm removal followed by bacterial enumeration by colony formation. This method is not useful for evaluation of the efficacy of enzymes against microbial biofilm, as the microbial cells left on a substratum after enzyme treatment are detached more easily than untreated cells. Furthermore, biofilm cells released from a substratum for enumeration are most often present in agglomerates, whereas biofilm cells released from an enzyme-treated substratum will be present as single cells or small clots as the extracellular polymers are degraded or partly degraded. Therefore, if bacterial enumeration by colony formation is used for the evaluation of enzyme activity against biofilm cells, then the enzyme efficacy will be underestimated. Using conductance measurements, the number of

[1] H. M. Ayres, D. N. Payne, J. R. Furr, and A. D. Russell, *Lett. Appl. Microbiol.* **27,** 79 (1998).
[2] C. Johansen, P. Falholt, and L. Gram, *Appl. Environ. Microbiol.* **63,** 3724 (1997).
[3] M. Johnston and M. V. Jones, *J. Microbiol. Methods* **21,** 15 (1995).
[4] H. M. Ayres, D. N. Payne, J. R. Furr, and A. D. Russell, *Lett. Appl. Microbiol.* **26,** 422 (1998).
[5] P. Connolly, S. F. Bloomfield, and S. P. Denyer, *J. Appl. Bacteriol.* **75,** 456 (1993).
[6] C. Johansen, T. Gill, and L. Gram, *J. Appl. Bacteriol.* **78,** 297 (1995).
[7] P. Silley and S. Forsythe, *J. Appl. Bacteriol.* **80,** 233 (1996).

0076-6879/99 $30.00

microorganisms left on substrata after enzyme treatment can be evaluated without removal of the cells from the substrata, simply by immersing the substrata in a conductance measuring cell. This article describes a method based on conductance measurements for evaluating the enzymatic degradation of microbial biofilm.

Growth of Biofilm

Biofilm is formed by the use of two types of equipment: the modified Robbins device[8] or a steel rack in a beaker. Using the modified Robbins device, the biofilm is made on steel tablets [stainless steel type AISI 304 with a No. 4 finish (polish grain 180)]. Prior to use, the steel tablets are cleaned in water with Deconex, followed by acetone before sterilization by autoclaving at 121° for 20 min. The biofilm culture is made either by growing tap water microorganisms or by using characterized strains; the choice of biofilm culture depends on the conceivable application of the enzymes that are to be evaluated.[2] The characterized strains are used either as mono or mixed cultures, and the precultures are grown at conditions relevant for the particular strain. The tap water preculture is made by mixing 200 ml tap water and 5 ml Tryptone® soya broth (TSB) (Oxoid) followed by incubation at 20° for 3 days.

The outgrowing culture is recycled through the modified Robbins device, and 1000 ml of TSB diluted 1:5 and with glucose (1 g/liter) is added continuously to the culture while the biofilm is formed during 5 days (20°).

The steel rack holds up to 20 small vertically clamped disks in an arrangement that allows the free circulation of liquid when immersed in culture medium. The disks are made of different materials (steel, plastic, hydroxylapatite, contact lenses, etc.) depending on the conceivable application of the particular enzymes that are to be evaluated.[2] Precultures are made as described earlier and inoculated [approximately 10^3 cfu (colony-forming units)/ml] in TSB diluted 1:5 with sterile water. Inoculated media are poured into the beaker covering the disks, and a biofilm is allowed to develop on both sides of the disks over 4 days while stirring (200 rpm). For some microorganisms, the growth medium has to be supplemented with carbohydrates (sucrose, glucose, etc.) before a proper biofilm is obtained.

Enzymes for biofilm degradation are evaluated either by recycling of an enzyme solution through the modified Robbins device or by immersing the disks individually into enzyme solutions.

[8] H.M. Lappin-Scott, J. Jass, and J. W. Costerton, "Microbial Biofilms: Formation and Control" (S. P. Denyer *et al.*, eds.). Blackwell Sci., Oxford, 1993.

Two modified Robbins devices are kept assembled, and the enzyme solution is recycled through one Robbins device; the pH and temperature of the enzyme solution and the time for the enzyme treatment are chosen dependent on the optimum of the particular enzyme system. Conditions such as 50°, pH 5–6, and 2 hr are often used. A control solution without enzyme is circulated through the other Robbins device using the same conditions as for the enzyme treatment. After enzyme treatment, all disks are removed from the modified Robbins device, and the enzyme effect is evaluated as described later.

Alternatively, disks are removed from the modified Robbins device or the steel rack and aseptically rinsed for 1 min in sterile phosphate buffer (0.067 *M*, pH 7) to remove poorly attached cells before incubation with enzymes at, e.g., 20° for 15 min without agitation, and the enzymes are diluted in a buffer. The buffer system, pH, temperature, and time are dependent of the activity of particular enzyme that is to be evaluated and the desired application of the enzyme. Sterile buffer with no enzymes added is used as control.

After enzyme treatment, the different substrata are gently rinsed once in sterile buffer prior to enumeration of bacteria by conductance measurements and microscopy.

Evaluation of Enzymatic Degradation of Biofilm

Conductance Assay

The number of living cells left on the substrata after enzyme treatment is determined by an indirect conductance measurement[9–11] using the Malthus instrument (Malthus Flexi 2000, Malthus Instrument Limited, UK). Substrata with biofilm cells after enzyme treatment are transferred to a Malthus glass tube containing 3 ml of growth medium. Growth of biofilm cells results in the evolution of carbon dioxide, which will diffuse into an inner tube containing 0.5 ml potassium hydroxide (0.1 *M*), placed inside the Malthus glass tube (Fig. 1). The neutralization of potassium hydroxide results in a change in conductance, which is measured by electrodes immersed in the alkaline solution (Fig. 2). The inoculated growth medium is in a separate

[9] F. J. Bolton, *J. Appl. Bacteriol.* **69,** 655, (1990).

[10] T. Dezenclos, M. Ascon-Cabrera, D. Ascon, J.-M. Lebeault, and A. Pauss, *Appl. Microbiol. Biotechnol.* **42,** 232 (1994).

[11] J. D. Owens, D. S. Thomas, P. S. Thompson, and J. W. Timmerman, *Lett. Appl. Microbiol.* **9,** 245 (1989).

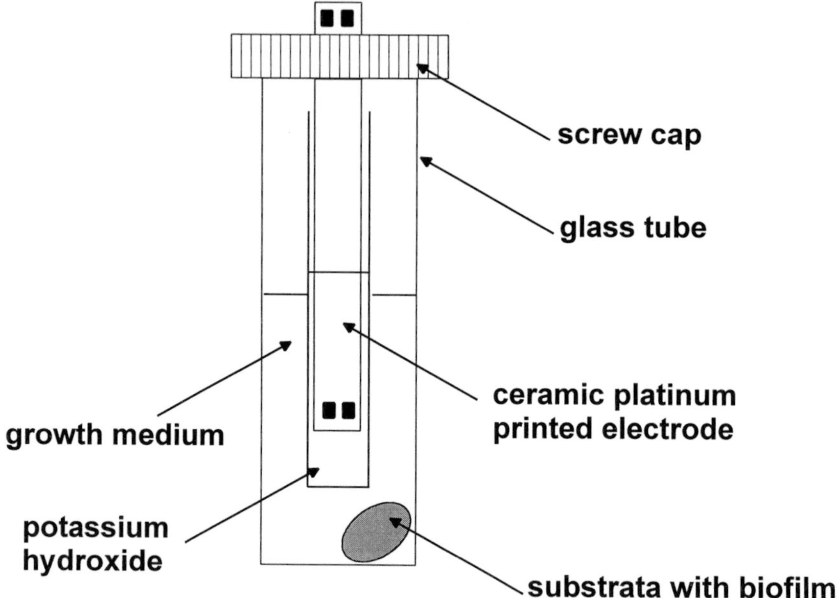

FIG. 1. Indirect Malthus cell. The substratum with biofilm is added to the growth medium, and carbon dioxide produced by the biofilm cells will diffuse into the inner tube containing potassium hydroxide, where a change in conductance is measured as the potassium hydroxide is being neutralized.

chamber and not in contact with the electrodes or potassium hydroxide. The Malthus glass tube is sealed tightly such that any carbon dioxide produced as a result of metabolism is absorbed by the potassium hydroxide.

The growth of some microorganisms is not measured easily by the indirect method, and biofilm with these organisms can be evaluated by the direct method,[3,12] where the electrodes are immersed directly in the growth medium containing the disks. As the microorganisms metabolize the growth medium, the conductance in the growth medium increases and the increase in conductance is determined by the Malthus instrument. However, when using the direct method, it is important that substrata do not touch the electrodes so as to avoid any interference with the conductance measurements. A significant interference with the conductance measurements was observed, particularly when using a metallic substrata in a direct Malthus cell.

The choice of Malthus growth medium will depend on the microorgan-

[12] R. Firstenberg-Eden and J. Zindulis, *J. Microbiol. Methods* **2,** 103 (1984).

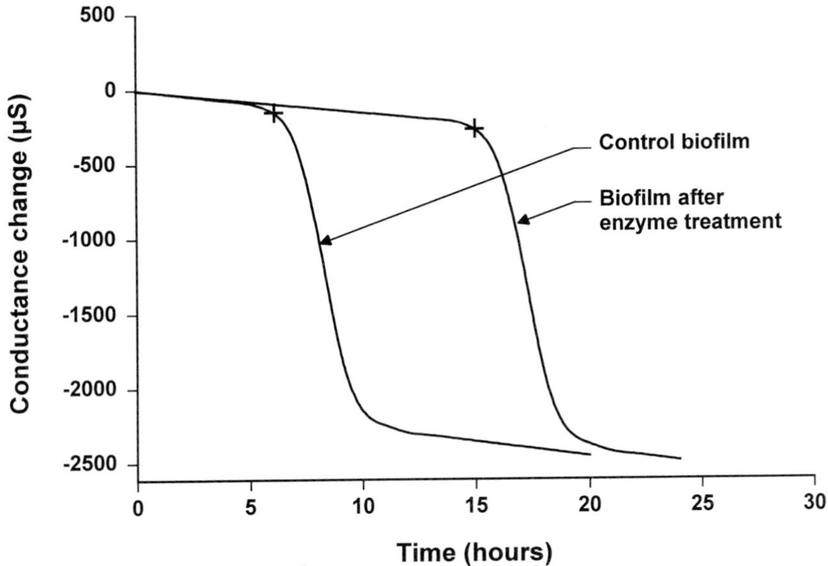

FIG. 2. Conductance measurement in indirect Malthus cells inoculated with *Pseudomonas fluorescens* AH2[14] biofilm disks. The biofilm was treated with an enzyme mixture of carbohydrases (Novo Nordisk A/S, Bagsvaerd, Denmark).

isms in the biofilm.[2,13] In general, TSB is used for the detection of biofilm with gram-negative bacteria and biofilm made from tap water, whereas brain heart infusion broth (BHI) (Oxoid) is used for the detection of gram-positive bacteria. However, other growth media are also useful for conductance measurements.

Calibration Curve

Changes in conductance were plotted against time and the detection time was determined as the time taken from the start of the measurement until a rapid change in conductance was detectable by the Malthus instrument (Fig. 2). The detection time can be related to the number of cells present at the start of the test by use of a calibration curve, which is to be constructed for each organism by inoculating Malthus tubes with a 10-fold dilution series of the planktonic cells from the biofilm culture.[3,6] Thereby it was assumed that there was a similar carbon dioxide metabolism of cells in the biofilm and planktonic cells from the biofilm culture. The carbon

[13] M. C. Easter and D. M. Gibson, *in* "Rapid Methods in Food Microbiology" (M. R. Adams and C. F. A. Hope, eds.), p. 57. Elsevier, Amsterdam/New York, 1989.

[14] L. Gram, C. Wedell-Nedergaard, and H. H. Huss, *Int. J. Food Microbiol.* **10,** 303 (1990).

dioxide metabolism can depend on the growth conditions of the cells before the conductance assay, thus a calibration curve made from a fresh liquid culture was significantly different from a calibration curve made from planktonic cells from the biofilm culture (Fig. 3). However, it was found that a calibration curve made from cells removed physically from the biofilm by ultrasonic treatment was not significantly different from a calibration curve made from planktonic cells of the same biofilm culture (Fig. 3). Therefore, the assumption that the carbon dioxide metabolism of biofilm cells is comparable to the carbon dioxide metabolism of the corresponding planktonic cells was considered to be acceptable.

The calibration curve for monocultures was reproducible, whereas calibration curves for mixed cultures are to be produced for each experiment. The choice of Malthus medium may result in a selective outgrowth from the mixed culture. Thus, it was assumed that if the medium was selective, it would select for outgrowth of the same microorganisms independent of whether these were inoculated into Malthus tubes as planktonic cells or as biofilm cells.

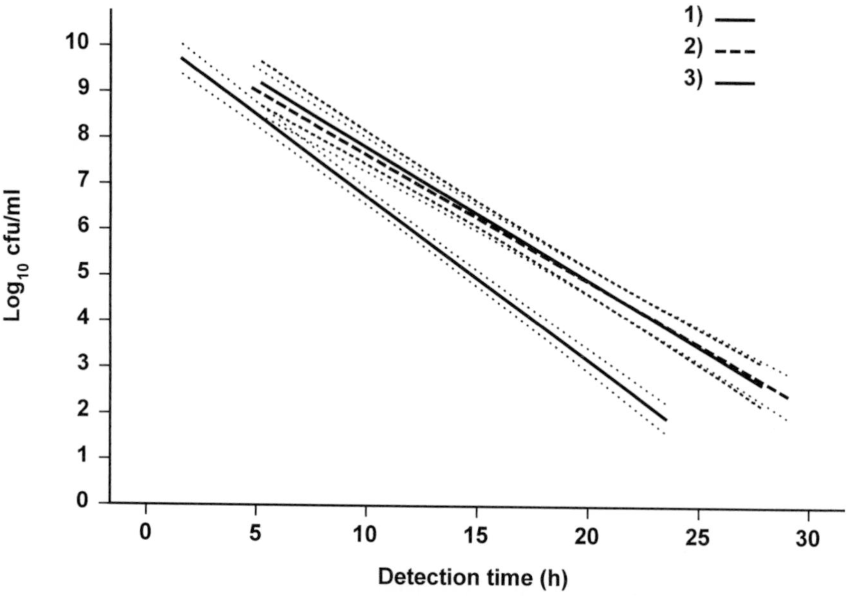

FIG. 3. Calibration curves for *Pseudomonas aeruginosa* ATCC 10148 (ATCC, Rockville, MD). (1) Cells grown as a liquid culture (20 hr, 25°). (2) Planktonic cells from the biofilm culture. (3) Biofilm cells removed from the substratum by ultrasonic treatment. A 95% confidence level for the regression curves is shown.

As control experiments, disks with biofilm but not treated with enzymes were incubated in Malthus tubes together with the different enzyme solutions to exclude interference of the enzymes with the carbon dioxide metabolism of the biofilm cells.

Biofilm degrading enzymes are, in general, nonbactericidal, and the enzymatic activity against biofilm will be an attack on the extracellular polymers, often by hydrolysis, and not a direct attack on the microbial cell. Using the Malthus method, it is not possible to distinguish between a bactericidal activity of the enzymes or an enzymatic removal of biofilm. Also, sublethally injured cells may prolong the detection time and result in underestimation of the number of living biofilm cells. Therefore, a decrease in living bacteria on the substrata has to be compared with the stimultaneous removal of biofilm from the substrata, estimated by fluorescent staining.

Fluorescence Microscopy

The effects of enzymes on biofilms were visualized by fluorescent staining of the microbial cells on substrata by a combined staining of total and respiring bacterial cells in the biofilm. The combined staining allows a quantification of living cells in the biofilm, and thereby a determination of possible bactericidal activity of the enzyme treatment. The tetrazolium salt 5-cyano-2,3-ditolyltetrazolium chloride (CTC) (Polysciences, Inc., Warrington, PA) was used as an indicator of cellular viability (functional electron transport).[15] The DNA-binding fluorochrome DAPI (4′,6-diamidino-2-phenylindole, Sigma, St. Louis, MO) was used as an indicator for the total cell number, and the biofilm cells were stained with DAPI after CTC to allow enumeration of total and respiring cells within the same preparation.[14] The stained cells were examined with the 100× oil immersion fluorescence objective on an Olympus Model BX50 microscope equipped with a 200-W mercury lamp. The filter combination used for viewing CTC-stained cells was a 480- to 550-nm excitation filter and a 590-nm barrier filter (Olympus cube Model U-MSWG). DAPI-stained cells were viewed with a 330- to 385-nm excitation filter and a 420-nm barrier filter (Olympus cube Model U-MWU).

The visual evaluation of enzymatic biofilm removal by fluorescence microscopy was particularly useful when a thick biofilm was treated with enzymes, whereas evaluations of the effects of enzymes on thin biofilms were more difficult. Therefore, the estimation of exact cell number was determined by conductance measurements.

[15] G. G. Rodriguez, D. Phipps, K. Ishiguro, and H. F. Ridgway, *Appl. Environ. Microbiol.* **58,** 1801 (1992).

Enumeration of Biofilm Cells

The number of microbial cells on substrata can be determined using a calibration curve. Using a linear calibration curve, the measured detection time (Dt) is converted to cfu/ml (N_C):

$$\log_{10}(N_C) = a\,Dt + b$$

where a and b are constants for one particular calibration curve (Fig. 3). The cell number N_C refer to the 10-fold dilution series of the planktonic cells from the biofilm culture, where 0.1 ml from each dilution was inoculated with 3 ml of growth medium in a Malthus cell. This cell number (N_C) can be converted to the number of cells present in that particular Malthus cell (N_M):

$$N_M = \frac{0.1\,N_C}{3}$$

The number of biofilm cells on a substratum was found by calculating the N_M for the measured detection time, whereas the number of cells per surface area is determined by dividing N_M with the total surface area for the substratum with biofilm.

Discussion

The described Malthus method was found useful for the evaluation of enzyme activity against microbial biofilm as the number of microorganisms left on the substratum after enzyme treatment can be evaluated and enumerated without removal of the cells from the substratum. Using the Malthus method, it was not possible to distinguish between a bactericidal activity of the enzymes and an enzymatic removal of biofilm. However, by combining the two described methods, Malthus and fluorescence microscopy, it is possible to evaluate both bactericidal activities against biofilm cells, as well as removal of microbial biofilm, and thereby it could be concluded whether the enzymes were bactericidal, able to degrade biofilm, or both. The Malthus/fluorescence staining is not limited to the evaluation of enzymes against biofilm cells, but is also useful for the evaluation of chemical biocides. Using this method, bactericidal activity can be evaluated against biofilm cells without removal of the cells from the substratum by simply placing the substratum with biofilm in an conductance measuring cell and immersing the electrodes. The determined antimicrobial activity may be more realistic, as the traditional methods involving biofilm removal and bacterial enumeration by colony formation overestimate biocide efficacy.[16]

[16] G. A. McFeters, F. P. Yu, B. H. Pyle, and P. S. Stewart, *J. Ind. Microbiol.* **15,** 333 (1995).

Furthermore, the standard deviation on cell density determined by conductance measurements was decreased compared to traditional plate counts. The described method was particularly useful for the evaluation of enzymatic degradation of biofilm, as cells left on the substratum after enzyme treatment was more detachable than untreated biofilm cells. The enzyme treatment can thereby interfere with the traditional evaluation methods involving biofilm removal followed by bacterial enumeration by colony formation, resulting in an underestimation of the enzyme efficacy.

The use of a calibration curve for determinating the number of biofilm cells involves the assumption that, in the Malthus tubes, the carbon dioxide metabolism of cells from the biofilm and cells from the planktonic suspension is comparable. This assumption was found to be acceptable, despite the indications of differences in the metabolism of biofilm and planktonic cells, respectively.[17] This can be explained by the fact that biofilm cells do not continue to grow as biofilm cells when transferred to Malthus tubes and thereby the actual detection will be on the outgrowth of planktonic cells in the Malthus medium.

[17] S. Møller, C. Kristensen, L. K. Poulsen, J. M. Carstensen, and S. Molin, *Appl. Environ. Microbiol.* **61,** 741 (1995).

[27] Evaluation and Quantification of Bacterial Attachment, Microbial Activity, and Biocide Efficacy by Microcalorimetry

By WOLFGANG SAND and HENRY VON RÈGE

Introduction

In the field of biofilm research, the interactions between microorganisms and the animate or inanimate surface are only partially understood. The primary process, which finally leads to a firm attachment of a microorganism to a substratum, is the focus of much attention. Up until now it remains impossible to predict which strain of a bacterial species will attach to a specific substratum/material.

Furthermore, once a cell has attached to a surface, the direct and continuous monitoring of its growth, multiplication, metabolic activity, and so on is very difficult. Most techniques allow only measurements at certain points (ATP, fatty acids, etc.), are destructive (electron microscopy, many electro-

0076-6879/99 $30.00

chemical techniques, etc.), or are only applicable for aerobic samples (O_2 electrode). In addition, special test rigs are required to expose sample coupons for consecutive detection and analyses.

Thus, there is a need for an on-line test system that allows for undisturbed, continuous measurements under aerobic as well as under anaerobic conditions. In addition, the test system should allow one to scrutinize the relevant microorganisms with respect to the relevant materials, to determine possible interactions such as attachment, and, in the case of unwanted effects such as biocorrosion, the optimization of countermeasures including biocide dosage.

To achieve these goals, the technique of microcalorimetry is the most suitable one. It is a nondestructive, highly sensitive technique that allows the detection on-line of a minimum of 10^4 metabolically active cells[1] without the need to disturb the system for/by sampling.

Description of Microcalorimetric Measurement System

Calorimetry has been applied as early as 1780 by Lavoisier for a determination of metabolic rates of mammals. From this first ice calorimeter it has been a long way to the development of present instruments.[2] Whereas the ice calorimeter used the amount of water, resulting from molten ice, as a measure of the heat evolved, present-day instruments are based on thermocouples, allowing for an accurate determination of the power output. Modern calorimeters, and here only the isothermal type of equipment will be referred to, allow the detection of thermal power quantities as low as 50 nW.

Measuring Principle

The principle on which modern instruments are based is given in Fig. 1. The measuring cup is the center of the system. It can take up closed ampoules of between 4 and 25 ml volume, microreactors of the same volume (open ampoules), and/or be surrounded by a coil of gold tubing, allowing for continuous flow experiments. The three systems are shown in Fig. 2. The flow system is available as a flow-through or as a flow-mix system. The latter allows the addition of a compound to the system under supervision immediately prior to the passage of a sample liquid through the thermocouples. Using this assembly short time reactions become detectable.

All metabolic reactions are accompanied by heat production or absorption (exergonic or endergonic reactions). In the microcalorimetric measure-

[1] J. P. Belaich, *in* "Biological Microcalorimetry" (A. E. Beezer, ed.). p. 1. Academic Press, London, 1980.

[2] J. Suurkuusk and I. Wadsö, *Chem. Scr.* **20,** 150 (1982).

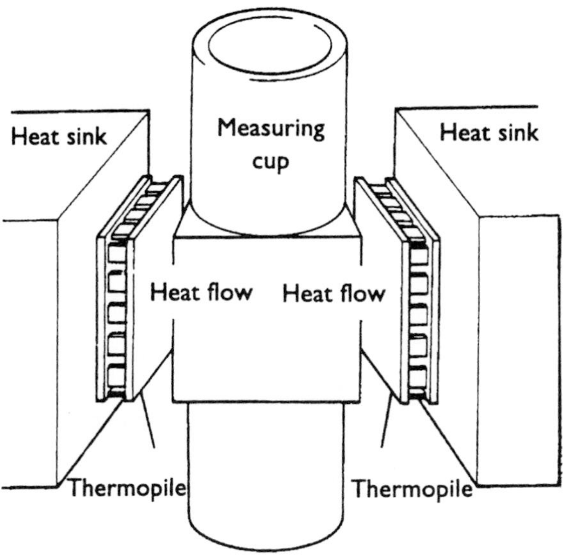

FIG. 1. Measuring unit of the microcalorimeter.

ment unit this heat is completely exchanged with a large, surrounding heat sink. This sink is a large water bath, accurately maintained at a constant temperature. When a thermal energy change occurs in a sample as a consequence of bacterial metabolism, a small temperature difference arises (relative to the heat sink), which forces a heat flow, until equilibrium is obtained again. This temperature difference is directly proportional to the heat flow. Highly sensitive thermopiles, located around the reaction vessel, are used to detect and quantify the temperature difference. The potential, generated

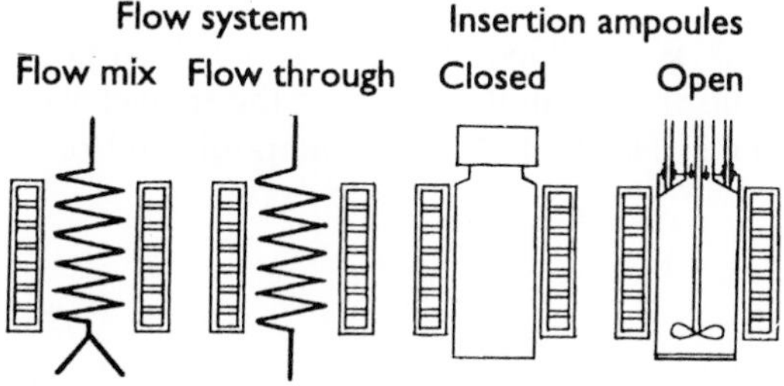

FIG. 2. Measuring systems of the microcalorimeter.

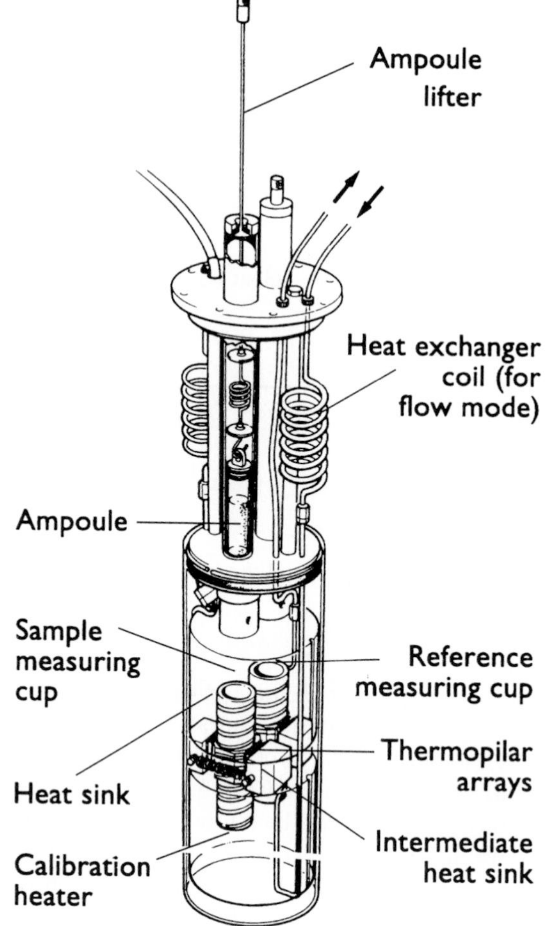

FIG. 3. View of a calorimetric unit.

by the thermopiles, is amplified and recorded as heat flow. Figure 3 gives the details of a calorimetric unit, displaying the spacial arrangement of the measuring elements. This unit allows for ampoule and/or flow-through experiments.

Measuring Units

Three general types of measuring units are available: (1) the 4-ml twin-ampoule/microreaction system unit, (2) the 20/25-ml twin-ampoule/microreaction system, and (3) the combined twin-ampoule/microreaction system plus flow or flow-mix installation unit.

Ampoule units are equipped with a twin detector for measuring and a reference cup. The ampoules are available with 4-, 20-, or 25-ml volumes. In all instances, either closed or open insertion ampoules (microreaction systems) may be used. The sensitivity of the system amounts to 50 nW (0.1 μW) for the 4-ml ampoules (microreaction system) or to 0.2 μW (0.4 μW) for the 20-ml units, respectively. Closed ampoules are generally used for stability or compatibility tests.[3] More specifically, they may be used for a quantification of biocide efficacy against biofilm samples on coupons or determinations of the microbial activity of cell suspensions with dissolved and/or particulate compounds.[4] Microreaction systems are equipped with a stirrer and gas/liquid inlets and outlets. Thus, experimental conditions can be manipulated during a run. Oxygen, nitrogen, or hydrogen gas may be added.

In addition, nutrients and/or trace elements as well as toxic compounds such as biocidal substances can be included in the experiment. These ampoules allow the study of ligand interactions, enzyme kinetics, binding, and conformational changes and, with living microorganisms, interactions between solids and liquids occurring as in biocorrosion systems. If test coupons of metallic or other materials are used, which contain a biofilm, material- and/or microorganism(s)-specific tests become possible. The response of the biofilm community to an addition of growth-enhancing or growth-inhibiting substances may be tested using planktonic, planktonic and attached, or only attached bacteria. Thus, effective measures for antifouling strategies with regard to a specific material can be developed or the course of biocorrosion be followed (see later).

Combined calorimetric units allow one to perform all of the previously described experiments plus continuous culture tests in the bypass mode. The solution or suspension under investigation may originate from an external fermenter and is pumped through the measuring coil for the quantification of the heat flow. The volume, available for a registration of the metabolic reactions, amounts to 0.6 ml. The whole system of a flow unit contains 1.8 (flow unit) or 2.1 (flow-mix unit) ml. The sensitivity of the system amounts to 0.1 μW for both types. Typical experiments, run in a flow unit, are assays involving the tests of drug effects on cell cultures or fermentation processes. Regarding biofilm, these flow units allow the detection of pure and mixed cultures, whether comprised of defined or

[3] K. Byström and A. B. Draco, "Thermometric Application note 22,004." Thermometric, Broma, Sweden.

[4] H. von Rége and W. Sand, *J. Microbiol. Methods* **33,** 227 (1998).

unknown (enrichment) microorganisms, the capability to attach to the surface of the tubing (in the measurement position). Because gold, an inert material, is used for the tubing, adhesion may be studied without any interfering effects resulting from corrosion. However, it needs to be pointed out that material-specific adhesion experiments are not yet possible. The advantage of the flow unit is that those microorganisms may be detected that attach readily to the surface of the tubing. They may even be detected in mixed cultures, allowing identification of the relevant ones for biofilm formation. In the case of pathogenic bacteria, to become established in potable water system biofilms, this possibility is of considerable importance. Up until now, no other test system is available that unequivocally allows for the detection of adhesion on line and, in addition, the test of countermeasures. The latter is achievable by adding antimicrobials before entering the measuring unit, i.e., outside of the instrument, to the pumped solution/ suspension in the case of the flow-type unit or by using the flow-mix type unit (mixing occurs inside the instrument) immediately prior to the measuring position with the pumped solution. The first possibility is used to measure long-term effects on pure cultures, attached, and/or planktonic ones, as well as on mixed cultures. The second possibility allows for tests of rapid reactions occurring immediately after the addition of toxic compounds. Other experiments that are possible using the flow-mix unit are useful for determinations of thermodynamic properties of reactions, such as the heat of dilution or ligand binding.

Practical Applications of Microcalorimetry

Following the previous explanation of the microcalorimetric system, specific experimental possibilities are described for an evaluation of adhesion, microbial activity of biofilms on materials, surfaces, and biocide efficacy.

Adhesion to Surfaces

Two examples are described: determining the possibility to detect the ability of bacteria to attach to a surface (of gold) and determining the amount of energy used for attachment and for biofilm formation.

Chemoorganotrophic bacteria of the species *Bacillus subtilis* were isolated from the river Elbe in Germany near Hamburg. The water samples originated from the surface and from deep water (0 to 12 m). From the water phase and from flocs found in the samples, strains of *B. subtilis* were isolated and subcultured. These were analyzed for their heat output and growth characteristics using flow-through measurement units (2277 type

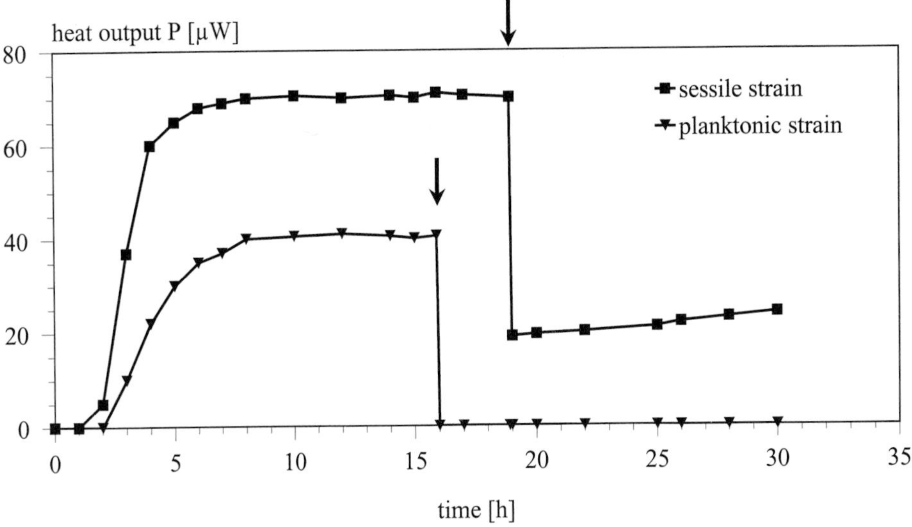

FIG. 4. Heat output of different strains of *Bacillus subtilis*. Arrows indicate the change from the culture suspension with bacteria to the sterile nutrient solution (M. Rudert, unpublished results).

TAM of Thermometric, Sweden, or C-3 Technik, Germany). The strains were cultivated aerobically and pumped through the flow-through unit with peristaltic pumps. Intermittently, air and solution at a rate of 20 ml each were pumped. The resulting power–time curves were recorded and registered. It has been shown previously that these power–time curves are strain specific.[5] Figure 4 gives a typical example of such a power–time curve. For evaluation, whether these strains had attached to the surface of the tubing, the flow from the external culture was interrupted and the solution was replaced by a fresh, sterile one. The consequences are shown in Fig. 4. In case of a strain able to attach, after the exchange of the nutrient solution, a measurable, constant heat output remained detectable. If growth occurs under these conditions, the heat output would increase. In contrast, in the experiment with the nonattaching strain the heat output decreased to almost the baseline after solution replacement. Obviously, all cells were washed out by the sterile solution, rendering the measuring position virtually free of bacterial cells. The ability for attachment was correlated with the origin of the strains. The attaching ones originated from flocs, whereas the planktonic ones did not attach. As a consequence, the calorimetric test allows an

[5] A. W. Schröter and W. Sand, *FEMS Microbiol. Rev.* **11,** 79 (1993).

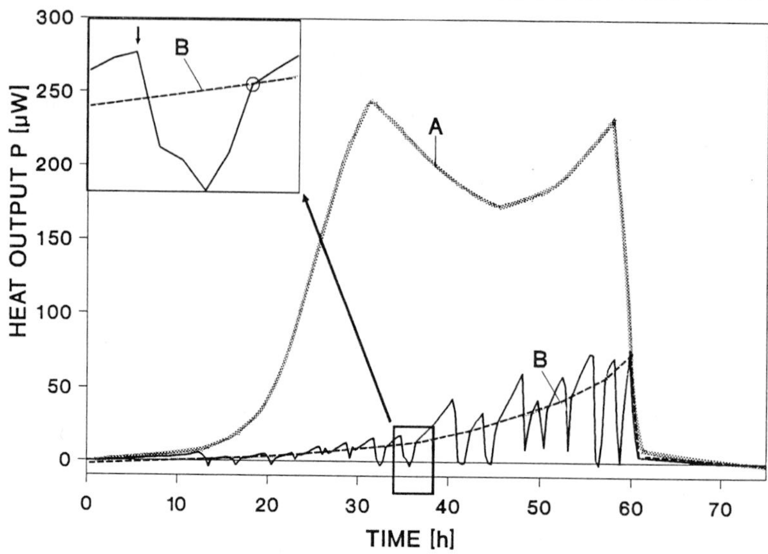

FIG. 5. Influence of adhering cells on the heat output of *Thiomonas intermedia* growing on thiosulfate. *A,* power–time curve without interruption by cleaning; *B,* power–time curve with cleaning intervals. Arrow, Start of cleaning procedure; circle, change from steep to moderate increase of heat output. Reprinted from S. Wentzien *et al., Arch. Microbiol.* **161,** 116 (1994), with permission.

explanation why some strains occurred in flocs and others only as planktonically growing ones.

In addition to chemoorganotrophic bacteria, autotrophic species are able to attach to surfaces and to form a biofilm. The following experiments were conducted with *Thiomonas intermedia* (previously *Thiobacillus intermedius*[6]) strain K12, which originally had been isolated from corroded sites of a concrete pipe in the Hamburg sewer system[7] and *Thiobacillus versutus* strain DSM 582. Both strains were cultured, using their respective growth media with thiosulfate as an energy source.[8] Microcalorimetric experiments were performed using a flow-through unit and a flow rate of 20 ml nutrient solution plus 20 ml air per hour each, achieved by means of a peristaltic pump. First growth characteristics were determined for both strains. The resulting power–time curves of typical experiments are given in Fig. 5. In the course of substrate degradation, *T. intermedia* exhibited a power–time curve with two peaks, accompanied by a large heat output.

[6] D. Moreira and R. Amils, *Int. J. Syst. Bacteriol.* **47,** 522 (1997).

[7] K. Milde, W. Sand, W. Wolff, and E. Bock, *J. Gen. Microbiol.* **129,** 1327 (1983).

[8] S. Wentzien, W. Sand, A. Albertsen, and R. Steudel, *Arch. Microbiol.* **161,** 116 (1994).

After exhaustion of the substrate, the heat output decreased rapidly to the baseline. In the experiment with a culture of *T. versutus,* the power–time curve increased gradually in the course of the experiment. After substrate consumption, the curve returned rapidly to baseline. In contrast to the experiment with *T. intermedia,* only a comparably small area, indicating a small total heat output, was registered for *T. versutus* in the course of substrate degradation. To demonstrate that the large heat output registered with *T. intermedia* resulted from attachment and biofilm formation, the experiment was repeated in the following way. From an external culture of *T. intermedia,* a suspension with growing, thiosulfate-oxidizing cells was split in two, each measured with an independent measuring unit of the microcalorimeter. In one unit the course of substrate degradation was recorded, as has been described previously. The same power–time curve was consequently recorded. In the other measuring unit, however, which was run under identical conditions, the flow of the culture solution was interrupted regularly and replaced by a cleaning solution [0.1 *M* NaOH/0.1% sodium dodecyl sulfate (SDS)] that had been shown previously to kill all attached bacteria and to remove most of the cell debris.[5] After circulating this latter solution for about 20 min, a thorough washing with distilled water followed to remove all traces of the cleaning solution. Afterward the culture solution was again pumped through this measuring unit. The resulting power–time curve is shown in Fig. 5. As a consequence of the application of the cleaning solution, the heat output decreased after each application almost to baseline. After reswitching to the culture solution, a rapid increase was registered. The heat output soon reached a point where it leveled off to a slight increase. By connecting these points, a power–time curve resulted, which resembled the one obtained with cells of *T. versutus.* As a consequence, by applying the cleaning solution the attached cells of *T. intermedia* were removed from the system, rendering it free from their heat output. The curve, resulting from the connected points, reflects the heat output of the planktonic population. The latter was almost identical to the curve of the planktonic cells of *T. versutus* (not shown). In addition, the area below the curve allows the determination of the amount of energy consumed by attachment and excretion, and biofilm formation as well. In the present case, about 85% of the total heat evolved had been used for attachment, and so on, whereas only 15% resulted from substrate oxidation.

Quantification of Microbial Activity

Most techniques require the removal of a biofilm from the sample surface for determinations of microbial activity. Because this alters the environmental conditions for biofilm microorganisms, a microcalorimetric procedure was developed that allows one to nondestructively determine microbial activity in biofilm samples. As a consequence, a realistic testing of a biocide treatment becomes possible. Furthermore, screening and optimization of other countermeasures against microbiologically influenced corrosion (MIC) and biofouling also become possible.

In the case of a biofilm on metal surfaces, biological and chemical processes such as corrosion produce a measurable heat. To determine the biological and the chemical contribution to the heat output, mild and stainless steel (AISI 304) coupons were incubated in a miniplant.[9,10] A mixed culture consisting of the biocorrosion-relevant bacteria *Desulfovibrio vulgaris* (NCIMB 8457), *T. intermedia,* and the chemoorganotrophic biofilm isolate *Ochrobactrum anthropi* were used as inocula (10^8/ml each). Medium[9] containing these bacteria was pumped continuously through ring columns of the miniplant to produce a biofilm on the metal surfaces. A continuous change between anaerobic and aerobic conditions within the columns was produced by alternatively gassing with nitrogen or compressed air during the incubation. As a result, high cell counts of all of these bacterial species were achieved. After 6 weeks of incubation, biofilm samples were withdrawn and analyzed for cell counts, for mass loss, and for microbial (biological) as well as chemical activity (heat output). Microcalorimetric measurements were performed with a 25-ml stainless steel ampoule (TAM-cylinder 2277-205). After sampling, the coupons (with biofilms) were put into the ampoule, together with 15 ml of sterile medium.[9] After stabilization of the heat output about 2 hr later, the value was registered. To differentiate between biological and chemical contributions to the heat output of biofilm samples, consecutive to the first measurement, a 24-hr incubation in 5% (w/v) formaldehyde was enclosed to kill all biofilm microorganisms. Afterward, the remaining (chemical) heat output was determined. The difference between the two values can be attributed to microbial activity.

The total heat output in this example, originating from biofilms on mild steel coupons, was 30 μW/cm^2. After formaldehyde incubation, a heat output of 14 μW/cm^2 remained, indicating that biological and chemical reactions contributed to about 50% each to the total value (Fig. 6). Cell counts in the biofilm samples were reduced by the formaldehyde treatment to 10^2 cells/cm^2 (not shown). Because this cell number is below the microcalorimetric limit of detectability, it was proven that the remaining output

[9] H. von Rége and W. Sand, *in* "Biodeterioration and Biodegradation, DECHEMA Monographs Vol. 133" (G. Kreysa and W. Sand, eds.), p. 325. VCH, Weinheim, Germany, 1996.

[10] U. Eul, H. von Rége, E. Heitz, and W. Sand, *in* "Microbially Influenced Corrosion of Materials" (E. Heitz, W. Sand, and H.-C.Flemming, eds.), p. 188. Springer-Verlag, Berlin, 1996.

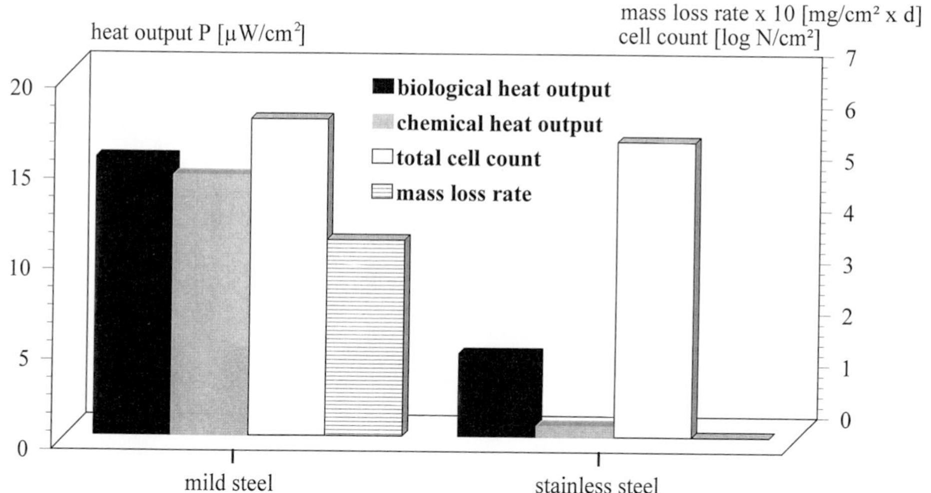

FIG. 6. Biological and chemical heat output, cell counts, and mass loss of mild or stainless steel coupons with biofilms consisting of *Thiomonas intermedia, Desulfovibrio vulgaris,* and *Ochrobactrum anthropi*. Coupons were incubated for 6 weeks under alternating aerobic or anaerobic conditions in a miniplant.

resulted from chemical activity only (possibly from the chemical oxidation of reduced sulfur compounds or from the iron compound oxidation). The microbial heat output of stainless steel coupons with biofilms was 4.4 μW/cm^2. Chemical heat output, measured after formaldehyde incubation, was only 0.6 μW/cm^2 (Fig. 6). Also, in this case, cell numbers in the biofilm after formaldehyde treatment were too low (not shown) to produce a detectable heat output. Thus, in contrast to the experiment with mild steel, the biological activity amounted to 86%. Biofilm samples on mild or stainless steel coupons incubated under fully aerobic or fully anaerobic conditions exhibited similar results (not shown). In the case of a biofilm on a mild steel coupon, the chemical activity generally contributed more to the total activity (due to corrosion, as indicated by the high mass loss observed) than in the case of a biofilm on a stainless steel coupon (without corrosion) where microbial activity was the main source for the heat output.

Summarizing, this technique allows a rapid detection and quantification of the ongoing processes in a biofilm on a steel surface.

Biocide Efficacy Testing

The effects of chemical agents, biocides, or cleaning procedures are easily quantifiable too, as has been shown earlier for the formaldehyde

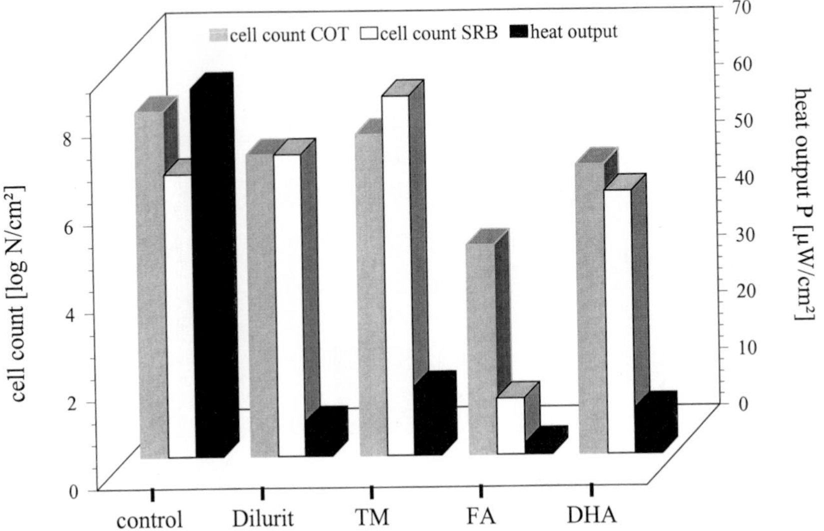

FIG. 7. Influence of four different biocides on microbial activity and cell counts of biofilm samples on mild steel. Coupons were incubated for 10 weeks with a mixed culture consisting of SRB and COT. Batch experiments with biocides lasted for 24 hr at 30°. Dilurit (500 mg/liter); TM (tetramethylammonium hydroxide, 500 mg/liter); DHA (1,8-dihydroxyanthraquinone, 4.8 mg/liter); FA (formaldehyde, 50,000 mg/liter).

treatment. Biocide application, to fight undesired biofilms, is still the most common practice in industry. Effective monitoring methods for an evaluation of biocide efficacy are lacking. Microcalorimetry offers a potential to fill this gap.

For a demonstration, the efficacy of biocides against biofilm microorganisms on mild steel coupons was scrutinized. Coupons were incubated anaerobically in the presence of a mixed culture containing sulfate-reducing bacteria (SRB) and chemoorganotrophic bacteria (COT).[4] After 10 weeks, identical coupons with biofilms were treated with several biocides to screen for the most effective one. The biocides were commercial products with glutaraldehyde as the active compound (Dilurit, BK Ladenburg), a quaternary ammonium compound (tetramethylammonium hydroxide), 1,8-dihydroxyanthraquinone (an ancoupler of ATP synthesis for SRB[11]), and formaldehyde. The latter compound was added in a high concentration in order to reduce the cell counts of living microorganisms in the biofilm samples

[11] F. B. Cooling, C. L. Maloney, E. Nagel, J. Tabinowski, and J. M. Odom. *Appl. Environ. Microbiol.* **62,** 2999 (1996).

to such a low amount that chemical contribution to the heat output could be determined.

As a result of a biocide application with a contact time of 24 hr, the microbial activity was in general reduced considerably (Fig. 7). The share of the chemical heat output only amounted to about 10% of the total heat output. Obviously, the biological activity was the dominating one. Cell counts of SRB remained stable in three of four experiments, despite biocide application, whereas those of COT were reduced slightly. Only in the case of the formaldehyde application were the cell counts of both groups reduced strongly. After an exchange of the medium against a biocide free one and 7 days of further incubation, regrowth occurred, but microbial activity did not reach the initial level (not shown). Microcalorimetry, consequently, enables differentiation between killing or inhibition of the microbiota as a result of a biocide application. Thus, a direct monitoring of biocide action for determining the time, when a further application becomes necessary, is possible. In industries where biocide use is unavoidable, as in paper and pulp manufacturing, this possibility is of considerable advantage.

On-line measurements of biocide action against biofilms by the flow-through technique offer additional possibilities. A biofilm established on the inner surface of the gold tubing (in the flow-through cylinder, described earlier) allows screening for the optimal dosage of a biocide. This was demonstrated by experiments with *Vibrio natriegens* (DSM 759). The bacte-

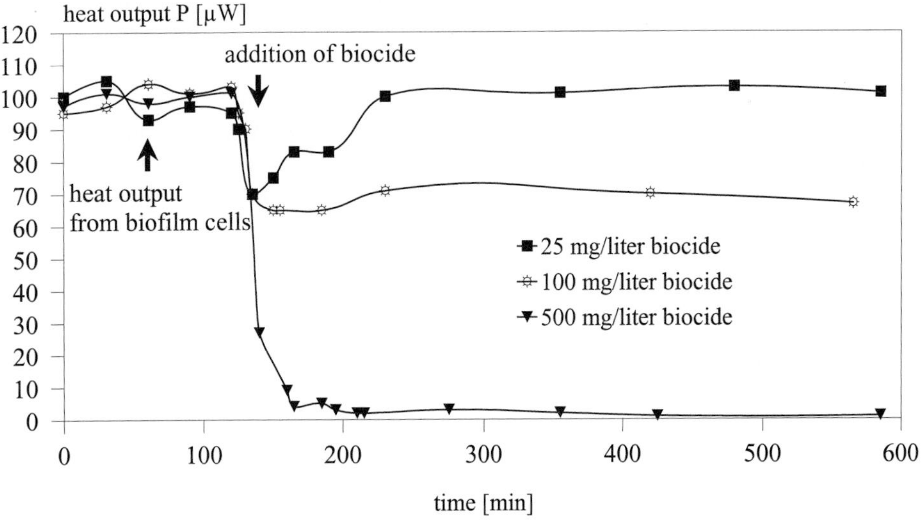

FIG. 8. Influence of biocide concentration on a *Vibrio natriegens* biofilm in continuous culture. Reprinted from H. von Rége and W. Sand, *J. Microbiol. Methods* **33,** 227 (1998).

rium formed a biofilm inside the gold tubing. After a stable heat output signal had been obtained, only sterile nutrient solution[4] was further pumped (20 ml/hr), followed by an addition of 25 mg/liter biocide to the medium. At this concentration only a slight and transient decrease of microbial activity was measured (Fig. 8). A concentration of 100 mg/liter had a more pronounced effect (see Fig. 8), although a considerable amount of microbial activity was still detectable. Only with a biocide concentration of 500 mg/liter was microbial activity almost totally abolished, remaining negligible until the end of the experiment. However, by replacing the biocide-containing medium against a biocide-free one (see earlier discussion), microbial activity increased within a 10-hr period. Obviously, some bacteria survived in the depth of the biofilm and were able to regrow. This technique allowed elaboration for this biocide. Furthermore, data demonstrated that some microorganisms remained alive within the (protecting) biofilm, inaccessible even to high biocide dosages. No other technique is known that allows such accurate determinations within such a short time. If it becomes possible to exchange the gold tubing by other materials, selective tests for adhesion and biocorrosion will allow for more insight into the interfacial processes governing biofilm formation and stability.

Summary and Outlook

Microcalorimetry is not yet a widely established technique in the field of biofilm research. This will change with time because of the potential the technique has to offer. Up until now, no other technique allows an on-line measurement of an attachment, which is also material specific. Furthermore, the possibility of measuring aerobic and/or anaerobic metabolism and differentiating between biotic and abiotic reactions is also unique. The comparably short times needed for an experiment are also a strong argument in favor of microcalorimetry, in addition to the simple handling required.

[28] Surface Analysis by X-ray Photoelectron Spectroscopy in Study of Bioadhesion and Biofilms

By YVES F. DUFRÊNE, CHRISTOPHE J. P. BOONAERT, and PAUL G. ROUXHET

Introduction

Various steps are involved in the formation of microbial biofilms; they are represented schematically in Fig. 1. In most natural environments, solid substrata are conditioned rapidly by adsorption of organic macromolecules (proteins, polysaccharides, lipids), which are likely to modify the substratum physicochemical properties. The cells may reach the substratum surface through various transport mechanisms: diffusion, sedimentation, convection, or active transport. Then, physicochemical interactions may lead to retention of the cells by the substratum, a process often referred to as initial adhesion. The production of extracellular macromolecules by the cells may both strengthen adhesion to the substratum and lead to cell–cell attachment. Finally, multiplication of the attached cells may lead to the formation of microcolonies and biofilms. Biofilm formation thus involves molecular interactions at cell–substratum and cell–cell interfaces.

Physicochemical interactions that have long been recognized to play a role in initial adhesion include long-range van der Waals and electrostatic interactions, on the one hand, and hydrophobic interactions, on the other hand. These interactions can be respectively described by the DLVO (Derjaguin, Landau, Verwey, and Overbeek) theory of colloid stability[1] and the balance of interfacial free energies involved in the creation of a new cell–substratum interface and in the destruction of the interface of each partner with water.[2] Van der Waals, electrostatic, and hydrophobic interactions depend on the surface properties of the interacting surfaces and therefore on their chemical composition.

The surfaces of microbial cells and substrata are not atomically smooth and rigid, as it is assumed in the approaches just described, but are coated with lyophilic macromolecules (Fig. 1) forming loosely structured layers with loops and tails protruding into the solution. This gives rise to macromo-

[1] P. R. Rutter and B. Vincent, *in* "Microbial Adhesion to Surfaces" (R. C. W. Berkeley, J. M. Lynch, J. Melling, P. R. Rutter, and B. Vincent, eds.), p. 79. Ellis Horwood, Chichester, 1980.

[2] H. J. Busscher, A. H. Weerkamp, H. C. van der Mei, A. W. J. van Pelt, H. P. de Jong, and J. Arends, *Appl. Environ. Microbiol.* **48,** 980 (1984).

0076-6879/99 $30.00

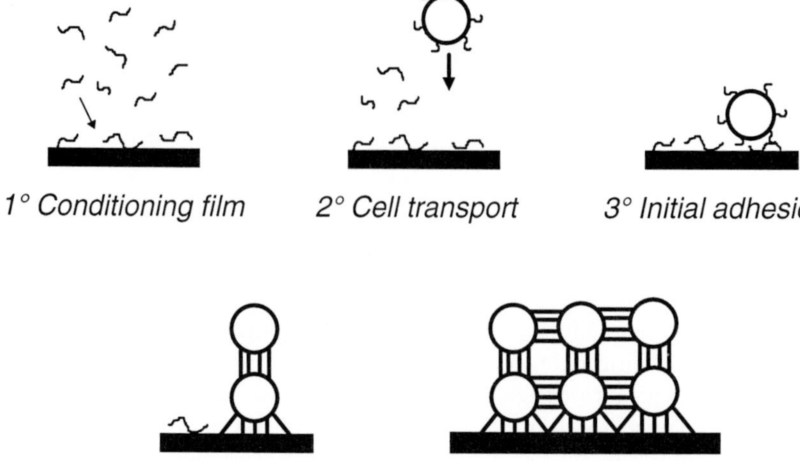

FIG. 1. Schematic representation of the various steps involved in the formation of biofilms by microbial cells.

lecular interactions[3] that can be either attractive (polymer bridging) or repulsive (steric hindrance) depending on the surface coverage and on the affinity of macromolecules for the solvent. Understanding interfacial interactions involved in bioadhesion and biofilm formation therefore requires a knowledge of the chemical composition of substratum surfaces, cell surfaces, and dissolved macromolecules.

X-ray photoelectron spectroscopy (XPS) provides a direct chemical analysis of solid surfaces, with an analyzed depth of about 5 nm.[4,5] Although well known in the field of material science, XPS has only been applied recently to microbial cells.[6] In contrast with classical (bio)chemical analysis, XPS provides information directly on the outermost surface rather than on the entire cell wall or on the bulk material, which makes it more relevant

[3] H. H. M. Rijnaarts, W. Norde, E. J. Bouwer, J. Lyklema, and A. J. B. Zehnder, *Colloids Surfaces B Biointerfaces* **4,** 5 (1995).

[4] B. D. Ratner and B. J. McElroy, *in* "Spectroscopy in the Biomedical Sciences" (R. M. Gendreau, ed.), p. 107. CRC Press, Boca Raton, Fl, 1986.

[5] P. G. Rouxhet and M. J. Genet, *in* "Microbial Cell Surface Analysis: Structural and Physicochemical Methods" (N. Mozes, P. S. Handley, H. J. Busscher, and P. G. Rouxhet, eds.), p. 173. VCH Publishers, New York, 1991.

[6] P. G. Rouxhet, N. Mozes, P. B. Dengis, Y. F. Dufrêne, P. A. Gerin, and M. J. Genet, *Colloids Surfaces B Biointerfaces* **2,** 347 (1994).

to the understanding of interfacial phenomena (adhesion, aggregation, biofilm formation).

The aim of this contribution is to present the various applications offered by XPS in bioadhesion and biofilm studies. This article includes (i) a presentation of the XPS technique (basic principles, spectroscopic aspects, and data interpretation), (ii) a description of a case study concerning the adhesion of the bacterium *Azospirillum brasilense* to model substrata, and (iii) an overview of the various applications of XPS and relevant methodologies in the study of bioadhesion and biofilms, based on experience developed over about 15 years in our laboratory.

X-Ray Photoelectron Spectroscopy

X-ray photoelectron spectroscopy involves irradiation of the sample by an X-ray beam, which induces the ejection of electrons. The kinetic energy of the emitted electrons is analyzed, and their binding energy in the atom of origin is determined. Due to inelastic scattering of electrons in the sample, the collected information concerns only the outermost molecular layers of the surface (2–5 nm). Each peak of the recorded spectrum is characteristic of a given electron energy level of a given element. Peaks are identified according to the binding energy, thus 1s designates electrons ejected from the first layer of the electronic configuration. Atomic concentration ratios and overall atomic fractions are calculated from the peak areas on the basis of acquisition parameters and of sensitivity factors. Detailed information about the basis of the technique and instrumentation can be found in the literature.[4,5,7]

Because the peak position is influenced by the chemical environment, further information on the functional groups present at the surface can be obtained by decomposing the peaks using least-squares curve fitting. This is illustrated in Fig. 2, which shows representative C_{1s} and O_{1s} spectra obtained for *A. brasilense* cells harvested in two different growth phases.[8] Attribution of the components of the carbon, oxygen, and nitrogen peaks can be made according to guidelines drawn from the analysis of polymers[9] and model biochemical compounds.[10] The carbon peak recorded on microbial cells and biochemical compounds can be decomposed into the following components (Fig. 2): (i) a component set at 284.8 eV due to carbon bound

[7] B. D. Ratner and D. G. Castner, *in* "Surface Analysis: The Principal Techniques" (J. C. Vickerman, ed.), p. 43. Wiley, Chichester, 1997.

[8] Y. F. Dufrêne and P. G. Rouxhet, *Can. J. Microbiol.* **42,** 548 (1996).

[9] G. Beamson and D. Briggs, "High Resolution XPS of Organic Polymers, the Scienta ESCA300 Database." Wiley, Chichester, 1992.

[10] P. A. Gerin, P. B. Dengis, and P. G. Rouxhet, *J. Chim. Phys.* **92,** 1043 (1995).

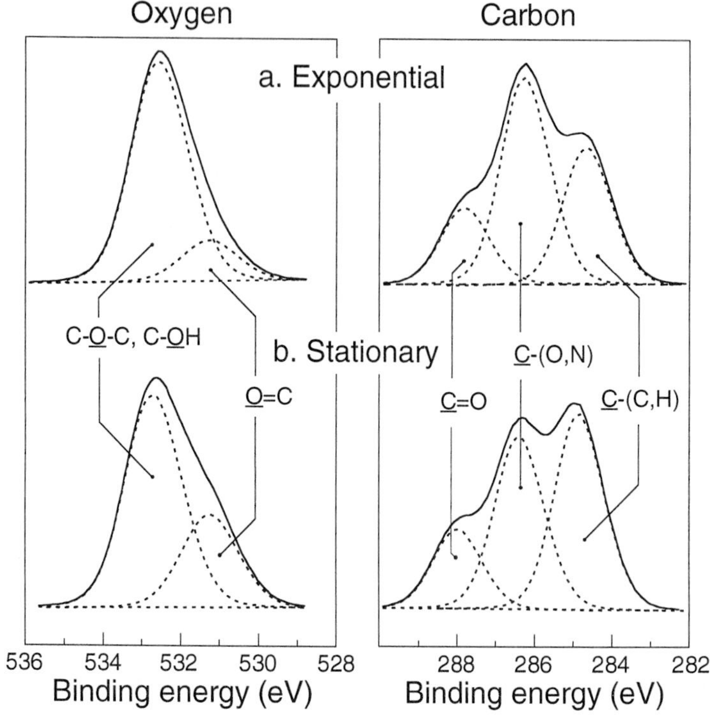

FIG. 2. Representative C_{1s} (right) and O_{1s} (left) peaks of *A. brasilense* cells harvested in the exponential growth phase (a) and in the early stationary growth phase (b).[8]

only to carbon and hydrogen $\underline{C}$—(C,H); this component is used as a reference to calculate the binding energy of all the other components/peaks; (ii) a component at about 286.4 eV due to carbon bound singly to oxygen or nitrogen $\underline{C}$—(O,N), including ether, alcohol, amine, and amide; (iii) a component at about 288.0 eV due to carbon making one double bond or two single bonds with oxygen $\underline{C}$=O, including amide, carbonyl, carboxylate, acetal, and hemiacetal; and (iv) sometimes a weak component (not observed here) found near 289.0 eV and attributed to carboxylic acid. The oxygen peak can be decomposed into two components: (i) a component at about 531.4 eV attributed to oxygen making a double bond with carbon ($\underline{O}$=C) in carboxylic acid, carboxylate, ester, carbonyl, or amide and (ii) a component at about 532.8 eV attributed to hydroxide (C—$\underline{O}$H), acetal, and hemiacetal (C—$\underline{O}$—C—$\underline{O}$—C). A nitrogen peak typically appears at about 400.0 eV due to unprotonated amine or amide functions.

Relationships between elements and functional groups provide insight into the nature of the molecular constituents present at the cell surface.

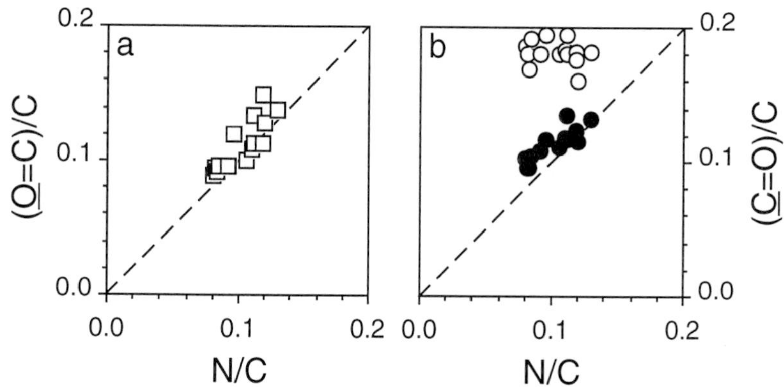

FIG. 3. Atomic concentration ratios, with respect to total carbon, (a) of oxygen doubly bound to carbon (O=C)/C and (b) of carbon making one double bond or two single bonds with oxygen (C=O)/C (○) and the same after deduction of the contribution of acetal (●), as a function of nitrogen; interrupted lines have unit slope and zero intercept. Results are obtained for *A. brasilense* throughout growth.[8]

For *A. brasilense* harvested at different culture times, Fig. 3 presents relationships among N/C, (O=C)/C, and (C=O)/C ratios.[8] For amide functions (O=C—N), a 1:1:1 proportion is expected among these ratios. Although a good agreement is obtained between (O=C)/C and N/C (Fig. 3a), the (C=O)/C ratio is clearly in large excess compared to the amount attributable to amide functions (Fig. 3b, open symbols). However, after subtraction of the contribution of acetal function present in polysaccharides from the (C=O)/C ratio, a good agreement is obtained with the N/C ratio (Fig. 3b, closed symbols). For *A. brasilense* cells, XPS data are thus consistent with the presence of proteins and polysaccharides.

Data can be further worked out to give quantitative information on the molecular composition of the surface.[6,8] This requires one to make assumptions about the nature and composition of the major surface constituents. For microorganisms, three classes of basic constituents must be considered: proteins (Pr), polysaccharides (PS), and hydrocarbon-like compounds (HC), which refer here to lipids and other compounds that contain mainly carbon and hydrogen. The chemical composition of model constituents is given in Table I and corresponds to $C_6H_{10}O_5$ for polysaccharides and to CH_2 for hydrocarbon-like compounds. The amino acid composition taken for the surface proteins is that of the major outer membrane protein of *Pseudomonas fluorescens* OE 28.3.[11]

The parameters deduced by XPS can be expressed as a function of the

[11] R. De Mot, P. Proost, J. Van Damme, and J. Vanderleyden, *Mol. Gen. Genet.* **231,** 489 (1992).

TABLE I

CHEMICAL COMPOSITION OF MODEL CONSTITUENTS CONSIDERED FOR DEDUCTION OF MOLECULAR COMPOSITION OF GRAM-NEGATIVE BACTERIA

Constituent	$\underline{C}-(C,H)/C$	$\underline{C}-(O,N)/C$	$\underline{C}=O/C$	O/C	N/C	Carbon concentration (mmol/g)
Protein[a]	0.428	0.293	0.279	0.325	0.279	43.5
Polysaccharide	0.000	0.833	0.167	0.833	0.000	37.0
Hydrocarbon	1.000	0.000	0.000	0.000	0.000	71.4

[a] Computed from R. De Mot *et al., Mol. Gen. Genet.* **231,** 489 (1992).

concentration of the model constituents using two independent schemes.[6,8] The first scheme is based on the three components of the carbon peak:

$$[(\underline{C}=O)/C]_{obs} = 0.279\,(C_{Pr}/C) + 0.167\,(C_{PS}/C)$$
$$[(\underline{C}-(O,N))/C]_{obs} = 0.293\,(C_{Pr}/C) + 0.833\,(C_{PS}/C)$$
$$[(\underline{C}-(C,H))/C]_{obs} = 0.428\,(C_{Pr}/C) + 1\,(C_{HC}/C)$$

whereas the second scheme is based on elemental concentration ratios:

$$[N/C]_{obs} = 0.279\,(C_{Pr}/C)$$
$$[O/C]_{obs} = 0.325\,(C_{Pr}/C) + 0.833\,(C_{PS}/C)$$
$$[C/C]_{obs} = (C_{Pr}/C) + (C_{PS}/C) + (C_{HC}/C) = 1$$

Solving the systems of equations provides the proportion of carbon associated with each molecular constituent: (C_{Pr}/C), (C_{PS}/C), and (C_{HC}/C). These proportions can then be converted into weight fractions, using the carbon concentration of each constituent (Table I). Comparison of the molecular compositions obtained by the two independent modeling approaches is useful in assessing not only the consistency of the compositions considered for the model constituents, but also the accuracy of the sensitivity factors and peak decomposition.

Using the just-described approach, the surface molecular composition of *A. brasilense* was shown to vary significantly during growth: the surface protein content increased from 30% (exponential phase cells) to 50% (stationary phase cells), whereas the polysaccharide content decreased from 60 to 35%.

Case Study: Adhesion of *A. brasilense* to Model Substrata

XPS has been applied to gain a better understanding of the adhesion mechanisms of *A. brasilense* to inert surfaces. The surface chemical composition of free cells was determined during growth in complex medium (Figs. 2 and 3).[8] Interpretation of data in molecular terms revealed that the surface protein content increased during growth, concurrently with a decrease in the polysaccharide content. There was a direct correlation among cell surface protein concentration, cell surface hydrophobicity, and cell adhesiveness, which pointed to the involvement of cell surface proteins and hydrophobicity in adhesion.

To examine the nature of macromolecules produced at interfaces during adhesion (Fig. 1), the surface chemical composition of the following samples was determined: cell sediments, polystyrene substrata after cell adhesion and detachment of the adhering cells, and substrata after contact with the liquid phase of a cell suspension.[12,13] Figure 4 presents C_{1s} and O_{1s} XPS peaks of bare polystyrene (a), of cell sediments obtained after 24 hr contact time at 30° (b) and 4° (c), and of polystyrene after cell adhesion (exponential phase, 24 hr contact time at 30°) and subsequent detachment of adhering cells (d). Spectra obtained for polystyrene conditioned with the liquid phase of a cell suspension were similar to those presented in Fig. 4d. Peaks were decomposed as for free cells, except that in some cases a component was present in the carbon peak at about 291.4 eV, due to an energy loss attributed to a $\pi \rightarrow \pi^*$ transition characteristic of the aryl moiety in polystyrene (shake-up component). Figure 4d shows that after adhesion and detachment of the adhering cells, the surface composition of polystyrene was rich in oxygen, nitrogen, and oxidized carbon, reflecting the presence of biological compounds. Strikingly, the prevailing component of the O_{1s} peak was due to $\underline{O}{=}C$, reflecting proteins, whereas for free cells (Fig. 2) and cell sediments (Figs. 4b and 4c) it was due to $C{-}\underline{O}H + C{-}\underline{O}{-}C$, reflecting essentially polysaccharides.

The surface composition was modeled in terms of proteins, polysaccharides, and hydrocarbon-like compounds originating from both polystyrene and biomolecules.[12] For cell sediments prepared at 4°, the major constituent was polysaccharide as observed with free exponential phase cells. However, for cell sediments prepared at 30°, the protein surface concentration increased and became similar to that of polysaccharides, as observed for free cells in the stationary phase.

The surface composition of substrata examined either after adhesion and cell detachment or after conditioning with the liquid phase of a cell suspension was clearly different from that of free cells and of cell sediments

[12] Y. F. Dufrêne, H. Vermeiren, J. Vanderleyden, and P. G. Rouxhet, *Microbiology* **142,** 855 (1996).

[13] Y. F. Dufrêne, C. J. P. Boonaert, and P. G. Rouxhet, *Colloids Surfaces B Biointerfaces* **7,** 113 (1996).

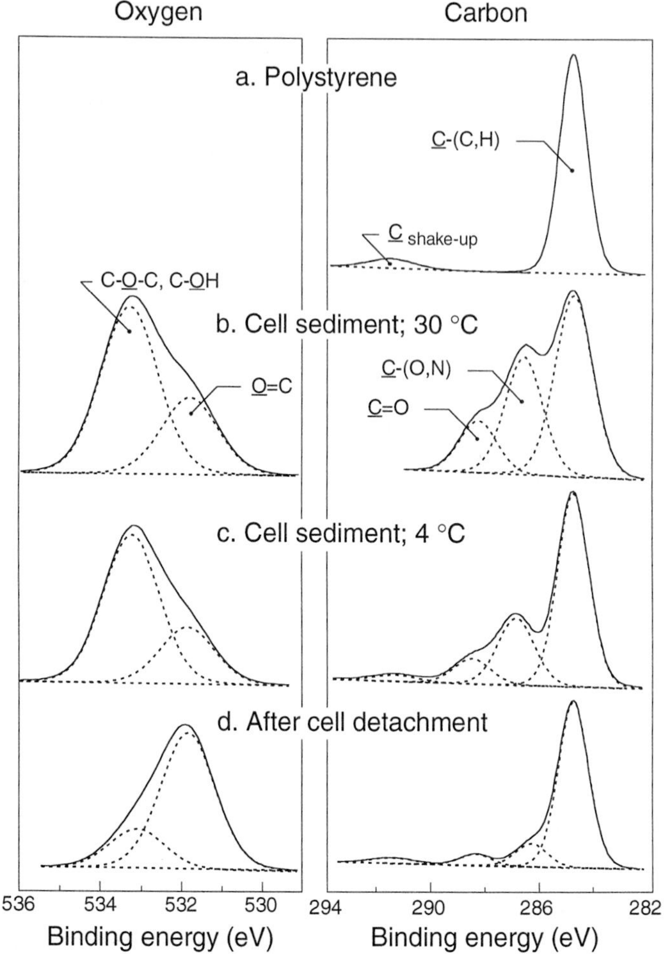

FIG. 4. Representative C_{1s} (right) and O_{1s} (left) peaks of bare polystyrene (a), of cell sediments (exponential phase cells, 24-hr contact time) prepared either at 30° (b) or at 4° (c) and of polystyrene after 24 hr contact time with exponential phase cells at 30° and detachment of the adhering cells (d).[12]

in that the protein content was much larger than the polysaccharide content. Lowering the contact time and performing adhesion under unfavorable metabolic conditions (4°) or in the presence of a protein synthesis inhibitor (tetracycline) resulted in a decrease of protein concentration at the substratum surface. Changes of the protein concentration, when varying experimental conditions, were directly correlated with differences in the density of cells adhering to polystyrene.

It thus appears that (i) for both cell sediments and free cells during growth, cell aging leads to an enrichment of the cell surface in proteins; (ii) proteins are the major constituent at the substratum surface after adhesion and cell detachment; and (iii) changes in experimental conditions affecting the protein concentration at the cell surface (growth phase, temperature) or at the substratum surface (time, temperature, addition of tetracycline) influence the adhesion density in the same direction, thereby providing evidence for the involvement of extracellular proteins in the adhesion to inert surfaces.

The role of extracellular proteins was investigated further by studying cell adhesion to polystyrene in a parallel plate chamber in the absence of flow, thus allowing cell transport by sedimentation.[13] The adhesion pattern on the bottom plate was found to be heterogeneous, and the density of adhering cells increased with the concentration of the suspension up to a plateau characterized by a low surface coverage (10%), indicating the influence of cell–cell interactions and of detachment of aggregates on rinsing.

Bottom and top plates were analyzed by XPS after the detachment of adhering cells. Analyses revealed the presence of proteins, the concentration of which was the same on the bottom and top plates. These results demonstrate that a direct contact between the cells and the substratum is not required for the accumulation of proteins at the substratum surface and that proteins are released progressively by the cells into the solution and adsorb at the substratum surface. The progressive release of extracellular proteins into the aqueous phase was demonstrated further by characterizing the supernatant of cell suspensions by UV-visible spectrophotometry and protein assay.

Substratum preconditioning by contact with a cell suspension and cell aging during 24 hr prior to the adhesion test, two treatments that cause an increase of the protein concentration at the cell–substratum interface, favored significantly adhesion after a short contact time (2 hr). Moreover, a prolonged contact (24 hr) between the cells and the substratum increased the adhesion density appreciably, pointing to the *in situ* excretion of anchoring proteins by the adhering cells.

It was therefore concluded that the role of proteins in the adhesion of *A. brasilense* to inert surfaces is twofold. In a first stage, proteins accumulate at the cell surface, are liberated into the solution, and adsorb onto the substratum; the increase of the protein concentration at the surface of cells and substratum promotes initial adhesion (Fig. 1, step 3). In a next stage, the prolonged contact between adhering cells and the substratum leads to the *in situ* excretion of proteins, which ensures cell anchorage to the substratum and cell–cell aggregation (Fig. 1, step 4). These results demonstrate that XPS is a powerful tool in microbial adhesion and biofilm studies in investigating the surface of cells and substrata as well as the composition of extracellular macromolecules at interfaces.

Applications and Relevant Methodologies

XPS may be applied to the surface of substrata, to surface active molecules in the liquid phase, and to the surfaces of cells for understanding better interfacial (cell–substratum, cell–cell) phenomena involved in bioadhesion and biofilm formation. The following section gives an overview of these applications and describes the relevant methodologies.

Surface of Substrata

Native Substrata. XPS can be used to determine the elemental and functional composition of the outermost surface of materials (polymers, metals, oxides) used as substrata in bioadhesion and biofilm studies. Analyses reveal that the surface composition of materials may be very different from that expected on the basis of bulk composition. It was shown that the surface of polyvinyl chloride is strongly depleted in chlorine and enriched in oxygen, presumably due to the migration of plasticizers from the bulk.[14] In a study of fungal spore immobilization to polycarbonate plates, no carbonate peak could be detected in the XPS carbon peak of the polycarbonate substrata; the carbon peak was similar to that of the protective foil that coated the substrata.[15] For metals and oxides, the surface is always contaminated by organic constituents. Hence, XPS is useful in comparing the surface composition of substrata with their bulk stoichiometry and in assessing the importance of surface contamination.

The use of XPS for studying materials is now well established.[4,9] For polymers, a detailed examination of the shape of the peaks provides a surface analysis in terms of functional groups.[16] For some polymers, a weak shake-up component is observed (Fig. 4a), typical of aromatic carbon and conjugated systems, which may be useful in assessing the spatial organization (thickness, surface coverage) of overlayers. Sensitivity factors can be either determined empirically using adequate standards or computed from photoionization cross sections, asymmetry parameters, inelastic mean free paths of photoelectrons, and analyzer transmission factors. For polymers,

[14] N. Mozes, F. Marchal, M. P. Hermesse, J. L. van Haecht, L. Reuliaux, A. J. Léonard, and P. G. Rouxhet, *Biotechnol. Bioeng.* **30,** 439 (1987).

[15] P. Gerin, M. N. Bellon-Fontaine, M. Asther, and P. G. Rouxhet, *Biotechnol. Bioeng.* **47,** 677 (1995).

[16] P. G. Rouxhet, A. Doren, J. L. Dewez, and O. Heuschling, *Progr. Organ. Coat.* **22,** 327 (1993).

a good agreement was found between the O/C ratios deduced with these two procedures and the ratios were close to values expected from bulk stoichiometry.[16]

Substrata after Surface Physicochemical Modification. Bioadhesion and biofilm formation may be influenced by modifying substratum surface properties. Electrostatic repulsion between cells and substrata can be reduced by modifying the substratum surface by various treatments, including coating with positively charged colloidal particles and adsorbing hydrolysable cations or polycations.[15,17,18]

Chemical modifications brought about by surface treatments may be determined by XPS. Changui *et al.*[18] studied the influence of oxidation by the sulfochromic mixture and of treatment by solutions of ferric nitrate on the surface properties of polycarbonate substrata. Changes of the surface composition detected by XPS were correlated with changes of hydrophobicity, electrical properties, and yeast cell adhesion. Furthermore, comparison of XPS analysis with chemical analysis revealed that iron was present at the surface in the form of particles of about 10 nm in size.

The surface composition of polycarbonate membranes modified by the sulfochromic mixture, nitration, oxygen and ammonia radio frequency glow discharge, and corona discharge was determined.[19] Although all treatments led to an increase of oxygen concentration, the presence of sulfur- and nitrogen-containing groups bound to the surface was demonstrated after application of the corresponding treatment. Analysis in terms of functional groups revealed that the decrease of hydrophobicity was related to the surface concentration of the fraction of oxygen bearing a high electron density.

Conditioned Substrata. The adsorption of conditioning films on substrata is a key step in the formation of biofilms (Fig. 1); therefore, knowledge of both their chemical composition and their spatial organization is of prime importance.

A first application of XPS is to examine the material remaining on the substratum in the form of "footprints"[20] after removal of the adhering cells. This was first demonstrated with *A. brasilense*,[12] where the surface composition of polystyrene substrata after cell detachment was found to be rich in oxygen, nitrogen, and oxidized carbon, reflecting the presence of proteins (Fig. 4d). The procedure of sample preparation involves pouring

[17] P. G. Rouxhet and N. Mozes, *Wat. Sci. Tech.* **22,** 1 (1990).

[18] C. Changui, A. Doren, W. E. E. Stone, N. Mozes, and P. G. Rouxhet, *J. Chim. Phys.* **84,** 275 (1987).

[19] J. L. Dewez, A. Doren, Y. J. Schneider, R. Legras, and P. G. Rouxhet, *Surf. Interface Anal.* **17,** 499 (1991).

[20] K. C. Marshall, R. Stout, and R. Mitchell, *J. Gen. Microbiol.* **68,** 337 (1971).

a cell suspension into a recipient containing the substratum, leaving the suspension undisturbed for a given time and temperature, and rinsing and detaching the adhering cells by ultrasonication in demineralized water until optical microscope observation ensures that all cells are detached.

Another application consists in investigating substrata after conditioning by the adsorption of macromolecules. After adhesion of *A. brasilense* in a parallel plate chamber,[13] the protein concentration determined by XPS was the same on the top and bottom plates, indicating that proteins adsorb at the substratum surface from the solution. The effect of preconditioning polystyrene substratum with milk constituents on the adhesion of *Lactococcus lactis* was investigated.[21] Proteins appeared to be the most abundant constituent on substrata conditioned with skimmed milk, caseinate, and whey protein isolate, whereas proteins, polysaccharides, and hydrocarbon-like compounds were present in similar proportions on substrata conditioned with whey permeate. Large variations of XPS data were observed when considering independent experiments, which were attributed primarily to the effect of drying performed prior to the analyses. No correlation was found between the gross molecular composition of the adsorbed films and the cell adhesion behavior, as only preconditioning with skimmed milk and whey permeate promoted adhesion.

Variation of the angle of photoelectron collection ("angularly resolved XPS" or ARXPS) allows to probe different depths at a sample surface and to provide information on the spatial organization of the near surface.[22] Using this approach, it was shown that the surface composition of polystyrene substrata after the detachment of *A. brasilense* cells is consistent with the presence of a discontinuous proteinaceous layer of 3 nm thickness, covering about 40% of the surface.[13]

Conditioning substrata by contact with adhering cells or by adsorption can be performed either in a recipient (petri dish, beaker)[12] or in a parallel plate chamber,[13,23] the latter being recommended because it offers a much better control of hydrodynamic conditions. XPS analysis requires the samples to be dehydrated before introduction into the spectrometer, which is under high vacuum. Thus, it also requires preliminary rinsing in order to avoid surface deposition of the substances present in the excess solution. Rinsing may provoke some desorption; however, operating quickly should avoid the removal of macromolecules, the adsorption of which is often highly irreversible. Drying is often performed in room conditions by holding

[21] C. J. P. Boonaert and P. G. Rouxhet, *Med. Fac. Landbouww. Univ. Gent* **63,** 1147 (1998).

[22] B. D. Ratner, T. A. Horbett, D. Shuttleworth, and H. R. Thomas, *J. Colloid Interface Sci.* **83,** 630 (1981).

[23] J. Sjollema, H. J. Busscher, and A. H. Weerkamp, *J. Microbiol. Methods* **9,** 73 (1989).

the sample vertically to allow drainage of the liquid film or by flushing the sample with a gentle nitrogen flow. In both cases, shear forces associated with the dewetting process may cause profound rearrangements of the adsorbed film. Using atomic force microscopy (AFM) in combination with XPS, air drying was shown to provoke marked reorganization of adsorbed protein films, thereby affecting both surface coverage and film thickness.[24,25] Similar observations were made with adsorbed phospholipid layers (unpublished data). As a result, the information provided by XPS on air-dried samples may be not representative of the *in vitro,* hydrated state, which is relevant to bioadhesion. Freeze drying the sample is therefore recommended to minimize reorganization of the adsorbed film. To this end, the wet substratum is immersed in liquid nitrogen immediately after rinsing, while avoiding dewetting, and freeze-dried according to a procedure described later.

Surface of Cells

Suspended Cells. During the last decade, XPS has been applied to study the surface of a variety of microbial cells, including bacteria, yeast cells, and filamentous fungi.[5,6,8,26–31] Representative carbon and oxygen peaks obtained on bacterial cells are presented in Fig. 2. It was shown earlier that peak decomposition provides information on the functional groups present at the surface and that XPS data can be converted into concentrations of model molecular constituents. Despite the fact that analyses are performed under high vacuum, the relevance of XPS to the probing of microbial surfaces has been supported by correlations between XPS data, on the one hand, and other interfacial properties (hydrophobicity, electrical properties, and cell behavior at interfaces)[6,8,32] or biochemical data obtained on isolated cell walls,[31] on the other hand.

[24] C. C. Dupont-Gillain, B. Nysten, and P. G. Rouxhet, *Polymer Int.* **48,** 271 (1999).

[25] Y. F. Dufrêne, T. G. Marchal, and P. G. Rouxhet, *Appl. Surf. Sci.* **144–145,** 638 (1999).

[26] D. E. Amory, N. Mozes, M. P. Hermesse, A. J. Léonard, and P. G. Rouxhet, *FEMS Microbiol. Lett.* **49,** 107 (1988).

[27] H. C. van der Mei, A. J. Léonard, A. H. Weerkamp, P. G. Rouxhet, and H. J. Busscher, *J. Bacteriol.* **170,** 2462 (1988).

[28] H. C. van der Mei, M. M. Cowan, M. J. Genet, P. G. Rouxhet, and H. J. Busscher, *Can. J. Microbiol.* **38,** 1033 (1992).

[29] P. A. Gerin, Y. Dufrêne, M. N. Bellon-Fontaine, M. Asther, and P. G. Rouxhet, *J. Bacteriol.* **175,** 5135 (1993).

[30] P. B. Dengis and P. G. Rouxhet, *Yeast* **13,** 931 (1997).

[31] Y. F. Dufrêne, A. van der Wal, W. Norde, and P. G. Rouxhet, *J. Bacteriol.* **179,** 1023 (1997).

[32] H. C. van der Mei, P. Brokke, J. Dankert, J. Feijen, P. G. Rouxhet, and H. J. Busscher, *Appl. Environ. Microbiol.* **55,** 2806 (1989).

The high vacuum imposed for the analysis requires preliminary dehydration of the samples. Protocols of sample preparation and analysis have been established, and a series of precautions aiming at minimizing contamination and rearrangement of the surface molecules have been described.[6,33,34] The recommended procedure involves washing the cells at least once and pouring a concentrated suspension into a vial precooled at liquid nitrogen temperature. The cooling rate obtained with this procedure is high enough to prevent migration of intracellular material to the surface. The freeze dryer shelf is precooled at −50°. The temperature is maintained at −50° for 3 hr, raised progressively to −5° in about 15 hr, maintained at −5° for 6–12 hr, and finally raised to 25°. Due to the risk of surface oxidation, the freeze-dried samples should not be stored for a prolonged period before analysis. XPS results are not affected significantly by details of the preparation procedure, which provides a satisfactory precision of the analysis and supports the fact that the analyzed surface is not altered by the migration of intracellular material. However, it must be kept in mind that the representativity of the analyzed surface with respect to the native surface in the hydrated state may be limited due to the removal of cell surface macromolecules or appendages during washing or to the rearrangement of the polymer network at the surface. Further details about the sample preparation procedure for XPS analysis of microorganisms can be found in the literature.[33,34]

Biofilms. The first attempt to apply XPS directly to the surface of biofilms has been described with *A. brasilense*[12]: a significant enrichment in proteins was detected for the surface of biofilms after 24 hr at 30° (Fig. 4b), suggesting the possible involvement of extracellular proteins in biofilm formation.

The recommended procedure for analyzing biofilms is as follows: the substratum carrying the biofilm (about 1 × 1 cm) is removed gently from the recipient, while avoiding dewetting; it is then immersed immediately in liquid nitrogen and freeze-dried according to the same procedure as for suspended cells.

Concluding Remarks

XPS can be applied to analyze the chemical composition of substratum surfaces, microbial cell surfaces, and surface active molecules in the liquid phase in terms of elemental, functional, and, to a certain extent, molecular composition. Correlations among XPS data, surface properties (hydropho-

[33] D. E. Amory, M. J. Genet, and P. G. Rouxhet, *Surface Interface Anal.* **11,** 478 (1988).
[34] P. B. Dengis and P. G. Rouxhet, *J. Microbiol. Methods* **26,** 171 (1996).

bicity, electrical properties), and interfacial behaviors provide a better understanding of bioadhesion and biofilm formation.

The high vacuum required for the analysis implies that samples must be dehydrated. For both adsorbed conditioning films and cells, the recommended procedure is freeze drying under controlled conditions, as air drying may cause reorganization of adsorbed films and migration of intracellular constituents, respectively. Although freeze drying is a critical step, it can be performed in a reproducible way, and the numerous relationships obtained between XPS data and interfacial properties indicate that the results are relevant to the native, hydrated state. However, for some microorganisms the representativity of the analyzed surface may be limited due to the removal of cell surface macromolecules or appendages during washing or to the rearrangement of the polymer network at the surface.

It should be kept in mind that XPS is a global technique supplying data on the overall surface with limited lateral resolution. Therefore, the potential of XPS is clearly amplified when combined with techniques capable of probing surfaces at high spatial resolution. Exciting new possibilities are now offered by AFM to investigate, *in situ,* the physicochemical properties of biosurfaces on the nanometer scale. With respect to conditioning films, AFM confirms the continuous or discontinuous character of the adsorbed film deduced from XPS data and reveals topographic features at a supramolecular scale (fibrillar structures, aggregates). A challenging goal for future research is to map the local surface properties (hydrophobicity, charged groups, distribution of surface macromolecules) of microorganisms and biofilms in physiological conditions.

Acknowledgments

The authors thank N. Mozes for fruitful discussion and M. J. Genet for pertinent discussion and technical assistance. The support of the National Foundation for Scientific Research (FNRS), of the Foundation for Training in Industrial and Agricultural Research (FRIA), and of the Federal Office for Scientific, Technical and Cultural Affairs (Interuniversity Poles of Attraction Program) is gratefully acknowledged.

Section VI

Physiology of Biofilm-Associated Microorganisms

[29] Evaluating Biofilm Activity in Response to Mass Transfer-Limited Bioavailability of Sorbed Nutrients

By RYAN N. JORDAN

Introduction

The rate of substrate metabolism (r, mole/liter-time), as described by the classical Michaelis–Menten model, depends on the concentration of the limiting nutrient (S_{aq}, mole/liter) that governs the rate of enzymatic reaction:

$$r = \frac{r_{max} S_{aq}}{K_s + S_{aq}} \tag{1}$$

In a well-mixed aqueous culture, S_{aq} is simply the nutrient concentration in the bulk liquid (aqueous) phase. It is commonly assumed that bacteria can only metabolize aqueous phase species. Thus, the aqueous phase shall be referred to as the *directly bioavailable* phase.

However, if the limiting nutrient exhibits a high degree of surface activity, such that it tends to associate with a solid–water interface, distribution of the nutrient between solid and aqueous phases might be described by a simple linear partitioning process:

$$K_p = \frac{S_s}{S_{aq}} \tag{2}$$

where S_s represents the sorbed nutrient concentration (moles of nutrient sorbed per unit area of solid–aqueous interface). Thus, sorption of hydrophobic compounds (i.e., those having a high partitioning coefficient K_p) tends to reduce their concentration in the directly bioavailable phase (i.e., S_{aq}). More sorption may reduce S_{aq} to the extent that Michaelis–Menten kinetics [Eq. (1)] are reduced to simple, first-order kinetics (by assuming that $K_s \gg S_{aq}$):

$$r = k_b S_{aq} \tag{3}$$

where k_b is a first-order biotransformation rate coefficient (1/time).

If the partitioning process [Eq. (2)] is "rapid" relative to the biotransformation process [Eq. (3)], then S_{aq} will govern the rate of biotransformation. This special case is referred to as *concentration-limited bioavailability.* However, if partitioning is "slower" than biotransformation, then metabolism will occur at a rate that is faster than the rate at which the directly bioavail-

0076-6879/99 $30.00

able pool can be replenished. Consequently, biotransformation proceeds at progressively slower rates until S_{aq} falls below the minimum substrate concentration required for metabolism (i.e., "S_{min}"). Slow mass transfer from sorbed to aqueous phases can often be approximated by a first-order mass transfer process:

$$\frac{dS_{aq}}{dt} = k_m(S_s - K_p S_{aq}) \tag{4}$$

where k_m is defined as the *first-order mass transfer rate coefficient* (1/time).

For the simple series desorption–biotransformation model described by Eqs. (2)–(4), the characteristic times for mass transfer and biotransformation are defined by Bouwer *et al.*[1]:

$$T_m = \frac{1}{k_m K_p R_{s/w}} \tag{5}$$

$$T_b = \frac{1}{k_b} \tag{6}$$

where $R_{s/w}$ is defined as the surface area of solid sorbent per unit volume of water in the system (i.e., the "solid-to-water" ratio, having units of area/volume).

The characteristic times defined by Eqs. (5) and (6) describe the *process time scales* for the mass transfer and biotransformation processes, respectively. Thus, concentration-limited bioavailability as described earlier (i.e., mass transfer is "rapid" relative to biotransformation) occurs when $T_m \ll T_b$. For $T_m \ll T_b$, bioavailability is *mass transfer limited* (i.e., mass transfer is "slow" relative to biotransformation), indicating that the rate of biotransformation is governed by the rate of replenishment (i.e., solid-to-aqueous phase mass transfer) of the directly bioavailable nutrient pool (i.e., the aqueous phase).

Previous studies have addressed both concentration- and mass-transfer limited bioavailability in solid–aqueous systems (e.g., soil slurries) where the substrate concentration was monitored,[1–4] but few studies[5] have addressed quantification of the growth of the microorganisms responsible for

[1] E. J. Bouwer, W. Zhang, L. P. Wilson, and N. D. Durant, *in* "Soil and Aquifer Remediation: Non-Aqueous Phase Liquids—Contamination and Reclamation" (H. Rubin, N. Narkis, and J. Carberry, eds.). Springer-Verlag, Heidelberg, 1997.

[2] T. N. P. Bosma, P. J. M. Middeldorp, G. Schraa, and A. J. B. Zehnder, *Environ. Sci. Technol.* **31,** 248 (1997).

[3] L. M. Carmichael, R. F. Christman, and F. K. Pfaender, *Environ. Sci. Technol.* **31,** 126 (1997).

[4] C. Fu, S. Pfanstiel, C. Gao, X. Yan, and R. Govind, *Environ. Sci. Technol.* **30,** 743 (1996).

[5] H. Harms, *Appl. Environ. Microbiol.* **62,** 2286 (1996).

substrate biotransformation. Further, most of the methods used in these studies were performed in batch systems, making it difficult to interpret the separate contributions of concentration and mass transfer processes to bioavailability limitations without extensive model calibration.

Thus, the objective of the methodology described herein is to evaluate biofilm activity in response to mass transfer limited bioavailability in a flowing system. Of key importance to the method are design parameters that govern mass transfer rates and reactor flows. System design will be based on an engineering process approach to modeling the mass transfer-biotransformation process kinetics as described earlier.

Method Description

While previous methods applied to bioavailability-limited systems have addressed the fate of the substrate, the goal of this method is to establish a biofilm on a surface where the only available substrate is present in the sorbed phase at the solid–water interface. Thus, this method is designed to investigate the fate of the biofilm in a bioavailability-limited system.

In general, an appropriate solid medium (i.e., a coupon or some other suitable substratum for biofilm growth, not to be confused with *microbiological* media used to provide nutrition for cell metabolism) is selected or manufactured such that sorption of the desired limiting nutrient at the solid–water interface is encouraged. The coupon is placed in a reactor, inoculated with a selected microbial culture, and exposed to a flowing nutrient stream for a predetermined period of time. Depending on the reactor configuration, the coupon is observed nondestructively (e.g., via microscopic techniques) or sampled destructively. (Some of these techniques are addressed later.)

As one example of applying this methodology, a system in which the desorption-limited bioavailability of soil-sorbed hydrophobic organic compounds (HOCs) governs biofilm activity will be described. However, this methodology is certainly applicable to monitoring biofilm growth on a variety of surfaces harboring low-solubility chemicals. Some applications might include assessing the efficacy of engineered surfaces (e.g., biomaterial-coated medical implants) with embedded antimicrobial agents or monitoring biofilm activity in response to growth on insoluble substrates (e.g., chitin).

Reactor Systems

A variety of flowing reactor systems exist for growing biofilms on solid media. Foremost among these systems are porous media (e.g., packed column) reactors, parallel plate (e.g., flow cell) reactors, and other coupon-based reactors (e.g., the Robbins device or the "rotating disk reactor" described by Zelver *et al.*[6]). Any of these systems could be incorporated into this method. Selection of the reactor system will depend primarily on limitations in the physical configuration of the coupon and the types of data to be collected from the system.

For the purpose of illustrating this method, a parallel plate (flow cell) coupon reactor will be described (Fig. 1). The flow cell is approximately 8 (length) by 6 (width) by 1 (depth) cm. The flow cell housing consists of two machined parallel plates (Fig. 1, A and B). A 1-mm (depth) flow channel [Fig. 1, C, 4 (length) by 2 (width) cm] is machined into a polycarbonate bottom plate (Fig. 1, A) having dimensions of 8 (length) by 6 (width) by 0.6 (thickness) cm. In turn, a 1.25 (diameter) by 0.5 (depth)-cm hole is drilled into the center of the flow channel to accommodate the biofilm coupon (Fig. 1, D). Pipe channels (Fig. 1, E and F, 3 mm diameter) are drilled in appropriate locations to introduce laminar flow and to allow effluent to exit the reactor. Inserted into the pipe channels on each end of the plate are sections of 3-mm (outside diameter) tubing (e.g., stainless steel). The top plate (Fig. 1, B) of the housing consists of an 8 (length) by 6 (width) by 0.3 (thickness)-cm sheet of aluminum or stainless steel. An observation window (Fig. 1, G) having dimensions large enough to accommodate a microscope objective [e.g., 4 (length) by 3 (width) cm] is cut from the center of the top plate. After the coupon is inserted into the flow cell, a gasket (Fig. 1, H, i.e., cut from a sheet of Viton rubber having the same dimensions as the top plate) and a glass coverslip (170 μm thickness), used as the observation window pane (Fig. 1, I), are sandwiched between the plates. The plates are held secure by a series of bolts with thumbscrews through holes (Fig. 1, J) drilled through the top plate, gasket, and bottom plate.

The biofilm coupon consists of a 1.25 (diameter) by 0.5 (thickness)-cm machined glass cylinder that is used as a substratum for depositing a reactive surface as described in the next section. (It should be emphasized that the choice of material used for the coupon is entirely up to the researcher—consideration must be given to the type of reactive surface being manufactured.) The flow cell is placed in-line with a nutrient delivery system consisting of a nutrient feed vessel, peristaltic pump, and waste collection container. Alternatively, the flow cell can be operated in a continuously recirculating batch mode, in which the waste stream simply feeds into the

[6] N. Zelver, M. Hamilton, B. Pitts, D. Goeres, D. Walker, P. Sturman, and J. Heersink, *Methods Enzymol.* **310** [45] 1999 (this volume).

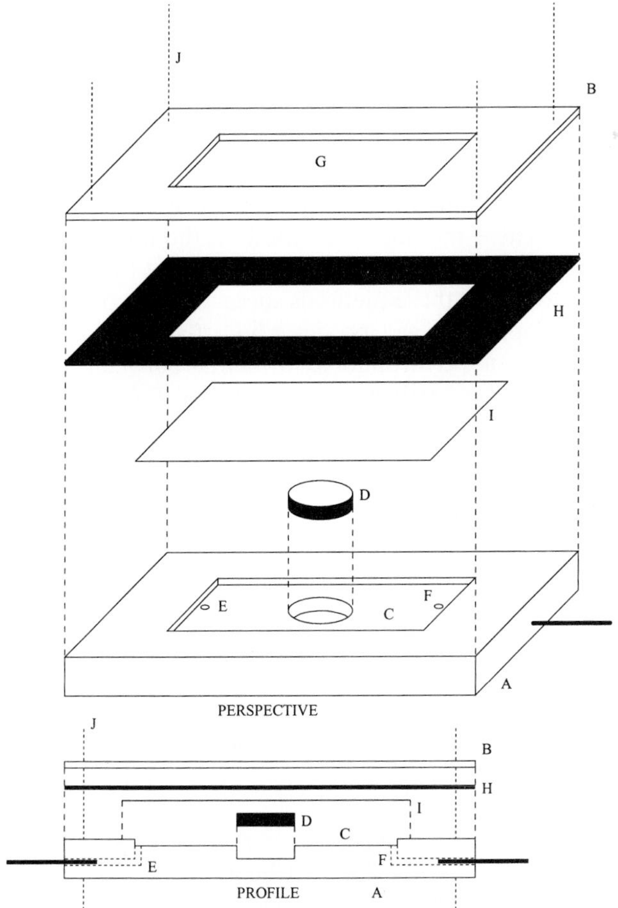

FIG. 1. Flow cell reactor configuration. See text for part descriptions.

inlet, with a pump placed in-line. This recirculating mode is quite helpful for initial inoculation.

Biofilm Coupon Manufacture

Although an engineered surface that mimics the chemical and physical complexity of a natural soil particle does not exist, a variety of analogs can be imagined. The most simple of these analogs is SiO_2 (glass). While SiO_2 is often the most dominant mineral in soils, it is an unrealistic model for recreating the chemical reactions that dominate the sorption of hydrophobic organic chemicals in the subsurface. Key among these sorption processes is the partitioning of HOCs with natural organic matter (NOM)[7,8] Further, NOM is often associated with the highly reactive mineral surfaces of iron oxides such as ferrihydrite, goethite, and hematite. Thus, manufacture of a coupon having a composite iron oxide (Fe_2O_3, hematite) and NOM (humic acid) sorbent phase is described.

A number of methods for depositing iron oxide on a silicon dioxide surface exist.[9,10] These methods are based on the heat-induced growth of α-FeOOH (goethite) from Fe nuclei present at the SiO_2–water interface. Although we have used these methods successfully in our own laboratory, a much simpler approach will be described that takes advantage of the strong electrostatic and hydrophobic attraction between the SiO_2 surface and Fe_2O_3 colloids during thermal dehydration. Following deposition of the iron oxide, the coupon is soaked in a humic acid solution and rinsed, leaving behind only humic acid molecules that are bound tightly to the iron oxide surface functionalities.[11] Finally, the oxide/NOM coupon is aged in an aqueous solution containing a selected HOC and rinsed again, leaving only strongly bound HOC. The result of this procedure is a coupon having an HOC-contaminated oxide/NOM surface that is chemically analogous to those found in HOC-contaminated soils. The procedure is outlined as follows.

Coupon Preparation. Select a glass coupon having dimensions appropriate for incorporation into the flow cell. Clean the coupon by soaking in chloroform (1 hr), 6 *M* H_2SO_4 (1 hr), rinsing with deionized water, and baking in a foil-covered glass beaker at 550° for 6 hr to oxidize residual organic contaminants.

Iron Oxide Deposition. Using a pipetter, place a 100-μl drop of a slurry containing 10 g/liter of Fe_2O_3 (Fisher Chemicals, Pittsburgh, PA; 99%) in deionized water in the center of the coupon. Using the pipetter tip, gently spread the drop so that it covers the entire coupon surface. Bake the coupon in a foil-covered glass beaker at 550° for 6 hr to dehydrate the oxide layer and remove organic contaminants in the oxide matrix.

NOM Deposition. Soak the coupon in a gently stirred solution containing 100 mg/liter of humic acid for 24 hr. Well-characterized and purified humic materials isolated from surface waters, soils, and peat can be obtained

[7] J. J. Pignatello and B. Xing, *Environ. Sci. Technol.* **30,** 1 (1996).

[8] J. J. Piatt and M. L. Brusseau, *Environ. Sci. Technol.* **32,** 1604 (1998).

[9] P. C. Rieke, B. D. Marsh, L. L. Wood, B. J. Tarasevich, J. Liu, L. Song, and G. E. Fryxell, *Langmuir* **11,** 318 (1995).

[10] Y. Chang and M. M. Benjamin, *J. Am. Water Works Assoc.* **89,** 100 (1997).

[11] E. M. Murphy, J. M. Zachara, S. C. Smith, J. L. Phillips, and T. W. Wietsma, *Environ. Sci. Technol.* **28,** 1291 (1994).

from the International Humic Substances Society.[12] Rinse the coupon in three successive solutions (1 hr per rinse) of gently stirred sterile buffer solution (SBS, 0.05*M* HCO_3^- buffer adjusted to pH 7 with HCl).

HOC Aging. HOC bioavailability is largely governed by its aging time with NOM[3,13] Thus, the researcher should select an aging technique with some caution. One simple technique involves soaking the coupon in a gently stirred solution of SBS (as described in the previous step) that contains a selected concentration of HOC (e.g., 1 mg/liter phenanthrene). Aging times ranging from 15 min to several months are suggested. After the aging procedure, the coupon should be rinsed in SBS for 1 hr.

Coupon Storage. Coupons should be used as soon as possible following the surface deposition process to minimize biological and/or chemical degradation of the organic constituents. Careful consideration must be given to coupon storage conditions, including atmospheric moisture, redox conditions, and preservation techniques. Selection of coupon storage criteria will be based largely on the chemical nature of the sorbent surface. At a bare minimum, coupons should be stored at 4°. For the oxide/NOM coupons described herein, it is recommended that they be placed into the reactor and used immediately following manufacture. Under no circumstances should the finished coupons be dried and exposed to the atmosphere so as to avoid chemical degradation of the organic compounds on the surface.

Reactor Start-Up and Inoculation

The objective of the inoculation step is to promote the initial adhesion of cells to the coupon surface. These cells will constitute the population base for further biofilm growth. To maintain repeatability of adhesion during inoculation, the cells should initially be grown in a stable chemostat under exponential-growth dilution rates as opposed to batch culture. (Batch cultures contain cells of different "ages" or growth rates, and consequently, they may possess different membrane characteristics that could contribute to differential adhesion behavior.) The cells should be centrifuged and washed (in a sterile salt buffer) to remove traces of growth media constituents that could interact with the coupon and provide nutrient sources, preventing the coupon-sorbed nutrient from limiting biofilm activity. The washed cells are then suspended in SBS at a density appropriate for inoculation (e.g., 10^6 cells/ml).

[12] http://www.gatech.edu/ihss/

[13] R. G. Luthy, G. R. Aiken, M. L. Brusseau, S. D. Cunningham, P. M. Gschwend, J. J. Pignatello, M. Reinhard, S. J. Traina, W. J. Weber, Jr., and J. C. Westall, *Environ. Sci. Technol.* **31,** 3341 (1997).

The simplest method for inoculating the reactor is to recirculate the resulting cell suspension in the flow cell for a predetermined period of time. For example, using the coupons described herein and a culture of *Pseudomonas aeruginosa* (grown in a chemostat with a 5-hr dilution rate on 300 mg/liter of LB media), an inoculation period of 5–15 min is usually sufficient to achieve a 5–10% surface coverage of cells on the coupon.

Reactor Operation

After the inoculation period, the flow configuration should be switched from batch recirculating mode to a once-through flow mode. Selection of the flow rate through the reactor (and the resulting fluid residence time) is the most important operating parameter when evaluating biofilm growth on desorbing nutrients. Thus, particular attention will be paid to this issue.

In addition to the characteristic times for desorption (mass transfer) and biotransformation defined by Eqs. (5) and (6), one must also consider the characteristic time for transport in the reactor (i.e., the fluid residence time). If the flow rate (Q, volume/time) and the reactor volume (V) are known, then the fluid residence time is defined as

$$T_Q = \frac{V}{Q} \tag{7}$$

Process dynamics (hydraulics, biotransformation, and mass transfer) in the reactor must satisfy the following criteria: (1) $T_Q \ll T_b$ and (2) $T_Q \ll T_m$.

Satisfaction of condition (1) is required to prevent the accumulation of active planktonic organisms in the bulk liquid. Condition (2) must be satisfied so any substrate that is desorbed (and not retained by the biofilm) will be washed rapidly out of the reactor. Satisfaction of condition (2) is the key operating principle of this design: it ensures that the biofilm has access *only* to that fraction of the limiting nutrient that is either (1) sorbed at the solid–water interface or (2) retained in the biofilm matrix. Consequently, this design is ideal for examining the effect of unique strategies that bacteria employ to facilitate desorption (e.g., by biosurfactant production) or otherwise increase the bioavailability of sorbed nutrients.

It should be noted that these conditions apply only for the case of mass transfer-limited bioavailability (i.e., $T_b \ll T_m$). If the reactor environment satisfies criteria (1) and (2) listed earlier, but bioavailability is concentration limited (i.e., $T_b \gg T_m$), then the substrate will simply desorb and be washed out of the reactor before the biofilm has the opportunity to metabolize it.

Order-of-magnitude estimates of T_b, T_m, and T_Q can be used to invoke the just-described analysis. Estimation of T_b can be performed in a traditional biokinetics experiment (e.g., planktonic batch culture or chemostat) by fitting experimental substrate degradation kinetics data to Eq. (3). T_m can be estimated in an abiotic desorption experiment in which the surface harboring the sorbed nutrient is placed in a batch vessel containing nutrient-free buffer solution. The concentration of the nutrient in the aqueous phase is then monitored over time and data are fitted to Eq. (4) to determine k_m. At equilibrium, the resulting values of S_{aq} and S_s can be used to compute K_p from Eq. (2). Finally, estimation of T_Q is determined by monitoring the flow rate through the reactor by measuring the amount of fluid collected from the effluent in a given period of time. These approaches are somewhat general; thus, it is advised to review the literature to identify more specific protocols for assessing process kinetics that are appropriate to the experimental system under investigation.

Nutrient Feed

The composition of the nutrient feed stream must be considered carefully. If it is assumed that the limiting nutrient will be initially present in the coupon-sorbed state (as is the HOC in this example), then this nutrient concentration must be zero in the feed. Further, to promote bacterial metabolism, the feed must be supplemented with other nutrients required for growth (nitrogen, phosphorus, trace metals, etc.). In the example presented herein, the sole source of carbon for energy and growth is the HOC sorbed to the coupon surface. A supplemental mineral salts medium with no carbon is then used as the nutrient feed.

Coupon Analysis

The methodology culminates, of course, with analysis of the biofilm. A variety of approaches can be employed and are limited only by the imagination of the researcher and the analytical tools available. Two approaches compatible with the flow cell methodology described herein are summarized briefly.

Real-Time Microscopy. The flow cell can be designed in such a way that it can be mounted onto a microscope stage with an observation window large enough to accommodate the desired objective. Consequently, the user brings the stage up to the objective and can view the biofilm during the experiment in real time. Combined with organisms that exhibit bioluminescence or fluorescence (e.g., green fluorescent protein, GFP), both light and epifluorescent techniques can be used to identify the biofilm while minimizing emission interference from the coupon surface. A word of caution is in order for examining biofilms attached to the coupon housed in the lower plate of the flow cell: the objective working distance must be at least as long as the observation window thickness (e.g., 170 μm) plus the vertical height of the flow channel (e.g., 1000 μm) so that the focal plane of the biofilm is within the working range of the objective. While objectives having working distances up to 2 mm are not uncommon, they are not usually standard issue in the 60–100× magnifications required for the useful observation of intact biofilms. Thus, it may be necessary to specify a microscope objective with an "ultralong" working distance from the microscope's manufacturer.

Biofilm Quantitation. It is not the objective of this article to address the art of quantifying (counting) biofilm bacteria. However, a variety of techniques may be employed by removing the coupon from the reactor at the end of the experiment and performing a desired quantitative analysis. These techniques span the range of sophistication from traditional "scrape-and-plate" methods to more elaborate molecular biology techniques.

Conclusion

Methods for growing and monitoring biofilms are not new. However, few methods address the growth and analysis of biofilms under flowing conditions where biofilm activity is controlled by the bioavailability of nutrients localized at a solid–water interface. The objective of the method described herein is to provide a standard protocol based on sound engineering principles that can be used to further investigate biofilm microbiology in these systems. Thus, a reactor-based methodology is introduced that affords the researcher a maximum amount of flexibility in reactor selection, coupon manufacture, and coupon analysis. It is expected that implementation of this methodology will provide particular insight into the physiology of microorganisms possessing unique nutrient scavenging strategies (e.g., biosurfactant production) that facilitate the bioavailability of sorbed nutrients. We are also investigating the applicability of the methodology and process analysis approach described herein to determine the bioavailability of surface-embedded antimicrobial agents in an effort to evaluate the efficacy of engineered antimicrobial surfaces.

Acknowledgments

The author acknowledges support from the National Science Foundation Engineering Research Centers Program (Cooperative Agreement EEC-8907039). This work resulted from a cooperative effort between the Center for Biofilm Engineering at Montana State University and the Pacific Northwest National Laboratory via the U. S. Department of Energy Office of Biological Research (Environmental Technology Partnerships Program). Special thanks are also addressed to Jessica Metcalf, a visiting undergraduate fellow sponsored by the National Science Foundation Research Experience for Undergraduates (REU) Program for contributions to the development of the methodology described herein.

[30] Bacterial Survival in Biofilms: Probes for Exopolysaccharide and Its Hydrolysis, and Measurements of Intra- and Interphase Mass Fluxes

By RONALD WEINER, E. SEAGREN, C. ARNOSTI, and E. QUINTERO

The biofilm is a complex ecosystem. Depending on the position of the organism in the film, levels of nutrients, waste products, predation, gases, toxins, degradative enzymes, and other factors can vary greatly.[1] The microorganism, however, is not without defenses. It is often constructive to consider factors that contribute to survival in biofilms whether it is to assess the efficacy of a medical treatment, the remediation rates in bioreactors, or the effects of antifouling measures. Briefly, these defenses come under the headings of cell orientation,[2] cell structures (e.g., S-layers), exopolymers, genetic variation (e.g., ability to regulate capsule deposition), and metabolic versatility. In other words, factors influencing survivability include the biofilm architecture itself, symbiosis, colonial characteristics, cellular fine structure, secretion of polymers, and, of course, physiology, including metabolism.

Two of the most important factors that dictate survivability in biofilms are obviously the ability to adhere there (form a matrix) and the ability to utilize available nutrients in the matrix. There have been several reports covering novel and emerging approaches to assess cell viability,[3] particularly in the biofilm population,[4,5] and the subject is discussed elsewhere in this volume. This article covers methods that probe the biofilm matrix, which determine whether its complex polysaccharides are being utilized and examine nutrient fluxes there.

The biofilm matrix is composed of substantial quantities of capsular exopolymeric substances (EPS), which in turn normally contain substantial quantities of capsular polysaccharide (CPS). The survival value of capsular

[1] K. C. Marshall, *ASM News* **58,** 202 (1992).

[2] S. L. Langille and R. M. Weiner, *Appl. Environ. Microbiol.* **64,** 2906 (1998).

[3] R. S. Burlage, *in* "Manual of Environmental Microbiology" (C. J. Hurst *et al.,* eds.), p. 115. ASM Press, Washington, DC, 1997.

[4] C. G. Rodriguez, D. Phipps, K. Ishiguro, and H. F. Ridgway, *Appl. Environ. Microbiol.* **58,** 1801 (1992).

[5] L. K. Poulson, G. Ballard, and D. A. Stahl, *Appl. Environ. Microbiol.* **59,** 1354 (1993).

0076-6879/99 $30.00

EPS has been reported.[6–8] Briefly, survival functions fall under the headings of (a) barrier, including protection from predation and resistance to toxins, antibiotics, and poisons; (b) cell–cell interaction and recognition, including symbiosis; and (c) adhesion and biofilm formation. Along with fimbriae, EPS may tether the organism to the surface, thereby serving as a primary adhesin.[2] It is also the secondary or permanent adhesin.

Probes for Exopolymeric Substances and Capsular Polysaccharides

Capsular polysaccharides can be probed with antibodies, lectins, and specific stains (Table I). *In situ* populations, colonies growing on agar, individual cells under light microscopy, sections of cells under electron microscopy, and fractionated EPS, separated by polyacrylamide gel electrophoresis, can each be specifically tagged and identified. The specificity can be broad (stains for acidic EPS) or absolute (monoclonal antibody for a specific epitope; Table I). This section uses the biofilm colonizer *Hyphomonas* as the example.[2,9]

Macroscopic: Colonies and Biofilmed Surfaces

Bacteria are plated on marine agar (MA) containing Congo red (150 μg/ml) or Calcofluor (4,4′-bis[4-anilino-6-bis(2-hydroxyethyl)amino-*s*-triazin-2-ylamino]-2,2′-stilbenedisulfonic acid) (80 μg/ml). The use of these dyes for the detection of EPS production is described elsewhere.[10,11] They have been reported to be specific for polysaccharides containing $(1 \rightarrow 3)$-β- and $(1 \rightarrow 4)$-β-D-glucopyranosyl units.[12] After incubating at optimum growth temperature until the colonies are well developed, they are examined for dye accumulation. In the case of Calcofluor, the plates are illuminated with ultraviolet (UV) light to score dye binding to EPS.

Light (Including Fluorescence) Microscopy: Cells and Biofilms

First, cells from broth cultures can be stained negatively with nigrosin and observed under a microscope to check for the expression of capsule.

[6] A. W. Decho, *Oceanogr. Mar. Biol. Annu. Rev.* **28,** 73 (1990).
[7] W. F. Dudman, *in* "Surface Carbohydrates of the Prokaryotic Cell" (I. W. Sutherland, ed.), p. 357. Academic Press, London, 1977.
[8] R. M. Weiner, S. Langille, and E. Quintero, *J. Ind. Microbiol.* **15,** 339 (1995).
[9] E. Quintero, K. Busch, and R. M. Weiner, *Appl. Environ. Microbiol.* **64,** 1246 (1998).
[10] F. Bastarrachea, M. Zamudio, and R. Rivas, *Can. J. Microbiol.* **34,** 24 (1988).
[11] D. Doherty, J. A. Leigh, J. Glazebrook, and G. C. Walker, *J. Bacteriol.* **170,** 4249 (1988).
[12] P. J. Wood, *Carbohydr. Res.* **85,** 271 (1980).

TABLE I
PROBES USED TO SPECIFICALLY BIND CPS AND EPS

Probe			
Class	Name	Specificity[a]	Use[b]
Dyes	Congo red	(1 → 4)-α-D-glucopyranosyl units and basic or neutral EPS	1
	Calcofluor	(1 → 3)-β and (1 → 4)-β-D-glucopyranosyl units	1, 2, 4, 5
	Cationic ferritin	Negatively charged EPS	3
	Alcian blue		3, 5
	Toluidine blue-O	Negatively charged or helical polysaccharide	2, 5
	Ruthenium red	Acidic EPS	3, 5
Lectins[c]	*Bauhinia purpurea* (BHA)	*N*-Acetyl-D-galactosamine	*d*
	Coral tree (CTL)	Galactose-β-1,4-*N*-acetylglucosamine	
	Sweet pea (SPL)	Mannose and *N*-acetylglucosamine	
Antibody	Polyclonal	Several different epitopes; 4–12 residues specific linkages and tertiary structure.	*d*
	Monoclonal	As above, but single epitope	

[a] Primary specificity (most affinity).

[b] Principal use, although may be adapted for additional approaches: 1, macroscopic (colonies and biofilmed surfaces); 2, light microscopy (including fluorescence) (cells and biofilms); 3, electron microscopy (capsular fine structure); 4, purified EPS and CPS; and 5, electrophoresis and blotting.

[c] Over 100 with varying specificity (generally specific for monosaccharides, e.g., SPL or disaccharides with specific linkages, e.g., CTL) are available (from, e.g., EY Lab., Inc., San Mateo, CA). Three lectins, specific for strains of *Hyphomonas* EPS, are included: MHS-3 polar capsule,[9] BHA; VP-6 capsule, SPL; VP-6; holdfast, CTL.[2]

[d] Primary or secondary antibody conjugated with fluorescein isothiocyanate (FITC), horseradish peroxidase (HRP), or other fluoresceins or indicator systems, respectively, or colloidal gold.

For more sensitive observations, acid-washed glass slides (75 × 25 mm) are placed in culture vessels containing appropriate growth medium. The slides are removed at different intervals and stained. Two different techniques can be used to stain biofilms following fixation and drying.[13] Briefly, after removal from the cultures, the slides are treated with 10 m*M* cetyl pyridinium chloride (to precipitate the EPS matrix in the biofilms), air dried, and heat fixed. They are then treated with a 2:1 mixture of saturated aqueous Congo red solution and 10% (v/v) Tween 80 (EPS stain), rinsed, and treated with 10% Ziehl carbol fuchsin (cellular stain).

[13] D. G. Allison and I. W. Sutherland, *J. Microbiol. Methods* **2,** 93 (1984).

The second method uses Calcofluor, without dehydrating the biofilm. Slides are removed from the cultures and placed in sterile fresh growth medium, containing Calcofluor, and incubated at 22° for 6–12 hr.

Labeling of Hyphomonas MHS-3 EPS Capsule with Lectins

A protocol similar to this one for *Hyphomonas* MHS-3 is followed. A wide array of lectins are normally tested. Examples include, but are by no means limited to, *Triticum vulgaris* lectin (wheat germ agglutinin) (WGA) and *Bandereira (Griffonia) simplicifolia* lectin (GS-II), which are specific for *N*-acetyl-D-glucosamine; concanavalin A (Con A), specific for terminal α-D-mannose and α-D-glucose; and *Ulex europaeus* agglutinin (UEA-I), which binds α-L-fucose. Those binding *N*-acetyl-D-galactosamine (GalNAc) include *Glycine max* lectin (SBA), *Maclura pomifera* lectin (MPA), *Bauhinia purpurea* lectin (BPA), and *Arachis hypogaea* lectin (peanut agglutinin) (PNA).[14] Primary screening is carried out as follows: a sample of a midexponential culture of cells is centrifuged at 16,000*g* for 5 min at 4°. The cell pellet is washed with 1× phosphate-buffered saline (PBS, 8.0 g NaCl, 0.2 g KCl, 1.44 g Na_2HPO_4, 0.24 g KH_2PO_4, to 1 liter with distilled H_2O, pH 7.2), then suspended in 1× PBS. Fluorescein isothiocyanate (FITC)-labeled lectins are added to a final concentration of 50 μg/ml and are incubated at 25° for 30 min. The cells are then centrifuged and washed with sterile 1× PBS and resuspended in 1× PBS. The extent of lectin binding is determined under epifluorescence microscopy.

Electron Microscopy: Capsular Fine Structure

Immunoelectron Microscopy

Cells are layered on Formvar-coated copper grids. The grids are "blocked" for 10 min with 5% skim milk in PBS and then inverted on a drop of appropriate dilution of IgG antibody in skim milk solution for 10 min. The dilution of the antibody is determined experimentally. The grids are rinsed by inverting on two successive drops of skim milk, placed on a drop of 1 : 10 gold-conjugated protein A in skim milk (colloidal gold particles of 15 nm), and incubated 10–30 min to allow protein A to bind IgG on the cell surface. Finally, the grids are rinsed by inverting on four to six successive drops of distilled H_2O (10–20 sec on each), and then observed

[14] A. M. Wu and S. Sugii, *Carbohydr. Res.* **213,** 127 (1991).

by transmission electron microscopy (TEM). This general technique is described elsewhere.[15,16]

Labeling of Capsule with Polycationic ferritin

Cells are fixed with glutaraldehyde, suspended in cacodylate buffer (0.1 *M* sodium cacodylate, pH 7.0), and allowed to react with polycationic ferritin (final concentration, 1.0 mg/ml) for 30 min at 25°.[17,18] The reaction is slowed by a 10-fold dilution with buffer. The cells are centrifuged and washed three times with cacodylate buffer, mixed with molten agar, imbedded in LR white resin, and thin sectioned.

Lectin Gold Labeling of Capsular EPS

Gold-labeled lectin (10-nm gold particles) (EY Laboratories, Inc., San Mateo, CA) can be used to visualize capsular EPS. Briefly, a drop of mid-exponential culture is placed on a Formvar-coated copper grid and incubated at room temperature for 1 min. It is blocked with a solution of 5% (w/v) bovine serum albumin (BSA, in 0.1 *M* PBS) for 5–10 min, incubated in a 1:10 dilution of the specific gold-conjugated lectin in 5% BSA for 10–15 min, and then rinsed five times with distilled water. Glutaraldehyde and skim milk normally interfere with the lectin–gold conjugate reaction.

Purified EPS and CPS

Purification

The culture is centrifuged at 16,000*g* for 20 min at 4° to separate cells from spent medium. The supernatant is precipitated with 4 volumes of ice-cold 2-propanol, and, for species that synthesize an integral capsule, the cell pellet is blended in a Waring blender with 10 m*M* EDTA, 3% NaCl for 1 min at 4° to shear off the capsular EPS. The suspension is again centrifuged, the cell pellet is discarded, and the sheared EPS in the supernatant is precipitated with 2-propanol as described earlier. The precipitate is suspended (or dissolved) in a minimum volume of distilled H_2O, dialyzed exhaustively against distilled H_2O, and lyophilized. The following steps are

[15] H. Levanony and Y. Bashan, *Curr. Microbiol.* **18,** 145 (1989).

[16] M. Wrangstahd, U. Szewzyk, J. Ostling, and S. Kjelleberg, *Appl. Environ. Microbiol.* **56,** 2065 (1990).

[17] M. Jacques and B. Foiry, *J. Bacteriol.* **169,** 3470. (1987).

[18] M. E. Bayer, *in* "Current Topics in Microbiology and Immunology" (K. Jann and B. Jann, eds.), Vol. 150, p. 129. Springer-Verlag, New York, 1990.

a modification of the purification procedure of Read and Costerton.[19] The crude EPS is dissolved in a minimum volume of 0.1 M $MgCl_2$. DNase and RNase are added to a final concentration of 0.1 mg/ml and are incubated at 37° for 4 hr. Protease K is added to 0.1 mg/ml and is incubated at 37° overnight. The residual protein is removed with a hot phenol extraction, followed by a chloroform extraction. The EPS solution is dialyzed exhaustively against distilled H_2O and freeze-dried. This partially purified EPS is redissolved in distilled H_2O and is purified further by gel permeation chromatography to separate the capsular EPS from contaminating lipopolysaccharide.

To purify the EPS further, solutions (1 mg EPS/ml) are chromatographed on a column (45 × 1.5 cm) of Sephacryl S-400-HR gel permeation resin (Pharmacia, Piscataway, NJ) using 50 mM ammonium acetate, pH 7.0 (containing 0.02% NaN_3), as the elution buffer at a flow rate of 0.5 ml/min. Three-milliliter fractions are collected and analyzed for total carbohydrate as described later. Solutions (1 mg/ml) of dextran molecular weight standards of 5,000,000, 2,000,000, 500,000, 70,000, 40,000, and 10,000 (Pharmacia; Sigma Chemical Co., St. Louis, MO) are used to standardize the column. Ion-exchange chromatography can also be used with a quaternary amine column, preferably in an FPLC (first protein liquid chromatography, Pharmacia) system. EPS solutions (500 μl of 2 mg/ml in 20 mM Tris buffer, pH 8.0) are loaded into the column. Samples are eluted with 5 ml of Tris buffer and then with a linear gradient of NaCl from 0.05 to 1.0 M (in buffer) at a flow rate of 1 ml/min. One-milliliter fractions are collected and analyzed for total carbohydrate.

Characterization

Lectin Specificity and EPS Purity by Hemagglutination Inhibition Assay (HAA). This type of assay has been used to determine the specificity and the degree of affinity of lectins for distinctive carbohydrate moieties.[14] Because lectins have two or more carbohydrate-binding sites per molecule, they can agglutinate cells that present specific sugar or oligosaccharides on their surfaces.[20] Inhibition of this agglutination by the addition of competing mono- or oligosaccharides is used to define the binding specificity of the molecule.[14]

The procedure is a modification of the Matsumoto and Osawa[21] hemagglutination inhibition assay. Briefly, 100 μl of a 3% suspension of human group O erythrocytes (Baxter Diagnostics, Inc.) is placed in individual wells of a 96-well microtiter plate with 100 μl of solutions of specific concentra-

[19] R. R. Read and J. W. Costerton, *Can. J. Microbiol.* **33,** 1080 (1987).

[20] H. Lis and N. Sharon, *Annu. Rev. Biochem.* **55,** 35 (1986).

[21] I. Matsumoto and T. Osawa, *Arch. Biochem. Biophys.* **140,** 484 (1970).

tions of sugar or polysaccharide prepared in 0.01 *M* PBS. Galactosamine, galactose, lactose, *N*-acetylgalactosamine, glucose and *N*-acetylglucosamine can be purchased from Sigma and 70,000 molecular weight dextran can be obtained from Pharmacia. The final density of red blood cells in each well is 1.5%, and the serial dilutions of the different carbohydrate solutions are from 10,000 to 1 μg/ml. Eight microliters of pure lectin (EY Laboratories) solution (1 mg/ml in 0.01 *M* PBS) is added to each well to obtain a final lectin concentration of 40 μg lectin/ml per well. Controls with no carbohydrate and no lectin are also included. The plate is incubated at room temperature and then the minimum carbohydrate concentrations that inhibited agglutination are scored.

Colorimetric Assays. NEUTRAL HEXOSE. The neutral hexose content of samples can be measured according to Dubois *et al.*[22] Briefly, 0.5 ml of an aqueous sample is mixed with 0.5 ml of phenol solution (5 g phenol in 100 ml distilled H_2O) in a glass test tube, and then 2.5 ml of sulfuric acid reagent (2.5 g hydrazine sulfate in 500 ml sulfuric acid) is rapidly added and vortexed. The tubes are incubated in the dark for 1 hr, when their absorbance at 490 nm (A_{490}) is measured. Glucose standards (10 to 100 μg/ml) are used to prepare the standard curve.

URONIC ACID. Uronic acids can be quantitated following the procedure of Blumenkrantz and Asboe-Hansen.[23] In iced glass test tubes, 0.2 ml of aqueous sample is mixed with 1.2 ml of sodium tetraborate reagent (12.5 m*M* sodium tetraborate in concentrated sulfuric acid). The mixture is vortexed, incubated at 100° for 5 min, and cooled on ice. Then, 20 μl of *m*-hydroxyphenyl reagent [0.15% *m*-phenylphenol (Eastman Kodak Co., Rochester, NY) in 0.5% (w/v) sodium hydroxide solution] is added, and within 5 min the absorbance of the samples is read at 520 nm (A_{520}). To correct for any carbohydrate interference, an additional tube is prepared for each sample and the same protocol is followed, but instead of adding *m*-hydroxyphenyl developing reagent, 20 μl of 0.5% NaOH is added. The absorbances of these control tubes are subtracted from those of their corresponding samples.

ACETYLATION. The presence of acetyl groups on the EPS is assayed using a modified colorimetric procedure.[24,25] Just prior to the test, a working reagent is prepared mixing equal volumes of 8.0 *M* hydroxylamine hydrochloride and glycine reagent (1.0 *M* glycine in 8.5 *M* NaOH). Two hundred

[22] M. Dubois, K. A. Gilles, J. K. Hamilton, P. A. Rebers, and F. Smith, *Anal. Chem.* **28,** 350 (1956).

[23] N. Blumenkrantz and G. Asboe-Hansen, *Anal. Chem.* **54,** 4841 (1973).

[24] S. J. Hestrin, *J. Biol. Chem.* **180,** 249 (1949).

[25] E. A. McComb and R. M. McCready, *Anal. Chem.* **29,** 819 (1957).

microliters of test sample is mixed with 400 μl of this reagent in a large glass test tube and is incubated at room temperature for 3 hr. Then, 2.5 ml of 1.0 *M* HCl and 6 ml of ferric chloride reagent (0.1 *M* $FeCl_3$ in 0.01 *M* HCl) are added. The absorbance is measured immediately at 540 nm (A_{540}), before a precipitate forms.

PYRUVILATION. Pyruvilation of the EPS is tested using an enzymatic assay.[26] Briefly, 1.5 ml of 0.08 *N* oxalic acid is added to 5 mg of EPS (in 1.5 ml distilled H_2O) and refluxed for 5 hr at 100° to hydrolyze pyruvate from EPS backbone. After cooling, 230 mg of calcium carbonate is added to each tube to neutralize the solutions, and the amount of free pyruvate present is assayed using lactate dehydrogenase (Sigma Chemical Co.), following the manufacturer specifications.

MONOSACCHARIDE ANALYSIS. High-performance anion-exchange chromatography (HPAE) is used to identify the major monosaccharide components of the EPS, following a protocol described elsewhere.[27] Briefly, 200 μg of purified EPS is hydrolyzed in 200 μl of 2 *N* HCl at 100° for 2 hr. The samples are dried under a stream of nitrogen and dissolved in 200 μl of distilled H_2O. The system used for HPAE consists of a BioLC gradient pump (Dionex Corp.) with a pulsed amperometric detector (PAD). A Carbopac PA1 (4 × 250 mm) pellicular anion-exchange column (Dionex Corp.) with a Carbopac guard column is used at a flow rate of 1 ml/min at room temperature. Two different eluants (degassed with helium) are used: eluant 1, 15 m*M* NaOH, useful for the analysis of neutral and amino sugars; and eluant 2, 100 m*M* NaOH and 150 m*M* sodium acetate, effective in the analysis of acidic monosaccharides. A monosaccharide standard solution (85 μg each sugar/ml, 20 μl injected) is run after each hydrolyzed EPS sample (injection volume, 20 μl) to identify the monomers.

DETECTION OF SULFATE GROUPS BY INFRARED SPECTROSCOPY (IR). One milligram of EPS is mixed with 100 mg of potassium bromide (IR grade) and ground with a mortar and pestle. Half of the fine powder is placed into a die and compressed into a translucent pellet. The pellet is placed in a FTIR (e.g., Perkin Elmer 1600 Series; Perkin-Elmer Co., Norwalk, CT), and a spectrum is obtained.[28] Chondroitin sulfate is the standard.

[26] M. Duckworth and W. Yaphe, *Chem. Ind.* **23,** 47 (1970).

[27] G. Reddy, U. Hayat, C. Abeygunawardana, C. Fox, A. Wright, D. Maneval, Jr., C. A. Bush, and J. G. Morris, Jr., *J. Bacteriol.* **174,** 2620 (1992).

[28] M. Matsuda, W. Worawattanamateekul, and K. Okutani, *Nippon Susian Gakkaishi* **58,** 1735 (1992).

Electrophoresed CPS and EPS

Electrophoresis

Purified EPS solutions are mixed with an equal volume of 2× treatment buffer [12.5 m*M* Tris, pH 6.8, 2% (v/v) sodium dodecyl sulfate (SDS), 10% glycerol, and 2% 2-mercaptoethanol] to a final concentration of 2 mg EPS/ml and boiled for 5 min. These EPS samples are stored at −20° and boiled for 1 min just prior to loading in gels. EPS is mobilized by discontinuous polyacrylamide gel electrophoresis with sodium dodecyl sulfate (SDS–PAGE; 4% stacking gel and a 10% running gel).[29] For many applications, the current is set at 35 mA/gel for 3–5 hr.

Staining

Several different carbohydrate-specific stains can be used to directly visualize the EPS in the gel. For example, toluidine blue-O (0.08% in 7% acetic acid) is used to stain alginates.[30] Alcian blue stains other EPS.[31,32] Solutions of Congo red and Calcofluor (0.03% in 1X PBS) are also used, staining overnight, and destaining with distilled H_2O.

In a different approach, the purified EPS is cross-linked with the fluorophore 8-aminonaphthalene-1,3,6-trisulfonic acid (ANTS) (Molecular Probes) by the reducing ends of the polymer prior to electrophoresis. This procedure had only been used previously to label mono- and oligosaccharide.[33] In our modification of the procedure for EPS, 600 μg of polymer is placed in a microcentrifuge tube and then mixed with 5 μl of 0.2 *M* ANTS solution (in acetic acid/water, 3:17, v/v) and 5 μl of 1.0 *M* sodium cyanoborohydride (Aldrich Chemical Co., Milwaukee, WI) (in dimethyl sulfoxide, DMSO). The mixture is vortexed and centrifuged briefly, incubated in the dark at 37° for 24 hr, and then brought to 200 μl with distilled H_2O and frozen at −70° (always kept in the dark). Between 100 and 250 μg is loaded in the gels. A UV transilluminator is used to detect the EPS band(s) after electrophoresis.

Electroblotting EPS (Eastern Blot)

Samples are electrophoresed as described earlier and electroblotted [e.g., between 60 and 90 min onto nitrocellulose membranes using a 2117-

[29] U. K. Laemmli, *Nature* **227,** 680 (1970).
[30] W. M. Dunne and F. L. A. Buckmire, *Appl. Environ. Microbiol.* **50,** 562 (1984).
[31] H. Min and M. K. Cowman, *Anal. Biochem.* **155,** 275 (1986).
[32] G. N. Misevic and M. M. Burger, *J. Cell. Biochem.* **43,** 307 (1990).
[33] P. Jackson, *Biochem. J.* **270,** 705 (1990).

250 Novablot Electrophoretic Transfer Kit (Pharmacia) for semidry blotting]. To probe for EPS after electroblotting, the nitrocellulose is blocked for 1 hr in 5% BSA in 0.01 *M* PBS buffer, followed by a 2- to 12-hr incubation in 50 ml horseradish peroxidase-labeled probe (appropriate lectin or antibody), normally diluted 1:500 in 5% BSA in 0.01 *M* PBS. The blot is washed three times with PBS and incubated for 20–60 min in peroxidase substrate buffer. Transferring to Zeta-Probe (nylon) blotting membranes (Bio-Rad Richmond, CA) also works well for some EPS.

Probes for Extracellular Enzymatic Hydrolysis of Polysaccharides

The ability to use macromolecular substrates in biofilms, as in other microbial communities, is closely linked to the capabilities of at least some members of a community to produce the extracellular enzymes required to hydrolyze organic macromolecules to sizes small enough to be transported across microbial membranes. The direct uptake limit for microbial substrates is on the order of 600 D,[34] approximately equivalent to a trisaccharide. Above this uptake limit, free or cell surface attached extracellular enzymes exhibiting the correct structural specificity are required to hydrolyze a substrate. The specificity and activity of microbial extracellular enzymes are therefore critical determinants of the nature and range of substrates available to biofilm bacteria.

A wide range of microbial extracellular enzymes have been isolated from pure cultures of bacteria, and these enzymes have been characterized to determine enzymatic specificities and activities (e.g., Refs. 35 and 36; previous volumes in this series). While essential for investigations of enzyme structure and catalysis, this approach provides no information about enzyme activity under *in situ* conditions. In addition, uncertainties about the qualitative and quantitative relationships between bacteria isolated successfully in pure culture and a naturally complex microbial consortia make *in situ* determinations of extracellular enzymatic activity a highly desirable goal.

To date, a limited number of low molecular weight proxies have been used to characterize the enzymatic capabilities of sedimentary and biofilm microbial communities (e.g., Refs. 37–39). This approach uses small sub-

[34] M. S. Weiss, U. Abele, J. Weckesser, W. Welte, E. Schiltz, and G. E. Schulz, *Science* **254,** 1627 (1991).

[35] C. A. White and J. F. Kennedy, *in* "Carbohydrate Chemistry" (J. F. Kennedy, ed.), p. 343. Clarendon Press, Oxford, 1988.

[36] M. W. W. Adams, *Annu. Rev. Microbiol.* **47,** 627 (1993).

[37] J. Marxsen and K. P. Witzel, *in* "Microbial Enzymes in Aquatic Ecosystems" (R. J. Chrost, ed.), p. 270. Springer-Verlag, New York, 1991.

[38] J. Marxsen and D. M. Fiebig, *FEMS Microb. Ecol.* **13,** 1 (1993).

[39] D. R. Confer and B. E. Logan, *Water Res.* **32,** 31 (1998).

strate analogs, commonly methylumbelliferyl–monosaccharide dimers (MUF substrates), which yield fluorescent methylumbelliferone on cleavage of the fluorophore–monosaccharide bond.[40–43] MUF substrates, however, have limited abilities to mimic the structural complexities of macromolecules. Structural features determined by linkage positions, heterogeneous monomer composition, and secondary or tertiary structure cannot be investigated with these substrates. Additionally, amino sugars, uronic acids, and sulfated monomers are not available as MUF substrates, hence, hydrolysis rates of these substances, common components of exopolymers and biofilms,[6,44] cannot be investigated with this approach. Experimental studies have demonstrated that MUF substrates may be poor proxies for their putative macromolecular counterparts.[45] In addition, because MUF substrates can be hydrolyzed within the periplasmic space of bacteria,[46] this approach measures some undefinable combination of periplasmic and extracellular hydrolysis.

A new approach, based on the use of fluorescently labeled (FLA) polysaccharides, has been developed to measure potential enzyme activities in environmental samples. With this method, polysaccharides, which include neutral and sulfated components, as well as uronic acids, can be utilized as hydrolysis probes. In principle, naturally occurring polysaccharides isolated from a wide variety of sources could serve as substrates. This approach has been tested particularly in marine sediments,[47,48] where biofilms may act as "cement," and are abundant coatings on sediment grains.[49]

General Procedures

Fluorescently labeled polysaccharides are synthesized and characterized as described later and are then added to a liquid sample or a slurry or injected into intact sediment cores. After incubation at *in situ* temperature, a sample is removed and filtered (sediments must be centrifuged to obtain core water) and then analyzed on a gel permeation chromatography (GPC) system attached to a fluorescence detector. The molecular weight distribu-

[40] H.-G. Hoppe, *Mar. Ecol. Prog. Ser.* **11,** 299 (1983).
[41] M. Somville, *Appl. Environ. Microbiol.* **48,** 1181 (1984).
[42] G. M. King, *Appl. Environ. Microbiol.* **51,** 373 (1986).
[43] L.-A. Meyer-Reil, *Appl. Environ. Microbiol.* **53,** 1748 (1987).
[44] K. D. Rinker and R. M. Kelly, *Appl. Environ. Microbiol.* **62,** 4478 (1996).
[45] E. Helmke and H. Weyland, *Kieler Meeresforsch. Sonderh.* **8,** 198 (1991).
[46] J. Martinez and F. Azam, *Mar. Ecol. Prog. Ser.* **92,** 89 (1993).
[47] C. Arnosti, *Geochim. Cosmochim. Acta* **59,** 4247 (1995).
[48] C. Arnosti, *Organic Geochem.* **25,** 105 (1996).
[49] A. W. Decho and G. R. Lopez, *Limnol. Oceanogr.* **38,** 1633 (1993).

tion of the hydrolysis products is determined via their elution profile, and potential hydrolysis rates are calculated as the minimum number of hydrolyses required to reduce the polysaccharide from its initial molecular weight to the molecular weight distribution observed in the chromatogram.

Synthesis and Characterization of Labeled Polysaccharides

Synthesis and characterization of fluorescently labeled polysaccharides are based on an adaptation of the method of Glabe *et al.*[50] Forty milligrams of highly soluble polysaccharides [such as pullulan, fucoidan, and laminarin (Sigma, Fluka, Ronkonkoma, NY)] are dissolved in 2.0 ml H_2O. Less soluble polysaccharides, such as xylan, are sonicated in 2.0 ml H_2O and then filtered (0.2-μm pore-sized filter) to remove insoluble material. Uronic acid-containing polysaccharides, such as alginic acid (Fluka), are dissolved (~7 mg/ml) in 0.25 *M* NaOH and are then filtered to remove any insoluble material. Activation of a polysaccharide is initiated by dissolving 30 mg CNBr in 350 μl H_2O and mixing the polysaccharide and CNBr solutions. Solution pH is monitored, and 25-μl portions of a 0.25 *M* NaOH solution are added to keep the pH above 9.5 for 6 min. The reaction mixture is then immediately desalted by injection onto a GPC column (18.0 × 1.0 cm, Sephadex G-25) with a mobile phase of 0.2 *M* $Na_2B_4O_7$, adjusted to pH 8.0, at a flow rate of 1.8 ml/min. The fraction corresponding to the void volume is collected in a vial containing 4.0 mg fluoresceinamine (isomer II; Sigma), and the reaction mix is incubated at room temperature overnight.

Neutral and sulfated polysaccharides are separated from unreacted fluorophore by injecting the reaction mix (typically in portions of 2–3 ml) onto a 26 × 1.0-cm column of Sephadex G-25 gel, with a mobile phase of 100 m*M* NaCl + 50 m*M* NaH_2PO_4/Na_2HPO_4, pH 8.0. The yellow band corresponding to the void volume of the column (=labeled polysaccharides) is collected; complete separation of labeled polysaccharide and unreacted tag can be checked by reinjecting a few microliters of the FLA–polysaccharide and monitoring the fluorescence profile (excitation: 490 nm; emission: 530 nm). Because uronic acid-containing polysaccharides interact strongly with the gel column under these conditions, the unreacted fluoropor is separated from the labeled polysaccharide by dialysis (Spectrapor CE membrane; 5000 molecular weight cutoff) against Milli-dried, Q–H_2O. The retentate (containing labeled polysaccharide) is freeze-dried, and complete separation of labeled polysaccharide from the free fluorescent tag can be checked by GPC (26 × 1.0-cm column of Sephadex G-25 gel, mobile phase of 1 *M* NaCl + 50 m*M* NaH_2PO_4/Na_2HPO_4, pH 8.0.)

[50] C. G. Glabe, P. K. Harty, and S. D. Rosen, *Anal. Biochem.* **130,** 287 (1983).

We have labeled polysaccharides with a variety of fluorophores other than fluoresceinamine, including dansyl ethylenediamine (excitation 334 nm; emission 512 nm), Texas Red sulfonylcadavarine (excitation: 590 nm; emission: 612 nm), and 5-aminomethylfluorescein (excitation: 492 nm; emission: 516 nm; all from Molecular Probes, Eugene, OR). In principle, any fluorophore containing a reactive primary amine group can be coupled to a CNBr-activated polysaccharide. Some fluorophores, such as fluorescein cadaverine (also Molecular Probes), which contain a thioureidyl group, are unstable in aqueous solution; the stability of a fluorophore should be tested thoroughly before use. Fluorophores that are insoluble in water, such as Texas Red sulfonylcadavarine (Molecular Probes), can be coupled to activated polysaccharides by first dissolving the fluorophore in 350–400 μl dimethylformamide (DMF). The activated polysaccharide collected from the borate column is mixed thoroughly with the DMF solution and is incubated overnight. Prior to on-column separation of unreacted tag from the labeled polysaccharide, the reaction mixture must be filtered to remove any possible precipitates. Note that the fluorescent properties of fluoresceinamine are particularly well suited for measurements in samples containing high concentrations of natural dissolved organic carbon, whose fluorescence maxima are typically centered at 350 nm (excitation) and 450 nm (emission).[51]

The labeling densities of the FLA–polysaccharides are calculated by measuring fluorophore absorbance against a series of standards and determining the carbohydrate concentration using the methods described earlier. Labeling densities can vary greatly for different polysaccharides. Although the general chemistry of the reaction is the same—conversion of polysaccharide cyanate esters to stable isourea linkages on reaction with primary amines[52]—the ultimate labeling density depends on the pH of the reaction mixture,[50] the number of potentially reactive hydroxyl groups, and the solution conformation of the polysaccharide. The presence of a fluorescent tag does not measurably affect biological activity of labeled polysaccharides.[50] The fact that relatively few monomers bear a fluorescent tag ensures little interference with polysaccharide activity and is also the principal reason that calculations made with this technique represent a lower limit on potential hydrolysis rates (see later).

The activated, derivatized, and purified FLA–polysaccharides we have tested are stable to repeated cycles of freezing and thawing, as well as a considerable temperature range. Labeled polysaccharides have been stored

[51] S. A. Green and N. V. Blough, *Limnol. Oceanogr.* **39,** 1903 (1994).
[52] J. Kohn and M. Wilchek, *Biochem. Biophys. Res. Commun.* **84,** 7 (1978).

in the dark at room temperature for days and frozen for months with no sign of deterioration.

Hydrolysis Measurements

FLA–polysaccharides have been used to measure potential hydrolysis rates in oxic and anoxic sediments, slurries, and seawater.[47,48,53,54] They could also be used in flow cells and mesocosms. Measurements in sediment cores have been made using the method of Jørgensen,[55] in which the substrate is injected into silicon-stoppered injection ports located at 1-cm intervals in a 2.6 × 20-cm core liner. The cores are incubated at *in situ* temperature, sectioned (typically in 2-cm intervals), and centrifuged to obtain core water samples, which are analyzed as described later. Substrate additions are in the pmol–nmol polysaccharide range; lower concentrations are used for seawater experiments. Potential hydrolysis rates are monitored and sampling intervals are optimized throughout the course of an experiment by periodically analyzing a sample on a GPC system and comparing the molecular weight distribution of fluorescence to the zero-time distribution pattern.

The molecular weight distribution of the polysaccharide is determined at each time point by filtering a sample (0.2-μm pore size) and injecting it onto a GPC column. Our current system consists of a Sephadex G-50 gel column (24.5 × 1 cm) connected to a Sephadex G-75 gel column (22 × 1 cm), with a mobile phase of 100 m*M* NaCl + 50 m*M* NaH_2PO_4/Na_2HPO_4, pH 8.0, at a flow rate of 0.8 ml/min. (The G-75 gel column has a 1-cm base layer of G-25 Sephadex gel to prevent crushing the G-75 gel into the column frit.) For analysis of uronic acid-containing polysaccharides, the mobile phase consists of 1 *M* NaCl + 50 m*M* NaH_2PO_4/Na_2HPO_4, pH 8.0. The column system is calibrated with FITC–dextran standards (Sigma) with molecular weights of 150, 70, 40, 10, and 4 kDa as well as FITC–glucose and free fluoresceinamine, and provides reasonable resolution in the molecular weight ranges of >10, 10, 3–4 kDa, monosaccharides, and free tag. Higher resolution could be achieved through the use of HPLC-GPC columns. The columns are connected to a fluorescence detector set to the excitation and emission maxima of the fluorophore (fluoresceinamine: excitation: 490; emission: 530 nm).

[53] C. Arnosti, B. B. Jørgensen, J. Sagemann, and B. Thamdrup, *Mar. Ecol. Prog. Ser.* **159,** 59 (1998).

[54] S. C. Keith and C. Arnosti, in preparation.

[55] B. B. Jørgensen, *Geomicrob. J.* **1,** 11 (1978).

Potential Hydrolysis Rate Calculations

The calculated rates are "potential" rates, as the substrates naturally present in the system are competing with added substrates for enzyme active sites. If saturating concentrations of substrate are added, enzyme kinetics should approximate zero order, dependent only on the characteristics of the enzymes themselves. Measuring the substrate concentration naturally present in many natural samples is difficult or impossible, as methods to determine the concentrations of polysaccharides of specific chemical structures in marine sediments, for example, have yet to be developed. An upper limit of polysaccharide concentration can be set by measuring total carbohydrate concentrations and/or concentrations of specific monomers.

The rate calculation model used here is intended to set a lower boundary on the potential hydrolysis rate. All of the fluorescence is initially associated with polysaccharides of known molecular weight. As the polysaccharides are progressively hydrolyzed, their elution profiles and the fluorescence signal detected in the GPC system, progressively change. By knowing the initial quantity of polysaccharide added to the system and quantifying the fluorescence that elutes in a specific molecular weight fraction at a specific time point, the minimum number of hydrolyses required to reduce the parent polysaccharide to the lower molecular weight products can be calculated. This calculation yields a conservative estimate: because only a small percentage of the monomers in a polysaccharide are labeled, only a fraction of the hydrolyses that occur are observable in an elution profile. Hydrolyses involving the unlabeled portion of a polysaccharide are equally probable, but unobservable. This factor also accounts for the progressive apparent decrease in the potential hydrolysis rate with incubation time: as a polysaccharide is increasingly hydrolyzed, the probability that a hydrolytic event will include a tagged fragment decreases.

Using Dimensionless Parameters to Study Substrate Fluxes

The metabolic activities in a biofilm are central to the survival of its inhabitants. Biofilm metabolic activity and accumulation are generally controlled by intra- and interphase transport processes, which, in turn, can be modified significantly by biofilm accumulation.[56] A useful tool for quantitatively studying the interactions between physicochemical mass transport processes and biokinetics is the application of dimensionless parameters in

[56] W. G. Characklis, M. H. Turakhia, and N. Zelver, *in* "Biofilms" (W. G. Characklis and K. C. Marshall, eds.), p. 265. Wiley, New York, 1990.

substrate mass balance and flux calculations. This section uses biofilms in subsurface porous media as an example for applying dimensionless parameters to examine the influence of internal and external mass transport resistance, and competing reaction sinks, on the substrate flux and reaction rate in the biofilm.

Modeling-Attached Growth in Porous Media

Conceptual Models

Conceptual models of bacterial growth in porous media include strictly macroscopic, microcolony, and biofilm models.[57] The strictly macroscopic model only assumes that bacteria are immobilized. Biokinetics are a function of the macroscopic biomass and bulk fluid substrate concentrations. The microcolony model[58] assumes the biomass is present as small colonies growing on the solid surfaces. External mass transfer resistance is accounted for by assuming that the substrate in the bulk solution must diffuse across a diffusion layer to reach the colony surface. The microcolonies are assumed to be thin enough that internal mass transfer resistance is negligible. The biofilm model[59,60] assumes that bacteria and their EPS are distributed as a continuous film with uniform cell density, X_f (M_x/L^3), and locally uniform thickness, L_f(L), giving on the macroscopic scale, a biomass per unit aquifer volume of X_fL_fa, where a is the total surface area per unit volume of porous medium. External mass transport resistance is represented by an effective diffusion layer of thickness, L (L). Three substrate concentrations (M_s/L^3) are important: (1) C, in the bulk liquid; C_s, at the biofilm surface; and C_f, within the biofilm, which is assumed to change only in the z direction, i.e., normal to the substratum.

Among other things, these models differ in terms of the impact of mass transport on the biokinetics in the biofilm. Because strictly macroscopic and microcolony models can be viewed as limiting cases of the biofilm model, the biofilm model kinetics are reviewed before examining the effects of mass transport on substrate utilization.

Biofilm Kinetic Fundamentals

Growth of attached bacteria is proportional to the substrate utilization rate, which is a function of the amount of attached biomass present, the

[57] P. Baveye and A. Valocchi, *Water Resour. Res.* **25,** 1413 (1989).

[58] F. J. Molz, M. A. Widdowson, and L. D. Benefield, *Water Resour. Res.* **22,** 1207 (1986).

[59] B. E. Rittmann and P. L. McCarty, *Biotechnol. Bioeng.* **22,** 2343 (1980).

[60] B. E. Rittmann and P. L. McCarty, *J. Environ. Engr.* **107,** 831 (1981).

biokinetics, and mass transport. Simultaneous reaction and molecular diffusion within the biofilm at steady state are represented as [60,61]:

$$D_f \frac{\partial^2 C_f}{\partial z^2} = q_m \frac{X_f C_f}{K + C_f} \tag{1}$$

where D_f is the molecular diffusion coefficient of the substrate within the biofilm (L^2/T) (typically within 80% of D, the molecular diffusion coefficient for the substrate)[62]; q_m is the maximum specific rate of substrate utilization (M_s/M_xT); and K is the half-maximum rate substrate concentration (M_s/L). The flux, J (M_s/L^2T), of substrate into the biofilm is described as

$$J = \eta q_m X_f L_f \frac{C_s}{K + C_s} \tag{2}$$

where η is the dimensionless effectiveness factor equal to the ratio of the actual substrate flux to that for a biofilm with no internal diffusion resistance. Solutions for η are available.[60,61]

In addition, external mass transport resistance from the bulk liquid to the biofilm surface is incorporated by assuming that external mass transport occurs across a diffusion layer of depth, L (L), representing the mass transport resistance from the bulk fluid to the biofilm. Thus, the substrate flux from the bulk liquid to the biofilm surface is defined by Fick's first law as[60]

$$J = \frac{D}{L}(C - C_s) \tag{3}$$

Intraphase Mass Transfer

Intraphase mass transport within the biofilm occurs primarily by molecular diffusion through the gel-like EPS matrix. As microorganisms in the biofilm consume substrates diffusing through the biofilm, substrate concentration gradients may develop, significantly reducing C_f and influencing microbial metabolic activity as well as population distributions within the biofilm.[62]

The influence of these diffusion gradients on the reaction rate can be evaluated using η. When $\eta = 1$ in Eq. (2), there is no internal mass transport resistance due to diffusion in the biofilm, and $C_f = C_s$ at all depths in the

[61] B. Atkinson and I. J. Davies, *Trans. Inst. Chem. Eng.* **52,** 248 (1974).

[62] B. E. Christensen and W. G. Characklis, *in* "Biofilms" (W. G. Characklis and K. C. Marshall, eds.), p.93. Wiley, New York, 1990.

biofilm. As the biofilm thickness increases, C_f becomes less than C_s and η becomes less than 1.0.

Biofilm modeling studies of porous media systems[63,64] found η values to equal 1, indicating no internal mass transport resistance. In the study of Odencrantz *et al.*,[63] typical biomass accumulations were averaged over the porous medium. Using a typical X_f and the geometrical a, the calculated biofilm thickness, L_f, was too thin to create internal mass transport resistance; thus, $\eta = 1$. This is consistent with the generally sparse microbial coverage observed on particles in natural porous media.[57]

The conditions for which $\eta = 1$ were defined by Suidan *et al.*[65] for low substrate concentrations ($C^* = C/K \ll 1$) using the dimensionless biofilm thickness, L_f^*:

$$L_f^* \leq \cosh^{-1}\left[\frac{L^*(1 + 0.98(L^{*2} - 1))^{1/2} - 1}{0.99(L^{*2} - 1)}\right] \qquad \text{for } L^* \neq 1.0 \qquad (4)$$

where L_f^* is $L_f\,(q_m X_f / D_f K)^{1/2}$ and L^* is the dimensionless thickness of the external diffusion layer $= L(q_m X_f / D_f K)^{1/2}\,(D_f / D)$. For $L^* = 1.0$, full penetration occurs when $L_f^* \leq 0.01$. If $X_f L_f$ becomes sufficiently small, the concept of a continuous biofilm becomes questionable. When the biofilm becomes noncontinuous, it can be modeled by decreasing a, L_f, or X_f without affecting the substrate flux, as long as η is not altered, as discussed elsewhere.[66]

Rittmann[66] analyzed experimental data for the colonization of porous media using the normalized surface loading:

$$\bar{J} = J/J_R \qquad (5)$$

where J is the actual substrate flux into the biofilm and J_R is the minimum substrate flux giving a steady-state biofilm that is deep, i.e., the internal diffusional resistance and biofilm thickness are sufficient that C_f approaches zero in the biofilm. This analysis demonstrated that the surface loading could be used to differentiate between conditions leading to continuous ($J/J_R > 1.0$) and discontinuous biofilm form (possible for J/J_R less than about 0.25), with an undefined transition in continuity. With high loading conditions ($J/J_R > 1.0$), biofilm accumulation is large and substrate removal is controlled by biokinetics and mass transport resistance. However, with low substrate loading ($J/J_R \ll 1.0$), e.g., in natural porous media, the sub-

[63] J. E. Odencrantz, A. J. Valocchi, and B. E. Rittmann, *in* "Proceedings of Petroleum Hydrocarbons and Organic Chemicals in Groundwater," p. 355. Am. Pet. Inst., Houston, TX, 1990.

[64] S. W. Taylor and P. R. Jaffé, *Water Resour. Res.* **26,** 2181 (1990).

[65] M. T. Suidan, B. E. Rittmann, and U. K. Traegner, *Water Res.* **21,** 491 (1987).

[66] B. E. Rittmann, *Water Resour. Res.* **29,** 2195 (1993).

strate removal is controlled by biomass accumulation, not internal mass transport.

Interphase Mass Transfer

External Mass Transport Resistance

Interphase mass transport processes are important for delivering substrates from the bulk liquid to the biofilm/liquid interface and for carrying away metabolic by-products. Quantitative criteria define when the biokinetics are slowed by external mass transfer resistance [Eq. (3)]. For the case of $\eta = 1$ and $C^* \ll 1$, external mass transfer resistance is unimportant when L^* is less than approximately 0.1.[65] If internal and external mass transport resistance are not important, the full biofilm model reduces to the macroscopic model and $C_f = C_s = C$. The substrate flux into the biofilm [(Eq. (2)] then becomes[66]:

$$J = q_m X_f L_f \frac{C}{K + C} \tag{6}$$

This is appropriate for groundwater situations with low substrate loads and substrate utilization kinetics limited by biomass accumulation, not mass transport.[66] Nevertheless, L_f^* and L^* criteria should be checked.

In a direct comparison of the macroscopic and biofilm models for realistic groundwater conditions, the two solutions converged for the organic substrate plume and the biomass distribution.[63] Thus, the added complexity of the biofilm model (and the microcolony model) is probably not needed for modeling the solute concentration for most groundwater situations. Further, the macroscopic model does not make assumptions concerning the specific spatial distribution of biomass, nor does it require difficult-to-determine biofilm parameters.[57,67] Finally, it avoids the conceptual disparity between the macroscale groundwater transport equations and the microscale biofilm model.

Other Physicochemical Limitations to Bioavailability

Although the mass transfer to the biofilm from the bulk phase, and within the biofilm, may not limit the biokinetics of subsurface attached biomass, other physicochemical processes may affect the bulk substrate concentration and, in turn, the biokinetics. Indeed, field and laboratory studies suggest that a large fraction of pollutants present in environmental

[67] T. P. Clement, B. S. Hooker, and R. S. Skeen, *Ground Water* **34,** 934, (1996).

systems are unavailable for microbial degradation.[68–70] Such observations indicate that the overall biotransformation rate depends not only on biokinetic factors, but also on physicochemical constraints that control pollutant bioavailability.[71]

For this discussion, mass transport and biodegradation are represented using a one-dimensional advection–dispersion reaction equation, with groundwater flow in the x direction, a sorptive porous media, and nonaqueous phase liquid (NAPL) present as a residual saturation of uniformly distributed immobilized blobs:

$$\frac{\partial C}{\partial t} = D_x \frac{\partial^2 C}{\partial x^2} - \frac{q_x}{nS_w}\frac{\partial C}{\partial x} + K_{\mathrm{ln}}(C_{\mathrm{eq}} - C) - K_{\mathrm{B1}}C + K_{\mathrm{ls}}(q/K_{\mathrm{D}} - C) \quad (7)$$

in which t = time $[T]$; D_x = longitudinal dispersion coefficient $[L^2/T]$; q_x = specific discharge $[L/T]$; n = porosity; S_w = water saturation; $K_{\mathrm{ln}} = a_{\mathrm{na}}k_{\mathrm{ln}}/\mathrm{n}nS_w$ = the lumped mass transfer rate coefficient for NAPL dissolution $[1/T]$, where $a_{\mathrm{na}} = A_{\mathrm{na}}/V$ = specific NAPL/aqueous phase interfacial area $[1/L]$, A_{na} = interfacial area between the NAPL and aqueous phases $[L^2]$, V = the total system volume $[L^3]$, k_{ln} = NAPL mass transfer rate coefficient $[L/T]$; C_{eq} = NAPL equilibrium concentration $[M/L^3]$; K_{B1} = lumped first-order biodegradation rate constant $[1/T]$; $K_{\mathrm{ls}} = a_{\mathrm{sa}}k_{\mathrm{ls}}/n_{Sw}$ = the lumped mass transfer rate coefficient for soil desorption $[1/T]$, where $a_{\mathrm{sa}} = A_{\mathrm{sa}}/V$ = specific soil/aqueous phase interfacial surface area $[1/L]$, A_{sa} = interfacial area between the NAPL and aqueous phases $[L^2]$, k_{ls} = the soil desorption mass transfer rate coefficient $[L/T]$; q = mass sorbate/mass solid $[M/M]$; and K_{D} = distribution coefficient $[L^3/M]$. Processes on the right-hand side of Eq. (7) represent, respectively, longitudinal dispersion, advection, and three reaction processes of interest: (1) mass transfer limited NAPL blob dissolution, e.g., Ref. 72; (2) biodegradation of dissolved contaminants, using macroscopic first-order biokinetics, e.g., Ref. 72; and (3) desorption from aquifer solids, using a simple linear driving force and lumped first-order mass transfer coefficient approach, e.g., Ref. 73. Equation (7) is easier to examine when transformed into a nondimensional form using the dimensionless parameters defined in Table II, which measure the relative rates of the interacting processes. Such dimensionless parameters have been used to succinctly capture the complexity of the system in a

[68] M. Alexander, *Environ. Sci. Technol.* **29,** 2713 (1995).

[69] A. J. Beck, S. C. Wilson, R. E. Alcock, and K. C. Jones, *CRC Crit. Rev. Environ. Sci. Technol.* **25,** 1 (1995).

[70] T. N. P. Bosma, P. J. M. Middeldorp, G. Schraa, and A. J. B. Zehnder, *Environ. Sci. Technol.* **31,** 248 (1997).

[71] A. Ramaswami, and R. G. Luthy, *in* "Manual of Environmental Microbiology" (C. J. Hurst *et al.,* eds.), p. 721. ASM Press, Washington, DC, 1997.

TABLE II

DIMENSIONLESS PARAMETERS Pe, Da_1, Da_2, Da_3, Da_4, AND Da_5[a]

Dimensionless parameters comparing mass transfer rates:[b]

$$Pe = \text{Peclet number} = \left(\frac{\text{advection rate}}{\text{dispersion rate}}\right) = \frac{q_x L}{D_x n S_w} = \frac{v_x L}{D_x}$$

Pe: $\gg 1$, dispersion is slower than advection; $\ll 1$, advection is slower

$$Da_1 = \text{Damköhler number 1} = \left(\frac{\text{NAPL mass-transfer rate}}{\text{advection rate}}\right) = \frac{k_{\text{ln}} a_{\text{na}} L}{q_x} = \frac{K_{\text{ln}} L}{v_x}$$

Da_1: $\gg 1$, advection is slower than NAPL dissolution; $\ll 1$, NAPL dissolution is slower

$$Da_4 = \text{Damköhler number 4} = \left(\frac{\text{soil mass-transfer rate}}{\text{advection rate}}\right) = \frac{k_{\text{ls}} a_{\text{sa}} L}{q_x} = \frac{K_{\text{ls}} L}{v_x}$$

Da_4: $\gg 1$, advection is slower than desorption from soil; $\ll 1$, desorption is slower

Dimensionless parameters comparing biodegradation and mass transfer:[b]

$$Da_2 = \text{Damköhler number 2} = \left(\frac{\text{biodegradation rate}}{\text{advection rate}}\right) = \frac{K_{\text{B1}} L}{v_x}$$

Da_2: $\gg 1$, slow advection limits biotransformation; $\ll 1$, slow biokinetics limit

$$Da_3 = \text{Damköhler number 3} = \left(\frac{\text{biodegradation rate}}{\text{NAPL mass-transfer rate}}\right) = \frac{Da_2}{Da_1} = \frac{K_{\text{B1}}}{K_{\text{ln}}}$$

Da_3: $\gg 1$, slow NAPL dissolution limits biotransformation; $\ll 1$, slow biokinetics limit

$$Da_5 = \text{Damköhler number 5} = \left(\frac{\text{biodegradation rate}}{\text{soil mass-transfer rate}}\right) = \frac{Da_2}{Da_4} = \frac{K_{\text{B1}}}{K_{\text{ls}}}$$

Da_5: $\gg 1$, slow desorption from soil limits biotransformation; $\ll 1$, slow biokinetics limit

[a] Adapted from Seagren *et al.* and Ramaswami and Luthy using the notation of Seagren *et al.* [E. A. Seagren, B. E. Rittmann, and A. J. Valocchi, *J. Contam. Hydrol.* **12,** 103 (1993); A. Ramaswami, and R. G. Luthy, *in* "Manual of Environmental Microbiology" (C. J. Hurst *et al.,* eds.), p. 721. ASM Press, Washington, DC, 1997.]

[b] L, characteristic length of the system [L]; v_x = average pore water velocity = q_x/nS_w.

quantitative framework.[70,72,74,75] A systematic comparison of the dimensionless numbers can be used to predict the rate-limiting phenomenon in the environmental system. Such an identification process is shown schematically in Fig. 1. The first three steps are used to identify the slowest mass transfer process and the fourth step compares that process rate with the biodegradation rate to determine the overall rate-limiting process. This approach is viable as long as the dimensionless parameters are significantly smaller or larger than unity because of the uncertainties associated with estimating mass transfer and biokinetic parameters.[71] Complementary ex-

[72] E. A. Seagren, B. E. Rittmann, and A. J. Valocchi, *J. Contam. Hydrol.* **12,** 103 (1993).

[73] M. T. van Genuchten and P. J. Wierenga, *Soil Sci. Soc. Am. J.* **40,** 473 (1976).

[74] E. A. Seagren, B. E. Rittmann, and A. J. Valocchi, *Environ. Sci. Technol.* **28,** 833 (1994).

[75] A. Ramaswami, S. Ghosal, and R. G. Luthy, *Water Sci. Technol.* **30,** 61 (1994).

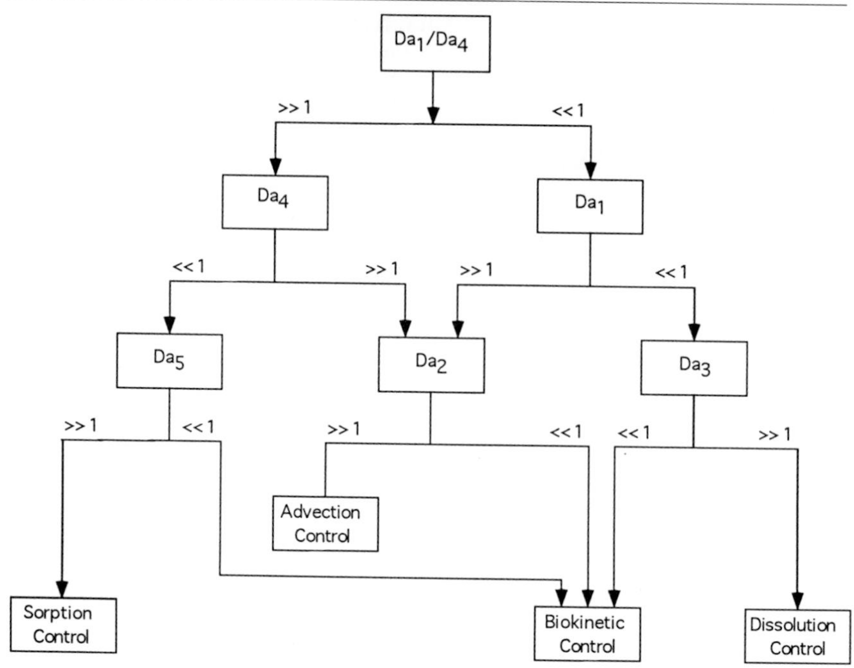

FIG. 1. Flow chart to predict rate-limiting phenomenon in the system [adapted from A. Ramaswami and R. G. Luthy, *in* "Manual of Environmental Microbiology" (C. J. Hurst *et al.*, eds.), p. 721. ASM Press, Washington, DC, 1997].

perimental protocols for evaluating the physicochemical limitations on biotransformation are described elsewhere.[71]

Using appropriate system boundary conditions, the governing equation [Eq. (7)] can be solved for the transient and steady-state solute concentration and flux as a function of the dimensionless parameters, and the influence of the various processes on the solute concentration can be investigated. For example, the effect of biodegradation on the solute concentration in this domain was investigated by Seagren *et al.*[72] for the steady-state solution of Eq. (7) without sorption, using a Pe value of 1.0 and three values of Da_1: 10^{-4}, 10^0, and 10^3. For the case $Da_1 = 10^3$, advection is the rate-limiting mass transport process, which allows the system to come to equilibrium with respect to NAPL/water interphase mass transfer ($C = C_{eq}$). Because $Da_1 \gg 1$, Da_2 is inspected next to compare the advection and biodegradation rates. If $Da_2 \ll 1$, biodegradation is the rate-limiting process (in fact, biodegradation is so slow relative to the advection rate that it has no impact on the solute concentration). If, however, $Da_2 \gg 1$, advection is the rate-limiting process and limits the substrate supply for biodegradation.

Bioenhancement of Mass Transport

As biofilms accumulate, they also can have significant impact on subsurface mass transport processes. For example, biofilm accumulation can influence solute transport through a porous medium by affecting the hydraulic conductivity, frictional resistance,[76] and hydrodynamic dispersion.[77] In addition, biofilm accumulation may result in a "bioenhancement" of interphase mass transfer. For example, biodegradation by attached microorganisms can also accelerate the interphase transfer processes of dissolution or desorption by increasing the concentration gradient at the interface. The phenomenon of NAPL pool dissolution flux bioenhancement was demonstrated experimentally by Seagren *et al.*[78] by comparing quasi-steady-state toluene dissolution fluxes determined using a ^{14}C mass balance approach with, and without, active solute biodegradation.

Seagren *et al.*[74] derived a dimensionless parameter for evaluating the bioenhancement of NAPL pool dissolution flux. This bioenhancement factor, E, is defined as the ratio of the total mass flux in the presence of biodegradation to the mass flux in the absence of biodegradation:

$$E = \frac{1}{2}\sqrt{\frac{\pi}{Da_2}}\left(\left(Da_2 + \frac{1}{2}\right)\operatorname{erf}\sqrt{Da_2} + \sqrt{\frac{Da_2}{\pi}}\exp(-Da_2)\right) \tag{8}$$

When Da_2 is less than 10^{-1}, E approaches 1 and no bioenhancement is observed; however, for large values of Da_2, the slope of the log E versus $\log(Da_2)$ curve becomes the constant limiting value of 0.5, illustrating increased bioenhancement at increased biokinetic rates (erf, error function).

Analogously, bioenhancement can also explain enhanced desorption of biodegradable organic compounds from solids[79–81] and the enhanced

[76] A. B. Cunningham, E. J. Bouwer, and W. G. Characklis, *in* "Biofilms" (W. G. Characklis and K. C. Marshall, eds.), p. 697. Wiley, New York, 1990.

[77] S. W. Taylor and P. R. Jaffé, *Water Resour. Res.* **26,** 2171 (1990).

[78] E. A. Seagren, B. E. Rittmann, and A. J. Valocchi, *in* "Proceedings of the Water Environment Federation 67th Annual Conference and Exposition," Vol. 2, p. 37. Water Environment Federation, Alexandria, VA, 1994.

[79] H. H. M. Rijnaarts, A. Bachmann, J. C. Jumelet, and A. J. B. Zehnder, *Environ Sci. Technol.* **24,** 1349 (1990).

[80] W. F. Guerin and S. A. Boyd, *Appl. Environ. Microbiol.* **58,** 1142 (1992).

[81] P. C. Wszolek and M. Alexander, *J Agric. Food Chem.* **27,** 410 (1979).

dissolution of solid organic compounds in the presence of aqueous phase microbial activity.[82]

Acknowledgments

We thank Gill Geesey, Montana State University; Barth Smets, University of Connecticut; Joseph Odencrantz, Tri-S Environmental, Inc.; and Neil Blough, University of Maryland whose valuable discussions contributed to this article. Funding was provided by NSF and EPA (CA), NOAA/Maryland Sea Grant, Oceanix Biosciences Corp., and MIPS (RW).

[82] J. M. Thomas, J. R. Yordy, J. A. Amador, and M. Alexander, *Appl. Environ. Microbiol.* **52,** 290 (1986).

[31] Biosurfactants Produced By *Lactobacillus*

By Gregor Reid, Christine Heinemann, Martine Velraeds, Henny C. van der Mei, and Henk J. Busscher

Introduction

Lactobacillus bacterial species have been shown to be important in the maintenance of a healthy gastrointestinal microbiota in men and women and to help protect against urogenital infections in women.[1] The critical components of importance in antagonizing pathogenic bacteria include an ability of the lactobacilli to adhere to surfaces, inhibit attachment and growth of the pathogens, resist spermicidal killing, and coaggregate with other members of the microbiota to produce a balanced microbial niche.[2] More recently, strains of lactobacilli were found to produce a biosurfactants collectively termed surlactin, characterized by the ability to reduce liquid surface tension, as measured by the axisymmetric drop shape analysis (ADSA)[3,4] and by the ability to inhibit the adhesion of enterococci and other uropathogens to polymer surfaces.[3–5] The active components of these

[1] G. Reid, A. W. Bruce, J. A. McGroarty, K.-J. Cheng, and J. W. Costerton, *Clin. Microbiol. Rev.* **3,** 335 (1990).

[2] G. Reid, A. W. Bruce, and V. Smeianov, *Int. Dairy J.* **8,** 555 (1998).

[3] M. C. Velraeds, H. C. van der Mei, G. Reid, and H. J. Busscher, *Appl. Environ. Microbiol.* **62,** 1958 (1996).

[4] M. C. Velraeds, H. C. van der Mei, G. Reid, and H. J. Busscher, *Coll. Surf. B Biointerfaces* **8,** 51 (1996).

[5] M. C. Velraeds, B. van der Belt, H. C. van der Mei, G. Reid, and H. J. Busscher, *J. Med. Microbiol.* **49,** 790 (1998).

0076-6879/99 $30.00

particular biosurfactants appear to be proteinaceous and carbohydrate. There are other examples of carbohydrate-containing biosurfactants, such as the rhamnolipid produced by *Pseudomonas aeruginosa*.[6] The heterogeneous nature of biosurfactants is illustrated further in *Bacillus licheniformis* with a fatty acid, lipopeptide, and polysaccharide content.[7] The intruiging part about the surlactin is its potential importance in helping to regulate the microbiota of the urogenital tract, making it potentially a critical factor in health maintenance of a balanced microbiota.

Methods Used to Identify Biosurfactants

Bacterial Strains

A number of lactobacilli strains tested have been found to express biosurfactants and thus a positive control strain should be included. In our experience, *Lactobacillus fermentum* RC-14, isolated originally from the urogenital tract of a healthy female, is an excellent marker strain. The organisms should be stored at −60° in MRS both (BDH, Darmstadt, Germany) and for culture they are thawed, streaked onto MRS agar, and incubated at 37° in 5% (v/v) CO_2.

Biosurfactant Production and Isolation

An 8.5-ml inoculum of a 24-hr RC-14 culture is put into 400 ml MRS broth and incubated for 18 hr. The stationary phase cells are harvested by centrifugation (10,000*g*, 10 min, 10°); (Sorvall RC-5B, SS-34 rotor, Du Pont Instruments, Newtown, CT), washed twice in demineralized water, and suspended in 67 ml phosphate-buffered saline (PBS: 10 m*M* KH_2PO_4/K_2HPO_4, 150 m*M* NaCl, pH 7.0). The lactobacilli are placed at room temperature for biosurfactant release with gentle stirring for 2 hr.

The bacteria and supernatant are separated by centrifugation (10,000*g*, 15 min, 10°), and the supernatant is removed, filtered (0.22 μm cellulose acetate, VWR Scientific Inc., West Chester, PA), and dialyzed against double demineralized water at 4° in a Spectrapor membrane tube (6000–8000 molecular weight cutoff, Spectrum Medical Industries Inc., CA). The retentate can be used at this state or freeze-dried at −10° and −5 μm Hg for 1–2 days, collected, and dried overnight in a Savant speed vacuum (RCT 4104). The collected biosurfactant powder is stored at −20°.

[6] M. E. Mercade, L. Monleon, C. de Andres, I. Rondon, E. Martinez, E. Espuny, and A. Manresa, *J. Appl. Bacteriol.* **81,** 161 (1996).

[7] K. Jenny, O. Kappeli, and A. Fiechter, *Appl. Microbiol. Biotechnol.* **36,** 5 (1991).

Axisymmetric Drop Shape Analysis by Pendant Drop Method (ADSA-P)

The ADSA-P is a technique used to determine liquid interfacial surface tension,[8] where the shape of the drop is fitted to a theoretical drop profile according to the Laplace equation of capillarity. The technique has been applied successfully to identify biosurfactant-producing organisms.[9,10] The profile of the droplet is fitted to the Laplace equation by deriving exact x and y magnification factors from profiles of a perfectly spherical stainless steel ball. ADSA-P is more suitable for surface tension measurements than dynamic methods such as ring tensiometry, which involves continuous destruction and creation of interfacial areas.

We have used various concentrations of biosurfactant suspension and principally 1 mg/ml PBS. The profile for each 100-μl droplet placed on fluoroethylene–propylene–Teflon in an enclosed chamber is monitored for 2 hr at room temperature and the surface tension is calculated as a function of time.

Adhesion Assays

The biological activity of the biosurfactants is measured with respect to the interference of adhesion of pathogens to surfaces. It is feasible that once more information is known about lactobacilli biosurfactants, alternative activity tests will be derived. Both static and flow techniques are used to determine microbial adhesion.

A polystyrene adhesion assay was developed from previously reported methods.[11–13] A suspension of 3×10^8/ml PBS-washed *Enterococcus faecalis* 1131, grown on BHI agar (BDH, Darmstadt, Germany), is added to wells of a flat, polystyrene microtiter plate (Corning Glassworks, Corning NY) that has been coated with biosurfactant for 18 hr at 4° on a rotating platform (2.5 rpm). Control wells comprise buffer without biosurfactant. A 200-μl suspension of enterococci is added and incubated in the wells for 4 hr at 4° at 2.5 rpm. Unattached organisms are removed by pipetting off suspending fluid and rinsing the wells gently three times with PBS. Adherent bacteria are then stained with crystal violet and optical density readings

[8] D. Y. Kwok, M. A. Cabrerizo-Vilchez, Y. Gomez, S. S. Susnar, O. del Rio, D. Vollhardt, R. Miller, and A. W. Neumann, *in* "Dynamic Properties of Interfaces and Association Structures" (V. Pillai, and D. Oshah, eds.), p. 278. AOCS Press, Illinois, 1996.

[9] H. J. Busscher, M. van der Kujil-Booij, and H. C. van der Mei, *J. Ind. Microbiol.* **16,** 15 (1996).

[10] W. Van der Vegt, H. C. van der Mei, J. Noordmans, and H. J. Busscher, *Appl. Microbiol. Biotechnol.* **35,** 766 (1991).

[11] S. Goldberg, Y. Konis, and M. Rosenberg, *Appl. Environ. Microbiol.* **56,** 1678 (1990).

[12] S. A. Klotz, D. Drutz, and J. E. Zajic, *Infect. Immun.* **50,** 97 (1985).

[13] M. Rosenberg, *FEMS Microbiol. Lett.* **25,** 41 (1984).

are taken at 595 nm. The change in adhesion with controls is calculated as percentage adhesion.

A second adhesion assay is performed whereby the organisms were allowed to adhere to the polystyrene and then are challenged with 200 μl biosurfactant for 4 hr at 4° to determine whether bacterial detachment occurred.

The first two assays represent a static situation similar to that found on devices (at the outer surface and tissue interface) and uroepithelial and vaginal cells. The assay can be carried out using vaginal epithelial cells with a view to simulating the *in vivo* situation where bacteria and their byproducts interact. In this third scenario, vaginal epithelial cells are recovered from a healthy woman either via a swab, which is then suspended in 1–5 ml saline, or by instilling 5 ml of saline and collecting the washings. The suspension is centrifuged at 1000*g* for 10 min and washed in saline or PBS (pH 5.5–7.1) and the supernatant is discarded. The sedimented cells will have indigenous bacteria attached, including lactobacilli. Cell suspensions of 100,000 per milliliter are incubated for various time periods up to 24 hr at 4° with biosurfactant. A control incubated with PBS is used. In order to determine whether the treatment caused bacterial detachment, 20 to 50 cells per triplicate sample are gram stained and examined under 1000× oil immersion and adherent bacteria counted. Of particular interest is the reduction in net numbers of gram-negative rods. This experiment also acts to determine the potential for the biosurfactant to actually treat bacterial biofilms on tissues.

A fourth method is designed to examine interfaces exposed to flow (urine, blood). For this purpose, a parallel plate flow chamber is used attached to a microscope and image analysis system.[3–5] A laminar flow is obtained in the middle of the chamber by increasing and decreasing gradually the depth and width of the inlet and outlet channels, respectively. The chamber is mounted on a stage of an Olympus Model BH-2 phase-contrast microscope equipped with a 40× objective having an ultralong working distance (Model ULWD-CD Plan 40 PL, Olympus). A charge-coupled device camera (CCD-MX High Technology, Eindhoven, The Netherlands) is linked to an image analyzer (TEA image manager, Difa, Breda, The Netherlands), which is installed on a personal computer. The system allows direct observation over a field of view of 0.011 mm^2. The top and bottom glass plates of the chamber (5.5 × 3.8 cm) and two Teflon spacers (0.06 cm thick) are cleaned ultrasonically in a 2% RBS surfactant solution in water (Omnilabo International BV, The Netherlands) for 10 min, rinsed thoroughly with warm tap water, methanol, and demineralized water, and finally secured in the flow chamber. These glass plates are wettable by water (contact angle 0°). The flow chamber is then filled with 10 ml biosurfactant and left at room temperature (approximately 25°) overnight for adsorption.

The pathogen (in our case, *E. faecalis, Escherichia coli, Klebsiella*, etc.) suspension (3×10^8 cells^{-1} in 250 ml PBS) is allowed to flow through the chamber at room temperature. A pulse-free flow (rate 0.034 ml/sec) is created by hydrostatic pressure, which produces a constant shear rate of 15 sec^{-1}, and the suspension is recirculated using a Multiperspex Model 2115 peristaltic pump. Images are grabbed during the experiment and stored on the computer for later analysis.

An initial deposition rate is calculated by an examination of the number of bacteria adherent per unit area with time by a linear least-square fitting procedure. After 4 hr, the number of adherent bacteria is again determined and the suspension is drained from the system, a process which allows an air–liquid interface to pass over the substratum. A final image is then taken. A comparison of pre- and postdrainage counts yields an indication of the strength of adhesion.

Expected Results and Clinical Relevance

ADSA-P will indicate whether biosurfactant activity is present. The drop in liquid surface tension of *Lactobacillus* spp. should be between 12 and 29 mJ/m^2 and possibly even higher (Fig. 1). A negative control, such as *E. faecalis* 1131, has a value of 4 mJ/m^2.

The biological activity is measured by the adhesion assays. Our experience with over several hundred experiments is that biosurfactant deposition should reduce subsequent adhesion by up to 98%. The clinical significance of these values has yet to be determined and so it is unclear how significant a reduction of say 23% in adhesion of *Klebsiella pneumoniae* 3a means (Fig. 2).[14] These reductions have been found in static and flow cell studies. Preferably, the effect should exist for a range of pathogens, but it is possible that only a select few species or strains are inhibited. In the urogenital tract, interference with binding of the main pathogens (generally gram-negative coliforms and gram-positive cocci) could imply clinical importance in keeping the pathogen colonization levels down.

It has been suggested that the biosurfactant can detach pathogens from vaginal epithelial cells, but our studies show that detachment does not occur to biomaterials in a flow cell. For clinical relevance, a detachment of 20 to 50% of pathogens from vaginal cells would be significant and would suggest that the agent could perhaps be applied as a douche-type treatment to

[14] M. Velraeds, Ph.D. Thesis, University of Groningen, The Netherlands, 1998.

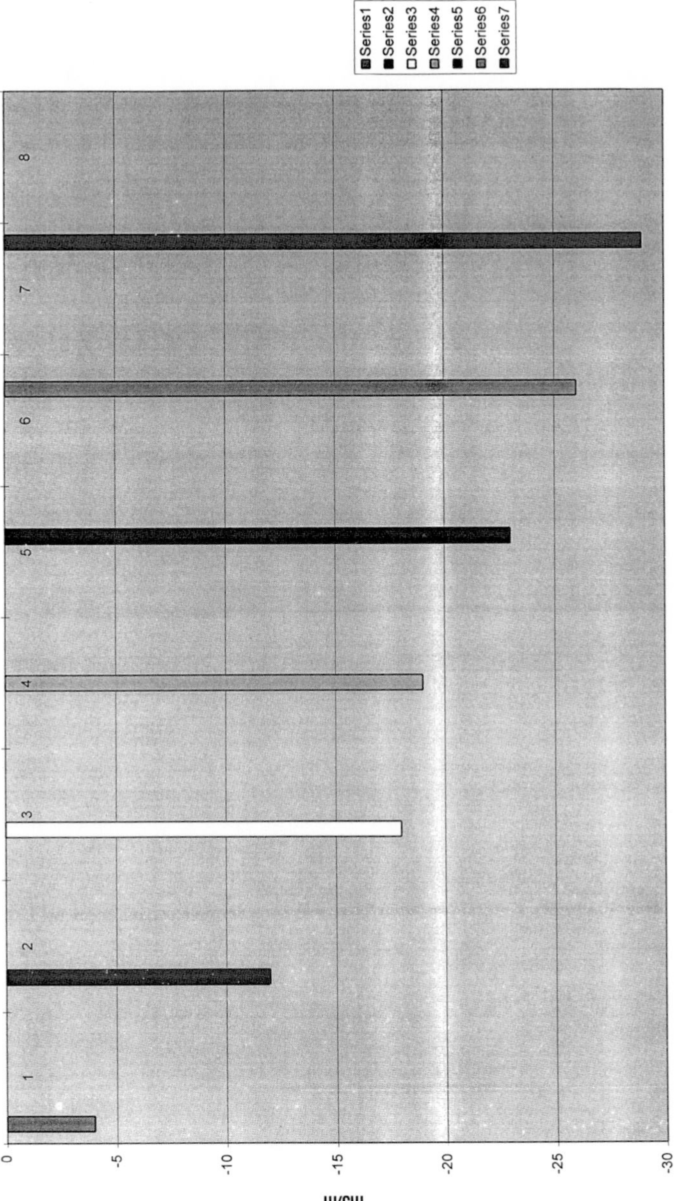

FIG. 1. The decrease in liquid surface tension as measured by ADSA-P found for the suspensions from the following strains: Series 1, *E. faecalis* 1131; series 2, *L. casei* subsp. *casei* 393; series 3, *L. acidophilus* T-13; series 4, *L. rhamnosus* GR-1; series 5, *L. plantarum* RC-6; series 6, *L. fermentum* RC-14; series 7, *L. fermentum* B-54.

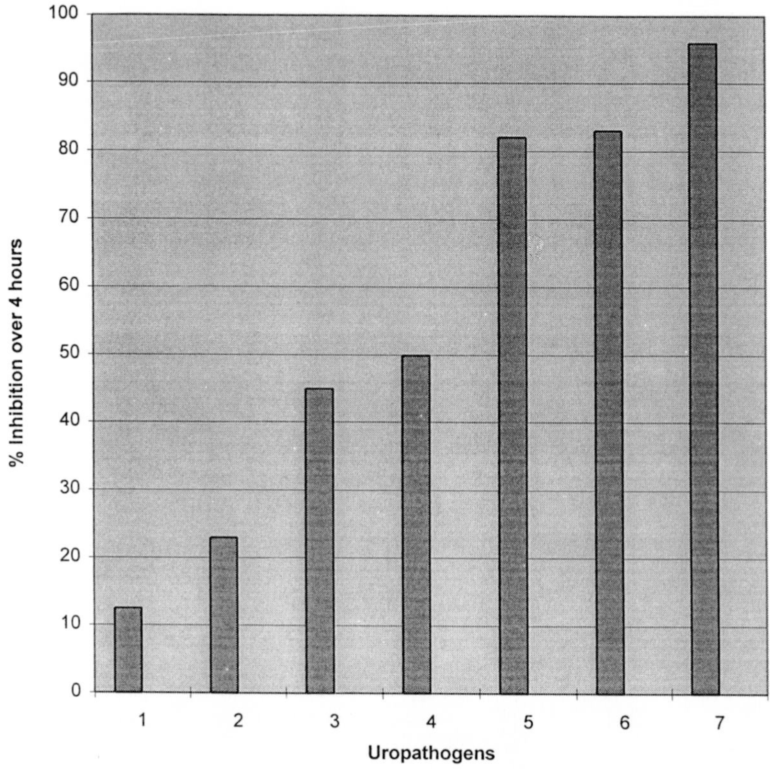

FIG. 2. Percentage inhibition of uropathogen adhesion to silicone rubber with an adsorbed layer of surfactant from *L. fermentum* RC-14 as studied in a parallel plate flow chamber after 4 hr. Uropathogens: (1) *Candida albicans* Urine 1, (2) *Klebsiella pneumoniae* 3a, (3) *Providencia stuartii* UHL 5292, (4) *Pseudomonas aeruginosa* ATCC 10145, (5) *Escherichia coli* Hu734, (6) *Staphylococcus epidermidis* 3081; and (7) *Enterococcus faecalis* 1131.

reduce the colonization levels of potential infectants. However, the theory needs to be proven *in vitro* as well as *in vivo*.

To date, a mutant *Lactobacillus*, which is a nonproducer of biosurfactant (as measured by ADSA-P) has not been isolated or found by our group. This is not to say that such strains do exist. Ideally, these experiments would be best carried out with a negative lactobacilli control. What has been found, however, is that some strains have less of a biological activity than others in inhibiting the adhesion of pathogens to substrata.

The importance of lactobacilli in intestinal and urogenital biofilms is now unquestioned. By producing biosurfactants, it appears that these organisms have a mechanism to manipulate the structure and composition of the microbial biofilms in which they inhabit. Given the slower growth rate of lactobacilli over many pathogens, such properties could be critical to the survival of the organism.

Acknowledgments

We thank Vladimir Smeianov, Dominique Lam, Rocio Navarro, and others who have helped to develop these techniques and improve the understanding of biosurfactants in relation to urogenital biofilms.

[32] Proteome Analysis of Biofilms: Growth of *Bacillus subtilis* on Solid Medium as Model

By BRIAN S. MILLER and MARIA R. DIAZ-TORRES

Introduction

A number of techniques have been developed to study gene expression in cells growing in biofilms. Typically, these are used to study expression from single genes such as by reporter gene fusions.[1] Fluorescent systems have been used to gauge gene expression by the detection of secreted gene products when genetic systems are unavailable.[2] Radiolabel assay techniques have been used to measure biofilm growth by total protein production.[3] We have moved to take advantage of recent technological advances to begin to study global gene expression during bacterial growth in biofilms.

Advances in DNA sequencing technology has led to the complete sequencing of the genomes of a number of microorganisms. The future promises an explosion of the number of complete microbial genomes available.[4] These developments have coincided with advances in protein analysis techniques for the rapid identification of proteins separated by two-dimensional (2D) polyacrylamide gel electrophoresis.[5] While still under development, these advances in 2D gel technology have made possible the serious study of the proteome for organisms with a sequenced genome. The proteome

[1] D. G. Davies and G. G. Geesey, *Appl. Environ. Microbiol.* **61,** 860 (1995).

[2] C.-T. Huang, K. D. Xu, G. A. McFeters, and P. S. Stewart, *Appl. Environ. Microbiol.* **64,** 1526 (1998).

[3] M. Hussain, C. Collins, J. G. Hastings, and P. J. White, *J. Med. Microbiol.* **37,** 62 (1992).

[4] E. Pennisi, *Science* **280,** 672 (1998).

[5] M. J. Dunn, *Biochem. Soc. Trans.* **25,** 248 (1997).

0076-6879/99 $30.00

concept involves nothing less than examination of the pattern of expression of all cell proteins under different conditions or by a specific cell type.[6] For those microorganisms with limited or no DNA sequence available, protein identifications can be made many times with high confidence using N-terminal Edman microsequencing and/or combinations of other identification techniques. The identifications are then made by sequence database search tools.[6]

In the course of our work we have isolated *Bacillus subtilis* mutant strains that show interesting phenotypes during growth on a solid medium compared to isogenic parental strains. For some of these strains, such phenotypic differences have been difficult to detect in liquid media. We have also found strains that are not readily amenable to genetic analysis. We therefore decided to see if we could radiolabel such a set of mutant and control cells on a solid medium of interest and reveal obvious differences using two-dimensional gel electrophoresis. While we were successful in this endeavor, we feel that expansion of the technique would be ideal for the general study of growth on a solid medium. *Bacillus subtilis* can serve as a model for biofilm studies as it is the most well-studied gram-positive organism and it has a completely sequenced genome.[7] As far as growth on a solid medium, *B. subtilis* has been reported to undergo differential gene expression during colony formation.[8] *Bacillus subtilis* has also shown solid media-dependent sporulation phenotypes and regulatory pathways.[9,10] Peptide cell signaling factors have been identified in *B. subtilis*, which stimulate various developmental pathways associated with the beginning of the stationary growth phase.[11]

Procedure for Radiolabeling of *Bacillus subtilis* Cells on Solid Growth Medium and Two-Dimensional Gel Electrophoresis

Prior to labeling plate cultures, strains are grown overnight on agar plates at 37°. The following day, cells are removed from the plate with a sterile loop and streaks are made of strains onto the solid medium plate of choice. After 16 hr of growth at 37°, 1 × 5 mm-long streaks are labeled with EXPRE^{35}S^{35}S (L-[^{35}S]methionine, L-[^{35}S]cysteine) protein labeling mix

[6] M. R. Wilkins, J.-C. Sanchez, A. A. Gooley, R. D. Appel, I. Humphery-Smith, D. F. Hochstrasser, and K. L. Williams, *Biotech. Genet. Engin. Rev.* **13,** 19 (1995).

[7] F. Kunst, N. Ogasawara, I. Moszer *et al., Nature* **390,** 249 (1997).

[8] B. Salhi and N. H. Mendelson, *J. Bacteriol.* **175,** 5000 (1993).

[9] D. Foulger, G. F. Parker, and J. Errington, *Microbiology* **141,** 1763 (1995).

[10] M. O'Reilly, K. Woodson, B. C. A. Dowds, and K. M. Devine, *Mol. Microbiol.* **11,** 87 (1994).

[11] B. A. Lazazzera, J. M. Solomon, and A. D. Grossman, *Cell* **89,** 917 (1997).

(NEN Life Science Products, Boston, MA) by directly applying 5 μl (40 μCi) of label to the streak. Plates are then incubated for various time periods at 37° at which time cells were scraped off with a sterile loop for processing.

Protein Sample Preparation

Bacillus subtilis cells are processed by a modification of a method used for labeling *Streptomyces* cells at Institute Pasteur, Paris, France.[12] Both radiolabeled and unlabeled cells harvested from solid media may be processed by the following method. Cells are first washed three times with cold 10 m*M* Tris, pH 8.0. The pellets are then suspended in 30 μl of 10 mg/ml lysozyme (in 10 m*M* Tris, pH 8.0), plus 10 μl phenylmethylsulfonyl fluoride (10 m*M* stock) and incubated for 30 min at 37°. Following incubation, 1 ml of SB1 [0.3% sodium dodecyl sulfate (SDS), 200 m*M* dithiothreitol (DTT), 28 m*M* Tris–HCl, and 22 m*M* Tris base] is added to the lysed cells and heated at 100° for 5 min. After cooling, 50 μl of SB2 (24 m*M* Tris base, 476 m*M* Tris–HCl, 50 m*M* $MgCl_2$, 1 mg/ml DNase I, and 0.25 mg/ml RNase A) is added, followed by incubation at room temperature for 5 min. Lysates are then heated to 100° for 1 min and then centrifuged at 12000 rpm for 10 min at room temperature. Supernatants are precipitated for 10 min with 10% (final concentration) trichloroacetic acid (TCA) at 4°. After centrifugation at 12,000 rpm for 5 min, the pellets are washed twice with acetone at 4° and air dried for 5 min. Fifty microliters of SB3 [9.9 *M* urea, 4% Nonidet P-40, 2.2% 3-10/2D Ampholytes (ESA Inc.), and 100 m*M* DTT] is then added to the pellets and allowed to solubilize for 30 min. The amount of radioactivity in the samples is then determined by liquid scintillation counting. In our model system we obtain similar numbers for total TCA precipitable counts from 1- and 4-hr labelings. The total counts in 50 μl of SB3 ranged from 13 to 26 $\times 10^6$ cpm.

Two-Dimensional Gel Electrophoresis

Two-dimensional gel electrophoresis can be performed using any of the available systems. We use the Investigator 2D electrophoresis system (Genomic Solutions Inc., Chelmsford, MA). 3-10/2D Ampholytes are used for the first-dimension IEF separation and a 12% gel is used for the second-dimension separation. A total of 3×10^6 cpm of protein sample is loaded per gel.

Completed gels are stained with Coomassie Blue R-250 to visualize the protein standards (Bio-Rad, Richmond, CA), dried on 3MM chromatography paper (Whatman, Clifton, NJ), and exposed to X-OMAT AR film

[12] B. S. Miller, A. K. H. Hsu, E. Ferrari, and M. R. Diaz-Torres, *Anal. Biochem.* **245,** 245 (1997).

(Kodak, Rochester, NY) for 8 days. Gels from unlabeled cell preps are silver stained to visualize the protein spots.[13]

Results

Protein Patterns Revealed by Two-Dimensional Gels

Figure 1 shows the type of result expected using the described labeling, sample preparation, and 2D gel electrophoresis techniques. Examination of the autoradiograms from the 4-hr plate culture labeling reveal excellent protein resolution and large differences in the protein synthesis pattern between an isogenic mutant and control (Figs. 1A and 1B). The cells labeled for 1 hr produced a similar pattern (not shown). These gels were compared to the pattern of protein synthesis for cells labeled during liquid culture at the beginning of the stationary phase. Liquid culture-labeled cells only showed a few differences in the pattern of induction between mutant and control. These differences included two proteins, which were strongly induced by the mutant strain (Figs. 1C and 1D). When the protein spots showing different induction patterns between the mutant and control labeled on plates were compared to the spots showing differences between the strains labeled in liquid medium, we were only able to find two obvious spots in common. These spots, labeled 1 and 2 in Fig. 1 at approximately 30,000 M_r, were strongly induced in the mutant strain compared to the control in both the solid and the liquid media. These data show the potential of the technique to reveal the correspondence of a specific phenotype seen on solid media to a specific protein expression pattern on two-dimensional gels and to allow one to make comparisons with cells labeled the more conventional way in a liquid medium.

Identification of Proteins Revealed by Two-Dimensional Gel Electrophoresis

The full potential of the proteomic approach is reached via the identification of proteins of interest. A number of techniques have been used to identify protein spots on two-dimensional gels, including N-terminal Edman microsequencing, amino acid composition, and peptide mass fingerprinting.[5] In the future, with the standardization of techniques, it should be possible to conduct experiments and identify the spots of interest from master

[13] T. Rabilloud, *Electrophoresis* **13,** 429 (1992).

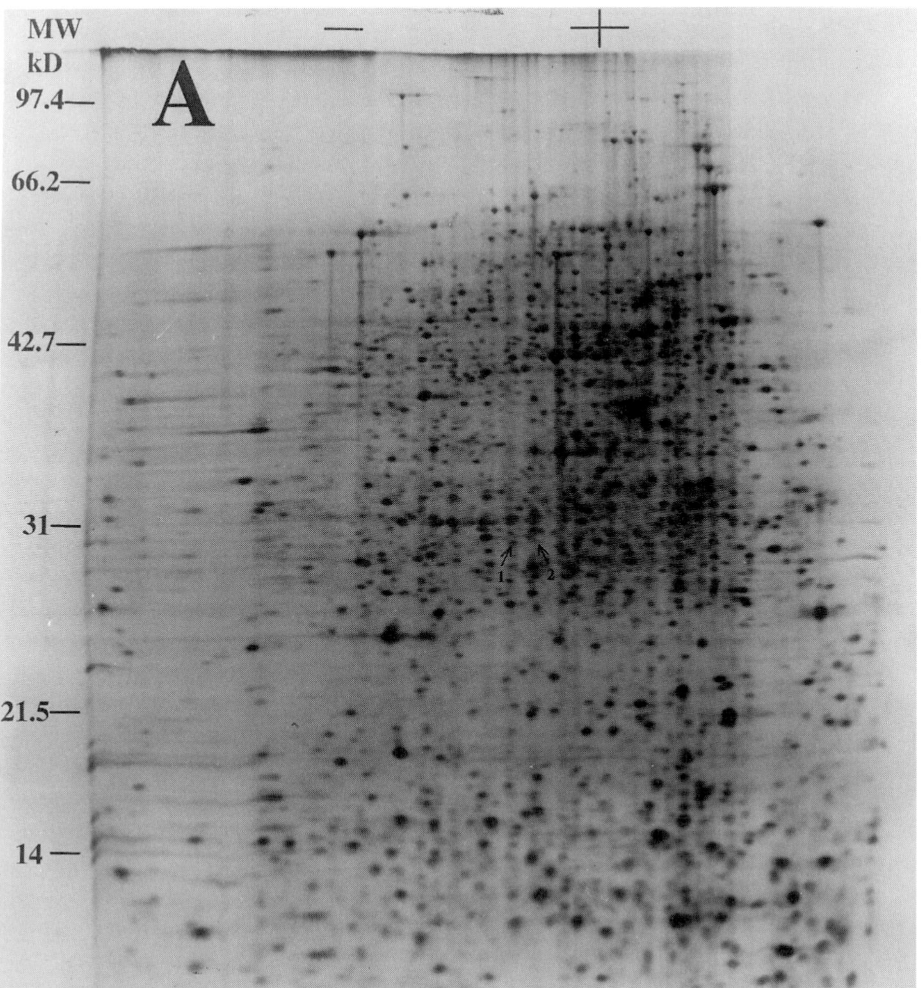

FIG. 1. Autoradiograms of two-dimensional gels of *B. subtilis* control and mutant proteins labeled on a solid medium for 4 hr (A and B, respectively) and liquid media (C and D, respectively). The orientation of the isoelectric focusing gel is indicated at the top of the gel by plus and minus symbols. Molecular weight standards are indicated on the left. Positions of the two proteins (1 and 2), which are induced in the mutant in both solid and liquid media, are indicated by arrows.

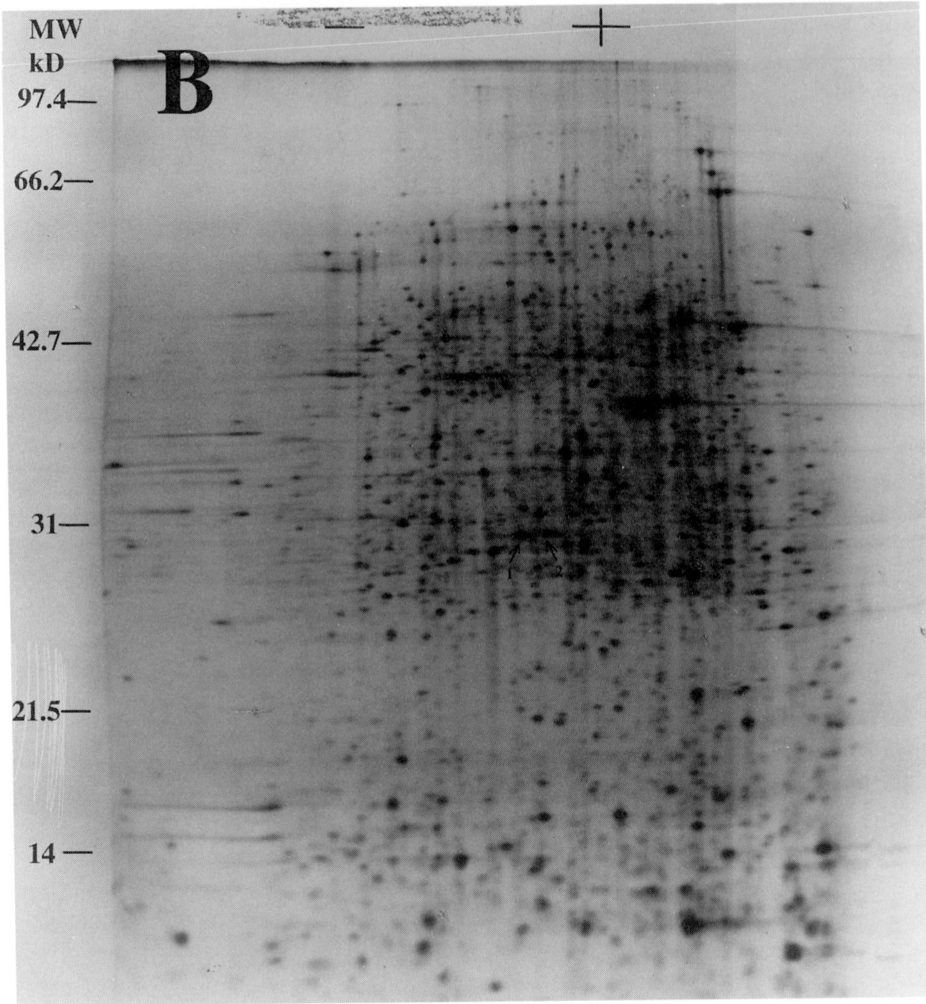

FIG. 1. (*continued*)

gels of the organism in question available on the World Wide Web.[5] The beginnings of such a site for *B. subtilis* 2D gels are now available.[14]

Discussion

In our first report of the method presented here we discussed one of the possible limitations of this type of analysis of biofilms, that being the

[14] R. Schmid, J. Bernhardt, H. Antelmann, A. Volker, H. Mach, U. Volker, and M. Hecker, *Microbiology* **143,** 991 (1997).

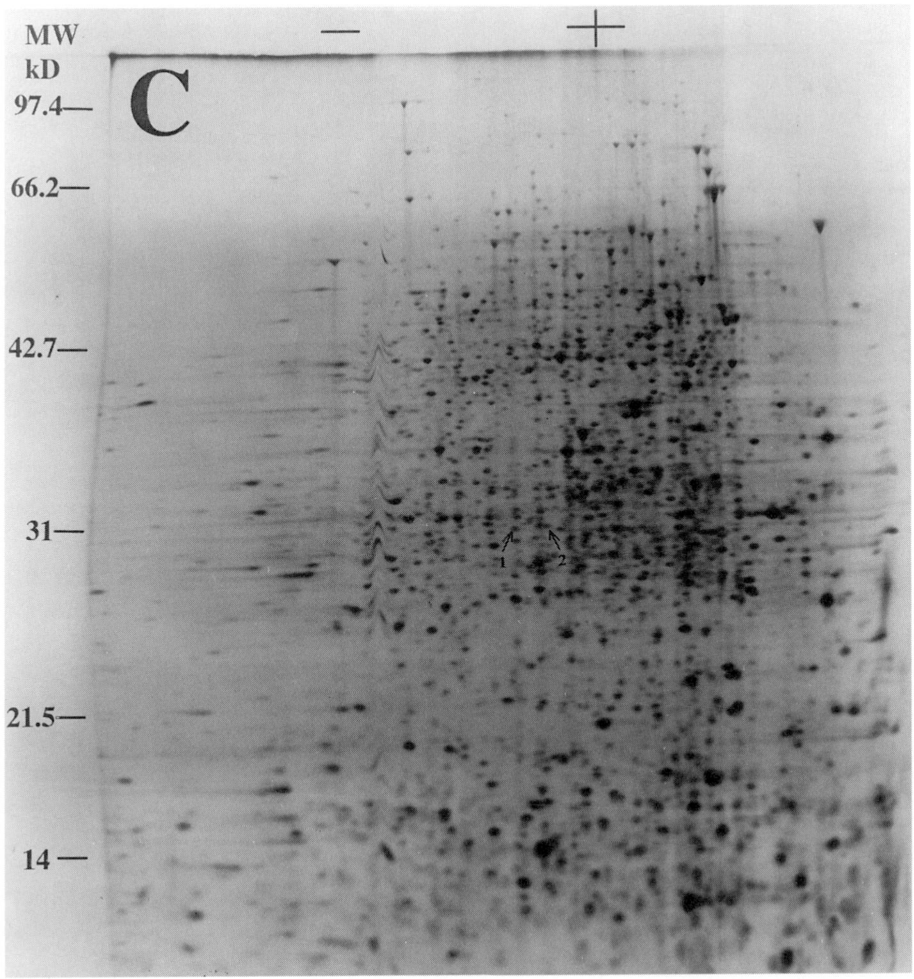

FIG. 1. (*continued*)

protein sample represents a heterogeneous population of cells.[12] This is in comparison to the more homogeneous populations represented by well-designed liquid medium experiments. Our results with solid medium labeling showed major differences between strains and media and provided valid data for study. However, we can envision modifications to how the cells are grown and harvested to sample subpopulations. This could include the plating of cells in films that are allowed to grow for short time periods before harvest to minimize heterogeneity. Of course much of the interest in biofilms involves the dynamics of interactions between cells in different

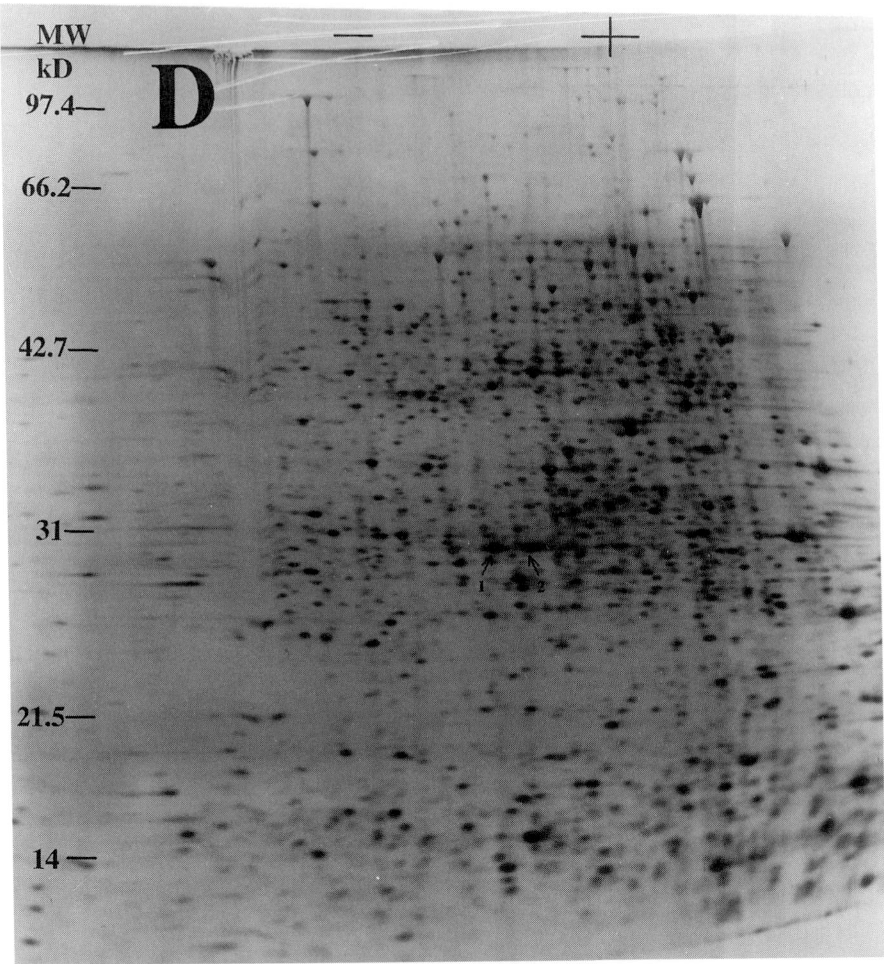

FIG. 1. (*continued*)

states of gene expression. Perhaps the sampling of cells from different zones of large colonies will be the next step for this type of proteomics approach. Mixed strain/species biofilms will require more creativity, but one could see a system that allows diffusion of small molecules while retaining separate populations for sample harvest. While we used radiolabeling to capture a picture of protein synthesis in a specific time period, this may be difficult in some situations. Analysis of proteins separated on 2D gels by silver stain would offer a look at the entire protein complement of the cell. Using *B. subtilis* as a model would begin to give an understanding of the many aspects of biofilm growth. Any number of questions can be envisioned such as the effect of medium composition, nutrient limitations, colony aging, and different environmental conditions. A database of the biofilm proteome responses should give a better picture of *B. subtilis* and other organisms in their natural habitat, rather than the useful but more artificial system of liquid media.

[33] Physiologic Homeostasis and Stress Responses in Oral Biofilms

By Robert A. Burne, Robert G. Quivey, Jr., and Robert E. Marquis

Biofilms, composed of more than 300 different species of bacteria, coat the hard and soft tissues of the human oral cavity. Nutritional requirements of the microorganisms are fulfilled by metabolism of the complex mixtures of nutrients that are provided continuously in secretions of the major and minor salivary glands and intermittently by constituents of the host diet. In most cases, the biofilms are benign and the integrity of the host tissue is maintained through a counterbalance between the detrimental forces of the enzymatic and metabolic products of the bacteria and the ameliorative forces of the biofilm metabolism and host defenses. Essentially all diseases of the oral cavity that are elicited by microorganisms arise as a result of perturbations of the normal physiologic homeostasis of oral biofilms, most frequently by changes in diet or alterations in normal host functions, such as changes in salivary flow rate or rather poorly understood changes in immunologic function. These perturbations can encourage emergence of a pathogenic microbiota with enhanced or additional metabolic and enzymatic activities that lead to the destruction, and loss of normal function, of host tissues. For example, the development of dental caries is fostered by high carbohydrate diets, which drive the emergence of an aciduric, acidogenic microbiota capable of demineralizing the tooth. Concurrently, organisms that are abundant in a "healthy" supragingival microbiota and have enzymatic activities that can generate alkali to neutralize plaque acids, such as certain sanguis streptococci, oral hemophili, and *Actinomyces* spp., are reduced in numbers or eliminated from the populations. Similarly, periodontal diseases result from rather poorly defined local changes that

0076-6879/99 $30.00

drive a shift in the subgingival microbiota from one that is rich in gram-positive, facultative organisms, to a predominantly gram-negative, anaerobic population.[1–3]

Interest in biofilms of oral microorganisms, mainly bacteria, derives in part from the desire to study the basic biology of the organisms in laboratory systems that approximate those of dental plaque in the animal body and from very practical needs to test the efficacy of agents developed to control oral infectious diseases in systems intermediate between suspension cultures in the laboratory and clinical trials in the field. A major challenge in the use of biofilms is to be able to monitor the same physiological processes that can be monitored more easily in suspension cultures. Knowledge of the physiology of bacteria in biofilms is increasing steadily, but at a relatively slow rate. Still, we have growing appreciation for general differences between biofilm populations and populations of planktonic bacteria. One of the major advantages of the biofilm mode of life is that environmental stresses are moderated because of high population densities and because of diffusion barriers. Thus, cells in biofilms generally have more time to adapt to a stress applied to the film than if the same stress were applied to cells in suspensions. The growing knowledge of the adaptive responses to environmental stresses in most, and probably all, microorganisms suggests that these responses were developed in relation to biofilm life. The adaptive responses can be slow and generally require synthesis of new proteins. In effect, the responses are to stresses that augment slowly, as do stresses for cells within biofilm communities. Importantly, these stressors are those that have also been well established to have major impacts on the gene expression patterns and physiologic responses of all bacteria. Stimuli that have been shown to be important in community composition or for the persistence of oral bacteria include acid and alkali, oxidative, osmotic, and salt stresses, as well as stresses imposed by starvation or chemical agents, such as fluoride, transition metals, and antiplaque agents. This article details a variety of methods that have been used to study the fundamental biochemical, physiologic, and genetic events that contribute to biofilm homeostasis or that allow for the emergence of pathogenic bacteria in adherent populations.

[1] R. A. Burne, *J. Dent. Res.* **77,** 445 (1998).

[2] W. F. Liljemark and C. Bloomquist, *Crit. Rev. Oral Biol. Med.* **7,** 180 (1996).

[3] G. H. W. Bowden, D. C. Ellwood, and I. R. Hamilton, *in* "Advances in Microbial Ecology" (M. Alexander, ed.) p. 135. Plenum Press, New York, 1979.

In Vitro Cultivation of Oral Biofilms: An Overview

Batch Cultivation

There are a wide variety of biofilm model systems based on batch cultivation, usually of single species, that are simple, inexpensive, and have multiple applications. Organisms can be grown in microtiter wells or, alternatively, cultivated on a variety of materials, including glass rods, wires, microscope slides, or even hydroxylapatite disks. Biofilms are grown batchwise, and if thicker films are desired, media can be replaced periodically until the biomass or biofilm depth is sufficient for analysis. Immobilized organisms grown in this fashion are useful for a wide variety of assays, such as testing for resistance to physical or chemical stresses, exploration of gene expression, or for physicochemical analyses. The most serious limitation of batch cultivation is that there is comparatively little control over environmental conditions. Other considerations are that microtiter well systems have a limited repertoire of readily available substrata and that the cell yields can be insufficient for assay systems lacking high levels of sensitivity. Batch systems in which the cells are grown on a material inserted into a culture tube or vessel can provide higher cell yields and allow for more variety in the substratum.

Continuous Flow Systems

Flow cells, stirred tank reactors, chemostat-based systems, and the continuous depth film fermentor have proven to be powerful tools for analyzing adherent populations of oral microorganisms. Each of these systems has distinct benefits and limitations. For example, flow cell technologies are the only reactors suitable for the real-time, nondestructive microscopic examination of biofilms. Chemostat-based systems appear to offer the best control over environmental conditions, whereas stirred tank reactors with large surface areas offer a reasonable level of control over the environment while providing substantial amounts of cell material. Detailed descriptions of the specific biofilm model systems and technical considerations have been reviewed elsewhere[4] and are covered in other articles in this volume, but there are factors worth considering that are germane to studies with oral bacteria. First, all areas of the mouth are exposed to low or essentially zero fluid shear. However, shear force and fluid flow across oral biofilms vary in a site-specific manner, and these factors should be accounted for in oral biofilm models. Second, when plaque is isolated from humans, it is rich in polysaccharides, presumably derived from (i) capsule synthesis by

[4] J. W. T. Wimpenny, *Adv. Dent. Res.* **11,** 150 (1997).

plaque microorganisms, (ii) sucrose-derived exopolysaccharides, and (iii) a variety of adsorbed polymers from the diet. Finally, although some very important findings have come from studies,[5] it is probably impossible to accurately mimic the nutritional composition of oral biofilms or to mimic the intermittent dietary intake in a fashion that is realistic and widely applicable. Arguably, then, experiments designed to probe physiology and gene expression patterns should be conducted in models that are well defined and offer as much control over environmental and nutritional parameters as possible.

Cultivation of Strains of Oral Streptococci in Commercially Available Biofilm Reactor

We originally conducted gene fusion and physiology studies of oral streptococci in the chemostat[6] and subsequently wanted to utilize a system that was not designed to be a direct mimic of the oral cavity, but instead was one that would allow us to contrast the behavior of chemostat-grown cells with that of cells immobilized on surfaces under conditions very similar to continuous culture. The goal was to control as many variables as possible so as to be able to distinguish genetic and physiologic differences between biofilm and suspended bacteria arising as a result of surface-specific phenomena versus environmental changes induced by stimuli created by diffusion limitation in a three-dimensional environment. We selected the Rototorque (Montana State University, Bozeman, MT), a commercially available, continuous flow biofilm reactor.[7] We had the manufacturer modify the vessel, removing the reflux loop, and we operate the system at low rpm's to facilitate thorough mixing without creating substantial shear. Details of the configuration and operation of the Rototorque for use with oral streptococci, along with alterations and dimensions, are available elsewhere.[8,9] We have used the Rototorque successfully for studying recombinant and wild-type strains of *Streptococcus mutans, S. sobrinus*, and *S. salivarius*.

Although continuous flow reactors can be operated in batch modes, operation in a continuous flow mode, similar to a chemostat, is the preferred method. One of the major advantages of a continuous chemostat is the ability to control multiple environmental conditions and bacterial growth

[5] D. J. Bradshaw, P. D. Marsh, H. Gerritsen, J. Vroom, G. K. Watson, and C. Allison, *J. Dent. Res.* **77,** 988 (1998).

[6] D. L. Wexler, M. C. Hudson, and R. A. Burne, *Infect. Immun.* **61,** 1259 (1993).

[7] W. G. Characklis and K. C. Marshall, *in* "Biofilms" (W. G. Charaklis, ed.), p. 55. Wiley, New York, 1990.

[8] R. A. Burne, Y.-Y. M. Chen, J. E. C. Penders, *Adv. Dent. Res.* **11,** 100 (1997).

[9] R. A. Burne and Y. Y. M. Chen. *Methods Cell Sci.* **20,** 181 (1998).

rates. Based on equations summarized elsewhere,[10] when cells achieve steady state in single-stage continuous culture, the dilution rate (D) is equal to the specific growth rate, μ. Thus, one can calculate the generation time (t_g) using equations derived elsewhere[10] and substituting D in place of μ (at steady state):

$$\frac{0.693}{\mu} = t_g$$

For example, cells growing at steady state with $D = 0.1\ \text{hr}^{-1}$ are dividing about every 6.9 hr. Of note, evidence suggests that cells in comparatively thin biofilms (<20–50 μm) are physiologically similar to planktonic phase cells.[8,11] Nonetheless, achieving true steady-state conditions in biofilms may not be possible. Consequently, cells are cultivated in a biofilm to a "quasi" steady state and the growth rate of bacteria in the films is approximated using equations derived for the chemostat. Growth rate control is not ideal as some variability will arise because of the diffusion limitation of nutrients and end products, but continuous flow systems still offer much more control than batch culture systems and probably more closely mimic the conditions commonly found in the human oral cavity.

The control of pH in a biofilm is far more challenging than in the planktonic phase. Conditions that favor biofilm formation usually result in the fouling of pH probes. The use of buffers is a practical, although less than ideal, alternative. We have had success controlling the pH of the liquid phase and of the biofilms of the Rototorque vessel using potassium phosphate buffers (50 mM) with sugar concentrations of 10–20 mM, although pH gradients appear to develop (Burne and Li, unpublished results, 1999). Obviously, when one is considering exploring bacterial response to stresses in biofilms, pH is just one example of the potential for confounding data arising from mass transport limitations. There is evidence that the diffusion of antimicrobials, oxygen, and nutrients is also restricted in biofilms,[12] so great care must be taken in the interpretation of data, and considerable attention needs to be given to the fact that films formed *in vitro*, like dental plaque, are extremely heterogeneous.

Specifics of Growth of Oral Streptococci in the Rototorque

Overnight cultures used for inoculation of the biofilm reactor vessel are grown at 37° in a 5% (v/v) CO_2, aerobic atmosphere in TY with 10 mM

[10] D. I. C. Wang, C. L. Cooney, A. L. Demain, P. Dunnill, A. E. Humphrey, and M. D. Lilly, "Fermentation and Enzyme Technology." Wiley, New York, 1979.

[11] G. H. W. Bowden and Y. H. Li, *Adv. Dent. Res.* **11,** 81 (1997).

[12] J. W. Costerton, Z. Lewandowski, D. E. Caldwell, D. R. Korber, and H. M. Lappin-Scott, *Annu. Rev. Microbiol.* **49,** 711 (1995).

glucose and with antibiotic if necessary. For cultivation of cells in biofilms, TY medium[13] supplemented with 10 m*M* sucrose (TYS) can be used. We have also found that TY base medium can be diluted two- to fourfold without compromising the efficiency of biofilm formation by oral streptococci. The Rototorque reactor is filled with about one-half the working volume of TYS, autoclaved for 30 min, equilibrated at 37°, and inoculated with 10 ml of an overnight culture of the strain to be studied. Fresh medium is introduced immediately after inoculation. Rotation of the inner drum is controlled at a speed of approximately 75 rpm. In the past, we have conducted analyses of biofilm cells after 48 hr or 7 days postinoculation[8] or at "quasi-steady state" following a calculated 10 generations of growth (Burne and Li, in preparation). Biofilms begin to become visible in as little as 12 to 24 hr.

The thickness of the biofilms formed by oral streptococci is strongly dependent on the sucrose concentration in the growth medium, the bacterial strain being used, the length of time the biofilms are allowed to form, and the pH. It is possible to get a high degree of reproducibility from one run to the next, but empirical determination of the growth conditions and media composition will need to be done for particular strains and research questions. We have found that 48-hr biofilms of *S. mutans* growing in 10 m*M* sucrose are generally about 20 μm in depth, whereas biofilms of *S. mutans* formed in 7 days achieve thicknesses of greater than 100 μm. It is with these thicker films that we have found that the gene expression patterns of the exopolysaccharide synthases, glucosyltransferase and fructosyltransferase, are altered as compared with their planktonic counterparts.[8] Strains of *S. sobrinus* can form biofilms that are much thicker, polysaccharide-rich biofilms under these conditions, whereas the soft tissue microorganism, *S. salivarius*, forms thinner films that are removed more easily from the polystyrene slides.

Dissecting responses of oral microbes to environmental stresses in adherent populations will be key to understanding the ecology, population dynamics, and virulence of oral biofilms. The following methods are used to explore physiology and gene expression in oral biofilms, using primarily environmental acidification and oxidative stresses as examples.

Acid–Base Physiology and Acid Adaptation in Biofilms

Methods for Measuring Acid Adaptation in Oral Streptococci

Oral streptococci face a severe challenge as a result of their metabolism of dietary sugars to lactic acid. In dental plaque, the pH can fall below a

[13] Burne, R. A., K. Schilling, W. H. Bowen, and R. E. Yasbin, *J. Bacteriol.* **169** 4507 (1987).

value of 4.0, and it is becoming evident that *S. mutans* and other oral streptococci[14] have evolved to take advantage of the acidification of plaque by expressing a more acid-resistant phenotype than many other plaque bacteria and by manifesting a significant acid tolerance response (ATR).[15] For example, *S. mutans* is able to grow and carry out glycolysis at pH values that prohibit these processes in other oral streptococci.[16] Also, like many other bacteria,[17] when *S. mutans* is grown at pH 5.0 as compared to growth at pH 7.0, there is a 3 to 5 log increase in the survival after incubation of the bacteria at pH 2.5 for 30 min. In enteric bacteria, it is established that the ATR involves a complex, multigene regulon with considerable overlap with systems that protect microorganisms from a wide variety of stressors. Acid adaptation in *S. mutans* requires both catabolism and protein synthesis, indicating a cellular requirement for additional proteins or increased proportions for some proteins.[18,19] For the quantitative assessment of acid adaptation by oral streptococci, the measurement of resistance to killing at low pH values is, at this time, the most reliable assay. A major advantage of examining acid-killing kinetics is that relatively small numbers of cells are sufficient to provide information about whether cells have become acid adapted. The following method is used for assessing adaptation for cells grown in biofilms.

Resistance to Acid-Mediated Killing as Measure of Adaptation

1. Biofilms are harvested from cultures grown at acidic or neutral pH values, and cells are removed from the substratum as detailed earlier and dispersed by sonication at 300 W for 30 sec at 4°. Typically, yields from the Rototorque, based on studies conducted with *S. salivarius*, are about 10^8 cells/cm^2 after 10 generations of growth at $D = 0.1$ hr^{-1}. Cells are harvested by centrifugation at 4° for 15 min at 15000 rpm.
2. The cell pellet is suspended in 0.1 *M* glycine, pH 2.5. While the cell suspension is stirred at room temperature, samples are removed at various time points, serially diluted in brain–heart infusion medium (BHI, Difco Laboratories, Detroit, MI), and plated for survival on solid BHI agar medium.

One important point to note is that it may not be appropriate to compare the kinetics of killing of biofilm bacteria directly with those of planktonic

[14] I. R. Hamilton and G. Svensäter, *Oral Microbiol. Immunol.* **13,** 292 (1998).
[15] W. A. Belli and R. E. Marquis, *Appl. Environ. Microbiol.* **57,** 1134 (1991).
[16] G. R. Bender, S. V. W. Sutton, and R. E. Marquis, *Infect. Immun.* **53,** 331 (1986).
[17] J. W. Foster, *Crit. Rev. Microbiol.* **21,** 215 (1995).
[18] K. Hahn, R. C. Faustoferri, and R. G. Quivey, Jr., *Mol. Microbiol.* **31,** 1489 (1999).
[19] Y. Ma, T. M. Curran, and R. E. Marquis, *Can. J. Microbiol.* **43,** 143 (1997).

cells for a variety of reasons, including (i) removal and dispersal of the biofilms can be difficult, (ii) subpopulations of adapted cells may exist in areas where the diffusion of acids from the biofilm is limited, (iii) the exopolysaccharide content of the biofilms and perhaps the cell-associated polysaccharides of cells derived from the biofilms are considerable higher than planktonic cells, and (iv) a diffusion limitation. Consequently, comparisons of survival of biofilms should be restricted to contrasting adherent populations with one another while the pH or other stressor is varied. Also, resistance to acid killing varies somewhat with temperature, e.g., resistance of *S. mutans* GS-5 is maximal at about room temperature.

ATPase Measurements

An enzyme that appears to be central to acid adaptation and pH homeostasis in *S. mutans* is the membrane-bound, proton-translocating ATPase (F-ATPase).[16,20] The role of the F-ATPase in oral streptococci is to pump protons out of the cytoplasm in order to maintain a ΔpH across the membrane,[16,21] preserving acid-sensitive glycolytic enzymes. As part of the adaptive repertoire of *S. mutans*, F-ATPase activity is known to upregulate as the organism begins to experience a downward shift in the pH value.[22] The mechanism by which the level of ATPase is regulated is not yet clear, nor is it well established how many additional gene products are necessary for, or participate in, the ATR of oral streptococci. However, given the clear connection between enhanced acid resistance and increased F-ATPase activity, measurement of ATPase activity in biofilm bacteria could be a relatively simple and efficient mechanism in assessing the acid adaptation and acid resistance mechanisms of adherent populations.

Determination of F-ATPase Activity in Streptococci

The most commonly used method to assay F-ATPase activity involves the colorimetric determination of phosphate release from ATP, generally in the presence of magnesium,[15,23] This method is rapid and inexpensive and it is valid if the assay is conducted over a short period of time. However, the phosphate release assay can be affected by buildup of ADP, which is inhibitory for F-ATPases. The most commonly used alternative method involves ATP regeneration to maintain the ATP supply constant and avoid the inhibitory effects of ADP. In the ATP-regenerating assay described by

[20] S. G. Dashper and E. C. Reynolds, *J. Dent. Res.* **71,** 1159 (1992).
[21] A. Casiano-Colon and R. E. Marquis, *Appl. Environ. Microbiol.* **54,** 1318 (1988).
[22] W. A. Belli and R. E. Marquis, *Oral Microbiol. Immun.* **9,** 29 (1994).
[23] C. H. Fiske and Y. SubbaRow, *J. Biol. Chem.* **66,** 375 (1925).

Pullman *et al.*,[24] ATP is regenerated by pyruvate kinase with phosphoenolpyruvate as the phosphate donor. The pyruvate produced can then be assayed by assessing changes in the absorbance of light of the 340-nm wavelength due to the oxidation of NADH associated with pyruvate reduction catalyzed by lactic dehydrogenase. The method has been used to assay F-ATPase activities of oral lactic acid bacteria.[25]

Because of the high polysaccharide content of bacterial biofilms, it is most appropriate to normalize data to total protein, rather than wet or dry weights. Again, as with acid killing, extreme caution should be exercised when comparing the results obtained from ATPase assays of biofilms cells with those from planktonic populations. Comparisons should be restricted to those between stressed and unstressed cells grown under otherwise identical conditions. Also, to ensure that differences in biochemical activities are not attributable to major differences in the proportions of live : dead cells under the various experimental conditions, we have also found it useful to enumerate populations to see if substantial portions of the cells in each test population are dead. We do this by dispersing cells and then determining viable counts by plating and counting cells in a Petroff–Hauser counting chamber.

Buffer Capacities of Biofilms

The importance of biomass concentration is clearly evident in relation to acid–base stress in biofilms. Basically, the biofilms in the laboratory or in plaque in the mouth have very high buffering capacities. Buffering capacity can be assessed directly by acid–base titration, as described by Marquis[26] for thick suspensions. Determinations are made of how much acid or base is required to change the pH by some amount per unit of biomass. An example is presented for *Streptococcus mutans* GS-5 in Fig. 1. In this titration, cells were permeabilized with 10% butanol prior to titration so that the total buffer capacity, that within the cells and that outside of the cell membrane, was assessed. Generally, data obtained with cells or biofilms not subjected to permeabilization are interpreted in terms of buffering only by constituents outside of the cells. However, the interpretation is not entirely straightforward because cells may have fairly high permeabilities to protons, and movements of protons across the cell membrane cannot be blocked readily. Biofilms tend to have high levels of extracellular or matrix polysaccharide. However, the buffer capacities of these polymers

[24] M. E. Pullman, H. S. Penefsky, A. Datt, and E. Racker, *J. Biol. Chem.* **235,** 3322 (1960).

[25] M. G. Sturr and R. E. Marquis, *Arch. Microbiol.* **155,** 22 (1990).

[26] R. E. Marquis, *in* "Cariology for the Nineties" (W. H. Bowen and L. A. Tabak, eds.), p. 309. University of Rochester Press, Rochester, NY, 1993.

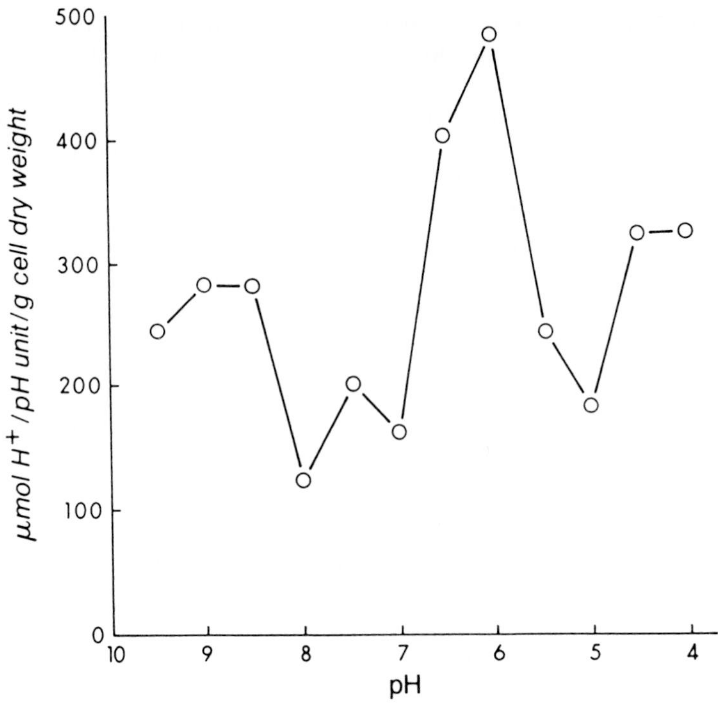

FIG. 1. Buffer capacities of butanol-permeabilized cells of *Streptococcus mutans* GS-5.

are generally low, even if they are charged. In essence, the matrix is highly hydrated with a low solids content per unit volume. Thus, the major buffering is by the cells themselves.

Movements of Protons into Biofilms

Movements of protons into biofilms can be assessed rather simply. In this case, biofilms developed on glass slides in batch culture, or in some other suitable biofilm system, are washed with a solution containing 50 m*M* KCl and 1 m*M* $MgCl_2$ to remove medium constituents and loosely adherent cells. The films are then placed in chambers with sufficiently wide mouths to accommodate the slide. Acid is added to the chamber to lower the pH value by a measured amount, say 0.2 pH units. Then, the rise in pH associated with the movement of protons into biofilm cells is monitored with a standard pH electrode. There is a limited surface between the biofilm and its bathing fluid across which protons can diffuse after an addition of acid to a biofilm. However, diffusion into the biofilm is generally rapid compared with the diffusion of protons across the membranes of metabolizing cells. Some of the protons that enter the film are "consumed," depending on the buffer capacity of the film, which is limited. This consumption occurs rapidly, and it is the excess protons not titrated by buffers that cause acidification. Permeability of cell membranes to protons is of a dynamic type in that as protons diffuse into the cytoplasm, they can be pumped back out, primarily by proton-translocating F-ATPases. Thus, the net permeabilities of biofilm cells to protons depend on the integrity of their cell membranes and also on their capacities to generate ATP to fuel F-ATPases.

One way to evaluate the barrier capabilities of cell membranes in biofilms is to treat the films with 5% butanol to damage the cell membranes, basically to eliminate their barrier functions. Then, the major barrier to proton movements should be the interface, which is less likely to be severely altered by agents such as butanol, especially if the solvent is washed away before assays of proton uptake. Many other membrane active agents can be used, ranging from highly damaging agents, such as toluene or gramicidin, to nondamaging agents, such as fluoride or other weak acids that act as proton carriers across the cell membrane. A common parameter for assessing average proton permeabilities of cells is the half-time for pH equilibration, i.e., the time required for the pH to rise halfway from the minimum value after acid addition to the maximum value at the time of pH equilibration across cell membranes. Generally, the latter value is estimated rapidly by adding butanol to induce rapid equilibration. Alternatively, plots of pH versus time can be extrapolated to obtain an asymptotic value.

There have been very few studies of proton permeabilities of biofilms of oral bacteria, and there is a need for detailed studies to obtain a reasonable view of proton dynamics in these systems in which metabolism may be limited by nutrient diffusion and the capacities of cells to maintain high values of ΔpH may be compromised.

pH Values within Biofilms

The methods used, mainly involving microelectrodes, for assessing pH within dental plaque can be applied to assess pH values in biofilms of oral bacteria. A major problem with the data obtained is in determining what was actually measured. The pH measured by the electrodes is that of the interstitial matrix and not that of the cells. In dental plaque, the pH value of the matrix is pertinent to caries development because the plaque fluid is in direct contact with pellicle, and the major flow of protons across the pellicle is likely to be from the plaque fluid or matrix fluid. Ultimately the protons come from the cells because the cytoplasm is the site of acid production by glycolytic and certain other enzymes, and the cell membrane is the site of respiration-linked proton extrusion for organisms with respiratory metabolism. However, the flow of protons across the cell membrane is highly regulated, and generally the cytoplasmic pH value is higher than the pH value of the matrix or plaque fluid. For oral biofilms matrix pH values may not be as informative as values for ΔpH between the interior and the exterior of cells, which are more indicative of the state of acid stress in the biofilms.

Measurements of ΔpH in Biofilms

Methods for assessing ΔpH across cell membranes, i.e., the difference between the pH value of the suspending medium and that of the cell cytoplasm, have been described in detail elsewhere.[27] In theory, they can be applied directly to biofilms. Thus, for example, the uptake of weak acids such as benzoate by biofilm cells can be assessed with use of radiolabeled compounds, and ΔpH can be calculated from knowledge of the difference in levels of appropriately chosen weak acids within cells in biofilms and in the bathing fluid. Calculation of ΔpH values requires knowledge of the volume of the biofilm external to the membranes of cells. Extracellular volumes of biofilms tend to be high because they include not only cell wall volumes but also intercellular matrix volume. The latter volume can be estimated in terms of biofilm volume available to large solutes such as dextrans, which are available in radiolabeled or dye-tagged forms. The cell wall volume can be estimated in terms of differences between the biofilm volume penetrated by small nonmetabolized solutes that cannot be transported across the cell membrane and that penetrated by large molecules such as dextrans. The total water volume, in matrix, walls, and cytoplasm, can be estimated by use of tritiated water. Because cells are highly permeable to water, tritiated water exchanges rapidly with nontritiated water in the cytoplasm as well as that outside of the cell membrane. All of the volumes should be related to some biofilm parameter, e.g., the dextran-impermeable volume per unit of dry weight or volume per unit of protein.

Methods for assessing ΔpH have many problems that lead to difficulties in interpreting the numbers obtained. Kashket[28] has presented a very incisive review of methods for assessing ΔpH and also for assessing $\Delta\psi$, the transmembrane electrical potential. The net conclusion is that the methods are useful indicators of changes in ΔpH (or $\Delta\psi$) but do not give absolute values. The methods have not yet been applied extensively to biofilms of oral bacteria, but could potentially provide valuable information about the

[27] R. E. Marquis, *in* "Methods for General and Molecular Bacteriology" (P. Gerhardt, editor-in-chief), p. 587. ASM Press, Washington, DC, 1994.

[28] E. R. Kashket, *Annu. Rev. Microbiol.* **39,** 219 (1985).

energetic status of the membranes of cells in biofilms relative to planktonic cells and of the effects of stresses on energized states of membranes of biofilm cells.

Alkali Production by Biofilms

Most of the focus in the study of biofilms in relation to dental caries is on acid production, acid tolerance, and acid adaptation. Biofilms of oral streptococci have been found to be active also in alkali production. For example, monoorganism biofilms of *S. sanguis* NCTC10904 were found[29] to be highly repressed for the arginine deiminase system when grown in tryptone yeast extract broth with 1% (w/v) sucrose, but could be fully induced/derepressed after transfer to the same medium containing arginine in place of sucrose. Small peptides as well as proteins, such as gelatin, also served to induce the system. Levels of the arginine deiminase system in biofilms were assessed by first permeabilizing the cells with toluene and two freeze-thaw cycles and then assaying the activity of arginine deiminase, the signature enzyme of the system, in terms of capacities to produce citrulline from arginine. In addition, arginine was shown to be protective against acid killing of cells in biofilms induced/derepressed for the system prior to acid exposure.

Oxygen Uptake by Biofilms, Oxidation–Reduction Potentials, and Oxidative Stress

Studies over many years with immobilized cells systems have shown that immobilization within a matrix commonly results in metabolism becoming diffusion limited. Diffusion limitation is to be expected for mature biofilms because the influx of nutrients for large and diverse populations must occur across a limited film surface in contact with the environment. Biofilms tend to have irregular surfaces, commonly with deep clefts that would increase the surface for diffusion. However, the entire film is also covered with an hydration layer and so there is some difficulty in estimating the functional surface area that limits the diffusional movement of solutes. Diffusion limitation also applies to movements of metabolic products out of the biofilm. If the products are inhibitory or toxic, a slow, diffusion-limited release can prolong the time of stress and resultant damage, but of course, because production is also generally substrate limited and the biomass concentration in the biofilm is high, the severity of the stress is moderated in rate and extent.

[29] T. M. Curran, Y. Ma, G. C. Rutherford, and R. E. Marquis, *Can. J. Microbiol.* **44,** 1078 (1998).

The commonly used relationship for describing the diffusion of solutes into and out of biofilms is the Fick equation:

$$ds/dt = PA\,\Delta C$$

where ds/dt is the net flow of solute per unit time, P is the permeability coefficient, A is the surface area, and ΔC is the difference in concentration of the solute in the bathing fluid and in the biofilm. Because it is often difficult to estimate A for biofilms, the terms P and A can be combined into a single constant.

Movement of O_2 into Biofilms

A key nutrient for biofilms containing aerobic organisms is oxygen. Biofilms are basically thin films, and the standard methods used in biochemical engineering to assess oxygen movements and oxygen demand apply. Oxygen demand in biofilms of oral bacteria is generally high because the organisms, even including anaerobes such as *Treponema denticola*, have vigorous oxygen metabolism.[30,31] For estimates of oxygen transfer in biochemical engineering studies, the combined PA term in Eq. (1) is designated $K_L A$, the volumetric mass transfer coefficient for oxygen.

The procedures we have used for assessing O_2 movements are similar to those for measuring the movement of protons into biofilms, only using standard oxygen electrodes. For these measurements, the biofilms were developed on glass slides, although biofilms generated in a number of ways can be used in these studies. The films were washed with a solution containing 50 mM KCl and 1 mM $MgCl_2$ to remove medium constituents and loosely adherent cells. The films were then placed in chambers and the chambers were filled with aerated media of known oxygen content. The volume of the chamber was known, and the volume occupied by the biofilm and support can be estimated by displacement with direct weighing of the volume of water displaced in an overflow vessel. Thus, the initial O_2 concentration and the total volume of added water can be determined to allow for calculation of the total O_2 in the vessel at zero time.

Each chamber was sealed quickly with a rubber stopper through which the oxygen probe extends. The glass–rubber interface can be sealed further with Parafilm. The insertion of the stopper–electrode assembly should be done without introducing an air bubble. Before use, the electrode was calibrated as suggested by the manufacturer. After insertion, the decline of O_2 concentration associated with biofilm metabolism can be recorded

[30] R. E. Marquis, *J. Indust. Microbiol.* **15,** 198 (1995).

[31] C. E. Caldwell and R. E. Marquis, *Oral Microbiol. Immunol.* **14,** 66 (1999).

at intervals or continuously. From a knowledge of biofilm characteristics, such as dry mass or protein content, the O_2 uptake per biomass unit per time unit can be calculated. Generally, it is most reasonable to estimate O_2 uptake rates per unit of biofilm protein rather than biomass because the polysaccharide in the interstitial matrix is not active in O_2 metabolism.

Oxygen levels in biofilms can be assessed using the same procedures used to assess O_2 levels in dental plaque with microelectrodes. However, more often, the required information is the rate of oxygen metabolism rather than the residual O_2 level in the films. As indicated, biofilms generally have very high oxygen metabolic capacities, and the determination of residual O_2 in the films generally only indicates the excess of influx compared with metabolism and is generally very low, less than, say, 10% of the value for air-saturated water.

A limitation of the usual method of assessing O_2 uptake is that the system becomes depleted of oxygen as the cells metabolize. Fortunately, the K_m for oxygen metabolism is generally in the micromolar range. Therefore, respiration becomes O_2 limited only at extremely low oxygen levels, and plots of change in the O_2 level with time are generally linear nearly until exhaustion of the supply. There are many ways to supply more oxygen to the cells. One simple way is to remove the biofilm to a new vessel with freshly aerated medium. It is also possible to gas the suspending fluid with pure O_2. However, many organisms are sensitive to this high a concentration of O_2, and there are problems in maintaining the level of gas during transfer from gassing vessels to respirometers. A more versatile way to provide additional O_2 is to supply the gas in a reservoir of fluorocarbon liquid, e.g., products such as FC-80 (3M Company, St. Paul, MN). O_2 is highly soluble in these liquids, which are not miscible with water. Therefore, as O_2 is metabolized from the aqueous phase, more of the gas diffuses from the reservoir. Tables of values for oxygen solubilities in various fluorocarbon liquids can be obtained from manufacturers. Fluorocarbons have been used for liquid breathing of animals and can supply sufficient O_2 to meet the respiratory needs of mice.[32]

Assessment of Oxidation–Reduction States of Cells in Biofilms

Oxidation–reduction potentials (E_h) of dental plaque biofilms have been assessed with microelectrodes.[33] Electrode methods have the problem of discerning what is actually being measured. For example, when pH electrodes are used, the measurement is generally considered to be indicative

[32] J. A. Klystra, R. Nantz, J. Crowe, W. Wagner, and H. A. Saltzman, *Science* **158,** 793 (1971).
[33] E. B. Kenny and M. M. Ash, Jr., *J. Periodontol.* **40,** 630 (1969).

of the pH value in the matrix outside of the cells in the film. Thus, even in biofilms, it is likely that bacteria maintain ΔpH across the cell membrane and that the electrode measures only the extracellular pH, which in an acidified biofilm would be lower than the pH within the cell. There is a less clear view of what is measured by E_h electrodes inserted into biofilms because there is relatively little information on ΔE_h between the interior and the exterior of cells.

One approach to assessing intracellular E_h values is to use redox dyes, for example, as described by Bradshaw *et al.*[5] for pH measurements. Their method involves use of two-photon excitation microscopy, which allows individual bacteria to be distinguished at depths in films up to 140 μm. The method is amenable to the use of redox dyes and the study of individual bacteria in films. However, it requires highly specialized equipment and skills.

A more direct way to assess the average oxidation–reduction of cells in biofilms is to use specific oxidation–reduction couples as indicators. Perhaps the most generally useful are the NAD/NADH or NADP/NADPH couples, which have a standard oxidation–reduction potential (E'_o) of −32 mV. These couples are generally good indicators because their E'_o value is appropriate for expected oxidation–reduction potentials (E_h) within cells of oral bacteria in biofilms and because the components of the couples are mainly unbound in the cytoplasm and are not excreted by the cells. Because the couples involve two-electron transfer, the equation for calculating E_h can be derived from the standard relationships.

$$NAD^+ + H^+ + 2e^- \rightarrow NADH$$
$$E_h = E'_o + (2.3\ RT/nF)\log[NAD^+][H^+]^b/[NADH]$$

where E_h is the oxidation–reduction potential, E'_o is the standard oxidation–reduction potential, R is the gas constant (8.31 volt-coulombs/mol-°K), T is the Kelvin temperature, n is the number of electrons transferred, F is the Faraday (96,496 coulombs/mol), [NAD^+] is the concentration (actually the activity) of NAD^+, [H^+] is the proton concentration, b is the number of protons involved in the reaction, and [NADH] is the concentration of NADH. The terms 2.3 RT/F are commonly combined into a single term z, which has a value of 59 mV at 25°.

A pH term can be separated (pH = log {1/[H^+]} = −log [H^+]) to give the equation

$$E_h = E'_o + (z/n)\log[NAD^+]/[NADH] - (z)(b/n)pH$$

For the NAD^+/NADH couple, n is 2 and b is 1.

The levels of NAD^+ and NADH in cells can be estimated by the methods used, for example, by Snoep *et al.*[34] with cells of *Enterococcus faecalis* NCTC 775. The methods involve differential extraction of NAD^+ and $NADP^+$ with acid at a pH of about 1.5 and extraction of NADH and NADPH with KOH solution at a pH of about 11.5. Levels of NAD^+ in the acid extract and NADH in the alkaline extract can be determined after the extracts are boiled, centrifuged, and neutralized by the recycling method of Bernofsky and Swan[35] with the use of alcohol dehydrogenase. $NADP^+$ and NADPH can be assayed using glucose-6-phosphate dehydrogenase. A recent modification of the cycling assay using tetrazolium salts has been described by Gibon and Larher.[36] These methods are highly sensitive. In addition, a recycling method for assaying acetyl-CoA/CoASH has been developed.[37]

The status of other oxidation–reduction couples in biofilms can be assessed, for example, $FMN/FMNH_2$ ($E'_o = -190$ mV), $FAD/FADH_2$ ($E'_o = -220$ mV), or pyruvate/lactate ($E'_o = -190$ mV). Acids of the pyruvate/lactate couple do not necessarily remain within the cells, and transmembrane transport, especially of lactate, can introduce a complication, although there is some question about whether there is a specific transport system for lactate in oral bacteria. Various oxidation–reduction dyes could also be used for E_h assessments, but they tend to upset metabolism and so potentially have undesirable effects.

Using Gene Fusions to Probe Gene Expression and Regulation in Biofilm Bacteria

At this time, virtually nothing is known about the expression and regulation of bacterial genes when cells are growing as biofilms. We have combined reporter gene technology with the exploration of gene expression in adherent population.[8] We have reported detailed methods for the examination of "quasi-steady state" gene expression and induction of genes using the Rototorque. The advantages to using the Rototorque are, as indicated earlier, the ability to tightly control environmental parameters and to extend observations made in continuous chemostat culture, the "gold standard" for exploring physiology of oral streptococci. Also, the ability to examine induction or repression of genes kinetically in response to an imposed

[34] J. L. Snoep, M. J. Teixeira de Mattos, P. W. Postma, and O. M. Neijssel, *Arch. Microbiol.* **154,** 50 (1990).

[35] C. Bernofsky and M. Swan, *Anal. Biochem.* **53,** 452 (1973).

[36] Y. Gibon and F. Larher, *Anal. Biochem.* **251,** 153 (1997).

[37] S. Chohnan, H. Furukawa, T. Fujio, H. Nishihara, and Y. Takamura, *Appl. Environ. Microbiol.* **63,** 553 (1997).

stimulus or stress by removing 1 or more of the 12 slides aseptically over a time course is another major advantage.

To measure reporter gene activity after biofilms have achieved the desired characteristics, slides are removed from the vessel and cells are mechanically dissociated from the slides using a rubber policeman, razor blade, or scalpel. The cells are scraped directly into a solution of ice-cold 10 m*M*/liter Tris–HCl (pH 7.0) containing 10 μg each of rifampicin and tetracycline per milliliter to arrest transcription and translation, respectively. The cells are immediately centrifuged at 8000*g* for 10 min at 4°. Cells are washed in ice-cold 10 m*M* Tris–HCl (pH 7.8) and the cell pellets are kept on ice until all samples are ready for processing.

Two relevant observations about reporter enzymes that we have made in our studies with *cat* and *lacZ* fusions in continuous culture and biofilms are that CAT has a short half-life in oral streptococci, whereas Lac Z appears somewhat more stable in streptococci. Thus, CAT probably gives a better picture of transcriptional initiation, whereas Lac Z tends to accumulate, so downregulation is not detected as easily. Also, CAT appears to be amenable to use in cells grown at low pH values, whereas this is not the case for Lac Z. These observations should be taken into consideration when evaluating data generated using these constructs in oral streptococci. For measuring reporter gene product activity in biofilms of oral streptococci:

1. Cells are prepared as described earlier.
2. Transfer the concentrated cell suspension (1 ml) to 2-ml screw-cap microfuge tubes containing 0.75 g (roughly one-third volume) of glass beads (0.1 mm average diameter, Sigma, St. Louis, MO).
3. Homogenize the mixture in a Bead Beater (Biospec Products, Bartlesville, OH) for 1 min at 4°, followed by chilling the tubes on ice for 2 min. Repeat.
4. Centrifuge the mixtures at 10,000 rpm for 10 min at 4° in a microcentrifuge. The protein content of each lysate is determined by using the Bio-Rad protein assay, based on the method of Bradford.[38] Bovine serum albumin serves as the standard. Cell extracts retain CAT activity when stored at −80° for at least 1–2 months, but, if possible, the extracts should be used immediately.
5. CAT activity is measured using the method of Shaw.[39] This is a kinetic assay and it is very temperature sensitive. All reagents and samples must be prewarmed to 37° and reactions must be carried out in a temperature-controlled cuvette chamber. We employ a Beckman

[38] M. M. Bradford, *Anal. Biochem.* **72,** 248 (1976).
[39] W. V. Shaw, *Methods Enzymol.* **43,** 737 (1979).

DU640 equipped with a Peltier effect temperature-controlled system with six cuvette holders and utilize the kinetics software package. With the six cuvette holder, we load three identical test samples and one blank. The instrument plots rates and automatically subtracts any background detectable in the absence of chloramphenicol.

We have also effectively used Lac Z as a reporter enzyme by adapting the protocol of Miller.[40] Briefly, cells are harvested as described earlier and washed twice with an equal volume of 10 mM $NaPO_4$, pH 7.0, 1 mM $MgCl_2$, and resuspended to 10% of the original culture volume in the same buffer. Samples of the cultures are then added to the assay medium (Z buffer[40]) to a final volume of 1 ml and the cells are permeabilized as described previously. From this point, the Miller method can be followed exactly. Data can be normalized to OD_{600}, or homogenates of the suspensions can be prepared as for the CAT assays and data can then be normalized to total protein. As is the case for CAT, cell dry weight is not a good method for normalizing data as the carbohydrate content of the films can exceed 95%.

Considerations

The only key considerations to take into account are that biofilm cells occur in an aggregated state and that the biofilms themselves can be particularly rich in polysaccharides. For these reasons, it is not recommended to utilize dry weights for the normalization of activities. Rather, quantification of protein is the preferred approach. In the case of *cat* gene fusions, we homogenize the material in the films as described earlier and simply measure protein by the Bradford assay. We have confirmed that the Bradford assay is suitable for this purpose by completely hydrolyzing the biofilms and conducting amino acid analysis and quantification of total polysaccharide. The results we have obtained using analytical chemistry correlate well with estimates of protein from the Bradford assay. For measuring Lac Z activity, we have normalized the assay to OD_{600}. In these cases, the cells are dispersed by sonication two times at 350 W with a microprobe for 30 sec, assays are performed, and optical density readings are obtained. Preparation of cell-free lysates for assaying of chloramphenicol acetyltransferase (CAT) activity.

Other Uses for Gene Fusions

Oral streptococci are amenable to analysis by microscopic techniques that have been used to study other biofilms. In conjunction with fluorescent dyes, optical sectioning of the films is achieved readily. Their distinctive

[40] J. H. Miller, "Experiments in Molecular Genetics." Cold Spring Harbor, NY, 1972.

morphology and resistance to lysis by low osmolality make oral streptococci ideal for confocal and real time microscopic monitoring of biofilms. Results of staining with vital dyes are consistent with results obtained by plating techniques, and fluorescent substrates for the Lac Z reporter gene work well with these cells. Therefore, *in situ* analysis of gene expression is achievable with these bacteria.

Summary

Studies performed since the early, 1970s have yielded tremendous amounts of information about the physiology, genetics, and interactions of oral bacteria. This pioneering work has provided a solid foundation to begin to apply the knowledge and technologies developed using suspended populations for studying oral bacteria under conditions that more closely mimic conditions in the oral cavity, in biofilms. Our current understanding of phenotypic capabilities of individual and complex mixtures of adherent oral bacteria is in its infancy. There is ample evidence that oral streptococci have different patterns of gene expression than planktonic cells, but we have little understanding of the basis for these observations. Even in biofilm-forming bacteria with very well-developed genetic systems it is only very recently that genetic loci involved in biofilm formation and responses to surface growth have been identified. A comprehensive study of the physiology and gene expression characteristics of adherent oral bacteria not only will enhance our abilities to control oral diseases, but it will provide critical information that can be applied to a variety of other pathogenic microorganisms.

Acknowledgments

This work was supported by Grants DE11549, DE12236 (R.A.B.), DE06127 (R.E.M.), and DE10174 (R.G.Q.) from the National Institute for Dental and Craniofacial Research.

Section VII

Substrata for Biofilm Development

[34] Biofouling of Membranes: Membrane Preparation, Characterization, and Analysis of Bacterial Adhesion

By HARRY RIDGWAY, KENNETH ISHIDA, GRISEL RODRIGUEZ, JANA SAFARIK, TOM KNOELL, and RICHARD BOLD

I. Background

A host of synthetic polymer membrane materials are currently employed for water purification applications.[1,2] These materials range from common substituted cellulose derivatives such as cellulose acetate (CA) or cellulose nitrate to more complex polymers with highly specialized properties such as aromatic cross-linked polyamides, polyether ureas, and polyethyleneimines (Fig. 1). Membranes constructed from these materials may be microporous in nature, as in microfiltration (MF) or ultrafiltration (UF) membranes, which are frequently employed in colloidal separations and wastewater reclamation applications. Other membranes may be essentially nonporous and semipermeable, as in nanofiltration (NF) or reverse osmosis (RO) membranes. The two most commonly employed RO membranes used in industrial and municipal water and wastewater purification applications are thin-film composite (TFC) polyamide (PA) membranes and asymmetric CA membranes (Fig. 1). Common MF and UF membrane materials include polypropylene, polyacrylonitrile, or modified (e.g., sulfonated) polyether sulfone. These polymer membranes can be packaged into various module (element) configurations, including spiral-wound, plate-and-frame, and hollow-fine fiber.[3]

A. Phenomenon of Membrane Biofouling

Biological fouling (biofouling) of the surfaces of synthetic polymer membranes used in water treatment applications is a commonly encountered problem that can dramatically diminish process efficiency (and cost effec-

[1] R. J. Petersen, *J. Membr. Sci.* **83,** 81 (1993).

[2] A. Allegrezza, *in* "Reverse Osmosis Technology: Applications for High Purity Water Production" (B. Parekh, ed.), p. 53. Dekker, New York, 1988.

[3] P. Aptel and C. A. Buckley, *in* "Water Treatment Membrane Processes" (J. Mallevialle, P. E. Odendall, and M. R. Wiesner, eds.), p. 21. McGraw-Hill, New York, 1996.

0076-6879/99 $30.00

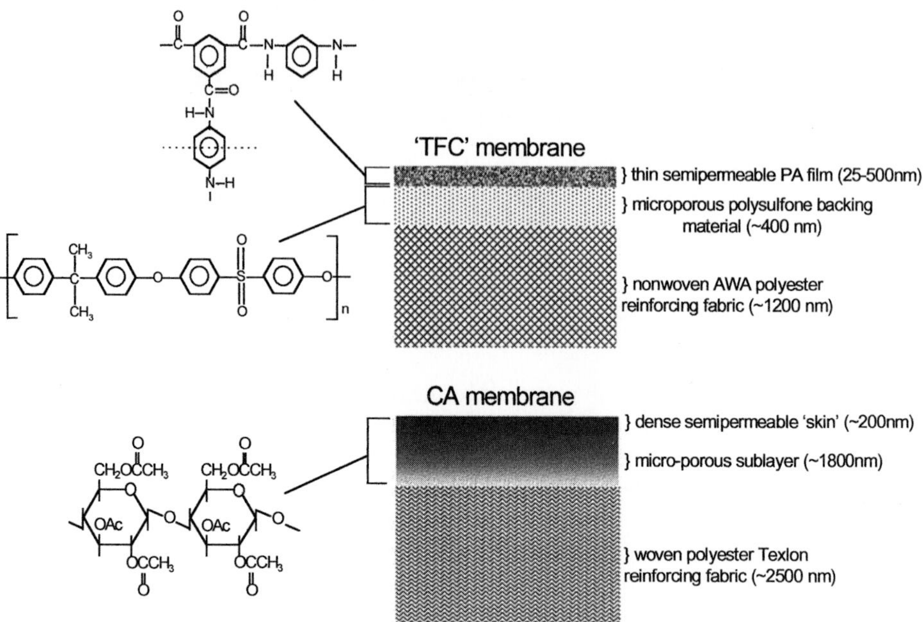

FIG. 1. Polymer compositions and structures of the two most common separation membrane designs; the TFC and asymmetric CA type RO membranes. The aromatic cross-linked PA polymer structure shown is currently the most widely employed TFC membrane. The degree of CA acetyl substitution is varied by manufacturers to achieve different flux and solute transport properties.

tiveness) in an engineered treatment system.[4–8] Membrane biofouling is initiated by the irreversible adhesion of one or more bacteria to the membrane surface followed by growth and multiplication of the sessile cells at the expense of feedwater nutrients (Fig. 2).[4,9,10] Given adequate nutrients and time, the initial sessile microbial population can eventually form a

[4] H. F. Ridgway and H.-C. Flemming, in "Water Treatment Membrane Processes" (J. Mallevialle, P. E. Odendaal, and M. R. Wiesner, eds.), p. 6.1. McGraw-Hill, New York, 1996.

[5] H.-C. Flemming, G. Schaule, R. McDonogh, and H. F. Ridgway, *in* "Biofouling and Biocorrosion in Industrial Water Systems" (G. G. Geesey, Z. Lewandowski, and H.-C. Flemming, eds.), p. 63. CRC Press, Boca Raton, FL, 1994.

[6] H.-C. Flemming, G. Schaule, and R. McDonogh, *Vom Vasser* **80,** 177 (1993).

[7] G. L. Leslie, R. P. Schneider, A. G. Fane, K. C. Marshall, and C. J. D. Fell, *Colloids Surf A Physicochem. Engin. Aspects* **73,** 165 (1993).

[8] H. F. Ridgway, *in* "Reverse Osmosis Technology: Applications for High-Purity Water Production" (B. S. Parekh, ed.), p. 429. Dekker, New York, 1988.

[9] H.-C. Flemming and G. Schaule, *Desalination* **70,** 95 (1988).

[10] H. F. Ridgway, M. G. Rigby, and D. G. Argo, *J. Am. Water Works Assoc.* **77,** 97 (1985).

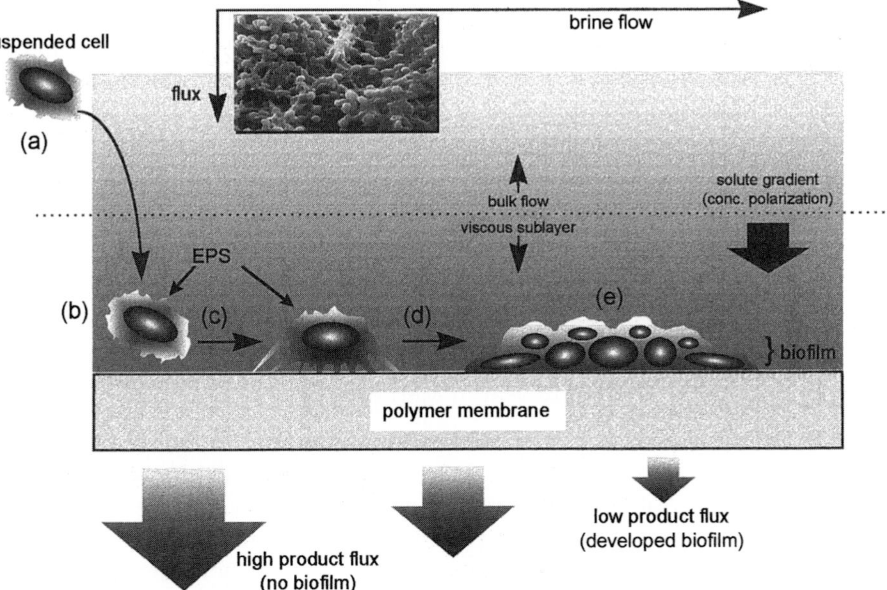

FIG. 2. Schematic of biofilm formation in membrane systems. Bacteria present in bulk flow (a) undergo transport to the membrane surface where initial (reversible) adhesion occurs (b). Irreversible adhesion is associated with biosynthesis of extracellular polymeric substances (EPS) (c) and a biofilm results from subsequent growth and multiplication at the expense of feedwater nutrients (e). Biofilm development is associated with reduced product flow and increased solute concentration in the viscous sublayer (concentration polarization). The latter phenomenon results in increased solute transport across semipermeable membrane and deterioration of product quality. (Inset) SEM image of biofilm on RO membrane treating activated sludge effluent. Diameter of rod-shaped cells $\cong 0.5\ \mu m$.

confluent lawn of bacteria (i.e., a biofilm) on the membrane surface.[11] Whereas the basic processes of bacterial adhesion[12,13] and biofilm formation[14,15] are similar in membrane systems and other natural and industrial systems, membranes are perhaps uniquely impacted by pressure-driven

[11] H. F. Ridgway, C. A. Justice, C. Whittaker, D. G. Argo, and B. H. Olson, *J. Am. Water Works Assoc.* **79,** 94 (1984).

[12] K. C. Marshall and B. L. Blainey, *in* "Biofouling and Biocorrosion in Industrial Water Systems" (H.-C. Flemming and G. G. Geesey, eds.), p. 29. Springer-Verlag, Berlin, 1991.

[13] K. C. Marshall, *in* "Bacterial Adhesion: Mechanisms and Physiological Significance" (D. C. Savage M. and Fletcher, eds.), 133. Plenum Press, New York, 1985.

[14] J. W. Costerton, Z. Lewandowski, D. DeBeer, D. Caldwell, D. Korber, and G. James, *J. Bacteriol.* **176,** 2137 (1994).

[15] J. W. Costerton, T. J. Marrie, and K. J. Cheng, *in* "Bacterial Adhesion: Mechanisms and Physiological Significance" (D. C. Savage and M. Fletcher, eds.), p. 3. Plenum Press, New York, 1985.

water and solute transport phenomena that serve to influence biofouling kinetics.[16] Typical symptoms of membrane biofouling include (i) a reduction in membrane water flux due to establishment of a gel-like diffusion barrier (i.e., the biofilm), (ii) an increase in solute concentration polarization accompanied by lowered solute rejection (in RO and NF membranes), (iii) an increase in the module differential pressure (Δp), (iv) biodegradation and/or biodeterioration of the membrane polymer or other module construction materials (e.g., polyurethane-based glue lines), and (v) establishment of concentrated populations of primary or secondary human pathogens on membrane surfaces.[4]

II. Membrane Biofouling Research Methods

Each of the problem categories just listed invokes important research questions, such as: Why do bacteria attach to synthetic membrane materials? What kinds of bacteria are responsible for early and late stages of membrane biofouling? How does biofilm architecture and community structure influence membrane performance? Do bacteria exhibit equal affinities for different polymer membrane materials? Can new membrane materials be designed with reduced biofouling potentials? How rapidly do membrane biofilms form? Can chemical agents be identified that interfere with bacterial adhesion, thereby retarding membrane biofouling? Is it possible to disrupt the biofilm and return the membrane to its original performance prior to biofouling? Associated with each of these queries are one or more appropriate research strategies or methodologies. This article provides descriptions of the primary protocols and methods currently used to explore and determine the mechanisms of biofouling on the surfaces of separation membranes. Whereas molecular genetic methods (e.g., 16S rRNA gene analysis) are currently employed to determine membrane biofilm community structure, these methods will not be described herein because they are covered in other articles in this volume.[16a] The use of microscopic techniques (e.g., fluorescence deconvolution microscopy) to observe and analyze membrane biofilm structure and function is described elsewhere in this volume.[16b] Most methods described in this article have a practical bias that has grown from a perception by the water treatment industry that solutions to membrane biofouling are urgently needed.

[16] M. R. Weisner and P. Aptel, *in* "Water Treatment Membrane Processes" (J. Mallevialle, P. E. Odendaal, and M. R. Weisner, eds.), p. 4.1. McGraw-Hill, New York, 1996.

[16a] *Methods Enzymol.* **361** [5] 1999 (this volume).

[16b] *Methods Enzymol.* **361** [14] 1999 (this volume).

A. Membrane Module Retrieval, Disassembly, and Biofilm Sampling

It is frequently necessary to disassemble membrane modules and visually inspect the membrane surfaces for evidence of biofouling. In addition, samples of the putative biofilm may be required for physicochemical analysis and for the isolation, identification, and characterization of representative fouling bacteria. Retrieval of membrane biofilm samples for analysis must be done with considerable care to preclude chemical and microbial contamination of samples from extraneous sources. This is especially important if molecular genetic techniques such as polymerase chain reaction (PCR) amplification of biofilm target sequences are going to be used to delineate biofilm community structure. Typically, membrane modules are removed from the system under investigation and partially drained of excess liquid, with care being exercised to maintain the interior of the module (and the membrane surfaces in particular) in a moist or wetted condition. After draining, the modules may be stored for transport to the laboratory by carefully sealing them in plastic wrap to preclude or retard drying. In order to maintain a humidified condition during membrane storage and transport to the laboratory, the modules may be first wrapped in dampened towels. Placement of the modules on ice is preferable if transit times are expected to exceed 24 hr, but this is not always practicable, especially if numerous large elements must be transported. Therefore, if possible, transport times to the laboratory should be kept to a minimum, e.g., <24 hr, and the modules should not be permitted to experience temperatures above about 20°.

Once in the laboratory, the membrane modules should be handled and dissected using aseptic techniques. However, it is frequently necessary to use power saws or other large cutting or prying tools to remove the durable fiberglass or plastic housings that commonly surround and protect the fragile membrane leaves. If this is the case, special precautions should be exercised to sample the interior regions of the membrane module that have not been exposed to the effects of the tools. Once the protective housing has been removed, it is generally possible to carefully unroll or otherwise expose the biofouled membrane surfaces on a disinfected table top or other appropriate venue. Specimens of the membrane can then be removed using sterile razors or scalpels and these samples can be placed (with little or no disturbance to the biofilm) into tared presterilized empty vessels (e.g., glass scintillation vials) or vessels containing physiological buffer solutions or chemical fixatives for electron microscopy (Fig. 3). Alternatively, for purposes of bulk biochemical analyses (e.g., total protein, carbohydrate, adenosine 5′-triphosphate [ATP]) or total genomic DNA extraction for PCR or direct gene probe analyses, the biofilm may be scraped from the membrane surface

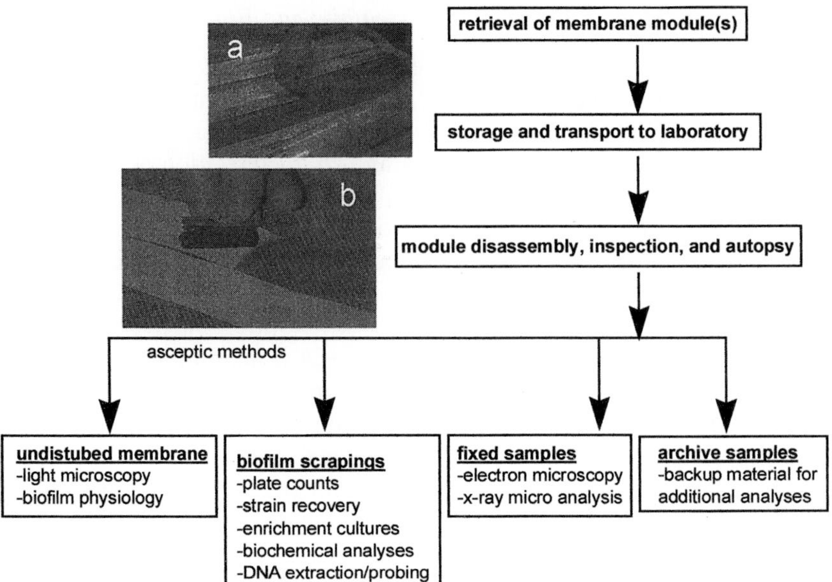

FIG. 3. Outline summary of typical biofilm sampling protocol for membranes. (Insets) Removal of biofilm samples using a sterile single edge razor.

using a sterile single edge razor blade or other suitable instrument. After a region of interest is scraped clean, that region should be measured carefully so that final analytical results may be quantitatively expressed on a "per unit area" basis (e.g., mg protein or colony-forming units per cm^2). It is also advisable to obtain macroscopic photographs of the biofouled membrane surface to indicate possible heterogeneity in biofilm distribution due to channeling or other fluid dynamic effects. Typically, a range of membrane and biofilm samples are collected from a single membrane autopsy. The samples may include (but are not limited to) the following: (i) one or more undisturbed membrane sections placed in sterile filtered water or a physiologic buffer solution for direct microscopic observations where cells must be maintained in a viable or metabolically active condition, (ii) samples fixed in 10 m*M* phosphate- or HEPES-buffered 2.5–5.0% (w/v) glutaraldehyde (pH 7.0) for scanning electron microscopy (SEM) or transmission electron microscopy (TEM), (iii) scraped biofilm samples for DNA or phospholipid fatty acid extraction for microbial community analysis, (iv) scraped biofilm samples for plate counts or enrichment cultures, (v) scraped material for biochemical analyses (e.g., total protein, ATP), and (vi) air-dried biofouled membrane samples for surface analytical methods such as attenuated total reflection Fourier transform infrared (ATR-FTIR) spectrometry (see Sections II,C,1 and II,D,2). Additional undisturbed membrane samples should be archived at −80° or colder to serve as a source of reserve material in the event that collateral or confirmatory analyses are required at a later time.

B. Preparation of Defined Polymer Membrane Substrata

It is tacitly assumed that adhesive interactions between the bacterial surface and the membrane substratum are the fundamental cause of early biofilm development in membrane systems. These adhesive interactions may be modified by the ionic properties of the aqueous environment[7,13] or by the presence of surface active compounds, biocides, etc.[10] One of the primary goals of membrane biofouling research is to improve understanding of the complex relationships between the physicochemical nature of the polymer membrane substratum and the adhesion behavior of bacteria involved in early biofilm formation. It is therefore essential in studies concerning the initial events of bacterial adhesion and biofilm formation on separation membranes that the substrata be precisely defined chemically and physically. If the substratum remains uncharacterized, it becomes difficult or impossible to establish meaningful correlations between measured biological parameters (e.g., cell surface properties, adhesion kinetics) and the nature of the membrane surface. In most instances it is not possible to use commercially available membranes as model substrata because manufacturers frequently modify their membranes by proprietary surface chemical treatments, such as polyvinyl alcohol to increase surface wettability and vinyl acetate/polyacrylic acid copolymers (Colloid 189) to improve solute rejection of RO membranes. Not only do such treatments alter membrane surface charge and chemistry, but they can also affect the microtopographic and mechanical (e.g., elastic, compressive) properties of the membrane. Because of this problem, it is generally advisable for critical work to prepare defined membrane substrata *de novo*. While the preparation and surface characterization of all membrane types are well beyond the scope of this article some generic examples of the preparation and characterization of the most common polymer membranes are provided.

1. Preparation of CA Membrane Substrata. Figure 4 outlines the basic steps involved in CA membrane preparation via a standard phase inversion process.[3] Many variations of method and scale of this basic approach are possible and routinely employed. Preparation of CA membranes involves initial solubilization of a well-characterized polymer (e.g., degree of acetylation, molecular weight, charge) in a suitable water-soluble organic solvent such as acetone, methanol, dioxane, pyridine, or some mixture of these

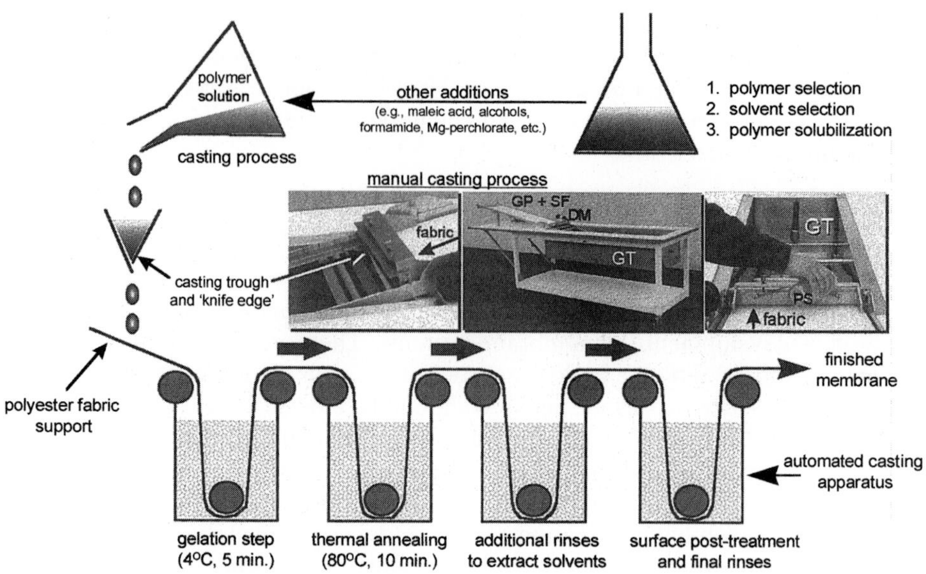

FIG. 4. Schematic of automated CA membrane casting process. (Photo insets) Manually operated laboratory glass plate casting apparatus. GP + SF, glass plate with support fabric attached; DM, drive mechanism; GT, gelation tank; PS, polymer solution; arrows indicate direction of glass plate/fabric movement during polymer casting; see text for details.

agents. In addition to the primary solvent(s), other compounds may also be present to affect changes in stability and performance (i.e., flux and solute rejection properties) of the finished membrane, such as detergents, salts, organic acids (e.g., maleic acid), and secondary charged or uncharged polymers (in the case of heteropolymer blend membranes). In the laboratory, the polymer solution may be cast directly onto a porous woven polyester support fabric (e.g., Texlon, Texlon, Inc., Los Angeles, CA) attached to a glass plate carrier, as depicted in Fig. 4 (photo insets). The glass plate carrier and attached support fabric (coated with the CA solution) are then immersed quickly into an initial aqueous ("gelation") bath followed by a series of additional temperature-controlled water baths, each containing proprietary ionic compositions. The immersion and rinse processes are typically performed at a controlled rate of several feet per minute. The CA polymer is physisorbed to the support material as no chemical reaction is involved and the film thickness is controlled by the temperature and concentration of the polymer solution, casting and immersion rates, and other process variables. During aqueous casting, the humidity and temperature must be controlled carefully as the solvent phase and any other chemical additives are dissolved rapidly into the aqueous phase in a controlled gelation step (phase inversion) in which the CA effectively precipitates onto the surface of the support fabric. The initial bath is typically ice-cold water, followed by a combined rinsing and annealing step at 80–90° for up to 10 min. Thermal annealing results in condensation and contraction of the polymer matrix only at the outer (semipermeable) surface and thus produces the characteristic asymmetric structure of the CA membrane (see Fig. 1) and determines the solute rejection and water transport (flux) properties of the finished membrane. Following thermal annealing, the membrane is cooled and rinsed extensively through additional water baths. It is also possible to cast the CA (dissolved in high vapor pressure solvents such as chloroform or dichloromethane) onto nonporous support materials such as glass, polycarbonate, or metals and then evaporate the solvent phase rapidly by means of a dry air stream (see Section II,B,3). The finished CA membrane is stored in a fully hydrated state to prevent surface defects (shrinking, cracking) and other damage associated with drying. The final membrane may be posttreated with wetting agents (e.g., glycerol), Colloid 189 (to improve solute rejection properties), or other compounds. Long-term storage of CA membranes is best done at 4° to discourage microbial growth (particularly fungi). Use of chemical biocides (e.g., glutaraldehyde, bisulfite, and EDTA) or sterilization by gamma irradiation is also possible. Clearly, even for CA membranes, which are considered among the easiest to manufacture, there are a multitude of interactive process variables that are difficult to control precisely from one membrane batch to the next. Thus, surface structural differences may occur among CA membranes supplied by different manufacturers or by the same manufacturer at different times.

2. *Preparation of Polyamide TFC Membrane Substrata.* Like the CA membranes described earlier, there are numerous and complex process variables involved in the preparation of TFC membranes.[1,3] As illustrated in Fig. 5, most commercial TFC membranes are synthesized via an interfacial condensation (polymerization) reaction involving an initial deposition of a water-soluble difunctional amine, such as *m*-phenylenediamine (MPD), onto a porous support layer, such as polysulfone (see Fig. 1). The polysulfone layer is itself supported further by a highly porous nonwoven polyester (AWA) backing material (Awa Paper Co., Tokyo, Japan), which provides structural integrity to the finished membrane. Immediately following deposition of the MPD layer, a hydrophobic (organic soluble) multifunctional acid chloride, such as trimesolyl chloride (TMC), is cast directly onto the MPD surface from the organic solvent phase (e.g., hexane). This initiates an extremely rapid interfacial polymerization and cross-linking reaction at the water/organic interface. Unreacted monomers, solvent, and other process chemicals are then exhaustively leached away from the nascent membrane in a series of typically proprietary water baths at controlled pH,

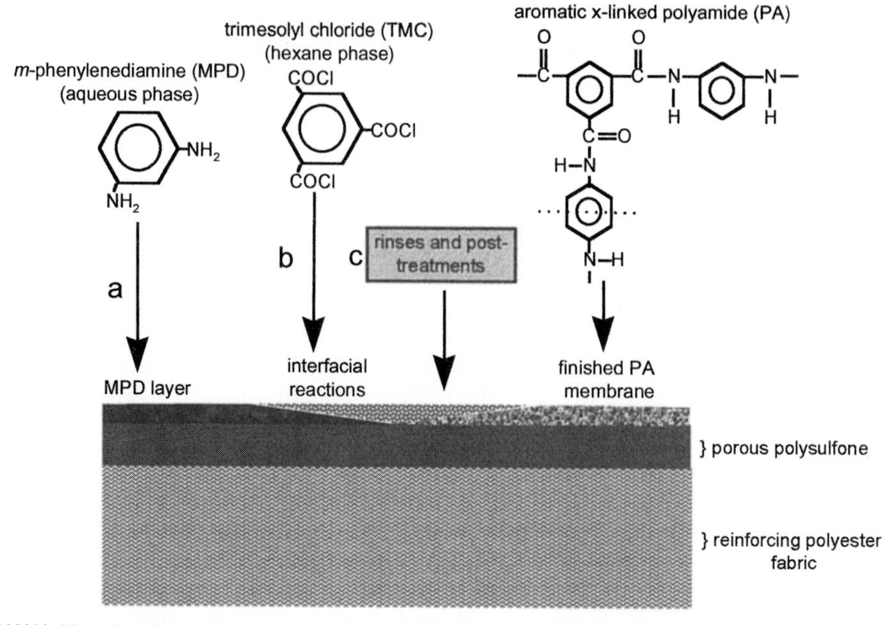

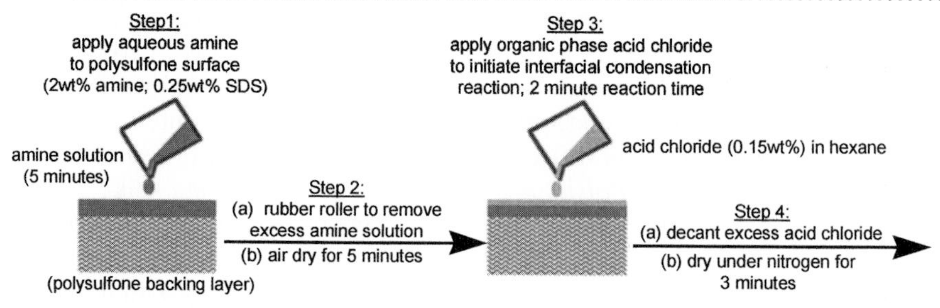

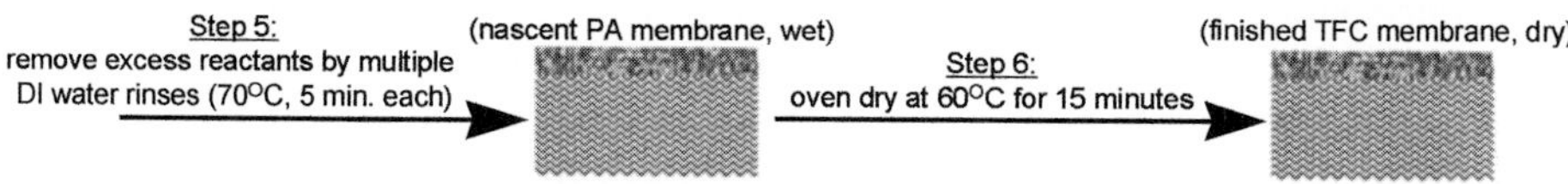

FIG. 5. (Top) Major steps in interfacial polymerization process involving deposition of MPD onto porous polysulfone support (a), addition of TMC from hexane phase to initiate interfacial polymerization (b), and rinsing steps to remove solvents and unreacted monomers. (Bottom) Detailed steps in lab-scale synthesis of TFC membrane carried out on porous polysulfone support.

temperature, humidity, and ionic strength. The entire process can be carried out readily in the laboratory on a small scale according to the specific steps outlined in Fig. 5. The finished membranes may be stored in a wetted condition (e.g., shrink-wrapped) in 50 m*M* sodium bisulfite solution or air-dried and placed under a nitrogen atmosphere and/or refrigerated at 4°. TFC membranes typically exhibit superior solute rejection and higher water fluxes compared to CA membranes. The greater flux of TFC membranes may be partly attributed to the much thinner (but more fragile) rejection layer that is produced by the interfacial polymerization reaction. Because of the great number of process variables involved in TFC membrane synthesis, the surface properties and performance of such membranes from different manufacturers may be quite different, even when the same basic polymer chemistries are employed. In addition, some types of PA membranes are quite unstable after preparation, resulting in spontaneous changes in their surface morphology and performance after storage in air for different periods. These considerations must be kept in mind when interpreting the results of membrane surface characterization work or bacterial attachment and biofouling studies.

3. Preparation and Casting of CA Thin Films on Internal Reflection Elements (IREs) and Glass Coupons. It is often useful to prepare defined model membrane surfaces that can be used to investigate organic or bacterial adsorption phenomena by direct microscopic methods (see later) or indirectly and noninvasively by methods such as ATR-FTIR spectrometry (see later). In either case, there is a requirement that the membrane polymer be cast onto the surface of an IRE or a suitable inert carrier or support material such as glass. Certain polymer systems, e.g., CA homopolymers or heteropolymer blends of CA comprising different degrees of acetylation are especially well suited for this purpose because they can be prepared as monophasic solutions that can be cast readily onto inert glass or metal support materials. The CA-coated coupons created in this manner can then be incorporated into special flow cells (see later), which permit the measurement of bacterial adsorption/desorption kinetics.

Parallelopiped (50 × 10 × 2 mm) Ge and (50 × 10 × 3 mm) ZnSe IREs are purchased from Harrick Scientific, Inc. (Ossining, NY). Use of ZnSe IREs for this purpose is recommended over Ge IREs because CA polymers adsorb more strongly to the former material. The Ge IREs are polished with 1.0-μm diamond paste (Buehler, Inc., Lake Bluff, IL), washed with detergent, rinsed with tap water, and then rinsed with 18 MΩ/cm deionized water (E-pure, Barnstead/Thermolyne, Dubuque, IA). ZnSe IREs are washed as described earlier, but not polished. Polymers of CA (approximately 100,000 MW) with varying levels of acetylation (~40–43.9 wt% acetyl content) are obtained from Eastman Kodak, Inc. (Rochester,

NY). The CA polymers are dissolved (at 0.5 wt%) in high purity dichloromethane (B&JGC, Burdick & Jackson, Muskegon, MI). Solutions are mixed with a Teflon stir bar and sonicated in a warm water bath (50–60°) until the CA is completely dissolved. Solutions are filtered through lens paper to remove insoluble fibrous material (i.e., undissolved CA). A Pyrex cylinder (20 × 6 cm outer diameter) is used in the coating process (Fig. 6). Compressed air passed through a dryer (Balston Model 75-20) and a 0.2-

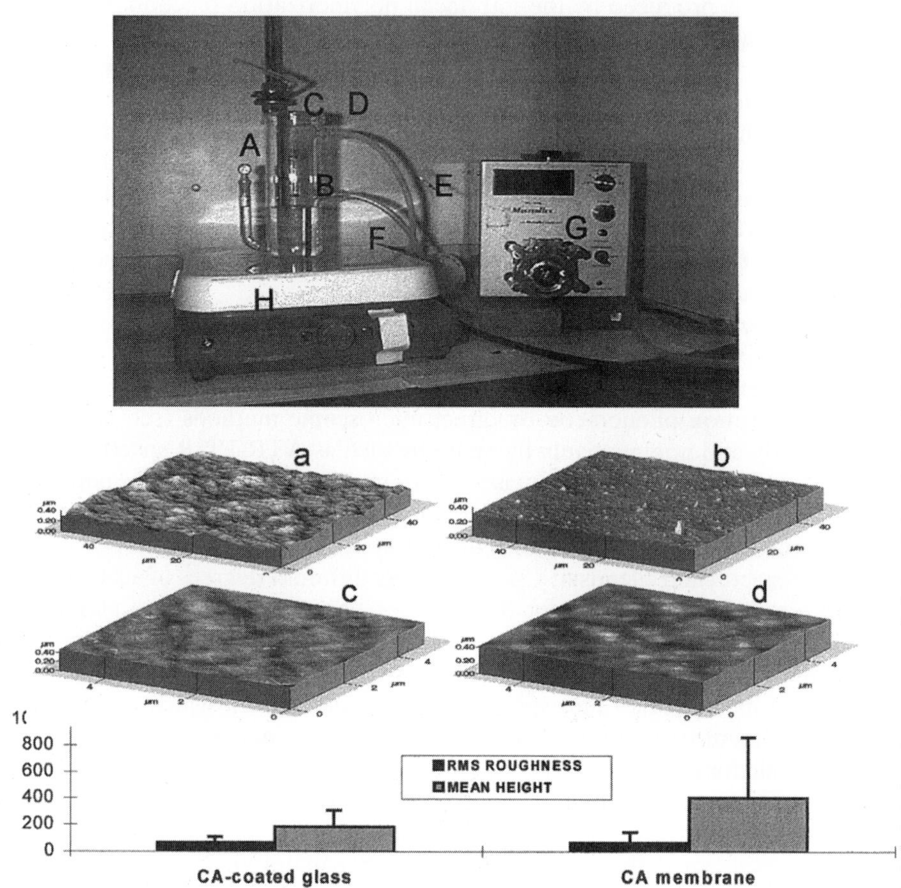

FIG. 6. (Top) Apparatus for casting CA thin films on IREs or glass coupons. (A) Coating cylinder, (B) ZnSe internal reflection element, (C) fine copper wire (connecting to IRE), (D) nylon monofilament, (E) brass swivels, (F) dry air inlet/outlet lines, (G) peristaltic pump, (H) stir plate. (Middle): AFM images of commercial CA membrane at low (a) and high (c) magnification, and CA thin film cast on glass at low (b) and high (d) magnification. (Bottom) AFM height and roughness data for CA-coated glass and commercial CA membranes.

μm polytetrafluoroethylene filter is used to continuously purge the cylinder of water vapor. Mixing of the CA solution is achieved with a Teflon stir bar; however, both stirring and air purge are stopped just prior to casting of the CA film to eliminate turbulence at the air–solution interface, which can result in the uneven deposition of the polymer. A peristaltic pump (Masterflex, Cole-Parmer Instrument Co., Chicago, IL) is used to withdraw the IRE from the polymer solution at a constant rate. Nylon monofilament (2-lb test) is tied to the drive shaft of the pump, and the IREs are secured to a fine copper wire with Teflon adhesive tape. The wire is attached to the nylon line with a brass swivel. The IREs are dipped and withdrawn from the CA solution at a rate of 1 cm/sec. The flow of air to the cylinder is reinitiated immediately after the IRE is withdrawn from the solution. The ZnSe IREs are dipped only once. The CA is removed from the end of the IRE with a cotton swab saturated with chloroform. Gravimetric analyses in which the weights of IREs are compared before and after coating with CA indicate that the film thickness on a ZnSe IRE dipped once in a 0.5 wt% CA solution is about 1500 Å, assuming a density of 1.3 g/cm^3. Precleaned glass microscope slides and coverslips are coated in the same manner as described earlier. A CA film thickness of 1260 $\pm$ 80 Å was calculated from gravimetric analyses. A similar thickness of CA is assumed to deposit on the Ge IREs. Analysis of CA-coated glass (not thermally annealed) and annealed commercial CA membranes by AFM reveals similar but not identical surface morphologies (see Fig. 6). Due to the lack of thermal annealing, the simulated CA "membrane" is presumed to be structurally symmetrical with high flux and poor solute rejection properties compared to commercial membranes.

C. Surface Property Characterization of Polymer Membranes

Characterization of the surface morphology and physicochemical attributes of polymer membranes is necessary in order to identify possible relationships with bacterial adhesion and early biofilm formation. Minimum requirements for membrane surface characterization include knowledge and confirmation of the polymer chemistry as well as quantitative measurements of (i) the microtopography of the hydrated membrane surface, including pore size distribution and geometry in microporous membranes, (ii) surface hydrophobicity, and (iii) surface charge. Specific methods designed to quantify each of the above surface properties are described in this section.

1. Confirmation of Surface Polymer Chemistry by ATR-FTIR Spectrometry. Absorption in the mid-infrared region (4000–400 cm^{-1}) can provide a convenient means of acquiring a unique spectroscopic fingerprint for each type of polymer separation membrane. Absorption spectra provide information on specific polymer functional groups (e.g., carbonyl, sulfonate, or amine groups) exposed at the membrane surface and are characteristic of the specific polymer surface chemistry (Fig. 7). Such IR absorbance spectra may be obtained readily using ATR-FTIR spectrometry. Small rectangular pieces of a dry membrane material are pressed against a ZnSe or Ge IRE (Fig. 7 top). The IRE sample holder is then placed on the ATR mirror assembly (see Fig. 15) in the sample compartment of an FTIR spectrometer (e.g., a Nicolet Magna 550, Nicolet Instrument, Madison, WI) equipped with either a liquid nitrogen-cooled mercury–cadmium–telluride (MCT) or a deuterated triglycine sulfate (DTGS) detector. Single-beam sample spectra are obtained by signal averaging multiple (256) scans at 4-cm^{-1} resolution and utilizing a Happ–Genzel apodization function. Each sample spectrum is ratioed against a bare IRE background spectrum and then converted to absorbance. Each absorbance spectrum is then corrected to

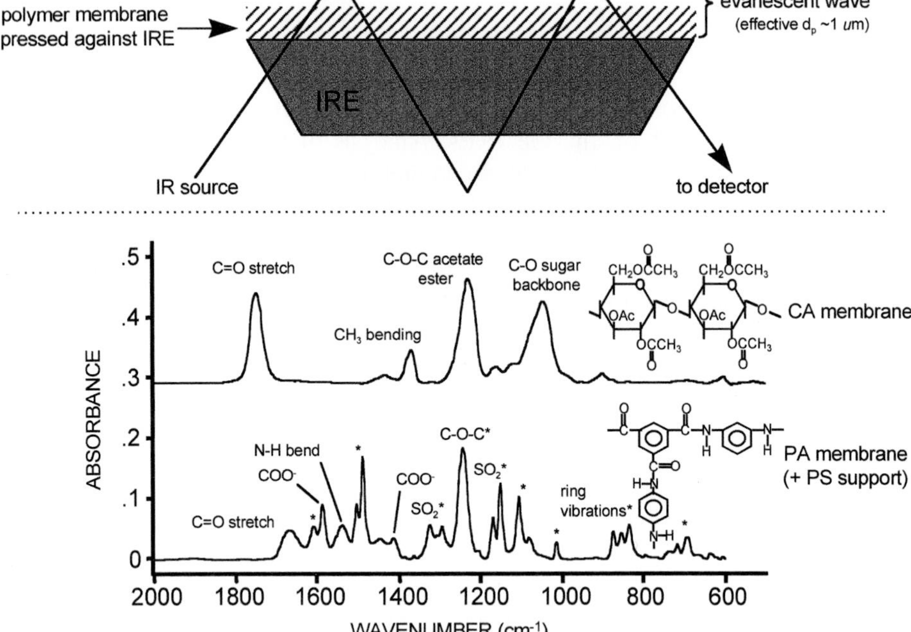

FIG. 7. Diagram of method for acquiring ATR-FTIR spectra of polymer membrane materials (top) and examples of mid-IR spectra for CA and PA membranes (bottom). The IR spectrum of the PA membrane is a combination of the PA and polysulfone (PS) support layer (functional groups denoted by *) as the evanescent wave passes through the thin polymer film. Spectra are corrected to account for the wavelength dependence of ATR spectrometry of bulk materials. Depth of penetration (d_p) of evanescent wave ≅1.0 μm.

account for the wavelength dependence of the depth of penetration of the evanescent wave.[17] Longer wavelength radiation penetrates farther into the adjoining sample medium; therefore, adsorption bands are relatively stronger and broadened on the long wavelength side. An Array Basic program provided with a spectral processing software package (GRAMS/32 version 5.10, Galactic Industries, Salem, NH) is used to make the correction and carry out other advanced spectral manipulations. The incidence angle of the IREs used is 45°, and the refractive index of the polymer membrane is assumed to be 1.5. The same correction can be done with software provided by the instrument manufacturer.

2. Determination of Membrane Surface Topography by Atomic Force Microscopy. The microscale surface features of polymer membranes influence colloidal and biofouling kinetics.[18] Therefore, knowledge of the surface morphology of membranes, especially in their hydrated states, is essential for understanding the mechanism(s) of bacterial adhesion and biofilm formation. The microscale topography and pore geometries of air-dried as well as fully hydrated (i.e., submerged) polymer membranes can be best characterized and mapped by intermittent ("tapping") or noncontact mode atomic force microscopy (AFM). Instruments such as Park Scientific Instrument's Model CP AutoProbe (Sunnyvale, CA) equipped with a noncontact/contact head and a 100-μm scanner operated in a constant force mode are suitable for this purpose. The wet membrane coupons are attached to a circular stainless steel sample holder using 12-mm carbon conductive tape (Ted Pella, Inc., Redding, CA). The holder with the attached membrane coupon is then mounted on the piezo scanner of the AFM, which is equipped with a liquid microcell. Before scanning and image acquisition, it is imperative that sufficient time is allowed to pass for the polymer membrane specimen to undergo complete hydration and swelling to equilibrium, a process that may require up to several hours at room temperature (about 23°). If scanning is attempted while membrane swelling is still in progress, a stable signal cannot be attained. Images are acquired utilizing silicon "ultralevers" (force constant = 0.24 N/m; Park Scientific), which are gold-coated cantilevers with integrated high-aspect ratio silicon nitride conical tips designed for maximum penetration into pores and other surface irregularities that are encountered frequently on the surfaces of polymer membranes, especially microporous membranes. Due to the extreme fragility of many polymeric membranes in the hydrated state, tapping mode AFM, which is similar to noncontact mode AFM, is generally employed to mini-

[17] N. J. Harrick, "Internal Reflection Spectroscopy," p. 13. Harrick Scientific Corporation, Ossining, New York, 1979.

[18] M. Elimelech, X. Zhu, A. E. Childress, and S. Hong, *J. Membr. Sci.* **127,** 101 (1997).

mize translational forces between the AFM tip and the polymer membrane surface. In the tapping mode, the AFM cantilever is maintained at some distance from the membrane substratum and is oscillated at a relatively high amplitude (on the order of 1000 Å) at or near its resonant frequency. The vibrating cantilever/tip is then moved closer to the sample surface until it just touches ("taps") the sample once during each oscillation. Thus, tapping mode AFM is less likely than contact mode AFM to cause friction-induced artifacts on the fragile membrane surface due the elimination of lateral mechanical forces between the tip and the sample. AFM images are acquired at a scan rate of 1.0–2.0 kHz with a minimum information density of 256 × 256 pixels. The root mean square roughness (RMS roughness) and mean surface height may be calculated for each membrane using Park Scientific software provided with the CP AutoProbe; however, virtually all AFM systems include similar software capabilities. For a transect containing N data points, the RMS roughness is given by the standard deviation of the individual height measurements (Park Scientific):

$$R_{\text{ms}} = \sqrt{\frac{\sum_{n=1}^{n} (z_n - \bar{z})^2}{N-1}}$$

where $\bar{z}$ is the mean z height. The mean height is given by the average of the individual height determinations within the selected height profile:

$$\bar{z} = \frac{1}{N} \sum_{n=1}^{N} z_n$$

where $\bar{z}$ is the mean z height. Representative tapping-mode AFM images of three different polymer membrane materials are presented in Fig. 8. Each membrane was imaged submerged in distilled water. The associated graph indicates the surface height and roughness measurements for each membrane.

3. Determination of Membrane Surface Hydrophobicity by Captive (Air) Bubble Contact Angle Measurement. The hydrophobicities of cells and inanimate substrata influence the strength and kinetics of microbial adhesion and early biofouling.[19] Therefore, the relative hydrophobicities of polymer membrane materials represent an important surface parameter in biofouling studies. The surface hydrophobicities of polymer membranes are best determined by captive (air) bubble contact angle measurements.[20,21]

[19] M. Rosenberg and R. J. Doyle, *in* "Microbial Cell Surface Hydrophobicity" (R. J. Doyle and M. Rosenberg, eds.), p. 1. Am. Soc. Microbiology, Washington, DC, 1990.

[20] J. Drelich, J. D. Miller, and R. J. Good, *J. Colloid Interface Sci.* **179,** 37 (1996).

[21] R. M. Prokop, O. I. del Rio, N. Niyakan, and A. W. Neuman, *Can. J. Chem. Engineer.* **74,** 534 (1996).

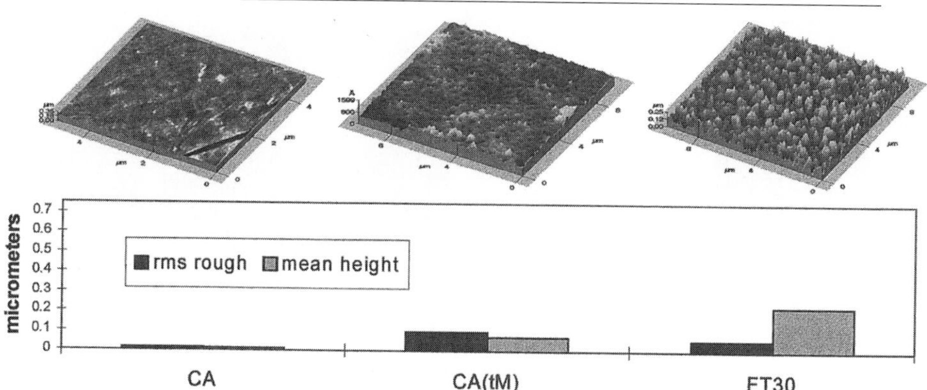

FIG. 8. Example of AFM images and associated morphometric data for three polymer membrane samples, including commercial CA, an experimental microporous CA-trimellitate membrane (CA[tM]), and a commerical PA membrane (FT30). All membranes were imaged fully hydrated in the intermittent contact (tapping) mode. Images are 10 × 10 μm.

Use of the captive bubble method instead of the conventional sessile water droplet method is preferable because many polymer membranes (e.g., CA) cannot be air dried without introducing significant surface artifacts, such as shrinkage and cracking. Furthermore, application of a water droplet to a dried membrane surface would result in immediate and continuous membrane swelling, which, in turn, would preclude accurate contact angle measurements. Finally, because polymer membranes are normally operated in a fully hydrated (i.e., submerged) condition, captive bubble determinations on wet membranes are more relevant for actual operating conditions. An example of a homemade captive bubble contact angle apparatus is depicted in Fig. 9. The main components of the system consist of a clear Plexiglas liquid reservoir, an aluminum sample support stage, a charge-coupled device (CCD) camera, an imaging lens, a *x-y-z* camera mount, and an illumination source (Fig. 9).

The major components mount on a flat sheet of aluminum (0.95 cm) equipped with threaded legs to maintain a level plane. The specimen mounting stage consists of a 10.16 × 5.08 × 2.54-cm block of 316 stainless steel with a 1-cm slot cut 3.8 cm deep down the middle. An alternate stage for more fragile membrane samples was also machined with a 0.5-cm wide slot. The stage is equipped with a stainless steel top plate to hold the sample flat. The sample stage is placed in 10.16 × 10.16 × 10.16-cm clear Plexiglas reservoir filled with 18 MΩ/cm distilled water. A thread-feed syringe with a Luerlock needle connection is angle mounted on the side of the reservoir. The syringe is equipped with a 7.62-cm, 22-gauge, 90° bevel stainless steel

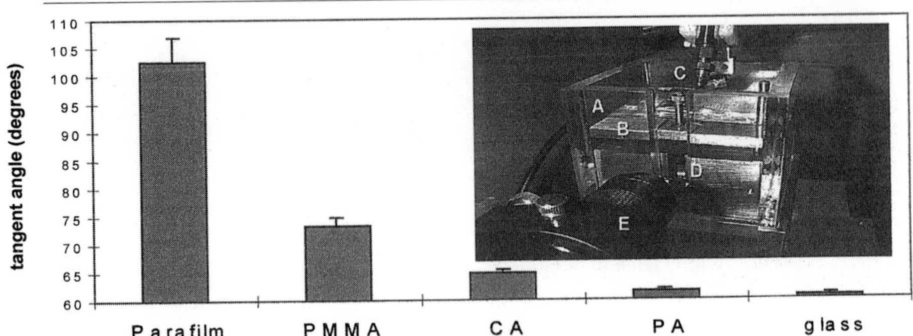

FIG. 9. A captive bubble contact angle apparatus and an example of contact angle data obtained using this device. A, clear Plexiglas housing; B, stainless-steel lid for holding membranes flat; C, angle-mounted syringe for delivering 10 μl air bubble to membrane sample; D, stainless-steel base plate with central vertical viewing channel; E, CCD camera and focusing lens. PMMA, polymethyl methacrylate.

needle (Hamilton Co., Reno, NV). Air bubbles discharged from the syringe are estimated at 7–10 μl. The syringe needle is reamed with a 0.025-cm nickel wire prior to the day's measurements to ensure needle diameter. A glass microscope slide is mounted in the wall of the reservoir, opposite the syringe, to enable capture of images. The CCD camera (COHU, Model 48155000 AL2D) is equipped with a 0.75 to 3.0× objective (Edmund Scientific, Barrington, NJ) and mounts on the x-y-z positioning stage. Images are captured and processed using CUE2 series image analysis software (Olympus America, Inc., Lake Success, NY). A Sobell filter is used to outline the circular perimeter of the bubble and the contact baseline. The CUE2 program generates the bubble height and diameter data for calculation of the height : diameter ($h : d$) aspect ratio. For purposes of comparison, Parafilm (National Can, Inc., Greenwich, CT), polymethyl methacrylate (PMMA), nonporous (i.e., dense) polypropylene, and glass microscope slides are analyzed in parallel with any membrane samples. Buoyancy carries the introduced air bubble upward and onto the hydrated membrane surface where it becomes trapped ("captive"). Departure of the air bubble from a perfect sphere with an aspect (diameter/height) ratio of 1.0 is related to the degree of spreading of the bubble over the membrane surface. Smaller aspect ratios ($\ll$1.0) correspond to greater bubble spreading and a more hydrophobic surface (larger contact angle), i.e., water is excluded from the bubble–membrane interface. Aspect ratios approaching 1.0 indicate a more hydrophilic membrane surface and a smaller contact angle. The following formulas may be used to convert bubble aspect ratio values to contact angles (Kruss, Inc., Charlotte, NC):

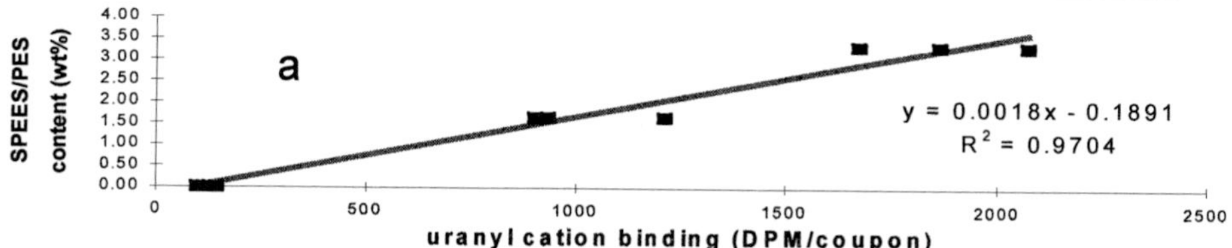

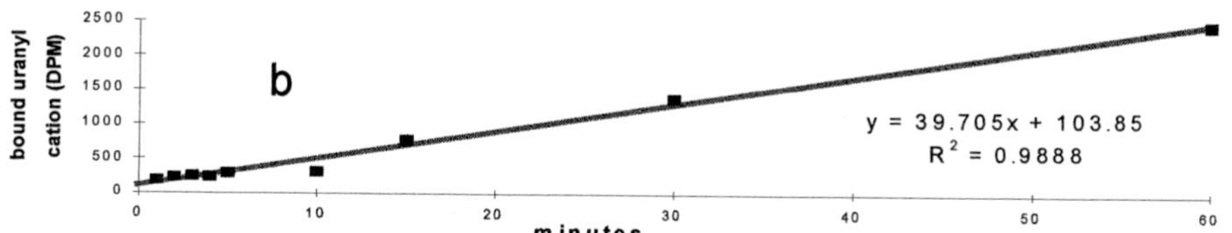

FIG. 10. UCB assay results for a microporous polyether sulfone membrane containing different amounts of a sulfonated polyether-ethersulfone/polyether sulfone (SPEES/PES) block coploymer (a). Assay was performed as described in the text. UCB kinetics (b) indicate a finite amount of uranyl cation binding at $t = 0$ (~104 dpm) reflecting surface-associated sulfonate groups.

$$\text{Contact angle} = 2\arctan(2h/d) \text{ for angles} < 90°$$
$$\text{Contact angle} = 2\arctan[(2d/h) - 1)] \text{ for angles} > 90°$$

where h is the bubble height and d is the bubble diameter.

4. Determination of Membrane Surface Charge by Uranyl Cation-Binding Assay. Most polymer separation membranes possess some degree of surface charge due to trace quantities of free carboxylate or sulfonate groups.[22,23] Relative membrane surface charge (due to the presence of free carboxylate or sulfonate groups in the membranes) may be conveniently determined by uranyl cation binding (UCB) according to a modification of the gravimetric procedure first described by Farrar *et al.*[24] Results from a modified UCB assay using polyether sulfone membranes with known amounts of sulfonation are presented in Fig. 10a. To perform the UCB assay, membrane coupons are secured in plastic reaction tubes identical to those described later for the disk bacterial adhesion assay (see Section II,D,1 and Fig. 12). Five milliliters of a 10 mM solution of uranyl acetate (Baker Chemical, Inc., Phillipsburg, NJ) in MS buffer (pH 5.0; Table 1) is added to reaction tubes to initiate the binding assay. The specific activity of the uranyl acetate as supplied by the manufacturer is approximately 1.459×10^4 Bq/g.

[22] A. E. Childress and M. Elimelech, *J. Membr. Sci.* **119,** 253, (1996).
[23] M. Elimelech, W. H. Chen, and J. J. Waypa, *Desalination* **95,** 269, (1994).
[24] J. Farrar, S. M. Neale, and G. R. Williamson, *Nature* **168,** 566 (1951).

After a 2-hr contact time at room temperature (about 23°), the membrane coupons are removed and rinsed at least 10 times in 18 MΩ/cm distilled water. The amount of bound uranyl cation is determined by liquid scintillation counting (LSC). All UCB assays are performed in triplicate.

Depending on the nature of the polymer system under investigation, it may be preferable to shorten the contact time in order to minimize diffusion of the uranyl cation into the polymer matrix. Thus, very brief contact times of perhaps only a few minutes may provide a more accurate reflection of the true polymer surface-binding capacity, whereas longer contact times probably result in a greater proportion of the counts originating from uranyl ions bound deeper within the polymer matrix. Surface-binding capacity can be estimated by back extrapolation of UCB kinetics through the *Y* axis (at time 0), thereby yielding a theoretical uranyl-binding activity at an infinitely short contact time (Fig. 10b).

D. Determination of Bacterial Adhesion to Polymer Membranes

It is possible to rapidly evaluate the potential of experimental or commercial polymer separation membranes to undergo biofouling (and therefore experience performance losses) by determining the relative affinities of the membranes for a "standard" set of surrogate fouling bacteria. Two bacteria that have been used for this purpose include a hydrophobic strain of *Mycobacterium cheloneae* (strain BT12-100) and a hydrophilic *Flavobacterium* species (strain PA-6), both of which were isolated from biofouled RO membranes used to demineralize municipal wastewater at Water Factory 21, a 15-mgd wastewater reclamation facility located in Fountain Valley, CA.[10,11,25] These and other suitable bacteria can be used in various laboratory adhesion assays to quantify and compare the "biofouling potentials" of polymer separation membranes of differing compositions and structures. Numerous adhesion assay methods have been developed over the years, but membrane systems have special requirements that distinguish them from other kinds of adhesion substrata. For example, it is frequently necessary to examine bacterial adhesion to only one surface of a membrane. Therefore, adhesion assays involving membranes must be designed to either restrict access of the test bacteria on only one surface of the membrane coupon or, if this is not done, the adsorbed bacteria on only one surface of the membrane must be determined by some appropriate means (e.g., by direct microscopic enumeration or autoradiography). In addition, polymer membranes are often quite fragile and may require thorough hydration before an adhesion assay can be performed. Although many methods

[25] H. F. Ridgway, M. G. Rigby, and D. G. Argo, *Appl. Environ. Microbiol.* **47,** 61 (1984).

are possible, four adhesion assay techniques used commonly in membrane biofouling research will be described. These methods include (i) a radiometric "disk" assay method in which radiolabeled bacteria are allowed to come into contact with one surface of a disk-shaped membrane coupon, (ii) autoradiographic scanning of radiolabled bacteria on membranes, (iii) ATR-FTIR spectrometry of bacterial adhesion to polymer-coated IREs held in flow cells, and (iv) use of microscope flow cells. The latter (microscope flow cell) method is described elsewhere in this volume.[26]

1. Determination of Bacterial Adhesion to Polymer Membranes by Radiometric Disk Assay. Four modifications of a rapid bacterial adhesion ("disk") assay described previously by Ridgway and co-workers[4,8,26a] are outlined in Fig. 11. In preparation for an adhesion assay, test bacteria are grown in a defined sterile mineral salts (MS) buffer containing 1.0 g/liter of mannitol, glucose, glycerol, acetate, or other suitable carbon and energy source (Table I). The MS medium is supplemented with 1.4 μCi/ml of sterile-filtered $Na_2{}^{35}SO_4$ (about 1.8×10^{-5} mM sulfate; Amersham, Inc., Arlington Heights, IL). Cultures are incubated on a rotary shaker at 200 rpm for 48–72 hr at 28°. Following incubation, uniformly radiolabeled bacteria are harvested by centrifugation (~10,000 rpm, 15 min, 4°), washed twice in sterile MS buffer (lacking the growth substrate; Table I), and resuspended in fresh MS buffer to yield a final cell density of approximately 5.0×10^9/ml, as determined by epifluorescence microscopy using the DNA-specific fluorochrome 4,6-diamidino-2-phenylindole (DAPI).[10]

In the "standard" version of the disk assay (Figure 11), sterile 16 × 125-mm plastic tubes with screw caps (Falcon 2025, Becton Dickinson, Lincoln Park, NJ) serve as the "reaction vessels" in the adhesion assay (Fig. 12). The bottom one-third of the tubes is cut off and a 1- inch-diameter coupon of CA, PA, or other polymer membrane is placed over the threaded end of the tube and oriented with the active (semipermeable) surface facing the tube interior. The tube interior diameter is ~1.6 mm, yielding ~2.0 cm^2 of available surface for bacterial adhesion. Prior to use in assays, membrane coupons are washed a minimum of 10 times in 18 MΩ/cm distilled water (Barnstead/Thermolyne, Inc., Dubuqe, IA) to remove preservative and/or wetting agents (e.g., bisulfite, glycerol, polyvinyl alcohol) or other surface chemical treatments. After tightening the screw cap, tubes are inverted and 5.0 ml of sterile MS buffer is added to each reaction vessel. Detergents and other test compounds suspected of influencing bacterial

[26] *Methods Enzymol.* **310** [12] 1999 (this volume).

[26a] H. F. Ridgway and J. Safarik, *in* "Biofouling and Biocorrosion in Industrial Water Systems" (H.-C. Flemming and G. G. Geesey, eds.), 81. Springer-Verlag, Berlin, 1991.

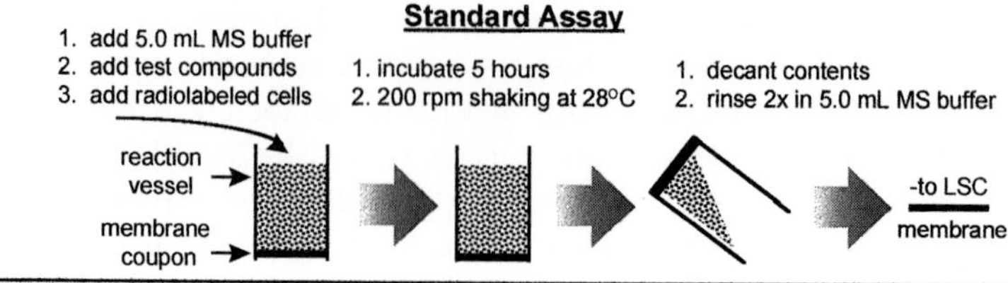

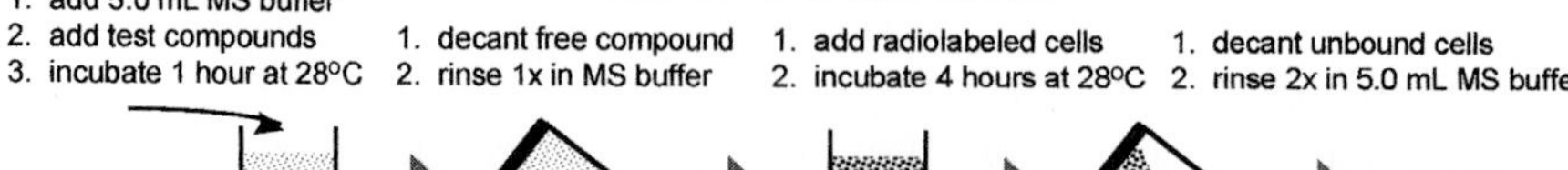

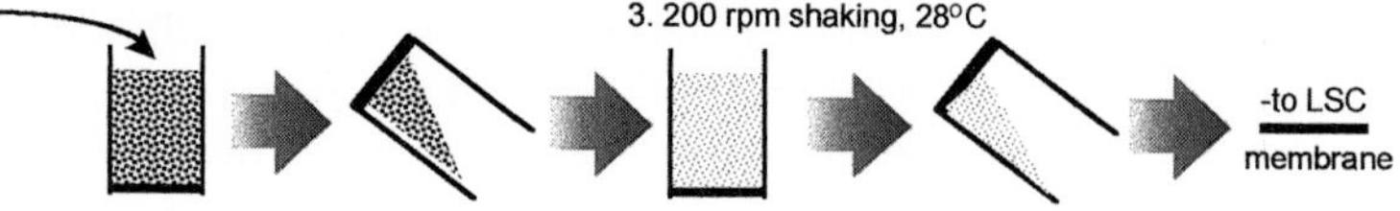

FIG. 11. Schematic outline of steps involved in the four versions of the disk adhesion assay used for determining bacterial affinities for polymer separation membranes. See text for details.

adhesion are typically added to reaction tubes at this point, with care being exercised to minimize overall volume changes. A total of 100 μl of the washed uniformly radiolabled cell suspension is added immediately to each reaction tube at zero time ($t = 0$) to initiate the adhesion assay (final cell density $\cong 1 \times 10^8$/ml). Identically prepared control vessels containing no detergents or other additives are routinely run with each group of test compounds. To account for the possibility of interexperiment variations, one or more "control" membranes (e.g., standard CA or PA membranes) are used as substrata with each experimental run or condition. Control and

TABLE I
COMPOSITION OF MS BUFFER AND MS GROWTH MEDIUM[a,b] USED FOR ADHESION ASSAYS

Component	Concentration
Na_2HPO_4	0.75 g/liter
K_2HPO_4	0.75 g/liter
NH_4Cl	1.0 g/liter
$MgSO_4 \cdot 7H_2O$	0.01 g/liter
$CaCl_2 \cdot 2H_2O$	0.011 g/liter
NH_4NO_3	0.14 mg/liter
$ZnCl_2$	0.0142 mg/liter
$MnCl_2 \cdot 4H_2O$	0.00114 mg/liter
$CuCl_2$	0.00028 mg/liter
$CoCl_2 \cdot 6H_2O$	0.00023 mg/liter
NaCl	0.00014 mg/liter
Ferric ammonium citrate pentahydrate	0.00159 mg/liter
pH 7.0–7.2 prior to sterilization	

[a] MS buffer containing 1.0 g/liter mannitol used to grow and radiolabel mycobacteria strains.
[b] MS buffer containing 1.0 g/liter glucose used to grow and radiolabel flavobacteria strains.

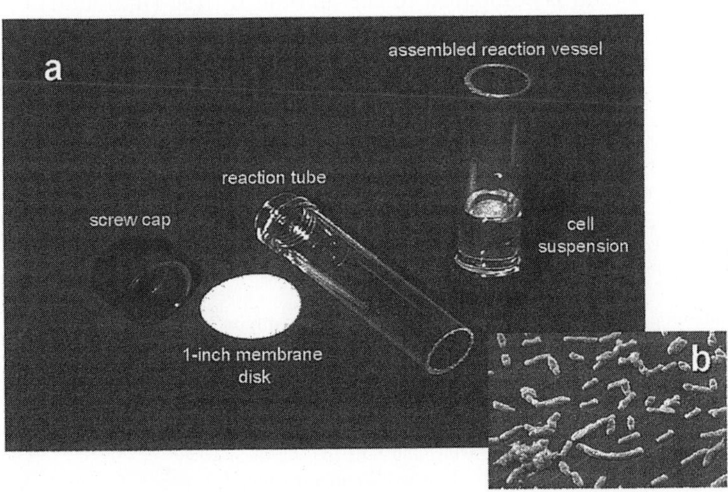

FIG. 12. (a) Disk adhesion assay "reaction tubes" and membrane (disk) coupon. Cell suspensions and other reagents may be added/removed from an assembled reaction tube by way of an open end. (b) SEM of *Mycobacterium cheloneae* strain BT 12-100 cells attached to CA surface following a 2-hr contact time in standard assay.

test vessels are typically prepared in replicates of 10; however, duplicate or triplicate replication of experiments is recommended for a total of N = 20–30. Previous research has demonstrated that equilibrium adhesion is reached in approximately 2–3 hr.[10,26] Therefore, reaction tubes are incubated on a rotary shaker at 200 rpm for 5 hr at 28° (see Table II). Following incubation, the vessel contents are decanted and membrane coupon surfaces are rinsed by two consecutive 5.0 ml volumes of MS buffer to remove loosely bound or unbound cells. The two rinse steps are performed by rapidly dispensing the MS buffer down the inside wall of the tube at a shallow angle (relative to the tube wall) so as to maximize turbulence at the surface of the membrane coupon without directly impinging the buffer stream onto the membrane surface. Rinsed membranes are drained thoroughly and removed from reaction vessels, placed in scintillation vials with 10 ml of cocktail (e.g., OptiFluor, Packard, Inc., Meridan, CT), and analyzed for bound radioactive cells using a liquid scintillation counter (LSC) with external quench correction (e.g., using ^{228}Ra). Using the labeling and adhesion regimen described earlier, several hundred to several thousand counts per minute (cpm) are typically associated with each membrane coupon, and counting is generally performed for a minimum of 20 minutes, routinely yielding maximum counting errors of <0.5%. Typically, LSC software is used to correct raw counts for chemiluminescence. The quantity of membrane-bound radioactive bacteria is reported in cpm or disintegrations per minute (dpm) and is compared to the amount of free (unbound) radiolabeled bacteria at t = 0. Absorption of radioactive counts by the membrane

TABLE II
CONDITIONS USED FOR DISK ADHESION ASSAY METHODS

Parameter	Description
Buffer system	Sterile MS buffer
Test bacterium	*Mycobacterium* strain BT12-100 or *Flavobacterium* strain PA-6
Radiolabeling	Cells labeled uniformly in MS medium with [^{35}S] sulfate
Growth and labeling medium	MS buffer + 0.1% (w/v) mannitol or glucose; cells grown 48 hr
Final cell density at t = 0	About 1×10^8 cells/ml
Incubation conditions	28°, gyratory shaking at 200 rpm
Contact time	Usually 5 hr for standard assay
Rinsing method	Two consecutive turbulent rinses, 5.0 ml each, MS buffer
Membrane pretreatment time	1 hr in MS buffer + test compound
Typical cell pretreatment time	30 min in MS buffer + test compound
Typical cell detachment time	1 hr in MS buffer + test compound

coupons is assumed to be constant across experiments; thus, cpms and dpms are directly proportional. The degree of bacterial adhesion to membrane coupons is expressed as the ratio of bound radioactive cells at the end of an experiment to free cells at $t = 0$ (B/F ratio), which allows direct comparison of cell-binding efficiency from one experiment to the next. The specific activity of bacteria (dpm/cell) is determined by dividing total radioactivity (dpm/ml) of the initial ($t = 0$) cell suspension by cell density (determined by epifluorescence microscopy). The number of membrane-bound bacteria is then estimated from dpm/coupon X cell/dpm. Student *t*-tests (or other appropriate statistical methods) may be used to determine whether cell adhesion data from experimental (i.e., detergent or biocide treated) and untreated control preparations are significantly different at $p \leq 0.05$.

In the "membrane pretreatment" assay, compounds are tested for their ability to affect bacterial adhesion after first having had the opportunity to adsorb to or otherwise interact with the virgin membrane material. In this assay, the active or semipermeable membrane surface is pretreated in reaction tubes for 1 hr at room temperature (about 23°) with a solution of the experimental compound prepared in MS buffer at the desired concentration. Following incubation, the test solution is decanted and the surface of the membrane coupon is rinsed once with 5.0 ml of MS buffer to remove unadsorbed compound. To initiate the adhesion assay, the reaction tube is subsequently filled with 5.0 ml of sterile MS buffer, and radiolabeled test bacteria (prepared as described earlier) are added at $t = 0$. The remainder of the assay is conducted as described earlier for the "standard" adhesion assay except that the incubation (contact) time is generally shortened to 4 hr (Fig. 11).

In the "cell pretreatment" assay, test compounds are evaluated for their ability to influence bacterial adhesion following pretreatment of the cells for a specified period. To conduct the assay, surrogate fouling bacteria are grown and radiolabeled as described earlier, spun down, and resuspended in an equal volume of sterile MS buffer containing the desired concentration of the experimental compound whose action is to be evaluated. After incubation under a prescribed set of conditions and time period (e.g., 30 min at 28°), bacteria are washed twice by centrifugation and resuspension in MS buffer and then used as described earlier in the standard version of the disk adhesion assay (Fig. 11).

In the "cell detachment" assay, experimental compounds or conditions are evaluated for their ability to remove sessile bacteria from a nascent monolayer biofilm. This version of the disk assay is undertaken by first adding a suspension of washed radiolabeled test bacteria in sterile MS buffer to reaction tubes and incubating on a gyratory shaker at 200 rpm for 4 hr to establish a nascent monolayer biofilm. Epifluorescence microscopy of membrane coupon surfaces typically reveals that the early biofilms are nonconfluent and that the spatial distribution of bacteria over the membrane is heterogeneous. Following the initial incubation, the contents of the reaction tubes are decanted and the membrane coupon surface is rinsed once in 5.0 ml of MS buffer (to remove unadsorbed or loosely adherent cells) followed by introduction of 5.0 ml of sterile MS buffer containing the test compound at the desired concentration. Control biofilms are prepared in an identical manner, but are exposed only to MS buffer. Control and experimental preparations are incubated with gyratory shaking (200 rpm) under prescribed conditions and time (e.g., 1 hr at 28°). The tube contents are then decanted and the membrane coupon surfaces are rinsed as described earlier with two consecutive (5.0 ml) changes of MS buffer (Fig. 11). The coupons are then removed and analyzed by LSC.

The disk assay methods described earlier are prone to errors caused by "edge" effects, i.e., accumulation of bacteria in the region where the surface of the membrane coupon and the lip of reaction tube come into contact. Adsorbed bacteria occupying this region of low fluid convection may not experience sufficient shear force during buffer rinses to be removed efficiently. Autoradiographic scans of membrane coupons following disk adhesion assays have indicated that approximately 40% of the total coupon counts are routinely associated with the peripheral region of the membrane (see Section II, D,3, Fig. 17). However, this value is effectively a provisional "constant" that can be ignored when comparing relative bacterial affinities across different membrane materials or conditions. However, estimations of the absolute amounts of bacterial adhesion to different membrane materials may require the subtraction of edge-associated counts from total coupon counts. Such a correction would be particularly important, for example, in calculating the theoretical number of available membrane-binding sites following equilibrium adsorption kinetics.

A low proportion of reaction tubes in disk adhesion assays (typically <5%) will contain membrane defects that allow excessive seepage or diffusion of radiolabeled bacteria from the semipermeable surface of the membrane coupon to the support fabric on the product water surface. Such defects may correspond to microscopic tears, holes, or other physical imperfections along the periphery of the coupon caused by excessive mechanical friction with the lip of the reaction tube during assembly. Because of the fibrillar nature of the polyester support fabrics (see Section I), bacteria can sometimes accumulate rapidly by means of capillary forces on the product surface, resulting in the appearance of inordinately high radioactive counts on these defective membranes. To detect potentially defective (i.e., leaky) membrane coupons, it is possible in any of the versions of the disk assay outlined earlier to place a small circular section of Whatman filter paper between the membrane coupon and the plastic tube cap during assembly. Following the assay, both the membrane coupon and the filter paper are analyzed separately by LSC. High radioactive counts on the filter paper relative to the membrane coupon (e.g., >10% of coupon-associated counts) indicate a possible leakage of radiolabeled cells to the back side of the membrane. An alternative approach for compensating for leaking membrane coupons is to omit from the assay results any coupon counts that exceed 95% confidence limits or that fall outside of two or more standard deviations from the mean value for a set of replicate samples.

Two examples of the application and results from the disk adhesion assay method are presented in Figs. 13 and 14. Data shown in Fig. 13

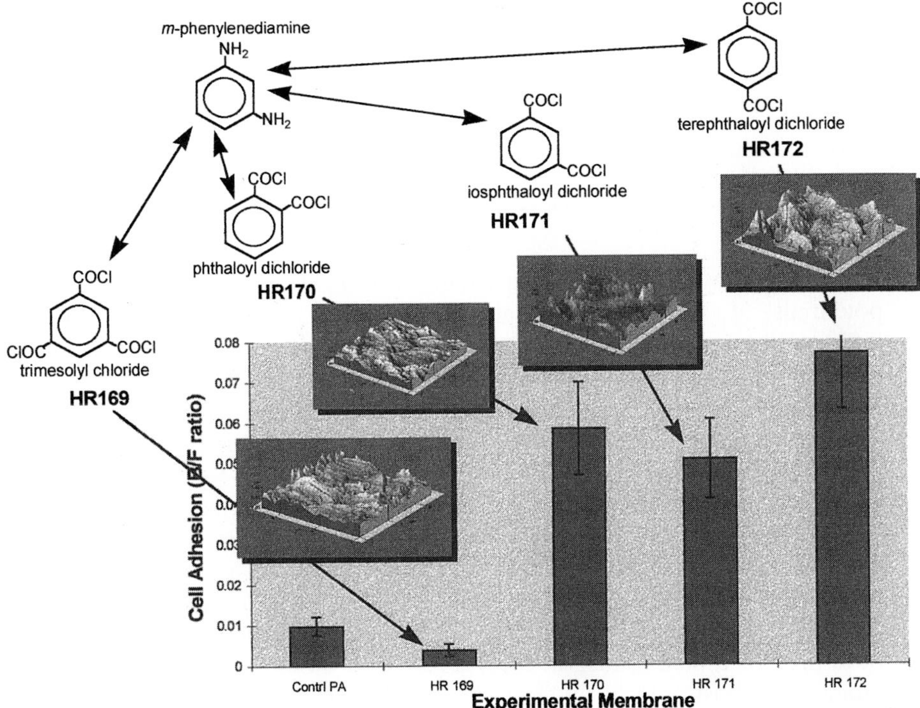

FIG. 13. Example of use of a standard disk adhesion assay for comparing biofouling potentials of experimental TFC polyamide RO membranes (HR169–HR172). Membranes were synthesized from the monomer pairs indicated by an interfacial condensation reaction carried out on a porous polysulfone support (see Section II,B,2). Control PA is commercial PA membrane (FT30) from Dow FilmTec, Inc. (Midland, MI). Results are averages and standard deviations from 10 replicate reaction tubes for each membrane. Data not corrected for edge effects (see text). AFM images indicate topographic features of each membrane.

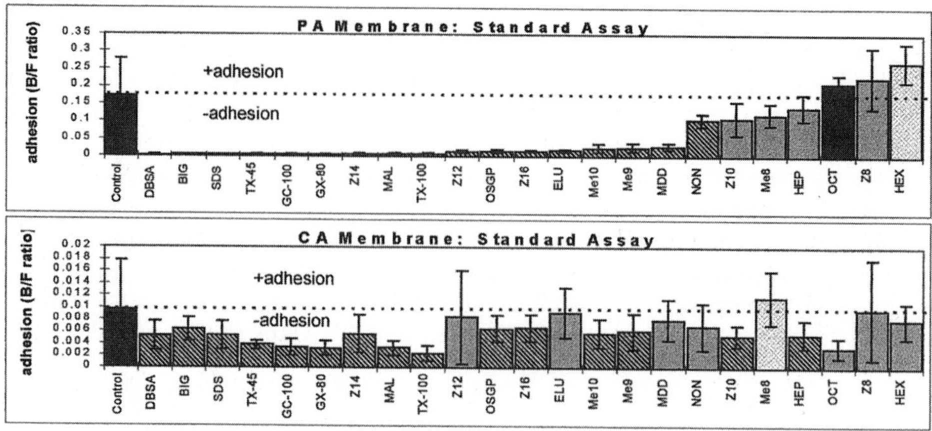

FIG. 14. Example of use of standard disk assay method for determining the influence of various surfactants on the adhesion of *Mycobacterium cheloneae* strain BT12-100 to CA and PA membrane substrata. All surfactants were used at 0.1 wt% final concentration in reaction tubes (MS buffer, pH 7.0). Contact time, 5 hr. Student *t* tests were used to evaluate statistical significance. Black filled bars, untreated controls; cross-hatched bars, inhibition of adhesion compared to controls ($p \leq 0.05$); gray filled bars, no statistical difference from controls; stippled bars, stimulation of adhesion compared to controls. (Data from Ref. 27.)

illustrate the use of the disk assay method in evaluating the "biofouling potentials" of several experimental PA membranes. In this experiment, four different PA membranes were synthesized via an interfacial condensation reaction according to the methods described earlier (Section II,B,2) using the MPD difunctional amine reacted with four different di- and trifunctional acid chloride monomers. Following surface characterization of each new membrane by AFM and other means, disk adhesion assays were performed to rank the membranes according to their affinity for a *Mycobacterium* species known to be involved in early membrane biofilm formation in wastewater systems.[10] Results indicated that the MPD–TMC reaction pair resulted in the TFC membrane with the lowest bacterial adhesion (and, hence, the lowest biofouling potential). The three difunctional acid chlorides produced membranes with relatively high biofouling potentials. Although not illustrated in this example, multivariate statistical techniques could be employed to search for possible correlations between one or more of the surface properties of the experimental membranes (e.g., charge, hydrophobicity, roughness) and the extent of bacterial adhesion observed in each case. Using such a strategy, it should be possible to design new membrane materials with reduced biofouling potentials.

[27] P. Campbell, R. Srinivasan, T. Knoell, D. Phipps, K. Ishida, J. Safarik, T. Cormack, and H. Ridgway, *Biotechnol. Bioeng.* **64,** 527 (1999).

Results presented in Fig. 14 illustrate use of the disk assay method for rapidly screening chemical agents (surfactant/biocides) for their ability to interfere with or otherwise influence bacterial adhesion in two different polymer membrane systems (CA and PA). Results indicate that (compared to untreated control preparations) the majority of the test compounds were significantly more effective as inhibitors of *M. cheloneae* strain BT12-100 adhesion in the PA membrane system than in the CA system. Moreover, compared to untreated controls, some compounds appeared to stimulate bacterial adhesion in both membrane systems. In such experiments, it is important to confirm that the surfactants do not result in extensive cell lysis during the course of the adhesion assay. Cell lysis would presumably result in the leakage of soluble radiolabeled substances (e.g., polypeptides, lipopolysaccharides, nucleic acids) into the medium (MS buffer), and possibly leading to an underestimation of cell adhesion. A convenient test for cell lysis may be carried out following the 5 hr contact time by filtering or pelleting cells from one or more reaction vessels in each detergent series and determining the distribution of radioactivity between the pellet and supernatant phases. Leakage of >20% of the total ^{35}S label into the supernatant fluid during the course of an assay (typically about 5 hr) is generally considered excessive and may lead to erroneous interpretation of the adhesion assay results. Fortunately, unlike most laboratory strains of *Escherichia coli,* the mycobacteria used in these assays are quite resistant to the potential lytic effects of most surfactants, presumably due to a robust cell envelope incorporating mycolic acids.[28] Nevertheless, it is advisable to independently determine the extent of cell leakage for each chemical compound and test bacterium evaluated.

2. Monitoring Bacterial Adhesion and Early Membrane Biofilm Formation by ATR-FTIR Spectrometry. The kinetics of bacterial adhesion and early biofilm development on simulated polymer separation membranes may be determined by ATR-FTIR spectrometry as illustrated in Fig. 15. The principal advantages of this approach are that biofilm formation may be quantified noninvasively, nondestructively, and virtually in real time. The primary disadvantages are the relatively low sensitivity of the ATR-FTIR instrument for detecting adsorbed bacteria (>10^5 cells are required to obtain a signal) and the lengthy time required for spectral processing following an experiment. The ATR-FTIR technique involves semicontinuous monitoring of one or more the mid-IR absorption bands that are characteristic of the developing biofilm. Typically the broad amide I (~1650 cm^{-1}) and/or amide II (~1550 cm^{-1}) bands that are indicative of cell pro-

[28] B. Bendinger, H. H. M. Rijnaarts, K. Altendorf, and A. J. B. Zehnder, *Appl. Environ. Microbiol.* **59,** 3973 (1993).

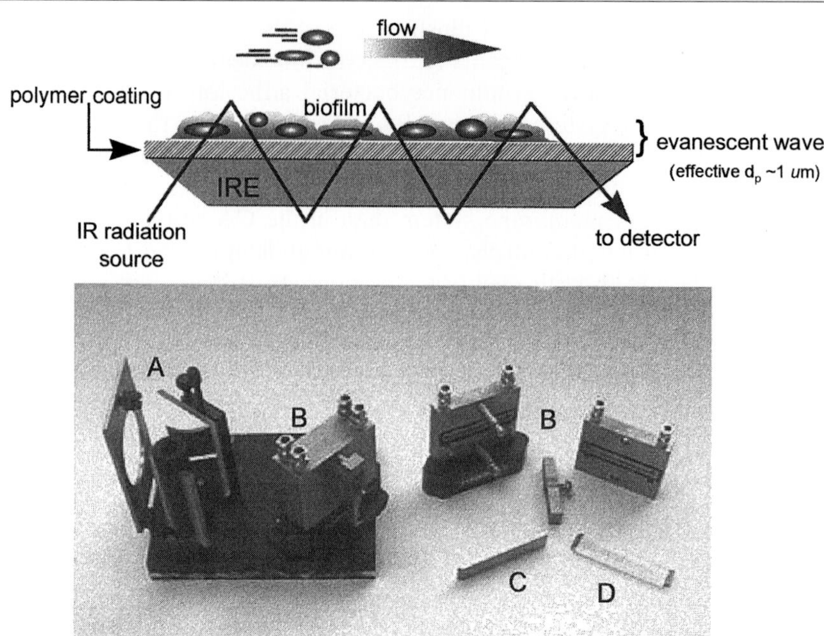

FIG. 15. Schematic of ATR-FTIR method for measuring bacterial adhesion and early biofilm formation on polymer-coated IREs (top) and ATR flow cell (B), mirror assembly (A), and coated paralellepiped IREs (C,D) (bottom). See text for details.

teins are used for this purpose,[29,30] but any convenient set of absorption bands that correlate with the presence of bacteria may be employed. A stringent requirement, however, is that the bacterial absorption bands selected must fall into a region of the IR spectrum that is largely or wholly unoccupied by absorption bands associated with the polymer coating on the surface of the IRE. The ZnSe or Ge IRE is coated with a polymer of choice (e.g., CA) to simulate an actual membrane surface as described in Section II,B,3 and is placed into an ATR flow cell apparatus as shown in Fig. 15. Although not depicted in Fig. 15, both sides of the IRE are coated with polymer and exposed to flowing solution. Prior to final assembly, surfaces of the coated IRE and the flow cell components may be disinfected by UV irradiation from a standard germicidal lamp.

To perform an ATR-FTIR experiment, the CA film on the IRE is first allowed to undergo complete hydration by pumping sterile MS buffer through the assembled ATR flow cell for a period of 30–60 min. During

[29] P. J. Bremmer and G. G. Geesey, *Biofouling* **3,** 89 (1991).
[30] G. G. Geesey and P. J. Bremmer, *Mar. Technol. Soc. J.* **24,** 36 (1990).

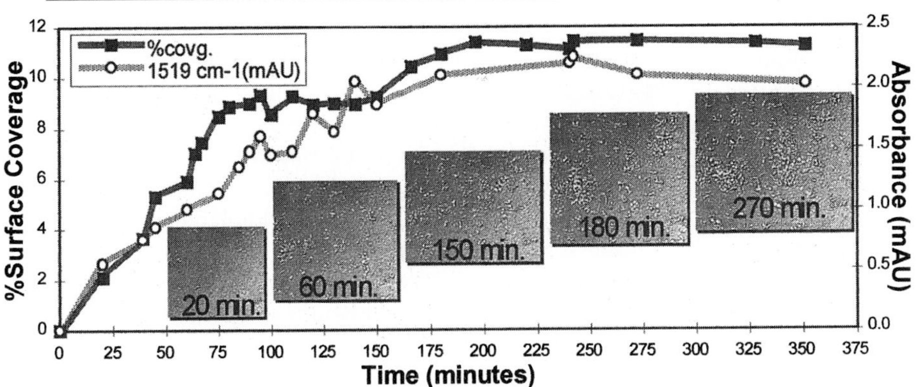

FIG. 16. Kinetics of *Mycobacterium cheloneae* strain BT12-100 adhesion to a CA-coated ZnSe IRE in MS buffer, pH 7.0. IR absorption at the 1519-cm^{-1} band was used to monitor cell adhesion. The percentage coverage of the CA film surface (in an identical sister flow cell) was calculated from Normarski DIC images. Flow velocity ≅8 ml/hr.

this time, IR spectra (characteristic of CA; see Fig. 7, Section II,C,1) are collected at short intervals (e.g., every few minutes) to monitor film stability and ensure that polymer swelling (hydration) has equilibrated completely. Polymer swelling leads to a reduction in polymer density within the depth of penetration of the evanescent wave. Thus, completion of membrane swelling is indicated by little or no further absorbance reductions in specific CA bands over time. Following hydration and swelling of the CA membrane, a cell suspension of *M. cheloneae* strain BT12-100 or other isolate of choice ($\sim 1.0 \times 10^8$ cells/ml) is introduced into the flow cell. Flow rates are typically on the order of ~10 ml/hr. Infrared spectra are then collected

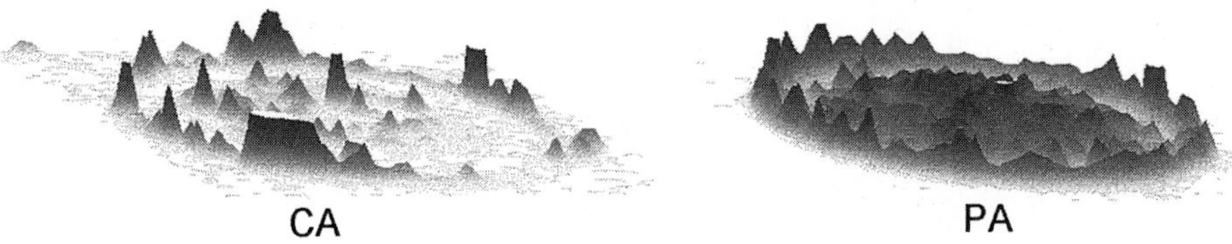

FIG. 17. ARS images of *Mycobacterium cheloneae* strain BT12-100 adhesion to CA and PA reverse osmosis membrane coupons used in the standard disk assay method (see Section II,D,1). Note that the spatial distribution of bacteria on the two membrane surfaces is nonuniform and that cell attachment to the PA, membrane is greater than to the CA membrane. Adhesion assay conditions: contact time, 5 hr at 28°; free (unbound) cell densities at start of both assays were identical ($\sim 1 \times 10^8$ cells/ml); MS buffer, pH 7.0. Scanning was performed at a spatial resolution of 400 μm. Scan time was approximately 18 hr. Note evidence of concentration of bacteria along the periphery of the membrane coupons ("edge effect").

at prescribed intervals (e.g., hourly for 24–48 hr) by computer macro software supplied with the spectrometer. The ATR-FTIR results presented in Fig. 16 serve to illustrate the kind of quantitative adhesion kinetic data that can be obtained using this technique.

3. Determination of Bacterial Adhesion and Distribution on Membrane Surfaces by Autoradiographic Scanning. Autoradiographic scanning (ARS) may be used to record the spatial distribution of radiolabeled bacteria on a polymer membrane surface following adhesion bioassays. An ARS instrument allows membrane disks or larger membrane sheets (e.g., 20 × 25 cm) to be scanned quantitatively for β emissions at spatial resolutions of $\geq$400 μm. Most ARS devices employ a planar gas ionization detector to register the two-dimensional locations of β particle emissions with energies in excess of about 0.1 MEV. Thus, tritium-labeled bacteria cannot be detected by this method. Another disadvantage of the ARS method is that bacterial detection limits are typically fairly high, often in excess of 10^5 cells/cm^2, but detection limits are increased proportionately when the adsorbed bacteria are distributed over progressively smaller areas. An important advantage of the ARS method (compared to ATR-FTIR, for example) is that both model (e.g., CA-coated IREs) and actual commercial membranes may be used as adhesion substrata with equal ease. As illustrated in Fig. 17, the method can be useful in quantifying and comparing the total amount of bacterial adhesion as well as the spatial distribution of cells on different membrane substrata.

[35] Simple Physical Model to Study Formation and Physiology of Biofilms on Urethral Catheters

By DAVID J. STICKLER, NICOLA S. MORRIS, and CAROLE WINTERS

Introduction

Indwelling urethral catheters are used in enormous numbers in modern medicine to relieve urinary retention and to manage long-term urinary incontinence.[1] Unfortunately, they also provide access for bacteria from a contaminated environment into a vulnerable body cavity. As a result, catheter-associated urinary tract infection is the most common of the infections

[1] J. Zimakoff, B. Pontoppidan, S. O. Larsen, and D. J. Stickler, *J. Hosp. Infect.* **24,** 183 (1993).

METHODS IN ENZYMOLOGY, VOL. 310
0076-6879/99 $30.00

that are acquired in hospitals and other health care facilities.[2] The risk of infection is related to the length of time the catheter is in place, and in the many patients undergoing long-term indwelling bladder catheterization, infection of the urinary tract is inevitable.[3] While the catheter remains in place, these infections are difficult to eliminate by antibiotic therapy[4] and it is common practice not to intervene with therapeutic agents unless clinical symptoms suggest that the bloodstream or the kidneys have become infected.[5] Patients on permanent catheterization generally have their catheters changed at 8- to 12-week intervals, so infected urine can be flowing through a catheter for periods of up to 3 months. Under these circumstances, substantial bacterial biofilms form on the lumenal surfaces of the catheter and can even completely block the flow of urine from the bladder.[6] The resistance of the biofilm cells to antibacterial agents also contributes to the difficulties in eliminating these infections by systemic antibiotics and antiseptic bladder washouts.[7]

Studies on the formation of biofilms in models such as the modified Robbins device, the constant film fermenter, and the flow cell have provided fundamental information about the general properties and development of biofilms on different types of material under various conditions. We felt that in the case of urethral catheter biofilms, it would also be useful to have a simple laboratory model to study their formation under conditions in which the design features of the catheter and the hydrodynamics of the catheterized bladder are considered.

The model does not constitute a system in which fine differences in the abilities of different biomaterials to support biofilm growth can be quantified and resolved. Nevertheless, it has proved useful in providing information about some important questions relating to the formation, physiology, and control of catheter biofilms. It has been used to examine which bacterial species are capable of producing catheter-blocking crystalline biofilms[8] and which of the many different types of catheters currently available for clinical use are capable of resisting colonization by crystalline biofilms.[9] It has permitted a study of the development of the crystal formation in catheter-

[2] J. W. Warren, *Med. Clin. N. Am.* **75,** 481 (1991).

[3] R. A. Garibaldi, J. P. Burke, M. R. Britt, W. A. Miller, and C. B. Smith, *N. Engl. J. Med.* **291,** 215 (1974).

[4] C. L. Clayton, J. C. Chawla, and D. J. Stickler, *J. Hosp. Infect.* **3,** 39 (1982).

[5] J. W. Warren, *Infect. Control Hosp. Epidemiol.* **15,** 557 (1994).

[6] D. J. Stickler, *Biofouling* **9,** 293 (1996).

[7] D. J. Stickler, C. L. Clayton, and J. C. Chawla, *J. Hosp. Infect.* **10,** 219 (1987).

[8] D. Stickler, N. Morris, M. Moreno, and N. Sabbuba, *Eur. J. Clin. Microbiol. Infect. Dis.* **17,** 1 (1998).

[9] N. S. Morris, D. J. Stickler, and C. Winters, *Br. J. Urol.* **80,** 58 (1997).

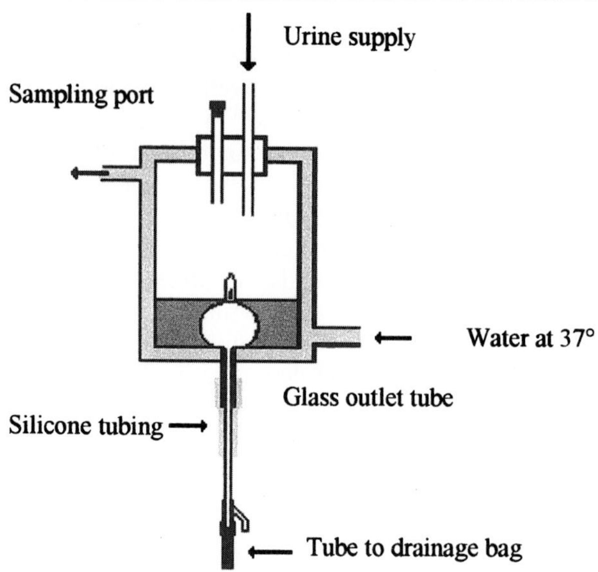

FIG. 1. Model of the catheterized bladder.

encrusting biofilms[10] and is used to test novel strategies for the control and prevention of catheter encrustation and blockage.[11] Most recently, it was used in the study that demonstrated that quorum-sensor signaling molecules are produced in biofilms.[12]

Bladder Model

The model consists of a glass vessel (200 ml) maintained at 37° by a water jacket (Fig. 1). The model is sterilized by autoclaving and then catheters (#14) are inserted aseptically into the vessel through a section of silicone tubing attached to a glass outlet at the base. The catheter balloon is inflated with 10 ml of water, securing the catheter in position and sealing the outlet from the "bladder." The catheter is then attached to a drainage tube and reservoir bag. Sterile urine is supplied to the bladder via a peristaltic pump. In this way a residual volume of urine (30 ml) collects in the

[10] C. Winters, D. J. Stickler, N. S. Howe, T. J. Williams, N. Wilkinson, and C. J. Buckley, *Cells Mater.* **5,** 245 (1995).

[11] N. S. Morris and D. J. Stickler, *Urol. Res.* **26,** 275 (1998).

[12] D. J. Stickler, N. S. Morris, R. J. C. McLean, and C. Fuqua, *Appl. Environ. Microbiol.* **64,** 3486 (1998).

bladder below the level of the catheter eyehole. As urine is supplied to the model, the overflow drains through the catheter to the collecting bag.

Growth Media

Pooled human urine or an artificial urine can be used as the growth medium for the test organisms. Human urine, collected from healthy volunteers with no history of urinary tract infection, is pooled, prefiltered through cotton wool, and sterilized by filtration through 0.8-μm and then 0.2-μm capsule filters (Sartorious AG, Goettingen, Germany). The artificial urine that we have used most commonly is based on that formulated by Griffith *et al.*[13] It contains calcium chloride (0.49 g/liter), magnesium chloride hexahydrate (0.65 g/liter), sodium chloride (4.60 g/liter), sodium sulfate (2.30 g/liter), trisodium citrate dihydrate (0.65 g/liter), disodium oxalate (0.02 g/liter), potassium dihydrogen phosphate (2.80 g/liter), potassium chloride (1.60 g/liter), ammonium chloride (1.0 g/liter), urea (25.0 g/liter), and gelatin (5.0 g/liter). The pH of the medium is adjusted to pH 6.1 prior to sterilization by filtration through a 0.2-μm capsule filter. Tryptone soya broth (Oxoid Ltd., Basingstoke, UK) is sterilized separately by autoclaving and is then added to the sterile basal medium to a final concentration of 10.0 g/liter. This artificial urine will support the growth of a wide range of species that have been isolated from the catheterized urinary tract.

We have used the model to produce catheter biofilms using a range of species, including *Escherichia coli, Pseudomonas aeruginosa, Klebsiella pneumoniae, Providencia stuartii, Morganella morganii, Proteus vulgaris,* and *Proteus mirabilis.* The inoculum for the bladder (10 ml of a young exponential phase batch culture of the test organism in the urine medium) is added to 20 ml of residual medium in the bladder and the culture is allowed to incubate in the vessel for 1 hr to establish itself before the supply of urine, usually at rates of 0.5–1.0 ml/min, is switched on. Under these circumstances, stable bladder populations of 10^7–10^8 cfu (colony-forming units)/ml are produced in the model for periods of up to 10 days. Sets of up to eight models have been used at a time in experimental work. At the end of the experimental period the supply of urine to the bladder is turned off, the balloon is deflated, and the catheter is removed through the base of the model. Formation of the biofilm on the lumenal surfaces of the catheters can then be assessed in several ways.

[13] D. P. Griffith, D. M. Musher, and C. Itin, *Invest. Urol.* **13,** 346 (1976).

Assessment of Biofilm Formation

Scanning Electron Microscopy

A very convenient way to visualize the biofilm involves the use of a scanning electron microscope with a low vacuum capability. Sections (1 cm long) of catheter are cut with a razor blade, mounted on adhesive carbon disks on aluminum stubs, and viewed directly at 20 kV. The low vacuum facility allows the direct examination of specimens that have not been fixed, stained, or treated in any way. The preparations are not permanent, however, they cannot be examined at a later date as specimen dehydration causes loss of morphological integrity. Examples of scanning electron micrographs of a *P. mirabilis* biofilm examined by this method are shown in Fig. 2.

An alternative procedure that produces permanent preparations involves freeze fracturing and freeze drying of the biofilms. Sections (3 cm) of the catheters are cut, plunged into liquid nitrogen-cooled liquid propane, and transferred to liquid nitrogen. Cross sections for examination are then produced by freeze fracturing the specimens in a custom-built copper block

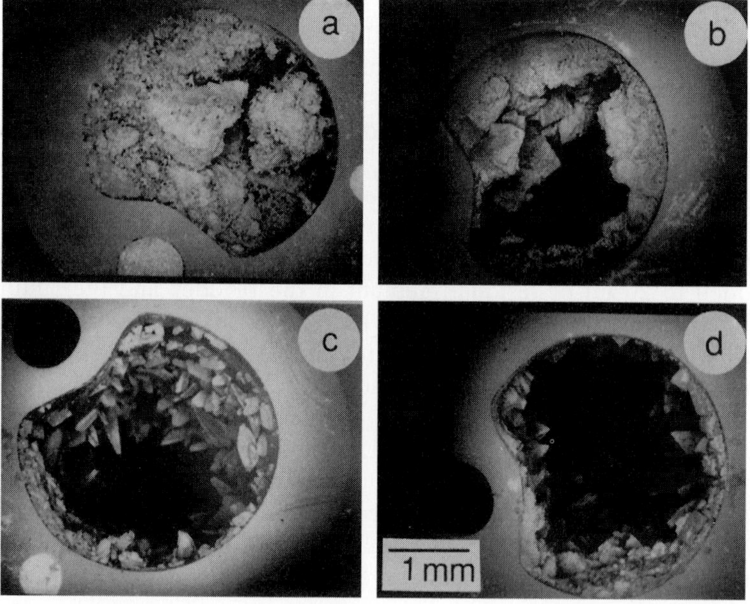

FIG. 2. Micrographs taken using the low vacuum facility on a scanning electron microscope showing serial cross sections of a catheter colonized by a *Proteus mirabilis* biofilm. The catheter had been incubated for 55 hr in the bladder model, and the sections were taken at distances (a) 1, (b) 4, (c) 12, and (d) 32 cm below the eyehole.

that holds both the catheter and a single-edged razor blade in position. The samples are then freeze dried for 24 hr at −80° in an Edwards-Pearse tissue drier (Model EP03, Edwards Ltd., Crawley, UK), mounted fractured surface uppermost onto adhesive carbon disks attached to aluminum stubs, using silver paint to provide extra conductivity. They are then sputtered with gold in a sputter coater (Model S150B, Edwards Ltd.) and examined in a scanning electron microscope operating at an accelerating voltage of 5–10 kV.

To observe the nature of the biofilm surfaces and visualize the cells in the biofilm, sections of catheter (approximately 1 cm long) are cut longitudinally into halves. They are fixed in 3% (v/v) glutaraldehyde in 0.1 *M* phosphate buffer (pH 7.4) for 1 hr and then washed overnight in the 0.1 *M* phosphate buffer. The samples are dehydrated in a graded series of aqueous ethanol solutions (30–100%) and then critically point dried in a Balzer Critical Point Drier (Model CPD030, Balzer, Furstentum, Liechtenstein) using liquid carbon dioxide as the substitution medium. Finally, the samples are mounted on aluminum stubs, sputtered with gold, and examined in the scanning electron microscope.

Transmission Electron Microscopy

Catheter sections (1 cm long) taken from below the retention balloon are placed in a solution containing glutaraldehyde (2.5%, v/v), ruthenium red (0.075%, w/v), and 50 m*M* lysine in 0.1 *M* phosphate-buffered saline (pH 7.4) for 30 min. This fixative solution is then decanted and replaced with fresh glutaraldehyde (3.0%, v/v) in the phosphate buffer and left for 2 hr at 4°. The catheter sections are then washed in the phosphate buffer overnight and postfixed with Millonig's phosphate-buffered (1%, w/v) osmium tetroxide[14] for 1 hr at 4°. Dehydration of the samples is performed in a graded series of ethanol solutions (30–100%), followed by embedding in LR White resin (Agar Scientific Co., Stansted, UK) in gelatin capsules (size 00). Anaerobic polymerization of the resin is allowed to occur by incubation of the sealed capsules at 60° for 24 hr. The gelatin capsules are stripped from the solidified block of resin and a longitudinal saw cut is made in the catheter wall to allow it to be peeled away, leaving the lumenal biofilm embedded in the resin core. Cross sections (approximately 60 nm thick) of the embedded biofilm are then cut using an ultramicrotome and mounted onto copper grids. After staining with aqueous uranyl acetate and Reynold's lead citrate,[15] the sections are examined in a transmission

[14] G. Milonig, *J. Appl. Phys.* **32,** 1637 (1961).

[15] D. G. Robinson, U. Ehlers, R. Herken, B. Herrman, F. Mayer, and F.-W. Schurman, "Methods of Preparation for Electron Microscopy," p. 69. Springer-Verlag, Berlin, 1987.

microscope operated at 80 kV. The treatment of these specimens with lysine and ruthenium red preserves and stains the polysaccharide matrix of the biofilm.[16] We have used this method to observe *P. aeruginosa* catheter biofilms[17] and the crystalline biofilms formed on catheters by *P. mirabilis.*[10] In the latter case the elongate crystals of struvite (magnesium ammonium phosphate) and the amorphous particles of apatite (calcium phosphate) can be seen together with the bacteria, distributed throughout the matrix of the biofilm. These biofilms are of particular interest as they encrust the surface of the catheter balloon and can rapidly obstruct the catheter lumen and seriously complicate the care of catheterized patients.[18]

X-ray microanalysis of freeze-substituted sections of *P. mirabilis* biofilms can also be used to identify and observe the formation of the encrustations. Pieces of encrusted catheter (1 cm in length) are plunged into liquid nitrogen-cooled liquid propane and then transferred to liquid nitrogen before storage in precooled and dried acetone at −70°. The temperature is slowly increased to −50° over 24 hr and the specimens are gradually impregnated with precooled Lowicryl K11M resin (Agar Scientific Co.) in increasing ratios with the acetone until a 100% resin is achieved.[19] After a change of 100% resin for 24 hr the specimens are transferred to fresh precooled resin in gelatin capsules that are then sealed to allow anaerobic polymerization of the resin by ultraviolet light. The capsules are kept in specially designed aluminum blocks in a bath of ethanol to dissipate the heat of polymerization and prevent it from affecting the specimens adversely.[20] Specimens are irradiated with UV for 5 days, after which the temperature is gradually increased and polymerization is completed at ambient temperature. As an alternative to these procedures, an automatic freeze-substitution system such as the Reichert AFS (Leica Vienna, Austria) can be used to prepare these specimens. Finally the gelatin capsules and the catheter wall are peeled away, leaving a central core of resin containing the freeze-substituted biofilm. Sections approximately 90 nm in thickness are cut using a diamond knife in an ultramicrotome. Spot chemical analyses of features of interest in the biofilm sections can be carried out with an X-ray detector and a multichannel X-ray microanalyzer (Link 860 series 11, Oxford Instruments, High Wycombe, UK) connected to a transmission electron microscope.

[16] T. A. Fassel, J. R. Sanger, and C. E. Edmiston, *Cells Mater.* **3,** 327 (1993).

[17] D. J. Stickler, N. S. Howe, and C. Winters, *Cells Mater.* **4,** 387 (1994).

[18] C. M. Kunin, *J. Clin. Epidemiol.* **42,** 835 (1989).

[19] B. Humbel and M. Muller, *in* "Science of Biological Specimen Preparation" (M. Muller, R. P. Becker, A Boyde, and J. L. Wolosewick, eds.), p. 175. Scanning Electron Microscopy Inc., Chicago, 1985.

[20] A. M. Glauert and R. D. Young, *J. Microsc.* **154,** 101 (1989).

Areas of the biofilm can be photographed in the microscope and then subjected to X-ray mapping techniques.[10]

Chemical Analysis

The extent of crystalline biofilm formation by *P. mirabilis* on the lumenal surfaces of catheters can be measured by chemical analysis for calcium and magnesium. Encrusted catheters are removed from the models and cut into sections (2 cm) that are then immersed in 10 ml of 4.0% (v/v) nitric acid (metal oxide semiconductor grade in double-deionized water) in universal containers. The encrustations are disrupted by placing the containers in a sonic cleaning bath for 5 min (Transsonic Water Bath, Camlab, UK). The crystalline suspensions are left to dissolve for 24 hr and then the calcium and magnesium contents of the resulting solutions are determined by atomic absorption spectroscopy.

Acknowledgment

Nicola Morris was supported by a postgraduate grant from the Welsh Scheme for the Development of Health and Social Research.

[36] Studying Initial Phase of Biofilm Formation: Molecular Interaction of Host Proteins and Bacterial Surface Components

By ATSUO AMANO, ICHIRO NAKAGAWA, and SHIGEYUKI HAMADA

Introduction

Planktonic cells in flowing water or protein-rich body fluids such as saliva and serum form sessile bacterial communities in response to the proximity of a surface and the juxtaposition of bacterial cells of the same or other species. During the complex process to biofilm maturation, the adhesion of sessile cells to the substratum triggers the alternation of a large number of gene expressions, and their phenotypes are changed dramatically in oligotrophic environments.[1] The initial microbial adhesion is affected by several physicochemical factors, such as van der Waals forces, electrostatic

[1] J. W. Costerton, Z. Lewandowski, D. E. Caldwell, D. R. Korber, and H. M. Lappin-Scott, *Annu. Rev. Microbiol.* **49,** 711 (1995).

0076-6879/99 $30.00

forces, hydrogen bonding, and Brownian motion forces.[2] Subsequent events for nonshedding bacteria are selective bindings to the substratum with remarkable avidities indispensable to biofilm accretion, and the interactions are usually mediated stereochemically by specific adhesins with a clear expression in hydrophobicity and charge. Bacterial surface components can recognize and bind to host adhesins such as salivary components on mucosal surfaces[3] and ligand components of other bacterial species.[4] Kinetic analyses between isolated adhesive molecules are advantageous in understanding the dynamics of the initial phase of biofilm formation.

For the assessment of protein interactions, various methods are used.[5,6] Polystyrene surfaces (e.g., microtiter plates or beads) are the most convenient and are widely used for protein immobilization to generate functional surfaces. Because protein immobilization on polystyrene is mainly dependent on hydrophobic interaction, the assay might be impeded by several factors, such as poor adsorption, random orientation, alteration of protein conformation, steric hindrance, and altered kinetics leading to a partial or complete loss of binding capacities.

An improved measurement system has been provided to examine interactions between different biomolecules.[7–10] Surface plasmon resonance-based biosensors are being used to define the kinetics of a wide variety of macromolecular interactions. This article describes several methods, including the surface plasmon resonance system, to study the specific interactions between biomolecules.

Materials

Bacterial Fimbriae as Adhesins

Porphyromonas gingivalis, a gram-negative anaerobic rod, has been recognized as a causative agent of periodontal diseases.[11] *Porphyromonas*

[2] H. J. Busscher and H. C. van der Mei, *Adv. Dent. Res.* **11,** 24 (1997).

[3] F. A. Scanapieco, *Crit. Rev. Oral Biol. Med.* **5,** 203 (1994).

[4] P. E. Kolenbrander, *Trends Microbiol.* **5,** 475 (1997).

[5] Å. Ljungh and T. Wadström, *Methods Enzymol.* **253,** 501 (1995).

[6] R. J. Doyle and M. Rosenberg, *Methods Enzymol.* **253,** 542 (1995).

[7] L. G. Fägerstam, Å. Frostell-Karlsson, R. Karlsson, B. Persson, and I. Rönnberg, *J. Chromatogr.* **597,** 397 (1992).

[8] B. Johnsson, S. Lofås, and G. Lindquist, *Anal. Biochem.* **198,** 268 (1991).

[9] R. Karlsson, A. Michaelsson, and L. Mattsson, *J. Immunol. Methods* **145,** 229 (1991).

[10] S. Löfås and B. Johnsson, *J. Chem. Soc. Chem. Commun.* **21,** 1526 (1990).

[11] T. J. M. van Steenbergen, A. J. van Winkelhoff, and J. de Graaff, *in* "Periodontal Disease: Pathogens & host immune responses" (S. Hamada, S. C. Holt, and J. R. McGhee, eds.), p. 41. Quintessence Publisher Co., Ltd., Tokyo, Japan, 1991.

gingivalis fimbriae function as an important adhesin so that the organism can bind to various host surface proteins, including saliva, serum, and extracellular matrix glycoproteins.[12] *Porphyromonas gingivalis* strain 381 is grown in 5 liters of half-strength (18 mg/ml) brain–heart infusion (BHI) broth supplemented with yeast extract (5 mg/ml), hemin (5 mg/ml), and menadione (0.2 mg/ml) and is buffered at pH 7.4 under anaerobic condition.[13] Organisms grown in half-strength medium and at the late exponential phase will give more fimbriation on the surface. Cells are washed and suspended in 50 m*M* Tris–HCl (pH 8.0). Fimbriae can be mechanically detached by repetitive pipetting with a 10-ml glass pipette. The supernatant is precipitated with 40% saturated ammonium sulfate and is centrifuged at 10,000*g* for 30 min at 4°. The material is suspended and dialyzed in 20 m*M* Tris–HCl (pH 8.0) and is applied to a column of DEAE-Sepharose CL-6B (25 by 5 cm, Pharmacia LKB Biotechnology, Piscataway, NJ). Fimbriae are eluted as a single peak containing purified materials by a linear gradient 0–0.3 *M* of NaCl in 20 m*M* Tris–HCl (pH 8.0).[14] Purified fimbriae are radiolabeled with ^{125}I using the Bolton–Hunter reagent [9.25 MBq (250 μCi), ICN, Costa Mesa, CA].[15] The specific activities of ^{125}I-labeled fimbriae are generally 12 $\pm$ 3 mCi/μmol.

Salivary Components

Human whole saliva (HWS) is collected in an ice-chilled container after paraffin stimulation and is clarified by centrifugation at 12,000*g* for 10 min at 4°. Citric acid-stimulated human parotid saliva (HPS) is obtained with a collecting device,[16] and human submandibular–sublingual saliva (HSMSL) is collected and processed.[17] To suppress protease activities in saliva, a solution (20 m*M* Tris–HCl buffer, pH 8.0) containing 2 m*M* Na_2EDTA, 10% 2-propanol, and 2 m*M* phenylmethylsulfonyl fluoride is added at a ratio of 1 : 9 immediately after collection. This saliva sample is concentrated by adding solid ammonium sulfate to 100% saturation. The pellet is dissolved and dialyzed in 20 m*M* Tris–HCl buffer (pH 8.0) containing 0.5 *M* NaCl and is applied to a Sephacryl S-200 gel filtration column (150 by 2 cm; Amersham Pharmacia Biotech., Uppsala, Sweden) equili-

[12] S. Hamada, A. Amano, S. Kimura, I. Nakagawa, S. Kawabata, and I. Morisaki, *Oral Microbiol. Immunol.* **13,** 129 (1998).

[13] A. Amano, H. T. Sojar, J.-Y. Lee, A. Sharma, M. J. Levine, and R. J. Genco, *Infect. Immun.* **62,** 3372 (1994).

[14] F. Yoshimura, K. Takahashi, Y. Nodasaka, and T. Suzuki, *J. Bacteriol.* **160,** 949 (1984).

[15] J. J. Langone, *Methods Enzymol.* **70,** 221 (1980).

[16] H. J. Keene, *J. Dent. Res.* **42,** 1041 (1963).

[17] N. Ramasubbu, M. S. Reddy, E. J. Bergey, G. G. Haraszthy, S.-D. Soni, and M. J. Levine, *Biochem. J.* **280,** 341 (1991).

brated with the same buffer.[18] Further purification is performed using high-performance liquid chromatography (HPLC) with a cation-exchange column (PolyCAT A column, PolyLC, Inc., Columbia, MD) with 20 m*M* potassium phosphate buffer (pH 5.5), and the adsorbed proteins are eluted with a linear gradient of the same buffer containing 1 *M* NaCl.[19] Other human proteins are purchased commercially: hemoglobin (Sigma Chemical Co., St. Louis, MO), fibrinogen (Kabi Vitrum, Stockholm, Sweden), fibronectin (Sigma), and laminin (Sigma).

Methods

Overlay Assay Using Nitrocellulose

Procedure

1. Host proteins are dissolved in 0.125 *M* Tris–HCl buffer (pH 6.8) containing 2% (w/v) SDS, 10% (v/v) glycerol, and 0.001% (w/v) bromphenol blue and then are incubated for 30 min at room temperature. Do not boil the samples. The samples are electrophoresed in 7–15% SDS–PAGE and are transferred to a nitrocellulose membrane (0.45 μm pore size, Bio-Rad Laboratories, Hercules, CA) by an electrophoretic blotting technique using Trans-Blot (Bio-Rad). Neutral or positively charged nitrocellulose membranes are used for Western transfer. Although the positively charged membrane has a very good protein-binding capability, it usually gives a higher background. Binding abilities of the immobilized proteins may be variable, depending on the membrane types used. Several types of membranes such as ProBlott (Applied Biosystems, Foster City, CA), Immobilon-P or -NC (Millipore Ltd., Bedford, MA), and PolyScreen PVDF (DuPont NEN, Boston, MA) should be tried to obtain good interactions.
2. The membrane is immersed for 1 hr at 4° in phosphate-buffered saline (PBS, pH 7.4) containing 1% (w/v) lipid-free bovine serum albumin (BSA, Sigma) as a blocking agent for unoccupied sites.
3. The membrane is incubated with a fimbrial solution (41 μg/ml, nonlabeled or ^{125}I-labeled fimbriae) and, if necessary, with inhibitors overnight at 4°. It is then washed three times with phosphate buffer containing 0.3 *M* NaCl and 0.1% Tween 20. This washing procedure

[18] N. Strömberg, T. Borén, A. Carlén, and J. Olsson, *Infect. Immun.* **60,** 3278 (1992).

[19] A. Amano, S. Shizukuishi, H. Horie, S. Kimura, I. Morisaki, and S. Hamada, *Infect. Immun.* **66,** 2072 (1998).

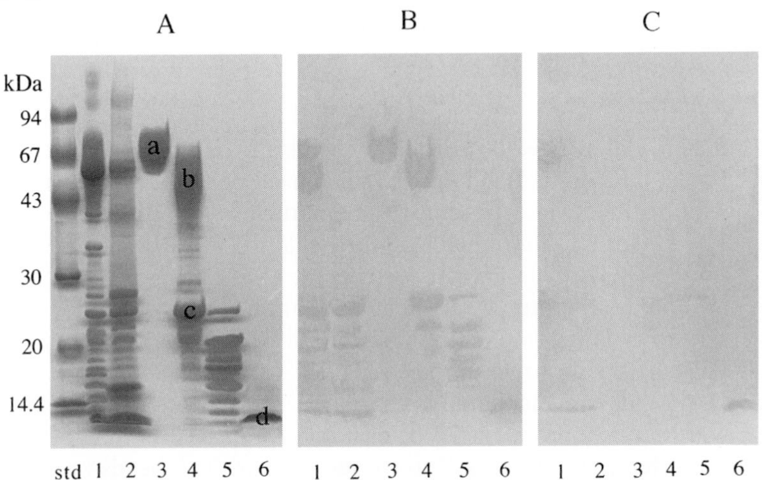

FIG. 1. Detection of salivary components capable of binding to fimbriae by the overlay assay. Fimbriae were probed with rabbit antiserum following the incubation with salivary proteins immobilized on the membrane (0.45-μm pore size). Four components (a, H-PRG; b, L-PRG; c, PRP; d, statherin) were found capable of binding to fimbriae. Lanes: std, molecular mass standard; 1, HPS; 2, HSMSL; 3, high molecular proline-rich glycoprotein (H-PRG); 4, low molecular (L-) PRG and proline-rich protein (PRP); 5, unknown components that bind to fimbriae; and 6, statherin. (A) SDS–PAGE (15%) profiles of salivary components by Coomassie brilliant blue staining. (B) Detection of salivary components binding to fimbriae. Fimbriae bound to salivary proteins on the membrane replica were visualized by goat antirabbit IgG–horseradish peroxidase conjugate (1:500). (C) The synthetic peptide (PRP-C) was found inhibitory in the fimbriae–PRP interaction.[20] The PRP-C peptide (100 nmol/ml) was added to the fimbriae solution (1 nmol/ml) prior to incubation with the membrane. The densities of the reaction bands became more pale than those in (B). From A. Amano *et al., Infect. Immun.* **66,** 2072 (1998), with permission of American Society for Microbiology.

should be optimized to obtain clearer binding and less background. Fimbriae bound to human proteins immobilized on the membranes are probed with rabbit anti-*P. gingivalis* fimbriae serum diluted 1:2000 in PBS (pH 7.4) or by autoradiography.[20] The experimental example is shown in Fig. 1.

Hydroxylapatite Beads Assay

Bacterial adhesion to the saliva-coated enamel surface is a prerequisite for dental plaque formation. Hydroxylapatite beads in calcifying buffer (KCl buffer, 50 m*M* KCl containing 1 m*M* KH_2PO_4, 1 m*M* $CaCl_2$, and 0.1

[20] K. Kataoka, A. Amano, M. Kuboniwa, H. Horie, N. Nagata, and S. Shizukuishi, *Infect. Immun.* **65,** 3159 (1997).

mM $MgCl_2$ buffered at pH 6.0-7.4) are used frequently as model enamel surfaces.[21,22]

Procedure

1. Spheroidal hydroxylapatite (HA) beads (2–10 mg, BDH Chemicals, Poole, UK) are washed once in siliconized borosilicate culture tubes with distilled water to remove fine particles and are equilibrated in 1.5–3 ml of KCl buffer with gentle oscillation in a Rototorque rotator (model 7637, Cole-Palmer Instrument Co., Chicago, IL) overnight at room temperature. Sufficient incubation time (>90 min) is necessary to obtain complete mineralization of the bead surfaces. Use a Pasteur pipette to wash the beads.
2. HA beads are incubated with 200 μl of each protein (1.0 mg/ml) and lipid-free BSA (A-7030) as a control in KCl buffer for 3 hr at room temperature in a Rototorque rotator. The beads are washed twice, and added with 200 μl of lipid-free BSA (50 μg/ml) to block the unoccupied sites, followed by washing once with KCl buffer.
3. Increasing amounts of ^{125}I-labeled fimbriae (0.125–1.25 nmol) are added in a total volume of 200 μl of KCl buffer to the tubes. The mixtures are incubated at 20 rpm in a rotator at room temperature for 1 hr. The mixtures are then washed once with KCl buffer containing 0.1 M NaCl and 0.01% Tween 20 and are washed twice with KCl buffer.
4. Amounts of ^{125}I-labeled fimbriae bound to HA beads are determined with a gamma counter. The binding level is calculated by subtracting the nonspecific binding in the presence of excess unlabeled fimbriae (50 nmol/tube). The experimental example is shown in Fig. 2.

The HA assay using whole bacterial cells was described previously.[22] For the assay of *P. gingivalis* cells, 100% Percoll (Sigma) is recommended to remove unbound, free cells because the cells autoaggregate quickly.

Polystyrene Microtiter Plate Assay

Multiwelled polystyrene microtiter plates (e.g., amino plate type A; Sumitomo Bakelite Co., Ltd., Japan, Maxisorp; Nalge Nunc International, Denmark) can be used for the passive coating of proteins. Solutions of host proteins (100 μl of 0.01–0.1 mg/ml in PBS, pH 7.4) are placed at room temperature for 2–3 hr. Coating with collagen is performed at 4° overnight.[5] Coating efficiency may be assessed by ^{125}I-labeled proteins, and the in-

[21] K. M. Schilling and R. J. Doyle, *Methods Enzymol.* **253,** 536 (1995).
[22] J.-Y. Lee, H. T. Sojar, G. S. Bedi, and R. J. Genco, *Infect. Immun.* **60,** 1662 (1992).

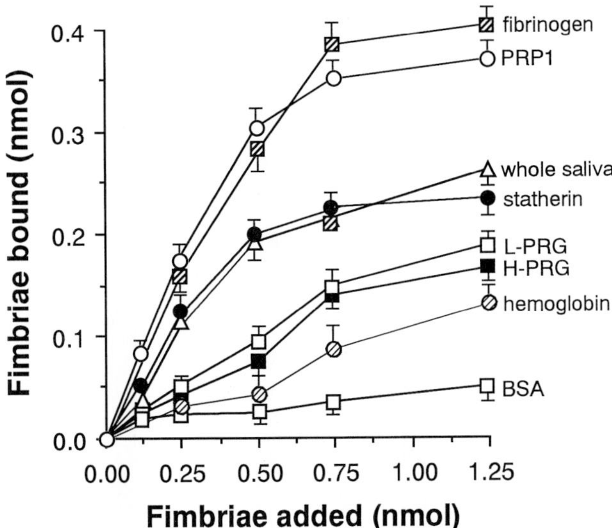

FIG. 2. Quantitive assay of binding of fimbriae to salivary proteins using HA beads. Increasing amounts of ^{125}I-labeled fimbriae were added to tubes containing host protein-coated HAP beads and incubated at room temperature for 1 hr. The specific binding level was calculated by subtracting the nonspecific binding level, which was obtained by the preincubation of HA beads with nonlabeled fimbriae (500 μl of 50 nmol/ml) at room temperature for 1 hr.

tactness of the uniform protein layer is also confirmed by antibodies. Lipid-free BSA or Block Ace, a casein solution prepared from homogenized milk (Snow Brand Co. Ltd., Sapporo, Japan), is used as a negative control.[20] After washing with 300 m*M* NaCl in phosphate buffer, nonspecific binding sites are blocked by incubation with BSA (100 μl of 1 mg/ml in PBS) at room temperature for 30 min. Following washing with 300 m*M* NaCl in phosphate buffer samples of fimbriae (5 μg/ml in PBS) and, if necessary, inhibitors are added to the wells and are incubated at room temperature for 1 hr. The wells are washed with 300 m*M* NaCl in phosphate buffer three times, and rabbit antifimbriae IgG (1:1000, v/v) is added to detect the amount of fimbriae bound to the wells. The experimental example is shown in Fig. 3.

Surface Plasmon Resonance Assay Using BIAcore System

A new measurement system enables the direct detection without labeling and real-time monitoring of biomolecular interactions. The biomolecular interaction analysis (BIA) system is based on a biosensor that uses

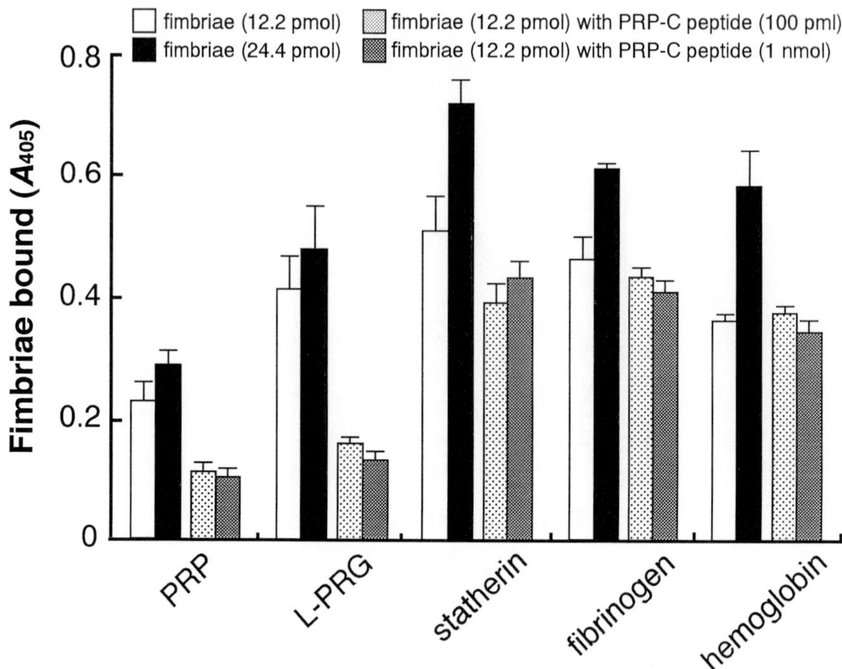

FIG. 3. Quantitive assay of binding of fimbriae to host proteins using a 96-well microtiter plate. Wells of 96-well microtitier plates were precoated with host proteins (100 μl of 0.1 mg/ml in PBS) at room temperature for 3 hr. After washing with 300 mM NaCl in phosphate buffer, the wells were coated with BSA. Following washing, samples of fimbriae (5 μg/ml in PBS) and, if necessary, PRP-C peptide as an inhibitor were added to the wells and were then incubated at room temperature for 1 hr. Rabbit antifimbriae IgG (1:1000) was added to detect the amount of fimbriae bound.

surface plasmon resonance (SPR).[5–8,23] In this system, one reactant (ligand) is immobilized covalently onto a sensory chip surface via the amino-terminal and ε-amino groups of the ligand protein.[24] The other reactant, referred as the analyte, flows over the sensory chip surface in solution. The technique measures small changes in refractive index at, or near to, the chip surface on which ligands are immobilized in a miniaturized flow system. Benefits of the SPR assay[25,26] are (i) direct and real-time observation of the interactions without any labeling of the reactants, (ii) kinetic analysis to provide rate

[23] M. Malmqvist and R. Karlsson, *Curr. Opin. Chem. Biol.* **1,** 378 (1997).

[24] B. Catimel, M. Nerrie, F. T. Lee, A. M. Scott, G. Ritter, S. Welt, L. J. Old, A. W. Burgess, and E. C. Nice, *J. Chromatogr.* **776,** 15 (1997).

[25] R. Karlsson and A. Fält, *J. Immunol. Methods* **200,** 121 (1997).

[26] D. G. Myszka, *Curr. Opin. Biotechnol.* **8,** 50 (1997).

and affinity constants of a one-to-one interaction, (iii) comparison of the binding properties of different analytes, such as other proteins and mutated recombinant proteins by point mutation or deletion, and (iv) screening of an unknown interactant in a crude sample. The analytic procedure using a Model 1000 BIAcore system (BIAcore AB, Uppsala, Sweden) is as follows.

Reagents: 100 m*M* HEPES, 0.15 m*M* NaCl, 3 m*M* EDTA, 0.005% (v/v) polysorbate 20 (or 0.05% Tween 20), 100 m*M* sodium acetate (pH is adjusted to 4.0 or 4.8 with 100 m*M* acetic acid), 100 m*M* HCl. Ready-to-use solutions of this mixture are prepared commercially by BIAcore AB.

Consumables (BIAcore AB): Sensory chips (five types); sensor chip CM5 (the most versatile chip), sensor chip SA (for capture of biotinylated peptides, proteins and DNA), sensor chip NTA (for capture of ligands via metal chelation), sensor chip HPA (for membrane-bound receptor study). Amine coupling kit; reagents for covalent immobilization of ligands carrying a primary amine group containing *N*-hydroxysuccinimide (NHS), *N*-ethyl-*N*′-[(3-dimethylamino)propyl]carbodiimide hydrochloride (EDC), and ethanolamine hydrochloride.

Interactants: purified ligand (host proteins, 1–10 μg/ml) and purified analyte (fimbriae, 10–700 μg/ml). Generally, the ligand must be purified, but a partially purified analyte can be used when the binding affinity is considerably high.[24–26] The interactants must be filtered through a 0.22-μm pore. Running buffer should be appropriate for the corresponding interactions.

Procedure

1. One of two interactants must be immobilized to the sensory chip. Generally, a more amplified signal can be obtained when a smaller molecular weight substance is immobilized. Acidic protein with $pI < 3.5$ is hard to be immobilized on the sensory chip CM5.
2. The analyte flows over the nontreated sensory chip. Survey optimal pH of running buffer and the concentration of the analyte solution. This step is called "preconcentration" to optimize the immobilization.
3. A mixture of NHC and EDC (1 : 1) is injected on the dextran matrix of the sensory chip to activate at a flow rate of 5 μl/ml at 25°, and the ligand (100 μg/ml) in 10 m*M* sodium acetate buffer (pH 4.0 or pH 4.8) is immobilized on the matrix. The pH of the immobilization buffer must be lower than the p*I* of the ligand. To equalize the amount (mol) of the immobilized proteins, the increase of resonance unit (RU) by immobilization is manually set to be 200–1000 × [molec-

ular mass of immobilized protein/molecular mass of fimbrillin (41 kDa)] RU. Excess active sites of the matrix are blocked with ethanolamine hydrochloride and washed with regeneration buffer (1 *M* NaCl in PBS). The flow pathway is rinsed with running buffer such as Tris–HCl buffer (pH 7.8) and PBS (pH 7.4).

4. Analytes (fimbriae) are injected at a flow rate of 10–100 μl/min at 25°, and the binding of fimbriae is monitored and presented as a sensorgram (a plot of RU versus time). The covalent nature of the bound ligand can be regenerated by washing out the bound analyte using 0.1 *M* HCl or 0.5–1 *M* NaCl, allowing repetitive measurements of the analyte binding using programmable robotics.

A 0.1° shift in SPR angle corresponds to 1000 RU, and this corresponds to a change in the surface concentration of 1 ng/mm^2.[7] For kinetic studies, fimbriae with increasing concentrations are injected over the sensory chip, and a dRU/dt versus RU plot is calculated from the sensorgram. The slopes at different fimbriae concentrations are replotted. The following constants are obtained from the following four equations:

$$\begin{aligned} dRU/dt &= k_{\text{ass}}\, C\, (RU_{\max} - RU) - k_{\text{diss}}\, RU \\ &= k_{\text{ass}} C RU_{\max} - (k_{\text{diss}}\, C + k_{\text{diss}})\, RU \end{aligned} \tag{1}$$

where k_{ass} is the association rate constant [1/*M*/sec (s)], k_{diss} is the dissociation rate constant (1/s), and C is the concentration of fimbriae.

$$\ln(Rt^1/Rt_n) = k_{\text{diss}}\, t \tag{2}$$

where Rt_1, RU at the time of the initial phase of the dissociation (t_1); Rt_n, RU at time t_n, $t = t_n - t_1$.

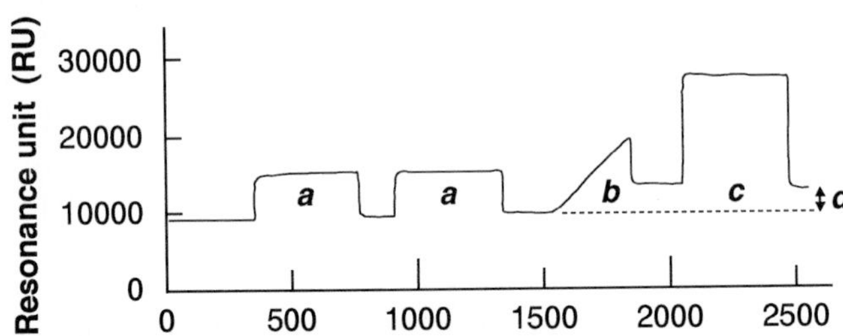

FIG. 4. Immobilization of ligand to the sensory chip CM5. Sensorgram of the immobilization of fibronectin to the sensor chip is shown as an example. A continuous Tris–HCl buffer (pH 7.8) flow of 5 μl/min is passed over the sensor chip. (a) A pulse of EDC/NHS was injected two times to activate the dextran matrix. After the pulse with a high RU had passed, the RU signal dropped to the level corresponding to the continuous buffer. The fibronectin solution was then injected (b) and ethanolamine hydrochloride was injected (c) to deactivate the surface. The amount, 3000 RU, of immobilized fibronectin is indicated (d). From M. Kontani *et al., Mol. Microbiol.* **24,** 1179 (1997), with permission.

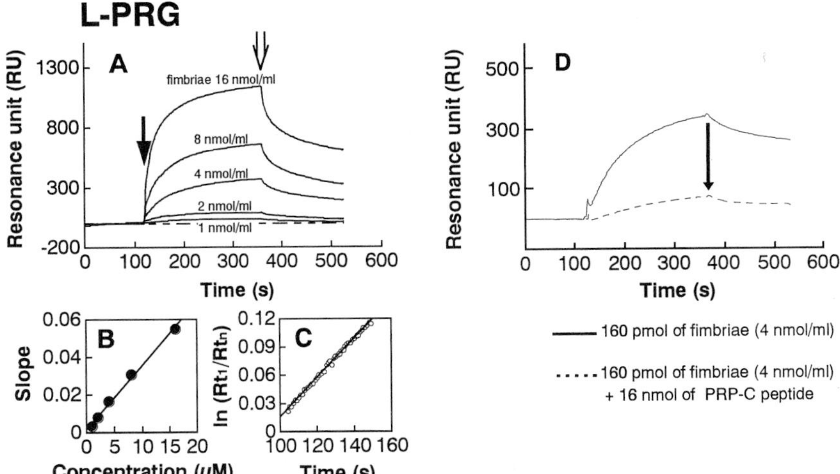

FIG. 5. Kinetic analysis of the interaction of fimbriae with L-PRG and inhibition assay of the inhibitory peptide using the BIAcore system. (A) For kinetic studies, the fimbrial solution was injected over L-PRG immobilized at different concentrations (1–16 nmol/ml) on the sensor chip surface (3 ng/mm^2). Arrows indicate the beginning and end of the injections at a flow rate of 10 μl/min at 20°, respectively. The association of fimbriae to L-PRG was observed as a rising curve steepened dependent on the amount of fimbriae bound. Denatured fimbriae (heat at 100°) showed negligible binding to the ligands. (B) A *dRU/dt* versus *RU* plot was calculated from the sensorgrams, and the slopes at different concentrations were replotted against the concentrations of fimbriae. Association rate constants (k_{diss}) can be obtained from Eq. (1). (C) The dissociation rate constant (k_{ass}) was determined directly from the dissociation phase. First-order kinetics is obtained by Eq. (2). (D) The potential inhibitor (PRP-C peptide, 16 nmol) against the interaction of fimbriae with L-PRG was injected simultaneously with fimbriae (160 pmol) to the sensory chip of BIAcore as described. The peptide markedly inhibited the association of fimbriae to immobilized L-PRG.

$$K_a = k_{\text{ass}}/k_{\text{diss}} \tag{3}$$

where K_a is the association constant.

$$K_d = k_{\text{diss}}/k_{\text{ass}} \tag{4}$$

where K_d is the dissociation constant.

Experimental sensorgram data are analyzed using a combination of BIAcore analysis programs and EXCEL 5.0. For sophisticated kinetic analyses using the programs, high flow rates (30–100 μl/min) and low amounts of immobilized ligands (20–100 RU) are recommended.[26] A new version of evaluation software, BIAevaluation 3.0, is now available commercially. This software enables appropriate analysis of various data under different

TABLE I
SPECIFIC CONSTANTS OF INTERACTIONS BETWEEN FIMBRIAE AND HOST PROTEINS

Host proteins	k_{ass} (1/*M*/sec)	k_{diss} (1/sec)	K_d (*M*)
Antifimbriae IgG	6.11×10^3	5.00×10^{-4}	8.19×10^{-8}
Hemoglobin	3.42×10^3	1.41×10^{-3}	4.12×10^{-7}
Fibrinogen	2.63×10^3	1.22×10^{-3}	4.63×10^{-7}
Fibronectin	3.41×10^3	1.65×10^{-3}	4.83×10^{-7}
Laminin	3.31×10^3	1.68×10^{-3}	5.07×10^{-7}
PRP	2.61×10^3	1.60×10^{-3}	6.13×10^{-7}
L-PRG	3.38×10^3	2.08×10^{-3}	6.15×10^{-7}
Statherin	2.49×10^3	1.68×10^{-3}	6.76×10^{-7}

conditions of setting up such as flow rates and immobilized ligand amounts compatible with several binding models and constants.[27] The following provisions are necessary to confirm that the binding is specific and not by "nonspecific" forces: (i) injection of denatured analyte, (ii) immobilization of known proteins without binding abilities to the ligand, (iii) injection of a known inhibitor(s) or free ligands in solution simultaneously with the analyte, and (iv) and observation of a dose-dependent effect of the inhibition.

Estimation of Fimbrial Binding Affinity to Host Proteins Using BIAcore System. Sensogram of the ligand immobilization to the sensory chip is shown in Fig. 4.[28] Kinetic analyses of the interactions of fimbriae with the host proteins, including hemoglobin, fibrinogen, fibronectin, laminin, proline-rich protein (PRP), low molecular proline-rich glycoprotein (L-PRG), and statherin, were performed. In Fig. 5, BIAcore sensograms of the binding of fimbriae to L-PRG are shown as an example.[29] An inhibition study using BIAcore was also performed (Fig. 5D). For these assays, PBS (pH 7.4) was used as the running buffer.

Specific constants of the interactions between fimbriae and host proteins are obtained by BIAcore assays and are summarized in Table I. The binding assay of fimbriae to antifimbriae rabbit IgG is also performed to know the affinity levels between fimbriae and host proteins. These results show that k_{ass} values of the interactions of fimbriae with the host proteins are very

[27] M. Fivash, E. M. Towler, and R. J. Fisher, *Curr. Opin. Biotechnol.* **9,** 97 (1998).
[28] M. Kontani, S. Kimura, I. Nakagawa, and S. Hamada, *Mol. Microbiol.* **24,** 1179 (1997).
[29] A. Amano, T. Nakamura, S. Kimura, I. Morisaki, I. Nakagawa, S. Kawabata, and H. Hamada, *Infect. Immun.* **67,** 2399 (1999).

close to that of antigen–antibody interactions and that the fimbriae–host protein interactions occur with a similar level of specificities (K_d values). These analyses suggest that the interactions with strong affinities might be crucial events in the initial phase of biofilm development *in vivo.*

[37] Studying Biofilm Formation of Mutans Streptococci

By SHIGETADA KAWABATA and SHIGEYUKI HAMADA

Introduction

Mutans streptococci (MS), including *Streptococcus mutans* (serotype *c, e,* and *f*) and *Streptococcus sobrinus* (serotype *d* and *g*), have been implicated as primary causative agents of dental caries in humans and experimental animals.[1,2] Dental plaque is a complex but typical bacterial biofilm that contains MS and other oral bacteria and their products. Dental plaque is formed through two different stages: initial and reversible cell-to-surface attachment and subsequent sucrose-dependent adhesion of the organisms, which is firm and "irreversible."

The former step is initiated by adsorption of organisms to acquired pellicle on the tooth surface. Salivary receptor molecules are more likely to promote bacterial adhesion. These receptors can influence the adhesion in several ways: (i) they can induce aggregation of oral bacteria,[3,4] (ii) such aggregation promotes adhesion of MS to other bacteria, but it can also facilitate the removal of oral bacteria by swallowing or flushing, (iii) cell surface protein antigen called PA I/II[5,6] or P1[7] of MS is presumed to participate in the initial attachment of the organisms to acquired pellicle on the tooth surface.

Glucosyltransferases (GTFs) are strongly involved in the latter step. These enzymes utilize sucrose as a substrate and yield fructose and glucan with predominant $\alpha(1 \rightarrow 3)$ and $\alpha(1 \rightarrow 6)$ bonds and fructose as the products. Several species of oral streptococci produce GTFs; however, only those of

[1] S. Hamada and H. D. Slade, *Microbiol. Rev.* **44,** 331 (1980).
[2] S. Hamada, T. Koga, and T. Ooshima, *J. Dent. Res.* **63,** 407 (1984).
[3] T. Ericson and J. Rundegren, *Eur. J. Biochem.* **133,** 255 (1983).
[4] C. G. Emilson, J. E. Ciardi, J. Olsson, and W. H. Bowen, *Arch. Oral Biol.* **34,** 335 (1989).
[5] N. Okahashi, C. Sasakawa, M. Yoshikawa, S. Hamada, and T. Koga, *Mol. Microbiol.* **3,** 221 (1989).
[6] M. W. Russell, L. A. Bermeier, E. D. Zanders, and T. Lehner, *Infect. Immun.* **28,** 486 (1980).
[7] H. Forester, N. Hunter, and K. W. Knox, *J. Gen. Microbiol.* **129,** 2779 (1983).

0076-6879/99 $30.00

MS cooperatively synthesize adhesive and water-insoluble glucan, resulting in firm adhesion of MS to the tooth surface. It is reported that *S. mutans* and *S. sobrinus* produce three and four kinds of GTFs, respectively.[8–10] However, fructose is used as an energy source for these organisms and it is metabolized to release lactic acid and other organic acids that serve as erosive agents in the cariogenic process. The biofilm keeps acids inside the structure, which may lead to localized decalcification of the enamel surfaces.

This article outlines methods used to analyze biofilm formation by MS.

Molecular Analysis of Cell Surface Components Involved in Adhesion

Hydrophobicity of bacterial cell surfaces has been considered to affect several biological phenomena, such as attachment of bacteria to host tissues, adhesion of bacteria to solid surfaces, and interactions between bacteria and phagocytes. Among others, PA of mutans streptococci is highly involved in cell surface hydrophobicity. PA-deficient mutants showed lower hydrophobicity and less adsorptive to saliva-coated hydroxylapatite beads (S-HA) than their parent strains.[11,12] Furthermore, isogenic mutants that synthesized larger amounts of cell-associated PA exhibited increased hydrophobicity as compared with their wild-type strains.[12,13]

Hydrophobicity Assay Using Hydrocarbon

For estimation of the hydrophobic nature of bacteria, a hexadecane method[14] is conveniently used. For this purpose, freeze-dried bacteria can be applied, and experimental time will be saved. Vortexing of rehydrated organisms is key in obtaining reproducible data.

Reagents

PUM buffer (pH 7.1): Prepare 17.0 g K_2HPO_4, 7.3 g KH_2PO_4, 1.8 g urea, 0.2 g $MgSO_4 \cdot 7H_2O$, and dissolve in a final volume of 1000 ml distilled water.

[8] T. Furuta, T. Koga, T. Nishizawa, N. Okahashi, and S. Hamada, *J. Gen. Microbiol.* **131,** 285 (1985).

[9] Y. Yamashita, N. Hanada, and T. Takehara, *J. Bacteriol.* **171,** 6265 (1989).

[10] S. Hamada, T. Horikoshi, T. Minami, N. Okahashi, and T. Koga, *J. Gen. Microbiol.* **135,** 335 (1989).

[11] S. F. Lee, A. Progulske-Fox, G. W. Erdos, D. A. Piacentini, G. Y. Ayakawa, P. J. Crowley, and A. S. Bleiweis, *Infect. Immun.* **57,** 3306 (1989).

[12] T. Koga, N. Okahashi, I. Takahashi, T. Kanamoto, H. Asakawa, and M. Iwaki, *Infect. Immun.* **58,** 289 (1990).

[13] I. Takahashi, N. Okahashi, and S. Hamada, *J. Bacteriol.* **175,** 4345 (1993).

[14] M. D. Rosenberg, D. Gutnick, and E. Rosenberg, *FEMS Microbiol. Lett.* **9,** 29 (1980).

Procedure

1. MS organisms are grown at 37° for 18 hr in brain–heart infusion (BHI) broth (Difco Laboratories, Detroit, MI). The organisms are washed twice with phosphate-buffered saline (PBS; pH 7.4) and freeze-dried. When live cells are used, the freeze-drying step can be omitted.
2. The organisms are suspended in PUM buffer to an optical density of 0.6 at 550 nm. Volumes (3 ml) of bacterial suspension are transferred into glass tubes (13 × 100 mm).
3. Hexadecane (0.3 ml) is mixed with the suspension using a vortex mixer for 1 min. The reaction mixtures are allowed to stand for 15 min at room temperature.
4. The optical density of the lower, aqueous phase is measured. Adsorption is calculated as the percentage loss in optical density relative to that of the initial cell suspension.

Adsorption of Mutans Streptococci to Salivary Glycoproteins

Bacterial adhesion is often the result of specific interaction between the carbohydrate portions of receptor glycoproteins and protein complexes termed adhesins on bacterial cell surfaces.[15] An example is the interaction between human salivary proteins and the A-repeat portion of PA of MS.[16,17] The assay for bacterial adsorption to S-HA described later[18] is sensitive.

Saliva Preparation

Whole saliva is collected in an ice-cooled container and is clarified by centrifugation (44,000*g*, 30 min, 4°). The collected samples are dispersed and frozen at −20° until use.[19]

Assay for Bacterial Adsorption to S-HA

Reagents

Buffered KCl solution: 50 m*M* KCl in 2 m*M* potassium phosphate buffer (pH 6.0)

Procedure

1. To prepare S-HA, spheroidal hydroxylapatite beads (20 mg, BDH Chemicals, Poole, UK) are coated with 1 ml of diluted whole saliva

[15] I. Ofek and A. Perry, *in* "Molecular Basis of Oral Bacterial Adhesion" (S. E. Mergenhagen and B. Rosan, eds.), p. 7. American Society for Microbiology, 1985.

[16] M. Nakai, N. Okahashi, H. Ohta, and T. Koga, *Infect. Immun.* **61,** 4344 (1993).

[17] P. J. Crowley, L. J. Brady, D. A. Piacentini, and A. S. Bleiweis, *Infect. Immun.* **61,** 1547 (1993).

[18] R. Eifert, B. Rosan, and E. Golub, *Infect. Immun.* **44,** 287 (1984).

[19] B. Appelbaum, E. Golub, S. C. Holt, and B. Rosan, *Infect. Immun.* **25,** 717 (1979).

(one part saliva to five parts KCl buffer) or KCl buffer for 1 hr at room temperature and are washed three times with KCl buffer.

2. Organisms are grown at 37° for 18 hr in a chemically defined medium[20] containing [*methyl*-^{3}H]thymidine [2.22 TBq (60 Ci)/mmol; ICN Radiochemicals, Costa Mesa, CA] at a final concentration of 370 kBq (10 μCi)/ml.
3. [^{3}H]Thymidine-labeled bacteria (5×10^7) are incubated with S-HA in 1 ml of KCl buffer for 2 hr at 37° with continuous rotation.
4. The S-HA beads are washed twice with KCl buffer to remove unattached cells.
5. The radioactivity associated with the S-HA beads is estimated by liquid scintillation spectroscopy. The number of bacteria adsorbed to the S-HA is estimated from the calculated specific radioactivity of the bacteria.

Role of GTFs Concerning Biofilm Development

The ability of *S. mutans* and other species of MS to adhere firmly to the tooth surface in the presence of sucrose is involved with glucosyltransferases (GTFs) of the organisms. Although initial attachment of MS to saliva-coated enamel surfaces occurs through the surface components of the organisms, the synthesis of water-insoluble, adhesive glucan from sucrose by GTFs is essential for biofilm and dental caries development. In recent years, a variety of purification methods of GTFs have been reported, and several GTF genes (*gtfB, C, D, I, S, T,* and *U*) have been sequenced from mutans streptococci.

Measurement of GTF Activity

GTF activity is estimated by [^{14}C]glucan synthesis from [^{14}C]sucrose.[21,22]

Reagents

[^{14}C-glucose]sucrose [1.85 MBq (50 μCi)/mmol, DuPont NEN Research Products, Boston, MA] Standard reaction mixture: 10 m*M* [^{14}C-glucose]sucrose with or without 20 μM dextran T10 in 20 μl of 0.1 *M* potassium phosphate buffer (pH 6.0)

[20] B. Terleckyj, N. P. Willett, and G. D. Shockman, *Infect. Immun.* **11,** 649 (1975).

[21] T. Koga, H. Asakawa, N. Okahashi, and S. Hamada, *J. Gen. Microbiol.* **132,** 2873 (1986).

[22] S. Hamada, T. Horikoshi, T. Minami, N. Okahashi, and T. Koga, *J. Gen. Microbiol.* **135,** 335 (1989).

Procedure

1. The standard mixture and GTF samples are mixed and incubated for 1 hr at 37° using a round-bottomed 96-well plate.
2. They are spotted on a filter-paper square (1.5 × 1.0 cm) and dried in air.
3. The squares are washed three times by stirring with methanol and then dried in air.
4. To determine the amount of [^{14}C]glucan synthesized, the filters are immersed in scintillation fluid and are quantitated by a liquid scintillation counter. One unit of GTF activity is defined as the amount of enzyme needed to incorporate 1 μmol of glucose residue from sucrose into glucan per minute under the conditions described earlier.

Sucrose-Dependent Cell Adhesion to Glass Surfaces

The number of adherent MS cells was determined turbidimetrically and was expressed as the percentage of the total cell mass (percentage cell adhesion) as follows.

Procedure

1. MS are grown at 37° at a 30° angle to the horizontal for 18 hr in 3 ml of BHI broth containing 1% sucrose. The use of 3 × 100-mm disposable glass tube (Corning, Corning, NY) is recommended for this assay.
2. Organisms that form biofilms on the surface of the tube are scraped with a rubber scraper and dispersed by ultrasonication (total cell mass).
3. The sample tubes are vibrated vigorously with a vortex mixer for 3 sec. Supernatant containing detached glucans and bacteria (unadhered cell mass) is ultrasonicated, and the absorbance at 550 nm is measured in a spectrophotometer.
4. Following measurement of A_{550} using BHI broth as background, the percentage of adhesion is defined at 100 × A_{550} (total cell mass-unadhered cell mass)/A_{550} (total cell mass).

Preparation and Purification of GTF from MS

Streptococcus mutans produces GTFs in cell-free (CF) and cell-associated (CA) states. CF-GTF synthesizes water-soluble glucan from sucrose, whereas CA-GTF almost exclusively forms water-insoluble glucan.[22] This CA-GTF is considered to be important in the pathogenesis of *S. mutans* because mutants lacking GTF do not induce experimental dental caries.[21]

Molecular biological studies[23–25] revealed that CA-GTF contains GTF-B synthesizing insoluble glucans as well as GTF-C synthesizing both insoluble and soluble glucans. Because GTF-B and GTF-C are extensively similar in terms of nucleotide sequences,[26] these enzymes have not been purified differentially from *S. mutans* yet.

However, *S. sobrinus* produces at least three kinds of GTFs: one GTF synthesizes water-insoluble glucan from sucrose and the other two GTFs make water-soluble glucan.[8] The organisms self-aggregate in the presence of primer dextran, and cells grown in sucrose-containing media bind GTFs, while the ability is not seen in the case of sucrose-free media. These GTFs can be isolated and purified from sucrose-free culture supernatants by chromatofocusing with PBE74 (Pharmacia, Piscataway, NJ).[8] The following section focuses on the purification of CA-GTF from *S. mutans.*

Procedure

1. *Streptococcus mutans* is grown in 8 liter of TTY broth culture[27] for 16 hr at 37°. If necessary, CF-GTF can be purified from the culture supernatant by 50% saturated ammonium sulfate precipitation, chromatofocusing on a Polybuffer exchanger PBE94 (Pharmacia), and subsequent HA chromatography (Bio-Rad, Richmond, CA) as described previously.[28]
2. Organisms are collected by centrifugation and washed three times with 50 m*M* sodium phosphate buffer (NaPB, pH 6.0).
3. The organisms are stirred with 8 *M* urea solution for 1 hr at room temperature to release CA-GTF.
4. The centrifuged supernatant is concentrated by 60% saturated ammonium sulfate precipitation, dialyzed against 50 m*M* NaPB (pH 7.5), and loaded onto a DEAE-Sephacel column (2.5 × 13 cm, Pharmacia).
5. After washing with 50 m*M* NaPB (pH 7.5), bound proteins are eluted with a linear gradient of 0 to 1.0 *M* NaCl. Fractions (15 ml) are collected, and GTF activity is determined as described earlier.
6. Fractions containing GTF eluted at about 0.6 *M* NaCl are pooled and dialyzed against 50 m*M* NaPB (pH 6.0), and GTF solution

[23] T. Shiroza, S. Ueda, and H. K. Kuramitsu, *J. Bacteriol.* **169,** 4263 (1987).

[24] N. Hanada and H. K. Kuramitsu, *Gene* **69,** 101 (1988).

[25] T. Fujiwara, S. Kawabata, and S. Hamada, *Biochem. Biophys. Res. Commun.* **187,** 1432 (1992).

[26] T. Fujiwara, Y. Terao, T. Hoshino, S. Kawabata, T. Ooshima, S. Sobue, S. Kimura, and S. Hamada, *FEMS Microbiol. Lett.* **161,** 331 (1998).

[27] S. Hamada and M. Torii, *Infect. Immun.* **20,** 592 (1978).

[28] S. Sato, T. Koga, and M. Inoue, *Carbohyd. Res.* **134,** 293 (1984).

TABLE I
PURIFICATION OF CELL-ASSOCIATED GTF FROM *S. mutans* MT8148[a]

Purification step	Total protein (mg)	Total activity (U)	Specific activity (U/mg)	Recovery (%)	Purification (-fold)
8 *M* urea extract	207	417	2.01	100	1
Ammonium sulfate (60% saturation)	119	260	2.18	62	1.1
DEAE-Sephacel	37.4	85.9	2.30	21	1.1
Hydroxylapatite	4.7	32.8	6.96	8	3.5

[a] Cells recovered from TTY culture (8 liter) were used for the extraction of cell-associated GTF.

is applied onto an hydroxylapatite column (1.0 × 13 cm, Bio-Rad).

7. The column is first eluted with 225 ml of 50 m*M* NaPB (pH 6.0) and is then eluted stepwise with 50 m*M* (90 ml), 0.2 *M* (165 ml), 0.26 *M* (135 ml), and 0.5 *M* (255 ml) NaPB (pH 6.0) at a flow rate of 2 ml/min.
8. The A_{280} and GTF activity of each fraction (15 ml) are measured. CA-GTF should be obtained by elution with 0.5 *M* NaPB. A summary of purification of GTF is described in Table I. Long-term storage is best achieved at −80° after the enzyme solution has been frozen by liquid nitrogen.

Mutagenesis of *gtf* Genes in *S. mutans*

To examine the role of each GTF in sucrose-dependent adhesion, the construction of *gtf*-isogenic mutants is essential. Assays for studying the *gtf* gene functions were reported by several research groups. For example, recombinant GTFs were expressed in *Escherichia coli* or *Streptococcus milleri,* purified, and characterized.[25,29,30] Sucrase and glucan-binding activities of GTF of MS were also determined by independent groups.[31–33] This

[29] N. Hanada and H. K. Kuramitsu, *Infect. Immun.* **56,** 1999 (1988).
[30] K. Fukushima, T. Ikeda, and H. K. Kuramitsu, *Infect. Immun.* **60,** 2815 (1992).
[31] C. Wong, S. A. Hefta, R. J. Paxton, J. E. Shively, and G. Mooser, *Infect. Immun.* **58,** 2165 (1990).
[32] H. Abo, T. Matsumura, T. Kodama, H. Ohta, K. Fukui, K. Kato, and H. Kagawa, *J. Bacteriol.* **173,** 989 (1991).
[33] C. Kato, Y. Nakano, M. Lis, and H. K. Kuramitsu, *Biochem. Biophys. Res. Commun.* **189,** 1184 (1992).

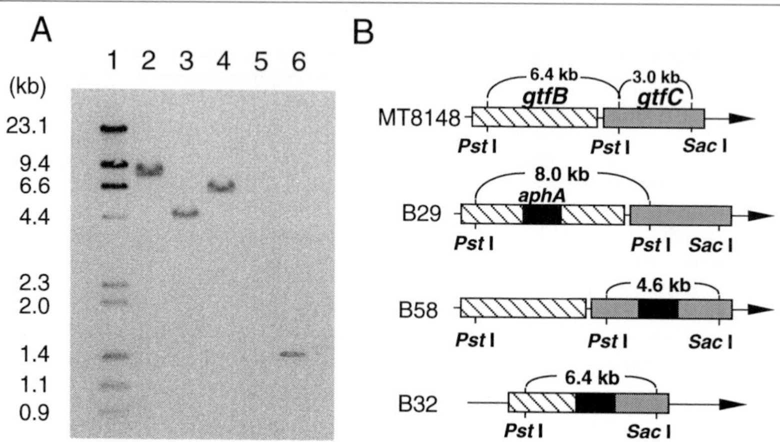

FIG. 1. Southern blot analysis of *S. mutans* MT8148 and its GTF-deficient mutants (A) and the location of the inserted *aphA* gene in GTF-deficient mutants (B). (A) Chromosomal DNA of the test organisms was digested and separated in a 0.8% agarose gel and then transferred onto a nylon membrane. The blotted membrane was hybridized stringently with the *aphA* gene fragment. Lanes: 1, DNA size marker; 2, mutant B29; 3, B58; 4, B32; 5, parent strain MT8148; and 6, the *aphA* gene. (B) The location of the inserted *aphA* gene (■) was determined by Southern blot analysis and the restriction enzyme mapping of *gtfB* (▧) and *gtfC* (▩) genes.

section describes inactivation and reconstruction of the *gtf* genes of *S. mutans.*

Transformation of DNA into S. mutans

Transformation of *S. mutans* is performed as described previously.[34]

1. *Streptococcus mutans* is cultured in Todd–Hewitt broth (Difco) supplemented with 10% heat-inactivated horse serum (GIBCO, Grand Island, NY) for 18 hr at 37°.
2. The overnight culture is diluted 1:40 into the broth (10 ml) and incubated for 1.5 hr at 37°, followed by the addition of donor DNA to a final concentration of 25 μg/ml.
3. It is further incubated for 2 hr at 37°, concentrated approximately 10-fold by centrifugation, and spread on a mitis-salivarius agar (Difco) plate containing appropriate antibiotics.
4. The plates are incubated at 37° for 2 to 3 days and possible transformants are harvested for further examination.

[34] T. Fujiwara, M. Tamesada, Z. Bian, S. Kawabata, S. Kimura, and S. Hamada, *Microb. Pathog.* **20,** 225 (1996).

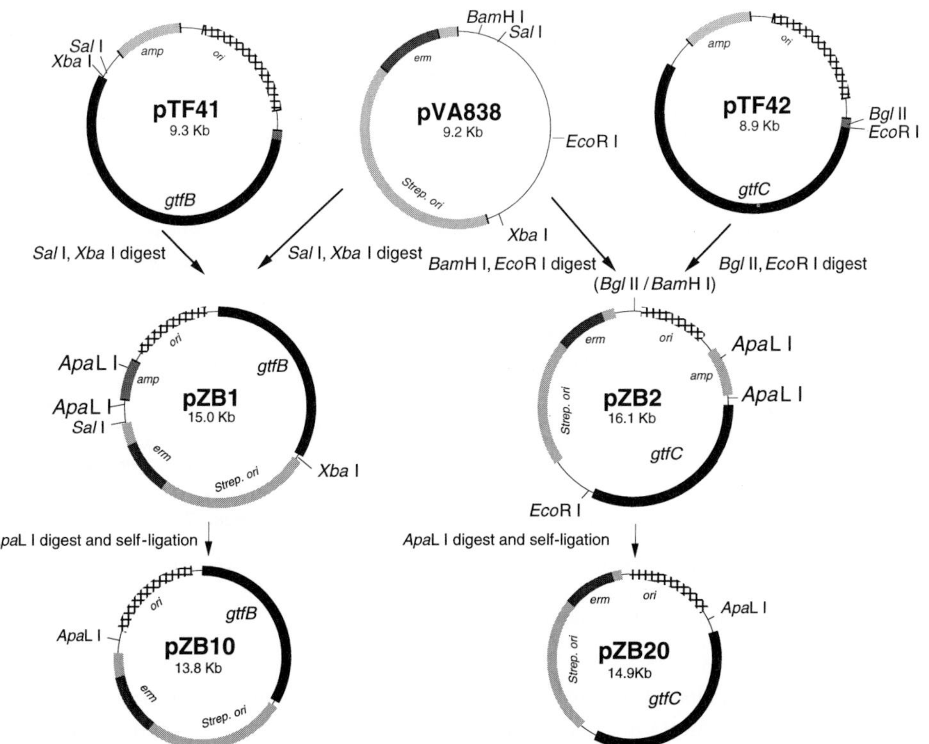

FIG. 2. Construction of *E. coli–S. mutans* shuttle plasmids carrying the *gtfB* or *gtfC* gene.

Insertional Inactivation of the *gtfB* and/or *gtfC* Gene in *S. mutans*

The 2-kb central portion of both *gtfB* and *gtfC* shows extensive similarity to the extent of 98%. Therefore, *gtfB*-, *gtfC*-, and *gtfBC*-inactivated isogenic mutants may be obtained simultaneously when a unique *Mlu*I site in the central region of the genes is inserted as an antibiotic marker and the resultant plasmid is transformed into *S. mutans*. According to this strategy, the *gtfB*-harboring plasmid, pSK6,[25] is cleaved with *Mlu*I and is ligated to a kanamycin resistance gene (*aphA*) from transposon Tn*1545*[35] to generate pTF55. Following linearization at a unique *Kpn*I site of the plasmid, the DNA fragment (10 μg) is introduced into *S. mutans* strain MT8148 by transformation to allow allelic replacement. Transformants are selected on MS agar containing 500 μg/ml kanamycin and are examined for their colony morphology on MS agar under a dissecting microscope.

[35] F. Caillaud, C. Carlier, and P. Courvalin, *Plasmid* **17,** 58 (1987).

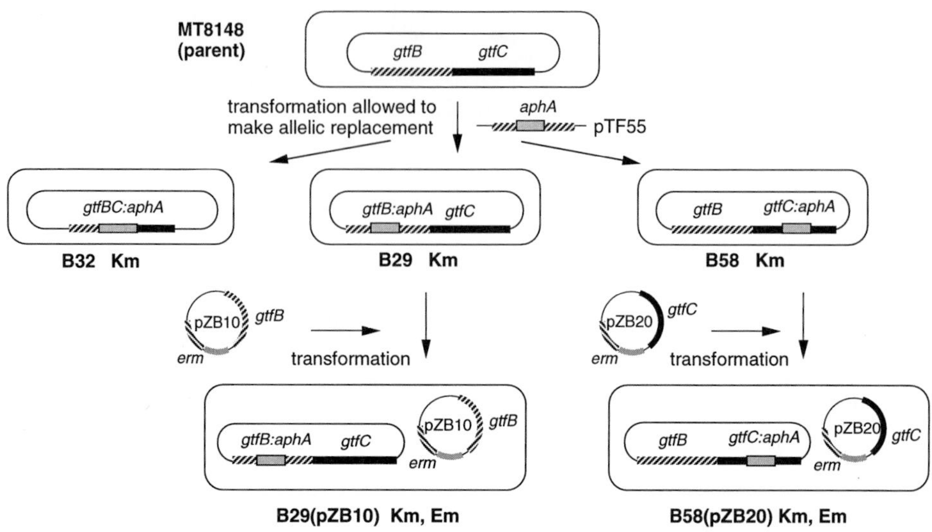

FIG. 3. Deletion and reconstruction of GTFs of *S. mutans* MT8148. *gtfB, gtfC,* or both genes were inactivated by insertional mutagenesis. Shuttle plasmids harboring the inactivated *gtf* genes were then transformed into the GTF-deficient mutant.

Streptococcus mutans grown on MS agar shows rough colonial morphology. Transformants are screened based on their different colonial appearances, and three kinds of mutant strains are obtained. Southern blot analysis using the 1.6-kb *aphA* gene as the probe reveals that strains B29, B58, and B32 are inactivated in *gtfB, gtfC,* and *gtfBC,* respectively (Fig. 1).

Reconstruction of *gtfB* and *gtfC,* and Reintroduction of GTF

A *Streptococcus–E. coli* shuttle vector pVA838[36] is digested with *Xba*I and *Sal*I to generate a shuttle plasmid carrying *gtfB* (Fig. 2). A 5.7-kb DNA fragment containing the genes of the origin of replication in *Streptococcus* and the erythromycin-resistant gene (*erm*) is isolated from an agarose gel. The fragment is ligated into pTF41, which is a derivative of pSK6 harboring the *gtfB* gene cleaved with the same enzymes, generating pZB1. To remove the *amp* gene, pZB1 is digested with *Apa*LI and self-ligated, resulting in pZB10.

The shuttle vector containing the *gtfC* gene, pZB2, is constructed by digesting pVA838 with *Eco*RI and *Bam* HI (Fig. 2). A 7.2-kb fragment is

[36] F. L. Macrina, R. P. Evans. J. A. Tobian, D. L. Hartley, D. B. Clewell, and K. R. Jones, *Gene* **25,** 145 (1983).

ligated with pTF42 digested with *Eco*RI and *Bgl*II. To remove the *amp* gene, pZB2 is digested with *Apa*LI and self-ligated, resulting in pZB20.

Both *E. coli* and *S. mutans* harboring pZB10 or pZB20 should be confirmed to be sensitive to ampicillin. To reintroduce deficient GTFs, pZB10 and pZB20 are transformed into GTF-deficient mutants B29 and B58, respectively (Fig. 3).

Remarks

This article outlines the protocol used to examine biofilm (dental plaque) development by mutans streptococci. Traditional and novel molecular biology techniques are required to understand the mechanism of the biofilm formation in the oral cavity. *Streptococcus mutans* and *S. sobrinus* apparently enjoy a favorite position among members of oral bacterial microbiota for the study of biofilms. It should be noted that similar biofilms are occasionally formed on heart valves, which eventually leads to bacterial endocarditis. In this case, MS as well as other oral and enteric streptococcal species are reported to be etiologically involved.

Acknowledgments

This work was supported by grants from Japan Society for the Promotion of Science and Ministry of Health and Welfare.

[38] Models for Studying Initial Adhesion and Surface Growth in Biofilm Formation on Surfaces

By BART GOTTENBOS, HENNY C. VAN DER MEI, and HENK J. BUSSCHER

Introduction

Biofilms can be considered as microecosystems in which different microbial strains and species efficiently cooperate in order to protect themselves against environmental stresses and to facilitate more efficient nutrient uptake. Most often, biofilms are unwanted and are related to diverse problems such as microbially induced corrosion of oil rigs and pipelines,[1] food and drinking water contamination,[2,3] dental caries and periodontal diseases,[4]

[1] J. T. Walker, K. Hanson, D. Caldwell, and C. W. Keevil, *Biofouling* **12,** 333 (1998).

[2] S. H. Flint, P. J. Bremer, and J. D. Brooks, *Biofouling* **11,** 81 (1997).

[3] S. Notermans, J. A. M. A. Dormans, and G. C. Mead, *Biofouling* **5,** 21 (1991).

[4] P. D. Marsh and M. V. Martin, "Oral Microbiology." Chapman and Hall, London, 1992.

0076-6879/99 $30.00

and a variety of biomaterial-centered infections in humans.[5] Biomaterial-centered infections in humans are especially troublesome, as biofilm organisms are protected against the host immune system and cannot be eradicated easily with antibiotics. Consequently, infection of a biomaterials implant will usually result in reoperation, osteomyelitis, amputation, or death.[6] Not all biofilms are unwanted, however, and in sewage treatment, biofilms are needed for the efficient degradation of xenobiotics,[7] while lactobacillus biofilms form part of the normal indigenous microbiota in humans and their maintenance is essential in the prevention of disease.[8]

Mechanism of Biofilm Formation

The formation of a biofilm in an aqueous environment is generally pictured to proceed in the following sequence[9] (see Fig. 1): (1) when organic matter is present, a conditioning film of adsorbed components is formed on the surface prior to the arrival of the first organisms; (2) microorganisms are transported to the surface through diffusion, convection, sedimentation, or active movement; (3) initial microbial adhesion occurs; (4) attachment of adhering microorganisms is strengthened through exopolymer production and unfolding of cell surface structures; (5) surface growth of attached microorganisms and continued secretion of exopolymers; (6) localized detachment of biofilm organisms caused by occasionally high fluid shear or other detachment forces. Localized detachment of biofilm organisms starts after initial adhesion, although the adhesion of individual microorganisms is frequently considered irreversible (whether justified or not) and increases with time as it is related to the number of microorganisms present in the biofilm.[10] Detachment of parts of a biofilm can occur by failure inside the bulk of the biofilm or by failure in the so-called linking film, involving detachment of the initially adhering organisms, cohesive failure in conditioning film, or interfacial rupture. Furthermore, as the number of biofilm organisms increases, growth rates will decrease due to nutrient and oxygen limitations and accumulation of organic acids, eventually leading to a sta-

[5] J. Dankert, A. H. Hogt, and J. Feijen, *CRC Crit. Rev. Biocompatibility* **2,** 219 (1986).

[6] A. G. Gristina, *Science* **237,** 1588 (1987).

[7] G. M. Wolfaardt, J. R. Lawrence, R. D. Robarts, S. J. Caldwell, and D. E. Caldwell, *Appl. Environ. Microbiol.* **60,** 434 (1994).

[8] G. Reid, A. W. Bruce, J. A. McGroarty, K. C. Cheng, and J. W. Costerton, *Clin. Microbiol. Rev.* **3,** 335 (1990).

[9] A. Escher and W. G. Characklis, *in* "Biofilms" (W. G. Characklis and K. C. Marshall, eds.), p. 445. Wiley, New York, 1990.

[10] M. J. Vieira, L. F. Melo, and M. M. Pinheiro, *Biofouling* **7,** 67 (1993).

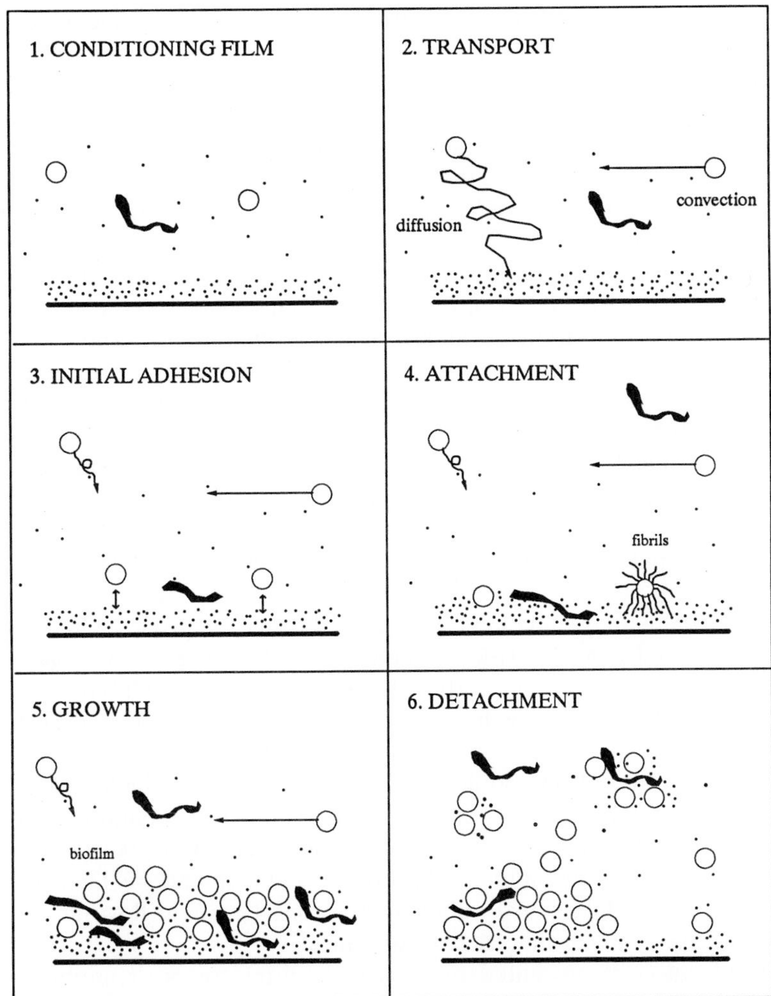

FIG. 1. Schematic, sequential presentation of the steps in biofilm formation.

tionary biofilm thickness, where adhesion and growth counterbalance detachment.

Prevention, Control, and Eradication of Biofilms

An obvious approach in the prevention of biofilm formation is the prevention of initial microbial adhesion. Microbial adhesion is mediated by specific interactions between cell surface structures and specific molecular groups on the substratum surface[11] or when viewed from an overall, physicochemical viewpoint by nonspecific interaction forces, including Lifshitz–van der Waals forces, electrostatic forces, acid–base interactions, and Brownian motion forces.[12] Specific interactions are in fact nonspecific forces acting on highly localized regions of the interacting surfaces over distances smaller than 5 nm, whereas nonspecific interaction forces have a long range character and originate from the entire body of the interacting surfaces. Upon approach of a surface, organisms will be attracted or repelled by the surface, depending on the resultant of the different nonspecific interaction forces. Lifshitz–van der Waals forces and Brownian motion usually promote adhesion, whereas electrostatic interactions can be either attractive or repulsive. Most organisms are negatively charged[5] and consequently a negatively charged substratum exerts a repulsive electrostatic force on the organisms. Control of the charge and hydrophobic properties of substratum surfaces is likewise a pathway to influence biofilm interaction with a substratum surface.

Another pathway that influences biofilm formation is through inhibition of growth. A possible approach is the design of (antibiotic)slow-release materials mediating direct kill on contact,[13] but such approaches are always temporary and bear the risk of inducing resistant strains. Alternatively, physicochemical surface properties also affect growth and it has been found that *Escherichia coli* growth is inhibited on positively charged surfaces through strong attachment.[14]

To control beneficial biofilms, conditions have to be adapted to obtain the optimal equilibrium of growth and detachment of biofilm organisms. The growth rate can be controlled by nutrient conditions, reactor design, oxygen, removal of metabolic end products, and temperature. Detachment can be influenced by substratum surface properties, fluid flow rate,[15] mechanical stress, and surface active substances.[16]

Eradication of unwanted biofilms with disinfectants or antibiotics is hampered because of the resistance of biofilm organisms against antimicrobial penetration through the biofilm.[17] The total removal of biofilms is only

[11] G. D. Christensen, L. M. Baddour, D. L. Hasty, J. H. Lowrance, and W. A. Simpson, *in* "Infections Associated with Indwelling Medical Devices" (A. L. Bisno and F. A. Waldvogel, eds.), p. 27. American Society for Microbiology, Washington, DC, 1989.

[12] C. J. Van Oss, "Interfacial Forces in Aqeuous Media." Dekker, New York, 1994.

[13] G. Golomb and A. Shpigelman, *J. Biomed. Mater. Res.* **25,** 937 (1991).

[14] G. Harkes, J. Dankert, and J. Feijen, *J. Biomater. Sci. Polym. Ed.* **3,** 403 (1992).

[15] B. E. Rittman, *in* "Structure and Function of Biofilms" (W. G. Characklis and P. A. Wilderer, eds.), p. 49. Wiley, New York, 1989.

[16] A. S. Landa, H. C. van der Mei, G. van Rij, and H. J. Busscher, *Cornea* **17,** 293 (1998).

[17] H. Anwar and J. W. Costerton, *ASM News* **58,** 665 (1992).

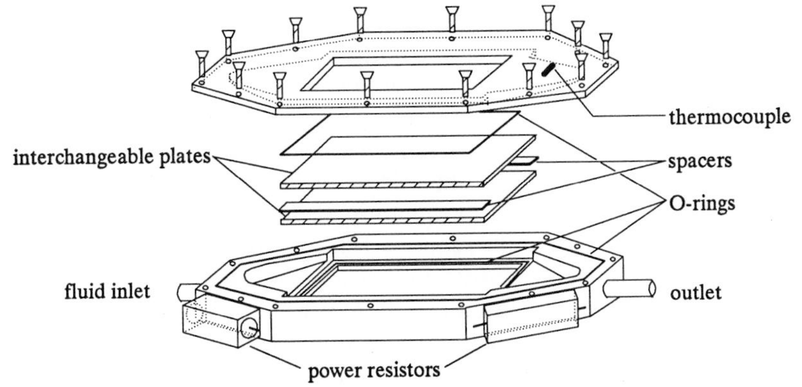

FIG. 2. Detailed view of the parallel plate flow chamber.

possible when biofilms are directly accessible, such as on exterior parts of the human body. Dental biofilms can be removed by mechanical cleansing, in combination with the use of surface active substances, as formed in dentifrices and mouth washes. Bacteria adhering on contact lenses can be removed by rubbing the lenses between the fingers in combination with cleansing solutions. The strength of biofilm adhesion to a substratum surface, i.e., the ease with which it can be removed, is greatly dependent on the strength with which the initially adhering organisms bind the substratum surface and cohesiveness of the conditioning film,[18] which makes initial microbial adhesion and surface growth an important issue of research.

This article summarizes the use of a parallel plate flow chamber model to study the initial microbial adhesion to surfaces and extends the use of flow chamber devices and data analysis to include surface growth of the initially adhering organisms.

Experimental Design

The parallel plate flow chamber and image analysis used in our laboratory is described accurately in an earlier volume of this series,[19] but will be repeated briefly for completeness. Figure 2 shows the chamber (external dimensions 16 × 8 × 2 cm, length by width by height), which is made of nickel-coated brass to allow sterilization. The internal dimensions of the chamber are 7.6 × 3.8 × 0.06 cm, although the height can be varied by

[18] H. J. Busscher, R. Bos, and H. C. van der Mei, *FEMS Microbiol. Lett.* **128,** 229 (1995).
[19] H. J. Busscher and H. C. van der Mei, *Methods Enzymol.* **253,** 455 (1995).

using different spacers. Microbial adhesion, surface growth, and detachment are observed directly, usually on the bottom plate using a microscope, thus biofilm formation can be followed *in situ* without any additional shear forces acting on the deposited microorganisms. The bottom plate can be of various materials but can be conveniently made of poly(methyl methacrylate) (PMMA) with a grove in the middle in which a substratum material of interest can be fixed. The top plate is made of glass. The application of phase-contrast microscopy requires transparent substratum materials. Nontransparent, reflective substrata can also be used when an ordinary metallurgical microscope, based on incident reflected light, is available.[20] To allow the detection of microorganisms on opaque materials, incandescent dark-field illumination is directed under a low angle onto the bottom plate of the channel by means of a lens supported on a slide.[21] Furthermore, it is possible to use fluorescence microscopy, provided it is ascertained that the required fluorescent dyes do not affect the adhesive properties of the microorganisms.

The microscope images can be recorded using a charge coupled device (CCD) camera and processed by an image analyzer (TEA, Image-Manager, Difa Breda, The Netherlands) in combination with dedicated image analysis programs.[22]

Flow is created by a roller pump and can be pulse free or a controlled pulsatile.[23] Temperature control using heating elements mounted on the sides of the chamber is optional. By means of a valve system it is possible to connect flasks containing, for example, buffer, reconstituted human whole saliva, bacterial suspension, growth medium, or detergent solutions, with the flow chamber without passing an air–liquid interface over the adsorbed conditioning film and/or adhering organisms.[24]

Initial Microbial Adhesion

Initial microbial adhesion experiments are typically done in buffer solutions without any additional nutrients to avoid complications caused by conditioning films or growth of microorganisms during deposition. Before each experiment all air bubbles are removed from the tubing and flow chamber and the buffer solution is perfused through the system for a predetermined time. Subsequently, flow is switched to a suspension of

[20] C. G. van Hoogmoed, H. C. van der Mei, and H. J. Busscher, *Biofouling* **11,** 167 (1996).

[21] J. Sjollema, H. J. Busscher, and A. H. Weerkamp, *J. Microbiol. Methods* **9,** 73 (1989).

[22] P. Wit and H. J. Busscher, *J. Microbiol. Methods* **32,** 281 (1998).

[23] T. G. van Kooten, J. M. Schakenraad, H. C. van der Mei, and H. J. Busscher, *J. Biomater. Sci. Polym. Ed.* **4,** 601 (1993).

[24] I. H. Pratt-Terpstra, A. H. Weerkamp, and H. J. Busscher, *J. Gen. Microbiol.* **133,** 3199 (1987).

microorganisms in buffer. During deposition, images of the bottom plate are recorded and the organisms present on the surface are counted using the image analysis program. From the number of microorganisms plotted versus time the initial deposition rate (j_0) is determined and through an iterative procedure the number of bacteria adhering at a stationary end point time (n_∞) is found.

As natural biofilms are not formed in plain buffer, it is relevant to study adhesion also in the fluids present in nature. For example, in studies on biofilm formation on teeth,[25] urinary catheters,[26] contact lenses,[16] or body implants,[27] buffer can be replaced by saliva, urine, tear fluid, or blood plasma. Here initial adhesion can be influenced by the composition of the conditioning film. In blood serum, for example, albumin reduces staphylococcal adhesion,[27] whereas fibronectin can promote the adhesion of certain staphylococcal strains.[28,29] The contribution of surface growth to the number of attached bacteria cannot be separated easily from that of adhesion, which complicates these types of experiments.

Surface Growth

The use of the parallel plate flow chamber can be extended to study the surface growth of sessile microorganisms.[30,31] For this purpose, the entire system is sterilized at 120°, except for the PMMA bottom plate and the substratum material of interest, which are sterilized partially by 70% (v/v) ethanol. During the experiment the flow chamber can be heated to 37° to get more relevant results for surface growth on biomaterials. The experiment starts with initial adhesion of the microorganisms during a short duration of time, as microorganisms tend to lose their viability in buffer. This can be avoided, however, by supplementing the buffer with growth medium. For *Pseudomonas aeruginosa,* for example, it was found that only 2% of the adherent bacteria was metabolically active after 4 hr of deposition in phosphate-buffered saline (PBS), whereas in a minimal (2%) growth medium, 67% appeared metabolically active.[32] In this case, addition of a minor amount of growth medium did not seriously complicate interpretation of the results.

[25] A. S. Landa, H. C. van der Mei, and H. J. Busscher, *Biofouling* **9,** 327 (1996).

[26] M. M. C. Velraeds, H. C. van der Mei, G. Reid, and H. J. Busscher, *Urology* **49,** 790 (1997).

[27] H. C. van der Mei, B. van de Belt-Gritter, G. Reid, H. Bialkowska-Hobrzanska, and H. J. Busscher, *Microbiology* **143,** 3861 (1997).

[28] Z. Zdanowski, E. Ribbe, and C. Schalen, *APMIS* **101,** 926 (1993).

[29] A. L. Cheung, M. Krishnan, E. A. Jaffe, and V. A. Fischetti, *J. Clin. Invest.* **87,** 2236 (1991).

[30] B. Gottenbos, H. C. van der Mei, and H. J. Busscher, *J. Biomed. Mater. Res.*, submitted (1999).

[31] A. J. Barton, R. D. Sagers, and W. G. Pitt, *J. Biomed. Mater. Res.* **32,** 271 (1996).

[32] M. B. Habash, H. C. van der Mei, G. Reid, and H. J. Busscher, *Microbiology* **143,** 2569 (1997).

After microorganisms are seeded on the substratum surface, the flow chamber is washed with adhesion buffer at the same flow rate as during deposition to remove the planktonic and loosely adhering organisms. Subsequently, flow is switched to growth medium at the same flow rate and is continued for a predetermined time. Images are recorded during surface growth, from which the number of organisms present is determined. With appropriate analysis, it is possible to follow individual microorganisms and determine their generation times.

Eradication of Biofilms

After microorganisms are deposited or grown in the flow chamber, various effects of environmental stress on the biofilm can be studied. To study the influence of high shear stress on the adhering organisms, we usually pass an air bubble over the substratum surface.[33] The passing of a liquid–air interface results in a detachment force on the adhering organisms of around 10^{-7} N,[19] which is much higher than the shear resulting from the flowing liquid. The number of organisms present before and after the passage of the liquid–air interface can be determined as a measure of the strength of microbial adhesion.

To determine the detergent-stimulated detachment of biofilm organisms, a surfactant solution can be led over the adhering microorganisms. Images are recorded before, during, and after the treatment, and the microorganisms are enumerated. In this way, the activity of, for example, mouth washes and contact lens cleansing solutions can be evaluated.

Examples of Use of Parallel Plate Flow Chamber for Studying Biofilm Formation and Eradication

Influence of Blood Plasma Conditioning Film on Initial Staphylococcal Adhesion

For six staphylococcal strains, staphylococcal adhesion was studied on silicone rubber with and without preadsorbed plasma proteins.[27] First, when appropriate, flow was switched to plasma for 1.5 hr to create a conditioning film. Thereafter, flow was switched for 30 min to buffer for removal of all remnants of plasma from the tubes and the chamber and then to the bacterial suspension, which was circulated through the system for 4 hr.

Figure 3 is an example of the deposition kinetics of one strain, *Staphylococcus epidermidis* 242. The presence of a conditioning film on silicone

[33] H. J. Busscher, G. I. Doornbusch, and H. C. van der Mei, *J. Dent. Res.* **71,** 491 (1992).

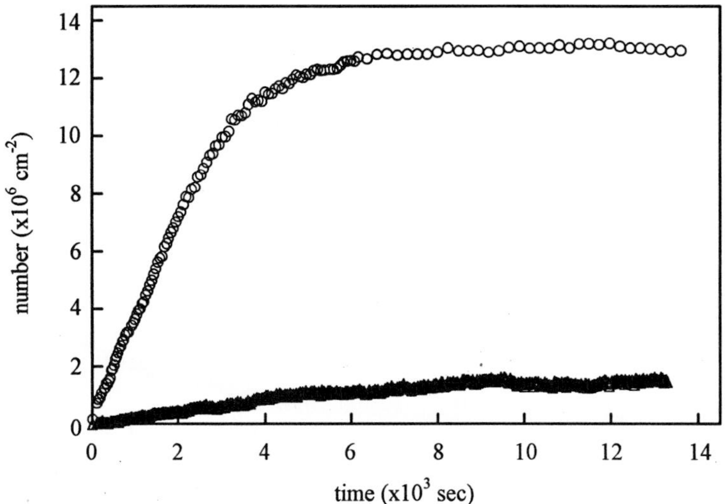

FIG. 3. Deposition kinetics of *S. epidermidis* 242 to silicone rubber with (△) and without (○) a plasma coating. For details, see van der Mei *et al.*[27]

rubber had a reducing effect on the adhesion of all strains studied. The reduction of the initial deposition rate, j_0, and the adhering numbers in a stationary end point varied between 77–97% and 55–98%, respectively, depending on the strain used.

Adhesion and Surface Growth of Staphylococcus epidermidis on Materials with Various Wettabilities

The adhesion and surface growth of *S epidermidis* HBH_2 102 were determined on different materials with varying water contact angles between 15 and 110°.[30] The flow chamber temperature was kept at 36° during the experiment. The initial adhesion rate was determined in phosphate-buffered saline (PBS, 10 m*M* potassium phosphate and 150 m*M* sodium chloride, pH 7.0) for 1 hr, and the flow chamber was washed with PBS for 15 min and was subsequently switched to 20 times diluted tryptone soya broth (TSB) in PBS. During 24 hr, surface growth was followed. Due to its grape-forming mode of growth, individual staphylococci are not counted easily and the numbers of adhering bacteria were derived from the measurement of the surface coverage of the biofilm. Figure 4 shows the number of bacteria on silicone rubber and glass. The desorption of biofilm bacteria was determined and expressed as the fraction of the biofilm that detaches per minute (k_{des}). From the biofilm doubling time (Δt), the generation time

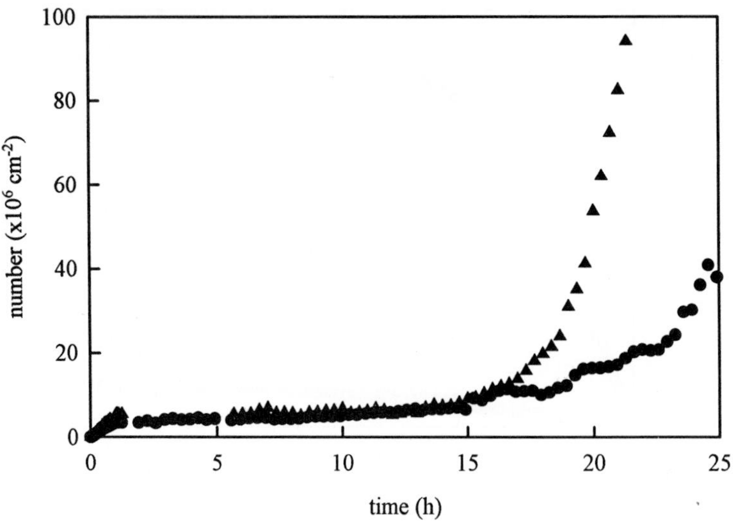

FIG. 4. The number of *S. epidermidis* HBH_2 102 during deposition and surface growth on silicone rubber (▲) and glass (●).[30]

(*g*) can be calculated using a modification of the mathematical growth model for microbial growth by Barton *et al.*[31]

$$g = \frac{\Delta t}{^2\log(2 + k_{des}\,\Delta t)} \tag{1}$$

Figure 5 gives the generation time of this staphylococcal strain on the different materials as a function of their water contact angle, revealing a relationship between substratum wettability and surface growth.

Detachment of Pseudomonas aeruginosa from Contact Lenses by Ophthalmic Solutions

To study the efficacies of two contact lens cleansing solutions, a biofilm was formed on a contact lens quarter and mounted in the parallel plate flow chamber.[16] For this purpose, a suspension of *P. aeruginosa* No. 3 in saline supplemented with 2% (w/v) TSB was circulated through the system for 20 hr to allow adhesion and surface growth. Then flow was switched to saline for 1 hr to remove nonadhering bacteria from the system, and subsequently 8 ml of an ophthalmic solution or saline (control), followed by 16 ml saline to clean the chamber of the solution, was perfused through the flow chamber. To study renewed bacterial adhesion, flow with saline was continued for 30 min, after which a new suspension of freshly cultured

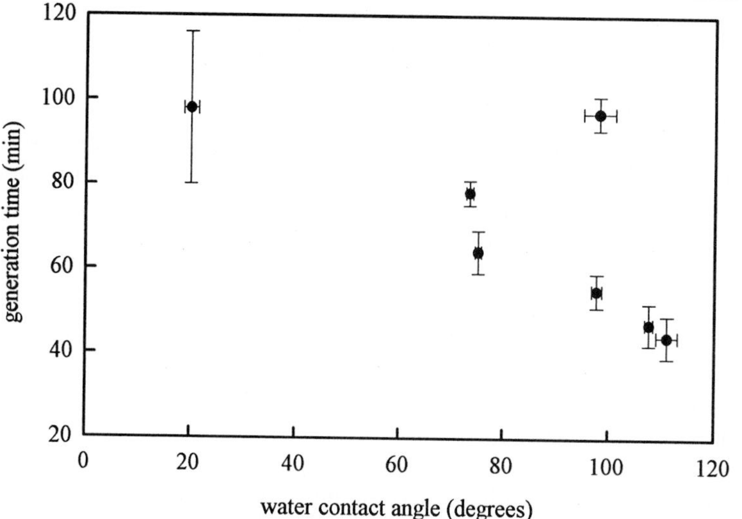

FIG. 5. The generation time of *S. epidermidis* HBH_2 102 during surface growth on different materials as a function of their water contact angle.[30]

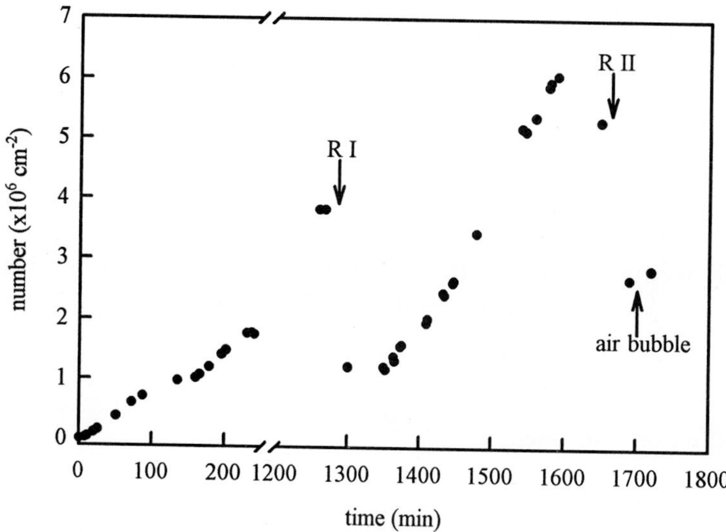

FIG. 6. Number of adhering *P. aeruginosa* No. 3 to a contact lens with an adsorbed tear film. After adhesion and surface growth, the surface was rinsed by a detergent mixture (R I), followed by a second adhesion phase and a second rinse (R II). Finally, an air bubble was passed over the surface. For details, see Landa *et al.*[16]

bacteria was circulated through the system for 4 hr. After removal of nonadhering bacteria with saline for 1 hr, another dose of ophthalmic solution was applied. Finally a liquid–air interface was led over the surface. Figure 6 illustrates the number of adhering *P. aeruginosa* on a contact lens with a tear film during one complete experiment. The ophthalmic solution clearly decreases the number of bacteria whereas the liquid–air interface does not yield a significant effect.

Advantages and Disadvantages of System

The major advantages of the system outlined are controlled shear and mass transport; a high data density in time; and the avoidance of air–liquid interface passages over the adhering microorganisms. Furthermore, the *in situ* observation offers the great advantage that all events in initial biofilm formation, including adhesion and growth, can be followed in time and that the fate of an individual microorganism in the biofilm can be studied.

A disadvantage of the system is that biofilm formation is only viewed in two dimensions, and consequently only initial biofilm formation can be studied. As the biofilm thickness extends to above one layer, the events are not clearly visible anymore. To study more mature biofilms, a three-dimensional viewing system is needed, as provided by scanning confocal laser microscopy combined with three-dimensional image analysis software.[34] Unfortunately, this method is invasive due to the fluorescent staining, and biofilm processes cannot be followed *in situ.*

[34] D. E. Caldwell, D. R. Korber, and J. R. Lawrence, *J. Appl. Bact. Symp. Suppl.* **74,** 52S (1993).

[39] Recovery and Characterization of Biofilm Bacteria Associated with Medical Devices

By MARC W. MITTELMAN

Introduction and Background

Device Evolution

Tremendous advances have been made in the biomedical device and biomaterials fields since the late 1960s. Where stainless steel and polymethyl methacrylate were once considered state of the art as engineered materials, titanium, fluoropolymers, and a variety of ceramic composites are now

0076-6879/99 $30.00

widely employed. Device designs have been radically changed and improved over this period. Devices that at one time were designed only for limited operating times, such as the total artificial heart, are now capable of *in situ* operations over a patient's lifetime. Energy sources have become more efficient, the devices smaller in size, and component parts more resistant to physiological corrosion activities. Improvements in surgical procedures, particularly those involving less invasive endoscopic-type procedures, have kept pace with engineering developments. Biocompatibility at the biomaterial–host interface has been improved greatly, particularly with the advent of what Ratner[1] and others have termed "biomimetic" surfaces.

The market for various implantable medical devices is more than $30 billion per year, with an estimated annual growth rate of 7–15% in North America and western Europe.[2] The demographics of an aging population in the Western world coupled with advances seen in both the device designs and associated surgical procedures suggest an increasing demand for implantable devices over the next several decades. Despite these advances in the biomaterials sciences and growth in the medical device market, the single greatest factor limiting the utility of existing devices, and those yet to be developed, is infection. Microbial adhesion to engineered surfaces is the essential first step in the development of device-related infections.

Infections and Implantable Devices

Bacteria colonizing medical devices exhibit a number of phenotypic and genotypic characteristics that are distinct from planktonic organisms and provide them with an adaptive advantage in an otherwise hostile environment replete with antibiotics and cellular and humoral antagonistic agents. The incidence of device-related bacterial infections is increasing with the more widespread application of implants in medicine. The sources of bacteria that colonize the vast majority of implant-associated infections include (1) perioperative contamination, (2) exit site contamination for percutaneous devices, or (3) hematogenous spread from locations distal to the implant area. Each of these sources has common and distinct characteristics in terms of species diversity, severity of infection, and predisposing host conditions. Infection problems linked to the contamination of devices labeled as "sterile" are extremely rare. This is due to the effectiveness of current steam, dry heat, ethylene oxide, and gamma irradiation sterilization processes.

[1] B. D. Ratner, *J. Biomed. Mater. Res.* **27,** 837 (1993).

[2] B. D. Ratner, A. S. Hoffman, F. J. Schoen, and J. E. Lemons, eds., "Biomaterials Science." Academic Press, San Diego, 1996.

The genesis of device-related infections and associated mechanisms has been reviewed previously.[3,4]

The ability of microorganisms to adhere to biomaterials is a key factor in the initiation of disease processes. However, local and systemic factors also play an important role in the host's response to a colonized foreign body. It is clear that bacterial colonization of biomaterials does not always lead to an intractable infection.[5] Although the mechanisms responsible for this type of "quiescent" colonization state are unclear, it is known that the immune response can be modulated by the presence of some kinds of biofilms. In addition to this altered host response to the presence of bacteria, device-associated bacteria possess an intrinsic resistance to topically or systemically applied antimicrobials.[6–8] The combination of these two factors—host-mediated immune modulation and antimicrobial resistance—accounts for the oftentimes severe consequences of device-related microbial infections. Bacteria (and fungi) in biofilms often exhibit a differential resistance to antimicrobial agents relative to their planktonic counterparts. This resistance is broad spectrum and includes antibiotics,[9] disinfectants,[10] and biocides.[11]

Clinical treatment failures are a frequent occurrence with device-related infections.[12–14] These treatment failures occur despite the relatively low minimum inhibitory concentration (MIC) and minimum biocidal concentration (MBC) values associated with planktonic isolates recovered from patient body fluids, tissues, and device surfaces. This observed poor predictive value is a result of the analytical methods employed for assessing antimicrobial efficacy. Namely, that planktonic organisms, instead of the putative agents of infection present within the biofilm, are challenged by the antibiotics of interest. The disparity between MIC or MBC and biofilm elimination

[3] M. W. Mittelman, *in* "The Molecular and Ecological Diversity of Bacterial Adhesion" (M. Fletcher, ed.), p. 89. Wiley, New York, 1997.

[4] A. L. Bisno and F. A. Waldvogel, eds., "Infections Associated with Indwelling Medical Devices." ASM Press, Washington, DC, 1994.

[5] C. Virden, M. K. Dobke, P. Stein, C. L. Parsons, and D. H. Frank, *Aesthetic Plastic Surg.* **16,** 173 (1992).

[6] E. A. Trafny, K. Kowalska, and J. Grzybowski, *J. Biomed. Mater. Res.* **41,** 593 (1998).

[7] H. Anwar and J. W. Costerton, *ASM News* **58,** 665 (1992).

[8] P. S. Stewart, *Antimicrob. Agents Chemother.* **38,** 1052 (1994).

[9] P. Vergeres and J. Blaser, *J. Infect. Dis.* **165,** 281 (1992).

[10] B. H. Pyle, S. K. Watters, and G. A. McFeters, *J. Appl. Bacteriol.* **76,** 142 (1994).

[11] V. J. Fraser, M. Jones, P. R. Murray, G. Medoff, Y. Zhang, and R. J. Wallace, *Am. Rev. Respir. Dis.* **145,** 853 (1992).

[12] J. C. Nickel and J. W. Costerton, *Can. J. Infect. Dis.* **3,** 261 (1992).

[13] S. Gander, *J. Antimicrob. Chemother.* **37,** 1047 (1996).

[14] C. Simon and M. Suttorp, *Support. Care Cancer* **2,** 66 (1994).

concentrations (BEC) has been demonstrated in bacteria as diverse as *Porphyromonas gingivalis* from periodontal infections[15] and *Staphylococcus epidermidis* from vascular infections.[16]

The methods described herein focus on both clinical and research laboratory techniques used for the recovery and evaluation of bacterial biofilms. Many of the techniques described elsewhere in this volume, including scanning confocal laser microscopy and Fourier transform infrared spectrometry (FTIR), have great potential for application to studies of device-associated biofilms. There are also a number of interesting surface analytical techniques that have been applied to the study of biofilms on various medical devices. These include X-ray photoelectron spectroscopy, energy-dispersive X-ray analysis, and static secondary ion mass spectrometry. A review of these techniques and some of their biomaterials applications has been published[17] and additional methods are presented elsewhere in this volume.

Recovery of Bacteria from Devices

Perioperative Sampling

Maki *et al.*[18] have described a semiquantitative culture technique for the recovery of putative infectious agents from intravenous catheters. The technique can also be modified for use with other types of percutaneous access devices and implanted devices. Maki *et al.*[19] have found this technique to be more specific in the diagnosis of catheter-related bloodstream infections than culture of the explanted catheter in broth. Direct (semiquantitative) plating of catheter sections or other devices onto agar surfaces or into broth medium shows significant correlation with the presence of bloodstream isolates.[19–21] A method has also been developed specifically for

[15] T. L. Wright, R. P. Ellen, J. M. Lacroix, S. Sinnadurai, and M. W. Mittelman, *J. Periodont. Res.* **32,** 473 (1997).

[16] T. M. Bergamini, T. M. McCurry, J. D. Bernard, K. L. Hoeg, R. B. Corpus, J. C. Peyton, K. R. Brittian, and W. C. Cheadle, *J. Surg. Res.* **60,** 3 (1996).

[17] G. Reid, H. J. Busscher, S. Sharma, M. W. Mittelman, and S. McIntyre, *Surf. Sci. Rep.* **7,** 251 (1995).

[18] D. G. Maki, C. E. Weise, and H. W. Sarafin, *N. Engl. J. Med.* **196,** 1305 (1977).

[19] D. G. Maki, C. E. Weise, and H. W. Sarafin, *N. Engl. J. Med.* **196,** 1305 (1977).

[20] D. R. Snydman, S. A. Murray, S. J. Kornfield, J. A. Majka, and C. A. Ellis, *Am. J. Med.* **73,** 695 (1982).

[21] A. McGeer and J. Righter, *Can. Med. Assoc. J.* **137,** 1009 (1987).

the recovery of bacteria from endoluminal surfaces of vascular access devices.[22]

Maki Roll Technique

1. Remove any antimicrobial preparations or blood from the exit site using an alcohol swab and then allow the contact site to air dry.
2. Withdraw the catheter using sterile forceps, taking care to avoid contact of the catheter with skin surfaces.
3. Aseptically cut the catheter into 5- to 7-cm segments, including a proximal segment beginning a few millimeters inside the former skin–catheter interface and the tip.
4. Transport the catheter segments in the appropriate transport medium contained within a sterile tube.
5. Place each catheter segment onto the surface of separate 100-mm agar plates [e.g., 5% (v/v) sheep blood agar].
6. Exert downward pressure on the segments using sterile forceps, then roll the segments across the agar four times. If the catheter segment (or other device) cannot be rolled, it should be smeared back and forth across the plate four times.
7. Remove the devices from the agar surfaces and then incubate plates at the appropriate temperature and time (e.g., 37° for 72 hr).
8. Resulting colony counts can be normalized by dividing the total count by the contact surface area.

Endoscopic Brush Technique

1. The endoscopic brush is available from FAS Medical (Leeds, UK). The tapered, nylon-bristled brush is 8 mm long on the end of a stainless steel introducer wire and is packaged as sterile. Attach the polyethylene sheath-brush assembly to the *in situ* catheter by means of the integral Luer lock-capped end piece.
2. Introduce the sterile brush through the catheter hub to the distal end of the catheter.
3. Withdraw the brush from the catheter and place it in its sterile polyethylene sheath for transport.
4. Use sterile scissors to cut the brush from the wire introducer and place in a test tube containing 1 ml of ice-cold phosphate-buffered saline (PBS), pH 7.4.
5. Sonicate the tube for 1 min at 44 kHz (Sonomatic, Jencons Scientific, Luton, UK).

[22] P. Kite, B. M. Dobbins, M. H. Wilcox, W. N. Fawley, X A. J. L. Kindon, D. Thomas, M. J. Tighe, and M. J. McMahon, *J. Clin. Pathol.* **50,** 278 (1997).

6. Vortex the tube for 15 sec at maximum speed.
7. Process the sonicated and vortexed suspension for bacterial characterization as described next.

Experimental Methods: Bacterial Biofilm Development in the Laboratory

In vitro bacterial adhesion and colonization experiments with implant materials can be conducted under a variety of conditions designed to simulate those of the *in situ* environment of interest. For example, evaluations of antimicrobial coatings can be tested under static culture conditions to simulate ureteral stent exposure to urinary stasis. Alternatively, flow cell experiments can be designed for once-through bacterial adhesion studies of vascular access devices or dental implants. Both methods have been employed for bacterial biofilm development on a variety of clinically relevant substrata.[23–27]

Semibatch Culture Biofilm Development

1. Dispense a 100-μl volume of an overnight culture of *Pseudomonas aeruginosa* or other strain of interest washed twice in PBS, pH 7.2, to 100 ml of sterile nutrient broth in 250-ml glass beakers. Bovine serum albumin (BSA) or complement-inactivated human serum can be added to a concentration of 5% (v/v) in the sterilized medium as required. Cover the beakers with sterile aluminum foil.
2. Add a minimum of five replicate catheter sections or other materials of interest to each beaker for each treatment set.
3. Place the inoculated beakers onto a rotary shaker set at 100 rpm and incubate at 37°.
4. At 12-hr intervals, carefully decant off one-half of the bacterial suspension and replace with sterile nutrient broth.
5. At the end of the colonization period, rinse each device section with a 10-ml volume of sterile PBS, pH 7.2, delivered via gravity through a 10-ml pipette.
6. Colonized surfaces can either be characterized directly as described later or be subjected to physical or chemical treatments to assess antimicrobial efficacy, also described later.

[23] F. Soboh, A. C. Zamboni, D. Davidson, A. E. Khoury, and M. W. Mittelman, *Antimicrob. Agents Chemother.* **39,** 1281 (1995).

[24] M. W. Mittelman, L. L. Kohring, and D. C. White, *Biofouling* **6,** 39 (1992).

[25] H. J. Busscher, H. Bialkowska-Hobrzanska, G. Reid, M. V. D. Kuijl-Booij, and H. C. van der Mei, *Coll. Surf. B Biointerfaces* **2,** 73 (1994).

[26] G. Reid, D. Lam, A. W. Bruce, H. C. van der Mei, and H. J. Busscher, *J. Biomed. Mater. Res.* **28,** 731 (1994).

[27] M. Cowan and H. J. Busscher, *Microbios* **73,** 135 (1993).

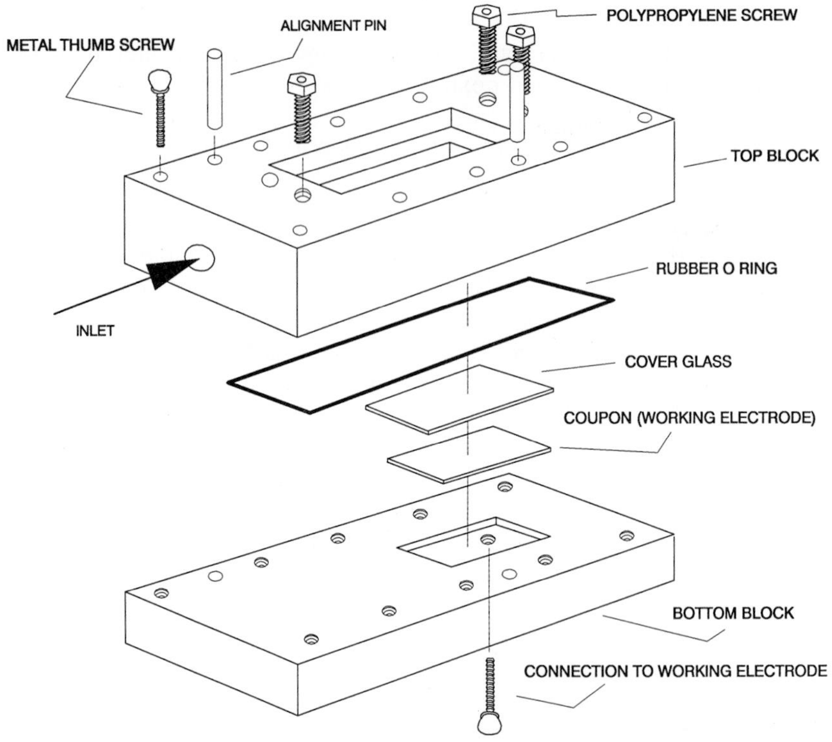

FIG. 1. Schematic of a laminar-flow adhesion cell for bacterial colonization studies.[24]

Continuous Culture Flow Cell Studies

1. Laminar-flow adhesion cells are used to develop biofilms on polymeric or metallic surfaces under dynamic flow conditions. The design for one type of device is shown in Fig. 1. Other adhesion cell designs have also been described.[28–30] Use of a Robbins device for the colonization of silicone,[31] polyurethane,[32] and other medical-grade plastics[33] has been described (see elsewhere in this volume).[33a]

[28] M. Cowan and H. J. Busscher, *Microbios* **73,** 135 (1993).
[29] H. M. W. Uyen, J. M. Schakenraad, J. Sjellema, J. Noordmans, W. L. Jongebold, I. Stokroos, and H. J. Busscher, *J. Biomed. Mater. Res.* **24,** 1599 (1990).
[30] A. D. Cook, R. D. Sagers, and W. G. Pitt, *J. Biomater. Appl.* **8,** 72 (1993).
[31] M. K. Dasgupta, K. Ward, P. A. Noble, M. Larabie, and J. W. Costerton, *Am. J. Kidney Dis.* **23,** 709 (1994).
[32] S. L. Fessia and M. J. Griffin, *Peritoneal Dialysis Int.* **11,** 144 (1991).
[33] G. D. Christensen, L. Baldassarri, and W. A. Simpson, *Methods Enzymol.* **253,** 477 (1995).
[33a] *Methods Enzymol.* **310** [27, 28] 1999 (this volume).

2. Fix the test surface to the bottom half of the flow cell using silicone sealant or other appropriate sealant.
3. Assemble the flow cell and then sterilize using ethylene oxide or other appropriate treatment. Cold treatments such as 3% (w/v, in deionized or distilled water) glutaraldehyde applied for 2 hr can also be used for sterilization. Cold sterilants should be rinsed thoroughly from the flow cells before the addition of bacterial inocula.
4. Bacterial cultures in continuous culture systems or from natural environments may be pumped through the flow cells using peristaltic pumps (e.g., Cole-Parmer, Chicago, IL) with recirculation or in a single-pass mode. A design for one type of experimental system is shown in Fig. 2.
5. Make T-type connections to 20 × 150-mm anaerobe tubes (Fisher Scientific) to eliminate the pulse flow, which is associated with peristaltic systems.
6. Flow rates of <1 ml/min have been found to produce consistent biofilms on various polymeric and metallic substrata.
7. Depending on the type of flow cell device used, adhesion and colonization can either be observed directly via optical microscopy and image analysis or cells can be quantitatively recovered and characterized as described next.

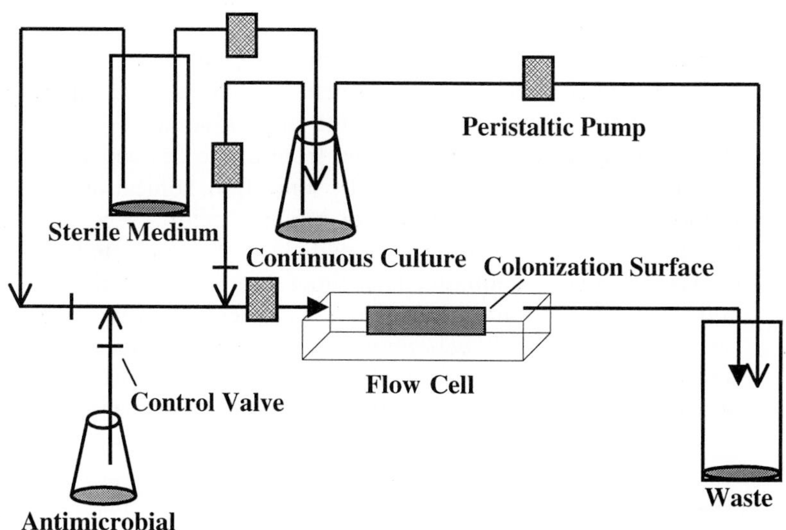

FIG. 2. Experimental setup for colonization and antimicrobial efficacy studies using flow cells.

Quantitative Recovery of Cells and Biomass Material from Surfaces

Flat, colonized surfaces can be quantitatively extracted for viable plate counts or microscopic direct cell counts (see later) using glass O-ring joints (Kontes Glass, Vineland, NJ) equipped with N-buna or Viton O rings. Various diameters are available, but a 1.2-cm-diameter providing an extraction area of 1.13 cm^2 has been found to work well with this technique.

1. Clamp the O-ring extractors onto the surface with "c clamps" (also available from Kontes Glass).
2. Add a 0.5- to 1.0-ml volume of ice-cold, sterile PBS, 50 m*M*, pH 7.2, or other appropriate solution to the extractor funnel.
3. Preclean a 3-mm-diameter sonicator probe (Heat Systems sonicator, Plainview, NJ) with an isopropanol wipe, followed by rinsing in 0.2-μm-filtered deionized or distilled water.
4. Apply three, 3-sec pulses at 20% power to the extraction medium inside the O-ring extractors to remove viable cells and biomass constituents from surfaces. Sonication cycles must be validated for other power settings or instruments to ensure the quantitative recovery of cells from surfaces.
5. Remove the sonicate from the O-ring extractors and add to sterile tubes. (If the sample is to be processed for protein, carbohydrate, or lipids, the tubes should be first washed in a suitable detergent, then cleaned with chloroform or hexane, followed by methanol. A final rinse in deionized or distilled water should be performed before steam sterilization of the tubes.)
6. Rinse the interior surfaces of the O-ring extractors and test coupons with an equivalent volume of extraction medium. Add this second aliquot to sterile tubes containing the original sonicate.
7. Alternatively, cells can be removed from entire devices or sections of devices by immersion, followed by sonication. Sonication has been found to be an effective means of biofilm bacteria recovery from explanted devices.[34] Sonication should always be conducted with ice-cold reagents using minimum pulse times. Use of a jeweler's sonicator and three, 30-sec sonication periods has been found to yield good recoveries of viable *P. aeruginosa* from catheter surfaces.[35] Teflon scrapers (Fisher Scientific) can also be used to quantitatively recover attached cells and biomass constituents from device surfaces. Before

[34] I. I. Raad, M. F. Sabbagh, K. H. Rand, and R. J. Sherertz, *Diagn. Microbiol. Infect. Dis.* **15,** 13 (1992).

[35] F. Soboh, A. C. Zamboni, D. Davidson, A. E. Khoury, and M. W. Mittelman, *Antimicrob. Agents Chemother.* **39,** 1281 (1995).

their use, the scrapers should be precleaned with glassware detergent, taken through a solvent series, and rinsed as described earlier.

8. Disrupt aggregates of biofilm bacteria recovered from device surfaces by adding 2-mm-diameter sterile glass beads to suspensions, followed by vortexing for 60 sec at high speed.

Evaluation of Biofilm Elimination Concentrations

Unlike MIC/MBC evaluations for planktonic bacteria,[36] there are no standardized clinical protocols available for assessing BEC values. Because bacterial challenge titers influence MIC/MBC values, it is likely that variations in colonization numbers will also affect BEC values. Because of the inherent variability associated with the development of biofilms, a minimum of three BEC replicates should be prepared for each treatment.

1. For semibatch culture experiments, place colonized and rinsed device sections into beakers containing nutrient broth as described earlier (with or without added serum) containing dilutions of the desired antimicrobial agent. Incubate at 37° on a rotary shaker set at 100 rpm. With flow cell experiments, antimicrobial agents can be dosed onto colonized surfaces by means of a T connection as shown in Fig. 2.
2. Time course experiments can be conducted to assess rate of kill for the agents of interest. Remove device sections at 2-hr time intervals and process for viable and/or total bacterial numbers as described later. A 24-hr exposure period may be used for comparison to MIC/MBC values obtained for corresponding planktonic organisms.
3. Inactivation of residual antimicrobial agents associated with colonized device surfaces is performed using the appropriate neutralizing reagents and/or media. Challenged devices are placed in neutralizing solutions for <1 hr before final rinsing and evaluation. Russell *et al.*[37] have provided information on neutralizing agents for a variety of antimicrobial agents.

The MBEC assay system (patent pending, MBEC Biofilm Technologies, Calgary) facilitates the production of multiple, reproducible biofilms for the study of antibiotic sensitivity, drug uptake studies by biofilms, or molecular expression studies. The MBEC system can also be used to study antibacterial coatings aimed at preventing biofilm formation. The device consists of

[36] B. A. Forbes and P. A. Granato, *in* "Manual of Clinical Microbiology" (P. Murray, E. S. Baron, M. A. Pfaller, F. C. Tenover, and R. H. Yolken, eds.), 6th ed., p. 265. ASM, Washington, DC, 1995.

[37] A. D. Russell, I. Ahonkhai, and D. T. Rogers, *J. Appl. Bacteriol.* **46,** 207 (1979).

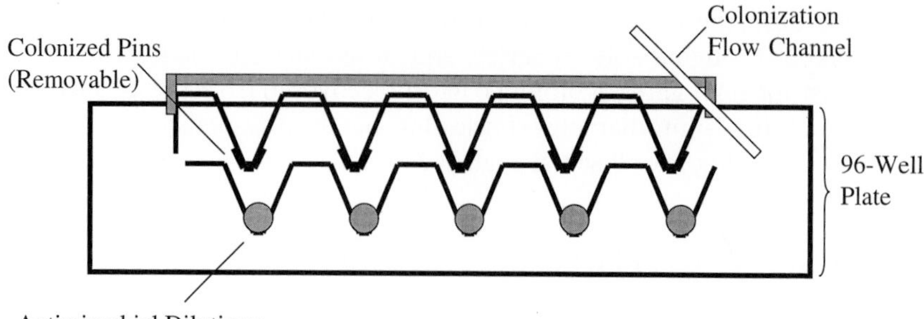

FIG. 3. Schematic of the MBEC device for biofilm bacteria antimicrobial susceptibility testing.[38]

two parts: a trough divided into channels and a lid, which has 96 pins that fit into the trough channels (Fig. 3).

1. Prepare a standardized bacterial inoculum for each of the MBEC microtiter plate wells.
2. Align the top of the device (containing pins for colonization) with the bottom half containing inoculum.
3. Incubate the apparatus on a tilt table to cause the flow of liquid along the troughs to induce equal shear force at each pin position. This arrangement eliminates the need for pumps, lines, and connections to produce the shear forces required to develop intact biofilms.
4. Following incubation, biofilms can be quantitated by breaking pins from the device and recovering adherent cells by sonication (as described earlier).
5. Once a biofilm is formed on the device pins, antimicrobial susceptibility (e.g., MBEC) can be evaluated by shifting the lid from the trough to a standard 96-well tray in which dilutions of antibiotics are prepared.
6. MBEC values (Fig. 4) can be derived by determining the concentration of antibiotic that eliminates the biofilm. This can be determined by a variety of end point assays, including sonicating bacteria from the pins after antibiotic challenge and performing plate counts to determine viable bacterial numbers on the pins. Alternatively, MBEC can be assessed by placing the treated pins into the appropriate neutralizing solution, rinsing, and then adding into fresh growth medium. Turbidity changes can then be used as an indication of viable bacteria remaining in the biofilm.

[38] H. Ceri, C. Stemick, M. E. Olson, R. R. Read, D. Morck, and A. Buret, Am. Soc. Microbiol. Ann. Meet., Atlanta, GA, 1998.

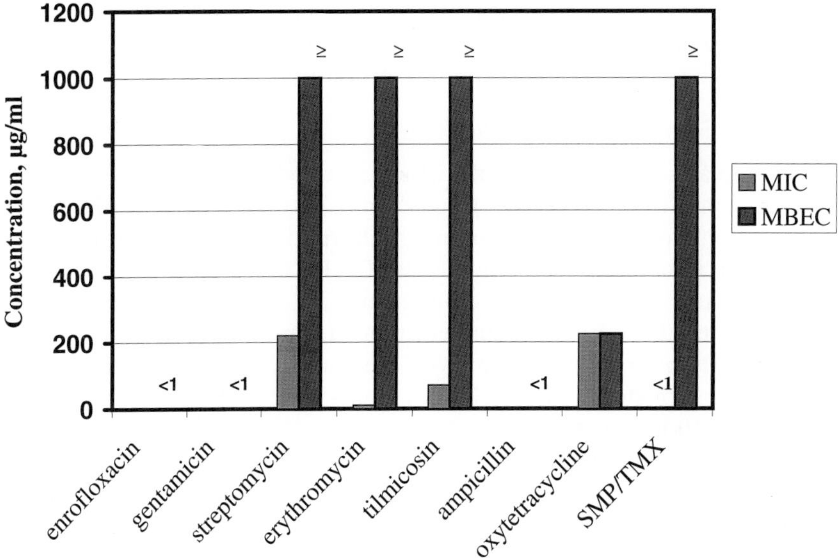

FIG. 4. MIC vs MBEC data for a clinical strain of *E. coli.* MBEC data were obtained with the MBEC device (courtesy of Dr. Howard Ceri, University of Calgary).

Examination of Device-Associated Biofilm Bacteria

Epifluorescence Microscopy

Direct observation of bacterial biofilms via epifluorescence microscopy is a simple, relatively inexpensive method for measuring bacterial attachment and desorption on metallic, polymeric, and composite (e.g., ceramic) substrata. The methods involve direct staining of biofilms with the appropriate fluorescent dye, observation, and characterization via epifluorescence microscopy. Acridine orange direct counts have been used for the direct, quantitative evaluation of organisms associated with explanted devices[39] and in laboratory studies of device-associated biofilms.[40] McFeters *et al.*[41] have reviewed applications of a wide range of fluorochromes for quantitation and characterization of bacteria. The techniques involve either direct staining of surfaces or sonication and removal of bacteria followed by filtration and staining. It is often difficult, however, to obtain a reliable count directly from surfaces with large numbers of adherent cells. Fixation

[39] J. Zufferey, B. Rime, P. Francioli, and J. Bille, *J. Clin. Microbiol.* **26,** 175 (1988).
[40] S. L. Fessia and M. J. Griffin, *Peritoneal Dialysis Int.* **11,** 144 (1991).
[41] G. A. McFeters, F. P. Yu, B. H. Pyle, and P. Stewart, *J. Microbiol. Methods* **21,** 1 (1995).

of biofilms before staining may be performed via immersion in 2.5% (w/v) glutaraldehyde in 100 m*M* PBS, pH 7.1. (If an assessment of cell viability is required, samples should not be fixed before staining.)

Total Direct Counts on Surfaces

1. Add a sufficient volume of 0.01 mg/ml acridine orange (AO) (Sigma Chemical, St. Louis, MO) in 100 m*M* phosphate buffer, pH 7.2, to cover the area to be examined. Alternatively, a solution of 1 μg/ml 4′,6-diamidino-2-phenylindole (DAPI; Polysciences, Warrington, PA) in 100 m*M* phosphate buffer, pH 7.2, may be used for staining.
2. Let stand for 5 min at ambient temperature.
3. Gently pour off the stain from the surface and rinse with a 10-ml volume of sterile PBS, pH 7.2. The PBS should be gravity-delivered via a 10-ml pipette.
4. Allow the stained and rinsed surface to air dry.
5. Place one drop of low fluorescing immersion oil (e.g., type B, Fisher Scientific) on the coupon surface. Cover with a coverslip.
6. Examine the coupon surface directly under epifluorescent illumination (e.g., Leitz Leico, Frankfurt, equipped with a 100-W mercury lamp, Leico PL Fluorstar 100×/1.3 na oil immersion objective) using the fluor clusters appropriate for AO or DAPI (Fig. 5).
7. A calibrated eyepiece reticule should be used for delineating the counting area. Sufficient fields should be counted such that a coefficient of variation (standard deviation/average) of <20% is obtained. Typically, 20–100 fields are scanned for the presence of attached organisms.
8. The number of cells per unit area desired is calculated by dividing the average count by the area of the counting reticule.

Total Direct Counts of Bacteria Recovered from Surfaces

1. Add 0.5-ml samples of dilutions prepared for viable counts (see sonication procedure, earlier) to 25-mm glass microfiltration units (Corning Corp., Acton, MA) containing black, 0.2-μm polycarbonate Nuclepore membrane filters (Corning Corp., Acton, MA). Note that the shiny side of the membranes should be oriented upward on the microfiltration funnel. Apply a vacuum to the microfiltration unit.
2. Remove the vacuum source and add an 0.5-ml volume of 0.01 mg/ml AO (Sigma Chemical, St. Louis, MO) in 100 m*M* phosphate buffer, pH 7.2, to the membrane surface. Alternatively, a 1-μg/ml solution of DAPI in 100 m*M* phosphate buffer, pH 7.2, may be used for staining.
3. Let stand 2 min at ambient temperature.

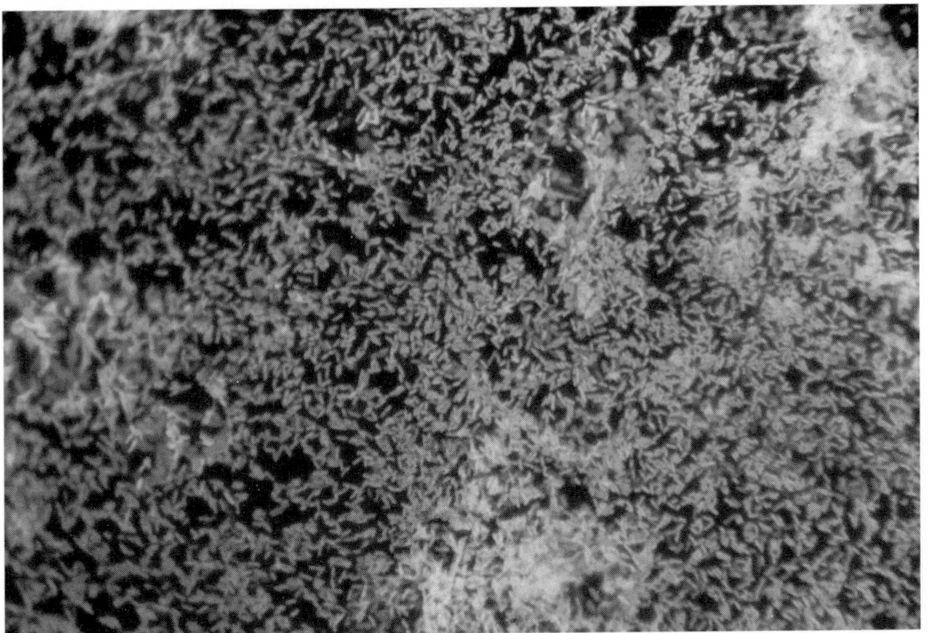

FIG. 5. Acridine orange preparation (black and white reproduction) of a *P. aeruginosa* biofilm on a silicone catheter surface. Magnification: ×1000.

4. Apply a minimal vacuum to remove the stain from the membrane surface, carefully remove the membrane, and let air dry for approximately 2 min.
5. Place the dried membrane onto the surface of a clean glass microscope slide (bacteria side up) containing a small drop of low-fluorescing immersion oil. Allow the oil to partially diffuse through the membrane and then add a coverslip to the surface of the membrane. Place a small drop of low-fluorescing immersion oil on the coverslip.
6. Examine the membrane surface under epifluorescent illumination as described earlier.
7. The number of cells/cm^2 of test surface may be calculated using the following formula:

$$a = \frac{xb}{mvde}$$

where a is cells/cm^2, x is average number of cells/eyepiece reticule area, b is effective filtration area (cm^2), m is area of eyepiece reticule (cm^2), v

is sample volume filtered (ml), d is dilution factor (1/ml), and e is area extracted (cm^2).

Viable Direct Counts. Counts of actively respiring biofilm bacteria (organisms with functioning electron transport systems) can be performed using the redox dye 5-cyano-2,3-ditolyltetrazolium chloride (CTC; Polysciences, Warrington, PA). Respiring organisms (aerobic and facultatively anaerobic bacteria) reduce the CTC to a fluorescent formazan salt, which can be detected under epifluorescent illumination. Incubation of the CTC can be performed *in situ*; however, heavy accretions of bacteria and biofilm constituents may impede the transport of CTC to cells at the biofilm–substratum interface.

1. Using the microfiltration system and membrane filters described earlier, filter 0.5–1.0 ml of the biofilm extract from viable-count dilution tubes.
2. Remove the vacuum source and add 0.5 ml of a 1.25 mM solution of CTC in pH 7.2, 100 mM phosphate buffer to the membrane surface.
3. Let stand at the *in situ* temperature (e.g., 37°) for 0.5–1.5 hr. The optimal incubation time must be determined for each type of biofilm and suite of experimental conditions.
4. Apply a minimal vacuum to the microfiltration apparatus.
5. Remove the vacuum and then add 0.5 ml of the DAPI solution to counterstain.
6. Let stand 2 min and then filter.
7. Process the membrane for epifluorescence microscopy as described earlier. The same fluor combination used for AO is appropriate for enumerating cells containing the reduced fluorescent formazan compound.
8. Actively respiring cells fluoresce orange, whereas inactive cells fluoresce blue-green.
9. Total viable numbers of bacteria are determined as described earlier.

Transmission Electron Microscopy

Transmission electron microscopy (TEM) methods have been described for characterization of bacteria associated with a number of implantable device types. These include peritoneal dialysis catheters,[42,43] vascular access

[42] M. K. Dasgupta, M. Larabie, K. Lam, K. B. Bettcher, D. L. Tyrrell, and J. W. Costerton, *Am. J. Nephrol.* **10,** 353 (1990).

[43] S. P. Gorman, C. G. Adair, and W. M. Mawhinney, *Epidemiol. Infect.* **112,** 551 (1994).

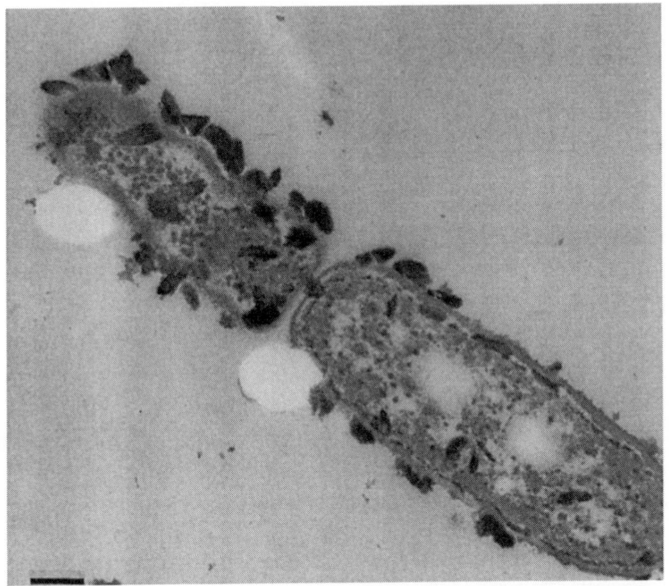

FIG. 6. TEM image of *Bacillus* spp. associated with copper plate. Bar: 250 nm.

devices,[44,45] orthopedic[46] and dental implant[47] materials, and silicone voice box prostheses.[48] Transmission electron microscopy can provide useful information on biofilm architecture, interactions of EPS with various substrata, and biofilm thickness (Fig. 6).

Although there are always variations in procedures used by different laboratories, the protocols and reagents employed are similar for different materials. Biofilm specimens are processed for TEM using a fixation, embedding, and thin sectioning procedure. This procedure is useful for examining biofilms associated with polymeric or metallic substrata.

1. Fix colonized specimens in 2.5% (v/v, in pH 7 PBS) EM-grade glutaraldehyde (Sigma) solution for 15 min at room temperature.

[44] R. Bambauer, P. Mestres, K. J. Pirrung, and P. Sioshansi, *Art. Org.* **18,** 272 (1994).

[45] S. E. Tebbs, A. Sawyer, and T. S. Elliott, *Br. J. Anaesthesia* **72,** 587 (1994).

[46] T. Arizono, M. Oga, and Y. Sugioka, *Acta Orthopaed. Scand.* **63,** 661 (1992).

[47] T. L. Wright, R. P. Ellen, J. M. Lacroix, S. Sinnadurai, and M. W. Mittelman, *J. Periodont. Res.* **32,** 473 (1997).

[48] T. R. Neu, G. J. Verkerke, I. F. Herrmann, H. K. Schutte, H. C. van der Mei, and H. J. Busscher, *J. Appl. Bacteriol.* **76,** 521 (1994).

2. Dehydrate the specimens through an ethanol drying series with pH 7 PBS as diluent: 25%, 5 min; 50%, 10 min; 75%, 10 min; 100%, 1 hr.
3. Immerse the fixed and dehydrated specimens in a 30:70 mixture (w/v, in 100% ethanol) of LR White medium-grade resin (Ted Pella Co., Redding, CA) for 30 min at room temperature. Disposable aluminum-weighing dishes (Fisher Scientific) can serve as embedding containers.
4. Immerse the specimens in a 70:30 mixture of the LR White resin for 30 min at room temperature followed by immersion in 100% LR White for 10 min at room temperature.
5. Remove the specimens to small glass jars and sparge with nitrogen for 5 min. Glass Ball canning jars are useful containers for this procedure.
6. Immediately seal the jars and place in a 65° oven for 12 hr.
7. At the end of the 12-hr curing period, a hard resin forms on the surface of the coupon. Separate the embedded biofilm from the

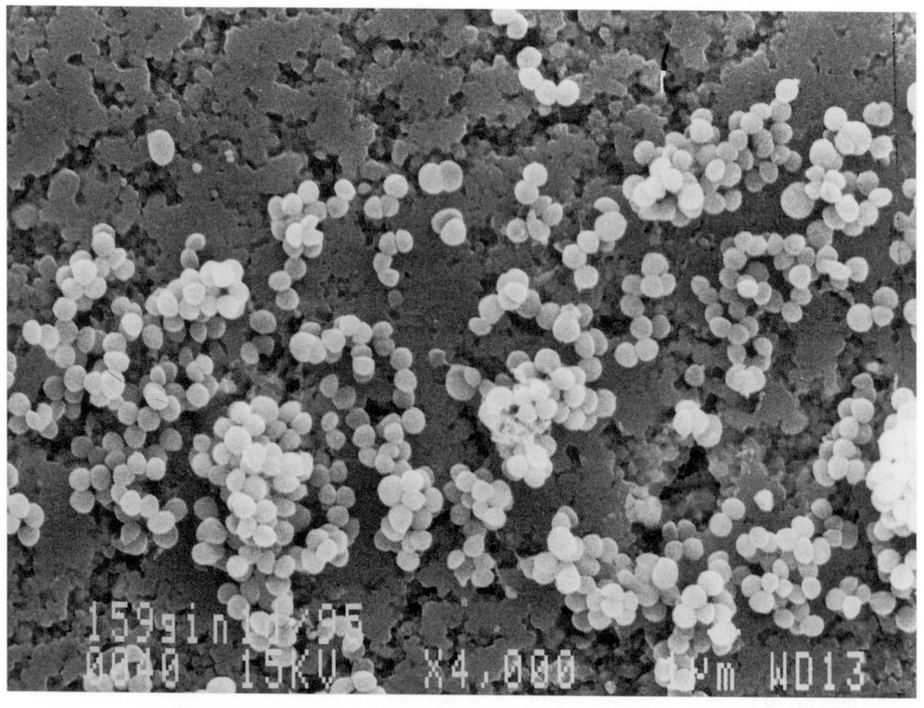

FIG. 7. SEM image of a *Porphyromonas gingivalis* biofilm associated with a hydroxylapatite coupon. Magnification: ×4000.

specimens by freezing at −80° for approximately 5 min, followed by immersion in room temperature deionized water.

8. Carefully pry off the resin from the specimens using a razor blade.
9. Thin section (1.2 μm sections) the resin on a microtome.
10. Stain the sections in 0.01 *M* lead citrate and saturated uranyl acetate solutions for 30 and 2 min, respectively.
11. Gently wash the sections in deionized water, affix to EM grids, and examine via TEM (e.g., Hitachi Model H-600, Tokyo).

Scanning Electron Microscopy

Significantly less work is required to prepare specimens for Scanning electron microscopy (SEM) examination. Embedding and thin sectioning—often tedious tasks—are not required with SEM. Specimens are fixed, critical point dried, sputter coated, and fixed to EM studs for examination. Scanning electron microscopy can provide useful information on relative numbers of biofilm bacteria associated with both polymeric and metallic substrata and on the influence of antimicrobials on bacterial and biofilm structure (Fig. 7).

1. Fix specimens in 2.5% (w/v) ice-cold glutaraldehyde (EM-grade) in 50 m*M* PBS, pH 7.2, for 1.5 hr.
2. Rinse specimens in three consecutive changes of distilled water.
3. Dehydrate specimens through an ethanol series as described earlier. Store, if required, at 4°.
4. Critical point dry.
5. Gold: palladium (60:40) sputter coat specimens to a thickness of about 200Å in a vacuum-coating device.
6. Examine under SEM (e.g., JSM-840, Jeol Ltd., Tokyo, Japan).

[40] Primary Adhesion of *Pseudomonas aeruginosa* to Inanimate Surfaces Including Biomaterials

By DONALD G. AHEARN, ROYA N. BORAZJANI, ROBERT B. SIMMONS, and MANAL M. GABRIEL

Introduction

The primary adhesion of bacteria to surfaces in governed by cell surface–surface and substrate–surface changes and hydrophobicities. Cells after initial contact may be desorbed from a surface by repulsive or shear forces

0076-6879/99 $30.00

or they may become irreversibly bound. The degree and strength of the "irreversible" bond appear to be strain-variable properties. Our studies have centered on "primary adhesion" to different "smooth" inanimate polymers, i.e., events of the first few hours of cell surface binding, by selected strains of bacteria. Cells at this time are mainly dispersed on the surface in a monolayer with occasional clumps. The deposition of exopolysaccharides and adhesion complex of coalesced cells in a glycocalyx has not yet formed, i.e., a mature biofilm with channels and "mushrooms" has not formed. Assessments of relative primary adhesion are of value in the selection of materials or coatings, particularly those containing bound or "insoluble" inhibitors.

Our studies have centered on a strain of *Pseudomonas aeruginosa* (GSU#3) that has been used in examining adhesion to contact lenses, intraocular lenses, and urinary catheters, as well as in assessment of the relative efficacies of various antimicrobials on adhered cells,[1–5] but any strain with consistent primary adhesion to materials such as Teflon, polyethylene, polyvinyl acetate, or silicone could be employed.

Methods

Resistance of Surfaces to Biofilm Formation

This protocol permits assessment of the relative susceptibility of surfaces to bacterial adhesion. The efficacy of antimicrobial coatings or bound inhibitors may be determined. The assays employ strains of gram-negative (e.g., *P. aeruginosa* or *Serratia marcescens*) and gram-positive (e.g., *Staphylococcus epidermidis* or enterococci) bacteria that have been documented for their ability to establish biofilms on standard surfaces (e.g., silicone, polyethylene, Teflon).

Bacteria (typically *P. aeruginosa* GSU#3 as a standard control) are grown overnight at 37° on Trypticase soy agar (BBL, Microbiology Systems, Cockeysville, MD). Cells from isolated colonies are inoculated into tryptic soy broth (Difco Laboratories, Detroit, MI) and incubated at 37° on a rotary shaker (150 rpm) for 12 to 18 hr. Cells are harvested by centrifugation at 2200*g* for 10 min and washed two times in phosphate-buffered saline

[1] D. G. Ahearn, L. L. May, and M. M. Gabriel, *J. Ind. Microbiol.* **15,** 372 (1995).

[2] M. M. Gabriel, A. D. Sawant, R. B. Simmons, and D. G. Ahearn, *Curr. Microbiol.* **30,** 17 (1995).

[3] M. M. Gabriel, M. S. Mayo, L. L. May, R. S. Simmons, and D. G. Ahearn, *Curr. Microbiol.* **33,** 1 (1996).

[4] M. J. Miller and D. G. Ahearn, *J. Clin. Microbiol.* **25,** 1392 (1987).

[5] D. L. Price, A. D. Sawant, and D. G. Ahearn, *J. Ind. Microbiol.* **8,** 83 (1991).

(PBS; 8.0 g NaCl, 0.2 g KCl, 1.44 Na_2HPO_4, 0.24 KH_2PO_4 in 1 liter distilled water) or in 0.9% saline. The cell pellet is adjusted in PBS or minimal broth to an optical density equivalent to a density of $\sim 2 \times 10^8$ cells/ml (in our laboratory we use an OD of 600 nm with a Sequaoia-Turner Model 340 spectrophotometer). The formula for the minimal broth is 1.0 g dextrose anhydrous, 7.0 g K_2HPO_4, 2.0 g KH_2PO_4, 0.5 g sodium citrate, 1.0 g $(NH_4)_2SO_4$, and 0.1 g $MgSO_4$ in 1 liter distilled water. The dextrose and $MgSO_4$ solutions are autoclaved separately. The cell suspension is incubated with shaking at 25° for 1 hr, and L-[3,4,5-^{3}H] leucine (NEN Research Products, Du Pont Company, Wilmington, DE) with a specific activity of >140 Ci/mM is added. The volume of leucine used is about 0.05% of the volume of the cell suspension (e.g., 20 μl leucine for 40 ml of cell suspension). After leucine addition, incubation is continued for 25 min. These cells are washed three times in 30 ml PBS or 0.9% saline and the pellets are suspended to an OD equivalent to 10^8 cells/ml.

The test samples (usually cylinders or cubes <10 mm in the largest dimension, including a "positive" silicone or polyethylene surface) are incubated with 3 ml of the radiolabeled cell suspension for 2 hr at 37° on a rotary shaker. The samples are removed from the cell suspension with forceps and immersed five times in each of three successive changes of >160 ml of saline (or deionized water for some microorganisms or surfaces). The samples are shaken free of excess saline and transferred to 20-ml glass scintillation vials. Ten milliliters Opti-Fluor scintillation cocktail (Packard Instrument Co., Downers Grove, IL) is added to each vial; the vials are agitated with a vortex mixer for 5 sec and counted in a liquid scintillation counter. A model dilution scheme is as follows: In PBS prepare the following dilution series on a sample of the radiolabeled cell suspension: 1:10, 1:50, 1:100, 1:500, and 1:1000. Transfer 100 μl of each dilution into scintillation vials containing 10 ml of Optifluor and measure the disintegrations per minute (dpm) of each with the scintillation counter.

Scintillation counts are converted to actual cell numbers with a calibration curve established from plate counts of dilutions of cell suspensions that may be subsequently labeled with leucine.

Viable Cell Counts. Prepare the following serial dilution of the cells suspended in PBS (a control dilution series could also be prepared with the radiolabeled cells): $1:10, 1:10^{-2}, 1:10^{-3}, 1:10^{-4}, 1:10^{-6}, 1:10^{-5}, 1:10^{-7}$, and $1:10^{-8}$. Spread plate or pour plate the last three dilutions onto TSA in duplicate and incubate for a time and at a temperature appropriate for the organism. Record the optical densities of the cell suspension. For most bacteria the cell densities are determined from plates with 30–300 colonies.

This protocol provides for comparison of the primary adhesion phase to inanimate "smooth" surfaces. The cells on these surfaces at 2 hr are not in a "biofilm" but are mostly single with small occasional clumps. Data obtained for these surfaces over short time intervals are similar for cells suspended in minimal medium or PBS. For the study of cells coalesced in a biofilm, we incubate the samples for extended time periods in either PBS or minimal broth inoculated with 10^8 cells/ml. Nonradiolabeled bacteria are allowed to adhere to the samples over specified time periods (4–24 hr) at 37° (temperature may be changed dependent on the bacterium under study). The samples are removed with sterile forceps and immersed three times in each of three successive changes (160 ml) of 0.9% saline or PBS. [^{3}H] Leucine (as described earlier) is added to fresh minimal broth in which the rinsed samples are incubated for an additional 30 min at 37° on a rotary shaker. After incubation, the samples are immersed five times in each of three successive changes of saline (160 ml). In most instances, only firmly attached cells remain on the test surface. If the adhesion phases are in excess of 4 hr with incubation in minimal broth, the firmly attached cells may include cells embedded in a glycocalyx. When substrates are incubated in PBS for 8–12 hr the cell numbers may increase only one- to threefold versus up to 3 logs in minimal medium. Samples with extended incubation periods are processed for liquid scintillation counting as described for the 2-hr primary adhesion assay. Surfaces exposed to the radiolabeled leucine without inoculation with microorganisms are needed as controls in all tests.

Statistical Analyses

For the 2- to 4-hr or longer time assays, repeat experiments are conducted each with five samples for each type of substratum to be compared. Adhesion data for the 2-hr assay may be converted from disintegrations per minute to colony-forming units (cfu) based on the standard calibration curves and expressed as cfu/mm^2 of surface area. For tests with extended incubation periods (>6 hr), particularly with minimal broth, data are expressed usually as dpm/mm^2. Once a glycocalyx has formed on a surface, the uptake of leucine may not be at the same rate as determined per cell for cells labeled prior to adhesion. Postlabeling of adhered cells provides relative data on surfaces (cell numbers in the attached biofilm may be estimated following direct counts of cells released following mild sonication in the presence of a nontoxic surfactant). All data from each test series are compared statistically. Typically we employ an unpaired "*t* test" and ANOVA (Sigma plot 4.0; Jandell Scientific, Sausalito, CA).

Scanning Electron Microscopy

Samples for scanning electron microscopy (SEM) are inoculated with the same inoculum preparation and are incubated under the same conditions as cells for the radiolabel studies. Samples are placed in glass scintillation vials and rinsed once (1 min) in sterile PBS prior to fixation. All fixing and rinsing solutions are filter sterilized immediately before use. Samples may be fixed in 3% glutaraldehyde in 0.1 *M* sodium cacodylate buffer for 90 min at room temperature and then rinsed three times (15 min each) in 0.1 *M* sodium cacodylate buffer. Postfixation is in 1% osmium tetroxide in 0.1 *M* sodium cacodylate buffer overnight at 4° (when appropriate). Samples are then rinsed in the same buffer three times (15 min each) and dehydrated in a graded ethanol series through 3× in 100%. Samples are rinsed with hexamethyldisilizane (HMDS, J. T. Baker, Phillipsburg, NJ) HMDS/ethanol, 1:1 for 1 min, HMDS (100%) 2 min [dimethoxypropane (Electron Microscopy Science, Ft. Washington, PA) may also be used at this step for the same dilutions and times], and allowed to air dry for 1 hr in glass petri dishes under a fume hood. The samples are transferred to a desiccation cabinet for storage. All samples are mounted on aluminum SEM stubs and sputter coated with 15–20 nm of gold palladium in a Denton Desk II sputter coater. SEM is carried out in a Leica Stereoscan S420 electron microscope operating at kV3-kV5. Images are stored as TIFF files in the microscope computer.

Model Experiment

Pseudomonas aeruginosa #3 adhered differentially to silicone with and without an antimicrobial surface coat. Data obtained in a representative experiment for a 2-hr adhesion study may be recorded as shown in the tabulation below.

Silicone (dpm)[a]	Antimicrobial-coated silicone (dpm)
11864	58741
19525	48274
17714	50038
15458	48857
16978	61034
Dilutions (cfu)[b]	**Dilutions (dpm)**
3.0×10^7	59933.500
1.5×10^7	30036.870
3.0×10^6	6092.412
1.5×10^6	2960.353
3.0×10^5	618.523

Slope $b(0) = -16845.9$
Intercept $b(1) = 500.6$
$y = b(1)\,x + b(0)$

Silicone (cfu)	Antimicrobial-coated silicone (cfu)
33093	163472
54401	134359
49362	139266
43089	135982
47316	169850

[a] Disintegrations per minute.
[b] Colony-forming unit.

These data may be graphed as in Fig. 1.

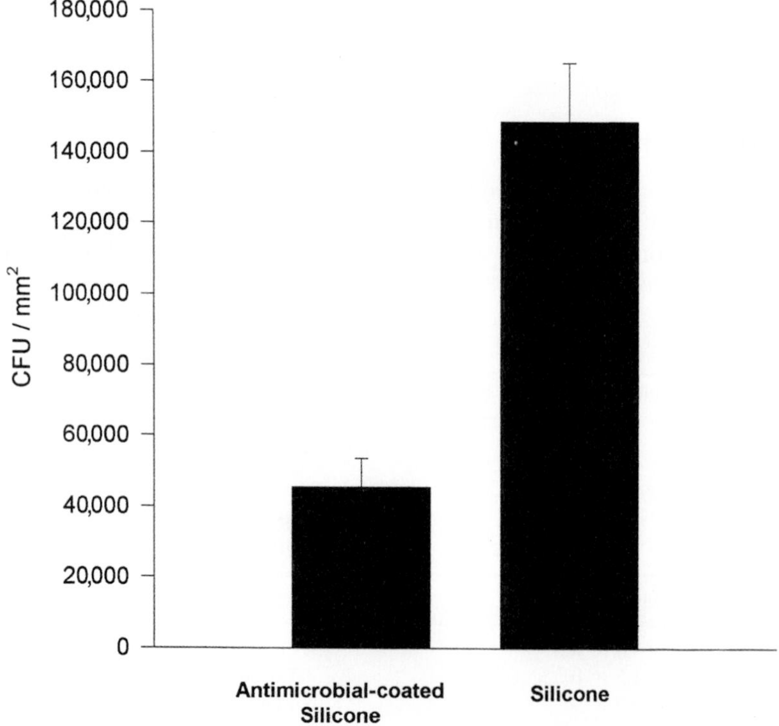

FIG. 1. Adhesion of *Pseudomonas aeruginosa* to biomaterials after 2 hr from PBS.

Discussion

We have examined at least 30 strains of *P. aeruginosa* in our studies of the primary adhesion of bacteria to inanimate materials. The strain, substratum, and other environmental conditions all interact in determining the degree of adhesion, with strain properties being dominant factors.

Interpretation of data for adhesion to most plastics and metals relates directly to reduced or enhanced adhesion versus a control surface. Interpretation of data for certain antimicrobial surfaces, however, requires consideration of mode of action of the inhibitor. Cells may selectively adhere to certain antimicrobial surfaces and a lag may occur before the antimicrobial properties are detected. In such instances, time studies with the described procedures may be required to detect the antimicrobial properties of delayed action coatings. The degree of solubility/mobility and the concentration of inhibitors embedded in the matrix of materials or coating are important considerations.[5] Some antimicrobial coatings with high solubilities in water provide short-term inhibition of adhesion and growth but lack long-term efficacy. These coatings give marked zones in agar diffusion tests. Coatings such as these may provide some efficacy during the implantation process for certain biomedical devices, but be essentially ineffective for other uses where long-term efficacy is required. Stability of the inhibitors in coatings and adsorption/desorption of additional inhibitory or neutralizing materials from the environment are also considerations for the evaluation of products from the field.

We have described a method with a 2-hr exposure period that permits comparison of five to seven separate materials with one microorganism by one individual in a day. Shorter times (5–10 min) of exposure of the surfaces to microorganism can be performed when fewer samples are handled. The shorter adhesion periods permit direct evaluations of materials from the field in the presence of other microorganisms. The preparation, experimentation, data collection, and interpretation for a single experiment involves 3–4 days.

The radiolabeled cell procedure permits reproducible relative comparisons of the affinity of microorganisms for surfaces. It also provides a method for assessing the relative activity of bound or low stability inhibitors. The test is predictive and may need supplementation with additional data from SEM observations, thin agar slurry challenges,[5] and recovery procedures, particularly for the evaluation of *in situ* materials.

[41] Biofilm and Biofilm-Related Encrustation of Urinary Tract Devices

By MICHAEL M. TUNNEY, DAVID S. JONES, and SEAN P. GORMAN

Introduction

Ureteral stents and urethral catheters fabricated from a range of polymeric biomaterials are used extensively to facilitate urine drainage in the upper and lower urinary tracts, respectively. Unfortunately, the ability of bacteria to colonize these biomaterial surfaces and form an adherent layer, or biofilm, composed of bacteria embedded in an organic matrix may be responsible for recurrent episodes of bacteriuria causing sepsis and fever.[1] Colonization of the biomaterial surface with urea-splitting bacteria such as *Proteus mirabilis, Proteus vulgaris,* and *Pseudomonas* spp. causes alkalinization of the urine, thus lowering the solubility of struvite (magnesium ammonium phosphate, $NH_4 \cdot MgPO_4 \cdot 2H_2O$) and hydroxyapatite [calcium phosphate, $Ca_{10}(PO_4)_6 \cdot H_2O$], which then become deposited on the device. The formation of encrusting deposits in association with bacterial biofilm is a recognized clinical occurrence. For example, of 40 retrieved stents examined by our group, 28% were found to have a profuse biofilm in association with encrustation.[2] The use of scanning electron microscopy (SEM) has also shown bacterial biofilms in close association with encrusting materials on both urethral catheters[3,4] and ureteral stents (Fig. 1).[5] The formation of these encrusting deposits may lead to obstruction or blockage of urine flow with the associated urine retention causing pain and distress to the patient. Management of encrustation formed on biomaterials in the urinary tract is difficult and, in many cases, necessitates removal of the device. Although encrusted catheters may be removed and replaced with relative ease, the removal of encrusted stents may require major surgery with associated trauma for the patient and increased cost.

The wide range of polymers used for the fabrication of urinary tract devices, including silicone, polyurethane, composite biomaterials, and a

[1] M. M. Tunney, S. P. Gorman, and S. Patrick, *Rev. Med. Microbiol.* **7,** 195 (1996).

[2] P. F. Keane, M. C. Bonner, S. R. Johnston, A. Zafar, and S. P. Gorman, *Br. J. Urol.* **73,** 687 (1994).

[3] M. Ohkawa, T. Sugata, M. Sawaki, T. Nakashima, H. Fuse, and H. Hisazumi, *J. Urol.* **143,** 717 (1990).

[4] D. J. Stickler, J. King, J. Nettleton, and C. Winters, *Cells Mater.* **3,** 315 (1993).

[5] M. M. Tunney, P. F. Keane, and S. P. Gorman, *Eur. J. Pharm. Sci.* **4,** S177 (1996).

0076-6879/99 $30.00

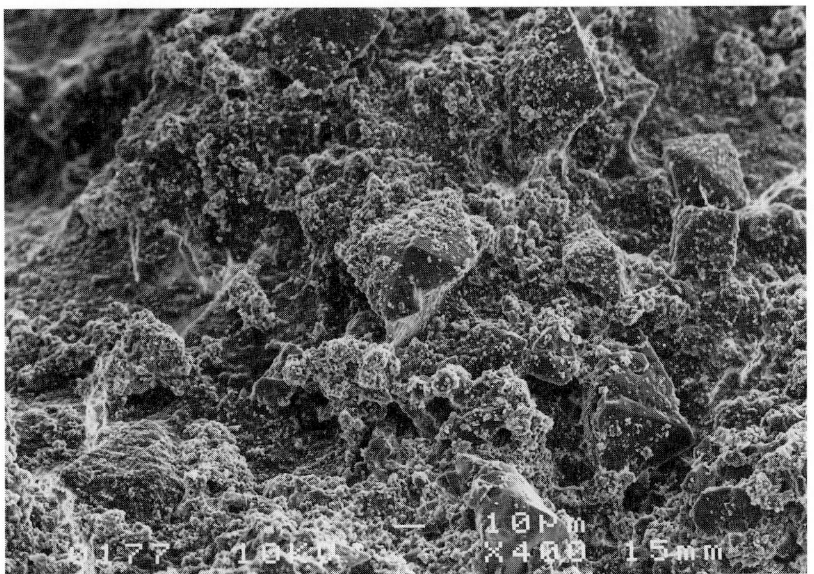

FIG. 1. Scanning electron micrograph showing large struvite crystals and small hydroxyapatite crystals formed in the presence of a bacterial biofilm on a retrieved ureteral stent.

range of hydrogel-coated biomaterials, suggests that no single biomaterial is more effective than the others and selection for clinical use appears to be on an *ad hoc* basis. Assessment of the ability of urinary tract biomaterials to resist encrustation is, therefore, beneficial when selecting biomaterials for use in the urinary tract and for evaluating the suitability of novel biomaterials for use in the urinary tract.

Encrustation of urinary tract biomaterials is assessed by the use of *in vitro* encrustation models based on either human or synthetic artificial urines. Human urine spiked with the urea-splitting microorganism *P. mirabilis,* which is frequently associated with catheter encrustation *in vivo,* has been used in several models.[6–8] However, conditions can be controlled more carefully and intra- and interexperimental variation eliminated by the use of an artificial urine with a constant composition. This article focuses on the development of artificial urine models that simulate *in vivo* encrustation resulting from device-related infection and can be used to assess bioma-

[6] M. J. Gleeson, J. A. Glueck, L. Feldman, D. P. Griffith, and G. P. Noon, *Trans. Am. Soc. Artif. Intern. Organs* **35,** 495 (1989).

[7] S. Sarangapani, K. Cavedon, and D. Gage, *J. Biomed. Mater. Res.* **29,** 1185 (1995).

[8] S. A. V. Holmes, C. Cheng, and H. N. Whitfield, *Br. J. Urol.* **69,** 651 (1992).

terial encrustation and the ability of biomaterials to resist intraluminal blockage.

Artificial Urine

The artificial urine used in all our artificial urine models consists of three solutions that are added separately to prevent the acidic precipitation of brushite.

Solution A: 0.76% (w/v) potassium dihydrogen orthophosphate, 0.36% (w/v) magnesium chloride hexahydrate, and 1.60% (w/v) urea

Solution B: 0.53% (w/v) calcium chloride hexahydrate and 0.20% (w/v) chicken ovalbumin

Solution C: 0.125% (w/v) jack bean urease Type IX

Urease is added to mimic the effects of urea-splitting bacteria, which increase the pH of the urine on release of ammonia from the hydrolysis of urea. The urine is maintained at a temperature of 37° and an atmosphere equilibrated to 5% (v/v) CO_2 to simulate physiological conditions in the urinary tract.[9]

"Static" Artificial Urine Model

This model involves a slight degree of content agitation relative to the other models described. The model is intended to mimic the urinary conditions prevailing in the bladder.

Reaction Vessel

The reaction vessel consists of a Perspex tank with a loose-fitting lid in which the artificial urine (5.16 liters; 2.5 liters solution A, 2.5 liters solution B, 160 ml solution C) is stirred by means of Teflon-coated metal stirrers. A plastic grid is positioned 80 mm above the base of the tank using Perspex columns attached to the walls of the tank and an aperture is cut from the center of the grid to allow solution exchange (Fig. 2). Using a siphon pump, 1 liter of urine is replaced on a daily basis with 500 ml each of solution A and B, and twice weekly 160 ml of urine is replaced with a similar volume of solution C.

Encrustation Development

Biomaterial sections (length, 50 mm) are suspended in the artificial urine using plastic-coated, color-coded paper clips, which allow several

[9] S. P. Denyer, M. C. Davies, J. A. Evans, R. G. Finch, D. G. E. Smith, M. H. Wilcox, and P. Williams, *J. Clin. Microbiol.* **28,** 1813 (1990).

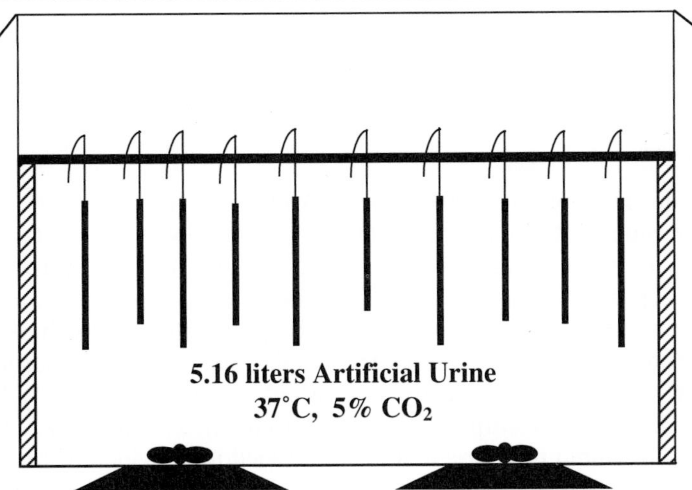

FIG. 2. Schematic diagram of the static artificial urine encrustation model.

biomaterials to be investigated concurrently. Sections are removed at 2, 6, and 14 weeks, rinsed gently in deionized water to remove albumin threads, and the type and quantity of encrustation then determined.

Encrustation Analysis

The type of encrustation produced on urinary tract biomaterials in this model is compared with that produced on stents *in vivo* and with the model encrustation components struvite, hydroxyapatite, and brushite using the following techniques:

Infrared Spectroscopy. One-milligram samples are ground with potassium bromide (20 mg; spectroscopic grade), compressed (10,000 kg) to a semitransparent disk, and the infrared spectra recorded in the wavelength range 4000 to 200 cm^{-1} using a Perkin-Elmer (Norwalk, CT) 983 spectrophotometer.

X-Ray Diffraction Spectroscopy. Samples are crushed into a fine powder, sprinkled onto a grease-smeared glass slide, and a diffraction scan obtained in the range 3° to 63° 2θ using a Siemens Diffraktometer D500.

Energy-Dispersive X-Ray Analysis. Samples are mounted on specimen stubs, coated using high vacuum carbon evaporation, and examined using a JEOL 733 Superprobe (JEOL Ltd, Akishima, Japan) to determine the concentrations of calcium, magnesium, and phosphorus present.

Analysis by these techniques has revealed that the encrustation produced using this model was composed primarily of struvite and hydroxyapatite and is similar in composition to that produced *in vivo*.[10]

Atomic Absorption Spectroscopy

Atomic absorption spectroscopy (AAS) is used to quantify the amount of magnesium and calcium deposited on biomaterials, which is taken as a measure of struvite and hydroxyapatite encrustation, respectively. Encrusted deposits are dissolved in acetic acid (8 ml, 1 *M*) using mild ultrasonication for a period of 4 hr. Each solution is made up to 50 ml with deionized water, filtered using a 0.20-μ filter, and the quantity of magnesium and calcium present determined by established techniques[11] using a Perkin-Elmer 2380 atomic absorption spectrophotometer. The quantity of magnesium and calcium present on a biomaterial section is divided by the surface area of that section to give the surface densities (in mg/cm^2) of magnesium and calcium present.

This model has been used to assess the encrustation resistance of five biomaterials currently used in the fabrication of ureteral stents and/or urethral catheters: silicone, polyurethane, hydrogel-coated polyurethane, Silitek™, and Percuflex™.[12] Silicone was least prone to struvite encrustation, followed by polyurethane, Silitek™, Percuflex™, and hydrogel-coated polyurethane in rank order. Silicone was also least prone to hydroxyapatite encrustation, followed by Silitek, polyurethane, Percuflex, and hydrogel-coated polyurethane in rank order.

Continuous Flow Artificial Urine Models

Continuous flow models representative of conditions in the urinary tract can also be used to evaluate the encrustation resistance of biomaterials when urine flow is continuous, as in the ureter, and to compare the ability of biomaterials to resist intraluminal blockage.

Modified Robbins Device Continuous Flow Model

The Robbins device (RD), which was first developed to study biofilm formation in industrial water systems and later modified (MRD) for biomedical experimentation, can be used to assess encrustation as part of a

[10] M. M. Tunney, M. C. Bonner, P. F. Keane, and S. P. Gorman, *Biomaterials* **17,** 1025 (1995).

[11] J. Bassett, R. C. Denney, G. H. Jeffrey, and J. Mendham, *in* "Vogels Textbook of Quantitative Inorganic Analysis," p. 837. Longman, London, 1978.

[12] M. M. Tunney, P. F. Keane, D. S. Jones, and S. P. Gorman, *Biomaterials* **17,** 1541 (1996).

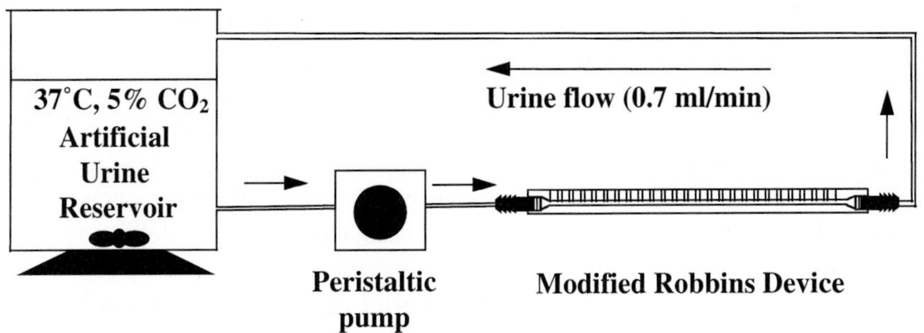

FIG. 3. Schematic diagram of the modified Robbins device continuous flow encrustation model.

continuous flow artificial urine model. The MRD is constructed of an acrylic block, 42 cm long with a 2 × 10-mm lumen. Twenty-five evenly spaced sampling ports are designed so that biomaterial disks attached to sampling plugs using a rubber backing sheet lie flush with the inner surface and do not disturb flow. Disks (diameter, 10 mm) of the biomaterials to be tested are attached to the sampling plugs, which are placed in the sampling ports. Artificial urine is pumped for the required period of time at a rate of 0.7 ml/min from a reservoir through the MRD and back to the reservoir, thereby simulating natural urine flow (Fig. 3). The disks are then removed, rinsed in deionized water, and the quantity of magnesium and calcium present in the encrusted deposits determined by AAS.

This model has been used to assess encrustation on silicone and polyurethane wherein less struvite and hydroxyapatite were deposited on the surface of silicone than on the surface of polyurethane.[13]

Dynamic Encrustation Model

The dynamic encrustation model consists of a series of glass reaction vessels and a pumping system to circulate the artificial urine around the vessels.[14] A 5-liter tank of artificial urine (as described) is used as a reservoir from which urine is pumped through silicone tubing (inner diameter, 7.7 mm). To set up the system the vessels are filled with urine from the reservoir using a plastic syringe. Gate clamps are used to temporarily seal the tubing leading from the vessels to prevent the backflow of urine from the vessels as they are being clamped into position. The inlet tube, pump manifold

[13] M. M. Tunney, P. F. Keane, and S. P. Gorman, *J. Biomed. Mater. Res.* (*Appl. Biomater.*) **38,** 87 (1997).

[14] C. P. Garvin, D. S. Jones, S. P. Gorman, and F. Quigley, *J. Pharm. Pharmacol.* **50,** 164 (1998).

tubing, and tubing leading to the vessels are filled with artificial urine from the reservoir by turning on the pump. The gate clamps are removed when the urine has been circulated to the clamps. A pump speed of 10 ml/min is required to maintain the urine level in the reaction vessel, therefore, the urine flow rate is monitored and the pump speed adjusted if required. A series of glass reaction vessels are arranged in parallel with each vessel clamped to a vertical steel pole. Each vessel has a separate inflow and outflow of urine and all outflowing urine is pumped back into the reservoir. Biomaterial sections to be tested are secured equidistant along each stainless steel mandrel (diameter, 0.95 mm) using cyanoacrylate adhesive with three sections of biomaterial per mandrel and three mandrels per vessel (Fig. 4). The mandrels are secured into size 27 rubber stoppers at each end of the vessel. Identification of each vessel and each mandrel may be by, for example, assignment of a letter A, B, C etc. and number 1, 2, or 3, respectively. The reaction vessels, pumping system, and reservoir are placed in an incubator at 37° to mimic physiological temperature. On a daily basis 1 liter of artificial urine is replaced. The operation of the model is checked daily to ensure that urine flows freely by collection of urine from the outlet tube over a 5-min period. Any crystalline material deposited in the tubing is also cleared daily to prevent blockages.

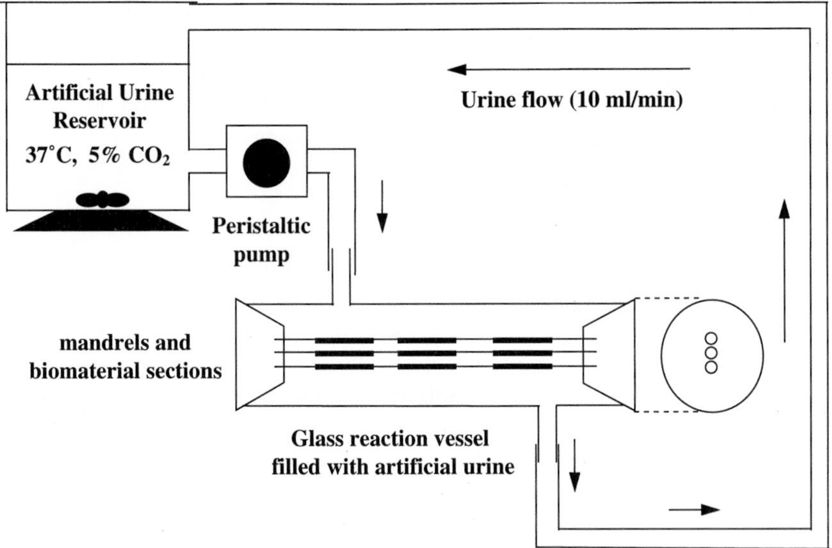

FIG. 4. Schematic diagram showing the reaction vessel and the urine pumping system in the dynamic encrustation model.

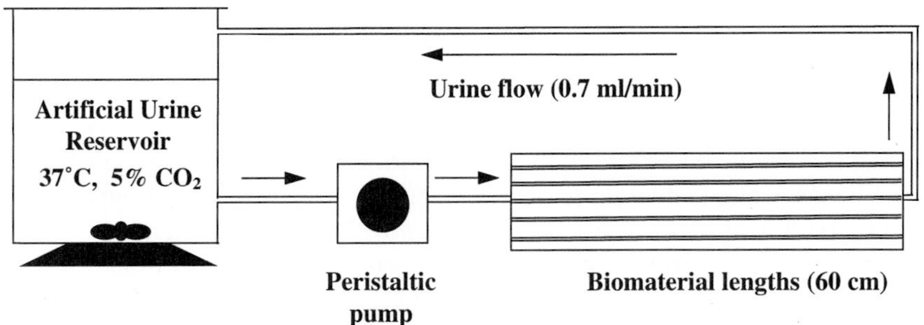

FIG. 5. Schematic diagram of the simulated urine flow encrustation model.

Validation of the models ensured that sections of the same biomaterial became similarly encrusted when placed in any position in any vessel.

Simulated Urine Flow Model

This model is similar in design to the MRD continuous flow model described previously with the artificial urine being pumped from a reservoir at a rate of 0.7 ml/min through 60-cm lengths of urinary tract biomaterials and back to the reservoir. As a control, deionized water is pumped at the same rate through similar lengths of the biomaterials (Fig. 5). The biomaterials are checked on a daily basis to ensure that urine is still free flowing, and when urine flow through a biomaterial is prevented by encrustation blocking the lumen, the biomaterial length and the corresponding control length are removed from the model. These biomaterial lengths are then dried for 7 days in a vacuum desiccator, after which the total weight of encrustation in each is calculated by subtracting the weight of the control section from the weight of the encrusted section. This model has been used to assess intraluminal encrustation and blockage of several urinary tract biomaterials. Polyurethane remained patent for the longest period of time followed by Percuflex™, silicone, Silitek™, and hydrogel-coated polyurethane in rank order.[15]

Conclusion

The extent to which different urinary biomaterials will encrust *in vivo* is a problem that needs to be addressed if new, less encrusting biomaterials are to be discovered. The optimal level of agitation within a model will be

[15] M. M. Tunney, P. F. Keane, and S. P. Gorman, *J. Pharm. Pharmacol.* **49,** 1102 (1997).

determined by the portion of the urinary tract in which the test biomaterial is destined to reside. For example, there will be more urine flow in the ureters than in the more stagnant portions of the bladder. Therefore, a biomaterial being tested for use as a ureteral stent should be tested in more dynamic conditions to mimic the flow of urine over the biomaterial surface *in vivo*. Models such as those described within, especially where urine flow is involved, will prove useful for testing novel biodegradable biomaterials for use as urinary prostheses. In practice, urine flow may encourage surface layers of the degradable polymer to be shed, thereby assisting the removal of encrusting deposits and adherent microorganisms. Such biomimetic biomaterials may have a role to play in the development of improved urinary tract devices.

[42] Detection of Prosthetic Joint Biofilm Infection Using Immunological and Molecular Techniques

By Michael M. Tunney, Sheila Patrick, Martin D. Curran, Gordon Ramage, Neil Anderson, Richard I. Davis, Sean P. Gorman, and James R. Nixon

Introduction

Total hip replacement is one of the most successful and cost-effective surgical operations ever devised, with over 50,000 hip replacements performed annually in the United Kingdom and 200,000 in the U.S. The majority of patients who undergo hip replacement experience dramatic relief of preoperative pain and restoration of satisfactory hip function.[1] A proportion (approximately 20% in Europe)[2] fail, and prosthesis removal and replacement (revision hip arthroplasty) is usually required with further trauma for the patient and increased cost to the Health Service.[3] Studies have reported that between 2[4] and 15%[5] of all revision operations result from infection of the implant. Unfortunately, the rate of infection is higher after revision hip arthroplasty than after primary procedures, with as many as 40% of revised hip joints failing due to infection.[6] It has been suggested

[1] R. H. Fitzgerald, *Orthop. Clin. N. Am.* **23,** 259 (1992).
[2] P. Christel and P. Djian, *Curr. Opp. Rheum.* **6,** 161 (1994).
[3] C. R. Dreghorn and D. L. Hamblen, *Br. Med. J.* **298,** 648 (1989).
[4] R. L. Barrack and W. H. Harris, *J. Bone Joint Surg.* **75,** 66 (1993).
[5] P. F. Lachiewicz, G. D. Rogers, and H. C. Thomason, *J. Bone Joint Surg.* **78,** 749 (1996).
[6] J. A. Dupont, *Clin. Orthop. Relat. R* **211,** 122 (1986).

0076-6879/99 $30.00

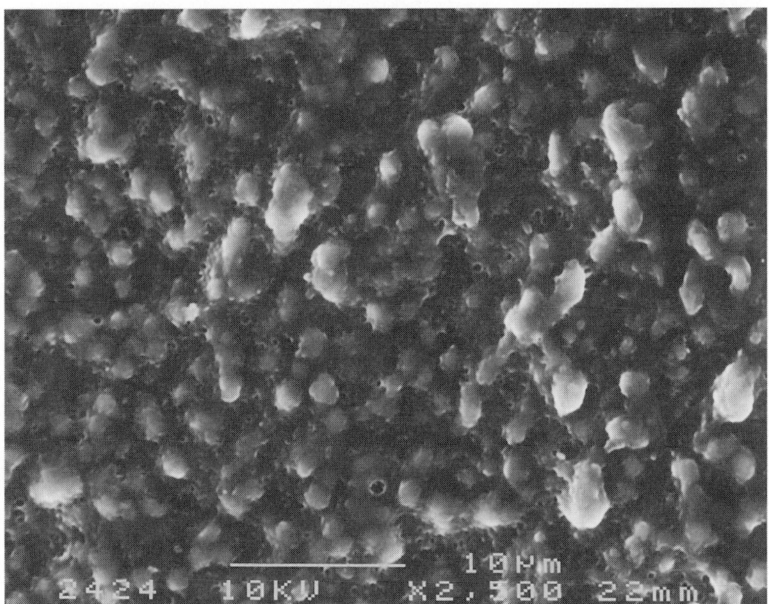

FIG. 1. Scanning electron micrograph showing bacterial biofilm on a retrieved orthopedic implant.

that this higher rate of infection postrevision may be due to unrecognized infection at the initial revision operation.[6] This may be because bacteria colonizing the surface of implanted biomaterials grow predominantly in adherent biofilms[7] (Fig. 1) and may not be detected by aspiration and routine culture techniques, which fail to examine the retrieved prostheses.[8] Additionally, given the well-proven incidence of anerobic bacteria in joint infection,[9] it is likely that this is also due to the fastidious culture requirements of anaerobic bacteria.

This article describes how the detection of bacteria by culture from revision hip prostheses can be improved by employing strict anaerobic techniques and by using mild ultrasonication to dislodge bacteria adhering to the surface of the retrieved implants. Nonculture techniques that can be used to further improve the detection of infection, including examination of the inflammatory response in implant-associated tissue and the detection of bacteria using immunological and polymerase chain reaction (PCR)-based molecular approaches, are also described.

[7] M. M. Tunney, S. P. Gorman, and S. Patrick, *Rev. Med. Microbiol.* **7,** 195 (1996).

[8] A. G. Gristina and J. W. Costerton, *J. Bone Joint Surg.* **67,** 264 (1985).

[9] I. Brook and E. H. Frazier, *Am. J. Med.* **94,** 21 (1993).

Improved Culture Technique for Detection of Infection

Current clinical laboratory practice for the detection of prosthetic hip infection involves the culture of bacteria from joint fluid aspirated peroperatively and from tissue samples removed at the time of surgery. A previous study has shown that mild ultrasonication of retrieved tissue, bone, and cement can increase the number of bacterial isolates cultured compared with the numbers cultured from joint fluid or swabs of excised tissue and prosthetic surfaces.[8]

Sample Collection

1. When removed from the patient, the femoral and acetabular components of the prosthetic hip are placed aseptically in separate sterile bags.
2. Tissue in contact with the implants is also removed and placed in sterile bottles.
3. The femoral and acetabular components of the prostheses and the tissue samples are placed immediately in an anaerobic jar (e.g.,Oxoid Anaerobic Jar HP11, Unipath Ltd., Basingstoke, UK) containing a catalyst (e.g., Oxoid Low Temperature Catalyst BR42, Unipath Ltd.) and a Gaspak (e.g., Oxoid Gas Generating Kit BR038B, Unipath Ltd.) is added immediately.
4. The jar is transported to the laboratory and placed in an anaerobic cabinet (e.g., Don Whitley Mk III, Don Whitley Scientific Ltd., Shipley, UK) containing an atmosphere of 80% N_2, 10% H_2, and 10% CO_2 (v/v/v), where it is opened.

Sample Processing

The two parts of the prostheses are processed independently.

Prostheses

1. Ringer's solution (25%, v/v) is prereduced by the addition of L-cysteine (1%, v/v) and by the subsequent removal of oxygen by either boiling at a temperature of 100° for 1 hr in a water bath or microwaving at full power (650 W) for 5 min. On cooling, the Ringer's solution is stored in the anaerobic cabinet prior to use.
2. Separate 100-ml volumes of prereduced Ringer's solution are added to the sterile bags containing the prostheses.
3. The prostheses are removed from the cabinet inside the sealed bags and any material attached to the prosthesis surface is dislodged into the Ringer's solution using mild ultrasonication (5 min in a 150-W

ultrasonic bath operating at a nominal frequency of 50 Hz). Ringer's solution is removed from the sealed bags and is placed in sterile bottles that are immediately transferred to the anaerobic cabinet. It is essential that this part of the procedure be performed quickly and efficiently to minimize exposure to air of any anaerobic bacteria present.

4. Total viable bacterial counts are performed by spreading 0.5-ml volumes of sonicate onto five blood agar [BA; 40 g/liter, Oxoid blood agar base No. 2, Unipath Ltd.; 5% (w/v) Oxoid horse blood, Unipath Ltd.] and five anaerobic blood agar (ABA; 47 g/liter, GIBCO anaerobic blood agar base, GIBCO Ltd., Paisley, UK) plates, which are stored in the anaerobic cabinet prior to use.
5. To enhance detection of bacteria that may be present in low numbers, two 5-ml volumes of each sonicate are concentrated using separate Millipore Sterifill units [Millipore (UK) Ltd., Watford, UK] and separate 0.22-μm filters (Supor-200 membrane filters, Gelman Sciences, Ann Arbor, MI), which are then placed on both a BA and an ABA plate.
6. The remaining 85 ml of each sonicate is centrifuged for 20 min at 10,000*g* and the pellet produced is resuspended in 3 ml Ringer's solution, which is then dispensed into 1-ml volumes that are stored at $-70°$.

Tissue

1. A known weight of tissue is homogenized for 3 min in 5 ml prereduced Ringer's solution using a mechanical homogenizer, and total viable bacterial counts are performed by spreading 0.5-ml volumes of homogenized tissue onto three BA and three ABA plates.
2. To detect bacteria present in low numbers, the remaining volume of homogenized tissue is added in equal amounts to tryptone soya broth (TSB; 30 g/liter, Oxoid tryptone soya broth, Unipath Ltd.) and cooked meat broth (CMB; 100 g/liter, Oxoid CM 81, Unipath Ltd.) for enrichment.

Bacterial Culture and Identification

BA and ABA plates are incubated at 37° aerobically and anaerobically, respectively, and examined after 1, 2, 4, and 7 days. The TSB and CMB are incubated at 37° aerobically and anaerobically, respectively, and subcultured onto both BA and ABA after 7 and 14 days. These plates are then incubated either aerobically or anaerobically at 37° for another 7 days.

All bacteria cultured are counted to allow direct quantitation of the number of infecting bacteria. Pure cultures of any bacteria isolated are gram stained and identified using commercially available biochemical test galleries (e.g., API20A, APIStaph, Biomerieux, Paris, France).

Nonculture Techniques for Detection of Infection

Histological Examination of Tissue

Assessment of histopathological reactions in tissue samples can be quantified easily and it has been shown that the presence of acute or chronic inflammation in tissue samples correlates well with culture results and can, therefore, be used to aid diagnosis of infection.[10–12]

1. Selected tissue samples are fixed in formal saline histological fixative (Gurr, BDH, Davidson and Hardy Ltd., Belfast, UK). Representative samples are then processed and embedded in paraffin wax, and 5-μm sections are cut and stained using a hematoxylin and eosin stain.
2. All slides are subsequently assessed microscopically without prior knowledge of culture results using the following technique. The hematoxylin and eosin stained sections are scanned at low power (×100 objective) and the areas of the slide containing the heaviest inflammatory infiltrate are selected for further examination at a higher magnification (×400 objective). If the sample is considered satisfactory, the number of inflammatory cells per high power field is assessed over five randomly selected fields.
3. The following parameters are assessed in each tissue sample:

   a. Acute inflammatory response (infiltration with polymorphonuclear leukocytes).

   b. Chronic inflammatory response (infiltration with lymphocytes and tissue macrophages).

For each type of inflammatory cell, the slide is graded according to the following scheme:

0 absent
1 1–10 cells/high power field
2 10–20 cells/high power field
3 20 or more cells/high power field

[10] T. K. Fehring and J. A. McAlister, *Clin. Orthop. Relat. R* **304,** 229 (1994).
[11] J. M. Mirra, H. C. Amstutz, M. Matos, and R. Gold, *Clin. Orthop. Relat. R* **117,** 221 (1976).
[12] W. J. Kraemer, R. Saplys, J. P. Waddell, and J. Morton, *J. Arthroplasty* **8,** 611 (1993).

Molecular Detection of Infection

The detection of bacterial 16S ribosomal RNA (rRNA) genes as an indicator of the presence of bacteria is a recognized technique that has been used for the detection of both environmental and medically important bacteria.[13,14] Because bacterial rRNA gene sequences contain regions that are conserved within the prokaryotes and regions that are unique to different bacterial genera and species, they provide an ideal tool for the specific detection and identification of bacteria. The availability of a growing database of rRNA sequences and the ability to amplify rRNA-encoding genes (rDNA) from the limited amount of material that is available for analysis via PCR further enhance the suitability of rRNA sequences for detection and subsequent identification of bacteria.[15]

DNA Extraction

All steps are carried out in 1.5-ml Eppendorf tubes.

1. Centrifuge a 1-ml amount of prosthesis sonicate at 10,000*g* for 15 min at room temperature. Discard the supernatant.
2. Lysozyme (1–2 mg/ml) in 10 m*M* Tris (pH 8.0) and 50 m*M* glucose can also be used to help in the extraction but attention to contamination with exogenous bacterial DNA must be taken into account.
3. Suspend pelleted cells by vortex mixing in 200 μl cell lysis buffer [containing 10 m*M* Tris (pH 8.0), 5 m*M* EDTA, 0.5% sodium dodecyl sulfate (SDS), and proteinase K (100 μg/ml)] and incubate at 55° for 3 hr.
4. Adjust the temperature to 37° and leave reaction mixture overnight.
5. Increase the temperature of reaction mixture to 55° for 1 hr.
6. Add an equal volume of saturated phenol/chloroform, vortex thoroughly, and centrifuge at 10,000*g* for 15 min at room temperature.[16] Repeat this step if necessary.
7. Remove the aqueous supernatant to a new tube, being careful to leave the interface behind.
8. Add ethanol (100%, 2.5 volumes), 3 *M* sodium acetate, pH 5.2 (0.1 volumes), and 2 μl of See DNA® (Pharmacia Biotech, Piscataway, NJ). Vortex briefly and centrifuge at 10,000*g* for 15 min at room temperature.

[13] K. Griesen, M. Loeffelholz, A. Purohit, and D. Leong, *J. Clin. Microbiol.* **32,** 335 (1994).

[14] M. J. Wilson, A. J. Weightman, and W. G. Wade, *Rev. Med. Microbiol.* **8,** 91 (1997).

[15] T. M. Schmidt and D. A. Relman, *Methods Enzymol.* **235,** 205 (1994).

[16] T. Maniatis, E. F. Fritsch, and J. Sambrook, "Molecular Cloning: A Laboratory Manual." Cold Spring Harbor Laboratory, Cold Spring Harbor, NY, 1982.

9. Pour off supernatant and wash the pelleted DNA with 1 ml of 70% (v/v) ethanol. Vortex briefly and centrifuge at 10,000*g* for 15 min at room temperature.
10. Remove the supernatant and dry the pellet in a vacuum dryer (DNA Plus, Heto Laboratory Equipment).
11. Dissolve extracted DNA in 50 μl of TE [10 m*M* Tris (pH 8.0), 1 m*M* EDTA] buffer and store at $-20°$.

Oligonucleotide Primer Design

Using DNAstar computer software (Megalign), alignments of eubacterial 16S rRNA gene sequences can be constructed. From these it is possible to improve the range of previously published broad-range PCR primer pairs[17] by adding in a number of degeneracies in the primers. An example of primer sequences that can be used are as follows:

D1: 5′-GAG GAA GGT RGG GAY GAC GT
D2: 5′-AGG CCC GGG AAC GYA TTY ACC G
R = AG, Y = CT.

Both primers flank a hypervariable region that aids in 16S rRNA microbial identification.

Polymerase Chain Reaction Amplification

The PCR mixture is made up to 50 μl in sterile water and contains 5 μl of 10× PCR buffer (Perkin-Elmer, Norwalk, CT), 5 μl $MgCl_2$ (25 m*M*), 200 μM of each deoxynucleotide triphosphate (Pharmacia Biotech, Piscataway, NJ), 20 p*M* of each primer, 3 units of Ampli*Taq* polymerase (Perkin-Elmer Corporation, UK), and 2 μl of lysate containing target DNA.

The typical PCR run profile is a 5-min denaturation of 96°, followed by 30 cycles of 1 min at 96°, 2 min at the annealing temperature of 55°, and 1 min at 72°. The final cycle is ended with a 5-min extension at 72°, and the reaction is then held at 15° until the tubes are removed from the thermocycler (Perkin-Elmer Gene-Amp PCR System 9600; Perkin-Elmer Corporation, UK). After amplification, 6 μl of the amplified product is run on a 1.5% (w/v) agarose gel in 1× Tris–borate–EDTA (TBE). DNA bands are detected by ethidium bromide staining and visualized by UV light photography.

The sensitivity of the extraction procedures should be investigated by extracting DNA from 10-fold serial dilutions of pure cultures and subse-

[17] M. N. Widjojoatmodjo, A. C. Fluit, and J. Verhoef, *J. Clin. Microbiol.* **33,** 2601 (1995).

quently determining the limits of detection following PCR amplification and agarose gel electrophoresis.

Single-Stranded Conformational Polymorphism Electrophoresis

Single-stranded conformational polymorphism (SSCP) is a simple method that can be used to detect nucleotide sequence changes in PCR products.[18] It is ideally suited for fingerprinting the hypervariable sequences of the 16S rRNA of eubacteria, providing a rapid preliminary identification for any PCR-positive samples. It also allows the detection of polymicrobial infections, as more than one SSCP profile will be apparent. The following is a brief outline of the procedure.

1. After thermal cycling, 4 μl of the PCR mixture is added to 11 μl of sequencing sample buffer [5 m*M* EDTA (pH 8.0), 0.25% (w/v) bromphenol blue, in deionized formamide] and heated for 5 min at 96°.
2. The denatured DNA is then placed directly on ice for 10 min before being applied to the gel in 10-μl volumes. Bacterial standards are also applied to the gel for direct comparison. The optimal gel composition is 0.5× mutation detection enhancement gel (Flowgen, Shenstone, Staffordshire, UK), 0.6× TBE, 0.04% (w/v) ammonium persulfate solution, and 0.004% (v/v) *N,N,N′,N′*-tetramethylethylenediamine.
3. Electrophoresis is performed at 23° on a Sequi-Gen, vertical gel electrophoresis apparatus (Bio-Rad, Richmond, CA) for 5 hr at 20 W constant power.
4. After electrophoresis, SSCP patterns are detected by silver staining according to the method of Bassam *et al.*[19] Briefly, the procedure is as follows.

a. Gels are fixed in 10% (v/v) glacial acetic acid for 30 min at room temperature and washed with deionized water four times for 2 min each.

b. Gels are then color impregnated for 30 min at room temperature with 0.1% (w/v) silver nitrate and 0.056% (v/v) formaldehyde.

c. Gels are then washed for 30 sec with deionized water prior to color development. Color development is for 2 to 10 min with a mixture of 30 g/liter sodium carbonate, 0.056% (v/v) formaldehyde, and 4 mg/liter sodium thiosulfate. The color reaction is stopped by the addition of 10% (v/v) glacial acetic acid.

[18] J. R. Kerr and M. D. Curran, *Clin. Mol. Pathol.* **49,** 315 (1996).

[19] B. J. Bassam, G. Caetano-Anolles, and P. M. Gresshoff, *Anal. Biochem.* **196,** 80 (1991).

DNA Sequencing

For definitive identification of bacteria, the PCR amplification products are sequenced directly by the dideoxynucleotide chain termination procedure using the PRISM Ready Reaction Terminator Cycle Sequencing Kit and read using a Model 373A automated sequencer (Applied Biosystems Inc.).

Immunological Detection of Infection

Direct immunological detection of bacteria in clinical samples can be achieved by the use of monoclonal antibodies (MAbs) and polyclonal antiserum prepared against the bacteria implicated in the clinical infection.[20]

Detailed methods describing the production of polyclonal antiserum and MAbs can be found in the laboratory manual of Harlow and Lane[21] and the textbook of Goding.[22] A brief description of both techniques follows.

Production of Polyclonal Antiserum

A New Zealand White rabbit is immunized with whole cells of bacteria. The rabbit is inoculated subcutaneously at four sites on the back with 0.1 ml of a bacterial suspension of 1×10^8 cfu (colony-forming units)/ml in 0.01 *M* phosphate-buffered saline (PBS). A further two inoculations of bacteria in PBS are made at approximately monthly intervals and the rabbit is test bled 2 weeks after the final booster dose. The blood is allowed to clot at 37° for 1 hr and contract at 4° overnight after which the serum is tested by immunofluorescence microscopy.

Production of MAbs

A BALB/c mouse is immunized by whole bacterial cells. The mouse is inoculated intraperitoneally with 0.2 ml of a bacterial suspension of 1×10^8 cfu/ml in 0.01 *M* PBS. A further inoculation of 0.2 ml is given 4 days before the mouse is killed. Spleen cells from the mouse are fused with P3X 63 Ag8-653 (NS-0/1) mouse myeloma cells by treatment with 50% polyethylene glycol 16000 (Sigma, Dorset, UK) in RPMI 1640 (Flow Labo-

[20] S. Patrick, L. D. Stewart, N. Damani, K. G. Wilson, D. A. Lutton, M. J. Larkin, I. Poxton, and R. Brown, *J. Med. Microbiol.* **43,** 99 (1995).

[21] E. Harlow and D. Lane, "Antibodies: A Laboratory Manual." Cold Spring Harbor Laboratory, Cold Spring Harbor, NY, 1988.

[22] J. W. Goding, "Monoclonal Antibodies: Principle and Practice." Academic Press, London, 1986.

ratories, Paisley, UK) using a modification of the method described by Galfre and Milstein.[23] Hybrid cell lines are selected with hypoxanthine–aminopterin–thymidine in RPMI 1640 medium (GIBCO, Grand Island, NY) containing 20% myoclone fetal calf serum (GIBCO). Culture supernatants are removed and screened by immunoblotting and immunofluorescence for IgG specific for the bacteria used for inoculation.

Immunofluorescence Microscopy

For further information on this technique, readers are directed to the detailed description of the technique provided by Patrick and Larkin.[24] The method in brief is as follows.

1. One-milliliter samples of prosthesis sonicate are centrifuged (10,000*g*, 15 min, room temperature) and the resulting pellets suspended in 100 μl PBS.
2. Samples (10 μl) are then applied in duplicate to multiwell slides. The slides are air-dried and fixed in methanol (100%, 10 min) at $-20°$, after which they may be stored for up to 6 months at $-20°$.
3. Samples are then examined by dual labeling using the following procedure:

a. Incubate the slides in a humidified box with undiluted MAb supernate for 45 min.

b. Wash slides briefly with a wash bottle containing PBS and then wash for 30 min in a bath containing PBS.

c. Incubate slides with polyclonal rabbit antiserum diluted in PBS for 45 min and repeat washing step.

d. Incubate slides for 45 min simultaneously with goat antimouse rhodamine conjugate (Sigma, Poole, UK) and sheep antirabbit fluorescein conjugate (Sigma), both diluted in PBS as recommended by the manufacturer.

e. After a final wash, mount slides with glycerol-PBS containing an antiphotobleaching agent (e.g., Citifluor, Agar Scientific Ltd., Essex, UK) and examine using a fluorescence microscope (e.g., Leitz fluorescence microscope) with filters suitable for examining fluorescein and rhodamine separately and in combination.

[23] G. Galfre and C. Milstein, *Methods Enzymol.* **73,** 1 (1981).

[24] S. Patrick and M. J. Larkin, *in* "Microbial Biofilms: Formation and Control" (S. P. Denyer and S. P. Gorman, eds.), p. 109. Blackwell Scientific Publications, Oxford, 1993.

4. For detection of bacteria on each well by immunofluorescence microscopy, a score is given of between 0 and 3 using the following criteria:

0	absent
1	1–10 bacteria/well
2	10–50 bacteria/well
3	50 or more bacteria/well

Concluding Remarks

The use of strict anaerobic techniques and mild ultrasonication described herein has resulted in bacteria being cultured from 26 of 120 (22%) retrieved implants examined.[25] Review of the notes from 18 of these 26 individuals revealed that infection prior to revision was only suspected in 6 cases and that in only 2 of these cases were bacteria cultured from preoperative aspirates or tissue removed at the time of surgery using conventional microbiological techniques.

We have also used the nonculture techniques described in this article to detect infection of retrieved prosthetic hip joint.[26] Results obtained using these techniques confirmed those obtained using bacterial detection by culture following mild ultrasonication of the implants, with all culture-positive samples immunofluorescence microscopy-positive and positive for the presence of bacterial 16S rRNA genes and associated tissue samples positive for the presence of inflammatory cells indicative of infection.[25] Their use also revealed the presence of bacteria in a large number of culture-negative samples, with 68% having detectable bacterial rRNA gene sequences, 53% having a positive immunofluorescence result, and 87% of associated tissue samples positive for the presence of inflammatory cells indicative of infection.[25] The use of these nonculture techniques to improve the detection of prosthetic joint biofilm infection coupled with appropriate postoperative antibiotic therapy should improve the clinical outcome for patients undergoing revision hip surgery.

[25] M. M. Tunney, S. Patrick, S. P. Gorman, J. R. Nixon, N. Anderson, R. I. Davis, D. Hanna, and G. Ramage, *J. Bone Joint Surg.* **80,** 568 (1998).

[26] M. M. Tunney, S. Patrick, M. D. Curran, G. Ramage, D. Hanna, J. R. Nixon, S. P. Gorman, R. I. Davis, and N. Anderson, *J. Clin. Microbiol.* **37,** in press (1999).

[43] *In Vitro* and *in Vivo* Models of Bacterial Biofilms

By HIROSHI YASUDA, TETSUFUMI KOGA, and TAKASHI FUKUOKA

Introduction

Bacterial biofilms are found in various places in nature and in industrial environments. To maintain coexistence between humans and bacteria, bacteria construct normal microbiota by forming biofilms in parts of the body such as the surface of the teeth. However, biofilm formation in the human body by pathogenic bacteria often causes serious problems. Extensive bacterial biofilms are found on inserted catheters and cannulas,[1–3] cardiac pacemakers,[4,5] wound drainage tubes,[2] artificial joints,[6,7] and infected tissues of the lungs, trachea, urinary tract, and other organs.[8–10] Figure 1 shows a typical biofilm mode of bacterial growth on the surface of a stent detained in the respiratory tract of a patient with lung cancer. Biofilms in a living body generally take on a more complex structure than those found in nature or industrial environments. This is partially because fibronectin that originates from the host, blood cells, and other materials participates together with bacterial exopolysaccharides (slime) in the biofilm formation.[11–13] Biofilm bacteria are a major concern for clinicians in the treatment of infections because of their resistance to chemotherapy[14–17] and also their

[1] J. C. Nickel, A. G. Gristina, and J. W. Costerton, *Can. J. Surg.* **28,** 50 (1985).

[2] G. Peters, R. Locci, and G. Pulverer, *Zentralbl. Bakterol. Mikrobiol. Hyg. I Abt. Orig. B* **173,** 293 (1981).

[3] J. H. Tenny, M. R. Moody, K. A. Newman, S. C. Schmpff, J. C. Wade, and J. W. Costerton, *Arch. Intern. Med.* **146,** 1949 (1986).

[4] T. J. Marrie and J. W. Costerton, *Circulation* **66,** 1339 (1982).

[5] T. J. Marrie and J. W. Costerton, *J. Clin. Microbiol.* **19,** 911 (1984).

[6] A. G. Gristina and J. W. Costerton, *Infect. Surg.* **3,** 655 (1984).

[7] A. G. Gristina and J. W. Costerton, *J. Bone Jt. Surg.* **67A,** 264 (1985).

[8] J. S. Lam, R. Chan, K. Lam, and J. W. Costerton, *Infect. Immun.* **28,** 546 (1980).

[9] T. J. Marrie and J. W. Costerton, *J. Clin. Microbiol.* **22,** 924 (1985).

[10] K. J. Mayberry-Carson, B. Tober-Mayer, J. K. Smith, and D. W. Lambe, *Infect. Immun.* **43,** 825 (1984).

[11] A. Buret, K. H. Ward, M. E. Orson, and J. W. Costerton, *J. Biomed. Mater. Res.* **25,** 865 (1991).

[12] P. E. Vaudaux, R. Suzuki, F. A. Waldvogel, J. J. Morgenthaler, and E. E. Nydegger, *J. Infect. Dis.* **150,** 546 (1984).

[13] H. Yasuda, Y. Ajiki, T. Koga, H. Kawada, and T. Yokota, *Antimicrob. Agents Chemother.* **37,** 1749 (1993).

[14] C. Chard, J. C. Lucet, P. Rohner, M. Herrmann, R. Auckenthaler, F. A. Waldvogel, and D. P. Lew, *J. Infect. Dis.* **163,** 1369 (1991).

0076-6879/99 $30.00

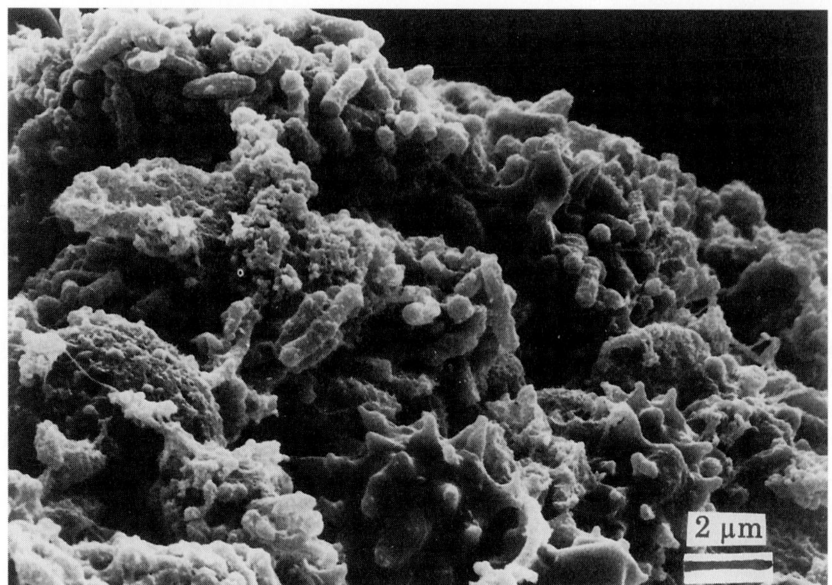

FIG. 1. Biofilm mode of bacterial growth on the surface of a stent detained in the respiratory tract of a patient.

resistance to clearance by humoral or cellular host defense mechanisms.[18–20] Many experimental efforts have been made to make bacterial biofilm models that reproduce those formed in the human body and to eradicate biofilm bacteria efficiently. In addition to giving some examples of inhibitory activities of clarithromycin (CAM), a macrolide, against slime production by biofilm bacteria, this article centers on several methods that have been developed in our laboratory to form bacterial biofilm models *in vitro* and *in vivo*, as well as several methods we have developed to investigate the nature of biofilm bacteria.

---

[15] D. J. Evans, D. G. Allison, M. R. W. Brown, and P. Gilbert, *J. Antimicrob. Chemother.* **27,** 177 (1991).

[16] B. F. Farber, M. H. Kaplan, and A. G. Clogston, *J. Infect. Dis.* **161,** 37 (1990).

[17] A. G. Gristina, C. D. Hobgood, L. X. Webb, and Q. N. Myrvik, *Biomaterials* **8,** 423 (1987).

[18] E. D. Gray, G. Peters, M. Verstegen, and E. Regelmann, *Lancet* **i,** 365 (1985).

[19] G. M. Jornson, D. A. Lee, W. E. Regelmann, E. D. Gray, G. Peters, and P. G. Quie, *Infect. Immun.* **54,** 13 (1986).

[20] P. E. Vaudaux, G. Zulian, E. Huggler, and F. A. Waldvogel, *Infect. Immun.* **50,** 472 (1985).

In Vitro Formation of Bacterial Biofilms on Membrane Filters[13,21]

Bacteria and Medium

Staphylococcus epidermidis 7646, a clinical isolate maintained in our laboratory and found to produce abundant slime, and *Pseudomonas aeruginosa* 1008, a nonmucoid clinical isolate maintained in our laboratory, are used.

Trypticase soy broth (TSB) is from Eiken-kagaku Co., Ltd. (Tokyo, Japan). Minimum medium (MM) is made by dissolving 10.5 g of K_2HPO_4, 4.5 g of KH_2PO_4, 1.0 g of $(NH_4)_2SO_4$, 0.1 g of $MgSO_4 \cdot 7H_2O$, 0.47 g of sodium citrate, and 4.0 g of glucose in 500 ml of water.

Membrane Filters

A cellulose membrane filter (filter type, GS; pore size, 0.22 μm; Nihon Millipore Kogyo K.K., Tokyo, Japan) and a glass microfiber filter (filter type, GF/F; Whatman International Ltd., Maidstone, England) are used.

Formation of Biofilms

Staphylococcus epidermidis and *P. aeruginosa* are precultured at 37° for 20 hr in 10 ml of TSB. Bacteria are washed with biological saline by centrifugation, resuspended in 5 ml of saline to 10^7 cfu(colony-forming units)/ml (*P. aeruginosa*) or 5 ml of MM to 10^7 cfu/ml (*S. epidermidis*), and transferred into a 3-cm-diameter plastic chamber with a cellulose membrane filter or a glass microfiber filter set inside. When quantitative analysis of hexose is performed in or on bacteria colonizing on a filter, a glass microfiber filter should be used. In these media, the number of viable *P. aeruginosa* or *S. epidermidis* gradually increases or decreases. Each chamber is kept at 37° for 10 days. When the activity of a drug (CAM is used in this procedure) to inhibit slime production or biofilm formation is investigated, the medium in the chamber is discarded and incubated at 37° for an additional 5 days in 5 ml of new medium with or without the drug. Figure 2 shows scanning electron micrographs (SEM) of biofilms of *P. aeruginosa* on the surface of membrane filters incubated in the medium with or without CAM. Figure 3 shows SEM of biofilms of *S. epidermidis* on the surfaces of membrane filters incubated in the medium with or without CAM.

[21] H. Yasuda, Y. Ajiki, T. Koga, and T. Yokota, *Antimicrob. Agents Chemother.* **38,** 138 (1994).

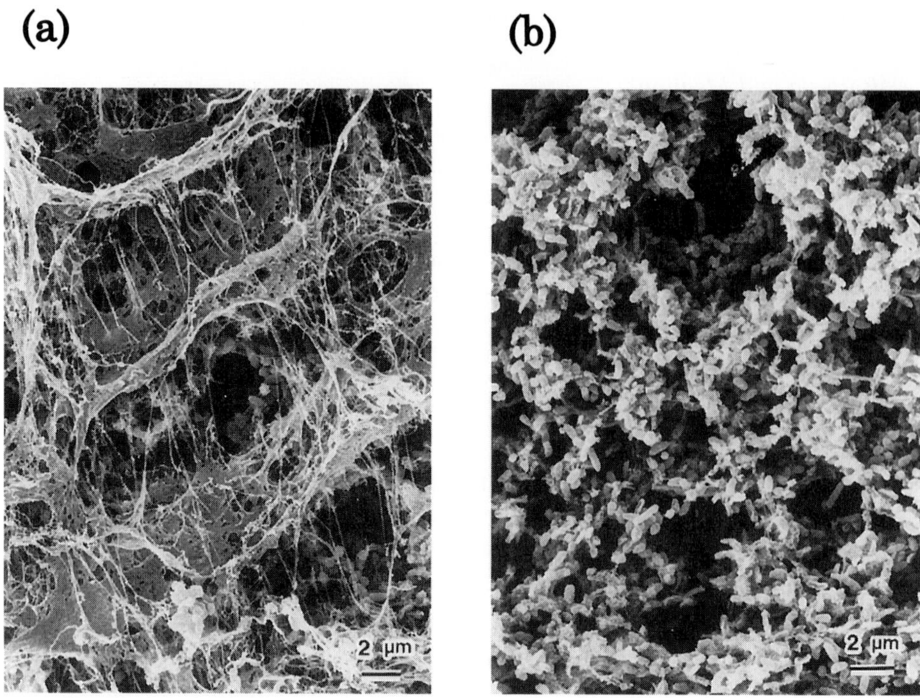

FIG. 2. Biofilm mode of growth of *P. aeruginosa* on the surface of membrane filters. (a) Control (without CAM) and (b) 10 μg of CAM per milliliter.

In Vitro Bacterial Biofilm Formation on Surfaces of Cotton Threads in Rat Subcutaneous Exudate[22]

Bacterial biofilms in the living body that result in chronic or refractory infections are probably constructed in tissues or on the surfaces of biomaterials by way of highly complicated mechanisms (refer to "*in vivo* models"). This method was developed to form bacterial biofilms *in vitro* in an environment reflecting an *in vivo* condition.

Preparation of Rat Subcutaneous Exudate[23]

An air pouch is formed on the back of a rat by injecting 10 ml of air subcutaneously with a 21-gauge needle after trimming the hair with clippers. After removal of the needle, the hole in the skin is sealed with an adhesive

[22] H. Yasuda, Y. Ajiki, J. Aoyama, and T. Yokota, *J. Med. Microbiol.* **41,** 359 (1994).

[23] Y. Ajiki, T. Koga, S. Ohya, T. Takenouchi, H. Yasuda, K. Watanabe, and K. Ueno, *J. Antimicrob. Chemother.* **28,** 537 (1991).

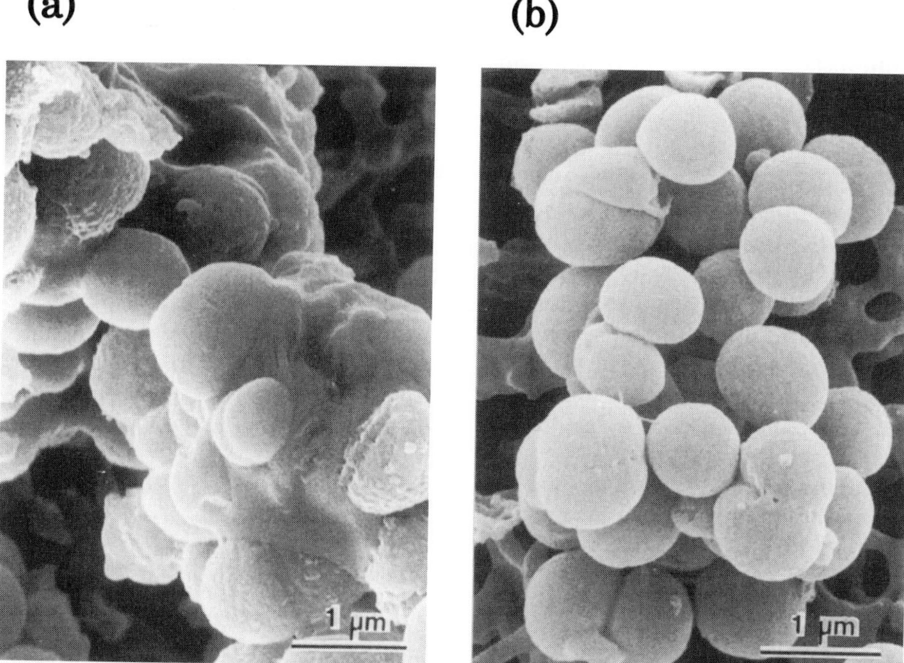

FIG. 3. Biofilm mode of growth of *S. epidermidis* on the surface of membrane filters. (a) Control (without CAM) and (b) 10 μg of CAM per milliliter.

agent. The next day, under anesthesia, the air in the pouch is aspirated and a carboxymethyl cellulose (CMC) pouch is formed by injecting 10 ml of sterilized CMC (Daiichi Pure Chemicals Co., Ltd., Tokyo, Japan) 1.5% (w/v) in saline. The pouch exudate is sampled by syringe 7 hr after the injection of CMC (about 7 ml of exudate could be sampled from each pouch).

Bacteria

Escherichia coli strain 704, a clinical isolate, is cultured in TSB at 37° for 20 hr before use.

Formation of Bacterial Biofilm on Surface of Cotton Threads

Sterilized cotton threads (No. 6, Suzuki-threads, Tokyo, Japan) measuring 1 mm in diameter are soaked in 5 ml of the pouch exudate, 5×10^6 cfu of *E. coli* is added, the mixture is incubated at 37° for 24 hr, and the threads are then transferred into fresh pouch exudate and incubated again.

The transfer is repeated several times. Figure 4 shows scanning electron micrographs of biofilms on the surface of cotton threads after zero and two transfer repetitions in pouch exudate.

Quantitative Analysis of Alginate and Hexose[13,21]

The exopolysaccharide secreted by *P. aeruginosa* is alginic acid (alginate), a linear copolymer of β-1,4-linked D-mannuronic acid and variable amounts of C-5 epimer L-guluronic acid. The exopolysaccharide secreted by *S. epidermidis* is known to contain hexose, but it has not yet been characterized in great detail.

We measured the quantities of alginate and hexose in or on the colonized bacteria on the membrane filters just described and in the medium in which bacterial biofilms had been formed.

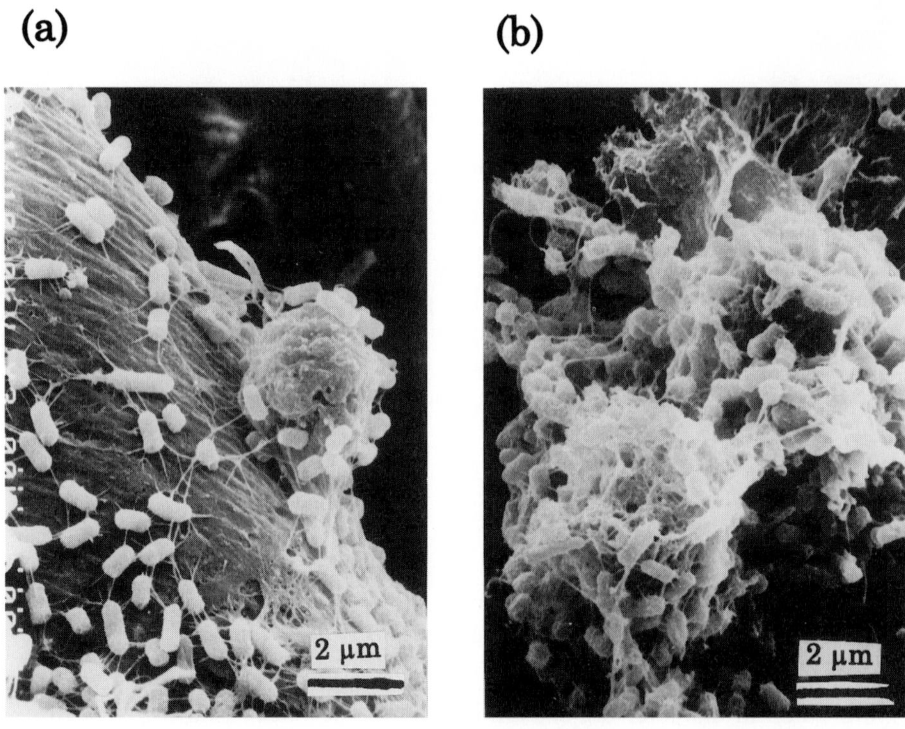

FIG. 4. Biofilm mode of growth of *E. coli* on the surface of cotton threads. After (a) zero and (b) two transfer repetitions in pouch exudate.

Preparation of Bacterial Suspension

To obtain bacteria colonized on the membrane filters, we place two colonized membrane filters (cellulose membrane filters and glass microfiber filters were used for the quantitative analysis of alginate and hexose, respectively) into 4 ml of saline and homogenize the mixture by a BT-10S homogenizer (Biotron). Homogenized samples are centrifuged at 1100*g* for 15 min at 4° to remove fragments of the filters.

Reagents

Copper–HCl reagent (CHR): 40 ml of concentrated HCl plus 1 ml of 2.5% copper sulfate solution plus 9 ml of water

Naphthoresorcinol reagent (NR): 100 mg of 1,3-dihydroxynaphthalene dissolved in 25 ml of water

Anthrone reagent (AR): A solution containing 0.05% anthrone, 1% thiourea, and 66% H_2SO_4 by volume

Quantitative Analysis of Alginate and Hexose

Quantitative analysis of alginate in or on *P. aeruginosa,* or in the saline in which the biofilm mode of colonization had occurred, is performed by the method of Toyoda *et al.*[24] with some modifications. One milliliter of the bacterial suspension or the saline is mixed with 3 ml of a 10% solution of copper sulfate. The reaction mixture is adjusted to pH 4.0 by 1 *N* HCl, kept at room temperature for 1 hr, and centrifuged at 1050*g* for 10 min at 4°. The precipitate is dissolved in 0.1 ml of 1 *N* NH_4OH and diluted with 0.9 ml of water. The sample (1 ml) is treated with 2 ml of CHR and 1 ml of NR and is kept in a boiling water bath for 40 min. The mixture is then chilled, mixed with 4 ml of butyl acetate, shaken well, and centrifuged to separate the butyl acetate layer. After one wash with 20% NaCl solution, the optical density at 565 nm is measured. The quantity of alginate is calculated on the basis of a standard curve made with alginate solution.

Quantitative analysis of hexose in or on *S. epidermidis,* or in the MM in which the biofilm mode of colonization of bacteria had occurred, is performed by the method of Roe.[25] Two milliliters of the bacterial suspension or MM is treated with 1 ml of 20% perchloric acid. The reaction mixture is kept in an ice bath for 40 min and then centrifuged for 10 min at 4° and 750*g*. The supernatant is neutralized by 2 *N* KOH. Each sample (0.3 ml) is treated with 3 ml of AR in an ice bath. The reaction mixture is

[24] M. Toyoda, C. Yomota, Y. Ito, and M. Harada, *J. Food Hyg. Soc. Jpn.* **26,** 189 (1985).
[25] J. H. Roe, *J. Biol. Chem.* **212,** 335 (1955).

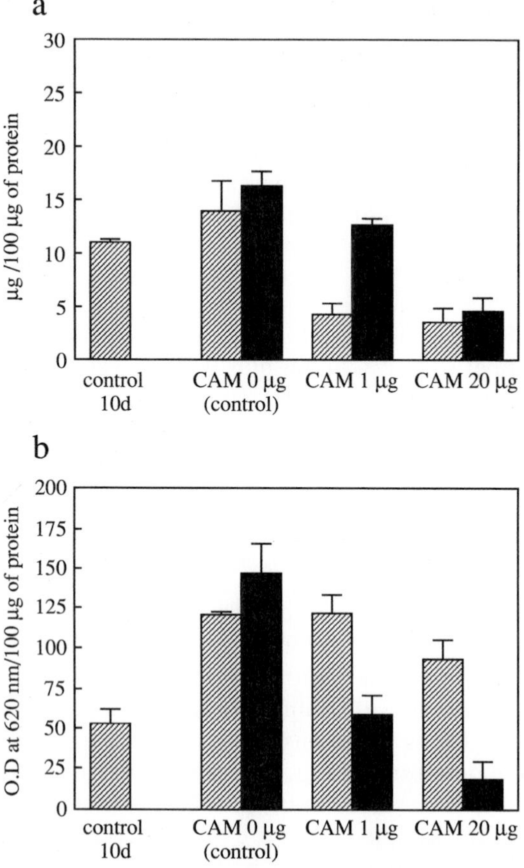

FIG. 5. Quantities of (a) alginate (*P. aeruginosa*) and (b) hexose (*S. epidermidis*) on or in bacterial colonies (▨) and in the environment (■). Control 10d indicates the value obtained using bacterial biofilm immediately before the addition of CAM. The quantity of alginate or hexose in the environment was negligibly small because it was measured just after the change of the medium. OD, optical density.

boiled at 100° for 15 min. After the mixture is chilled, its optical density at 620 nm is measured with a spectrophotometer (U-1000; Hitachi-Seisakusho, Tokyo, Japan).

Figure 5 shows the quantities of alginate and hexose on or in bacterial colonies and in the environment with or without CAM. Quantitative analysis of protein is performed by the biuret method.[26]

[26] L. H. Stickland, *J. Gen. Microbiol.* **5,** 698 (1951).

Measurement of Penetration of Antibacterial Agents through Bacterial Biofilms[13,21]

Biofilm bacteria have been known to be resistant to chemotherapy for a number of reasons. One of the reasons is that an exopolysaccharide secreted by the bacteria acts as a barrier against the penetration of antibacterial agents.

The cellulose membrane filters that had bacterial biofilms formed by *S. epidermidis* or *P. aeruginosa* on their surfaces (the first paragraph) are set in the plastic test systems illustrated in Fig. 6a (Fig. 6b is a photograph of the actual test systems). For the control, fresh membrane filters without

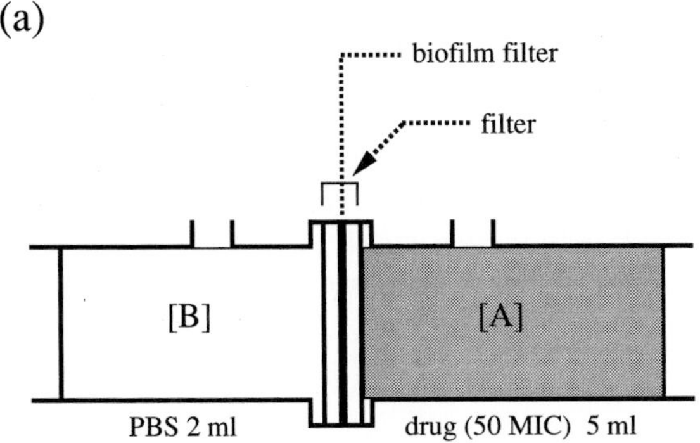

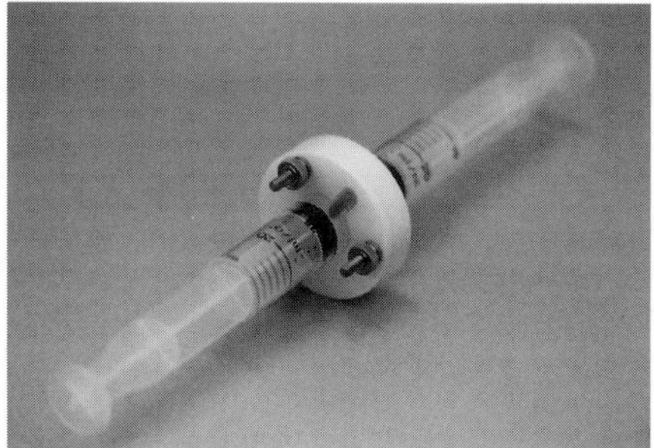

FIG. 6. The plastic test system for a measurement of penetration of antibacterial agents through bacterial biofilms formed on the surfaces of membrane filters. (a) Illustration of the system and (b) actual plastic test system.

bacterial biofilms are used in place of those with biofilms. A drug that has been tested for penetration is put into chamber A at a concentration 50 times the minimum inhibitory concentrations (MICs). The systems are kept at 37°. After 2 and 4 hr, the concentration of antibiotics in chamber B is measured by bioassay using *Bacillus subtilis* as the indicator. Figure 7 shows the permeability of ofloxacin, a new quinolone, through bacterial biofilms formed in the medium with or without CAM.

Measurement of Sensitivity of Biofilm Bacteria to Phagocytosis, Phagocytic Killing, and Killing Activity of H_2O_2[22]

Preparation of Biofilm Bacteria

The cotton threads with biofilms on their surfaces (the second paragraph: after two transfer repetitions) are rinsed once with TSB, soaked in fresh TSB, and incubated at 37° for 30–60 min in a shaking apparatus. The critical incubation time is determined according to the turbidity of the TSB to control the number of planktonic bacteria (PB) released from the biofilm. The PB are used as biofilm bacteria in the following experiments. As a control in each experiment, we use bacteria that had been cultured in TSB at 37° for 72 hr.

Human Serum

Human serum is obtained from fresh blood of a healthy volunteer. The serum is heated at 56° for 40 min before use.

Preparation of Human Polymorphonuclear Leukocytes

Human polymorphonuclear leukocytes (PMNL) are harvested and purified according to the method of Zimmerli *et al.*[27] Heparinized blood (60 ml) from a healthy volunteer is mixed with an equal volume of phosphate-buffered saline (PBS) containing 3.5% dextran T500 (Pharmacia AB, Uppsala, Sweden) and left at room temperature for 45 min. The white blood cell layer is collected and centrifuged at 1300 rpm for 5 min at 4°. The cell fraction is washed once with 10 ml of Eagle's basal medium (EBM; Nissui-seiyaku Co., Ltd., Tokyo, Japan) by centrifugation at 4°. Cells are resuspended in 4 ml of EBM. The suspension is added gently to 4 ml of Ficoll–Paque solution (Pharmacia AB) and centrifuged at 1300 rpm for 30 min at room temperature. The PMNL fraction is collected; 90% of the cells in the fraction are neutrophils.

[27] W. Zimmerli, P. D. Lew, and F. A. Waldvogel, *J. Clin. Invest.* **73,** 1191 (1984).

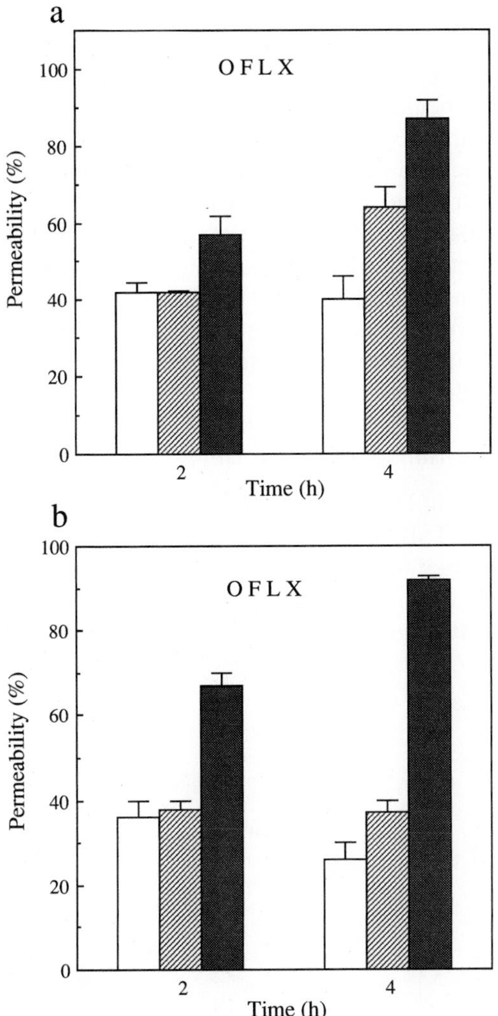

FIG. 7. Permeability of ofloxacin through bacterial biofilms of *P. aeruginosa* (a) and *S. epidermidis* (b). □, control (without CAM); ▨, 1 μg of CAM; and ■, 10 μg of CAM per milliliter. The vertical axis indicates the penetration rates, with the value for the control (without biofilms) assumed to be 100%.

Measurement of Sensitivity of PB to Phagocytic Killing and Phagocytosis

TSB (0.25 ml), containing about 3×10^7 cfu of PB/ml, is mixed with 0.55 ml of EBM, 0.2 ml of human serum, 0.5 ml of EBM containing 4×10^7 PMNL/ml, and 0.5 ml of EBM containing 4% (w/v) gelatin (Wakojunyaku Industry Co., Ltd., Tokyo, Japan). For the reaction mixture without PMNL, 0.5 ml of EBM is added in place of the PMNL suspension. The reaction mixtures are incubated at 37° in a rolling incubator (7 rpm). At 1 and 2 hr after the beginning of incubation, the number of viable bacteria in each reaction mixture is counted on heart infusion agar (Eiken-kagaku).

For measurement of the sensitivity of PB to phagocytosis, the reaction mixtures are centrifuged at 1300 rpm for 5 min at 4°, 15 min after the beginning of incubation. PMNLs are suspended in a small amount of human serum and smeared onto glass slides. Phagocytosis is observed by microscopic examination of Wright-stained cell smears. One hundred neutrophils on each smear are examined, and the percentage of phagocytosing neutrophils and the mean number of bacteria/mean number of phagocytosing neutrophils are calculated.

Measurement of Sensitivity of PB to Killing Activity of H_2O_2

A small amount of H_2O_2 (31%) solution (Iwai Kagaku Co., Ltd., Tokyo, Japan) is added to TSB containing about 3×10^6 cfu of PB/ml to make a H_2O_2 concentration of 0.01 or 0.001%. The reaction mixtures are incubated at 37° in a shaker. After 1 hr, the number of viable bacteria is counted on heart infusion agar. Figure 8 shows the susceptibility of biofilm bacteria

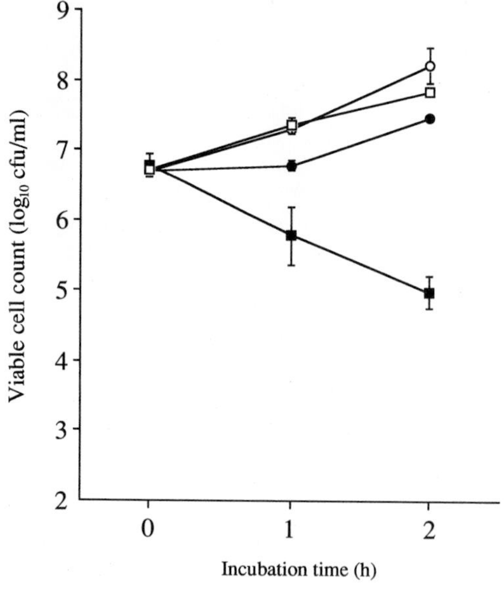

FIG. 8. Susceptibility of biofilm bacteria (PB) to killing by human PMNL. ○, Control bacteria; ■, control bacteria + PMNL; □, PB only; and ●, PB + PMNL.

TABLE I
SUSCEPTIBILITY OF BIOFILM BACTERIA (PB) TO PHAGOCYTOSIS BY PMNL[a]

Experiment No.	PB		Control	
	A	B	A	B
1	35	2.6	43	3.4
2	33	2.4	40	3.4
3	22	2.8	25	2.9
4	20	2.8	19	3.2
5	27	4.9	48	5.4
Mean (SE)	29 (3.5)	3.1 (0.46)	35 (5.5)	3.7 (0.44)

[a] A, number of phagocytosing cells/100 PMNL; B, mean bacterial number/phagocytosing cell.

(PB) to killing by human PMNL. Table I shows the susceptibility of biofilm bacteria (PB) to phagocytosis by neutrophils. Figure 9 shows the susceptibility of biofilm bacteria (PB) to killing by H_2O_2.

In Vivo Infection Models Manifesting Biofilm Mode of Bacterial Growth: Tissue Cage Infection Model[28] in Rats

Tissue Cage

The tissue cage infection system used in this experiment was originally developed by Zimmerli *et al.*[28] for a foreign infection model. Multiperforated [about 120 regularly spaced holes (diameter, <1 mm) and sealed at each end with a cap made from an identical material] polytetrafluoroethylene (Teflon) tissue cages (internal and external diameters of 8 and 10 mm, respectively; length, 32 mm) containing a Teflon slip (7 × 26 mm) (Fig. 10) are used.

Animals

Male Wistar–Imamichi rats weighing 220 to 240 g are from the Imamichi Institute for Animal Reproduction (Saitama, Japan).

[28] W. Zimmerli, F. A. Waldvogel, P. Vaudaux, and U. E. Nydegger, *J. Infect. Dis.* **146,** 487 (1982).

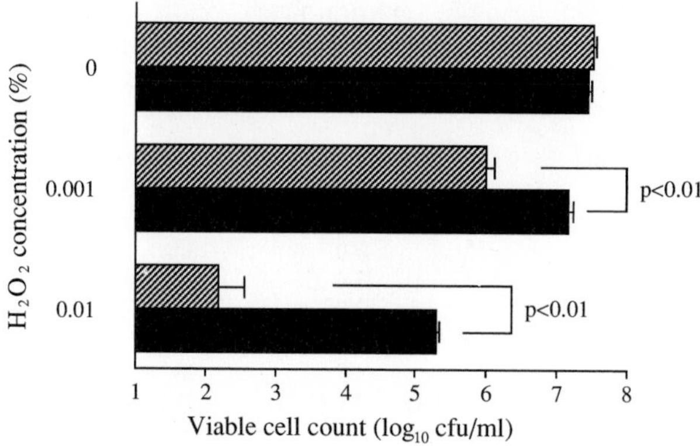

FIG. 9. Susceptibility of biofilm bacteria (PB) to killing by H_2O_2. ▨, control bacteria; and ■, PB.

Bacteria

S. epidermidis 871, a clinical isolate maintained in our laboratory, is used.

Induction of Infection in Tissue Cages and Therapy by Antibiotics

After trimming the hair from the backs of rats with a hair clipper, a 3-cm incision (parallel to the backbone) is made using an aseptic technique, and the subcutaneous space is dissected bluntly. One gas-sterilized tissue cage is implanted in each flank, and the skin is closed with metal clips. The clips are removed 5 days after surgery, and the tissue cage fluid is collected by percutaneous aspiration to confirm the sterility of the fluid 12 days after surgery. Fourteen days after surgery, 2.4×10^6 cfu/tissue cage of bacteria

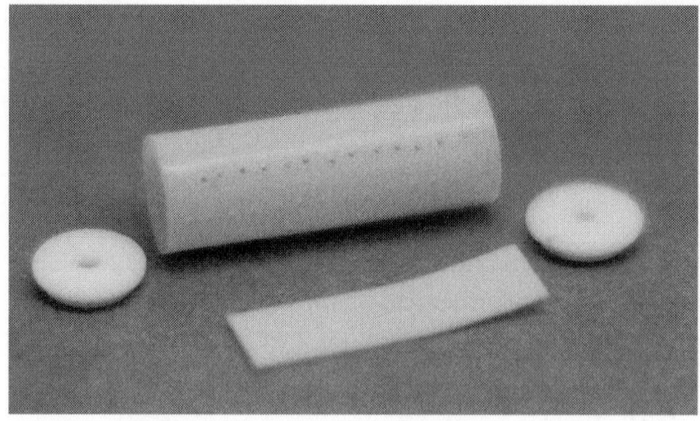

FIG. 10. A tissue cage.

is injected by syringe directly into the tissue cage. Drug administrators (orally) are initiated 1 day after the bacterial injection and continued twice a day for 8 days. The exudate in the tissue cage is sampled by syringe at 1 (just before the first administration of drug), 5, 7, and 9 days after bacterial infection. The numbers of viable bacteria in the tissue cage exudates are counted on Mueller–Hinton II agar (MHII; Becton-Dickinson, Cockeysville, MD).

Figure 11 shows the biofilm mode of growth of *S. epidermidis* on the surfaces of Teflon slips that had been set in the tissue cages. Figure 12 shows the changes of viable numbers of *S. epidermidis* in the tissue cages with or without therapy.

In Vivo Infection Models Manifesting Biofilm Mode of Bacterial Growth: CMC Pouch Infection Model in Rats[13,23]

Animal

Male Wistar–Imamichi rats weighing 140 to 150 g are from the Imamichi Institute for Animal Reproduction.

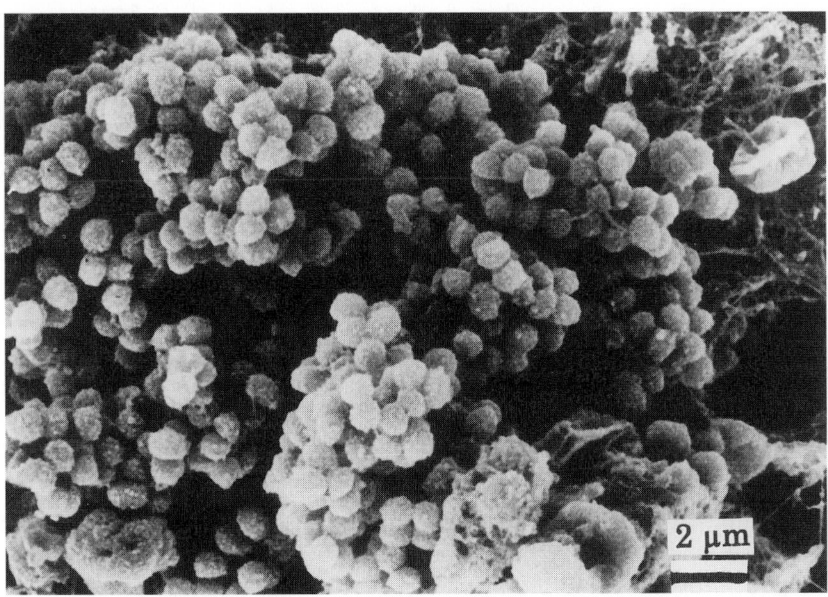

FIG. 11. Biofilm mode of growth of *S. epidermidis* on the surfaces of Teflon slips that had been set in the tissue cages.

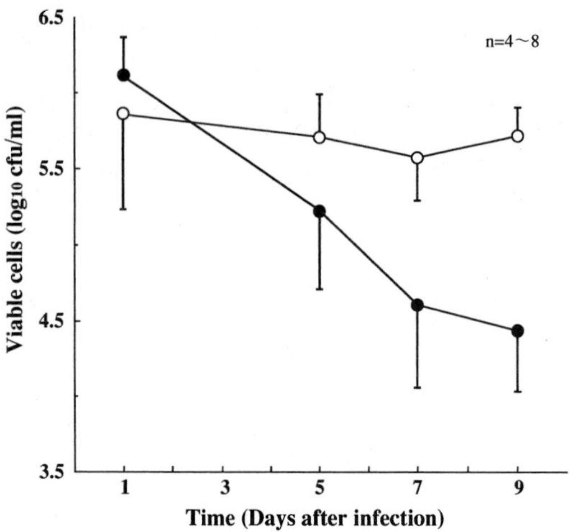

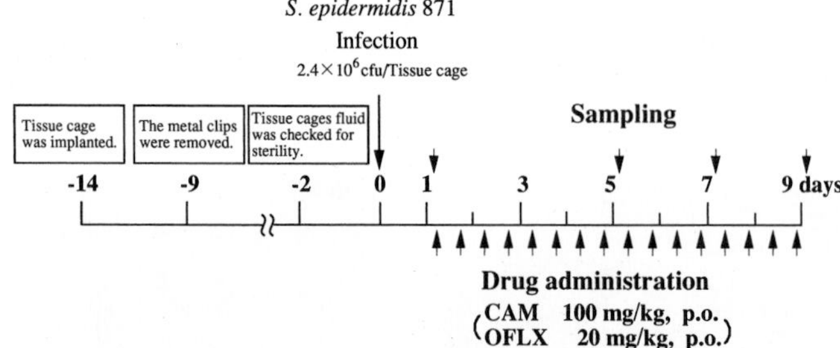

FIG. 12. Experimental schedules and the changes of variable numbers of *S. epidermidis* in the tissue cages with (●) or without (○) therapy.

Bacteria

Pseudomonas aeruginosa 1008 is precultured in TSB at 37° for 20 hr before use.

Induction of Infection in Pouches on Backs of Rats and Therapy by Antibiotics

Methods used to form CMC pouches on the backs of rats have been described earlier. An infection is induced by inoculating 10^6 cfu of bacteria per pouch together with the injection of CMC.

Drug administrations (orally) are initiated 4 days after the induction of the infection and continued twice a day for 5 days. Pouch exudates are sampled by syringe once a day for 6 days after the beginning of therapy. The numbers of viable bacteria in the pouch exudates are counted on MHII. Figure 13 shows the development of the biofilm mode of growth over time after the infection of *P. aeruginosa* in CMC pouches of rats. Figure 14 shows the changes of viable numbers of *P. aeruginosa* in rat CMC pouches with or without therapy.

Conclusions

In vitro experimental models of bacterial biofilms have been formed by many investigators on the surfaces of disks of silicone latex catheter material, cellulose acetate membranes, and the surfaces of other devices by culturing bacteria in an artificial medium, artificial urine, or serum. Ideal properties sought in an *in vitro* bacterial biofilm model are uniform film formation, uniform reproducibility, and the ability to be investigated quantitatively. However, the ability to mimic a highly complicated *in vivo* condi-

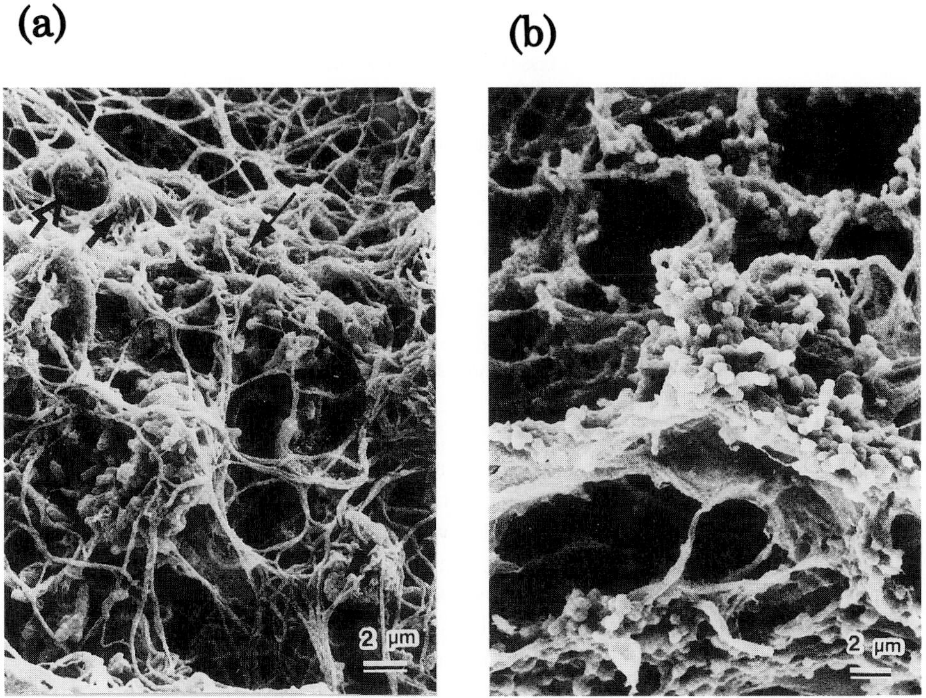

FIG. 13. Biofilm mode of growth of *P. aeruginosa* on subcutaneous tissues of rats 1 day (a) and 2 days (b) after infection. ➔, erythrocyte; ⇾, another type of cell; and →, fibrous structure.

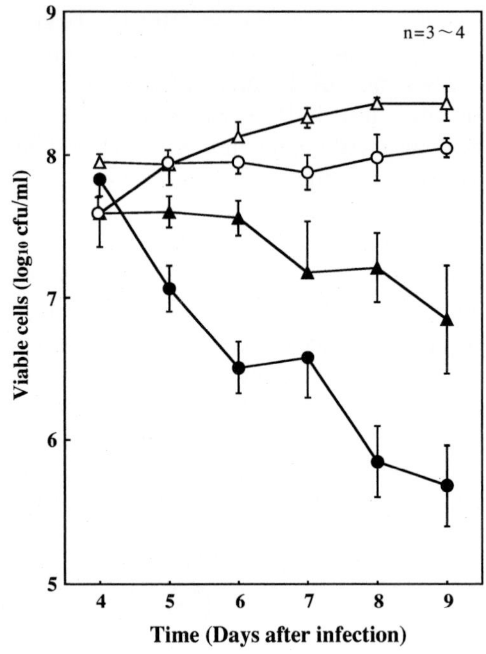

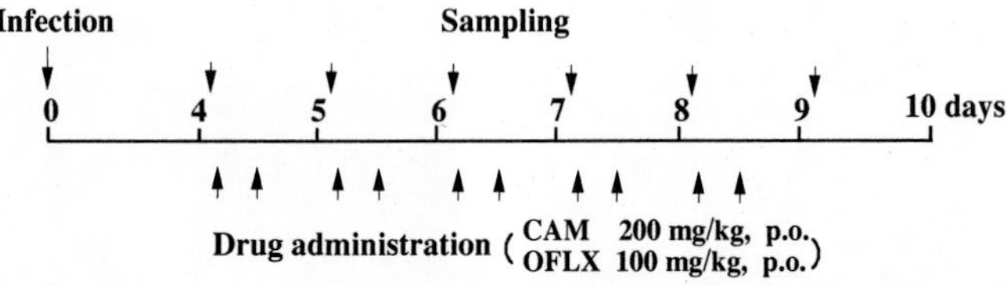

FIG. 14. Experimental schedules and the changes of viable numbers of *P. aeruginosa* in rat CMC pouches. ○, without therapy; △, CAM; ▲, ofloxacin (OFLX); and ●, CAM + OFLX.

tion is desirable in an *in vitro* model. We have not been able to apply a strict definition to "bacterial biofilm" as such a biofilm formed in a living body is a diverse and vague matrix, a bacterial living region, composed of bacteria, bacterial exopolysaccharides, various insoluble proteins, and blood cells. For this reason, researchers planning studies on bacterial biofilms need to choose models that suit their specific purposes, although none of the currently available models are always completely reliable and reproducible. By considering the advantages and disadvantages of the existing models and the reliability of the experimental results they yield carefully, researchers will be able to design new models using their own ideas or by combining well-established methods.

There have been relatively few reports on *in vivo* models of biofilm growth. One of the most difficult challenges to overcome is to devise a durable infection model. Generally, it takes several days for infected bacteria to form biofilms in an infected tissue or organ, and it takes several days longer to investigate the therapeutic effect of a drug after the biofilm mode of growth of the infected bacteria has been established. Ideally, the infection should continue in a stable condition for more than 7 or 10 days. Respiratory tract infections, urinary tract infections, infective endocarditis, foreign body infections, and so forth are typical infectious diseases in which the biofilm mode of bacterial growth is sometimes characteristic. Therefore, these diseases should be taken into account in making a useful *in vivo* experimental biofilm infection model.

Acknowledgment

The authors thank Mrs. Yohko Ajiki for efforts in the development and characterization of these methods.

Section VIII

Antifouling Methods

[44] *Pseudomonas aeruginosa* Biofilm Sensitivity to Biocides: Use of Hydrogen Peroxide as Model Antimicrobial Agent for Examining Resistance Mechanisms

By DANIEL J. HASSETT, JAMES G. ELKINS, JU-FANG MA, and TIMOTHY R. MCDERMOTT

Introduction

In nature, bacterial biofilms are complex, yet highly structured communities that are often composed of multiple genera. Biofilms are ubiquitous, found on various substrata, and can be problematic in environmental, industrial, and clinical settings. When occupying such diverse niches, one unifying hallmark of biofilms is their relative resistance to a myriad of biocides and antibiotics, especially when compared to planktonic organisms. In industrial and environmental settings, oxidizing agents such as hydrogen peroxide (H_2O_2) and sodium hypochlorite (NaOCl) are used for biofilm control. However, killing of biofilm organisms with these agents is typically incomplete, with regrowth commonly occurring rapidly. As *Pseudomonas aeruginosa* is often associated with biofilms, it has been used by various laboratories as a model organism for studying biofilm resistance to these oxidizing biocides.

If available, *P. aeruginosa* preferentially uses molecular oxygen (O_2) as a terminal electron acceptor. However, aberrant electron flow occurring in respiratory metabolism can lead to the production of reactive oxygen intermediates (ROIs). These include superoxide (O_2^-), H_2O_2, and the hydroxyl radical (HO·). Although such apparent metabolic infidelity is common in aerobic organisms, ramifications of ROI production include cell damage, mutations, or death. Relief of ROI stress is provided by antioxidant defenses, including superoxide dismutase (SOD), catalase, peroxidase, DNA repair enzymes, and free radical scavenging agents.[1,2]

Pseudomonas aeruginosa is well equipped to deal with oxidative stress. The organism possesses an iron- and a manganese-cofactored SOD,[3,4]

[1] W. Beyer, J. Imlay, and I. Fridovich, *Prog. Nucleic Res.* **40,** 221 (1991).

[2] J. Imlay and S. Linn, *J. Bacteriol.* **169,** 2967 (1987).

[3] D. J. Hassett, W. A. Woodruff, D. J. Wozniak, M. L. Vasil, M. S. Cohen, and D. E. Ohman, *J. Bacteriol.* **175,** 7658 (1993).

[4] D. J. Hassett, H. P. Schweizer, and D. E. Ohman, *J. Bacteriol.* **177,** 6330 (1995).

0076-6879/99 $30.00

whose function is to remove O_2^-.[5] To dispose of H_2O_2, *P. aeruginosa* possesses three catalases, designated KatA,[6,7] KatB,[6] and KatC.[8] KatA is the predominant catalase and is produced during all growth stages.[6,7] In planktonic cells, KatB has not been detected in exponential phase cells, but is observed at very low levels in stationary phase cells. However, when *P. aeruginosa* is exposed to exogenous H_2O_2 or the redox-cycling antibiotic paraquat, KatB levels are increased greatly due to the transcriptional activation of *katB*. Unlike KatA and KatB, little is known about KatC or its biological role in *P. aeruginosa.* Together, however, these catalases arguably constitute the major first-line defense mechanism in protecting *P. aeruginosa* against H_2O_2.

Because our knowledge of this organism's response to oxidative stress is reasonably well grounded genetically and physiologically,[3,4,7,9–11] we are in a position to begin asking specific questions regarding mechanisms by which biofilm bacteria resist oxidizing biocides. In studying bacterial biofilm behavioral responses to these antimicrobials, one can gain insight into the physiological processes of biofilm bacteria and how they differ from planktonic cells. This article describes methods for measuring biofilm responses to H_2O_2 in terms of cell survival and defense mechanisms. The basic method described for the quantification of cell resistance to H_2O_2 can be applied to study the effects of any biofilm biocide. To demonstrate basic utility of the approach in terms of assessing biofilm physiology, we also include spectrophotometric and activity gel-staining techniques for measuring catalase and SOD.

Growth of Planktonic Cultures and Biofilms

Planktonic cultures and biofilms are grown aerobically in a medium of choice. In the methods describe here we used either a defined media such

[5] J. M. McCord and I. Fridovich, *J. Biol. Chem.* **244,** 6049 (1969).

[6] S. M. Brown, M. L. Howell, M. L. Vasil, A. Anderson, and D. J. Hassett, *J. Bacteriol.* **177,** 6536 (1995).

[7] D. J. Hassett, L. Charniga, K. A. Bean, D. E. Ohman, and M. S. Cohen, *Infec. Immun.* **60,** 328 (1992).

[8] The *Pseudomonas* Genome Project, Pathogenesis Corporation and the Cystic Fibrosis Foundation, http://www.pseudomonas.com.

[9] D. J. Hassett, P. Sokol, M. L. Howell, J.-F. Ma, H. P. Schweizer, U. Ochsner, and M. L. Vasil, *J. Bacteriol.* **178,** 3996 (1996).

[10] D. J. Hassett, M. L. Howell, P. A. Sokol, M. Vasil, and G. E. Dean, *J. Bacteriol.* **179,** 1442 (1997).

[11] D. J. Hassett, M. L. Howell, U. Ochsner, Z. Johnson, M. Vasil, and G. E. Dean, *J. Bacteriol.* **179,** 1452 (1997).

as *Pseudomonas* basal mineral (PBM) medium[12] or complex media such as L-broth or Trypticase® soy broth (TSB) diluted 1 : 10 for batch planktonic cultures (1 : 10 TSB) or 1 : 100 for biofilms (1 : 100 TSB). The temperature used to culture planktonic cells should match that used for biofilms, and in the experiments described in this article, all cultures were grown at 25°. Planktonic cultures are grown to stationary phase (overnight) with shaking at 300 rpm and used to inoculate biofilm coupons or are collected for the preparation of cell extracts (described later).

Biofilms are grown in a multiple chamber drip flow reactor as described previously.[13] Briefly, medium is dripped over sterile coupons held in parallel polycarbonate chambers.[13] Each coupon (1.3 × 7.6 cm; composed of material relevant to the environment or substratum of interest), resting horizontally in the polycarbonate chamber, is inoculated aseptically with 1 ml of stationary phase culture and 10 ml of fresh media. The reactor cover is closed and bacteria are allowed to attach in a static environment for a 6- to 18-hr period. Longer attachment incubations will allow for better colonization of the coupon, resulting in cell replication and greater biofilm biomass. The entire reactor is then inclined at 10° and the nutrient flow (50 ml/hr) initiated. The medium drips onto the coupon at the raised edge, flows down lengthwise over the coupon, and then out an effluent port at the chamber base. Biofilms are grown for 24–72 hr, depending on the relative richness of the medium and required biofilm thickness.

Comparison of Planktonic Cell and Biofilm Cell H_2O_2 Resistance

Planktonic bacteria are grown overnight at room temperature and then diluted to the desired extent in the same medium. A sample is then diluted serially in phosphate-buffered saline, pH 7.0 (PBS), with appropriate dilutions spread onto R2A agar media (Difco, Franklin Lakes, NJ). The remaining cell suspension is treated with H_2O_2 at experimentally relevant concentrations (50 m*M* H_2O_2 in the experiments described here). After predetermined exposure periods, samples of the cell suspension are again diluted serially, except the dilution blanks contain 0.2% (w/v) sodium thiosulfate to neutralize the H_2O_2. Aliquots of several dilutions are plated immediately on R2A agar medium. Colony-forming units (cfu) are enumerated after a 24- to 48-hr incubation at 37°, with the $\log_{10}$ reduction in viability calculated as the comparison of final viable counts determined after H_2O_2 treatment versus initial viable cell counts of cells taken just prior to H_2O_2 exposure.

[12] R. M. Atlas, *in* "Handbook of Microbiological Media" (R. M. Atlas and L. C. Parks, ed.), 2nd ed., p. 1143. CRC Press, Ann Arbor, 1997.

[13] C.-T. Huang, K. D. Xu, G. A. McFeeters, and P. S. Stewart, *Appl. Environ. Microbiol.* **64,** 1526 (1998).

Biofilms are treated by continuous flow of media containing H_2O_2. The H_2O_2 concentration and exposure period are again varied relative to the planktonic culture treatment(s) and according to experimental interests. Following H_2O_2 treatment, biofilms on coupons are scraped into 50 ml of PBS containing 0.2% (w/v) sodium thiosulfate (to neutralize H_2O_2) and homogenized using a Brinkman homogenizer (Model PT 10/35, Brinkman Instruments, Westbury, NY). Homogenized biofilms are analyzed for viable bacteria by serial dilution and plating as for planktonic cultures. The suspended biofilms are also analyzed for total cell numbers by 4′,6-diamidino-2-phenylindole dihydrochloride (DAPI; Molecular Probes, Inc., Eugene, OR) direct counts. The same assays are performed on untreated control biofilm coupons to verify the accuracy and precision of the direct count method. Log_{10} reduction of viable biofilm bacteria is calculated based on the survival fractions defined as the ratio of colony forming units to direct microscopic counts with DAPI-stained cells. This approach factors out the detachment of cells occurring in biofilm experiments that is not a true measurement of disinfection. All experiments are repeated at least three times.

Measurement of Catalase and Superoxide Dismutase

Preparation of Cell Extracts

Following H_2O_2 or control treatments, planktonic cultures are collected by centrifugation at 5000*g* for 10 min at 4°, washed twice in ice-cold 50 m*M* potassium phosphate, pH 7.0 (KP_i), resuspended in 0.5 ml of the same buffer, and transferred to an Eppendorf tube for sonication. Biofilms are scraped and homogenized as described earlier. Bacteria are collected by centrifugation (5000*g* for 10 min at 4°), washed, suspended in 0.5 ml of ice-cold KP_i, and transferred to an Eppendorf tube for sonication as for planktonic cells. Cells are disrupted by sonication at an appropriate power setting (consult manufacturer guidelines) using a microtip. Thorough cell disruption is achieved by 5- to 30-sec sonications in an ice water bath. The sonicate is centrifuged at 13,000*g* for 10 min at 4°, and the supernatant is removed for protein estimates and enzyme assays. Protein concentration in the cell-free extracts is estimated by the method of Bradford[14] using bovine serum albumin (BSA) fraction V (Sigma, St. Louis, MO) as standard.

Spectrophotometric Measurement of Superoxide Dismutase and Catalase

Catalase activity is monitored by following the decomposition of 18 m*M* H_2O_2 in KP_i at 240 nm at 25°.[6,9,15] Briefly, an appropriate amount of

[14] M. M. Bradford, *Anal. Biochem.* **72,** 248 (1976).

[15] R. F. Beers, Jr., and I. W. Sizer, *J. Biol. Chem.* **195,** 133 (1952).

cell-free extract is added to 3 ml of KP_i/H_2O_2 in a quartz cuvette, mixed thoroughly, and H_2O_2 decomposition followed for 1 min. Specific activity is calculated as in Eq. (1).

$$\text{Specific activity} = \frac{\Delta OD_{240}/\text{min} \times 3000}{43.6\ (M\ \varepsilon) \times \text{mg protein}} \tag{1}$$

Superoxide dismutase activity is monitored by following the autoxidation of pyrogallol at 320 nm,[16] a modification of the original method described by Marklund and Marklund.[17] This method is considerably less tedious and reagent intensive than the original SOD-inhibitable cytochrome *c* reduction assay described by McCord and Fridovich.[5] A 10 m*M* solution of pyrogallol is prepared in 10 m*M* HCl. To a 1-ml quartz cuvette, add 1 ml of 50 m*M* Tris–HCl/1 m*M* EDTA (pH 8.2) that has been kept on a stir plate with constant stirring. Add 5 μl of pyrogallol solution and mix thoroughly. Record the change in absorbance at 320 nm for 1 min. Adjust the volume of pyrogallol added until the change in OD_{320} is 0.02 ± 0.002. The amount of cell-free extract that causes a 50% reduction in pyrogallol autoxidation (e.g., $\Delta OD_{320} = 0.01 + 0.001$) constitutes 1 U of activity. Specific activity is then calculated as U/mg protein.

Measuring Superoxide Dismutase and Catalase Activity in Native Gels

For detection of SOD activity in native gels, cell-free extracts (typically 40–80 μg protein) are loaded onto 10% polyacrylamide minigels and separated by electrophoresis at 140 V (constant) at 4°. As soon as the bromphenol blue-tracking dye reaches the bottom of the gel, the gel is removed and submerged in a small tray containing 80 ml of 50 m*M* potassium phosphate (pH 7.8), 0.1 m*M* EDTA, 2 mg riboflavin, and 16 mg nitro blue tetrazolium (NBT). Finally, 400 μl of TEMED is added to initiate NBT reduction and the gel is incubated in the dark for 30–45 min at 37°. The gel is removed, placed on a sheet of Saran wrap, and laid on a fluorescent light box. Within a few minutes, the gel turns purple, indicative of O_2-catalyzed reduction of NBT, whereas SOD activity bands are white. Once the color has developed sufficiently, the gel is wrapped in Saran wrap surrounded by aluminum foil to prevent further light-catalyzed NBT reduction and is photographed or scanned as desired.

For catalase activity gel staining, 5% polyacrylamide gels are used because of the larger size of the *P. aeruginosa* catalases (170 and 228 kDa) relative to SODs (42 and 44 kDa). Before loading the samples, add 0.1

[16] J. R. Prohaska, *J. Nutr.* **113,** 2048 (1983).

[17] S. Marklund and G. Marklund, *Eur. J. Biochem.* **47,** 469 (1974).

mg/ml sodium thioglycolate to the top buffer and prerun the gel at 20 mA constant current for 30 min at 4°. Thioglycolate prevents APS-catalyzed radical formation that can cause catalase activity banding artifacts. Because *P. aeruginosa* possesses such high catalase activity (~500–4000 U/mg[6,7]), load only 10–20 μg of cell-free extract per lane. As little as 5 μg can be used from organisms pretreated with H_2O_2. Load 50–100 μg of cell-free extract if working with organisms that produce significantly less catalase activity (e.g., *Escherichia coli,* 8–10 U/mg). Apply a constant current of 10–20 mA until the bromphenol blue reaches the bottom of the gel. At this point, the gel is ready to be removed from the casings. Use extreme care in dislodging the gel from the glass casings as 5% polyacrylamide gels are extremely fragile. After the top gel plate is removed, the gel and the bottom plate are placed in a small tray of distilled water. Using a flat metal spatula, carefully insert the spatula between the gel and the plate in short rapid motions. Slowly, the gel will become dislodged from the bottom plate. The water in the tray is removed by aspiration and replaced with 100 ml of distilled water containing 4 m*M* H_2O_2. The gel is incubated at room temperature for 10 min, the H_2O_2 solution is removed by aspiration, and the gel is washed with 100 ml of distilled water. The gel is then soaked in a solution containing 1% (w/v) ferric chloride and 1% (w/v) potassium ferricyanide. As soon as the gel turns a dark green, the ferric chloride/potassium ferricyanide solution should be removed and the gel rinsed with a light stream of distilled water to prevent overdevelopment. Finally, once all the dye is removed, the gel can be photographed immediately or, if necessary, stored in a tray of distilled water for a few days at 4° prior to photography.

Experimental Examples and Comments

As discussed earlier, one of the important features of biofilms is their innate resistance to biocides. The results depicted in Fig. 1 illustrate a typical response of biofilm *P. aeruginosa* to H_2O_2; H_2O_2 is significantly less effective at killing biofilm bacteria than highly vulnerable planktonic cells (Fig. 1A). In these experiments, planktonic bacteria were exposed to a single dose of 50 m*M* H_2O_2 for 60 min, while biofilms were exposed to a constant stream of 50 m*M* H_2O_2 for 60 min. Even after prolonged exposure, biofilm cell viability is high, assuring biofilm survival and regrowth subsequent to biocide treatment. These results suggest that biofilm bacteria are well equipped to resist H_2O_2.

As mentioned earlier, *P. aeruginosa* has two major catalases, KatA and KatB, that are expressed differentially, depending on exposure to oxidative stress, and two SODs, Fe-SOD and Mn-SOD. To illustrate the application

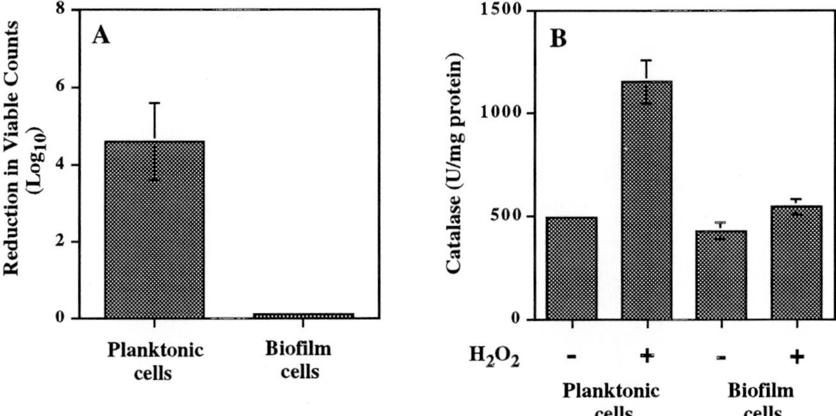

FIG. 1. (A) Sensitivity of planktonic cells and biofilm cells of *P. aeruginosa* wild-type strain PAO1 to H_2O_2 Data are shown as the $\log_{10}$ reduction of viable cells after exposure to 50 m*M* H_2O_2 for 1 hr. (B) Catalase-specific activity of planktonic and biofilm cells without (−) and with (+) H_2O_2 treatment. Both cell types were grown in PBM prior to H_2O_2 treatment.

of the previous methods in the detection and measurement of these different enzymes, experiments were performed with planktonic and biofilm bacteria. As measured with spectrophotometric enzyme assays, overall total catalase levels in planktonic cells grown in PBM medium are maintained at approximately 500 U/mg, but increase significantly after only a brief exposure to an external oxidizing agent such as H_2O_2 (Fig. 1B). Under these conditions, catalase activity in biofilm bacteria treated with H_2O_2 is only about half of that observed with planktonic cells, inducing poorly in response to H_2O_2 treatment. Therefore, catalase activity per se may not explain the increased resistance of biofilm bacteria to H_2O_2.

As can be seen in Fig. 2, native gel analysis can be used to show that KatA is expressed constitutively in both cell types, serving to detoxify H_2O_2 synthesized during normal aerobic metabolism, and upregulated weakly in cells exposed to H_2O_2 (Fig. 2). Small amounts of KatB are seen in stationary phase planktonic cells, but the complement of this isozyme increases dramatically after treatment with H_2O_2. KatB is rarely observed in untreated *P. aeruginosa* biofilms and its induction in response to H_2O_2 is significantly less than with planktonic cells (Fig. 2).

Superoxide dismutase expression in *P. aeruginosa* is at least in part governed by iron availability.[3,4,9–11] When iron is not limiting, only Fe-SOD is expressed, but under conditions of iron starvation, Mn-SOD is induced. This type of regulation is not coincidental as Fe-SOD requires iron for its active site, as Mn-SOD requires manganese. Therefore, *P. aeruginosa* has

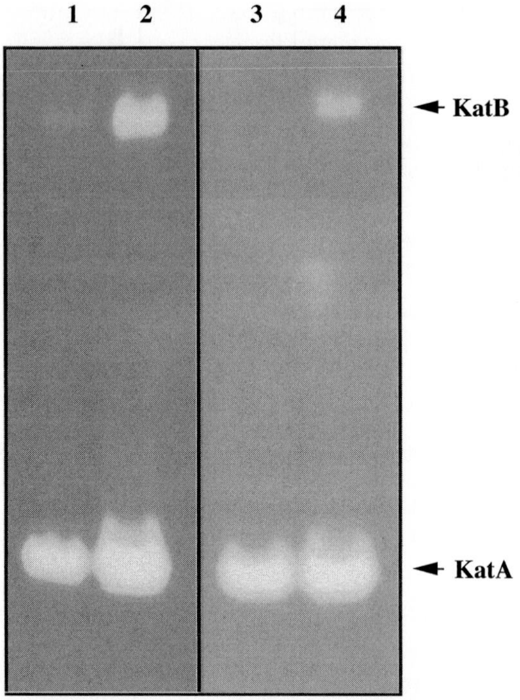

FIG. 2. Native gel catalase activity staining of cell-free extracts of *P. aeruginosa* wild-type strain PAO1 before and after treatment with H_2O_2; 15 μg protein loaded per lane. (*Left panel*) Extracts from a 24-hr batch planktonic culture: Lane 1, no H_2O_2 treatment; and lane 2, six doses of 1 m*M* H_2O_2 every 10 min for 1 hr. (*Right panel*) Extracts from 72-hr biofilms: Lane 3, no H_2O_2 treatment; and lane 4 after constant exposure to 50 m*M* H_2O_2 for 1 hr. Gels were scanned using a Hewlett-Packard ScanJet IIcx, composed with Adobe Photoshop 3.0 for Macintosh, and printed using an Epson Stylus Color 600 printer.

evolved a mechanism to guarantee synthesis of SOD and thus protection against ROIs, regardless of micronutrient availability. An analysis of SOD activity in the different cell types grown in TSB medium reveals that SOD expression appears dissimilar. In planktonic cells, Fe-SOD is expressed constitutively, whereas Mn-SOD activity is induced only after exposure to the hydrophobic iron-specific chelator 2,2′-dipyridyl (Fig. 3). In contrast, biofilm bacteria express both SODs without the addition of an iron chelator (Fig. 3), suggesting either that biofilm cells have an iron requirement in excess of planktonic cells, which is met by the iron in the medium used, or that the *fatA-fumC-orfX-sodA* operon is subject to other regulatory mechanisms that are imposed during the biofilm mode of growth. This is

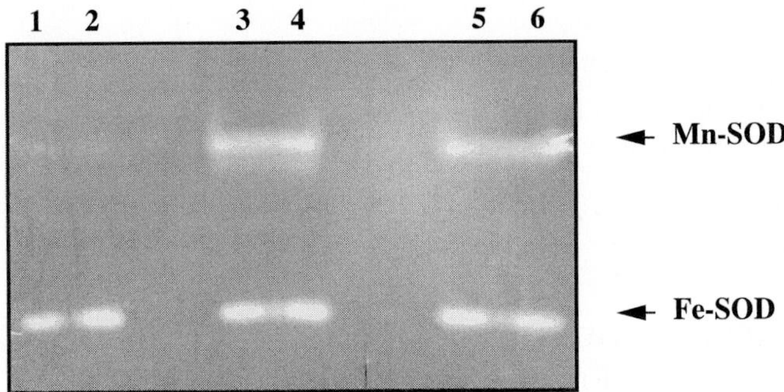

FIG. 3. Native gel Mn- and Fe-SOD activity stains of protein extracts of *P. aeruginosa* wild-type strain PAO1. All results are shown in duplicate, with 60 μg protein loaded per lane. Lanes 1 and 2: protein extracts from planktonic cultures grown in 1 : 10 TSB without the iron-specific chelator 2,2′-dipyridyl; lanes 3 and 4: protein extracts from planktonic cultures grown in 1 : 10 TSB with 500 μM 2,2′-dipyridyl; and lanes 5 and 6: protein extracts from biofilm cells grown in 1 : 100 TSB without 2,2′-dipyridyl. Gels were scanned using a Hewlett-Packard ScanJet IIcx, composed with Adobe Photoshop 3.0 for Macintosh, and printed using an Epson Stylus Color 600 printer.

one of few examples that demonstrate physiological differences between biofilms and their planktonic counterparts.

Summary

The biofilm mode of bacterial growth may be the preferred form of existence in nature. Because of the global impact of problematic biofilms, study of the mechanisms affording resistance to various biocides is of dire importance. Furthermore, understanding the physiological differences between biofilm and planktonic organisms ranks particularly high on the list of important and necessary research. Such contributions will only serve to broaden our knowledge base, especially regarding the development of better antimicrobials while also fine-tuning the use of current highly effective antimicrobials. Using H_2O_2 as a model oxidizing biocide, we demonstrate the marked resistance of biofilm bacteria relative to planktonic cells. Because many biocides are good oxidizing agents (e.g., H_2O_2, HOCl), understanding the mechanisms by which genes involved in combating oxidative stress are activated is important in determining the overall efficacy of such biocides. Future studies will focus on determining mechanisms of oxidative stress gene regulation in bacterial biofilms.

Acknowledgments

Work in the laboratory of D.J.H. was supported by grants from the National Institutes of Health (A.I. 40541) and the Cystic Fibrosis Foundation (HASSETPO97), and work in the laboratory of T.R.M. was supported by the National Science Foundation Center for Biofilm Engineering.

[45] Measuring Antimicrobial Effects on Biofilm Bacteria: From Laboratory to Field

By NICK ZELVER, MARTY HAMILTON, BETSEY PITTS, DARLA GOERES, DIANE WALKER, PAUL STURMAN, and JOANNA HEERSINK

Introduction

Biofilm organisms typically exhibit a high resistance to antimicrobial agents compared to their planktonic counterparts.[1–3] Chen and Stewart[4] and Xu *et al.*[5] showed that reactive biocides such as hypochlorite and hydrogen peroxide had limited penetration into a biofilm. Other types of antimicrobials, such as antibiotics, penetrate quickly into the biofilm[6] but may still have limited efficacy compared with application of antibiotics to planktonic cells. Studies, in which cells bearing reporter genes were observed directly by confocal scanning laser microscopy, have shown that a large number of genes are upregulated as planktonic cells adhere to a surface and form biofilms.[7] Such genetic transformations from the planktonic to the sessile state may also play a role in the antimicrobial resistance of biofilms.

The resistance of biofilms to antimicrobials suggests that new standard analytical methods should be developed for evaluating the efficacy of antimicrobials specifically against biofilms. To be accepted by regulators and industry, these new methods must be unbiased, repeatable, reproducible, and practical. This article discusses these key issues in the development of new assays for biofilms and provides a prototype method for growing and

[1] C. T. Huang, K. D. Xu, G. A. McFeters, and P. S. Stewart, *Appl. Environ. Microbiol.* **64,** 1526 (1998).
[2] P. S. Stewart, *Antimicrob. Agent Chemother.* **40,** 2517 (1996).
[3] M. R. W. Brown and P. Gilbert, *J. Appl. Bacteriol. Symp. Suppl.* **74,** 87 (1993).
[4] X. Chen and P. S. Stewart, *Environ. Sci. Technol.* **30,** 2017 (1996).
[5] X. Xu, P. S. Stewart, and X. Chen, *Biotech. Bioeng.* **49,** 93 (1996).
[6] J. D. Vrany, P. S. Stewart, and P. A. Suci, *Antimicrob. Agent Chemother.* **41,** 1352 (1997).
[7] D. G. Davies and G. G. Geesey, *Appl. Environ. Microbiol.* **61,** 860 (1995).

0076-6879/99 $30.00

evaluating biofilms in the laboratory. We include two case studies to demonstrate the application of this methodology for evaluating antimicrobial efficacy of biofilms in (1) the household and (2) in an oil production field.

Methods for Growing Laboratory Biofilms

Characklis[8] has presented a review of laboratory reactors and analytical methods used to grow biofilms under controlled environmental conditions. The typical reactor he described requires a continuous supply of nutrients at a short residence time and a hydraulic shear at the biofilm test surface. Roe and colleagues[9] described two commonly used systems, the annular reactor, which uses a rotating cylinder for flow shear, and the recycle loop, which pumps nutrient fluid through a pipe to produce flow shear. The advantage of these configurations is that they enable the researcher to conduct a material balance across the system to determine the utilization of nutrients and production of biomass that can be related to biofilm growth. Characklis[10] also described the use of a capillary tube to grow biofilm for direct, on-line viewing under a microscope. More recent studies at the Center for Biofilm Engineering included the use of a similar approach but with more defined dimensions, machined into a small flow cell specifically designed for use in a confocal scanning laser microscope.[11] Vidal *et al.*[12] and Genevaux *et al.*[13] monitored biofilm growth microscopically in microtiter plates. Das *et al.*[14] monitored biofilm growth in microtiter plates by light transmittance. Microtiter plates offer the advantage of conducting biofilm assays on numerous test surfaces (e.g., a 96-well plate).

Despite the history of developing systems for growing and evaluating biofilms in the laboratory, no standard regulatory-approved method exists for evaluating antimicrobial efficacy specifically against biofilms. Hoffert[15] reported that, "An EPA scientific advisory ... in September (1997) ... recommended that better methods of testing are needed to simulate condi-

[8] W. G. Characklis, *in* "Biofilms" (W. G. Characklis and K. C. Marshall, eds.), p. 55. Wiley, New York, 1989.

[9] F. L. Roe, E. J. Wentland, N. Zelver, B. K. Warwood, R. Waters, and W. G. Characklis, *in* "Biofouling and Biocorrosion in Industrial Water Systems" (G. G. Geesey, Z. Lewandowski, and H. C. Flemming, eds.). Lewis Publishers, 1993.

[10] W. G. Characklis, *in* "Biofilms" (W. G. Characklis and K. C. Marshall, eds.), p. 66. Wiley, New York, 1989.

[11] A. C. Camper, M. A. Hamilton, K. R. Johnson, P. Stoodley, G. J. Harkin, and D. S. Daly, *Ultrapure Water* **11,** 26 (1994),

[12] O. R. Vidal, R. Longin, C. Prigent-Combaret, C. Dorel, M. Hooreman, and P. Lejeune, *J. Bacteriol.* **180,** 2442 (1998).

[13] P. Genevaux, S. Muller, and P. Bauda, *Fed. Eur. Microbiol. Soc. B.V.* PII S0378, 1097 (1996).

[14] J. R. Das, M. Bhakoo, M. V. Jones, and P. Gilbert, *J. Appl. Microbiol.* **84,** 852 (1998).

[15] S. P. Hoffert, *Scientist* **12,** 11 (1998).

tions commonly encountered in the real world that can inhibit the activity of antimicrobial products. For example, microbes on household and industrial surfaces often develop biofilms, which inhibit the penetration of antibacterials and protect pathogens.''

At present, the only standard antimicrobial assays accepted by regulatory agencies are either dried-surface tests or suspension tests. The typical dried-surface test consists of the following steps applied to the test surface (carriers):

1. Place the carrier in a culture broth for a specified time.
2. Extract and dry the carrier to bind the microbes to the surface as a dried organic film.
3. Treat the carrier by immersing it in the test antimicrobial (or an inactive control).
4. Transfer the microbes from the surface to a suspension and use standard culturing techniques to count the number of viable organisms.
5. Calculate the log reduction (LR) between antimicrobial-treated carriers and controls.

Studies are needed to determine the extent to which the results of the dried-surface tests are applicable to biofilms. The only parallel test results known to us show that antimicrobial chemicals are less effective against biofilms than against dried-surface bacteria.[16]

Criteria for Unbiased, Repeatable, and Reproducible Antimicrobial Assays

Although biofilms present a new challenge in evaluating the efficacy of antimicrobials, the statistical and theoretical bases of current standard antimicrobial assays provide important benchmarks. Key benchmarks for any regulatory-approved standard antimicrobial assay are that the method be unbiased, repeatable, and reproducible.

Unbiased Assays. For established antimicrobial tests, the evidence that a method would produce unbiased estimates of the LR was based primarily on expert opinion. Experts reviewed the steps that made up the laboratory method, and if they were convinced that there was no reason to suspect bias, the method was proposed for acceptance as a standard. However, experts have since discovered biases in at least one existing standard method. Thirty-five years since the use dilution method (UDM) was ac-

[16] S. Bloomfield, personal communication.

cepted as a standard surface test for antimicrobials, a series of well-focused investigations uncovered potential biases in the UDM.[17–22]

To ensure that our biofilm assays are unbiased, we are developing a "gold standard" for assessing the efficacy of an antimicrobial agent against biofilm bacteria. The gold standard will consist of directly measuring viable cell numbers on the surface using microscopy (e.g., using labels or stains) and activity of attached bacteria using respirometry.[23] The gold standard is relatively labor-intensive and requires expensive equipment. Thus, we expect its use to be limited to determining that our prototype antimicrobial assays for biofilms are unbiased. We are in the early stages of developing such a gold standard and do not have a proven example to provide in this article.

Repeatability. With the inherent variability associated with microbiological experiments, it is improbable that replicate experiments will produce exactly the same LR values. Consequently, it is necessary to choose a statistical definition of repeatability. The conventional approach is to calculate the standard deviation of the LR values across replicate experiments, i.e., the repeatability standard deviation, denoted by s_r.[24] Small repeatability standard deviations indicate good repeatability.

There is not a single s_r for a method because s_r is dependent on the antimicrobial agent and the microbial species/strain used in the experiment. Typically, the value of s_r is larger for partially active antimicrobial agents than for inactive or very effective agents. A review of the literature showed that established standard antimicrobial assays have associated repeatability standard deviations that range from 0.2 to 1.0.[25]

Reproducibility. For regulatory purposes, different laboratories must arrive at approximately the same result when testing an antimicrobial agent. There are examples where convincing antimicrobial efficacy test results were presented in a successful registration application, but subsequent testing could not reproduce those results.[26] Some organizations that referee

[17] J. M. Ascenzi, R. J. Ezzell, and T. M. Wendt, *Appl. Environ. Microbiol.* **51,** 91 (1986).

[18] E. C. Cole, W. A. Rutala, and J. L. Carson, *J. Assoc. Official Anal. Chem.* **70,** 903 (1987).

[19] E. C. Cole and W. A. Rutala, *J. Assoc. Official Analyt. Chemists,* **71,** 9 (1988).

[20] E. C. Cole, W. A. Rutala, and G. P. Samsa, *J. Assoc. Official Anal. Chem.* **71,** 187 (1988).

[21] E. C. Cole, W. A. Rutala, J. L. Carson, and E. M. Alfano, *J. Assoc. Official Anal. Chem.* **71,** 288 (1988).

[22] E. C. Cole, W. A. Rutala, J. L. Carson, and E. M. Alfano, *Appl. Environ. Microbiol.* **55,** 511 (1989).

[23] J. L. Czekajewski, L. Nennerfelt, H. Kaczmarek, and J. F. Rabek, *Acta Polmer.* **45,** 369 (1994).

[24] Anonymous, in "Official Methods of the Association of Official Analytical Chemists" (K. Helfrich, ed.), p. 681. AOAC, Arlington, VA, 1990.

[25] N. Tilt and M. A. Hamilton, *J. Assoc. Official Anal. Chem.,* in press (1999).

[26] W. A. Rutala and E. C. Cole, *Infect. Control* **8,** 501 (1987).

and certify new methods, notably the Association of Official Analytical Chemists International, require a collaborative study involving at least eight laboratories for purposes of evaluating the reproducibility of a new method. The results of the collaborative study are described by the reproducibility standard deviation, denoted by s_R, and the laboratory-to-laboratory standard deviation, denoted by s_L.[24] Small values of s_L and s_R are desirable. The reproducibility standard deviation measures the total variability, including within laboratory and laboratory-to-laboratory variability, as is indicated by the formula $s_R = (s_r^2 + s_L^2)^{1/2}$. A statistical analysis of variance is typically used to calculate the standard deviations from collaborative study data.[27] Established standard antimicrobial assays have associated reproducibility standard deviations that range from 0.3 to 1.5.[25] It is not uncommon for s_L to be as large, or somewhat larger, than s_r. The notion that performance standards for quantitative antimicrobial tests should take account of the s_L and s_R is new, and no such performance standards exist at present.

Realistically, only a limited amount of data will be available at the time an antimicrobial chemical is submitted for registration. There will probably be LR values from only two or three laboratories. It is possible that, due to chance variation, the submitted LRs happen to be at the upper end of the normal spread of values. For this reason, the performance standards must take into account the risk that an inefficacious antimicrobial (one that truly does not meet the target) mistakenly passes the registration criterion. Such an error occurs because of the inherent statistical variation in the assay method. Although performance standards could be formulated so that errors are impossible, they would be impractical, either requiring a massive amount of testing or requiring the LRs to be very high relative to the target. For these reasons, there will be some technical, statistical work involved in creating performance standards that are practical, properly focused on the chosen definition of the target LR, and guarantee that the risk of an error is no larger than 0.05 or some other specified value.

Practical and Repeatable Assay for Biofilms

Any new standard method for evaluating antimicrobials against biofilms must meet the criteria of being unbiased, repeatable, and reproducible to meet regulatory approval. Proving that the method is acceptable will require data from an extensive set of demonstrations in multiple laboratories and with a range of antimicrobial categories. The method must also be practical such that screening of large numbers of antimicrobials can be conducted

[27] E. H. Steiner, *in* "Statistical Manual of the Association of Official Analytical Chemists" (by W. J. Youden and E. H. Steiner eds.). AOAC, Arlington, VA, 1975.

with a minimal investment of time and equipment. Under our strategy for ultimately developing recognized standard methods for evaluating the efficacy of antimicrobials against biofilms, we have first focused on developing a practical protocol for growing, sampling, and analyzing a biofilm and determining the repeatability of these steps.

Figure 1 summarizes our strategic approach to developing practical, relevant, and repeatable methods for evaluating antimicrobials against biofilms. The strategy includes taking research methods and conducting adequate field validation and laboratory tests to develop the methods as standard protocol. Ultimately, these methods will be published and databases of results developed to be available to industry, the biofilm research community, and regulatory agencies. We believe both industry and regulatory agencies urgently need such standardized approaches in order to advance antimicrobial technology in all areas where biofilms pose a problem or risk.

Disk Reactor Method for Growing and Analyzing Biofilms

Figure 2 shows the rotating disk reactor (RDR) system for growing biofilm. The RDR vessel is a 1-liter glass beaker fitted with a drain spout. The bottom of the vessel contains a magnetically driven rotor with six 1.27-cm-diameter biofilm test-surface coupons. The rotor is constructed from a

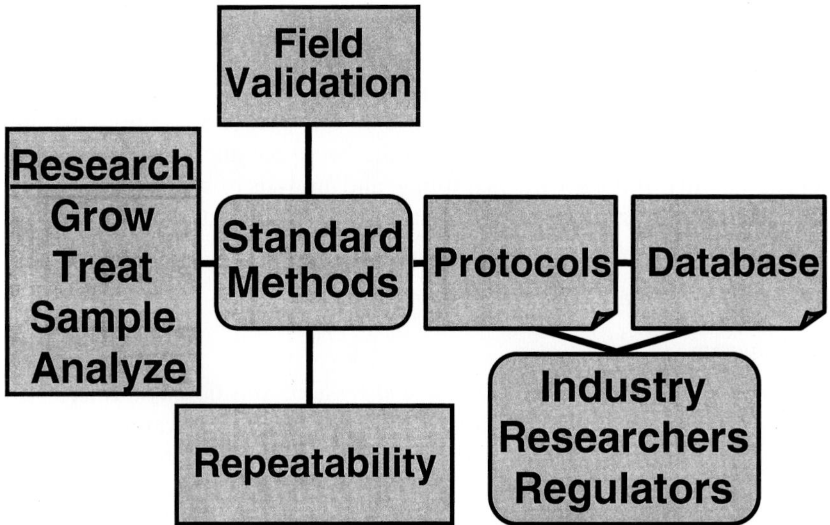

FIG. 1. Strategic process for developing standard methods for evaluating antimicrobials against biofilms.

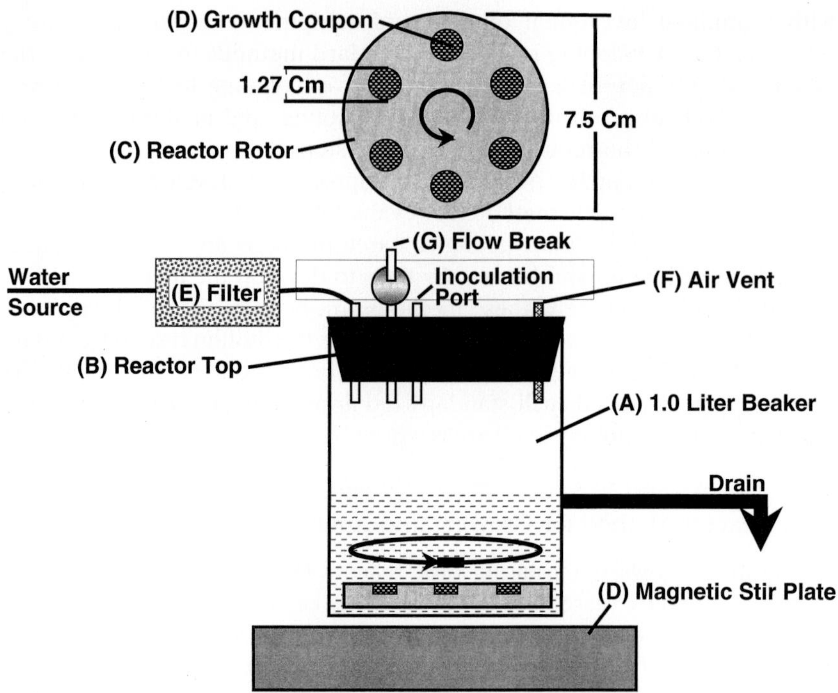

FIG. 2. Rotating disk reactor for laboratory biofilm assays.

star-head magnetic stir bar to which a Teflon and silicone rubber disk is attached that holds the coupons. The coupons rotate continuously to provide fluid shear. Biofilm growth nutrients are pumped continuously into the vessel. A series of control RDR studies were conducted to demonstrate reproducibility of the growth, sampling, and analysis of biofilm. The following RDR biofilm growth protocol was used for these studies.

Rotating Disk Reactor Biofilm Growth Protocol

Reagents

1. Batch nutrient broth. Dissolve 0.18 g of tryptic soy broth (TSB, Difco Inc., Detroit, MI) into 600 ml of reagent grade water. Autoclave for 20 min at 121°. Allow to cool to room temperature before inoculating.
2. Continuous flow nutrient broth. Dissolve 5 g of TSB into 500 ml of reagent grade water. Autoclave for 20 min at 121°. Autoclave 10 liters of reagent grade water in a suitable vessel (10 liter carboy with

at least two barbed bulk head fittings plumbed into the lid) for 3 hr at 121°. Aseptically pour the 5 g/500 ml sterile TSB into 10 liter sterile reagent grade water by flaming flask openings and working close to a Bunsen burner flame.

3. R2A agar plates (Difco).
4. Buffered water for serial dilutions.[28]

Apparatus Setup

1. Rotating disk reactor: Assemble parts A–H according to the schematic (Fig. 2). Tape the reactor stopper to the reactor vessel to ensure stability during autoclaving. Keep tubing connected to carboys, but disconnected from the reactor vessel. Foil and tape the ends of tubing that will connect to the nutrient and water supplies. Autoclave the reactor, tubing, and filter combination at 121° for 20 min. After all the parts have cooled, complete tubing connections by removing foil from the ends of tubing and disinfecting with 95% (v/v) ethanol. While completing all these connections, avoid excessive handling of sterile parts.

Parts A–H (Fig. 2)

A. Reactor vessel: 1-liter spoutless Pyrex beaker with a glass barbed drain spout added at the 500-ml mark. The spout can be attached at an angle of 90° or greater with the beaker wall. Connect a piece of tubing to the spout and clamp it prior to autoclaving.

B. Reactor top: Bore three holes into a size 15 rubber stopper. Push 7- to 8-cm lengths of glass tubing through the stopper. Fire polish the ends of the glass tubing before plumbing the stopper. Lubricate glass tubing with glycerol to ease through the rubber stopper. The tubing will form the nutrient, air, and water inlet ports. An inoculating port consisting of a glass tube fitted with an autoclavable rubber septum can also be added to the stopper.

C. Reactor rotor. This is constructed from a Teflon disk (7.63 cm diameter), a Viton disk (7.63 cm diameter), and a star-head magnetic stir bar. The rotor is held together with nylon screws. Six evenly spaced holes holding growth surfaces (1.27 cm diameter) are bored through the Teflon and Viton.

D. Growth coupons. Polycarbonate (or alternative test-surface) coupons (1.27 cm diameter, 0.4 cm thick) are inserted into corresponding holes bored into the rotor.

E. Supor cap sterile in-line filter (Gelman Sciences, Inc., Ann Arbor,

[28] Anonymous, *in* "AHPA, Standard Methods for the Examination of Water and Waste Water" (A. D. Eaton, L. S. Clesceri, and A. E. Greenberg, 19th ed., 9–17 (1995).

MI). Attach in-line water filter to tubing exiting the water port in reactor stopper.

F. Bacterial air vent. A bacterial air vent is attached to one of the three glass tubes exiting the reactor stopper and allows for sterile air exchange between the reactor volume and the environment.

G. Glass flow break. Install glass flow break in nutrient supply line to prevent back contamination of nutrient supply vessel.

2. Magnetic stir plate. Reactor sits on a stir plate and the stirring action spins the rotor during the experiment.
3. Ring stand three-prong extension clamp. Stabilizes tubing above the reactor.
4. Pumps, pump heads, and controller (Masterflex, Cole-Parmer, Vernon Hills, IL). One pump with a single pump head, set at 1 ml/min, for pumping nutrients into reactor. One pump with two pump heads, set at 16 ml/min, used to pump filtered water into reactor and pump waste out of waste vessel (20 liter carboy).
5. Reactor supports. Four L-shaped devices are evenly spaced and glued onto the stir plate to raise the reactor (~0.5 cm) above the plate face. This protects the reactor from heat produced by the stir plate. Placement of the supports is dependent on centering the individual disk rotor.
6. Portable sampling surface. A piece of 1.27-cm-thick polycarbonate (about 20 × 28 cm) can be used as a portable work surface. A thin cross section of norprene rubber sheeting [dimensions: 3.8 × 1.9 × 0.63 cm with a hole (~1.3 cm diameter)] is cut in the center with a leather bore. The strip is glued to the polycarbonate surface; this serves as a holder in which coupons can be immobilized during scraping.
7. Wooden applicator sticks. These are used to scrape coupons. Place several 15-cm wooden applicator sticks in a test tube and autoclave at 121° for 20 min.
8. Stainless steel dissecting tools. These are used to pull the rotor from a reactor and to pry coupons from rotors.
9. Empty sterile petri dish. This provides a sterile surface to place the rotor in after it is removed from the reactor.
10. Stainless steel scissors clamp. This is used to hold the coupon during scraping.
11. Vortex. For sample mixing during serial dilutions and plating.
12. Digital Pipetman (Rainin digital EDP2 pipettor, Emeryville, CA). Used for plate inoculation.
13. Tissue homogenizer (Tekmar-Dohrmann, Cincinnati, OH). For disaggregating biofilm bacteria.

Reactor Operation

1. Batch inoculation. Aseptically pour sterile batch TSB (described earlier) into the sterile reactor vessel just up to drain spout. Retrieve 1 ml frozen stock of *Pseudomonas aeruginosa* (or alternative biofilm forming test organism) at a bacterial density of approximately 10^8 cfu/ml from a $-80°$ freezer and let thaw. Disinfect the tubing attached to the bacterial air vent close to the point where it is connected to the glass tubing exiting the stopper with 95% ethanol. Aspirate the thawed culture with a sterile, disposable, 1-ml syringe and needle. Insert the needle into the disinfected tubing, pushing the needle down into the glass tubing. Quickly inject the thawed culture into the reactor. Inject inoculum through the inoculation port if it is available. Turn on the stir plate, allowing the rotor to turn at a medium speed. The reactor should remain in batch mode for 24 hr. Unclamp the drain spout before starting a continuous flow mode of operation. Operate with a continuous flow to provide a 30-min residence time. With a relatively short residence time of 30 min, most suspended organisms will be washed out before being able to grow. Thus, the bacterial growth and utilization of nutrients will be primarily in the attached or biofilm state.
2. Continuous flow operation. Also referred to as CSTR mode. After 24 hr in batch mode, continuous flow conditions will be implemented. Nutrients are pumped into the reactor through a pump set at 1 ml/min. Dilution water is pumped into the reactor via a pump set at 16 ml/min. The drain spout on the reactor allows overflow to occur, maintaining a constant nutrient concentration of 30 mg/liter TSB in the reactor during CSTR mode.

Treatment Protocol

1. Single application treatment protocols are under development.
2. Testing of biocidal agents under regrowth conditions (continuous application of treatment beginning after inoculation) can be implemented by the addition of antimicrobial agents to the input water supply.

Sampling Protocol. Coupons can be sampled (always in duplicate) at any time point after 24 hr under continuous flow (CSTR) conditions.

Sampling Steps

1. Disinfect area. Disinfect portable sample surface and surrounding lab bench with 95% ethanol. Flame sterilize stainless steel dissecting tools and clamp and set aside until needed.

2. Remove rotor and coupon. Turn off the stir plate. Carefully lift off the top of the reactor and insert a sterile dissecting tool into the extraction device. Hook the rotor and lift it out of the reactor. Place it in an empty sterile petri dish. Hold onto the rotor with a sterile clamp and carefully pry out a coupon using a sterile dissecting tool, taking care not to disturb the biofilm on the coupon surface. After the coupon is removed, carefully set the rotor back into the reactor with a sterile dissecting tool and turn the stir plate back on.
3. Scrape biofilm. Hold onto the coupon with a sterile clamp. Using a sterile applicator stick, thoroughly scrape the surface of the coupon for about 1 min, stirring occasionally into 9 ml of sterile buffered water. After sufficient scraping, rinse the coupon surface with 1 ml of sterile buffered water. The final volume in the sample test tube will be 10 ml.
4. Homogenize. Autoclave the homogenizer probe for 20 min at 121°. Insert the probe into the homogenizer. Homogenize the sample tube at ~20,500 rpm for 30 sec for bulk fluid samples and 60 sec for scraped biofilm samples. Remove sample tube and homogenize a sterile dilution blank for 30 sec at 20,500 rpm. Remove the probe from the homogenizer and place in a tube containing 95% ethanol. Let the probe sit in this tube for about 1 min. While the probe is still in ethanol, insert it back into the homogenizer and homogenize at 20,500 rpm for ~15 sec. Remove the probe from the homogenizer and shake it to remove excess ethanol. Flame sterilize the probe. Insert the probe and homogenize another sterile dilution blank for 30 sec at ~20,500 rpm.
5. Complete serial dilutions. Perform 1 : 10 serial dilutions on the original samples.
6. Plate. Plating in this protocol can be done using the drop plate method for pure cultures and the spread plate method[29] for mixed cultures. Prepare plates for the drop method by dividing each plate into four quadrants, allowing one quadrant per dilution. Prepare these plates in duplicate. Set an automatic pipettor to pick up 100 μl of sample and to expel it in 10-μl volumes. Vortex sample. Expel sample in five evenly spaced 10-μl drops per dilution, on duplicate plates. When drops have dried, invert the plates and incubate at 35° for 17–20 hr. For mixed cultures, use the spread plate method.

[29] A. L. Koch, *in* "Methods for General and Molecular Bacteriology" (P. Gerhardt, ed.), p. 255. ASM Press, Washington, DC, 1994.

Analytical Protocols

1. Viable counts. Also referred to as plate counts. Colony counts using the drop plate method are done by picking the dilution that gives 3–30 colonies per 10 μl drop of sample dispensed. Viable areal cell counts are expressed as cfu(colony-forming units)/cm^2.

Results of Rotating Disk Reactor Control Studies

The control biofilm growth experiment described earlier has been repeated 13 times. The repeats involved four technicians operating independently. A visual impression of the repeatability of the results is shown in Fig. 3. Densities among coupons in the same reactor (same experiment) are shown by points aligned vertically. The variability among experiments by the same technician can be discerned by averaging the densities in each vertical line and comparing those averages. Similarly, the variability among technicians can be discerned by averaging all results for a technician and comparing those four averages. The overall average of the $\log_{10}$ density results is 7.6. The range of log densities is about 1.0 (cf. vertical axis of Fig. 3), indicating good repeatability of results (note, the repeatability in this

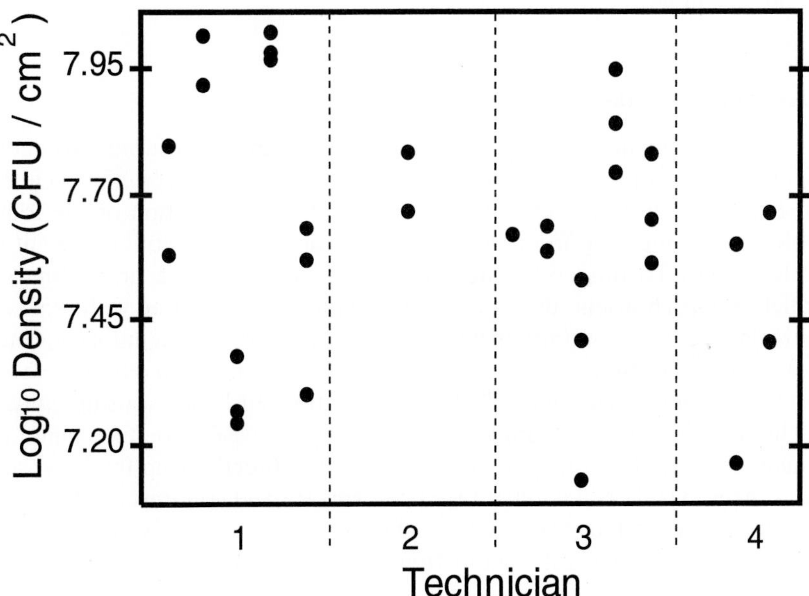

FIG. 3. Rotating disk reactor control biofilm experiment repeated 13 times.

context pertains to an untreated biofilm only, not to the repeatability of log reduction values).

The visual information conveyed by Fig. 3 can be made quantitative by conducting a statistical analysis of variance. The analysis shows that the repeatability standard deviation of log densities for control biofilms in the rotating disk reactor is 0.27. The analysis of variance shows that the differences among log densities is 25% attributable to differences among coupons in the same reactor (same experiment) and 75% attributable to differences among experiments. There was no discernable difference among technicians other than what could be attributable to coupons and experiments. These results are encouraging because they indicate that the protocol is transferable and removes the potential influence of technician subjectivity.

Field Studies

Each field site requires a unique methodology and validation to ensure that laboratory results can be properly scaled-up to represent the field conditions. The following case studies describe the use of the rotating disk reactor to represent field conditions in a household disinfection environment and in an oil production field.

Case Study: Household Disinfection in Toilet Bowls

The rotating disk reactor was used in an intermittent flow environment to evaluate the efficacy of antimicrobials against toilet bowl biofilm bacteria. The first step in development of this method is examination of the toilet bowl environment in order to characterize the bacterial biofilm present in toilet bowls. To this end, a field study was conducted, incorporating six toilets in local households as field systems. Each toilet is fitted with removal porcelain ceramic coupons on which biofilm is allowed to accumulate. Once a week (for a duration of 7 weeks) coupons are retrieved in duplicate from each study toilet and analyzed for viable, attached bacteria using culture techniques as detailed in Pitts *et al.*[30] In addition, weekly biofilm sampling includes duplicate scrapes of attached bacteria directly from the bowl surface. Areal viable cell counts from bowl surface and coupons are found to be strongly correlated (r = 0.85). The field phase of the study provides biofilm accumulation rates and extents for all six field systems, and these

[30] B. Pitts, P. S. Stewart, G. A. McFeters, M. A. Hamilton, A. Willse, and N. Zelver, *Biofouling* **13,** 19 (1998).

are used to construct an average toilet bowl fouling profile for this particular set of toilets and ambient conditions.

The next step in the method development involves incorporating some of the defining characteristics of the toilet bowl growth environment into a laboratory rotating disk reactor. Features identified as necessary in the laboratory system include intermittent nutrient provision and water flow, ambient temperature and tap water influent supplying approximately 1 mg/liter chlorine residual (as measured at the tap), and a continuous low-level inoculum of drinking water microorganisms. At start-up, laboratory growth reactors are also given an extra inoculation with organisms isolated from 3- to 7-week-old toilet bowl biofilms.

Once per hour, RDR vessels experience a simulated toilet bowl use cycle, in which sterile tryptic soy broth (TSB, Difco Inc., Detroit, MI) is added to the 400-ml working volume to produce a final concentration of 1 g/liter. After a 5-min residence time, the volume is stirred and drained. When the vessel is empty, a fresh supply of refill tap water is delivered to 400 ml. This cycle, controlled by a programmable timing module (Chrontrol Corp., San Diego, CA), is repeated once an hour for the duration of the experiment. A series of experiments lasting from between 1 week and 11 days is run and, as with the field systems, the areal viable cell counts collected allow completion of a fouling profile for each reactor.

Finally, antimicrobial efficacy testing is conducted with both laboratory and field systems. Chlorine is chosen as the biocidal agent and is used in the form of sodium hypochlorite, as household bleach. The chlorine is applied continuously in both toilets and reactors at 0, 1, 3, and 9 mg/liter. In toilets, bleach is dispensed automatically per flush to give the desired concentration in the filled bowl, and in reactors, bleach is added to the tap water refill supply to give the same final concentration of chlorine in the reactor volume.

The methodology is validated in three stages. First, measures are established by which biofilm growth and biocide efficacy could be evaluated. The following quantities were selected: biofilm accumulation at end point, biofilm accumulation at day 2, specific fouling rate, and log reduction. Second, it was required that the laboratory growth system be repeatable. This is shown graphically in Fig. 4. Repeatability standard deviations, which measure total variability between experiments, were low. Log reduction values were as repeatable in this biofilm growth system as in standard suspension and hard surface disinfection tests.

Third, it was necessary that the measures described previously could be calibrated reliably between the laboratory system and the field (actual toilets). In this investigation, biofilm accumulation curves from both lab and field were roughly parallel, where the time scale is days for reactors

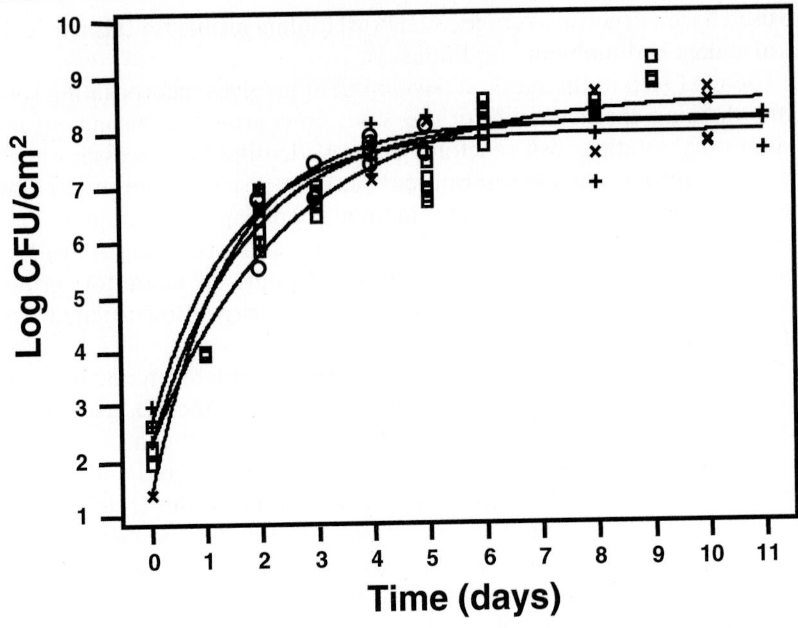

FIG. 4. Repeated control biofilm growth experiments in the toilet-simulating rotating disk reactor.

and weeks for toilets, as shown in Fig. 5. Based on criteria described earlier, we expected that a ranking of biocidal agents as used against laboratory-grown biofilms versus field-grown biofilms would be the same. The same ranking was achieved in the laboratory and in the field with three of the four measures, using three different concentrations of sodium hypochlorite. In summary, this reactor system effectively simulated the intermittent flow environment represented by a toilet. Results demonstrated that this biofilm growth system produced repeatable biofilm accumulation curves and provided repeatable assays of the efficacy of antimicrobial agents against those biofilms.

Case Study: Antimicrobial Efficacy Testing for Anaerobic Oil Field Application

A rotating disk reactor was used to evaluate nitrite inhibition of sulfide reducing bacteria (SRB) as a proposed method to control souring (hydrogen sulfide production) in oil field operations. The RDR was modified to operate under anaerobic conditions and to represent an oil dehydrator environment.

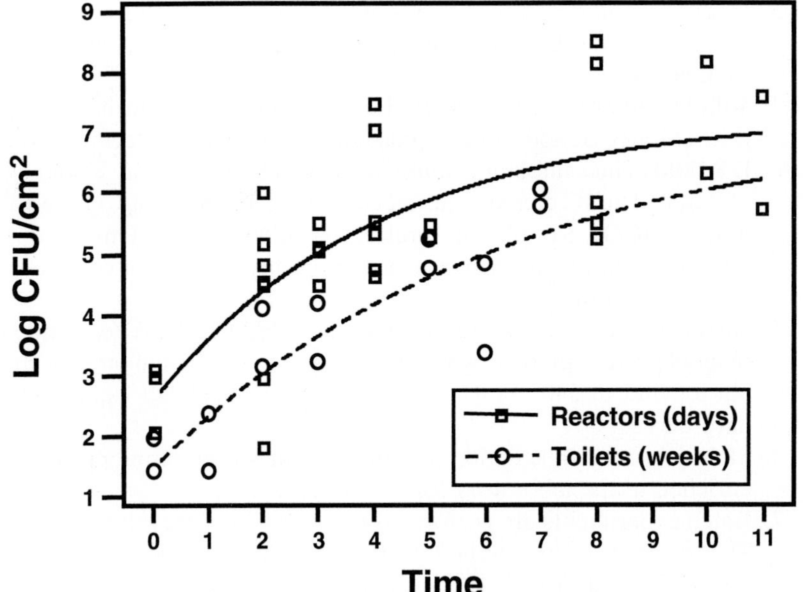

FIG. 5. Comparison of control biofilm growth experiments in the toilet-simulating rotating disk reactor (days) versus the actual toilet (weeks); both systems were treated with continuously applied chlorine at 9 mg/liter.

Two nitrite dosing methods were applied to determine the effectiveness of nitrite at SRB inhibition: (1) a high nitrite pulse dose was injected directly into the reactor and (2) a small continuous dose was supplied to the reactor via the media.

Methods

1. Growth media: 10 liters of modified Postgate medium C[31] is autoclaved for 3 hr and allowed to cool under nitrogen. Media are modified to include 200 mg S/liter as SO_4^{-2} (SO_4^{-2}-S/liter), 200 mg C/liter as lactate (lactate-C/liter), and the reducing agents are omitted. In addition, 2% (w/w) sodium chloride is included.
2. A mass balance approach is taken to evaluate SRB activity by sulfate utilization and hydrogen sulfide production.

[31] J. R. Postgate, *in* "The Sulfate-Reducing Bacteria," 2nd Ed., p. 31. Cambridge Univ. Press, Cambridge, UK, 1984.

3. Hydrogen sulfide in the aqueous phase is measured using the methylene blue method.[32] The samples are fixed in 1% zinc acetate prior to analysis.
4. Sulfate and nitrite are measured using a Dionex ion chromatograph (IC) (Model Al-450) with a pulse electrochemical detector (Model DX 300). The sampling column is an AS4A-SC that has a pore size of 2 mm. The IC eluant solution consists of 1.8 m*M* Na_2CO_3 and 1.7 m*M* $NaHCO_3$. Samples are pretreated with a Dionex OnGuard-Ag sample pretreatment cartridge to remove any chloride or cells before being run on the IC.
5. Nitrite concentrations are verified using HACH NitriVer 3 nitrite reagent powder pillows. Absorbance is read after 10 min on a spectrophotometer at 546 nm. Concentrations are calculated using a standard curve.
6. SRB consortia inoculation, obtained from an operating oil field, is enriched in the laboratory for lactate utilization.
7. Batch experiments are performed to determine a growth rate for the SRB consortium in modified Postgate medium C. Batch experiments are performed in triplicate over a 5-day period. Three enrichment vials with 30 ml modified Postgate medium C are inoculated with 1 ml sample of SRB bacteria labeled. One control vial remains uninoculated. All vials are incubated at ambient temperature (20°) and sampled daily for 5 days. A 0.05 hr^{-1} growth rate is determined for the SRB consortium. For experiments in the 500-ml rotating disk reactor, a medium flow rate of 0.8 ml/min provides a residence time of 10 hr to ensure that planktonic SRB washed out of the system and that measurements made during experimentation are based on biofilm growth.

Anaerobic Rotating Disk Reactor System

Growth of Relevant Biofilm

Nitrogen gas is supplied to the oil field RDR (Fig. 6) to maintain an anaerobic environment. The 90% nitrogen flows through a 400° furnace, which encases a glass tubing filled with copper filings to scavenge any residual oxygen. From there, the N_2 sparges the medium, creating a positive head space that forced the gas into the reactor. The effluent is then collected

[32] Anonymous, *in* "Standard Methods for the Examination of Water and Wastewater" (A. D. Eaton, L. S. Clesceri, and A. E. Greenberg, eds.), 19th ed., p. 4. American Public Health Association, Washington, DC, 1995.

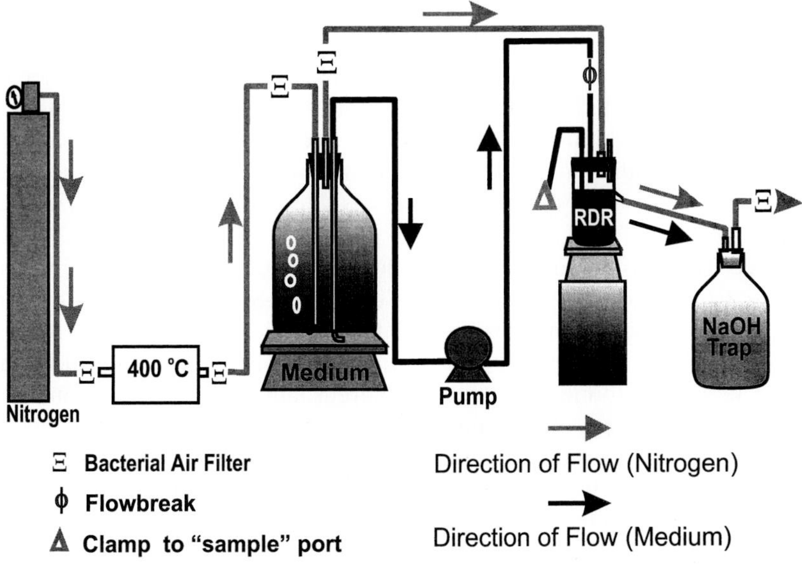

FIG. 6. Rotating disk reactor system representing an oil production field.

in a sodium hydroxide (NaOH) trap to keep the sulfide in solution and prevent its escape to the atmosphere. With the exception of the nitrogen tank and furnace, the remainder of the system is kept under a fume hood for safety reasons.

Prior to initiating the experiment, the disk reactor is sealed with a butyl rubber stopper affixed with plastic tie downs. The system is tested for air tightness by pressurization with nitrogen and checked for leaks. The RDR and tubing are then autoclaved for 20 min. After cooling, the RDR is attached to the medium container and filled via a peristaltic pump. The pump is then switched off, the medium tubing to the reactor is clamped shut, and approximately 30 ml SRB is injected into the reactor to grow in batch for 48 hr. Continuous flow could begin once sulfide levels are greater than 20 mg/liter, indicative of a well-established biofilm. After 72 hr in batch mode, however, sulfide levels remain below 20 mg/liter. The reactor is therefore reinoculated with an additional 30 ml SRB and, within 24 hr, 33 mg/liter H_2S-S is measured. At this point, the pump is turned on and continuous flow started. Sulfide and sulfate levels are measured daily until an apparent steady state is reached (as determined by reactor sulfide and sulfate concentrations). Upon initiation of continuous flow, sulfide and sulfate levels remain relatively stable at approximately 50 mg/liter H_2S-S and 130 mg/liter SO_4^{2-}-S. Sulfide and sulfate stability is taken as an indication that the SRB has reached a steady state.

Pulse Treating. Two hundred milligrams per liter NO_2^--N is injected directly into the disk reactor after a biofilm is established. Sampling is performed throughout the 10-hr reactor residence time. Predose samples are taken for sulfide and sulfate. After the injection with nitrite, samples are drawn every 2 hr throughout one residence time for sulfide, sulfate, and nitrite.

Predose sampling indicates the following concentrations of sulfide and sulfate: 63 mg/liter H_2S-S and 130 mg/liter SO_4^{2-}-S. Following nitrite injection into the reactor, time zero concentrations are measured to be 57 mg/liter H_2S-S, 135 mg/liter SO_4^{2-}-S, and 201 mg/liter NO_2^--N. Sulfide and nitrite levels are found to decrease, whereas sulfate increases throughout one residence time. Approximately 13 hr later, another sample is taken, which indicates a residual nitrite concentration of 8 mg/liter NO_2^--N, a decrease in sulfide to 11 mg/liter H_2S-S, and a corresponding increase in sulfate to 213 mg/liter SO_4^{2-}-S. Once all the nitrite is washed out of the system, sulfide levels return to pretest concentrations (Fig. 7).

To determine whether results are showing an actual SRB inhibition or depicting an oxidation–reduction reaction between hydrogen sulfide and

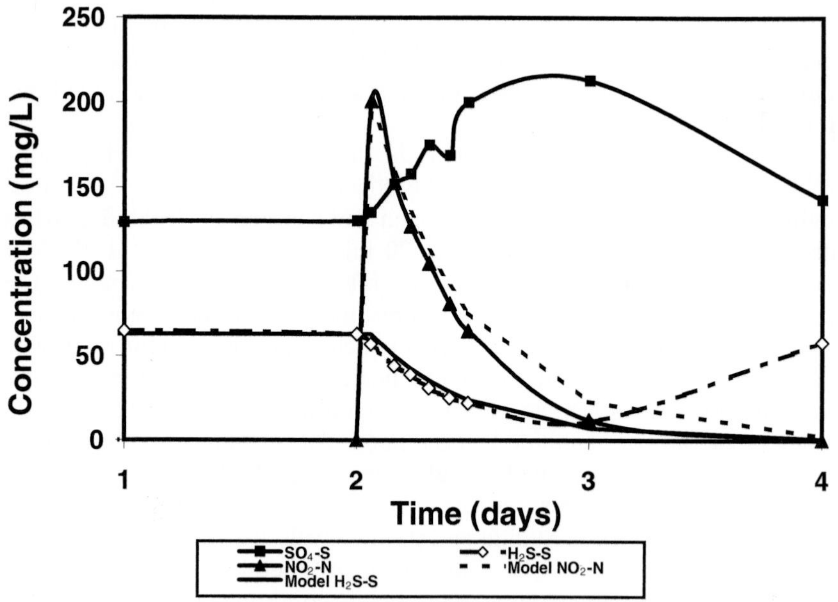

FIG. 7. Pulse dosing of nitrite.

nitrite, a mathematical model of a pulse tracer added to a continuous-flow stirred-tank reactor (CFSTR) is calculated based on the following equation for a pulse tracer concentration of C_o:

$$C = C_o(e^{-t(Q/V)})$$

where C is the concentration of tracer in reactor (mg/ml), V is the reactor volume (ml), Q is the volumetric flow rate (ml/min), C_o is the initial concentration of tracer (mg/ml), and t is time (min).

The modeled tracer curve for sulfide as well as nitrite nearly mimic the actual curves, indicating SRB inhibition by nitrite (Fig. 7). If scavenging has occurred, the actual curves for each compound would have fallen significantly below the tracer models.

Continuous Treating. The second treatment protocol includes continually applying a low dose of nitrite to the reactor and recording the inhibition of SRB activity over time. The modified Postgate medium C is prepared to include 10 mg/liter NO_2^--N. When continuous dose sampling begins, no nitrite is detected in the reactor (Fig. 8), signifying that a scavenging effect is occurring. Throughout the sampling period when scavenging is observed, sulfide levels average 50 mg/liter H_2S-S. A stock solution of nitrite is prepared that could be injected into the medium until a 10-mg/liter NO_2^--N

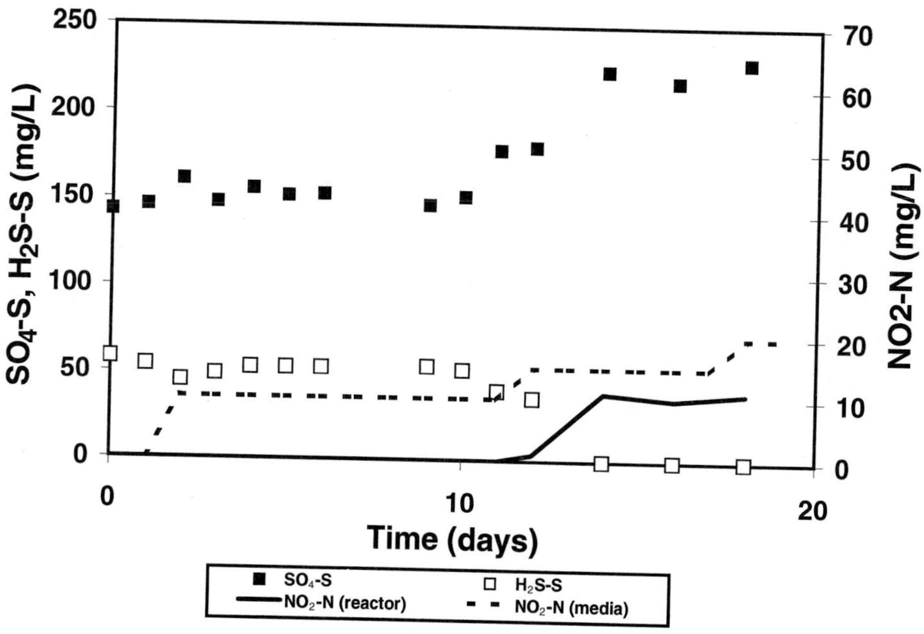

FIG. 8. Continuous dosing of nitrite.

residual is measured in the reactor. Once the nitrite residual is achieved, sulfide concentrations immediately drop to zero.

Field trials, conducted in a mature and soured North Sea oil-producing field, have demonstrated SRB inhibition with the application of nitrite. A qualitative comparison of the field demonstration to the oil dehydrator simulation in the disk reactor system shows the potential for the laboratory system to represent the field. This case study demonstrates the utility of using a mass balance for noninvasive monitoring of biofilm activity, particularly with a difficult-to-access anaerobic system. Further work is needed to validate the laboratory system in terms of repeatability and field relevance.

Conclusions

Our approach to developing acceptable screening tests is to start with an unbiased and repeatable laboratory protocol and then systematically improving the method to reduce the expense and time and developing a practical method without sacrificing the unbiased and repeatable. We have shown repeatability of a prototype laboratory biofilm growth, sampling, and analytical protocol using the rotating disk reactor and demonstrated how the RDR can be used to simulate various field applications. We will continue to generate data needed to promote this and other methodologies for screening antimicrobial efficacy against biofilms. Because biofilm organisms appear to have greater resistance than their planktonic counterparts, we believe that new standard analytical methods developed specifically for evaluating antimicrobials against biofilms are an essential need in the effort to control biofilm-related problems.

Acknowledgments

The authors acknowledge support from the National Science Foundation Engineering Research Centers Program (Cooperative Agreement #EEC-8907039). Special thanks are also addressed to the following: Montana State University researchers for their contribution to the toilet bowl case study; Philip S. Stewart, Department of Chemical Engineering; Gordon A. McFeters, Department of Microbiology; and Alan Willse, Department of Mathematical Sciences.

[46] Quantitative Assessment of Biocide Control of Biofilms Including *Legionella pneumophila* Using Total Viable Counts, Fluorescence Microscopy, and Image Analysis

By J. T. WALKER, A. D. ROBERTS, V. J. LUCAS, M. A. ROPER, and R. G. BROWN

Introduction

A study was undertaken to evaluate the control of *Legionella pneumophila* in biofilms of domestic water systems typical of that found in a 50 person occupancy building within the United Kingdom.[1,2] The tests were carried out in three full-size identical test rigs constructed at the Building Services Research and Information Association (BSRIA).

Legionnaires' disease[3] has attracted attention since the 1980s due to a number of high profile outbreaks.[4] This disease is contracted through inhalation of aerosols from water systems containing *L. pneumophila,* particularly *L. pneumophila* sero group 1 Pontiac, and can result in a mild flu-like disease or a severe life-threatening pneumonia.[5,6]

Legionella pneumophila has been shown to be ubiquitous in the environment,[7] in man-made water systems,[8] and has been recovered from biofilms on surfaces in water systems.[9,10]

A number of different biocides have been used to control the presence of *L. pneumophila* both in the laboratory and in actual water systems.[3,11–13]

[1] N. L. Pavey and M. Roper, Technical Note TN 2/98. ISBN 0 86022 486. 4. Bourne Press, 1998.

[2] N. L. Pavey and M. Roper, Technical Note TN 6/96. ISBN 086022 4384. Bourne Press, 1996.

[3] Anonymous, Health and Safety Booklet HS (G) 70. H. M. Stationery Office, London, 1995.

[4] Committee of Enquiry Comnd 256. H. M. Stationery Office, London, 1987.

[5] A. W. Pasculle, J. C. Feeley, R. J. Gibson, L. G. Cordes, R. L. Myerowitz, C. M. Patton, G. W. Gorman, C. L. Carmack, J. W. Ezzel, and J. N. Dowling, *J. Infect. Dis.* **141,** 727 (1980).

[6] A. Baskerville, R. B. Fitzgeorge, M. Broster, P. Hambleton, and P. J. Dennis, *Lancet* **ii,** 1389 (1981).

[7] C. B. Fliermans, W. B. Cherry, L. H. Orrison, S. J. Smith, D. L. Tison, and D. H. Pope, *Appl. Environ. Microbiol.* **41,** 9 (1981).

[8] J. E. Stout, V. Yu, and M. G. Best, *Appl. Environ. Microbiol.* **49,** 221 (1985).

[9] G. Bezanson, S. Burbridge, D. Haldance, and T. Marrie, *Can. J. Microbiol.* **38,** 3328 (1992).

[10] J. S. Colbourne, D. J. Pratt, M. G. Smith, S. P. Fisher-Hoch, and D. Harper, *Lancet* **i,** 210 (1984).

[11] G. M. Schofield and R. Locci, *J. Appl. Bacteriol.* **58,** 151 (1985).

[12] J. T. Walker, J. Rogers, and C. W. Keevil, *Biofouling* **8,** 47 (1994).

0076-6879/99 $30.00

TABLE I
CONFIGURATION OF TEST RIGS DURING INFECTION TRIAL

Test rig no.:	1	2	3
Water	Hard	Hard	Soft
Hot water (infection)	35°	35°	35°
ClO_2	None	None	None
Hot water (disinfection)	60°	35°	35°
ClO_2 (disinfection)	None	0.5 ppm	0.5 ppm

Microbial populations, including an avirulent strain of *L. pneumophila,* were used to inoculate the water systems. The aim of this study was to investigate the efficiency of disinfectants to control bacteria in the water system as well as examine the biofouling of copper pipe and glass-reinforced plastic (GRP) storage tanks. Microbial counts were obtained from the surfaces and image analysis was performed to determine a quantitative assessment of biofilm covering the surface.

Test Rig Protocol

The trial was composed of a contamination and biofouling phase for 8 weeks (Table I) followed by a 16-week disinfection phase in which each test rig (Fig. 1) operated according to a specific disinfection protocol.

Composition of Microbial Inoculum

Microorganisms contained in the inoculum are *Pseudomonas vesicularis, P. xylosoxidans, P. testosteroii, P. aeruginosa, P. acidovorans, P. diminuta, P. fluorescens, P. maltophilia, P. mendocina, P. paucimobilis, P. pseudomonas, P. stutzeri, Actinomycetes* sp., *Aeromonas* sp., *Alcaligenes* pp., *Flavobacterium* sp., *Methylobacterium* sp., *Klebsiella* sp., *Acinetobacter* sp., and an avirulent strain of *L. pneumophila* sero group 1 Pontiac.[14] The cultures are grown on suitable agar media, and a suspension is prepared using sterile distilled water (SDW) containing the appropriate density of cells to inoculate each test rig at 1×10^6/ml.

---

[13] J. T. Walker, C. W. Mackerness, D. Mallon, T. Makin, T. Williets, and C. W. Keevil, *J. Indust. Microbiol.* **15,** 384 (1995).

[14] S. B. Surman, D. Goddard, L. H. G. Morton, and C. W. Keevil, *in* "Biofilms." Community Interactions and Control" (J. Wimpenny, P. Handley, P. Gilbert, H. Lappin-Scott, and M. Jones, eds.), p. 269. Bioline, Cardiff, 1997.

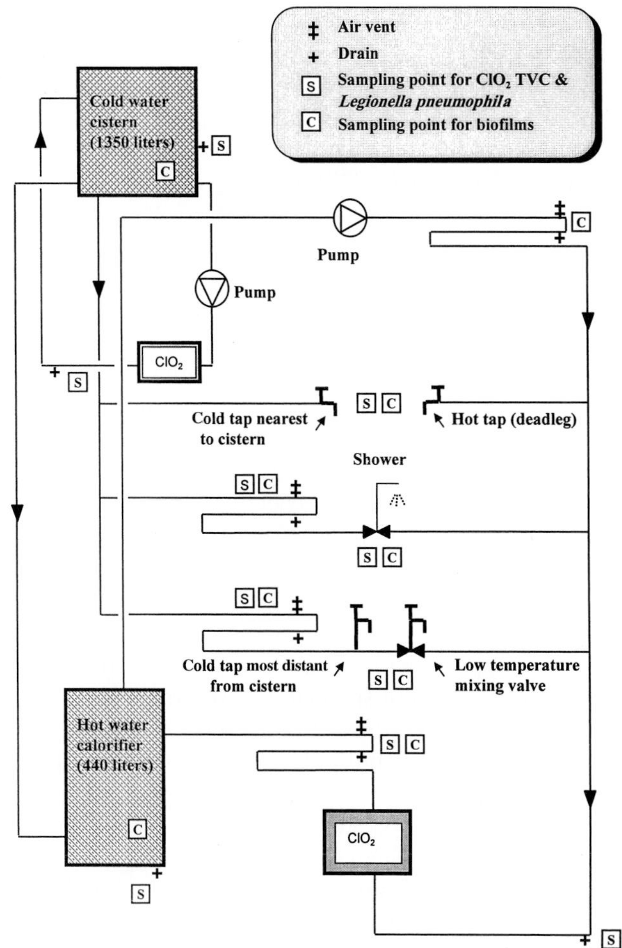

FIG. 1. Test rig for treatment.

Coupon Samples for Biofilm Assessment

Copper pipe samples are sealed at both ends to keep them hydrated with source water and forwarded to the laboratory. Glass-reinforced plastic sections from the cisterns are placed in universals containing source water.

Microscopy and Image Analysis

Copper pipe and GRP coupons are sectioned into duplicate 1-cm^2 coupons. One set is used for microscopy (flattened in a vice) and the other for microbial analysis.[15,16] Each specimen is stained for 1 min with 50 μl propidium iodide [Sigma, UK, 1 mg/ml stock in sterile distilled water, prefiltered (Sartorious, UK, 0.2 μm) before use] and is then rinsed in nonflowing sterile distilled water (twice). Coupons are visualized using a Nikon Labophot 2 microscope with episcopic fluorescence. Images are relayed to 35-mm transparencies (Nikon F801) and saved as computer files (*.tif) for image analysis (Optimus, Datacell, UK) of percentage coverage.

Microbial Analysis

Duplicate coupons are initially scraped with a sterile dental probe to remove biofilm and then vortexed to prepare a dilution series in SDW and EDTA (0.1 mM to chelate metal ions and prevent injury to microorganisms during incubation) from which 0.1 ml is used for plate counts. For total viable counts (TVC) of heterotrophic bacteria, nutrient agar is incubated at 22° and 37° for 7 days. BCYE and selective GVPC are used for growth of *L. pneumophila* and incubated at 37° for 10 days.

Biofouling Analysis of Plastic Cistern Material from Test Rig 1

Test rig 1 was supplied with hard water and was the control test rig with only the hot water system subjected to thermal control at 60° during the disinfection stage of the study (from day 0). Therefore, results from cold water biofilms represent biofouling in the absence of any disinfection mechanism.

During the study, the TVC at 22° from the plastic biofilm surfaces was found to be consistently high [>log 4.0 (cfu/ml)] (Fig. 2). *Legionella pneumophila* numbers were found to be greater than log 3 (cfu per liter). Correlating with the high numbers of TVC and *L. pneumophila,* biofouling was found to be initially greater than 25% and latterly greater than 15% coverage.

Biofouling Analysis of Plastic Cistern Material from Test Rig 2

Test rig 2 was supplied with hard water and was subjected to disinfection with chlorine dioxide up to 0.5 ppm.

Biofilm results indicated that using the disinfectant in hard water resulted in a log 2.0 decrease in the TVC (cfu/ml) at 22° and a log 3.0 decrease

[15] J. T. Walker, A. B. Dowsett, P. J. Dennis, and C. W. Keevil, *Int. Biodeterior.* **27,** 121 (1991).

[16] S. B. Surman, J. T. Walker, D. T. Goddard, L. H. G. Morton, C. W. Keevil, W. Weaver, A. Skinner, and J. Kurtz, *J. Microbiol. Methods* **25,** 57 (1996).

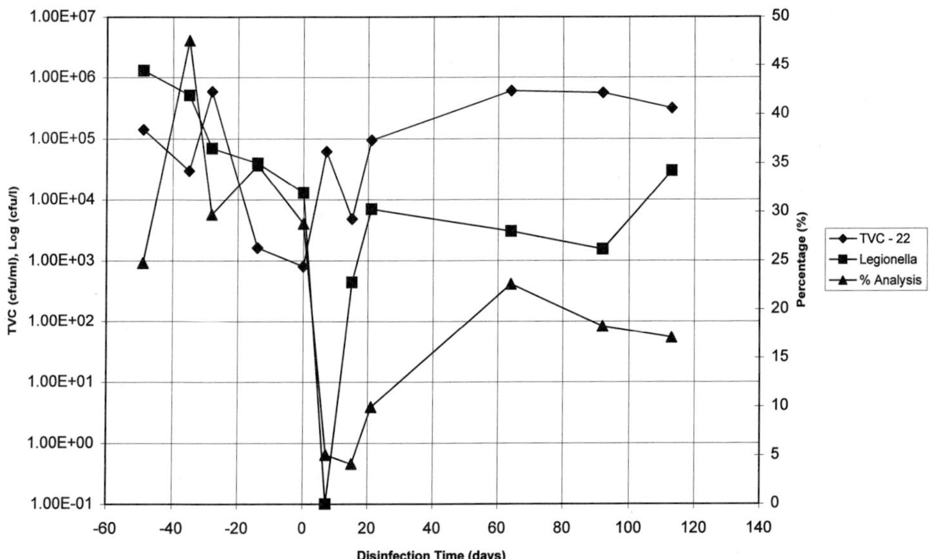

FIG. 2. Microbial analysis of biofilms from cisterns of test rig 1.

in the numbers of *L. pneumophila* (Fig. 3). Approximately 40% of the surface was fouled prior to disinfection and only after 20 days' treatment was the biofouling reduced to less than 10%.

Biofouling Analysis of Plastic Cistern Material from Test Rig 3

Test rig 3 was supplied with soft water that was treated with chlorine dioxide up to 0.5 ppm.

In the presence of disinfectant in soft water, biofilm TVC at 22° was reduced from log 5 to less than log 3 (Fig. 4). Within 20 days of applying disinfectant, *L. pneumophila* numbers were reduced significantly and were no longer recoverable. Similarly the biofilm coverage correlated with a significant reduction in the percentage coverage of the surface.

Comparison of Biofouling on Copper and Plastic Materials

A direct comparison of biofouling results within the cold water systems of rig 1 indicated that the copper was fouled to a lesser degree than the plastic materials (Fig. 5).

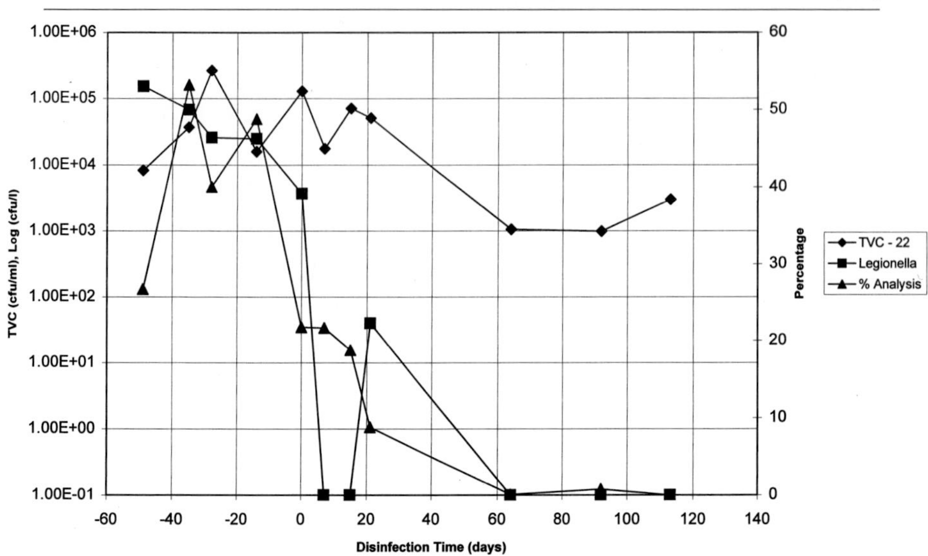

FIG. 3. Microbial analysis of cistern coupons from hard water test rig 2.

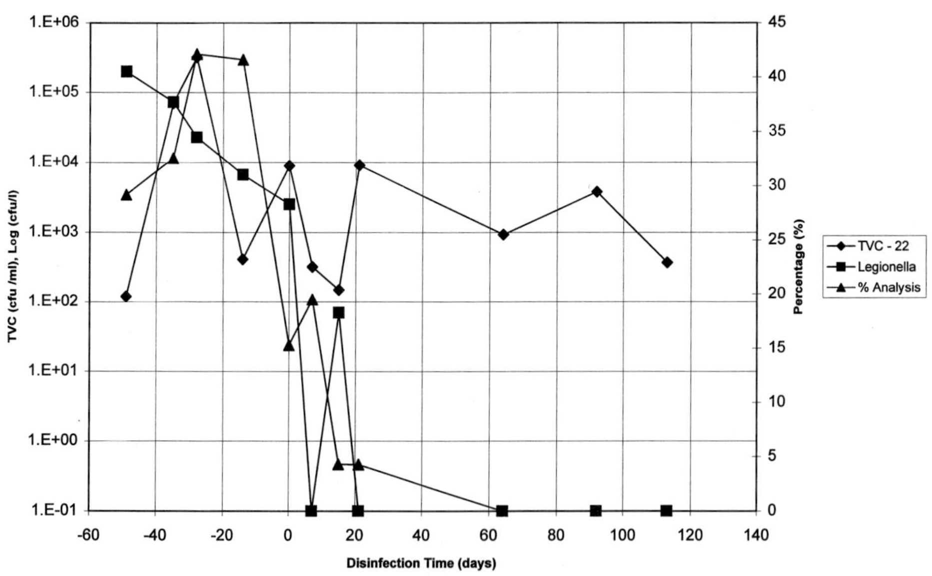

FIG. 4. Biofilm analysis of plastic material from rig 3.

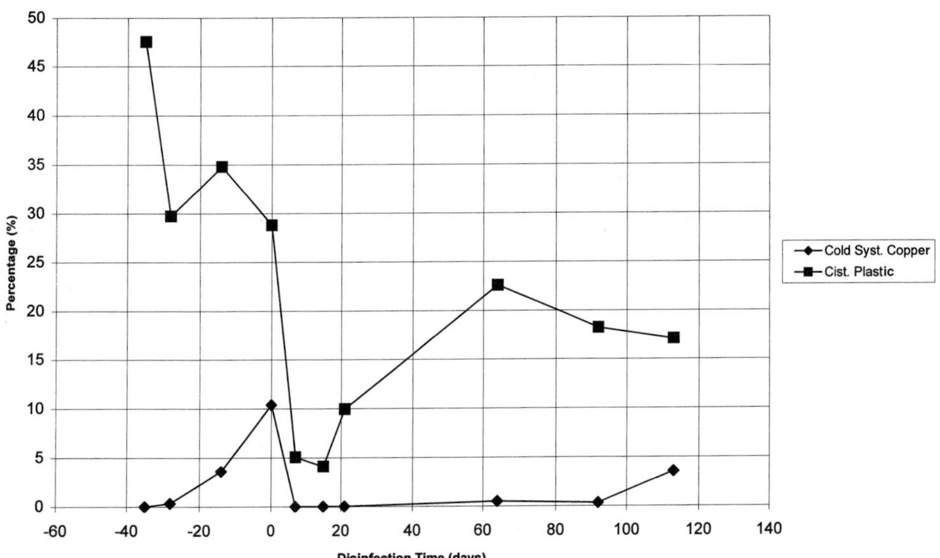

FIG. 5. Comparison of biofouling of copper (◆) and plastic (■).

Discussion and Conclusions

The assessment of biofilms within plumbing systems presents investigators with a complex situation, often resulting in the removal of tube sections for analysis. Such studies have been carried out in distribution mains,[17] hospital water systems,[18] and dental unit working stations.[19,20]

In this study, a full-size test rig, typical of that found in a 50 person occupancy building within the United Kingdom, was designed to allow the removal of sections of copper tubing from both the hot and the cold systems and also from the GRP cistern tanks.[1,2] Sections of materials were analyzed for the recovery of microorganisms from the surfaces by traditional culture techniques as well as direct visualization of the biofouling on the surfaces.

There was a correlation between the presence of both *L. pneumophila* and biofilm percentage coverage during the trial (Figs. 2–4).

[17] J. C. Block, K. Haudidier, J. L. Paquin, J. Miazga, Y. Levi, and J. D. Bryers, *Biofouling* **6,** 333 (1993).

[18] Z. Liu, J. E. Stout, M. Boldin, J. Rugh, W. F. Diven, and V. L. Yu, *Clin. Infect. Dis.* **26,** 138 (1998).

[19] J. F. Williams, J. A. Molinari, and N. Andrews, *Compend.-Contin.-Educ.-Dent.* **17,** 538 (1996).

[20] H. N. Williams, C. M. Paszko-Kolva, C. Shahamat, C. Palmer, and C. Pettis-Kelley, *J. Am. Dent. Assoc.* **127,** 1188 (1996).

When recovery of TVC at 22° was at log 3.0 (cfu/ml) or below, the biofilm percentage coverage was found to be very low or negligible.

Total viable counts provide important information on the actual numbers of bacteria present and on microbial speciation,[21] but results were not available for 48–72 hr, with identification of the *L. pneumophila* taking up to 10 days.

Biofilm percentage coverage results were available on the same day as removal of the materials, providing a correlative supplementary technique to viable culture.

The reduced fouling on copper could perhaps be due to the residual biocidal effect of the copper surface preventing buildup of the biofilm, but this was not investigated further.[22] However, one of the main considerations between the two surfaces was in the actual visualization of the biofilm on the surface. Being relatively smooth, the plastic surfaces presented a consistently flat surface on which to measure biofouling. The copper pipe presented a rougher surface, compounded by the effects of curvature and corrosion, making it more difficult to reliably observe the biofilm. To alleviate some of these problems, other researchers have treated the copper surface with abrasive paper to provide a smoother surface on which to visualize the microorganisms.[23]

A number of different microscopy techniques can be used to study biofouling,[16] including Hoffman modulation contrast, differential interference contrast, fluorescence,[24] scanning electron microscopy (SEM), environmental SEM, transmission EM,[25] atomic force microscopy, and scanning confocal laser microscopy.[26]

Studies have produced data suggesting that the direct microscopy assessment of biofilms provides more reliable and significant results than traditional culture using TVC.[27]

The microscopy techniques described earlier are direct and relatively rapid methods (no culture required) of examining biofilms on surfaces providing results that correlate with TVC.

[21] N. R. Ward, R. L. Wolfe, C. A. Justice, and B. H. Olson, *Adv. Appl. Microbiol.* **31,** 293 (1987).

[22] G. G. Geesey, P. J. Bremer, W. R. Fischer, D. Wagner, C. W. Keevil, J. T. Walker, A. H. L. Chamberlain, and P. Angell, *in* "Microbial Biofilms" (G. G. Geesey, ed.), p. 243. Lewis Publishing, New York, 1994.

[23] P. J. Bremmer, G. G. Geesey, and B. Drake, *Curr. Microbiol.* **24,** 223 (1992).

[24] J. T. Walker, D. Wagner, W. Fischer, and C. W. Keevil, *Biofouling* **8,** 47 (1994).

[25] J. T. Walker, J. B. Kurtz, and C. W. Keevil, *in* "Proceedings of the 9th International Biodeterioration and Biodegradation Society Symposium Leeds" (R. Edyvean, A. Bauscher, and M. Chandra, eds.), p. 17. Institute of Chemical Engineers, 1995.

[26] J. T. Walker, K. Hanson, D. Caldwell, and C. W. Keevil, *Biofouling* **12,** 333 (1998).

[27] J. T. Walker, D. J. Bradshaw, P. D. Marsh, and B. Gangnus, *J. Dent. Res.* **77,** 948 (1998).

Acknowledgments

The authors acknowledge the Department of Environment, Environment, Transport and the Regions (DETR), Partners in Technology Programme, and BSRIA for financially supporting this work.

[47] Quantifying Effects of Antifouling Paints on Microbial Biofilm Formation

By JOSEPH J. COONEY and RUEY-JING TANG

Much attention has been given to macroorganisms that foul submerged or intermittently submerged surfaces, but the first colonizers in such ecosystems are aquatic microorganisms. Microorganisms on such surfaces can produce toxins, corrosion, discoloration, alter the electrical conductivity of the material, and physically block the transport of chemicals to and from the surface.[1,2]

One approach to controlling microbial and/or macrobial growth on such surfaces is to coat the surface with a layer of material containing a biocidal or biostatic agent. Organotin compounds, particularly tributyl- and triphenyltins, are among the most effective additives. Copper is less effective than tributyltin (TBT) but is less toxic to many nontarget organisms than is TBT. There is extensive literature on the effects of TBT on macro-[3] and microorganisms.[4] We have used TBT and copper to inhibit microbial growth on various surfaces[5] and report here on methods we employ to quantify biofilm development.

There are three principal components of a biofilm: living cells, dead cells, and intercellular polysaccharide. We estimate living cells by plating cells from the biofilm on the surface of a suitable medium. Dead cells are quantified by counting the total number of cells in a sample from the biofilm and subtracting from it the number of viable cells. Extracellular polysaccharide is estimated by quantifying it colorimetrically in material removed from the biofilm.

[1] J. W. Costerton, Z. Lewandowski, D. E. Caldwell, D. R. Korber, and H. Lappin-Scott, *Annu. Rev. Microbiol.* **49,** 711 (1995).

[2] J. Wimpenny, P. Handley, P. Gilbert, H. Lappin-Scott, and M. Jones, "Biofilms." BioLine, Univ. Wales, Cardiff, UK, 1997.

[3] K. Fent, *Crit. Rev. Toxicol.* **26,** 1 (1996).

[4] J. J. Cooney, *Helgolander Meeresuntersuchungen* **49,** 663 (1995).

[5] R.-J. Tang and J. J. Cooney, *J. Indust. Microbiol. Biotechnol.* **20,** 275 (1998).

0076-6879/99 $30.00

TABLE I
EFFECT OF STANNIC CHLORIDE ON SEDIMENT MICROORGANISMS[a]

Tin (mg liter^{-1})[b]	Colony-forming units (cfu) per g ($\times 10^4$)[c]	Control (%)
0	74.0 ± 2.47	100
5	67.2 ± 1.48	91
10	52.7 ± 6.34	71
25	50.0 ± 3.62	68
50	33.3 ± 2.80	45
75	12.0 ± 0.32	16
100	5.8 ± 0.21	8
200	0.2 ± 0.04	<1

[a] Modified from L. E. Hallas and J. J. Cooney, *Appl. Environ. Microbiol.* **41,** 466 (1981), with permission.
[b] Concentration of Sn as $SnCl_4 \cdot 5H_2O$ in medium.
[c] Mean ± standard error.

Materials and Methods

In nature, most biofilms contain more than one microorganism. When estimating the number of living microorganisms it is necessary to use a medium on which the microorganisms will grow. In some cases, more than one medium will be necessary to quantify different organisms. We will demonstrate the methods using a single biofilm-forming bacterium, *Pseudomonas aeruginosa* PAO-1.

It is often necessary to estimate how much of a given antibacterial agent to incorporate in a surface coating. We do this by determining the concentration of the chemical that will inhibit growth of 80–90% of the bacteria found in the ecosystem in which the surface will be immersed.[6] For example, the amount of stannic chloride that decreased the number of viable bacteria in sediment samples from Chesapeake Bay by that amount was 75 mg liter^{-1} (Table I).

The surface on which a biofilm develops can be as simple as a slide immersed in a container of water. In the laboratory we have used a variety of chambers[5,7] that permit control and manipulation of a number of variables. One such chamber, a modified Robbins device (MRD), is diagrammed in Fig. 1. The coupon material can be varied and the effects of various coatings on biofilm development can be determined.[5] The liquid flowing through the chamber will vary with the questions being posed experimentally, e.g., distilled water, buffer, water from a natural body, or

[6] L. E. Hallas and J. J. Cooney, *Appl. Environ. Microbiol.* **41,** 466 (1981).
[7] M. M. Doolittle, J. J. Cooney, and D. E. Caldwell, *J. Indust. Microbiol.* **16,** 331 (1996).

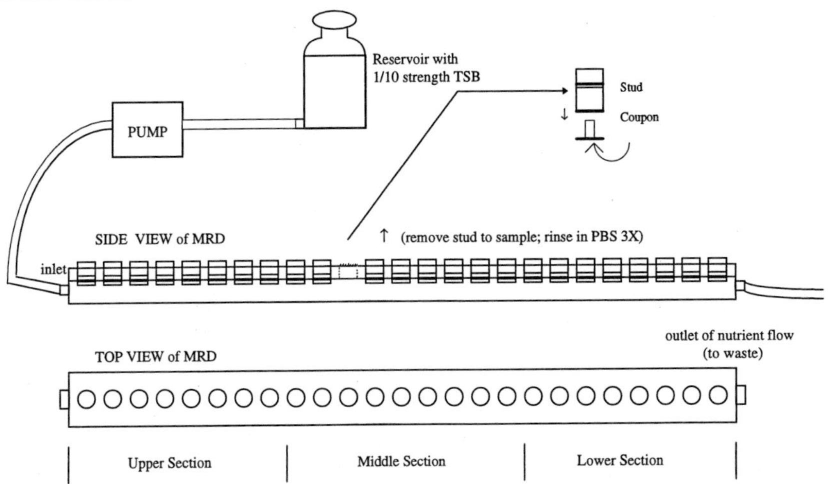

FIG. 1. Modified Robbins device (MRD). The nutrient, 1/10 strength tryptic soy broth (TSB), was pumped in by a peristaltic pump at 150 ml/hr. At each sampling time, one stud was taken from each of the three sections and biofilms from them were combined. PBS, phosphate-buffered saline. From R.-J. Tang and J. J. Cooney, *J. Indust. Microbiol. Biotechnol.* **20,** 275 (1998), with permission.

a laboratory medium. Preliminary experiments are usually necessary to establish appropriate flow rate(s) and strength of medium.

During biofilm development the MRD is placed with the studs down to avoid bubbles at the surface of the studs. Twenty minutes before each sampling time the chamber is placed so that the studs are on the upper surface. At each sampling time one or more studs are removed and replaced with blank studs. Biofilm development occurs at the same rate on all studs.[5] Studs removed are rinsed by swirling them in phosphate-buffered saline (PBS) to remove loosely attached cells. This swirling is repeated twice. Each coupon is then placed in 1.0 ml of sterile PBS or of a sterile medium in an Eppendorf tube. Tubes are kept in an ice bath.

The coupon is sonicated in the Eppendorf tube using an ultrasonic cleaner (Model 1210, Bransonic Co., Danbury, CT) for 8 min and vortexed for 10 sec to remove biofilms from the coupons. For some samples, e.g., early in biofilm formation when the film contains few cells, cells from more than one coupon may be combined.

For viable counts of *P. aeruginosa,* appropriate dilutions are plated on tryptic soy agar. Total cell counts are made using acridine orange (AODC),[8]

[8] J. E. Hobbie, R. J. Daley, and S. Jasper, *Appl. Environ. Microbiol.* **33,** 1225 (1977).

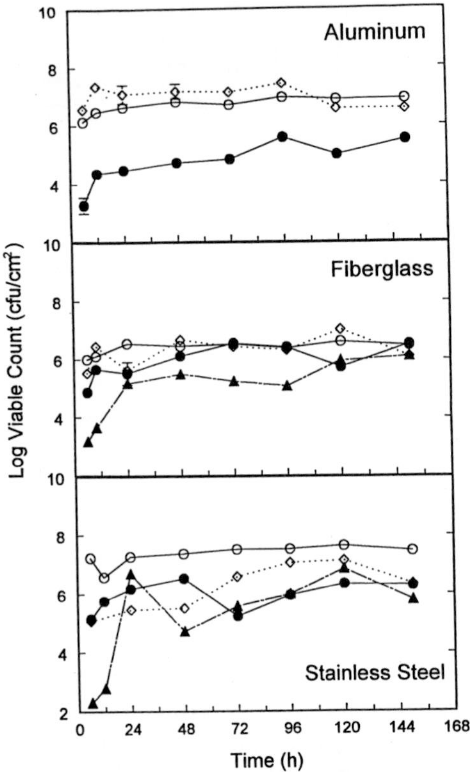

FIG. 2. Viable counts from biofilms grown on the surface of aluminum, fiberglass, or stainless steel. ○, no paint; ◇, painted with marine paint VC-18; ▲, VC-18 paint with copper added; ●, VC-18 with TBTF added. Paint with copper was not tested on aluminum. In this and subsequent figures, error bars indicate one standard deviation. When no error bar is shown, the error bar fell within the symbol. From R.-J. Tang and J. J. Cooney, *J. Industr. Microbiol. Biotechnol.* **20,** 275 (1998), with permission.

and extracellular polysaccharide (EPS) is estimated by the phenol–sulfuric acid method.[9]

Results

Viable counts (Fig. 2) and total counts (Fig. 3) on three materials with and without coatings show that numbers were 10- to 100-fold higher for total counts than for viable counts, suggesting that a large number of cells

[9] M. Dubois, K. A. Gilles, J. K. Hamilton, P. A. Rebers, and F. Smith, *Anal. Chem.* **28,** 350 (1956).

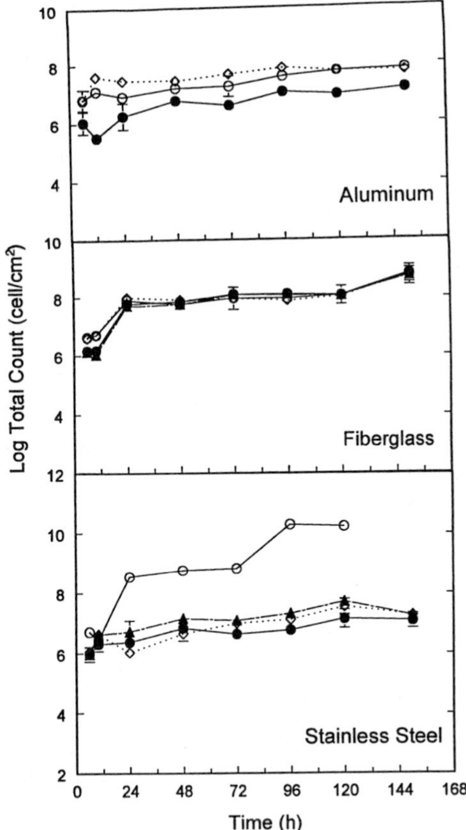

FIG. 3. Direct counts from biofilms grown on the surface of aluminum, fiberglass, or stainless steel. ○, no paint; ◇, painted with marine paint VC-18; ▲, VC-18 paint with copper added; ●, VC-18 paint with TBTF added. From R.-J. Tang and J. J. Cooney, *J. Indust. Microbiol. Biotechnol.* **20,** 275 (1998), with permission.

in the film were dead or viable but nonculturable. Viable counts were higher on noncoated stainless steel than on noncoated aluminum or fiberglass. The viable counts on stainless steel were almost 10-fold higher than those on aluminum and fiberglass. After biofilms reached equilibrium, the differences between viable counts and total counts between stainless steel and the other two surfaces were significant at the $\alpha = 0.01$ level. Thus, a more extensive biofilm developed on noncoated stainless steel than on aluminum or fiberglass. In our hands, EPS values (Fig. 4) showed considerable variation during the incubation period.

The addition of a coat of marine paint had little effect on the rate of

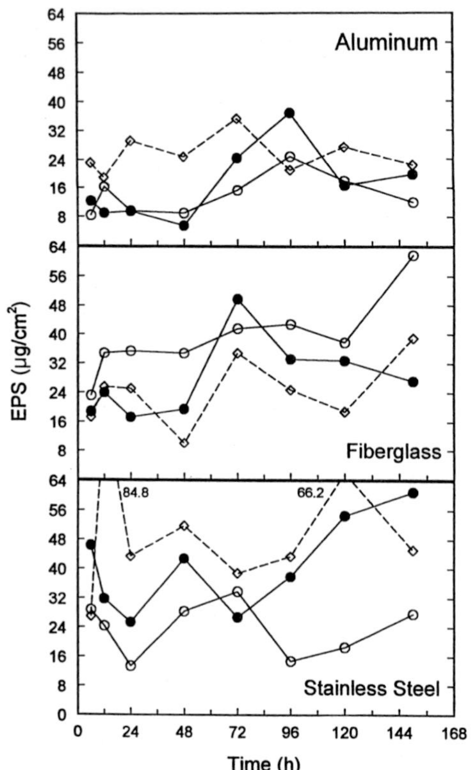

FIG. 4. Exopolysaccharide (EPS) content of biofilms grown on the surface of aluminum, fiberglass, or stainless steel. ○, no paint; ◇, painted with marine paint VC-18; ●, C-18 paint with TBTF added. Error bars are not shown because each point represents a single determination from three coupons that were combined to constitute one sample. From R.-J. Tang and J. J. Cooney, *J. Indust. Microbiol. Biotechnol.* **20,** 275 (1998), with permission.

biofilm development or the number of viable cells in films on aluminum or fiberglass, but viable counts were markedly lower on painted stainless steel than on nonpainted stainless steel (Fig. 2). Total cell numbers (Fig. 3) showed similar results but were 10- to 100-fold higher than viable counts, similar to results on noncoated surfaces. The EPS content of biofilms varied during the 150-hr period (Fig. 4), but EPS values tended to be higher on painted aluminum and stainless steel than on the respective noncoated coupons. Biofilms on fiberglass contained equivalent amounts of EPS.

Aluminum coupons with marine paint with TBT had significantly lower numbers of viable bacteria (Fig. 2) and of total cells (Fig. 3) than coupons with paint alone, but TBT had relatively little effect on viable counts on fiberglass or stainless steel. Approximately 1–10% of the cells from all three surfaces were able to form colonies. TBT added to paint applied to aluminum decreased the number of viable cells by approximately 1000-fold early in biofilm development and by 10- to 100-fold later. Values for EPS varied during film development (Fig. 4).

Paint containing copper had biofilms that developed more slowly and contained fewer viable cells than surfaces coated with paint alone (Fig. 2). As with TBT, copper inhibited viable cell numbers about 1000-fold early in biofilm development but only 10-fold later in development. Total cell numbers were not affected (Fig. 3). These results suggest that the same number of cells attached in the presence or absence of copper but the cells were later inhibited or killed. EPS was not measured for biofilms on copper-coated surfaces because copper interfered with the phenol–sulfuric acid method.

Discussion

There are a number of methods for quantifying microbial growth on surfaces. In addition to the methods employed here, one can use transmission or scanning electron microscopy, confocal laser microscopy, substrate uptake, or measurement of a chemical specific to the cellular or EPS component of the biofilm. The EPS component of a biofilm-forming organism can vary considerably[10,11] and one must define the conditions under which the biofilm occurs if comparisons are to be made with the work of others.

On stainless steel surfaces coated with paint containing copper, copper-resistant cells made up 100% of the viable population in the first 48 hr and then decreased to less than 10% of the population by 108 hr. Similarly, on stainless steel coated with TBT-containing paint, TBT-resistant cells constituted more than 50% of the population early in biofilm development and 10% or less later in development.[5] Thus, both copper and TBT inhibited development of a biofilm of *P. aeruginosa* in its early stages but not in its later stages. After 5–6 days, viable counts were equivalent on all painted surfaces, perhaps due to inactivation of the agents by the EPS. It is also possible that resistant cells may develop early in biofilm formation and form a layer that shields later developing cells from the agents.

Addition of TBT to a marine paint decreased the numbers of total cells and viable cells in biofilms on aluminum but not on the other two surfaces

[10] B. Ruiz, A. Jaspe, and C. San Jose, *in* "Biofilms" (J. Wimpenny, P. Handley, P. Gilbert, H. Lappin-Scott, and M. Jones, eds.), p. 47. BioLine, Univ. Wales, Cardiff, UK, 1997.

[11] I. W. Sutherland, *in* "Biofilms" (J. Wimpenny, P. Handley, P. Gilbert, H. Lappin-Scott, and M. Jones, eds.), p. 33. BioLIne, Univ. Wales, Cardiff, UK, 1997.

examined. Similarly, addition of copper to a marine paint decreased the number of viable cells in biofilms of *P. aeruginosa* initially, but after 6 days numbers were equivalent in films on paint with or without copper. These results suggest that a surface coated with paint can interact with materials in the paint in affecting microbial activities.

Because microorganisms on a submerged surface can increase the number of shellfish larvae on the surface,[12,13] inhibition of a microbial biofilm might decrease barnacle growth on the surface. Our results suggest that antifouling paints with copper or TBT are not likely to act primarily by inhibiting development of a microbial film.

[12] N. J. O'Connor and D. L. Richardson, *J. Exp. Mar. Biol. Ecol.* **206,** 69 (1996).
[13] R. M. Weiner, M. Walch, M. P. Labarre, B. D. Bonar, and R. R. Colwell, *J. Shellfish Res.* **8,** 117 (1989).

[48] *Candida* Biofilms and Their Susceptibility to Antifungal Agents

By George S. Baillie and L. Julia Douglas

Introduction

Of the yeasts known to be pathogenic for humans, the most versatile are a handful found in the genus *Candida.* These organisms can cause a variety of superficial and deep-seated mycoses that are distributed worldwide.[1] All are opportunistic pathogens, liable to attack immunocompromised hosts or those debilitated in some other way. The principal pathogen of the genus is the thrush fungus, *C. albicans,* which may grow either as oval budding yeasts or as continuous septate hyphae; both morphological forms are usually seen in infected tissue. In recent years, *Candida* species have been identified as potentially lethal agents of hospital-acquired infection.[2] Their emergence as important nosocomial pathogens is related to specific risk factors associated with modern medical procedures, notably the use of immunosuppressive and cytotoxic drugs, powerful antibiotics that suppress the normal bacterial microbiota, and implanted devices of various kinds. The majority of nosocomial septicemias caused by *Candida*

[1] F. C. Odds, "Candida and Candidosis." Bailliere Tindall, London, 1988.
[2] S. K. Fridkin and W. R. Jarvis, *Clin. Microbiol. Rev.* **9,** 499 (1996).

0076-6879/99 $30.00

species derive from biofilm formation on intravascular catheters.[3,4] Urinary catheters, prosthetic heart valves, cardiac pacemakers, silicone voice prostheses, endotracheal tubes, and cerebrospinal fluid shunts are also strongly associated with *Candida* infections.[1]

The role of adhesion in microbial pathogenesis is now widely recognized and, as a result, *Candida* adhesion to epithelial cells, endothelial cells, and various inert surfaces has received increasing attention since the late 1970s.[5–7] Much of this work has focused on adhesion mechanisms, with the ultimate goal being the development of measures to inhibit *Candida* adhesion *in vivo* and hence prevent infection. More recently, these studies have been extended to an investigation of biofilm formation *in vitro* and to possible mechanisms by which *Candida* biofilms on medical implants might resist the action of antifungal agents. A number of model biofilm systems have been devised,[8–14] all of which have been adapted from methods reported previously for bacteria. Three of these systems are described here.

Candida species differ from bacteria in several important characteristics, which are particularly relevant to any investigator planning to work with biofilms. *Candida albicans* is a dimorphic fungus and will produce biofilms consisting of yeast cells, hyphae, pseudohyphae (which are chains of elongated yeast cells), or a mixture of morphological forms depending on the growth conditions and the nature of the substratum. Other species, such as *C. parapsilosis,* will form pseudohyphae but not true hyphae.[1] All of these cell types are considerably larger than bacteria, and consequently they sediment more easily and are dislodged more readily from a surface. Moreover, the presence of hyphal forms can complicate any estimate of biofilm viability following drug treatment. Finally, it must be remembered

[3] D. A. Goldmann and G. B. Pier, *Clin. Microbiol. Rev.* **6,** 176 (1993).

[4] D. G. Maki, *in* "Infections Associated With Indwelling Medical Devices" (A. L. Bisno and F. A. Waldvogel, eds.), p. 161. American Society for Microbiology, Washington, DC, 1989.

[5] L. J. Douglas, *in* "The Yeasts" (A. H. Rose and J. S. Harrison, eds.), Vol. 2, p. 239. Academic Press, London.

[6] R. A. Calderone and P. C. Braun, *Microbiol. Rev.* **55,** 1 (1991).

[7] M. K. Hostetter, *Clin. Microbiol. Rev.* **7,** 29 (1994).

[8] S. P. Hawser and L. J. Douglas, *Infect. Immun.* **62,** 915 (1994).

[9] H. J. Busscher, C. E. Deboer, G. J. Verkerke, R. Kalicharan, H. K. Schutte, and H. C. Van der Mei, *Int. Biodeterior. Biodegrad.* **33,** 383 (1994).

[10] S. P. Hawser and L. J. Douglas, *Antimicrob. Agents Chemother.* **39,** 2128 (1995).

[11] H. Nikawa, H. Nishimura, T. Yamamoto, T. Hamada, and L. P. Samaranayake, *Microb. Ecol. Health Dis.* **9,** 35 (1996).

[12] S. P. Hawser, G. S. Baillie, and L. J. Douglas, *J. Med. Microbiol.* **47,** 253 (1998).

[13] G. S. Baillie and L. J. Douglas, *Antimicrob. Agents Chemother.* **42,** 1900 (1998).

[14] G. S. Baillie and L. J. Douglas, *Antimicrob. Agents Chemother.* **42,** 2146 (1998).

that the chemical composition and overall architecture of the *Candida* cell surface are quite different from that of bacteria.[15]

Biofilm Formation on Catheter Disks

The simplest model system[8,10,12] involves forming *Candida* biofilms on the surfaces of small disks cut from catheters. The method has been adapted from one used previously by Prosser and co-workers[16] for *Escherichia coli* and other gram-negative bacteria. Growth of the biofilms is monitored quantitatively using dry weight, colorimetric, or radioisotope assays and can be visualized readily by scanning electron microscopy.

Catheter Materials

Different types of catheters, made from various plastics, can be used for biofilm formation. These include central venous catheters [composed of polyvinyl chloride (PVC) or polyurethane] and urinary Foley catheters (composed of latex, silicone, or silicone elastomer-coated latex). Disks (surface area, 0.5 cm^2; diameter, 0.8 cm) are cut using a metal punch and sterilized with ethylene oxide. Alternatively, disks may be cut aseptically from sterile catheters. The disks are slightly curved, with the degree of curvature depending on the diameter of the catheter used. It is more convenient to use a wide catheter, such as a PVC Faucher tube (French gauge 36; Vygon, Cirencester, UK). The exact chemical composition of catheter materials, including plasticizers, is difficult to ascertain, but seems to vary from manufacturer to manufacturer and may affect the nature of the biofilm formed.

Preparation of Microorganisms

Cells are grown in a liquid medium consisting of yeast nitrogen base (Difco) containing an appropriate carbon source such as 50 m*M* glucose. Batches of medium (20 ml, in 100-ml Erlenmeyer flasks) are inoculated from fresh culture slopes and incubated at 37° in an orbital shaker at 60 rpm. Almost all *Candida* species and strains grow exclusively in the budding yeast phase under these conditions. Cells are harvested after 24 hr and washed twice in 0.15 *M* phosphate-buffered saline (PBS), pH 7.2. Before use in biofilm assays, all washed cell suspensions are standardized to an optical density of 0.8 at 520 nm.

[15] M. G. Shepherd, *Crit. Rev. Microbiol.* **15,** 7 (1987).

[16] B. T. Prosser, D. Taylor, B. A. Dix, and R. Cleeland, *Antimicrob. Agents Chemother.* **31,** 1502 (1987).

Growth of Biofilms

Two stages are involved: (i) adhesion of the organism to the catheter surface and (ii) biofilm formation following submersion of the disk in growth medium. A standardized cell suspension (80 μl) is applied to the surface of each catheter disk housed in a well of a 24-well Nunclon tissue culture plate (Nalge Nunc International, Rochester, NY). Disks are incubated at 37° for 1 hr (adhesion period). Nonadherent organisms are removed by gentle washing with PBS (5 ml) and the disks are submerged in 1-ml portions of growth medium (yeast nitrogen base with 50 m*M* glucose) predispensed in the wells of a fresh plate. They are then incubated at 37° for 24–48 hr (biofilm formation). Control disks incubated in growth medium without the prior addition of cells should always be included. Ideally, all biofilm and control assays are carried out three times in triplicate.

Quantitative Analysis of Biofilm Growth

Measurement of biofilm growth by means of viable counts can give variable results if hyphal forms are present. Dry weight estimations are reliable, although rather time-consuming. Methods involving quantitation of some metabolic activity can be very convenient for comparative studies, and the procedures described here involving leucine uptake and tetrazolium reduction give an excellent correlation with biofilm dry weight.[8]

Determination of Dry Weight. After biofilm formation, each disk is removed carefully from its well using forceps and washed gently in PBS (5 ml) to remove nonbiofilm cells. Biofilm organisms are scraped from the disk with a sterile scalpel into PBS (5 ml) and vortexed gently for 3 min. The disk is then washed thoroughly in more PBS (5 ml) to remove any remaining cells. All organisms are collected on a preweighed cellulose nitrate filter (0.45 μm pore size; 25 mm diameter) and given three washes with water (5 ml). The filter is dried to constant weight at 80°, and the dry weight of cells is calculated.

Incorporation of [^{3}H]Leucine. The method is a modification of one used for labeling biofilms of *Escherichia coli.*[17] After biofilm formation, L-[4,5-^{3}H]leucine (Amersham Life Science Ltd., Little Chalfont, Buckinghamshire, UK) in PBS is added to the growth medium in each well (final concentration, 1 μCi/ml), and the plates are incubated for a further 4 hr at 37°. Uptake of the radioactive label by the biofilm cells is linear during this time. After labeling, organisms are removed from the catheter surface by incubating each disk in a well containing 0.5 ml of 5% (w/v) sodium

[17] B. A. Dix, P. S. Cohen, D. C. Laux, and R. Cleeland, *Antimicrob. Agents Chemother.* **32,** 770 (1988).

dodecyl sulfate (SDS) for 3 hr at 37°. The SDS solutions, with cells, are then transferred to tubes containing 0.5 ml of 10% (w/v) trichloroacetic acid (TCA) supplemented with 1% (w/v) casamino acids (Difco Laboratories, Detroit, MI). All TCA precipitates are filtered through prewetted cellulose nitrate filters (0.45 μm pore size; 25 mm diameter). The filters are washed three times with 5 ml of 5% (w/v) TCA containing 1% (w/v) casamino acids and dried. Radioactivity is counted after the addition of 4 ml of Opti-Fluor O scintillant (Packard Instrument Company Inc., Downers Grove, IL).

Tetrazolium Reduction Assays. Two tetrazolium salts, MTT and XTT (Sigma, St. Louis, MO), can be used in colorimetric determinations of biofilm formation.[8] These salts are reduced by mitochondrial dehydrogenases to colored tetrazolium formazan products (violet for MTT and brown for XTT), which are then determined spectrophotometrically. Similar assays have been used to measure *Candida* viability[18] and adhesion to plastic.[19,20] XTT is easier to use than MTT because it yields a water-soluble formazan product and there is no requirement for dimethyl sulfoxide as a solubilizing agent. However, addition of the electron coupling agent menadione (Sigma) is necessary.

After biofilm formation, XTT or MTT dissolved in prewarmed PBS is added to each well to a final concentration of 250 or 50 μg/ml, respectively. Menadione solution (1 mM in acetone; 1 μl) is also added to the XTT assay wells (final concentration, 1 μM). After incubation at 37° for 5 hr, the medium plus MTT is removed and the wells are washed three times with PBS (2 ml) to remove all traces of MTT. Dimethyl sulfoxide (1 ml) is then added to solubilize the MTT formazan crystals that have formed on the catheter disk. The solution is transferred to a microfuge tube, centrifuged at top speed (13,000 rpm) for 3 min at room temperature to pellet any cells present, and MTT formazan in the supernatant is measured at 540 nm using a spectrophotometer. In the XTT assay, the medium containing XTT formazan is simply clarified by centrifugation and formazan production is measured at 492 nm.

Perfused Biofilm Fermenter

Growth rate can affect both the cell envelope composition of microorganisms and their susceptibility to antimicrobial agents. A low growth rate is characteristic of biofilms in natural environments where there is often a deficiency of nutrients crucial to microbial survival. The perfused biofilm

[18] S. M. Levitz and R. D. Diamond, *J. Infect. Dis.* **152,** 938 (1985).
[19] S. Hawser, *J. Med. Vet. Mycol.* **34,** 149 (1996).
[20] S. Hawser, *J. Med. Vet. Mycol.* **34,** 407 (1996).

fermenter represents a model system in which the growth rate of biofilms can be controlled accurately.[21] It has been used with a variety of gram-positive and gram-negative bacteria,[22,23] particularly in relation to biofilm resistance to antibiotics, and it is described in detail elsewhere in this volume.[24] The same fermenter vessel can be used without modification for the growth of *Candida* biofilms.[13] With this apparatus, a biofilm is established on the underside of a cellulose membrane, and a steady state develops in which the rate of perfusion with fresh medium controls the rate of biofilm growth. By varying the rate at which medium is supplied, biofilms can therefore be produced over a range of different growth rates. Chemostat-grown, planktonic populations are normally used as control cultures. This allows a direct comparison of the physiological properties (e.g., antimicrobial susceptibility) of biofilm cells with those of planktonic cells grown at an identical rate in a chemostat. Properties associated with the adherent phenotype of biofilm cells can thus be distinguished from those simply resulting from a low growth rate.

Preparation of Microorganisms

Cells are grown in yeast nitrogen base liquid medium without amino acids (pH 5.4; Difco) prepared from individual constituents. For the production of glucose-limited biofilms of *C. albicans*, glucose is incorporated at a concentration of 4 m*M*. This concentration allows batch growth of the organism to a stationary-phase optical density of approximately 1.3 at 540 nm. An exponential-phase culture is required as the inoculum for the perfused biofilm fermenter. This can be prepared by adding a portion (10 ml) of an overnight shake culture to fresh, prewarmed medium (40 ml) and incubating at 37° in an orbital shaker at 60 rpm for 3 hr. At this stage the culture should have an optical density of approximately 1 at 540 nm. The cells (4.5×10^8) are collected by pressure filtration on a cellulose acetate membrane (0.2 μm pore size; 47 mm diameter; Whatman, Clifton, NJ) and the membrane is inverted into the base of the fermenter. Care must be taken when manipulationg the filter to ensure that its underside remains sterile. Identical exponential-phase cultures can be used as inocula for the production of glucose-limited planktonic cells grown at different rates in a chemostat.

[21] P. Gilbert, D. G. Allison, D. J. Evans, P. S. Handley, and M. R. W. Brown, *Appl. Environ. Microbiol.* **55,** 1308 (1989).

[22] I. G. Duguid, E. Evans, M. R. W. Brown, and P. Gilbert, *J. Antimicrob. Chemother.* **30,** 803 (1992).

[23] D. J. Evans, D. G. Allison, M. R. W. Brown, and P. Gilbert, *J. Antimicrob. Chemother.* **27,** 177 (1991).

[24] D. Allison and P. Gilbert, *Methods Enzymol.* **310** [19] 1999 (this volume).

Growth of Biofilms

After insertion of the membrane into the fermenter, fresh growth medium (yeast nitrogen base containing 4 m*M* glucose) is passed into the vessel at controlled flow rates (18 to 138 ml/hr) via a peristaltic pump. A hydrostatic head develops above the membrane filter and, under steady-state conditions, perfuses the filter at the rate of medium addition to the fermenter. Initially, the filter is perfused at a rate of 1.12 ml/min. The eluate passing through the filter is collected at intervals, and viable counts are made by serial dilution in PBS and plating in triplicate on Sabouraud's dextrose agar.

Loosely attached cells are dislodged from the membrane by the perfusing medium for up to 80 min after the initiation of flow and more than 90% of the cells added to the filter are removed during this period.[13] After the initial cell loss, organisms are eluted from the filter at a constant rate. These represent newly formed daughter yeast cells budding from the biofilm. This steady state can be maintained for approximately 30 hr,[13] after which *Candida* hyphae appear to grow through the cellulose acetate support, presumably via thigmotropism[25] or contact guidance.

Increasing the medium flow rate, and hence the availability of the limiting nutrient, results in a greater yield of newly formed daughter cells. This can be demonstrated up to a flow rate of 1.7 ml/min (the critical medium flow rate), above which the production of daughter cells decreases and growth rate control under steady-state conditions is lost.[13] Growth rates (divisions/hr) are calculated by dividing the number of daughter cells produced per hour at steady state by the estimated adherent cell population (determined by obtaining the viable counts of the suspended biofilms).

Biofilm Formation on Cylindrical Cellulose Filters

Although the perfused biofilm fermenter is the only model system that permits the full control of biofilm growth rates, it suffers from a number of disadvantages. It requires specialized equipment, is fairly labor intensive, and, due to the two-dimensional nature of the biofilm support, supplies a relatively small number of cells for experimental purposes. Moreover, its operating time is restricted to about 30 hr because of the ability of *Candida* hyphae to penetrate the membrane filter. A much simpler model system involves biofilm formation within small, cylindrical, cellulose filters that are perfused with culture medium. This system was originally developed for

[25] J. Sherwood, N. A. R. Gow, G. W. Gooday, D. W. Gregory, and D. Marshall, *J. Med. Vet. Mycol.* **30,** 461 (1992).

use with *Pseudomonas aeruginosa* and *Staphylococcus aureus*[26] but also allows the formation of *Candida* biofilms[14] at reproducible, low growth rates. A modified apparatus, described here, lacks stainless steel components and so facilitates the production of iron-limited biofilms in addition to glucose-limited ones.

Preparation of Iron-Limited Microorganisms

To minimize iron contamination, disposable plasticware is used wherever possible; glassware is washed with Extran (BDH Laboratory Supplies, Poole, Dorset, UK), soaked in 1% (v/v) HCl for at least 48 hr, and rinsed extensively with distilled, deionized water. The growth medium[27] is a modification of yeast nitrogen base (Difco), prepared from individual constituents, in which monobasic potassium phosphate (a major source of contaminating iron) is replaced by the dibasic salt. Glucose (50 m*M*) is added as the carbon source. Deferration of the medium is accomplished by passage through columns of Chelex 100 ion-exchange resin (Bio-Rad, Richmond, CA). A stock solution containing most of the medium components (except glucose, $MgSO_4$, $CaCl_2$, $CuSO_4$, $MnSO_4$, $ZnSO_4$, and $FeCl_3$), made up in distilled, deionized water, is passed through a column (2.5×50 cm) of sodium-form resin. Glucose, as a 1 *M* solution, and the distilled, deionized water to be used for diluting the concentrated medium are passed through separate columns of sodium-form resin. Stock solutions of $CaCl_2$ and $MgSO_4$ are passed through columns of calcium-form and magnesium-form resin, respectively. Solutions of zinc, copper, and manganese salts are not treated with Chelex 100 because they contain insignificant amounts of iron. The various stock solutions are combined to give a 10-fold concentrate of medium, which is sterilized by filtration and stored at 4° until use. When diluted, the medium has an iron content of $<0.036\ \mu M$ as determined by graphite furnace atomic absorption spectrometry.[14] This concentration limits the growth of *C. albicans*[27] and results in a stationary-phase optical density of 1.3 at 540 nm in batch culture.

An exponential-phase culture is required as the inoculum for biofilm growth on a cylindrical filter. This is prepared by adding a portion (10 ml) of an overnight culture in iron-limited medium to fresh, prewarmed medium (40 ml) and incubating at 37° with shaking for 3 hr.

Preparation of Glucose-Limited Microorganisms

The medium used for glucose-limited cultures is a yeast nitrogen base made up from individual constituents without deferration and containing

[26] A. E. Hodgson, S. M. Nelson, M. R. W. Brown, and P. Gilbert, *J. Appl. Bacteriol.* **79,** 87 (1995).

[27] S. P. Sweet and L. J. Douglas, *J. Gen. Microbiol.* **137,** 859 (1991).

4 mM glucose. Exponential-phase inocula for biofilms are prepared by the addition of samples (10 ml) of an overnight culture in this medium to fresh batches (40 ml) of prewarmed medium and incubation at 37° with shaking for 3 hr.

Growth of Biofilms

Biofilms are grown on cylindrical filters consisting of compacted cellulose fibers (Gilson safety filters, 22 by 8 mm; Anachem, Luton, UK). Each filter is inserted into a section of silicone tubing attached to the bottom of a disposable syringe body (2 ml) from which the plunger has been removed. Medium is pumped directly into the vertically clamped syringe body via silicone tubing. The stainless steel needle, which was used to direct medium onto the top of the filter in the original procedure, is omitted in this modification.[14] Instead, the medium simply trickles down the silicone tubing onto the filter.

Filters are prewetted with sterile saline (5 ml) and then inoculated with an exponential-phase batch culture (10 ml) grown under glucose-limiting or iron-limiting conditions. After perfusion of the saline and inoculum, the filters are perfused with the appropriate medium at a flow rate of 0.87 ml/min. The eluate passing through the filter is collected at intervals and viable counts are made by serial dilution in PBS and plating in triplicate on Sabouraud's dextrose agar. Growth rates of biofilms (divisions/hr) are calculated by dividing the number of daughter cells produced per hour at steady state by the estimated adherent cell population (determined by viable counts of suspended biofilms).

Elution profiles differ from those obtained with glucose-limited populations maintained in a perfused biofilm fermenter. The number of cells eluted decreases over the first hour, but this is followed by a period of biofilm growth during which the adherent culture attains an optimal density. Iron-limited biofilms take longer to reach this stage (24 hr) than glucose-limited biofilms maintained at the same flow rate (6 hr), although both populations shed daughter cells at a constant rate thereafter.[14] In contrast, glucose-limited biofilms in a perfused biofilm fermenter achieve a steady state after only 80 min (see earlier discussion). This is presumably a reflection of the different nature of the filter supports as well as the different inoculation procedures used in these two model systems.

Scanning Electron Microscopy of Biofilms

Biofilm development in all three model systems is monitored most easily by scanning electron microscopy. Various procedures are available for sample preparation, and the choice of procedure will be governed by convenience and by any requirement to preserve the biofilm matrix. With *C. albicans* biofilms on catheter disks, the amount of matrix material visible by scanning electron microscopy depends not only on the preparative techniques used, but also on incubation conditions during biofilm development. Under static incubation conditions, the synthesis of matrix material appears to be minimal. However, matrix production increases dramatically when developing biofilms are subjected to a liquid flow, such as that resulting from incubation with gentle shaking.[12] Under these conditions, biofilm cells may be largely obscured by an extensive canopy of matrix material.

Routine Monitoring of Biofilm Development

Samples are fixed in 2.5% (v/v) glutaraldehyde in PBS for 1 hr at room temperature and immersed in 1% (w/v) osmium tetroxide for 1 hr. They are then washed three times with distilled water (3 ml), treated with 1% (w/v) uranyl acetate for 1 hr, and washed again with distilled water (3 ml). Dehydration of the samples is accomplished by immersion in a series of ethanol solutions ranging, in 10% increments, from 30% (v/v) ethanol in distilled water to dried absolute ethanol. The final step before gold coating is drying. Simple air drying overnight in a desiccator is adequate for many purposes and does not lead to significant sample collapse or wrinkling. Critical point drying is more labor intensive but results in superior conservation of the extracellular matrix.

Preservation of Biofilm Structure

The drying procedure used markedly affects the extent to which the biofilm structure, particularly the matrix of extracellular polymeric material, is retained. Air drying, or chemical drying with Peldri II or hexamethyldisilizane, results in little or no preservation of the matrix structure.[12] Critical point drying is more effective, but consistently better matrix preservation can be achieved using a freeze-drying technique. Samples are fixed with 2.5% (v/v) glutaraldehyde in 0.1 *M* cacodylate buffer (pH 7.0), washed gently three times in distilled water, and then plunged into a liquid propane–isopentane mixture (2 : 1; v/v) at −196° before freeze-drying under vacuum (10^{-6} Torr). With this procedure, even biofilms grown statically can be seen to contain cells clearly linked together by strands of extracellular polymer, whereas biofilms subjected to a liquid flow during incubation display copious amounts of matrix material.[12]

Morphological Forms in Biofilms

The nature of the morphological forms present in *C. albicans* biofilms depends on the model system used. Biofilms grown on catheter disks[8,10,12] comprise a mixture of yeasts, hyphae, and pseudohyphae (Fig. 1A). The perfused biofilm fermenter also produces a mixture of morphological forms[13] (Fig. 1B). In contrast, *C. albicans* biofilms grown on cylindrical cellulose filters consist exclusively of yeast cells[14] (Fig. 1C). The reason for the absence of hyphae is not clear. It could be related to a possible oxygen deficiency inside the cylindrical filter, although such conditions usually favor hyphal development. Alternatively, if morphogenesis in biofilms is dependent on contact-induced gene expression, as has been suggested,[8] the precise nature of the surface may be a crucial determinant in triggering the response.

Susceptibility of *Candida* Biofilms to Antifungal Agents

Like bacterial biofilms, *Candida* biofilms are resistant to a range of antimicrobial agents.[10,13,14] Drug susceptibility of biofilms can be assayed using any of the three model systems described here.

Drug Susceptibility of Biofilms on Catheter Disks

Amphotericin B (Sigma), flucytosine (5-fluorocytosine; Sigma), fluconazole (Pfizer Ltd., Sandwich, Kent, UK), itraconazole (Janssen Research Foundation, Beerse, Belgium), and ketoconazole (Janssen) have all been tested against 48-hr biofilms of *C. albicans* using this model system.[10] Stock solutions of the drugs in dimethyl sulfoxide (amphotericin B, itraconazole, and ketoconazole), sterile distilled water (flucytosine), and dimethylformamide (fluconazole) are prepared immediately prior to use and diluted in growth medium buffered to pH 7 with 0.165 *M* morpholinopropanesulfonic acid (MOPS) buffer (Sigma). Following incubation of catheter disks for 48 hr at 37°, the growth medium is removed from each well and replaced with 1 ml of buffered medium containing the test antifungal agent at concentrations ranging from 0 to 250 μg/ml. The biofilms are incubated for a further 5 hr at 37° and then washed gently in 5 ml of PBS. Biofilm activity is assessed by the [^{3}H]leucine incorporation and tetrazolium reduction assays described earlier. The effect of an antifungal agent is measured in terms of the percentage inhibition of [^{3}H]leucine incorporation or formazan formation by biofilms compared with values obtained for control biofilms incubated in the absence of the agent.[10] These results, in the form of dose–response curves, can then be used to calculate the drug concentration that causes 50% inhibition of [^{3}H]leucine incorporation or formazan formation.

Drug Susceptibility of Biofilms on Cellulose Filters

The drug susceptibility of biofilm cells grown on cylindrical cellulose filters[14] or on cellulose acetate membranes in the perfused biofilm fer-

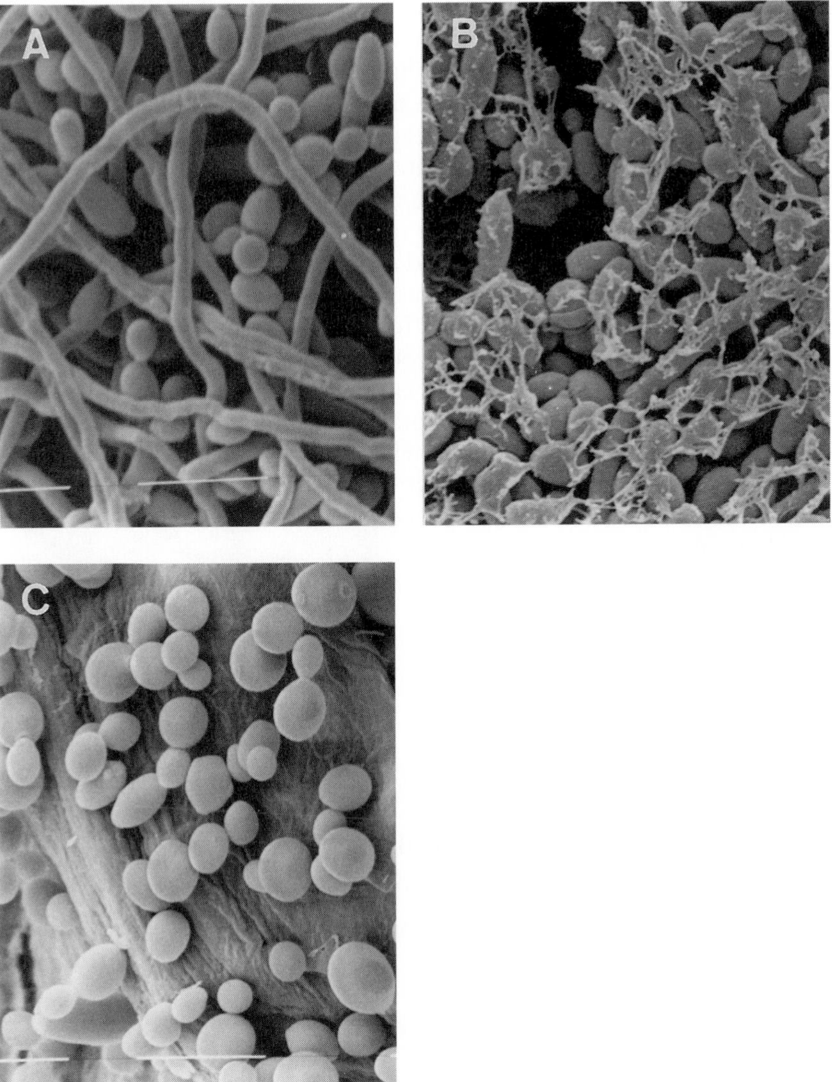

FIG. 1. Scanning electron micrographs of *C. albicans* biofilms grown on a PVC catheter disk (A), a cellulose acetate filter in a perfused biofilm fermenter (B), and a cylindrical cellulose filter perfused with glucose-limited medium (C). Biofilms in A and C were air dried overnight; the biofilm in B was freeze dried. Bar: 10 μm.

menter[13] can be tested using a protocol originally devised by Evans *et al.*[22] The method compares intact biofilms, suspended biofilm cells, and daughter cells newly budded from the biofilm. When the steady state has been reached in either model system, the filter is removed from the apparatus and cut in half (longitudinally for the cylindrical filter). One-half of the filter is immersed in drug solution (10 ml) for 1 hr at 37° and the adherent cells are then suspended. For the cylindrical filter, this is done by breaking up the filter with a sterile aluminum rod, followed by vortexing for 1 min; the flat cellulose acetate filter is shaken vigorously for 10 min. Cells on the other halves of the filters are first suspended in sterile water (10 ml) and then incubated with the drug at 37° for 1 hr. Samples of perfusate (1 ml) containing newly formed daughter cells are also incubated similarly at an identical final drug concentration. Viable counts are made for all samples by serial dilution and plating in triplicate on Sabouraud's dextrose agar. Values for percentage survival are calculated by using counts for untreated control samples processed similarly. Colony counts for control samples before and after the 1-hr incubation period show only very small increases in cell numbers.[13,14]

Acknowledgments

The authors' work was supported by Grants 92/3A and 94/22A from the Sir Jules Thorn Charitable Trust.

[49] Enhanced Bacterial Biofilm Control Using Electromagnetic Fields in Combination with Antibiotics

By B. R. McLeod, S. Fortun, J. W. Costerton, and P. S. Stewart

Bacterial Biofilms

Bacterial biofilms, two rather innocent looking words, describe an entity that has become the focus of a very large field of investigation involving researchers around the world. Even a casual glance at the titles of the articles of this volume conveys both the breadth and the depth of the studies being carried out on this persistent, opportunistic form of bacterial growth. Over the past few years, practitioners in the medical field have become aware of the extent of clinical infection problems that are caused or sustained by bacterial biofilms and how relatively ineffective antibiotics are in controlling them.

0076-6879/99 $30.00

Many of the research efforts are directed toward studying the physiology of the bacteria in a "biofilm environment," identifying the methods by which biofilms form, or studying the polysaccharide matrix in which bacteria reside. This article will, however, focus on a unique means of killing the bacteria in a biofilm and on establishing the physical parameters that produce "total kill."

Electromagnetic Fields and Biological Systems

There has been a certain amount of interest in using electromagnetic fields to produce observable changes in biological systems since about 1774 when Galvoni used short pulses of current to make an isolated frog leg twitch. Although controversy still abounds when this subject is discussed, at least one area has become established clinically in which electromagnetic fields play an important role in obtaining patient recovery. That area is the healing of problem bone fractures (more specifically, nonunion bone fractures). A large body of literature exists concerning this clinical practice.[1] As this clinical method of electromagnetic-enhanced bone healing has emerged, it is not surprising that results from experiments involving bacterial biofilms and electromagnetic fields also began to appear in the literature. In a paper published in 1992, Blenkinsopp *et al.*[2] coined the name "bioelectric effect" to describe data that small dc electrical currents could be used to enhance the efficacy of biocides against *Pseudomonas aeruginosa* biofilms. Today, one can access a number of other papers that describe work involving the bioelectric effect.[3–7]

The papers cited earlier[2–7] each reported varying degrees of success in enhancing the kill achieved with an antibiotic when a dc current was added as part of the biofilm treatment, but each laboratory used an exposure system, bacteria, antibiotic, and biofilm growth system that was designed to address their particular interest. Most of this work concentrated on holding the electrical parameter (e.g., the dc current) fixed and studying

[1] A. A. Pilla, *in* "Electricity and Magnetism in Biology and Medicine" (M. Blank, ed.), p. 17. San Francisco Press, San Francisco, 1993.

[2] S. A. Blenkinsopp, A. E. Khoury, and J. W. Costerton, *Appl. Environ. Microbiol.* **58,** 3370 (1992).

[3] J. Jass, J. W. Costerton, and H. M. Lappin-Scott, *J. Indust. Microbiol.* **15,** 234 (1995).

[4] N. Wellman, S. M. Fortun, and B. R. McLeod, *Antimicrob. Agents Chemother.* **40,** 2012 (1995).

[5] J. W. Costerton, B. Ellis, K. Lam, F. Johnson, and A. E. Khoury, *Antimicrob. Agents Chemother.* **38,** 2803 (1994).

[6] A. E. Khoury, K. Lam, B. Ellis, and J. W. Costerton, *ASAIO Trans.* **38,** M174 (1992).

[7] P. Stoodley, D. de Beer, and M. M. Lappin-Scott, *Antimicrob. Agents Chemother.* **41,** 1876 (1997).

the result of varying the biology (such as the level of antibiotic). The following methods and results resulted from work designed to hold the biology constant and vary the current. We wanted to establish a dc current versus an increased bacterial biofilm killing dose–response curve and to develop an experiment and an experimental protocol that could be reproduced readily in other laboratories. The dose–response curve is presented later. Also, a sufficiently high level of confidence in experimental repeatability was achieved for a current density of about 360 $\mu A/mm^2$ in the experimental chamber to allow the exploration of several paths that could lead to an explanation of the mechanism of the bioelectric effect. A summary of these results is also presented.

Experimental Approach

Biofilm Development

A strain of *P. aeruginosa* (maintained in the Montana State University-Bozeman (MSU) Center for Biofilm Research as ERC-1) is used for these experiments. The biofilm is grown in a growth chamber (described later) that is connected to two peristaltic pumps (MasterFlex Model 6-600 rpm and 1-100 rpm, Cole-Parmer Instrument Co., Chicago, IL) with peristaltic pump heads (MasterFlex Model, Cole-Parmer Instrument Co.). The growth chamber is stirred magnetically and is connected to the substrate, the buffer, and the dilution water by MasterFlex tubing. Dilution water is stored in a plastic container (34 gal) that is suffused continuously with air using an aquarium pump. The air pump is a Model Elite 802 (Rolf C. Hagen Corp., Mansfield, MA 02048). The growth chamber, along with pumps, support containers, and flexible tubing, is shown in Fig. 1.

The chamber itself is a cylindrical, straight-sided glass beaker that is modified slightly in the glass-blowing shop at Montana State University. The modification is the addition of a glass overflow pipe that is set to be just above the level of the growth suspension in which the biofilms are formed. A second, thin-walled polycarbonate cylinder with a slightly smaller diameter is made that slips into the glass beaker but whose diameter is chosen so it has a snug fit to the inner wall of the glass beaker. Slots are cut into the polycarbonate cylinder to support the slides (see Fig. 1) on which the biofilms are grown. Eight such slides can be inserted in the growth chamber in this arrangement, which gives sufficient biofilms to have four controls and four "experimentals" with each experimental run. To grow the biofilms, the sterile growth chamber (Fig. 1) is filled with the following buffer and substrate–mineral solutions at 30 times the concentrations listed,

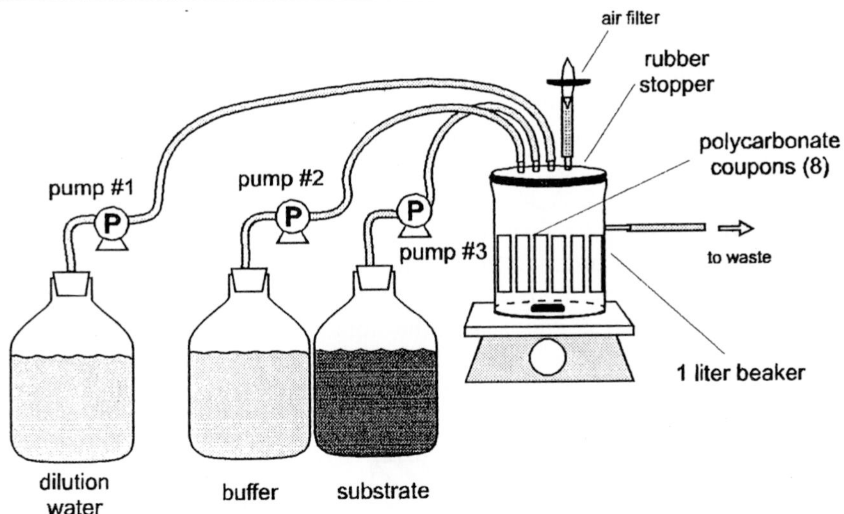

FIG. 1. Growth chamber used to grow the biofilms. The chamber consists of a straight-sided glass beaker with a glass overflow pipe added just above the level of the biofilm slides (the overflow pipe is marked "to waste" in the figure and the biofilm slides are marked as "polycarbonate coupons"). The polycarbonate slides are held close to but not touching the glass wall of the cylinder by a second, thin-walled (0.41 cm thick) polycarbonate cylinder in which slots have been cut. The biofilm slides slip into these slots and are held in position as the biofilm is being grown.

and 1 ml of the frozen *P. aeruginosa* culture is thawed and inoculated into the chamber. The buffer solution consists of (per liter) NaH_2PO_4, 454 mg and KH_2PO_4, 219 mg. The substrate solution consists of (per liter) KNO_3, 14.5 mg.; $MgSO_4 \cdot 7H_2O$, 1 mg.; $CaCO_3$, 1 mg.; $N(CH_2COOH)_3$, 427 μg; $(NH_4)_6Mo_7O_{24} \cdot 4H_2O$, 1.5 μg; $ZnSO_4 \cdot 7H_2O$, 151 μg; $MnSO_4 \cdot H_2O$, 12.2 μg; $CuSO_4 \cdot 5H_2O$, 3 μg; $Na_4B_4O_7 \cdot 10H_2O$, 1.5 μg; $Co(NO_3)_2 \cdot 6H_2O$, 1.79 μg; $FeSO_4 \cdot 7H_2O$, 170 μg; and glucose, 21.3 mg. This "batch phase" continues for 24 hr with constant stirring; 24 hr after inoculation, buffer, substrate, and dilution water are fed continuously into the reactor. The dilution water flow rate (the water is sterilized using two filters in series) is 30 ml/min and the substrate flow rate is 1 ml/min. After approximately 72 hr of growth, the polycarbonate slides, with the attached bacterial biofilm, are transferred aseptically to the experimental chambers for the dc electrical current experiments. All the growth and the experiments are carried out at room temperature (22° ± 2°). Our experience indicates that temperature is not critical in this protocol.

Experimental Chamber

An experimental chamber is machined from a solid, rectangular block of polycarbonate with outside dimensions of 30 × 40 × 84 mm. An end mill is used to remove the material in a volume measuring 16 × 34 × 70.5 mm (approximately 38 cm^3) as shown in Fig. 2. The fluid flow connectors are also shown in Fig. 2. A 34 × 88 mm, rectangular piece of polycarbonate, 5.7 mm thick, is used to form the lid for each experimental chamber. The lid is made thick in order to give solid support to the 22-gauge stainless steel wires (type 316, McMaster-Carr Supply Co., Santa Fe Springs, CA) that form the electrical contacts to the liquid in the chamber. Two holes are drilled through the lid, one at each end, such that the 22-gauge wire pushes easily through the holes and extends down into the exposure cham-

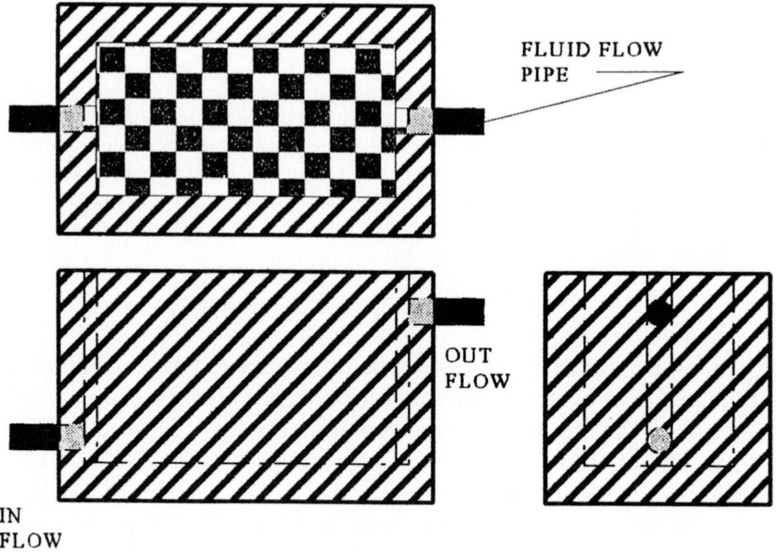

FIG. 2. Top, side, and end view of a polycarbonate exposure chamber. Note the small grooves at each end of the exposure chamber with which the biofilm support substrate (a 0.72-mm-thick, 18 × 73.5-mm strip of polycarbonate) is held upright and centered in the chamber. The 22-gauge stainless steel electrodes also fit into these grooves and are cut to clear the bottom of the chamber by about 1 mm. The chamber inside dimensions are 70.5 × 16 × 34 mm deep. Each end groove is 2 mm deep. The fluid input and outflow ports are discussed in Fig. 3. A flat cover of polycarbonate measuring 88 × 34 × 5.7 mm thick and fitted with a soft rubber gasket is held on the exposure chamber by six screws (three on each long side) during each experiment. There are two holes drilled in the cover through which the electrodes are pushed in order to make the electrical connections to the chamber. The holes provide a snug fit for the 22-gauge wire, and the gasket acts as a septum through which the electrodes are pushed. This provides a seal against liquid leakage during the experiment.

ber. The diameter of the holes is chosen so that the wires fit snugly in the holes. To further prevent leakage around the electrodes, a thin sheet of sterilized soft rubber is placed under the entire bottom side of the lid (e.g., a rubber sealing gasket) and the electrodes are pushed through the rubber at the start of each experiment. The holes for the electrodes are positioned so that the electrodes extend down the distal end of the grooves holding the biofilm substrate centered and upright in the experimental chamber (see Fig. 2). The wire electrodes are cut to a length that exposes 32 mm of electrode in the chamber (e.g., extending from the lid to near the bottom of the chamber), plus approximately 3 cm of wire extending outside the box for the purpose of connecting the electrode to the power supply circuit.

The lid and sealing gasket are then fastened to the chamber with six screws spaced equally along the long sides of the chamber (three on each side). With the lid mounted to the chamber in this way, the slowly flowing support liquid fills the entire chamber during each experiment, the fluid level remains constant, and there are no leaks from the experimental chamber. It should be pointed out that no data have been taken that indicate that the exact size or shape of the exposure chamber is important. It is considered important to keep the level of the support medium constant in order to have a constant cross section of conducting medium through which the current is flowing (e.g., to have a known and constant current density in the chamber).

Treatment of Biofilm

The antibiotic used for all of the experiments is tobramycin (Apothecon, Bristol-Myers Squibb Co., Princeton, NJ) at a concentration of 5 mg/liter [five times the minimum inhibitory concentration (MIC) of tobramycin for planktonic cells of this strain of *P. aeruginosa*]. The antibiotic is obtained in 2-ml vials (concentration 40 mg/ml) stored, when not in use, as per the manufacturer's directions and used well before the label expiration date. As a further check on the efficacy of the antibiotic, a planktonic control is run with each experiment, i.e., tobramycin is tested against planktonic bacteria at the just-mentioned concentration to be sure that the drug is still potent. The equipment used for this part of the experiment is shown in Fig. 3, and a nutrient support medium (pH 6.8 to 7), consisting of the buffer and support solution (without the salts) in the concentrations listed earlier, is prepared in 1000-ml flasks and autoclaved along with all of the experimental chambers, tubing, and connectors (anything that comes in contact with the biofilm is heat or filter sterilized). All material is left overnight to cool and then salts, glucose, and the antibiotic are added to the growth medium flask in a sterile procedure using sterile syringes, nee-

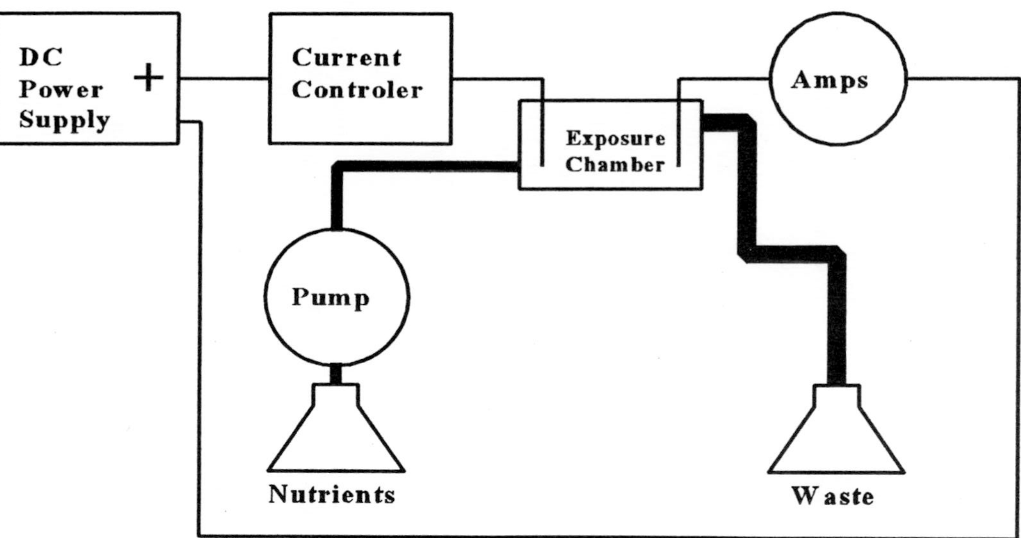

FIG. 3. A block diagram of the electrical circuit used in the experiments. Also shown is the fluid flow circuit. Note that the inflow and the outflow pipes are of different size. The outflow pipe was made larger simply to allow the fluid to exit freely. The input stainless pipe has dimensions of 1.7 mm o.d. and 1.21 mm i.d., and the outflow pipe has dimensions of 4.77 mm o.d. and 3.21 mm i.d. Both pipes are push fitted tightly into the experimental chamber.

dles, and syringe filters. The tubing is connected to the flasks and to the experimental chambers through peristaltic pumps, and the nutrients are allowed to flow to the chambers for about 1 hr in order to fill them. Once the experimental chambers are full, the transfer of the slides is completed using aseptic techniques, and the nutrient flow rate is continued at 2.7 ml/hr. Currents are set at the desired value in each individual chamber and the experiment is carried out for 24 hr. The original count (the cell density in the biofilm on one of the polycarbonate slides immediately after the slides are removed from the growth reactor and immediately before an experiment is initiated) is obtained using the same procedure used to obtain the experimental counts (described later). Tobramycin, at a concentration of 5 mg/liter, is then added to the original biofilm suspension, which is placed in a stirred water bath for 24 hr, serially diluted, and plated. This part of the protocol produces data about the original density of cells as well as information about the level of killing of the planktonic (free floating) bacteria by the antibiotic.

After 24 hr, each slide is removed from its experimental chamber, and the biofilm is scraped, using sterile techniques, into a sterile beaker from the polycarbonate slide using tweezers and a stainless steel spatula. Nine milliliters of sterile buffer solution is added to the beaker. An additional 1 ml of sterile buffer is added to the beaker after the first 9 ml containing the biofilm has been poured back into a sterile test tube. This extra 1 ml of buffer solution rinses out any biofilm left in the beaker and is then added to the 9 ml in the test tube. Eight serial dilutions are performed on the suspended cells using a Rainin 100-ml electronic digital pipette (edp 2 by Rainin), and the planktonic culture with antibiotic is also diluted and plated. The petri dishes are incubated at 35° and, after 17 to 18 hr, colonies are counted and data are recorded.

Electric Circuit

Because one of the goals of the work reported in this article is to develop a current versus killing dose response, it is important to be able to set the current at the desired value and know that the set value does not vary during the 24 hr during which the experiment is performed. Therefore, a current controller commonly used to control current to laser diodes (Model LDD200-1M, Wavelength Electronics, Bozeman, MT), is placed in the circuit as shown in Fig. 3. It is placed between the experimental chamber and the dc power supply in order to be sure that the set value of current is flowing to the chamber. The dc ammeter (shown in Fig. 3 in the circle labeled "Amps") is placed in the circuit to measure the amount of current flowing through the chamber. The controller has the capability of producing up to a constant 200 mA of current (when used in conjunction with a 12-V dc source). It should be pointed out that having exactly the same electronics in every circuit is not important as long as the current is controlled and measured carefully. There are five identical circuits used in each experimental run so that the desired current level can be set and maintained in individual chambers. The desired current in each chamber is set at the initiation of each experiment and the current controller keeps that value constant (desired value ± 5%) independent of any resistance changes in the chamber.

Evaluation of Results

As stated earlier, the initial work was designed to produce a dose–response curve for the current needed to produce increasing levels of killing of the bacteria in the biofilm. As the work progressed, the scope was enlarged to investigate the role of electrolysis products in mediating the bioelectric effect. Table I shows a summary of the general design of the experiment as well as the names attached to the various sets of data.

Because the biofilm growth reactor was designed to allow a biofilm to grow on eight polycarbonate slides, each experiment was designed to make use of all eight of the slides. One slide was removed from the reactor, the biofilm was scraped into a sterile beaker using a stainless steel scraper, and suspended in 10 ml of phosphate buffer. Antibiotic at the concentration used in these experiments (5 mg/liter) was then added and the suspension was placed in a stirred water bath for 24 hr. At the end of this period, eight serial dilutions were drop-plated into R2A agar and incubated for 18 hr at 35°, and then the plates were counted. These data indicated that the efficacy of the tobramycin at 5 mg/liter against the planktonic form of the bacteria was a mean log reduction of 4.27 ± 1.3.

Three of the slides for each experiment were used for the sham control (SC), the antibiotic control (C), and the current control (CC) as explained in Table I. The remaining four slides were exposed in the experimental

TABLE I
GENERAL EXPERIMENTAL DESIGN

Current	Antibiotic	Data name
No	No	Sham control (SC)
No	Yes	Antibiotic control (C)
Yes	No	Current control (CC)
Yes	Yes	Experiment (E)

chambers to set levels of current flow (recall that the antibiotic was present in the slowly flowing support medium as described in "Treatment of the Biofilm") with data being recorded as "experiment" (E). The (E) data from each experiment and from different experiments were treated as independent data sets because each circuit consisted of a complete set of independent equipment. All seven of these slides were treated in identical experimental boxes. The untreated sham control data exhibited a mean cell count of 8.7×10^8 cfu with a mean cell density of 7.8×10^7 cfu/cm^2.

When the biofilm was treated with antibiotic alone [log(C/SC)], the mean log reduction was 2.88 ± 0.66 ($p < 10^4$), and treatment with current alone [log(CC/SC) with the current level set at 2 mA], the log reduction was 0.65 ± 0.42 ($p = 0.0016$). At the 2 mA current level, the electrical enhancement of the antibiotic efficacy was calculated by comparing the combined effect of current and antibiotic with antibiotic alone [log(E/C) with the result being a mean log reduction of 2.75 ± 0.95]. If the combined current (2 mA) and antibiotic treatment was compared to the untreated biofilm [log(E/SC)], the result was a mean log reduction of 5.77 ± 0.26. Dose–response data shown in Fig. 4 is plotted as current plus antibiotic in the exposure chamber versus treatment with antibiotic alone [log(E/C)] and as current plus antibiotic in the chamber versus no treatment [log(E/SC] in order to illustrate both the increase in effectiveness of the antibiotic as the current flow is increased and also to allow one to observe the enhancement due to the bioelectric effect.

In addition to these results, data were also obtained from a series of experiments that were designed to investigate hypotheses concerning the mechanism by which the current enhances the antibiotic efficacy. In four experiments, the electrodes were placed outside the treatment chamber and a potential was applied across the electrodes. The electric field established in the chamber was adjusted to be the same as that developed in the normal experiment but, of course, there was no current flow. There was no enhancement of bacterial killing. In another series of experiments, oxygen was sparged into a treatment chamber that was receiving antibiotic but no current. There was a significant ($p = 0.027$) enhancement of the efficacy of the antibiotic of about 1.8 log in these experiments. For the same experiments with hydrogen replacing oxygen, there was no enhancement.

It was observed that a large change in pH occurred when the electric current was applied to the suspension in the experimental system. For example, pH values in the SC, the C, the CC, and the E at 2 mA were 7.16, 7.18, 4.52, and 4.74, respectively. A series of experiments was then performed in which the buffer strength was increased. This reduced the pH change that occurred when the current was flowing in the chamber, but the increased buffer strength also reduced the antibiotic efficacy. A

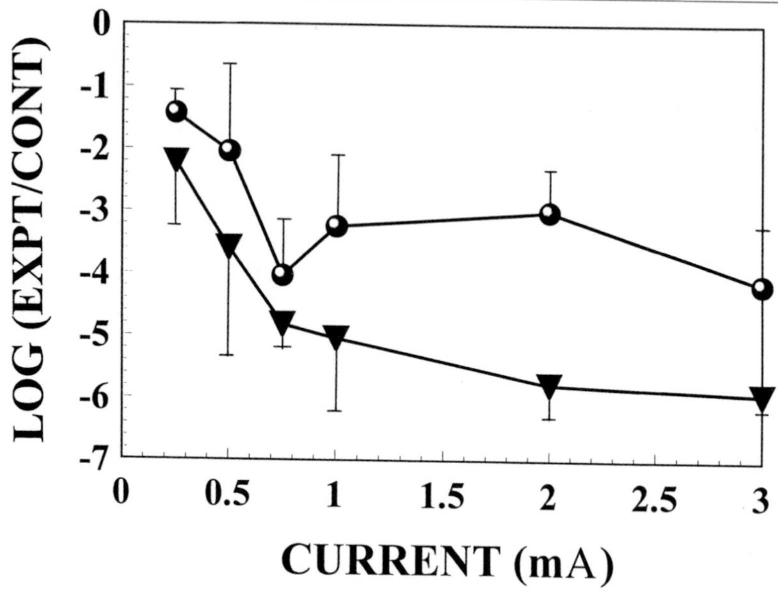

FIG. 4. Data from Table III plotted to show the results of antibiotic plus current compared to antibiotic with no current (log[E/C], ●) and for antibiotic plus current compared to the sham control (log[E/SC], ▼). Recall that control data are obtained with the experimental chambers using antibiotic but no current in the support medium. The sham control is obtained with no antibiotic and no current in the support medium. Therefore, in the upper curve (●) one can see the effect of adding current to an antibiotic treatment for a biofilm. The lower curve (▼) indicates the combined effect of adding both current and antibiotic to treat a biofilm. One-half of the error bars are shown upward for (●) data and downward for (▼) data. The ANOVA analysis of the data points is given in Table III.

further test of the role of pH in the bioelectric effect was conducted by artificially forcing a pH change by altering the relative proportions of the two buffer constituents. The phosphate buffer was reformulated to have a pH of 5.0 with the same total phosphate concentration. When this buffer was used in the experiments (with no current flow), the forced reduction in pH reduced the antibiotic efficacy instead of enhancing it.

Other possible mechanisms were investigated, such as a temperature change in the suspension in the experimental chamber, the possibility that active oxygen intermediates such as peroxide were being generated, and that ions from the salts in the support medium were involved in causing the bioelectric effect. The temperature increase with 2 mA flowing in the experimental chamber for 24 hr was measured and found to be 0.18 ± 0.05°, which is not sufficient to account for the effect. To test the possibility that oxygen intermediates were responsible for increasing the efficacy of the antibiotic, sodium thiosulfate was added to the medium, first at 1 g/liter then at 10 g/liter. At the lower concentration, the thiosulfate did not abolish the bioelectric effect nor did it affect the efficacy of the antibiotic alone. At the higher concentration, the efficacy of the antibiotic alone decreased, but the electrical enhancement of killing of the biofilm increased. When a skeletal medium consisting of only glucose and the two phosphate buffer compounds was used (e.g., no salts), the electrical enhancement remained the same. A summary of the various treatments of the biofilm is given in Table II.

Discussion of Bioelectric Effect on Biofilms

The material contained in this article has focused on the bioelectric effect and the electrical enhancement of antibiotic efficacy against a biofilm, which, in this case, was a model system of *P. aeruginosa* and tobramycin. Data indicate that close to a 3 log reduction in colony-forming units (when the treatment of current plus antibiotic is compared to the treatment of antibiotic alone) can be achieved when the current through the experimental chamber is 2 mA. This corresponds to a current density of 3.7×10^{-6} A/mm^2. If the enhancement of current plus antibiotic is compared to a biofilm with no treatment (the sham control), the result is more than a 5.5 log reduction in the colony-forming unit. This level of current has been established in the experimental system described in this article as the amount of current that is required to give a nearly complete kill of the

TABLE II
COMPARISON OF BIOELECTRIC ENHANCEMENT OF ANTIBIOTIC EFFICACY WITH OTHER TREATMENTS[a]

Treatment	Mean log(E/C)	n	p	
			Mean = 0	Mean = 2 mA
2 mA	−2.75	15	<0.0001	na
E field, no current	0.33	4	0.082	<0.0001
2 mA, 3× buffer	−2.3	3	0.077	0.27
No current, pH 5.0	2.41	3	0.008	<0.0001
2 mA, 1 g/liter, thiosulfate	−2.67	3	0.074	0.93
2 mA, 10 g/liter, thiosulfate	−5.56	3	<0.0001	<0.0001
2 mA, salts omitted	−2.87	3	0.044	0.87
No current, oxygen	−1.83	3	0.027	0.07
No current, hydrogen	0.51	3	0.11	<0.0001

[a] E denotes the viable cell count in the presence of both antibiotic and electric current and C denotes the viable cell count when treated with antibiotic alone.

biofilm. More sensitive tests are now available that could be used to establish 9 or 12 log reductions in the number of viable colonies, but that was not the point of this work. The two curves presented in Fig. 4 are quite suggestive of dose–response curves, even though the error bars are rather wide. ANOVA analysis (see Table III) indicated that the data points on the two curves were significantly different at 1 and 2 mA and close to being significantly different at 3 mA.

The mechanism by which the electric current augments the efficacy of the antibiotic has not been established. Because of this, a number of experiments were run using the exposure chambers and the 2-mA current level to eliminate some of the possible explanations for the bioelectric effect. It has been established that current must flow in the chamber containing the biofilm for the effect to be operable, i.e., a non-time varying (direct current or dc) electric field by itself does not increase the killing of the biofilm.

TABLE III
ANOVA ANALYSIS[a]

	Data points			
Condition	*p*	F	Critical F	Significant at 95%?
Antibiotic + I/current control				
2 to 0.5 mA	0.0693	3.79	4.494	Near
2 to 1 mA	0.5629	0.3461	4.3512	No
2 to 3 mA	0.052	4.4545	4.5431	Near
E/sham control				
2 to 0.5 mA	0.001	15.94	4.49	Yes
2 to 1 mA	0.0563	4.1053	4.3512	Near
2 to 3 mA	0.6162	0.262	4.5431	No
Between curves (E/SC to E/CC)				
0.5 mA SC to 0.5 mA CC	0.276	1.4355	5.9814	No
1 mA SC to 1 mA CC	0.0143	7.8074	4.6001	Yes
2 mA SC to 2 mA CC	1E-11	131.37	4.222	Yes
3 mA SC to 3 mA CC	0.0651	6.3708	7.7086	Near
Current value				Number of data points
0.5 mA				4
1 mA				8
2 mA				14
3 mA				3

[a] Data plotted in Fig. 4 are examined to determine if points on the same curve are significantly different at the 95% confidence level (E/CC or E/SC). Then the same analysis is used to determine if points at the same current level but on different curves are significantly different (E/SC to E/CC).

Data presented here also indicate that electrolytically generated changes in pH, an increase in antibiotic efficacy due to a current-mediated rise in the temperature of the medium, and the generation of reactive oxygen intermediates can be ruled out as mechanisms. We established that virtually the same electrical enhancement of the killing could be obtained with the current flowing in either direction (but not reversing during any individual experiment) in the exposure chamber, which eliminates enhanced convective transport via electrically driven contraction and expansion of the biofilm[7] as a mechanism. When all of the salts were removed from the medium, leaving only two phosphate buffer components, the bioelectric effect persisted. This suggests that the electrochemical generation of an inhibitory ion, such as nitrite from nitrate, is not likely to be the explanation for the electrical enhancement of antibiotic efficiency.

One possible explanation for the bioelectric effect did emerge from these experiments. It is possible that there is increased delivery of oxygen to the biofilm due to oxygen generation by *in situ* electrolysis. Calculations indicate that the flow of current at the levels established in these experiments was sufficient to saturate the aqueous medium with oxygen. When gaseous oxygen was bubbled into the treatment chamber during exposure to tobramycin but in the absence of current, a 1.8 log enhancement of the killing of the biofilm was recorded. We also noted that oxygen applied without antibiotic decreased the biofilm accumulation compared to the sham control (no current, no antibiotic) by about 0.47 log, which mimicked the effect(s) of direct current alone (which resulted in a reduction of about 0.65 log). The "bioelectric effect mechanism" for this set of experiments was linked to oxygen because a similar set of experiments with hydrogen replacing oxygen being applied to the biofilm resulted in no enhancement of killing of the biofilm. This line of reasoning is currently being further investigated, focusing on two hypotheses. It is possible that when oxygen reaches toxic levels in the biofilm, it weakens the bacterial cells and they may become more susceptible to the antibiotic. Another hypothesis is that the increased delivery of oxygen could enhance the growth in the biofilm, which would negate the reduced susceptibility of the bacteria in the biofilm associated with slow growth.[8] It has been shown that *P. aeruginosa* biofilms are readily oxygen limited, which leads to zones of slow or no growth within the depths of the biofilm.[9,10] It would follow then that if biofilm resistance

[8] P. Gilbert and M. R. Brown, *in* "Microbial Biofilms" (H. M. Lappin-Scott and J. W. Costerton, eds.), p. 118. Cambridge Univ. Press, Cambridge, 1995.

[9] C. T. Huang, K. D. Xu, G. A. McFeters, and P. S. Stewart, *Appl. Environ. Microbiol.* **64,** 1525 (1998).

[10] K. D. Xu, P. S. Stewart, F. Xia, C. T. Huang, and G. A. McFeters, *Appl. Environ. Microbiol.* **64**(10), 4035 (1998).

to antibiotics is due to slow growth in the biofilm, the augmenting of the concentration of the limiting nutrient could actually make the biofilm more susceptible.

Conclusions

This work has established a protocol and a test system for investigating the result of adding a dc current flow in conjunction with the antibiotic tobramycin to sharply increase the kill of the bacterium *P. aeruginosa* that is growing in a biofilm. The dose–response curves contained in Fig. 4 and the design of the experimental exposure chambers allow one to estimate the current densities that are needed to reach a desired level of killing. This has several implications with respect to the design of devices that would accomplish the sterilization of medical devices. Data that were developed to help define the mechanism of interaction are of equal importance as once again, if one desires to develop a device that will translate these results from the laboratory to a clinical application, the design of the device will depend on an understanding of the means by which the antibiotic efficacy is augmented. There appears to be no doubt that the bioelectric effect can be used to strikingly increase the ability of tobramycin to be effective against *P. aeruginosa* when the concentration of the antibiotic is about 5 MIC.

Author Index

Numbers in parentheses are footnote reference numbers and indicate that an author's work is referred to although the name is not cited in the text.

A

B

C

D

E

F

G

H

I

J

K

L

M

N

O

P

Q

R

S

T

U

V

X

Y

Z

Subject Index

A

B

C

D

E

G

H

I

J

L

M

N

O

P

Q

R

S

T

U

X